Energy Storage Systems and Power Conversion Electronics for E-Transportation and Smart Grid

Energy Storage Systems and Power Conversion Electronics for E-Transportation and Smart Grid

Special Issue Editors

Sergio Saponara
Lucian Mihet-Popa

MDPI • Basel • Beijing • Wuhan • Barcelona • Belgrade • Manchester • Tokyo • Cluj • Tianjin

Special Issue Editors
Sergio Saponara   Lucian Mihet-Popa
Università di Pisa   Østfold University College
Italy   Norway

Editorial Office
MDPI
St. Alban-Anlage 66
4052 Basel, Switzerland

This is a reprint of articles from the Special Issue published online in the open access journal *Energies* (ISSN 1996-1073) (available at: https://www.mdpi.com/journal/energies/special_issues/e_transportation_smart_microgrid).

For citation purposes, cite each article independently as indicated on the article page online and as indicated below:

LastName, A.A.; LastName, B.B.; LastName, C.C. Article Title. *Journal Name* **Year**, *Article Number*, Page Range.

ISBN 978-3-03936-425-1 (Hbk)
ISBN 978-3-03936-426-8 (PDF)

Contents

About the Special Issue Editors

Sergio Saponara is an Italian scientist, engineer and entrepreneur, active in the fields of electronics and automotive engineering. He is Full Professor of Electronics at the University of Pisa, Italy, where he is also President of the BSc and MSc degrees in Electronic Engineering. He got his Master and Ph.D. degrees cum laude in Electronic Engineering in 1999 and 2003, respectively. He started his research activities in the field of electronics in 1999, first with a research grant from the National Research Council, and then a grant from STMicroelectronics. In 2002, he was a Marie Curie Research Fellow at IMEC, Belgium, Leuven. Since 2005, he has been responsible for the teaching of courses in Automotive Electronics and Electronic Systems for Robotics and Automation at the University of Pisa, where he contributed to the foundation of the MSc degree in Cybersecurity, being responsible for HW and Embedded Security. Since 1999, he has also taught Electronics at the Italian Naval Academy. As an entrepreneur, he co-founded the company IngeniArs in 2014, and is the winner of several innovation prizes, such as the H2020 SME Instrument. He founded the Summer School in Enabling Technologies for Industrial IoT in 2016, and is also its director. The organisation was awarded by the IEEE CAS society in 2017 and 2018. In 2020, he was appointed director of the Automotive Electronics and Powertrain Electrifications specialization course for retraining in the field of vehicle electrification for the Vitesco Technologies (Continental) group. He co-organized more than 150 conferences and he is Associate Editor of the journals *Energies, Sensors, Designs, IEE Electronic Letters, SpringerNature Journal of Real Time Image Processing, IEEE Canadian Journal of Electrical and Computer Engineering, IEEE Consumer Electronics Magazine* and *IEEE Vehicular Technology Magazine*. He currently coordinates for the University of Pisa in the H2020 European Processor Initiative, where he is WP leader for security HW implementation and a member elected of the steering committee. He also has on-going projects with MIT relating to power converters and e-drives for electric/hybrid vehicles. He has co-authored about 300 scientific articles, four books, 10 journal special issues, and 20 patents. In technology transfer, he has collaborated with Magneti Marelli, STMicroelectronics, Renesas, Intel, Huawei, Infineon, Ericsson, P2P, Solari, Trenitalia, AMS, RiCo, INTECS, IDS, Sitael and Leonardo. He has worked as an evaluator for scientific projects for the governments of Italy, Romania, Kazakhstan and the European Commission. He has been selected as an expert for the AI4people initiative of the European Institute for Science Media and Democracy.

Lucian Mihet-Popa received his Habilitation (2015) and Ph.D. degree (2002) in Electrical Engineering, a Master's degree (2000) in Electric Drives and Power Electronics and a Bachelor's degree (1999) in Electrical Engineering, from the Politehnica University of Timisoara-Romania. Since 2016, he has worked as Full Professor in Energy Technology at Oestfold University College in Norway. From 1999 to 2016, Professor Mihet-Popa has been with the Politehnica University of Timisoara, Romania. He has also worked as a Research Scientist with the Danish Technical University (2011–2014) and Aalborg University (2000–2002) in Denmark and as a Post Doc with Siegen University in Germany in 2004. Dr. Lucian Mihet-Popa has published more than 120 papers in national and international journals and conference proceedings, and 10 books. Since 2017, he has been a Guest Editor for five Special Issues for the MDPI Energies and Applied Sciences Journals, for Maljesi and for the Advances in Meteorology Journal. He has served as Scientific and Technical Programme Committee Member for many IEEE Conferences. Professor Mihet-Popa has participated in more than 15 international grants/projects, such as FP7, EEA and Horizon 2020, and has been awarded more than 10 national research grants. He is also the head of the Research Lab "Intelligent Control of Energy Conversion and Storage Systems" and is one of the Coordinators of the Master Program in "Green Energy Technology" at the Faculty of Engineering. His research interest include modelling, simulation, control and testing of Energy Conversion Systems, Distributed Energy Resources (DER) components and systems, including battery storage systems-BSS (for electric vehicles and hybrid cars and vanadium redox batteries (VRB)), energy efficiency in smart buildings and smart grids. Professor Mihet-Popa was invited to join the Energy and Automotive committees by the President and the Honorary President of Atomium European Institute, working in close cooperation with—and under the umbrella—of the EC and EU Parliament, and he was appointed the Chairman of AI4People 2020, Energy Section.

Editorial

Energy Storage Systems and Power Conversion Electronics for E-Transportation and Smart Grid

Sergio Saponara [1],* and Lucian Mihet-Popa [2]

[1] Dip. Ingegneria della Informazione, Università di Pisa, via G. Caruso 16, 56122 Pisa, Italy
[2] Faculty of Engineering, Østfold University College, Kobberslagerstredet 5, 1671 Fredrikstad, Norway; lucian.mihet@hiof.no
* Correspondence: sergio.saponara@unipi.it

Received: 25 December 2018; Accepted: 1 February 2019; Published: 19 February 2019

Abstract: The special issue "Energy Storage Systems and Power Conversion Electronics for E-Transportation and Smart Grid" on MDPI *Energies* presents 20 accepted papers, with authors from North and South America, Asia, Europe and Africa, related to the emerging trends in energy storage and power conversion electronic circuits and systems, with a specific focus on transportation electrification and on the evolution of the electric grid to a smart grid. An extensive exploitation of renewable energy sources is foreseen for smart grid as well as a close integration with the energy storage and recharging systems of the electrified transportation era. Innovations at both algorithmic and hardware (i.e., power converters, electric drives, electronic control units (ECU), energy storage modules and charging stations) levels are proposed.

Keywords: electric vehicles (EVs); hybrid EV (HEV); plug-in EV (PEV); intelligent transport systems; renewable energy sources; battery; energy storage; electric machines; electric drives; electronic control unit (ECU); fast charging; smart grid; automotive electronics; power converters; Internet of Energy; Internet of Things (IoT)

---

1. Introduction

The proposed special issue has invited submissions related to energy storage, power converters and e-drive systems for electrified transportation and smart grid [1–20]. The particular topics of interest include:

- New emerging energy storage technologies;
- Ageing mechanisms of power converters and energy storage devices;
- Electronic control units (ECU) for energy storage system monitoring and management;
- Online estimation of state-of-charge and state-of-health;
- Power conversion electronics for renewable energy sources;
- Fast chargers and smart chargers for electric-vehicles;
- Integration of charging infrastructures in the smart grid for E-transportation;
- Predictive diagnostic for renewable energy sources and energy storage systems;
- Methods for design and verification of hardware and software for energy storage and renewable energy systems;
- Integration of Internet of Things (IoT) into E-transportation.

Research and technology transfer activities in energy storage systems, such as batteries and super/ultra-capacitors, are essential for the success of electric transportation and to foster the use of renewable energy sources. The latter are intermittent in nature and are not directly matched with users' requirements. Energy storage systems are the key technology to solve these issues and to increase the adoption of renewable energy sources in the smart grid. However, major challenges have still to be solved such as the design of high performance and cost-effective energy storage systems, the on-line estimation of state-of-charge/state-of-health of batteries and super/ultra-capacitors, the estimation of aging effects, the design and optimization of fast chargers and the integration within the smart grid of the charging infrastructure for electrified transportation. The strategic interest for this R&D activity is proved by the rise of initiatives such as the Battery 2030+ initiative or the European Battery Alliance [21] where a mixed effort of the European Commission, industries and research organizations aims at developing an innovative, sustainable and competitive battery "ecosystem" in Europe, from raw material to cell and battery manufacturing to electric vehicle (EV) manufacturing to recycling.

Power converters and electric drives need also optimization in terms of increased efficiency and implementation of predictive diagnostic features. Beside the hardware parts, also the role of the software is increasing and new design and verification methods have to be investigated to achieve high functional safety levels. Due to the increasing role of information and communication technology (ICT) in smart grid and electrified transportation, towards an Internet of Energy scenario, cybersecurity is also becoming a key issue.

The main objective of the 20 works published in the Special Issue is, hence, to provide timely solutions for the design and management of energy storage systems, of renewable energy sources and of the relevant power electronics conversion systems. Proposed works addressed these issues from the low component level, up to the Integration of all these sub-systems within the smart grid for e-transportation and smart/green cities.

2. Review of the Contributions

The special issue includes, after a strict review process, 20 published works of which 18 are original research papers [1–18], while papers [19,20] are survey papers.

Reference [1] entitled "Hybrid PV-Wind, Micro-Grid Development Using Quasi-Z-Source Inverter Modeling and Control—Experimental Investigation", by a group of authors, Priyadarshi et al. from India, Denmark, USA and Norway, dealt with the modeling and control of a hybrid photovoltaic-wind micro-grid system using a quasi Z-source inverter (QZsi). A single ended primary inductance converter (SEPIC) module used as DC-DC switched mode converter was employed for maximum power point tracking (MPPT) functions, while a modified power ratio variable step (MPRVS) based perturb and observe (P&O) method had been proposed, as part of the MPPT action for the photovoltaic system. The dSPACE real-time hardware platform had been employed to test the proposed micro grid system under varying wind speed, solar irradiation, load cutting and removing conditions. The results confirmed the performance of the proposed system for a standalone micro grid, which is used specifically in rural places.

Reference [2], entitled "An Overview of Energy Scenarios, Storage Systems and the Infrastructure for Vehicle-to-Grid Technology", by a group of authors, Harighi et al. from Turkey, Denmark, Norway and USA, presented important issues about energy scenarios, storage systems and the infrastructure of the grid related to vehicle-to-grid (V2G) technology. The scenarios, policies and targets of the governments and agencies of the world for lower greenhouse gases (GHG) emissions and high-energy efficiency had also been suggested. One of the conclusions was that the batteries needed to be further developed to comply better with the charge/life cycle and to provide more safety. All mentioned problems forced the infrastructure of the grid to work with poor quality. Today, after the development of the lithium type of batteries and the power electronics systems, including battery management unit, the battery-to-grid technologies will provide fast charging operations. With those benefits, the grid could work properly, so the lack of energy and temporary storage are solved. The power system infrastructure should be designed according to some parameters such as accessibility, reliability and being able to be developed. The result of this action was that the grid can be updated with some technologies such as V2G and the subset technologies. The grid should support the energy traffic, which was achieved by newer technologies such as V2G or energy storage systems.

Reference [3], entitled "Maximum Power Point Tracking for Brushless DC Motor-Driven Photovoltaic Pumping Systems Using a Hybrid ANFIS-FLOWER Pollination Optimization Algorithm", by a group of authors, Priyadarshi et al. from India, Denmark, and Norway, deals with a hybrid artificial neural network (ANN)-fuzzy logic control (FLC) tuned flower pollination algorithm (FPA) as a maximum power point tracker (MPPT), which was employed to amend root mean square error (RMSE) of photovoltaic (PV) modeling. Moreover, Gaussian membership functions had been considered for fuzzy controller design. Experimental results certify the effectiveness of the suggested motor-pump system supporting diverse operating states. Performed experimental responses revealed that, compared to different bio-inspired, swarm-intelligence and classical MPPT techniques reviewed in the literature, the ANFIS-FPA had superior power tracking ability, fast convergence velocity and accurate system response.

Reference [4] entitled "Boundary Detection and Enhancement Strategy for Power System Bus Bar Stabilization—Investigation under Fault Conditions for Islanding Operation", by a group of authors, Pouryekta et al., from Malaysia and Denmark, proposed a novel scheme for the detection of island boundaries and stabilizing the system during autonomous operation. In the first stage, a boundary detection method was proposed to detect the configuration of the island. In the second stage, a dynamic voltage sensitivity factor (DVSF) was proposed to assess the dynamic performance of the system. In the third stage, a wide area load shedding program was adopted based on DVSF to shed the load in weak bus-bars and stabilize the system. The proposed scheme was validated and tested on a generic 18-bus system using a combination of EMTDC/PSCAD and MATLAB software.

Reference [5], entitled "An Original Transformer and Switched-Capacitor (T & SC)-Based Extension for DC-DC Boost Converter for High-Voltage/Low-Current Renewable Energy Applications: Hardware Implementation of a New T & SC Boost Converter", by a group of authors, Padmanaban et al., from Denmark, Qatar, India and Slovakia, proposed a new transformer and switched capacitor-based boost converter (T & SC-BC) for high-voltage/low-current renewable energy applications. The proposed T & SC-BC was an original extension of the DC-DC boost converter, which is designed by utilizing a transformer and switched capacitor (T & SC). PV energy was a fast emergent segment among the renewable energy systems. The proposed T & SC-BC combines the features of the conventional boost converter and T & SC to achieve a high voltage conversion ratio. The proposed T & SC-BC topology was compared with the recently addressed DC-DC converters in terms of number of components, cost, voltage conversion ratio, ripples, efficiency and power range. Simulation and experimental results were provided, which validated the functionality, design and concept of the proposed approach.

Reference [6] by a group of authors, Hossain et al., from South Africa, Malaysia, Denmark, Norway and USA, "Sliding Mode Controller and Lyapunov Redesign Controller to Improve Microgrid Stability: A Comparative Analysis with CPL Power Variation", dealt with a storage-based load side compensation technique, which is used to enhance the stability of microgrids. Besides adopting this technique, sliding mode controller (SMC) and Lyapunov redesign controller (LRC), two of the most prominent nonlinear control techniques, were individually implemented to control microgrid system stability with desired robustness. Constant power load (CPL) is then varied to compare robustness of these two control techniques. This investigation revealed the better performance of the LRC system compared to SMC to retain stability in the microgrid with a dense CPL load. The simulation results have been validated on the MATLAB/Simulink software package for authentic verification. Reasons behind inferior SMC performance and ways to mitigate that were also discussed. Finally, the effectiveness of SMC and LRC systems to attain stability in real microgrids was verified by numerical analysis.

Reference [7], entitled "Minimization of Load Variance in Power Grids—Investigation on Optimal Vehicle-to-Grid Scheduling", by a group of authors, Tan et al. from Malaysia, South Africa, Denmark and Norway, dealt with an optimal scheduling of V2G using the genetic algorithm to minimize the power grid load variance. This was achieved by allowing EV charging (grid-to-vehicle) whenever the actual power grid loading is lower than the target loading, while conducting EV discharging (V2G) whenever the actual power grid loading is higher than the target loading. The performance of the proposed algorithm under various target load and EVs' state of charge selections were analyzed. The effectiveness of the V2G scheduling to implement the appropriate peak load shaving and load levelling services for the grid load variance minimization was verified under various simulation investigations. A performance index was also introduced in this paper to provide an excellent indication on the overall performance of the proposed V2G optimization algorithm.

Reference [8], co-authored by Bhaskar et al. from South Africa, India, Norway, Denmark and Malaysia, entitled "Hardware Implementation and a New Adaptation in the Winding Scheme of Standard Three Phase Induction Machine to Utilize for Multifunctional Operation: A New Multifunctional Induction Machine", presented a new distinct winding scheme that is used to utilize three phase induction machines for multifunctional operation. It can be used as a three-phase induction motor (IM), welding transformer and phase converter. The proposed machine design also worked as a single-phase IM at the same time it worked as a three-phase to single-phase converter. The proposed motor provided an operative solution for agricultural as well as industrial purposes because of rugged construction and less maintenance needs. The proposed concept was verified by designing a motor by modifying the windings of an old IM and the proposed motor was well tested to find its efficiencies and the experimental results are provided in the article to validate the design and construction.

Reference [9], entitled "Development of Sliding Mode Controller for a Modified Boost Ćuk Converter Configuration" from a group of authors, Padmanaban et al. from South Africa, Slovakia and Denmark, introduced an SMC-based equivalent control method to a novel high output gain Ćuk converter. An additional inductor and capacitor improves the efficiency and output gain of the classical Ćuk converter. An SMC-based equivalent control method, which achieved a robust operation in a wide operation range was also proposed. Switching frequency is kept constant in appropriate intervals at different loading and disturbance conditions by implementing a dynamic hysteresis control method. Numerical simulations conducted in MATLAB/Simulink confirm the accuracy of analysis of high output gain modified Ćuk converter. In addition, the proposed equivalent control method was validated in different perturbations to demonstrate robust operation in a wide operation range.

Reference [10], by Tiwari et al., a group of authors from India and South Africa, entitled "Coordinated Control Strategies for a Permanent Magnet Synchronous Generator Based Wind Energy Conversion System", proposed a novel coordinated hybrid MPPT-pitch angle based on a radial basis function network (RBFN) for a variable speed, variable pitch wind turbine. The proposed controller was used to maximize output power when the wind speed is low and optimize the power when the wind speed is high. The proposed controller provides robustness to the nonlinear characteristic of wind speed. It used wind speed, generator speed, and generator power as input variables and utilizes the duty cycle and the reference pitch angle as the output control variables. The duty cycle was used to control the converter so as to maximize the power output and the reference pitch angle was used to control the generator speed in order to control the generator output power in the above rated wind speed region. The effectiveness of the proposed controller was verified using MATLAB/Simulink software.

Reference [11], by Ganesan et al., a group of authors from India, South Africa and Norway, entitled "Study and Analysis of an Intelligent Microgrid Energy Management Solution with Distributed Energy Sources", proposed a robust energy management solution which will facilitate the optimum and economic control of energy flows throughout a microgrid network. This study enabled precise management of power flows by forecasting renewable energy generation, estimating the availability of energy at storage batteries, and invoking the appropriate mode of operation, based on the load demand to achieve efficient and economic operation. The predefined mode of operation was derived out of an expert rule set and schedules the load and distributed energy sources along with the utility grid. A robust control methodology had been developed and demonstrated in a deterministic way to operate the microgrid network in a sustainable mode. Faster communication topologies were deployed to achieve a better response time for the control commands at local as well as centralized controllers. This work can be enhanced by interlinking multiple microgrid networks with a more complex source and load system. The load management and control could be further improved by using artificial intelligence and optimization techniques as future work.

Reference [12], entitled "A Modular AC-DC Power Converter with Zero Voltage Transition for Electric Vehicles", by Ramirez-Hernandez et al., from Mexico and Chile, presented a study of the fundamental of operation of a three-phase AC-DC power converter that uses zero-voltage transition (ZVT) together with space vector pulse width modulation (SVPWM). The proposed converter was basically an active rectifier divided into two converters: a matrix converter and an H bridge, which transfer energy through a high-frequency transformer, resulting in a modular AC-DC wireless converter appropriate for plug-in EVs (PEVs). The principle of operation of this converter considered high power quality, output regulation and low semiconductor power loss. The circuit operation, idealized waveforms and modulation strategy were explained together with simulation results of a 5 kW design. The target application was suitable for light EVs (e-scooters, small city cars) or for hybrid vehicles where the internal combustion engine was supported at low speed by an electric motor.

Reference [13], entitled "Interconnecting Microgrids via the Energy Router with Smart Energy Management" by Liu et al. from China, presented a novel and flexible interconnecting framework for microgrids and corresponding energy management strategies. The proposed solution was presented in response to the situation of increasing renewable-energy penetration and the need to alleviate dependency on energy storage equipment. The key idea was to establish complementary energy exchange between adjacent microgrids through a multiport electrical energy router, according to the consideration that adjacent microgrids may differ substantially in terms of their patterns of energy production and consumption, which can be utilized to compensate for each other's instant energy deficit. Based on multiport bidirectional voltage source converters (VSCs) and a shared direct current (DC) power line, the energy router served as an energy hub, and enabled flexible energy flow among the adjacent microgrids and the main grid. The analytical model was established for the whole system, including the energy router, the interconnected microgrids and the main grid. Various operational modes of the interconnected microgrids, facilitated by the energy router, were analyzed, and the corresponding control strategies are developed.

Reference [14] by Brenna et al. from Italy and Canada, entitled "Modelling and Simulation of Electric Vehicle Fast Charging Stations Driven by High Speed Railway Systems" aimed at the analysis of the opportunity introduced by the use of railway infrastructures for the power supply of fast charging stations located in highways. Actually, long highways were often located far from urban areas and electrical infrastructure, therefore the installations of high power charging areas could be difficult. Specifically, the aim of the investigation in this paper was the analysis of the opportunity introduced by the use of railway infrastructures for the power supply of fast charging stations located in highways. This paper was focused on fast-charging electric cars in motorway service areas by using high-speed lines for supplying the required power. Economic, security, safety and environmental pressures were motivating and pushing countries around the globe to electrify transportation, which currently accounted for a significant amount, above 70% of total oil demand. Electric cars required fast-charging station networks to allowing owners to rapidly charge their batteries when they drive relatively long routes. In other words, this meant the infrastructure towards building charging stations in motorway service areas and addressing the problem of finding solutions for suitable electric power sources. A possible and promising solution was proposed in the study that involves using the high-speed railway line, because it allowed not only powering a high load but also it can be located relatively near the motorway itself. This paper presented a detailed investigation on the modelling and simulation of a 2×25 kV system to feed the railway. A model had been developed and implemented using the SimPowerSystems (Simscape/Specialized Power System) tool in MATLAB/Simulink to simulate the railway itself. Then, the model had been applied to simulate the battery charger and the system as a whole in two successive steps. The results showed that the concept could work in a real situation. Nonetheless if more than twenty 100 kW charging bays were required in each direction or if the line topology is changed for whatever reason, it cannot be guaranteed that the railway system will be able to deliver the additional power that is necessary.

Reference [15], entitled "Real-Time Analysis of a Modified State Observer for Sensorless Induction Motor Drive Used in Electric Vehicle Applications" by Krishna et al., from India, South Africa and Norway, proposed an adaptive sliding mode Luenberger state observer with improved disturbance rejection capability and better tracking performance under dynamic conditions. The sliding hyperplane was altered by incorporating the estimated disturbance torque with the stator currents. In addition, the effects of parameter detuning on the speed convergence are observed and compared with the conventional disturbance rejection mechanism. The entire drive system was first built in the MATLAB-Simulink environment. Then, the Simulink model was integrated with real-time (RT)-Lab blocksets and implemented in a relatively new real-time environment using OP4500 real-time simulator. Real-time simulation and testing platforms had succeeded offline simulation and testing tools due to their reduced development time. The real-time results validated the improvement in the proposed state observer and also correspond to the performance of the actual physical model. The real-time results also validated the improvement in the disturbance rejection capability for the different test cases presented and also provided more credibility as compared to other offline simulated results.

Reference [16], entitled "Control Strategy for a Grid-Connected Inverter under Unbalanced Network Conditions—A Disturbance Observer-Based Decoupled Current Approach", by Ozsoy et al., a group of authors from Turkey, Norway, Slovakia, and South Africa, presented a new approach on the novel current control strategy for grid-tied voltage-source inverters (VSIs) with circumstances of asymmetrical voltage conditions. A standard grid-connected inverter (GCI) allowed the degree of freedom to integrate the renewable energy system to enhance the penetration of total utility power. This paper proposed a proportional current controller with a first-order low-pass filter disturbance observer (DOb). The proposed controller established independent control on positive, as well as negative, sequence current components under asymmetrical grid voltage conditions. A numerical simulation model of the overall power system was implemented in a commercial software tool and the results showed that double-frequency active power oscillations were suppressed by injecting appropriate negative-sequence currents. The simulation results matched the developed theoretical background for its feasibility. The proposed current controller seemed to be a valid alternative solution for GCIs under unbalanced conditions.

Reference [17] by Vavilapalli et al., a group of authors from India, South Africa, Norway, entitled "Power Balancing Control for Grid Energy Storage System in Photovoltaic Applications—Real Time Digital Simulation Implementation" presented a Power Balancing Control (PBC) method for a grid energy storage system for PV applications containing three different power sources, PV array, battery storage system and the grid, was proposed to operate the system in three different modes of operation. Control of a dual active bridge (DAB)-based battery charger which provides a galvanic isolation between batteries and other sources is explained briefly. Various modes of operation of a grid energy storage system are also presented. Hardware-in-the-loop (HIL) simulation is carried out to check the performance of the system and the PBC algorithm. A power circuit (comprised of the inverter, DAB based battery charger, grid, PV cell, batteries, contactors, and switches) is simulated and the controller hardware and user interface panel are connected as HIL with the simulated power circuit through real time digital simulator (RTDS). The PBC technique is implemented on a TMS320F2812 processor-based controller card and tested. Dynamic responses of the inverter and battery charger system are verified by applying a step change in the reference values and satisfactory results are obtained.

Reference [18] by Chandramohan et al., a group of authors from India, South Africa and Norway, entitled "Grid Synchronization of a Seven-Phase Wind Electric Generator Using d-q PLL", presented the development of a comprehensive model of the wind turbine driven seven-phase induction generator (7PIG) along with the necessary power electronic converters and the controller for grid interface. The dynamic model of the system was developed in MATLAB/Simulink and the system response is observed for various wind velocities. The effectiveness of the seven phase induction generator was demonstrated with the fault tolerant capability and high output power with reduced phase current when compared to the conventional three-phase wind generation scheme phase model. The performance of the synchronous reference frame phase-locked loop (SRF-PLL) incorporated in the grid connected seven-phase wind electric generator was analyzed for various operating grid conditions. The use of multiphase machines along with the PLL synchronization of the grid increased the reliability of the wind electric generator.

After the first 18 research papers the special issue also includes two excellent and comprehensive survey papers.

The group of authors in reference [19], from Bangladesh, South Africa, USA and Norway, proposed "A Comprehensive Study of Key Electric Vehicle (EV) Components, Technologies, Challenges, Impacts, and Future Direction of Development", which presented a comprehensive study about EVs, including battery EV (BEV), hybrid EV (HEV), plug-in HEV (PHEV) and fuel cell EV (FCEV). This paper was focused on reviewing all the useful data available on EV configurations, battery energy sources, electrical machines, charging techniques, optimization techniques, impacts, trends, and possible directions of future developments. Its objective was to provide an overall picture of the current EV technology and ways of future development to assist in future research in this sector. The authors concluded that the EVs have great potential of becoming the future of transport while saving this planet from imminent calamities caused by global warming. They were a viable alternative to conventional vehicles that depend directly on the diminishing fossil fuel reserves. The impacts EVs cause in different sectors had been discussed as well, along with the huge possibilities they hold to promote a better and greener energy system by collaborating with smart grid and facilitating the integration of renewable sources. Limitations of current EVs had been listed along with probable solutions to overcome these shortcomings. The current optimization techniques and control algorithms had also been included. A brief overview of the current EV market had been presented. Finally, trends and ways of future developments had been assessed followed by the outcomes of this paper to summarize the whole text, providing a clear picture of this sector and the areas in need of further research.

The last contribution, reference [20], was a survey by the editors Mihet-Popa and Saponara, entitled, "Toward Green Vehicles Digitalization for the Next Generation of Connected and Electrified Transport Systems". This survey paper reviewed recent trends in green vehicle electrification and digitalization. First, the energy demand and emissions of EVs were reviewed, including the analysis of the trends of battery technology and of the recharging issues considering the characteristics of the power grid. Solutions to integrate EV electricity demand in power grids were also proposed. Integrated electric/electronic architectures for HEVs and full EVs were discussed, detailing innovations emerging for all components (power converters, electric machines, batteries, and battery-management-systems). 48 V HEVs were emerging as the most promising solution for the short-term electrification of current vehicles based on internal combustion engines. The increased digitalization and connectivity of electrified cars was posing cyber-security issues that were discussed in detail, together with some countermeasures to mitigate them, thus tracing the path for future on-board computing and control.

3. Conclusions

The special issue entitled "Energy Storage Systems and Power Conversion Electronics for E-Transportation and Smart Grid" presented 18 original research works and 2 comprehensive survey papers, with a worldwide group of authors, related to the emerging trends in energy storage, power converters and e-drives. The works, addressing both algorithmic-level and hardware-level aspects, were focused on applications such as transportation electrification (Full EV and HEV, mainly for automotive and railway scenarios) and the evolution of the electric grid to a smart grid, with massive use of renewable energy sources. Moreover, a close integration is foreseen between the smart electric grid and the electrified vehicles due to the need of an efficient and fast recharging infrastructure.

Energies **2019**, *12*, 663

Therefore, the proposed special issue provides an overview, for the energy and power electronic aspects, of the evolution and trend towards electrified, automated and connected vehicles.

Funding: This research was partially supported by the PRA2017-2018 E-project from University of Pisa.

Acknowledgments: The author is grateful to the MDPI publisher for the possibility to act as guest editor of this special issue and wants to thank the editorial staff of Energies for their kind co-operation, patience and committed engagement.

Conflicts of Interest: The authors declare no conflict of interest.

References

1. Priyadarshi, N.; Padmanaban, S.; Ionel, D.; Mihet-Popa, L.; Azam, F. Hybrid PV-Wind, Micro-Grid Development Using Quasi-Z-Source Inverter Modeling and Control—Experimental Investigation. *Energies* **2018**, *11*, 2277. [CrossRef]
2. Harighi, T.; Bayindir, R.; Padmanaban, S.; Mihet-Popa, L.; Hossain, E. An Overview of Energy Scenarios, Storage Systems and the Infrastructure for Vehicle-to-Grid Technology. *Energies* **2018**, *11*, 2174. [CrossRef]
3. Priyadarshi, N.; Padmanaban, S.; Mihet-Popa, L.; Blaabjerg, F.; Azam, F. Maximum Power Point Tracking for Brushless DC Motor-Driven Photovoltaic Pumping Systems Using a Hybrid ANFIS-FLOWER Pollination Optimization Algorithm. *Energies* **2018**, *11*, 1067. [CrossRef]
4. Pouryekta, A.; Ramachandaramurthy, V.K.; Padmanaban, S.; Blaabjerg, F.; Guerrero, J.M. Boundary Detection and Enhancement Strategy for Power System Bus Bar Stabilization—Investigation under Fault Conditions for Islanding Operation. *Energies* **2018**, *11*, 889. [CrossRef]
5. Padmanaban, S.; Bhaskar, M.S.; Maroti, P.K.; Blaabjerg, F.; Fedák, V. An Original Transformer and Switched-Capacitor (T & SC)-Based Extension for DC-DC Boost Converter for High-Voltage/Low-Current Renewable Energy Applications: Hardware Implementation of a New T & SC Boost Converter. *Energies* **2018**, *11*, 783. [CrossRef]
6. Hossain, E.; Perez, R.; Padmanaban, S.; Mihet-Popa, L.; Blaabjerg, F.; Ramachandaramurthy, V.K. Sliding Mode Controller and Lyapunov Redesign Controller to Improve Microgrid Stability: A Comparative Analysis with CPL Power Variation. *Energies* **2017**, *10*, 1959. [CrossRef]
7. Tan, K.M.; Ramachandaramurthy, V.K.; Yong, J.Y.; Padmanaban, S.; Mihet-Popa, L.; Blaabjerg, F. Minimization of Load Variance in Power Grids—Investigation on Optimal Vehicle-to-Grid Scheduling. *Energies* **2017**, *10*, 1880. [CrossRef]
8. Bhaskar, M.S.; Padmanaban, S.; Sabnis, S.A.; Mihet-Popa, L.; Blaabjerg, F.; Ramachandaramurthy, V.K. Hardware Implementation and a New Adaptation in the Winding Scheme of Standard Three Phase Induction Machine to Utilize for Multifunctional Operation: A New Multifunctional Induction Machine. *Energies* **2017**, *10*, 1757. [CrossRef]
9. Padmanaban, S.; Ozsoy, E.; Fedák, V.; Blaabjerg, F. Development of Sliding Mode Controller for a Modified Boost Ćuk Converter Configuration. *Energies* **2017**, *10*, 1513. [CrossRef]
10. Tiwari, R.; Padmanaban, S.; Neelakandan, R.B. Coordinated Control Strategies for a Permanent Magnet Synchronous Generator Based Wind Energy Conversion System. *Energies* **2017**, *10*, 1493. [CrossRef]
11. Ganesan, S.; Padmanaban, S.; Varadarajan, R.; Subramaniam, U.; Mihet-Popa, L. Study and Analysis of an Intelligent Microgrid Energy Management Solution with Distributed Energy Sources. *Energies* **2017**, *10*, 1419. [CrossRef]
12. Ramirez-Hernandez, J.; Araujo-Vargas, I.; Rivera, M. A Modular AC-DC Power Converter with Zero Voltage Transition for Electric Vehicles. *Energies* **2017**, *10*, 1386. [CrossRef]
13. Liu, Y.; Fang, Y.; Li, J. Interconnecting Microgrids via the Energy Router with Smart Energy Management. *Energies* **2017**, *10*, 1297. [CrossRef]
14. Brenna, M.; Longo, M.; Yaici, W. Modelling and Simulation of Electric Vehicle Fast Charging Stations Driven by High Speed Railway Systems. *Energies* **2017**, *10*, 1286. [CrossRef]
15. Krishna, S.M.; Daya, J.L.F.; Padmanaban, S.; Mihet-Popa, L. Real-Time Analysis of a Modified State Observer for Sensorless Induction Motor Drive Used in Electric Vehicle Applications. *Energies* **2017**, *10*, 1077. [CrossRef]

16. Ozsoy, E.; Padmanaban, S.; Mihet-Popa, L.; Fedák, V.; Ahmad, F.; Akhtar, R.; Sabanovic, A. Control Strategy for a Grid-Connected Inverter under Unbalanced Network Conditions—A Disturbance Observer-Based Decoupled Current Approach. *Energies* **2017**, *10*, 1067. [CrossRef]

17. Vavilapalli, S.; Padmanaban, S.; Subramaniam, U.; Mihet-Popa, L. Power Balancing Control for Grid Energy Storage System in Photovoltaic Applications—Real Time Digital Simulation Implementation. *Energies* **2017**, *10*, 928. [CrossRef]

18. Chandramohan, K.; Padmanaban, S.; Kalyanasundaram, R.; Bhaskar, M.S.; Mihet-Popa, L. Grid Synchronization of a Seven-Phase Wind Electric Generator Using d-q PLL. *Energies* **2017**, *10*, 926. [CrossRef]

19. Un-Noor, F.; Padmanaban, S.; Mihet-Popa, L.; Mollah, M.N.; Hossain, E. A Comprehensive Study of Key Electric Vehicle (EV) Components, Technologies, Challenges, Impacts, and Future Direction of Development. *Energies* **2017**, *10*, 1217. [CrossRef]

20. Mihet-Popa, L.; Saponara, S. Toward Green Vehicles Digitalization for the Next Generation of Connected and Electrified Transport Systems. *Energies* **2017**, *11*, 3124. [CrossRef]

21. Policies, Information and Services/European Battery Alliance. Available online: https://ec.europa.eu/growth/industry/policy/european-battery-alliance_en (accessed on 9 October 2018).

Article

Hybrid PV-Wind, Micro-Grid Development Using Quasi-Z-Source Inverter Modeling and Control—Experimental Investigation

Neeraj Priyadarshi [1], **Sanjeevikumar Padmanaban** [2], **Dan M. Ionel** [3], **Lucian Mihet-Popa** [4,*] and **Farooque Azam** [1]

[1] Department of Electrical Engineering, Millia Institute of Technology, Purnea 854301, India; neerajrjd@gmail.com (N.P.); farooque53786@gmail.com (F.A.)

[2] Department of Energy Technology, Aalborg University, 6700 Esbjerg, Denmark; san@et.aau.dk

[3] Power and Energy Institute Kentucky (PEIK), Department of Electrical and Computer Engineering, University of Kentucky, 689 FPAT, Lexington, KY 40506-0046, USA; dan.ionel@uky.edu

[4] Faculty of Engineering, Østfold University College, Kobberslagerstredet 5, 1671 Kråkeroy-Fredrikstad, Norway

* Correspondence: lucian.mihet@hiof.no; Tel.: +47-9227-1353

Received: 2 June 2018; Accepted: 21 August 2018; Published: 29 August 2018

Abstract: This research work deals with the modeling and control of a hybrid photovoltaic (PV)-Wind micro-grid using Quasi Z-source inverter (QZsi). This inverter has major benefits as it provides better buck/boost characteristics, can regulate the phase angle output, has less harmonic contents, does not require the filter and has high power performance characteristics over the conventional inverter. A single ended primary inductance converter (SEPIC) module used as DC-DC switched power apparatus is employed for maximum power point tracking (MPPT) functions which provide high voltage gain throughout the process. Moreover, a modified power ratio variable step (MPRVS) based perturb & observe (P&O) method has been proposed, as part of the PV MPPT action, which forces the operating point close to the maximum power point (MPP). The proposed controller effectively correlates with the hybrid PV, Wind and battery system and provides integration of distributed generation (DG) with loads under varying operating conditions. The proposed standalone micro grid system is applicable specifically in rural places. The dSPACE real-time hardware platform has been employed to test the proposed micro grid system under varying wind speed, solar irradiation, load cutting and removing conditions etc. The experimental results based on a real-time digital platform, under dynamic conditions, justify the performance of a hybrid PV-Wind micro-grid with Quasi Z-Source inverter topology.

Keywords: PV; MPRVS; Quasi Z-source inverter; MPP; SEPIC converter

---

1. Introduction

Micro-grid comprises the combination of interconnected loads and distributed energy resources (DER), including energy storage devices and several active loads/prosumers which work as a controlled unit to deliver the electric demand for miniature location. It supplies power generation with tremendous reliability as well as an affirmation to varying loads [1–3]. Fossil fuels and nuclear sources are treated as the traditional energy sources, which provide electricity and are not located closer to the load point. As the conventional energy sources are not environmentally friendly and due to the long-distance transmission, there are considerable power losses that can occur. Therefore, nowadays, renewable energy sources have been given more attention by the researchers and industry to generating alternative power [4–12]. Distributed generating (DG) source such as solar, wind, fuel cell, hydro, tidal, etc. are considered as the main renewable technology, which is highly flexible, expandable and has environmentally friendly behavior. The maximum power point tracking (MPPT) is the important constituent needed to achieve the maximum power point (MPP) as an operating point which enables the utmost power extraction for renewable sources [13,14]. Several MPPT techniques, including Perturb & Observe (P&O), Incremental Conductance (INC), Fuzzy logic control (FLC), Artificial Neural Network (ANN), Particle swarm optimization (PSO), Ant Colony Optimization (ACO), Artificial Bee Colony (ABC), Firefly Algorithm (FA) etc. reviewed in the literature were unable to detect global peak point with partial shade situations [15–27].In this work, Modified Power Ratio Variable Step (MPRVS) based on the P&O technique is proposed without the proportional-integral(PI) controller utilization, which reduces power oscillation near to MPP in comparison to a conventional P&O algorithm and also provides the prevention to battery charging from voltage fluctuation.

To avoid multi reversal generation occurrence in a micro-grid system, in the current research, a Quasi Z-Source inverter is employed [28–33]. The DC-DC converter is a vital interface to achieve a peak power generation from PV modules. In this work, a high-quality tracking behavior is achieved by employing single ended primary inductance converter (SEPIC), which provides high voltage gain with better buck/boost performance compared to other dc-dc switched power converters [34]. In this paper, an additional dc-dc converter (SEPIC converter) is used because it comprises buck/boost capabilities. Moreover, QZsi combines a boost converter and an inverter. The MPRVS based P&O MPPT is controlled through the SEPIC converter which provides MPP achievement and works effectively under varying sun insolation and wind velocity. Moreover, the SEPIC converter works as an impedance adapter between the PV panel and Z-source inverter. Jain et al. [35] have implemented QZsi based grid PV system using a predictive controller in which the active and reactive power have been regulated. However, this work is discussed only for the PV system which utilized the traditional INC MPPT with a classical PI controller as a dc bus regulator. Liu et al. [36] have discussed QZsi based multilevel inverter for grid PV power system, which provides precise MPPT and dc-link voltage regulation at the unity power coefficient. However, during practical justification, voltage/current sensors and bulk resistor models are required, which has a high cost. Nevertheless, this work only explains the performance of QZsi based multilevel inverter for only the PV systems rather than the hybrid system. Amini et al. [37] have discussed the cloud computing applications in micro grid clusters. A real time digital simulator is employed for the physical interpretation of power routing which can be utilized for electrical grid utility with the communication system. However, the application of the proposed scheme with hybrid PV-Wind micro grid systems is missing in this research work. Ali et al. [38] have conferred game theory structure for improvement of smart grid efficiency in which the Femtocell communication system is employed. However, the main disadvantage of this proposed communication system is interference in cross layer. Furthermore, the proposed game theory application with hybrid PV-Wind micro grid system has not been discussed in this research work. Vignesyn et al. [39] have discussed the hybrid micro grid for standalone/Grid mode operation with Z-source inverter. This paper discusses the behavior of micro grid under varying loading conditions, solar insolation and wind speed using simulation environment (MATLAB) only. The real time implementation is missing in this research work. In this research work, to reduce multiple reverse conversions and for improving the efficiency of the micro grid, hybrid PV-Wind with Quasi Z-source inverter has been implemented. Furthermore, SEPIC converter acts as a dc link interface with MPPT functioning. This research work is organized under 3 main sections. Section 1 discusses the micro-grid system with an extensive literature review of MPPT techniques, dc-dc converters with benefits of Z-source inverter. Section 2 presents the complete structure of the hybrid PV-Wind micro-grid system. It explains the PV generator modeling, wind turbine model, MPRVS based MPPT algorithm, design specifications of SEPIC converter, battery model as well as the modes of operations of the Quasi Z-source inverter. Section 3 presents the experimental results which validate the performance of the proposed hybrid PV-wind micro-grid system. The novelty of this research paper is MPRVS based advanced MPPT algorithm have neither been dis-coursed nor been utilized before for the hybrid PV-wind micro-grid with Quasi Z-source inverter experimentally.

2. Hybrid PV-Wind Micro Grid Structure

The proposed structure of the PV-Wind micro grid system is shown in Figure 1. The micro grid system contains a PV generator, a Wind Turbine, a battery system and the power electronic converter topologies. To analyze the proposed system, the equivalent circuit with two diode models for the PV generator has been used because of its better power extraction capability when compared with the single diode model. The rotor of the wind turbine is mechanically tied to a generator to produce electrical power. A wind turbine is a complex system, but a reasonably simple representation is possible by modeling the aerodynamic torque or power based on turbine characteristics (non-dimensional curves of the power coefficient). A battery solution is also necessary to balance the stochastic fluctuations of photovoltaic (PV) power and wind power injected to the grid/load. In this section, a short description about how these main components of the proposed micro grid system have been modeled are presented.

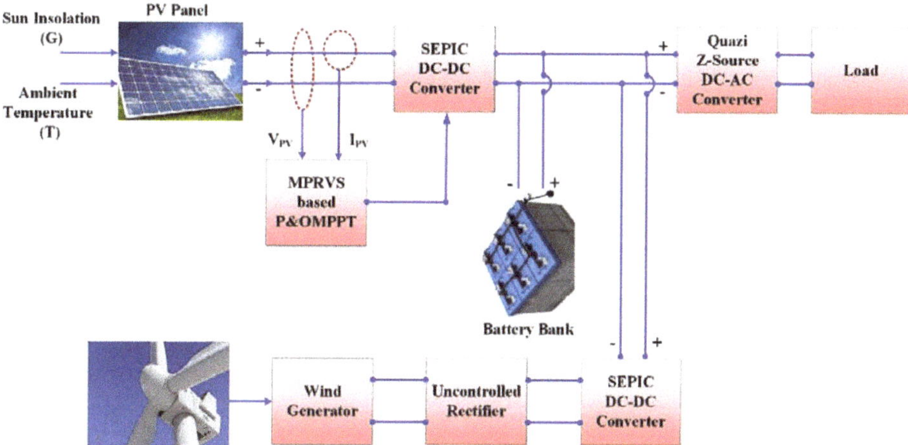

Figure 1. A block diagram with the structure of Hybrid PV-Wind micro grid system.

2.1. PVG Mathematical Model

Figure 2 illustrates the basic PV cell schematic diagram, which is responsible for the transformation of the solar energy into electric power using photoelectric effect which comprises numerous cells. In this paper, the two-diode model is considered to deliver better accuracy compared to the single diode model.

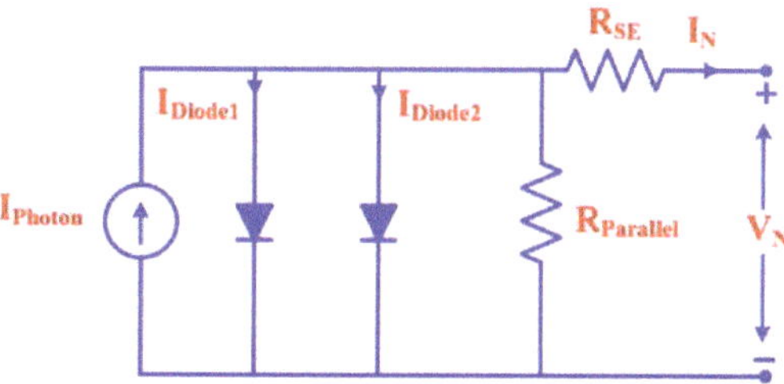

Figure 2. Equivalent circuit model of a PV cell with double diodes and a series and parallel resistance.

The PV cell output current is expressed mathematically as:

$$I_N = I_{Photon} - I_{Diode1} - I_{Diode2} - \left(\frac{V_N + I_N R_{SE}}{R_{Parallel}} \right) \tag{1}$$

Also, Photon current is evaluated mathematically as:

$$I_{Photon} = [I_{Photon\_STC} + K_S(T_C - T_{STC})] \times \frac{G}{G_{STC}} \tag{2}$$

Diode saturation current can be expressed as:

$$I_{Diode1} = I_{Diode2} = \frac{I_{Short\_STC} + K_S(T_C - T_{STC})}{exp^{\frac{[(V_{open\_STC} + K_{VL}(T_C - T_{STC}))]}{V_{Thermal}}} - 1} \tag{3}$$

2.2. Wind Turbine Modeling

A wind turbine is essentially a machine that converts the kinetic energy first into mechanical energy at the turbine shaft, and then into electrical energy. The wind turbine power generation depends mainly on wind velocity in which the rotors are mechanically linked to a generator. A simple model can be achieved by using the power coefficient (CPR) as a function of tip speed ration and the blade pitch angle. CPR (Performance/power coefficient) Vs tip speed ($\lambda_{T.S}$) curve is plotted for different $\beta_{P.B}$ (Pitch blade angle) in Figure 3.

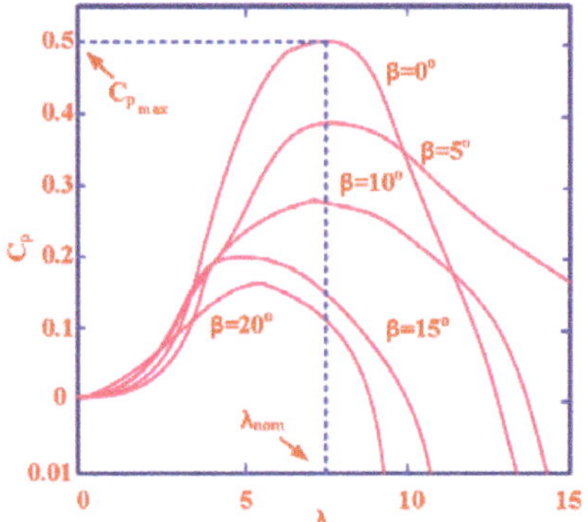

Figure 3. CPR (Performance coefficient) Vs tip speed ($\lambda_{T.S}$) curve is plotted for different $\beta_{P.B}$ (Pitch blade angle).

Generated mechanical power output from the wind turbine can be written using Equation (4) which is depending on wind velocity (V_{Wind}), R_T (Turbine radius) and C_{PR} (Performance coefficient) as:

$$P_{Mechanical} = \frac{1}{2}C_{PR}\pi R_T^2 \rho_{a.d} V_{Wind}^3 \tag{4}$$

Also, the ratio of tip speed ($\lambda_{T.S}$) can be described mathematically which is correlated with an angular velocity of the blade ($\omega_{A.V}$), V_{Wind} and R_T as:

$$\lambda_{T.S} = \frac{\omega_{A.V} \times R_T}{V_{Wind}} \tag{5}$$

And coefficient of performance is expressed with $\lambda_{T.S}$ and $\beta_{P.B}$ (Pitch blade angle) as:

$$C_{PR}(\lambda_{T\,S}, \beta_{P\,B}) = 0.72 \left[\frac{150}{\lambda_j} - 2 \times 10^{-3}\beta_{P.B} - 131 \times 10^{-1} \right] e^{\frac{-185 \times 10^{-1}}{\lambda_j}} \tag{6}$$

where,

$$\frac{1}{\lambda_j} = \frac{1}{(\lambda_{T.S} + 8 \times 10^{-2}\beta_{P.B})} - \frac{35 \times 10^{-3}}{1 + \beta_{P.B}^3} \tag{7}$$

$$\lambda_{T.S} = \frac{\omega_G \times R_T}{V_{Wind} \times \eta_{gear}} \tag{8}$$

$$\eta_{gear} = \frac{\omega_{GM} \times R_T}{\lambda_{T.S} V_{Wind}} \tag{9}$$

2.3. Electric Equivalent Circuit of the Battery Model

A battery is a vital component for a hybrid system which provides the solution under fluctuating action of renewable energy sources. In this work, the electric circuit-based battery model is employed, which provides better dynamics for a state of charge operation mode. It comprises a voltage source (ideal) with a series of internal resistance which evaluates the battery behavior as depicted in Figure 4.

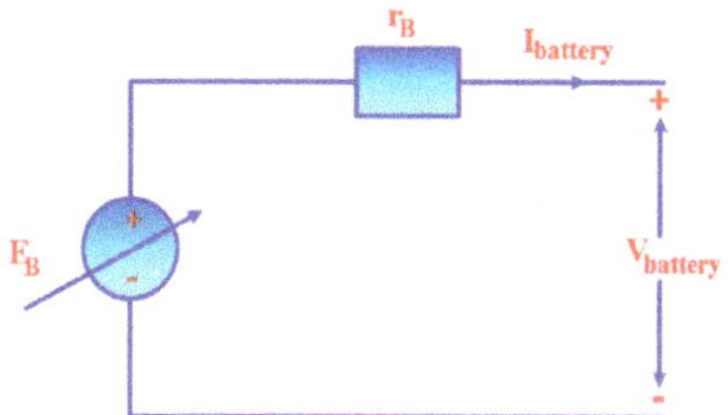

Figure 4. Electric equivalent circuit-based battery model.

Final voltage controlled is obtained mathematically as:

$$V = E_B - \frac{V_{PO} \times Q_{Bat}}{Q_{Bat} - \int I_{Battey}dt} + A_{exp}.e\left(B_{exp}\int I_{Battery}dt\right) \tag{10}$$

3. Power Electronic Converters used to Control the Proposed Micro Grid System. Description and Mathematical Modelling

The power converters have been developed to manage the maximum energy harvesting and power processing for the hybrid solution with Photovoltaic (PV) and wind power generators. The topologies involved in this study contains two topologies of power electronic converters: a SEPIC converter and a Quasi Z-Source Inverter (QZsi). The MPPT method for QZsi is introduced based on the P&O method to minimize the voltage stress on the inverter. Moreover, it prevents overlapping between Shoot-Through (ST) duty ratio and modulation index using DC-Link voltage controller. The output current is regulated using the stationary frame current controller, achieving lower Total Harmonic Distortion (THD) as much as possible. The SEPIC based soft switching for MPPT action is controlled through an advanced MPRVS based P&O MPPT. A quasi Z-source inverter with the common grounding characteristics is employed to get high voltage gain. Employed inverter operates in two modes of operation as the shoot through and the non-shoot through the states.

3.1. SEPIC Converter Model

Single ended primary inductor converter (SEPIC) is considered as an impedance adapter between the PV module and the Z-source inverter as it provides high gain throughout the operation, better voltage performance and high voltage rating for lower/higher power requirements. When boost converter combines with the additional inductor and the capacitor, a SEPIC converter is developed. In contrast with the buck boost converter, the polarity of SEPIC is kept positively as it is depicted in Figure 5. Table 1 portrays the employed SEPIC converter parameters during an implementation.

$$V_{output} = V_{supply} \times \frac{D_{duty}}{1 - D_{duty}} \tag{11}$$

$$L_A = \frac{V_{supply} \times D_{duty}}{\Delta I_{L_A} \times f_{switching}} \tag{12}$$

$$L_B = \frac{V_{supply} \times D_{duty}}{\Delta I_{L_B} \times f_{switching}} \tag{13}$$

$$C_A = \frac{V_{output} \times D_{duty}}{R_{Load} \times \Delta V_o \times f_{switching}} \tag{14}$$

$$C_B = \frac{V_{output} \times D_{duty}}{R_{Load} \times \Delta V_o \times f_{switching}} \tag{15}$$

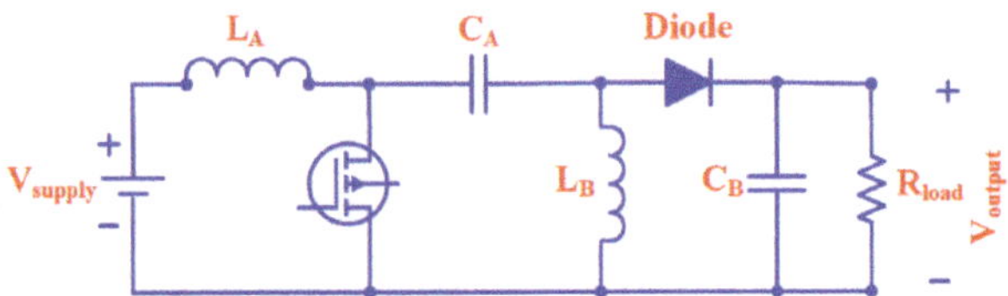

Figure 5. SEPIC converter equivalent circuit.

Table 1. SEPIC converter parameter.

SI. No.	Parameters	Value
1.	Inductors ($L_A = L_B$)	0.42 mH
2.	Capacitors ($C_A = C_B$)	3.5×10^{-3} μF
3.	Current ripple ($\Delta I_{L_A} = \Delta I_{L_B}$)	0.5 A
4.	Voltage ripple (ΔV_o)	1×10^{-3} V
5.	Switching frequency ($f_{switching}$)	20 Hz

3.2. Modified Power Ratio Variable Step Based P&O MPPT

Figure 6 demonstrates the working model of MPRVS based P&O technique for optimal PV power extraction from solar modules. The generation of gating pulses to the SEPIC converter is possible without the action of the PI controller, which makes the reduction of power oscillation nearer to MPP and forces operating point close to the MPP. It also prevents the battery charging system from over voltage. The instantaneous power obtained through PVG [$P_{PV}(N)$] at SEPIC output terminal is calculated as:

$$P_{PV}(N) = V_0(N) \times I_{PV}(N) \tag{16}$$

Also, the previous instantaneous power is mathematically described as:

$$P_{PV}(N-1) = V_{PV}(N-1) \times I_{PV}(N-1) \tag{17}$$

And if

$$\Delta P_{PV}(N) = P_{PV}(N) - P_{PV}(N-1) > 0,\ S = -1 \tag{18}$$

$$\& P_{PV}(N) - P_{PV}(N-1) < 0,\ S = +1 \tag{19}$$

Again,

$$D(N) = D(N-1) + S \times \Delta D \tag{20}$$

ΔD = Step perturbation of duty ratio = $K \times dT$

dT = Fixed step size

K = Variable power ratio

$$K = \frac{P_{PV}^{max} - P_{PV}(N)}{P_{PV}(N)} \tag{21}$$

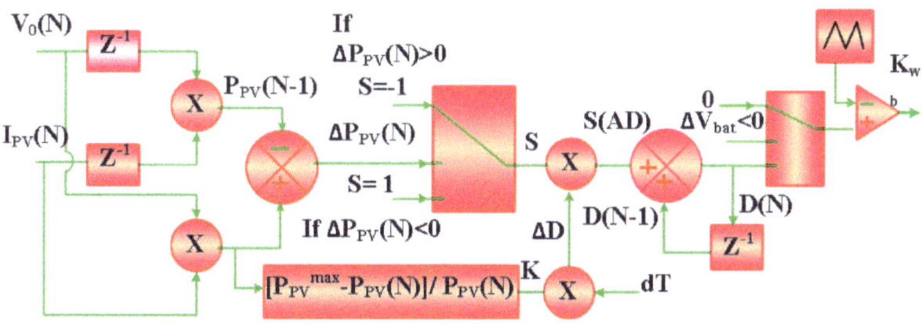

Figure 6. Working model of MPRVS based P&O technique.

3.3. Quasi Z-Source Inverter Mathematical Modeling

Figure 7 presents the equivalent power circuit of Quasi Z-source inverter which comprises of L_A, L_B, C_A, C_B components with impedance circuit. The considered Z-Source Quasi inverter has no filter requirement, better buck/boost characteristics, able to regulate the phase angle output, less size, continuous conducting mode working, less harmonic content, high efficiency and with better power performance over the conventional inverter as major advantages. The Quasi Z-source inverter operates in two modes of operation. In the non-shoot mode, the equivalent circuit has 6 active states with 2 zero states. The T_S is the total switched inverter with T_A and T_B as the shoot through the state and the non-shoot through state, respectively. The duty ratio D_{duty} of SEPIC converter is mathematically written as:

$$D_{duty} = \frac{T_A}{T_S} \tag{22}$$

Mode I: The equivalent model of Quasi Z-source inverter is depicted in Figure 8 and mathematical equations governing non-shoot through the state is expressed as:

$$V_{L_A} = V_{IN} - V_{C_A}$$

$$V_{L_B} = -V_{C_B} \tag{23}$$

$$V_{DIODE} = 0$$

Mode II: Figure 9 illustrates the equivalent model of Quasi Z-source inverter in shoot through the state mode with the mathematical expression as:

$$V_{L_A} = V_{IN} + V_{C_A}$$

$$V_{L_B} = V_{C_B} \tag{24}$$

$$V_{DIODE} = V_{C_A} + V_{C_B}$$

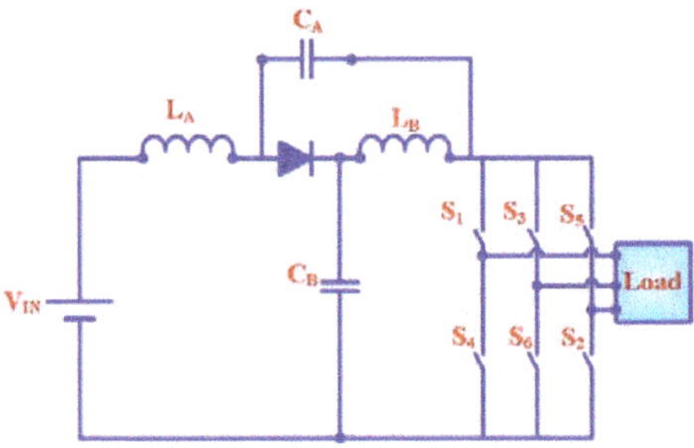

Figure 7. Equivalent power circuit of Quasi Z-source inverter.

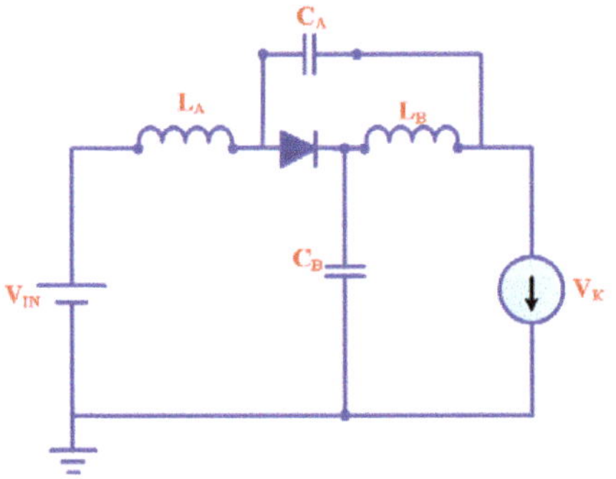

Figure 8. The equivalent model of Quasi Z-source inverter governing non-shoot through the state.

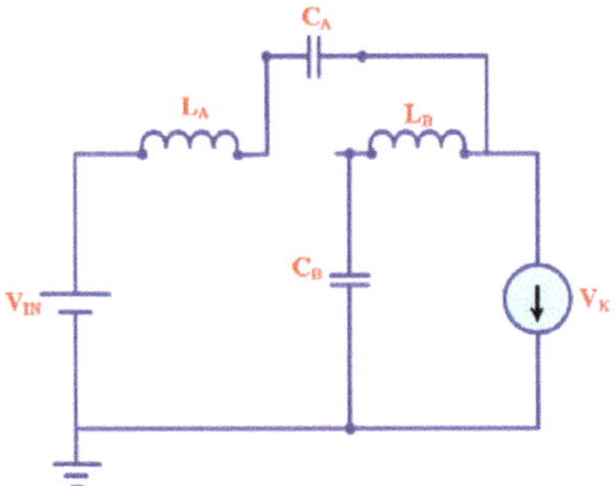

Figure 9. Equivalent model of Quasi Z-source inverter in shoot through the state.

Under the steady condition, the average inductor voltage becomes zero.

$$V_{L_A} = \left[\frac{(V_{IN} + V_{C_B})T_A + (V_{IN} - V_{C_A})T_B}{T_S} \right] = 0 \tag{25}$$

$$V_{L_B} = \left[\frac{V_{C_A}T_A + (-V_{C_B})T_B}{T_S} \right] = 0 \tag{26}$$

On solving the above equations, capacitor voltage (V_{C_A} & V_{C_B}) is calculated mathematically as:

$$V_{C_A} = \left(\frac{T_B}{T_B - T_A}\right) \times V_{IN} \tag{27}$$

$$V_{C_B} = \left(\frac{T_A}{T_B - T_A}\right) \times V_{IN} \tag{28}$$

$$\text{Maximum voltage across DC-link} = V_{C_A} + V_{C_B} \tag{29}$$

Putting Equations (26) and (27) in (28) we get

$$\text{Maximum DC-link voltage} = \left|\frac{1}{1 - 2\frac{T_A}{T_S}}\right| V_{IN} = K \times V_{IN} \tag{30}$$

4. Experimental Setup Description and Results

4.1. Description of the Experimental Setup

The considered hybrid PV-Wind micro grid is tested using MPRVS based P&O MPPT with employed Z-source inverter. Figure 10 depicts the developed practical structure of the proposed hybrid micro grid based on a real-time platform, dSPACE. The SEPIC converter is controlled through the MPRVS based P&O based MPPT, in which LV-25P and LA-25P, current and voltage sensors are employed for measuring the PV panel parameters, V_{PV} and I_{PV} respectively. The power factor coefficient and THD are evaluated using the power quality analyzer (FLUKE 43B), considering the main components of the converter: IGBT (IRG4PH50U), diode (Freewheel RHRG30120), driver circuit (HCPL 3120) etc. permanent magnet synchronous generator (PMSG) based wind emulator system is employed as the wind turbine generator and is mechanically coupled with the DC-motor. The switched mode power converter makes the wind turbine to have varying wind speed which produces the required mechanical torque by controlling wind turbine characteristics.

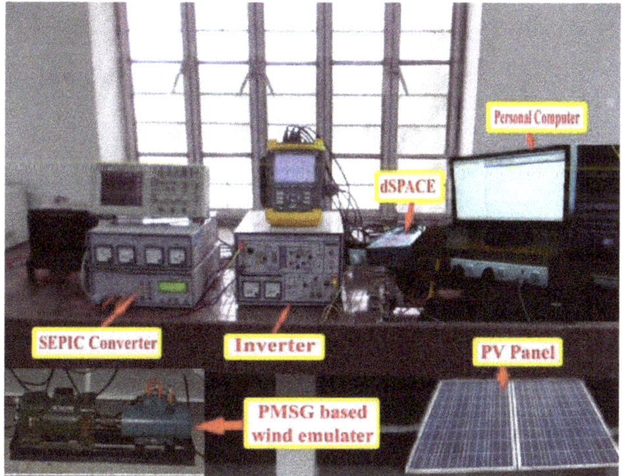

Figure 10. Developed experimental setup of the proposed hybrid micro grid system based on a real-time digital simulator-dSPACE platform.

4.2. Experimental Results and Scenarious Development

The accuracy of the proposed MPRVS based P&O MPPT has been tested with changing wind operating condition depicted in Figure 11a. The employed controller works in MPP area and provides optimal tracking of wind power under the sudden changes of wind velocity shown in Figure 11b.The corresponding duty ratio of SEPIC converter is shown in Figure 11c. Furthermore, the capability of proposed MPPT tracker is examined under the first scenarios with step varying solar irradiation. Figure 12 demonstrates that the PV array has obtained parameters under the step-changes in solar irradiation and the propped system has proved high accuracy and effective PV tracking in MPP region. The obtained experimental results in Figure 13a illustrate that the performance of the proposed hybrid micro grid under the second scenarios by varying wind velocity and constant solar irradiation. Also, Figure 13b demonstrates the behavior responses of the hybrid micro grid under varying solar irradiance and constant wind velocity with MPRVS based P&O MPPT employed. The performance of the hybrid micro grid is also tested under the third proposed scenarios in the absence of wind velocity and during this operation: the load is connected/disconnected to the utility grid, which is shown in Figure 14a under the load cutting condition, and in Figure 14b, under the load removing conditions. The performance of the wind generator is evaluated under disconnecting/reconnecting operating conditions to the micro grid, which are depicted in Figures 15 and 16 and reveal that the accurate performance of the proposed hybrid micro grid in varying operating situations (disconnecting operating conditions to the micro grid and reconnecting operating conditions to the micro grid), respectively.

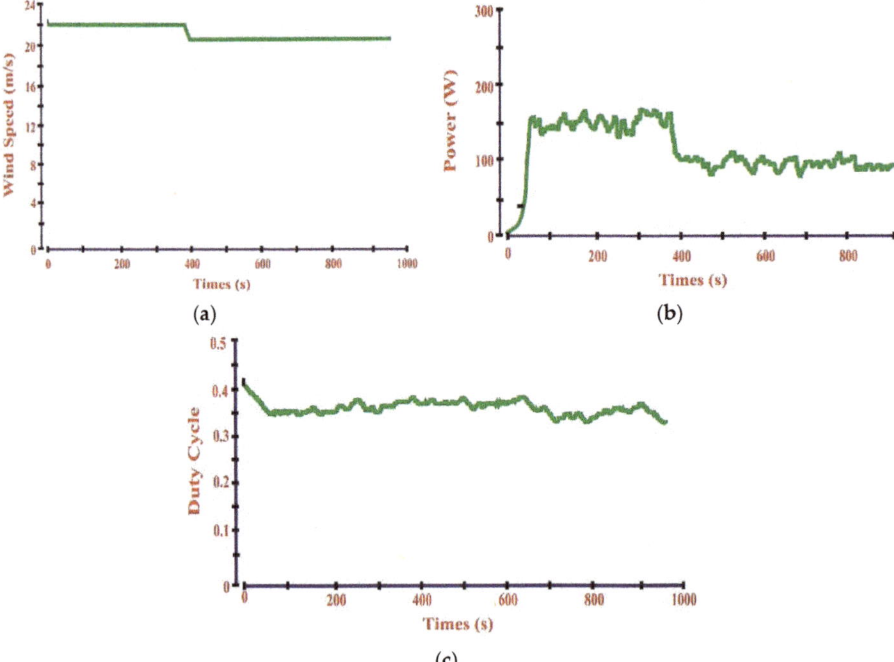

Figure 11. Experimental results (**a**) during a step-changed in wind speed; (**b**) wind power; and (**c**) Duty cycle of Cuk converter.

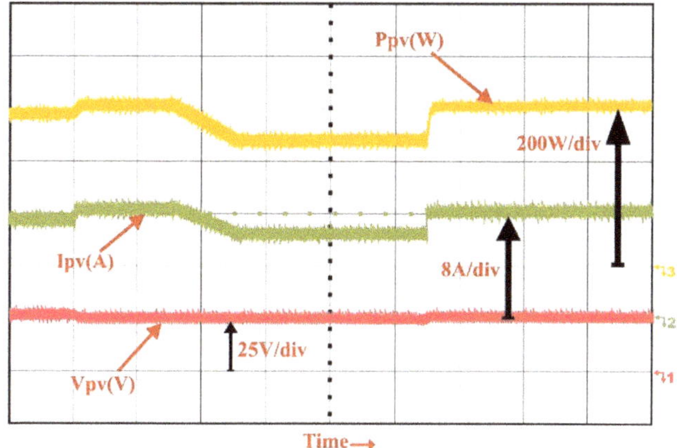

Figure 12. PV system responses under step-changes in solar irradiation.

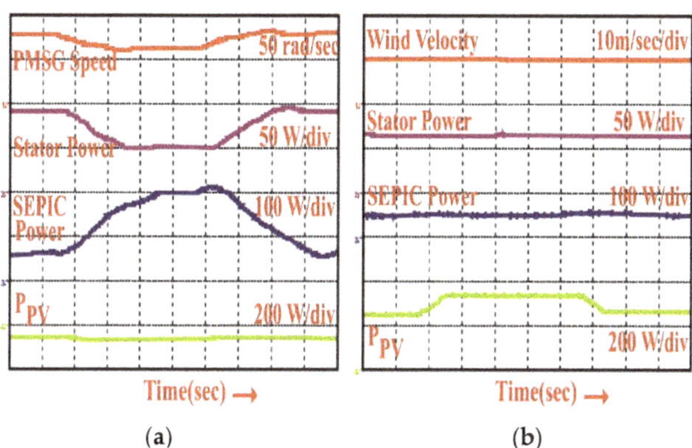

Figure 13. (**a**) Capability of the proposed hybrid micro grid under varying wind velocity and constant solar irradiation; (**b**) Behavior responses of the hybrid micro grid under varying solar irradiance and constant wind velocity.

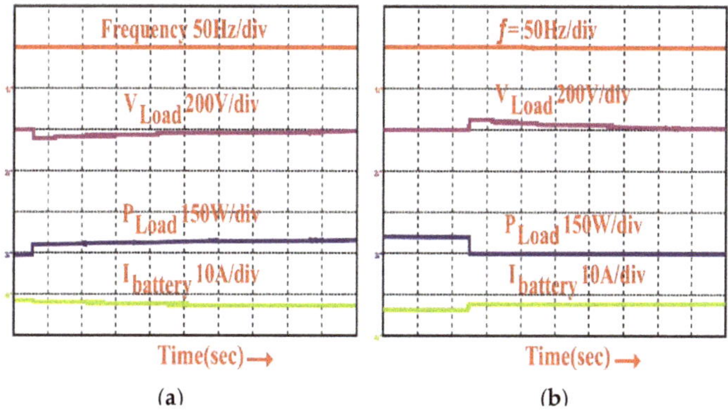

Figure 14. The performance of the hybrid micro grid (**a**) load cutting condition; (**b**) Load removing condition.

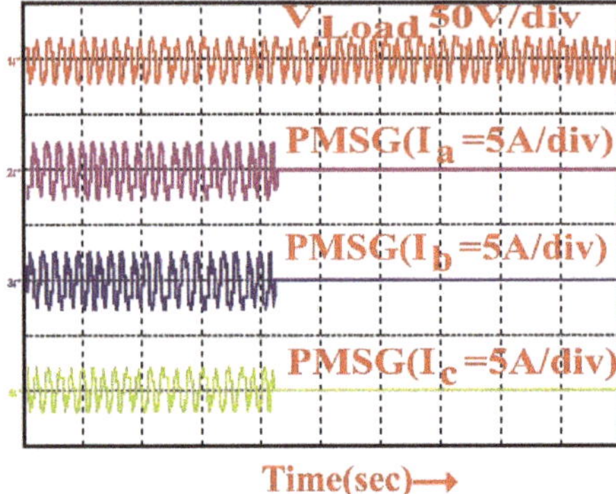

Figure 15. The performance of the wind generator is evaluated under disconnecting operating conditions to the micro grid.

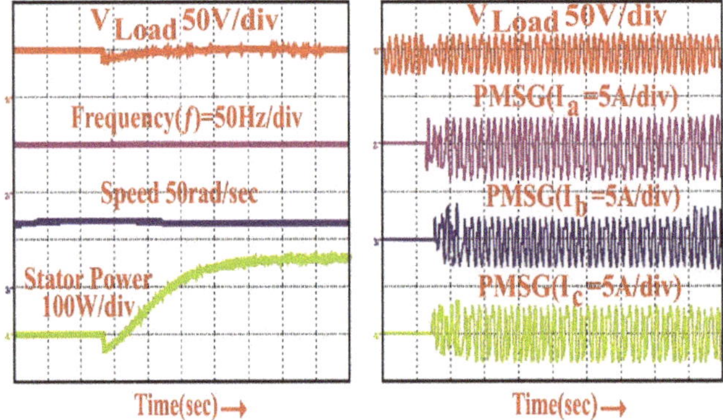

Figure 16. The performance of the wind generator is evaluated under reconnecting operating conditions to the micro grid.

5. Conclusions

The proposed hybrid PV-Wind micro-grid system using Quasi Z-source inverter is established practically and tested with the Real-time digital simulator dSPACE (DS 1104) platform.

The point wise findings that have been included in this section are as follows:

(i) The MPRVS based P&O MPPT performance with SEPIC converter has been validated effectively, which delivers MPP achievement with low power oscillation for the PV system.

(ii) The performance of the Quasi Z-source inverter has been evaluated experimentally as having better buck/boost characteristics with fast dc-link voltage regulation under different operating conditions.

(iii) The proposed QZsi topology for a hybrid PV-Wind Turbine application in a micro grid enhanced reliability, good output power quality and efficiency improvements.

(iv) Experimental results under dynamic conditions, such as step-changed in wind speed or solar irradiation, reveal that optimal power has been tracked through the PV-Wind renewable sources and proved the validity of the proposed solution.

(v) The two-diode model-based PV Generator provides high power extraction when compared to the single diode model.

As a future work, the paper can be extended by using the multilevel inverter with the application of advanced intelligent MPPT algorithms viz. Jaya DE, hybrid ANFIS-ABC methods.

Author Contributions: All authors contributed equally and formulated the research work to present in current version as full research article.

Funding: No funding addressed to this research activities.

Conflicts of Interest: The authors declare no conflict of interest.

Nomenclature

R_{SE}	Resistance in series
$R_{Parallel}$	Resistance in parallel
T_{STC}	Temperature at STC (Standard Test Condition)
G_{STC}	Solar irradiance at STC
K_S	Coefficient of short circuit current
I_{Photon_STC}	Photo current at STC
T_C	Ambient temperature
G	Solar irradiation
I_{Short_STC}	Short circuit current at STC
V_{open_STC}	Open circuit current at STC
$V_{Thermal}$	Diode thermal voltage
K_{VL}	Voltage temperature coefficient
$\rho_{a,d}$	Air density $\rho_{a,d}$
ω_G	Speed of generator
ω_{GM}	Peak allowed generator speed
η_{gear}	Gear ratio
E_B	Battery fixed voltage
V_{PO}	Polarized voltage
Q_{Bat}	Capacity of battery
$I_{Battery}$	Battery current
A_{exp}	Amplitude of exponential zone
B_{exp}	Inverse time constant exponential zone
$\Delta I_{L_A} = I_{L_B}$	Current ripple
ΔV_0	Ripple voltage
$f_{switching}$	Switched frequency
$V_0(N) \& I_{PV}(N)$	Sensed voltage and current
ΔD	Step perturbation of duty ratio
dT	Fixed step size
K	Variable power ratio
PI	Proportional Integral

References

1. Tiwari, S.K.; Singh, B.; Goel, P.K. Design and Control of Micro-Grid fed by Renewable Energy Generating Sources. *IEEE Trans. Ind. Appl.* **2018**, *54*, 2041–2050. [CrossRef]
2. Mi, Y.; Zhang, H.; Tian, Y.; Yang, Y.; Li, J.; Li, J.; Wang, L. Control and Operation of Hybrid Solar/Wind Isolated DC Micro grid. In Proceedings of the IEEE Conference and Expo Transportation Electrification Asia-Pacific (ITEC Asia-Pacific), Beijing, China, 31 August–3 September 2014; pp. 1–5.
3. Shastry, A.; Suresh, K.V.; Vinayaka, K.U. Hybrid Wind-Solar Systems using Cuk-Sepic Fused Converter with Quasi-Z-Source Inverter. In Proceedings of the IEEE Power, Communication and Information Technology Conference (PCITC), Bhubaneswar, India, 15–17 October 2015; pp. 856–861.
4. Kassem, A.M.; Zaid, S.A. Optimal Control of a Hybrid Renewable Wind/Fuel Cell Energy in Micro Grid Application. In Proceedings of the Nineteenth International Middle East Power Systems Conference (MEPCON), Shibin El Kom, Egypt, 19–21 December 2017; pp. 84–90.

5.   Tiwari, S.K.; Singh, B.; Goel, P.K. Design and Control of Micro-Grid fed by Renewable Energy Generating Sources. In Proceedings of the IEEE 6th International Conference on Power Systems (ICPS), New Delhi, India, 4–6 March 2016; pp. 1–6.

6.   Ahmed, J.; Salam, Z. An Enhanced Adaptive P&O MPPT for Fast and Efficient Tracking under Varying Environmental Conditions. *IEEE Trans. Sustain. Energy* **2018**, *9*, 1487–1496.

7.   Vavilapalli, V.; Umashankar, S.; Sanjeevikumar, P.; Ramachandramurthy, V.K. Design and Real-Time Simulation of an AC Voltage Regulator based Battery Charger for Large-Scale PV-Grid Energy Storage Systems. *IEEE Access* **2017**, *5*, 25158–25170. [CrossRef]

8.   Hussain, S.; Alammari, R.; Jafarullah, M.; Iqbal, A.; Sanjeevikumar, P. Optimization of Hybrid Renewable Energy System Using Iterative Filter Selection Approach. *IET Renew. Power Gen.* **2017**, *11*, 1440–1445. [CrossRef]

9.   Dramohan, K.; Padmanaban, S.; Kalyanasundaram, R.; Bhaskar, M.S.; Mihet-Popa, L. Grid Synchronization of a Seven-Phase Wind Electric Generator Using d-q PLL. *Energies* **2017**, *10*, 926. [CrossRef]

10.   Szeidert, I.; Prostean, O.; Filip, I.; Vasar, C.; Mihet-Popa, L. Issues regarding the modeling and simulation of wind energy conversion system's components. In Proceedings of the International Conference on Automation, Quality & Testing, Robotics (AQTR 2008), Cluj-Napoca, Romania, 22–25 May 2008; pp. 225–228.

11.   Mihet-Popa, L.; Bindner, H. Simulation models developed for voltage control in a distribution network using energy storage systems for PV penetration. In Proceedings of the 39th Annual Conference of the IEEE Industrial Electronics Society, Vienna, Austria, 10–13 November 2013; pp. 7487–7492.

12.   Maheswaran, G.; Hidayathullah, M.; Ismail, B.C.; Mihet-Popa, L.; Sanjeevikumar, P. Energy Management Strategy for Rural Communities' DC Micro Grid power system structure with Maximum Penetration of Renewable Energy Sources. *Appl. Sci.* **2018**, *11*, 585. [CrossRef]

13.   Priyadarshi, N.; Kumar, V.; Yadav, K.; Vardia, M. An Experimental Study on Zeta buck-boost converter for Application in PV system. In *Handbook of Distributed Generation*; Springer: Cham, Switzerland, 2017; pp. 393–406.

14.   Priyadarshi, N.; Anand, A.; Sharma, A.K.; Azam, F.; Singh, V.K.; Sinha, R.K. An Experimental Implementation and Testing of GA based Maximum Power Point Tracking for PV System under Varying Ambient Conditions Using dSPACE DS 1104 Controller. *Int. J. Renew. Energy Res.* **2017**, *7*, 255–265.

15.   Kumar, N.; Hussain, I.; Singh, B.; Panigrahi, B.K. Framework of Maximum Power Extraction from Solar PV Panel using Self Predictive Perturb and Observe Algorithm. *IEEE Trans. Sustain. Energy* **2017**, *9*, 895–903. [CrossRef]

16.   Elgendy, M.A.; Zahawi, B.; Atkinson, D.J. Assessment of the Incremental Conductance Maximum Power Point Tracking Algorithm. *IEEE Trans. Sustain. Energy* **2013**, *4*, 108–117. [CrossRef]

17.   Zamora, A.C.; Vazquez, G.; Sosa, J.M.; Rodriguez, P.R.M.; Juarez, M.A. Efficiency Based Comparative Analysis of Selected Classical MPPT Methods. In Proceedings of the IEEE International Autumn Meeting on Power, Electronics and Computing, Ixtapa, Mexico, 8–10 November 2017; pp. 1–6.

18.   Abu-Rub, H.; Iqbal, A.; Ahmed, S.K.M.; Peng, F.Z.; Li, Y.; Baoming, G. Quasi-Z-Source Inverter-Based Photovoltaic Generation System with Maximum Power Tracking Control Using ANFIS. *IEEE Trans. Sustain. Energy* **2013**, *4*, 11–20. [CrossRef]

19.   Mohamed, A.A.S.; Berzoy, A.; Mohammed, O. Design and Hardware Implementation of FL-MPPT Control of PV Systems Based on GA and Small-Signal Analysis. *IEEE Trans. Sustain. Energy* **2017**, *8*, 279–290. [CrossRef]

20.   Wang, L.; Singh, C. Population-Based Intelligent Search in Reliability Evaluation of Generation Systems with Wind Power Penetration. *IEEE Trans. Power Syst.* **2008**, *23*, 1336–1345. [CrossRef]

21.   Koad, R.B.A.; Zobaa, A.F.; El-Shahat, A. A Novel MPPT Algorithm Based on Particle Swarm Optimisation for Photovoltaic Systems. *IEEE Trans. Sustain. Energy* **2017**, *8*, 468–476. [CrossRef]

22.   Priyadarshi, N.; Sharma, A.K.; Azam, F. A Hybrid Firefly-Asymmetrical Fuzzy Logic Controller based MPPT for PV-Wind-Fuel Grid Integration. *Int. J. Renew. Energy Res.* **2017**, *7*, 1546–1560.

23.   Sundareswaran, K.; Sankar, P.; Nayak, P.S.R.; Simon, S.P.; Palani, S. Enhanced Energy Output from a PV System Under Partial Shaded Conditions Through Artificial Bee Colony. *IEEE Trans. Sustain. Energy* **2015**, *6*, 18–209. [CrossRef]

24.   Kalaam, R.N.; Muyeen, S.M.; Al-Durra, A.; Hasanien, H.N.; Al-Wahedi, K. Optimisation of controller parameters for grid tied photovoltaic system at faulty network using artificial neural network-based cuckoo search algorithm. *IET Renew. Power Gen.* **2017**, *11*, 1517–1526. [CrossRef]

25. Priyadarshi, N.; Padmanaban, S.; Mihet-Popa, L.; Blaabjerg, F.; Azam, F. Maximum Power Point Tracking for Brushless DC Motor-Driven Photovoltaic Pumping Systems Using a Hybrid ANFIS-FLOWER Pollination Optimization Algorithm. *Energies* **2018**, *11*, 1067. [CrossRef]
26. Priyadarshi, N.; Padmanaban, S.; Maroti, P.K.; Sharma, A. An Extensive Practical Investigation of FPSO Based MPPT for Grid Integrated PV System Under Variable Operating Conditions with Anti-Islanding Protection. *IEEE Syst. J.* **2018**, *PP*, 1–11. [CrossRef]
27. Priyadarshi, N.; Padmanaban, S.; Bhaskar, M.S.; Blaabjerg, F.; Sharma, A. A Fuzzy SVPWM Based Inverter Control Realization of Grid Integrated PV-Wind System with FPSO MPPT Algorithm for a Grid-Connected PV/Wind Power Generation System: Hardware Implementation. *IET Electr. Power Appl.* **2018**, *12*, 962–971. [CrossRef]
28. Alanisamy, R.; Mutawakkil, A.U.; Selvakumar, K.; Karthikeyan, D. Modelling and simulation of z source inverter based grid connected PV system. In Proceedings of the 2014 IEEE International Conference on Computational Intelligence and Computing Research, Coimbatore, India, 18–20 December 2014; pp. 1–4.
29. Kumaran, B.A.; Sekhar, C.S.A.; Bala, V. PV Powered Quasi Z-Source Inverter for Agricultural Water Pumping System. In Proceedings of the Third International Conference on Science Technology Engineering & Management (ICONSTEM), Chennai, India, 23–24 March 2017; pp. 544–549.
30. Haji-Esmaeili, M.M.; Babaei, E.; Sabahi, M. High Step-Up Quasi-Z Source DC-DC Converter. *IEEE Trans. Power Electron.* **2018**, 1. [CrossRef]
31. Zhou, Y.; Liu, L.; Li, H. A High-Performance Photovoltaic Module-Integrated Converter (MIC) Based on Cascaded Quasi-Z-Source Inverters (qZSI) Using eGaN FETs. *IEEE Trans. Power Electron.* **2013**, *28*, 2727–2738. [CrossRef]
32. Kayiranga, T.; Li, H.; Lin, X.; Shi, Y.; Li, H. Abnormal Operation State Analysis and Control of Asymmetric Impedance Network-Based Quasi-Z-Source PV Inverter (AIN-qZSI). *IEEE Trans. Power Electron.* **2016**, *31*, 7642–7650. [CrossRef]
33. AsSakka, A.O.; Hassan, M.A.M.; Senjyn, T. Small signal modeling and control of PV based QZSI for grid connected applications. In Proceedings of the 2017 International Conference on Modern Electrical and Energy Systems (MEES), Kremenchuk, Ukraine, 15–17 November 2017. [CrossRef]
34. Tey, K.S.; Mekhilef, S.; Seyedmahmoudian, M.; Horan, B.; Oo, A.T.; Stojcevski, A. Improved Differential Evolution-based MPPT Algorithm using SEPIC for PV Systems under Partial Shading Conditions and Load Variation. *IEEE Trans. Ind. Inform.* **2018**, 1. [CrossRef]
35. Jain, S.; Shadmand, M.B.; Balog, R.S. Decoupled Active and Reactive PowerPredictive Control for PV Applications using a Grid-tied Quasi-Z-Source Inverter. *IEEE J. Emerg. Sel. Top. Power Electron.* **2018**, 1. [CrossRef]
36. Liu, Y.; Ge, B.; Abu-Rub, H. Modelling and controller design of quasi-Z-sourcecascaded multilevel inverter-based three-phase grid-tie photovoltaic power system. *IET Renew. Power Gen.* **2014**, *8*, 925–936. [CrossRef]
37. Amini, M.H.; Broojeni, K.G.; Dragičević, T.; Nejadpak, A.; Iyengar, S.S.; Blaabjerg, F. Application of Cloud Computing in Power Routing for Clusters of MicrogridsUsing Oblivious Network Routing Algorithm. In Proceedings of the 19th European Conference on Power Electronics and Applications (EPE'17 ECCE Europe), Warsaw, Poland, 11–14 September 2017; pp. P.1–P.11.
38. Mohammadi, A.; Dehghani, M.J.; Ghazizadeh, E. Game Theoretic Spectrum Allocation in Femtocell Networks for Smart Electric Distribution Grids. *Energies* **2018**, *11*, 1635. [CrossRef]
39. Vigneysh, T.; Kumarappan, N. Operation and Control of Hybrid Microgrid Using ZSource Converter in Grid Tied Mode. In Proceedings of the 2nd International Conference on Applied and Theoretical Computing and Communication Technology (iCATccT), Bangalore, India, 21–23 July 2016; pp. 318–323.

Article

An Overview of Energy Scenarios, Storage Systems and the Infrastructure for Vehicle-to-Grid Technology

Tohid Harighi [1], Ramazan Bayindir [1], Sanjeevikumar Padmanaban [2,*], Lucian Mihet-Popa [3,*] and Eklas Hossain [4]

1   Department of Electrical and Electronics Engineering, Gazi University, Ankara 06500, Turkey; tohidharighi@gmail.com (T.H.); bayindir@gazi.edu.tr (R.B.)
2   Department of Energy Technology, Aalborg University, 6700 Esbjerg, Denmark
3   Norway Faculty of Engineering, Østfold University College, Kobberslagerstredet 5, 1671 Kråkeroy-Fredrikstad, Norway
4   Department of Electrical Engineering & Renewable Energy, Oregon Tech, Klamath Falls, OR 97601, USA; eklas.hossain@oit.edu
*   Correspondence: san@et.aau.dk (S.P.); lucian.mihet@hiof.no (L.M.-P.); Tel.: +45-716-820-84 (S.P.); +47-922-713-53 (L.M.-P.)

Received: 21 June 2018; Accepted: 13 August 2018; Published: 20 August 2018

Abstract: The increase in the emission of greenhouse gases (GHG) is one of the most important problems in the world. Decreasing GHG emissions will be a big challenge in the future. The transportation sector uses a significant part of petroleum production in the world, and this leads to an increase in the emission of GHG. The result of this issue is that the population of the world befouls the environment by the transportation system automatically. Electric Vehicles (EV) have the potential to solve a big part of GHG emission and energy efficiency issues such as the stability and reliability of energy. Therefore, the EV and grid relation is limited to the Vehicle-to-Grid (V2G) or Grid-to-Vehicle (G2V) function. Consequently, the grid has temporary energy storage in EVs' batteries and electricity in exchange for fossil energy in vehicles. The energy actors and their research teams have determined some targets for 2050; hence, they hope to decrease the world temperature by 6 °C, or at least by 2 °C in the normal condition. Fulfilment of these scenarios requires suitable grid infrastructure, but in most countries, the grid does not have a suitable background to apply in those scenarios. In this paper, some problems regarding energy scenarios, energy storage systems, grid infrastructure and communication systems in the supply and demand side of the grid are reviewed.

Keywords: vehicle-to-grid; grid-to-vehicle; electric vehicles; batteries; harmonic distortion; IEEE Bus standards

---

1. Introduction

The world population is growing rapidly, so the outcome is greenhouse gas (GHG) emission and energy consumption increase year by year. There is not a traditional fuel for transportation systems at hand that is both clear and efficient (mostly fossil fuels), while on the other hand, electric fleet systems can work with lower GHG emissions and energy losses. Therefore, changing fuels seems the best idea to get the best result here. To make this happen, in the first step, the electric grids that are used must be smart (as is mostly the case in North America, Europe and pacific Asia) [1–3]. The Electric Vehicle (EV) is one of the electric transportation technologies; therefore, EV and the smart grid have been integrated to execute our plan. EVs should connect to the smart grid in the form of Vehicle-to-Grid (V2G) or Grid-to-Vehicle (G2V). In V2G technology, the EV and grid share energy from the vehicle to the grid, and vice versa in G2V. Hence, it could be said that EVs are the subdivision of electric fleet systems and grids. The result of this integration is to have critical specific features such as having high storage and low GHG emissions. According to some existing options, technologies and disadvantages, all the energy arena actors have developed energy strategies. They have planned to change the transportation system to an electric fleet system to meet the targets made by IEA 2030 and 2050. The Paris agreement (UNFCCC, 2015a) has contributed to decrease GHG emissions in the world, so this agreement declares that countries should come up with plans to decrease the global average temperature up to 2 °C. The IEA 2DS trajectory sets the goal, which is reducing GHG emissions from 33 GtCO$_2$ approximately to 15 GtCO$_2$ in 2050, which is roughly 45% of the CO$_2$ that was emitted in 2013. Besides, in the 6DS IEA trajectory, GHG emission is approximately 55 GtCO2. Some predictions of IEA and UNFCCC until 2030 and 2050 on the EV plan are as follows [1]:

- About one billion electric vehicles, comprising above 40% of the total LDV stock, which is trajectory 2DS.
- More than 400 million electric two-wheeler vehicle will be produced in 2030.
- All of the cars will be electric two-wheeler vehicles by 2050.
- The EVI members are comprised of 16 governments today.
- Between 2014 and 2015, new enrolment of EV (BEV, BEV) increased by 70% (more than 550 K sold worldwide)
- The annual sale list of 2015 in comparison to 2014, increased more than 75% EVs in these countries: France, Germany, Korea, Norway, Sweden, The UK and India.
- The cost of the PHEV batteries decreased from USD 1000/kWh in 2008 to USD 268/kWh in 2015, and the target for 2022 is USD 125/kWh.
- The density of the PHEV batteries increased from 60 Wh/L in 2008 to 295 Wh/L in 2015, and the target for 2022 is 400 Wh/L.

Some countries have taken actions to reach the IEA 2030 and 2050 targets. For example, Ireland created a roadmap from 2011–2050. In this plan, 800K tons of oil and 4 million tons of CO$_2$ emissions will be reduced by 2050 per annum [4]. Renewable energies such as wind and solar are very important because they produce energy with zero GHG emissions, and the grids supplied by both energy types are more flexible than grids supplied by only fossil energy [5–7]. All types of renewable energies are only available in special situations because they are not stable energy sources. However, all of them can be constructed at any size and everywhere. V2G, the smart grid and micro grid are supplementary to each other and also can expand one another. All of the renewable energy conversion procedures include AC to DC or the inverse. The DC type of energy supplies batteries, and the EVs' energy storage system plays a big role in the grid when it needs energy exchange. EVs can save energy when the demand side of the grid strongly decreases, for example a decrease between 23:00 and 05:00. Hence, reducing the energy level immediately in power plants (in every condition) is not economical. Fossil ICE efficiency in the best condition and with the latest technology is 18~20%. However, the fossil power plant efficiency is 38~40%, and CHP energy efficiency is 60~75% [8]. This clearly shows that EVs' benefit is not limited only to having zero GHG emissions. Thus, the EVs are supplied with electric energy, which is generated as 38~75%, which depends on the energy generating condition, and all mentioned benefits (EVs are supplied by high efficiency energy and reduce GHG emissions) can be improved by V2G technology or similar technologies. This is based on a bidirectional energy transmission system, and some targets are easily reachable, such as peak reduction, stability and reliability of energy. EVs' batteries, the quantity of the EVs, the time of charge and the power electronic systems' topologies are forced to use an energy storage system in charging stations, and this is sensible in grids that are used in the DC line. In this situation, energy is conveyed to the grid in critical situations such as peaks and down time by the stable energy storage in charging stations and the temporary energy storage in EVs [9–14].

This paper covers discussions about energy scenarios, storage systems and the infrastructure of the grid related to V2G technology. The scenarios, policies and targets of the governments and agencies of the world for lower GHG emissions and high-energy efficiency have been suggested. They change the grid target and duty, but the collection of infrastructures cannot respond to the requirements mentioned in the scenarios. According to the scenarios (created for grid upgrades), storage systems (such as batteries and chargers) and the infrastructures of the grid should be replaced by the latest technologies in each section.

2. Energy Scenarios

Each energy plan has some scenarios. Energy efficiency and GHG emission are targets for energy efficiency and GHG emission for each energy scenario in the world, and scenarios for V2G technology depend on their own location, so different requirements and energy stocks exist all over the world. The road maps created by IEA give the result for general scenarios for 2050. Table 1 gives the number of EVs and PHEVs that will be sold in 2050. It is clear that EVs have lower GHG emissions than PHEVs. According to the 2050 road map, the North American, European and Pacific countries will have less PHEVs than China and India. This means that their fossil energy consumption in the transportation sector will be lower than India and China.

Table 1. Electric vehicles that will be sold in 2050 according to the IEA scenario.

Location	EV	PHEV
North America	8800 K	3800 K
Europe	6400 K	3100 K
China	9400 K	11,400 K
India	8600 K	9600 K
Pacific	2400 K	1300 K

For example, China's government has tried to reduce GHG emission more than other countries. However, the mentioned countries updated their grid from the cyber security, smart and micro grid side of the grid [15]. Energy efficiency is possible, if only the mentioned parameters are used together. Today, energy providers provide energy in different forms such as oil, gas (LPG, CNG and LNG) or electricity. The energy usage method is very important, and it must be observed in all parts of the energy consumption. Energy use percentages, in some places, could be as follows: electricity 30%, heat 40% and transport 30%. In addition, they can be supplied with renewable energy by the CEESA 2050 (Research project Coherent Energy and Environmental System Analysis (CEESA) financed by the Danish Council for Strategic Research.) scenario [16]. Control of supply and demand is the primary parameter of energy quality, so reduction of the GHG emissions and the efficiency of energy are possible by assuming supply and demand accurately [17–19]. The control of the mentioned items results in facilitates integrating the renewable energy network with the grid and EVs, but it requires data from metrology, the supply and demand of the grid, the infrastructure of the grid and some additional algorithms to provide good and predictable data [20–22]. V2G technology should be economical, so all of the mentioned items should have a reasonable relation between technical benefits and the economic condition of the investor. The performance of the V2G technology depends on the grid infrastructure and the type of electric vehicle storage system. Their quality and size directly influence the performance of the grid [23,24]. Hence, not all of them are completely controllable, but they are predictable. Predictable means that the management team can modify the demand time schedule [25–27]. Here, some scenarios are reviewed one by one, so all of them are issued for a special location and condition, also any changes in the main scenario can generate different results. According to the scenario (can be different in each location), the design of the grid have to be more developable, flexible and manageable than what grids use currently, so grids after design, will be more complex and comprehensive. Scenarios of the grid have some algorithms to use in normal and critical situations and all sections of the grid have a connection with another section of the grid, which is very important to control the grid and one of the grid designing protocols [28–30].

3. Storage Systems in V2G Technology

3.1. Batteries Use on the Electric Vehicles

Energy storage types are different in each situation. For example, the hydro power plant uses the pumping of water to save energy. The type of energy storage is categorized by some environmental and advanced parameters, such as GDR or technical aspects, and both must consider each other [31,32]. The type depends on the rating of power, charge and discharge, the density of power and energy, response time, efficiency, self-discharge and lifetime. EVs batteries must be solid and have the mentioned parameters [33]. Today, four types of batteries are better used in electric vehicles, such as Li-ion, NiCd, NaS and ZnBr. However, the specifics are different for each material. The properties of EVs' batteries are given in Table 2 [34]. The cost of each type of battery is different, but the technical side of this issue clearly shows that Li-ion batteries are the best choice for EVs in every situation. Li-ion batteries with some alloy such as Fe and Mn give the best performance, so they boost the battery capacity and safety. In the latest measurements, EVs can travel between 250 and 350 miles with Li-ion batteries. Generally, $LiFePO_4$, $LiCoO_2$ and $LiMn_2O_4$ types of batteries are used in EVs. All of them are produced as anode and cathode types [31,34,35]. Table 3 illustrates the type of Li battery and its specifications.

Table 2. Type of batteries and their specifications [34].

Battery Parameters	NiCd	NaS	ZnBr	Li-ion
Power rating (MW)	0~40	0.05~8	0.05~2	0~0.1
Discharge time	S~h	S~h	S~10 h	Min~h
Power density (W/I)	75~700	120~160	1~25	1300~10,000
Energy density (Wh/I)	15~8	15~300	65	200~400
Response time	<S	<S	S	<S
Efficiency (%)	60~80	70~85	65~75	65~75
Lifetime in years	5~20	10~15	5~10	5~100
Lifetime in cycles	1500~3000	2500~4500	1000~3650	600~1200
Cost $ (kW)	500~1500	1000~3000	700~2500	1200~4000
Cost $ (kW/h)	800~1500	300~500	150~1000	600~2500

Table 3. Some type of Li batteries and their specifications [35].

Type of Li-ion	Practical Energy Density (Wh/kg)	Cycle Life	Safety
$C/LiCoO_2$	110~190	500~1000	Poor
$C/LiMn_2O_2$	100~120	1000	Safer
$C/LiFePO_4$	90~115	>3000	Very safe
$LTO/LiCoO_2$	70~75	>4000	Extremely safe
$LTO/LiFePO_4$	~70	>4000	Extremely safe

The batteries' performance depends on the alloy used, so to know the grade of battery quality, manufactures of EVs or users of the Li battery check the result of tests performed on it. Current, voltage, mechanical strike and temperature are important parameters to be considered in all tests on the batteries, as these parameters fluctuate during the day.

3.1.1. Mechanical Strike Influence

Mechanical strike is an unavoidable problem, whether the EVs are in the normal or an abnormal condition. In the normal condition, the EV battery is abused by the mechanical strike effect of the road, but the EV batteries have an unreliable behavior in the case of an accident. Thus, it must pass some tests such as the T4 and FMVSS 305 mechanical absorb tests or the battery should conform to some standards such as SAE j2464 (Safety and Abuse Testing) [36,37]. The battery is strongly sensitive, when the SOC ratio is up to 80%. Figure 1 shows how important mechanical strike and temperature are. They influence the life and all other aspects of the battery [38–44].

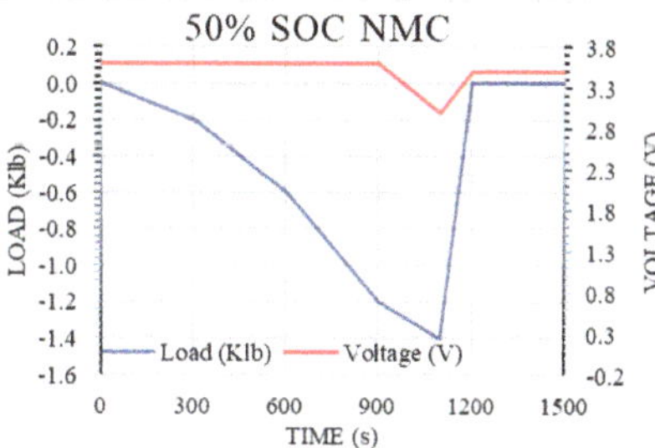

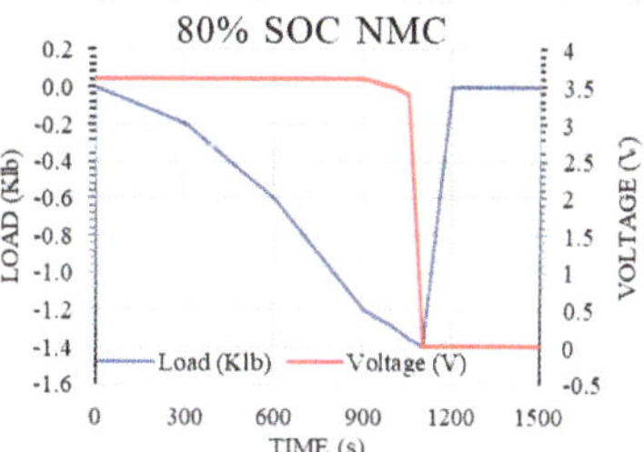

Figure 1. Effects of NMC battery mechanical strike [43].

For example in 2013, during flight JA829J, Japan Airline's (JAL) Boeing 787 APU (Auxiliary Power Unit) battery presented thermal runaway after 52,000 flight hours. Until grounding JA829J, their battery type was lithium cobalt oxide ($LiCoO_2$). According to the U.S. National Transportation Safety Board (NTSB) report, this problem was a result of T5 and T6 mechanical absorbance in the APU battery, and finally, a fire broke out [45].

3.1.2. Temperature Stability Significance

Temperature stability should be considered in all situations, as the battery has different behaviors at each temperature. It does not work perfectly at very extreme temperatures such as $-20°C$ or $120–130\ °C$, in this situation, the runaway threat is greater than in the normal condition. Hence, the EV battery must be protected against temperature (battery heating and cooling system) and the mechanical strike issue [46–48].

3.1.3. Control on the Storage Systems

Grid integration contributes to the grid's power efficiency in each branch of the electric grid. The storage system needs more control than other parts of the grid, so the health of the storage systems is critical and sensitive [49]. The storage control system is not limited to grid storages: EV batteries are part of the grid storage, and thereby, it can control all EVs' storage and the grid storage system, because EVs represent a short-term energy storage for the grid. Therefore, local governments can achieve a storage system without any payment. This is one of the economic benefits of V2G technology for the grid. The control system and EVs' storage systems protect the grid from shocks when the demand strongly decreases or increases. For the infrastructure of the grid response control for V2G technology and some environmental conditions, the control of the grid requires scenarios, plans and data about the grid's future, so the grid should predict data based on old data, plans and scenarios [49–52].

3.1.4. Longevity of the EVs' Batteries

The lifetime of the Li-ion batteries is a complex issue, and it depends on the mentioned parameters such as charge, discharge, thermal condition and some other parameters. Hence, the electric grid can assume batteries' life and capacity by controlling the grid fully [53,54]. EVs' batteries are used in AM and SM; however, they are used for both sides of the grid. This option encourages the economics and eventually the cost of the electric power and energy system to change in each location [55–62]. Ageing is one of the important problems in V2G technology and bidirectional energy systems. According to battery aging, calendar ageing depends on standing time and SOC, and temperature and cycling ageing depends on cycle number, DOD and charging rate. For V2G technology, it is recommended to use $LiFePO_4/C$ cell battery and avoid $LiNiCoAlO_2/C$-based batteries [63,64].

3.2. Charging System

The charging system in the electric grid is a bidirectional system, which supports V2G technology. The charging system quality has a special aspect: it is almost determined as the efficiency of the V2G system. The charging system has almost covered the core of V2G. Charging systems are changed in each grid, and this depends on the infrastructure of the grid. Some charging systems support only DC or AC or both of them. Therefore, the chargers produce the considered infrastructure of the grid with various power electronic parts. IEC 62196-2 Type 1, 2, hybrid, SAE J1772 (SAE Electric Vehicle and Plug in Hybrid Electric Vehicle Conductive Charge Coupler) Type 1, 2, Combo and CHAdeMO are types of the charger connectors. All of the connectors work in some critical scenarios. For example, CHAdeMO only works in a DC system [65,66]. Table 4 gives some charger connectors and types of charge models.

Table 4. EV charger connectors' power types [65,66].

Type of Port	AC	DC
IEC 62196-2	*	Hybrid version
SAE J1772	*	Combo version
CHAdeMO		*

*: its availability depends on the condition.

The bases of energies, like power plants or renewable energy, have to convert it to DC type of energy, if the grid has a connection to the storage network. When increasing the number of DC convertors, the ratio of the THD is automatically increased on the grid. The creation of DC lines in the infrastructure of the grid is beneficial to the grid, which are used in smart grids and V2G technology; for example, the DC line improves charging stations' performance in public places with solar energy or using a wireless charging system. Power electronic converters always change energy alongside the loss of energy, especially when changing from DC to AC. The infrastructure of the electric grid and THD issues are critical problems, and nowadays, countries are trying to reduce both problems. Table 5 mentions that the balance of the energy is a big future challenge, and a number of EVs, chargers, populations and spaces of the location have a crucial role in the load spread balance. All of them are important for control and modelling of the grid. Computing their data accuracy contributes to predicting the future of the grid. Nowadays, energy consumption should be efficient. Hence, heating, cooling and transportation systems are eager to use the electric type of energy as much as possible. The solutions of the grid modelling are different in each location. Creating energy nodes in the urban environment and redistributing energy and DC reserve lines to supply DC base energy equipment are ways to reinforce the infrastructure of the grid [67–71]. The support of the infrastructure and EVs has induced positive opinions in people's minds, so people are looking at all the facilities of V2G technology [72].

Table 5. Some numerical location data related to V2G technology.

EV and Population Information	US	China	Japan	UK	The Netherlands
EV (k)	404	312	126	49.67	87.53
Chargers (k)	28.15	46.65	16.12	8.716	17.78
Fast chargers (k)	3.524	12.1	5.99	1.158	0.465
Population (m)	324.6	1373.5	126.8	65.11	17.10
Population per square km	35	145	346	255	412
Area in square km	9,833,520	9,596,961	377,972	242,495	41,543

Potentials to Build Charging Stations for Renewable Energies

Having a sensible effect on people's life is important. It is one of the effective factors. Charging systems must be available at EV drivers' homes or offices. In fact, the charging system converts energy and transfers it from EV or grid to the storage system. The storage side of V2G technology (home, office, parking and some public places) might be one of the Virtual Power Plants (VPPs). VPPs can support the grid, and their efficiency is at a medium–high level, such as WPP, solar energy and CHP. Renewable energies can generate energy everywhere and at every scale [73]. Table 6 shows the major VPPs. This means that EVs complete the energy circle by using V2G technology. In this circle, EVs perform the transmission line and temporary storage duty. If both the home and office are equipped with VPP and V2G technology and personal EVs could always be connected to grid, all of them would provide some benefits to the energy sector, and these are mentioned below [74–77]:

✓   Low GHG emissions on the supply and demand side
✓   Strong reduction of grid shock
✓   Reduction of energy cost to help the economy of the home and office
✓   Reduction of fossil, coal and nuclear energy contribution

Table 6. Accessible VPP distribution by place and source type.

	Home	Office	Parking	Urban
Solar PV	*	*	*	*
Micro turbine	*	*		*
Regular turbine		*1		
CHP		*1		
Battery	*	*	*	*

*: available/executable condition; *1: its availability depends on the condition.

Some projects related to V2G have been accomplished, and all mentioned parameters are considered in projects; for example, Taiwan, the U.S. (Florida), The Netherlands and China [78–81]. The priority of the source to supply the charging station is local energy production. Electric storage systems and their charging systems in V2G technology, as well as similar technologies (for which, the energy transfer between them is bidirectional) are important part of the grid, so the reason for having this technology is the electric storage systems, which have good energy efficiency. According to the mentioned advantages and disadvantages, it seems that the LiFePO$_4$ type of Li batteries is the best choice between various types of batteries; it is very safe in comparison to other types of batteries.

4. Infrastructure

4.1. Strategy

The strategy of V2G technology works according to some plans and targets. Besides, the targets have zero GHG emissions and energy efficiency, such as stability and reliability. Some parameters cause changes in the plan. Infrastructure, energy sources, budget and some other parameters in each country are determined by the energy plan, which is necessary to develop V2G technology. Energy sources and the ability to generate renewable energy are determined with the benefits and market share of V2G technology. Renewable energy developments, V2G technology and the smart grid are impelled alongside [82,83].

4.1.1. Pricing of Energy

The energy sector pricing system is different from other sectors. The supply/demand and determining the cost of the energy by time and other grid priorities give balance and discipline to all of the grid. However, in the energy systems, the supply/demand balance is the second important factor. Balancing is a means to distribute energy correctly in the electric grid. The density of the population is not constant in each location, but the specification of the population is clear for governments. Hence, the location and density of the population, the time of supply/demand, the measure of supply/demand and the balancing of supply/demand [84–86] influence the price of the energy.

The location and density of the population determines the average grid demand in all electric grids. There is a high population density mostly located in the downtown of cities or the capitals of counties. In this condition, the density of the transmission lines is increased, so the ratio of the loss energy is increased. For example, when energy demands are at a high level and the energy generation is low, the cost of energy in these locations is at the highest level [87–90]. The time of the electric energy consumption is important; sometimes, all customers of the grid use energy. In this condition, the grid works at full capacity, and this means that it is prone to overload. Hence, the electric supervisors encourage people to use electric energy from the grid, except for the demand peak time (18:00–22:00) [91–93] considering the peak/down time; in some countries, the time zones differs between two and six hours across the country. Figure 2 gives some countries' populations in 2016 and 2050 [94]. Today, everyone can generate energy in their own house or office by using renewable energies, but this depends on the environmental condition. On the other hand, by using V2G technology, one can share energy with customers or the grid. Generating energy contributing to the grid and supervising of the pay cache for energies generated from customers [95], supervisors have to balance energy on the grid, as an unbalanced grid can damage the grid. They use some methods such as building switching centers to balance energy in the grid [96].

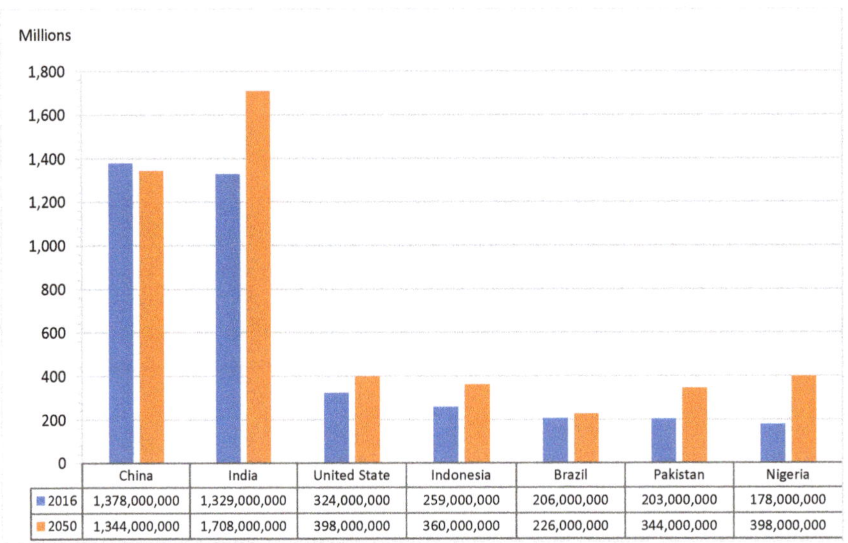

	China	India	United State	Indonesia	Brazil	Pakistan	Nigeria
2016	1,378,000,000	1,329,000,000	324,000,000	259,000,000	206,000,000	203,000,000	178,000,000
2050	1,344,000,000	1,708,000,000	398,000,000	360,000,000	226,000,000	344,000,000	398,000,000

Figure 2. Population from 2016–2050 in some countries [95].

This clearly shows that the mentioned issues influence the electric grid directly. Social infrastructures, such as the economy, psychology, applied and cognitive science, should be prepared to use EVs with people. People of the world have to use EVs, as fossil energies are going to deplete and people who use the transportation system with fossil energies would increase if the world population continues the present energy consumption method in the transportation system. For other types of energy, consumption differs in each area depending on sunrise and sunset, where in some countries' grids work synchronously. The management group must organize this considering local time and other local abilities [97–100] V2G technology helps to reduce the mentioned problems, and governments can create prefect charging and energy-sharing designs and pricing and grid infrastructure control plans with the challenging problems mentioned for the background and future targets. A good control system must be tested in international sample electric grids, and each part of the grid must be tested with international grid patents such as IEEE 33 Node, IEEE 34 Node and IEEE 300-Bus Network.

Grid coordination under the mentioned standards provides a wide facility for development, such as grid behavior prediction, charging scheduling and risk management [101–105].

4.1.2. Control System Communications

One of the control systems duties is online observation, as the EVs are mobile devices, and they can be on the supply or demand side of the grid (charge/discharge). This provides some favorably strong aspects for people. EVs users can see online the existing charging centers and their status. In the online system, all variable parameters change in each condition. This gives facility to provide fine pricing and a managing system via control system communications such as WAN, HAN, NAN, SCADA, DSL, GSM, satellite and GPS [106]. In fact, they have a chain connection with other parts of the grid. They are illustrated in Figure 3. This also depends on the location and situation, for example using GSM, DSL and satellite as the connection method to connect some data centers with other parts of the grid, and all of them are available everywhere [107–110].

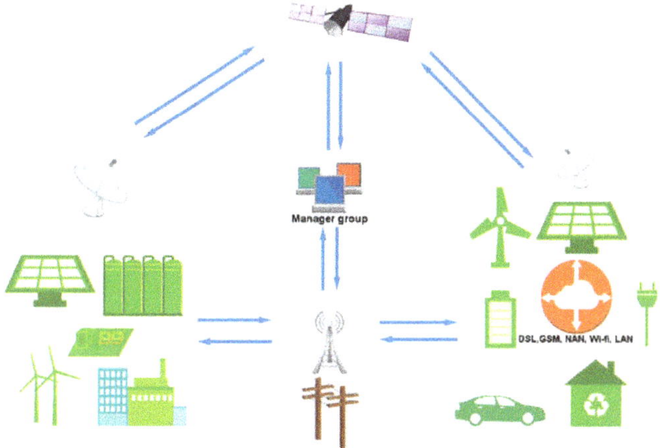

Figure 3. Communication system in the electric grid.

Data of the EVs are sent to the EVs' controller operator to control all EVs in each location. However, in some situations, the grid works in uncertain conditions, and uncertain conditions result in different behaviors of the grid. Some methods contribute to calculating the energy cost in uncertain conditions such as the ENTRUST (Energy trading under uncertainty in smart grid systems) algorithm [111], and some other algorithms estimate the active demand load on the grid; this means that the grid has high reliability and is ready for uncertain threats and conditions.

4.2. Grid Modelling

The infrastructure of the model of the grid is based on V2G technology. The mentioned topics are impelled toward the best efficiency and GHG emission targets. Thus, reforming some sections of grid is too hard and needs to be investigated deeply. Fossil power planets, renewable energies and V2G-G2V (subsets of renewable energies) are the inputs of grids. Today, the policy and planning of countries are to require grids to be supplied by renewable energy, and it is subsets to consume green and efficient energy. All energies injected to the grid should be managed, so all types of energies have particularities. For example, convertors of the renewable energies such as DC to AC and reverse generated THD or energies on the storage system should be controlled on the supply/demand side. Electric distribution is the last node of grid electrification. In this section, the grid supplies all demand side of the grid, and the energy distribution condition depends on the location. Grid modelling with

V2G technology is the same as the smart grid, so the grid in both systems works unidirectionally. The grid can supply both AC and DC electric energies [112]. Most of the electric devices work with the AC type of electric power. In this situation, energy providers have to provide the AC type of energy and the DC type of energy achieved by DC/AC convertors. The different types of energies are some of the causes of the energy loss. Today, having one of the DC lines from the source to the consumer is compulsory. In such a condition, the AC type of energy has the THD problem. The DC type of energy can respond to a wide range of customers of the grid. Different chargers, different convertors and different sources are examples of the energy type problems. AC and DC customers are also connected, and energy exchange will result in loss of energy and increased THD; however, the DC line has some critical advantages such as reducing energy loss, the response of the local renewable energy source and the other sources that work with the DC type of energy, which can work in a hybrid manner. For example, Egypt and India have tested DC lines in some projects [113–122].

4.2.1. Frequency Control

Electric power frequency is one of the most important parameters in the grid. The grid frequency is not controllable when supply and demand are not balanced; usually, the demand side of the grid overcomes the supply. In V2G technology, the charge and discharge influences the grid frequency, especially in the charge condition. EVs have batteries with a large capacity, and the quantity of EVs is high, which means overall energy exchange is homogenized; however, this is not true all the time. All of this has a negative effect on the grid frequency. Hence, the solution of the frequency issue is to control all parameters and equipment of the grid and demands such as the quantity of the EVs that are active or the EVs' energy capacity. Power systems must be controlled with special control nodes depending on the area, demand and sources [123–127].

4.2.2. System Integration

A smart and unidirectional power system must be an integrated system. Therefore, all it is parts should be connected and work synchronously. The process of the integration needs to provide the capacity and limit some factors depending on the grid design. Each part of the materials mentioned in this paper plays a big role in the power system integration. Why is power system integration important? Because V2G technology disrupts the power system and V2G works correctly when the supply/demand condition is fine. For example, the status of the charging station, the battery status of the EVs and the grid supply/demand power are some aspects of V2G technology related to power system integration. Figure 4 gives these facts clearly [128]. The power system is integrated to provide low energy loss and high efficiency, so with this technology, the gap of the supply/demand is reduce and the power of the control system is increased; For example, home or office grid integration to generate energy by some local renewable energy sources [129–133].

The power system infrastructure should be designed according to some parameters such as accessibility, reliability and being able to be developed. Result of this action is that the grid can be updated with some technologies such as V2G and the subset technologies. The grid should support the energy traffic, which is achieved by newer technologies such as V2G or energy storage systems. The grid designers must care about communication between the grid and people. Therefore, in emergency situations, the grid management group may increase or decrease the supply/demand according to people. This is only a recommendation, which would work with the grid infrastructure.

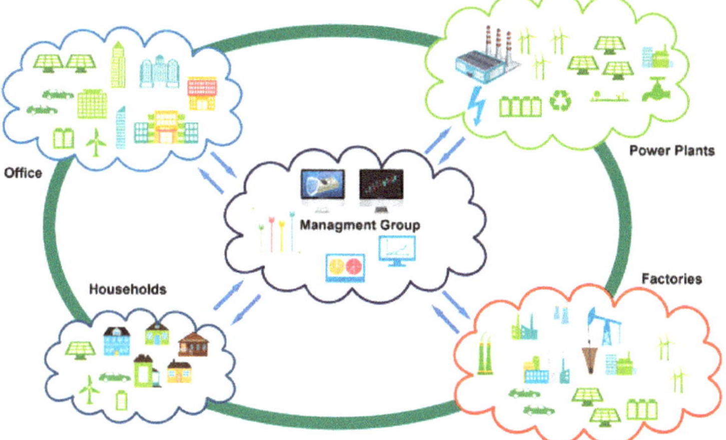

Figure 4. All grid energies connect to other sources and customers [128].

5. Conclusions

In the infrastructure of the world's electric grid, some revisions have been made. In addition, all energy scenarios after increasing GHG emissions in the world have changed according to some agreements and conferences such as UNFCCC 2015a in Paris. The infrastructure of each system should be changed when the enrolment scenarios change. The ratio of revisions depends on old infrastructure conditions and future targets, which are mentioned in context. Storage systems are some of the sensitive parts of the latest grid models such as micro and smart grids, but their development does not satisfy and respond to the grid requirements. Hence, batteries need to be further developed to comply better with the charge/life cycle and to provide more safety. At a glance, all mentioned problems force the infrastructure of the grid to work with poor quality. Today, after the development of the lithium type of batteries and power electronic systems, the researchers who are working on V2G technologies can easily take some actions so that the battery and power electronic systems will provide fast charging, on the supply and demand side of the grid. With those benefits, the grid could work properly, so the lack of energy and temporary storage are solved.

- ✓ The life/charge cycle, energy density and safety problems are solved with some polymer types of lithium batteries.
- ✓ The problems of the infrastructure of the grid can be detected quickly with a synchronized communication system.

The V2G technology has received great acceptance from people, but a greater market share is needed to develop this technology. The psychology of people is important, so people must have satisfaction with and a great viewpoint of EVs. Hence, officials have to achieve people's acceptance and create encouragement plans. For example, in some countries the tax of EV is zero and in some countries the benefits of V2G technology using Li batteries, mobile energy stations and ultra-capacitors are explained for people so that they use them more.

Author Contributions: All authors contributed for bringing the manuscript in its current state. Their contributions include detailed survey of the literatures and state of art, which were essential for the completion of this paper.

Funding: This research received no external funding.

Conflicts of Interest: The authors declare no conflict of interest.

Nomenclature

GHG	Green House Gas
ICE	Internal Combustion Engine
EV	Electric Vehicle
PHEV	Plug-in Hybrid Electric Vehicle
HEV	Hybrid Electric Vehicle
V2G	Vehicle-to-Grid
G2V	Grid-to-Vehicle
GDR	Generalized Demand-side Resources
IEA	International Energy Agency
UNFCCC	United Nations Framework Convention on Climate Change
IEC	International Electro technical Commission
SAE	Society of Automotive Engineers
GTCO2	Giga Tonnes of Carbon dioxide
2DS	2 °C Scenario
6DS	6 °C Scenario
W h/L	Watt hour per liter
TWh	Terra Watt hours
KWh	Kilo Watt hour
EVI	Electric Vehicles Initiative
AC	Alternating Current
DC	Direct Current
CHP	Combined Heat and Power
CHAdeMO	CHArge de Move
SOC	State of Charge
AM	Automotive Mode
LNG	Liquefied Natural Gas
LPG	Liquefied Petroleum Gas
IEEE:	Institute of Electrical and Electronics Engineers
WAN	Wide Area Network
HAN	Home Area Network
NAN	Neighborhood Area Network
SCADA	Supervisory Control and Data Acquisition
DSL	Digital Subscriber Line
GSM	Global System for Mobile Communications
GPS	Global Positioning System
THD	Total Harmonic Distortion
VPP	Virtual Power Plant
WPP	Wind Power Plant
LDV	Light-Duty Vehicle
Fe	Iron
Li	Lithium
Li-ion	Lithium-ion
NiCd	Nickel–Cadmium
NaS	Sodium–Sulfur
ZnBr	Zinc–Bromine
$LiFePO_4$	Lithium Iron Phosphate Oxide
$LiCoO_4$	Lithium Cobalt Oxide
$LiMn_2O_4$	Lithium ion Manganese Oxide
NMC	$LiNiMnCoO_2$
DOD	Depth of Charge
SM	Storage Mode
CNG	Compressed Natural Gas

HAN	Home Area Network
NAN	Neighborhood Area Network
SCADA	Supervisory Control and Data Acquisition
DSL	Digital Subscriber Line
GSM	Global System for Mobile Communications
GPS	Global Positioning System

References

1. International Energy Agency, Global EV Outlook. Available online: https://www.iea.org/publications/freepublications/publication/Global_EV_Outlook_2016.pdf (accessed on 28 May 2016).
2. Erdogan, N.; Erden, F.; Kisacikoglu, M. A fast and efficient coordinated vehicle-to-grid discharging control scheme for peak shaving in power distribution system. *J. Mod. Power Syst. Clean Energy* **2018**, *6*, 555–566. [CrossRef]
3. Kisacikoglu, M.C.; Erden, F.; Erdogan, N. Distributed Control of PEV Charging Based on Energy Demand Forecast. *IEEE Trans. Ind. Inform.* **2018**, *14*, 332–341. [CrossRef]
4. Sustainable Energy Authority of Ireland. Available online: http://www.seai.ie/Publications/Statistics_Publications/SEAI_2050_Energy_Roadmaps/Electric_Vehicle_Roadmap.pdf (accessed on 27 May 2016).
5. Energy Efficiency & Renewable Energy. The Transforming Mobility Ecosystem: Enabling an Energy-Efficient Future. Available online: https://www.energy.gov/sites/prod/files/2017/01/f34/The%20Transforming%20Mobility%20Ecosystem-Enabling%20an%20Energy%20Efficient%20Future_0117_1.pdf (accessed on 29 January 2017).
6. Energy Scenarios for 2030. Available online: https://www.energinet.dk/-/media/Energinet/Analyser-og-Forskning-RMS/Dokumenter/Analyser/Energy-Scenarios-for-2030-UK-Version.PDFpdf (accesssed on 4 October 2016).
7. Baumhefner, M.; Hwang, R. *How Utilities Can Accelerate the Market for Electric Vehicles*; Technical Report for The Natural Resources Defense Council; NRDC: New York, NY, USA; Washington, DC, USA; Los Angeles, CA, USA; San Francisco, CA, USA; Chicago, IL, USA; Beijing, China, 2016.
8. Combined Heat and Power Basics. Available online: https://energy.gov/eere/amo/combined-heat-and-power-basics (accessed on 29 November 2017).
9. Shirazi, Y.; Carr, E.; Knapp, L. A cost-benefit analysis of alternatively fueled buses with special considerations for V2G technology. *Energy Policy* **2015**, *87*, 591–603. [CrossRef]
10. Irean. *Road Transport: The Cost of Renewable Solutions*; Irena Publication: Abu Dhabi, UAE, 2013.
11. Electric Power Research Institute. Environmental Assessment of Plug-In Hybrid Electric Vehicles. Available online: https://www.energy.gov/sites/prod/files/oeprod/DocumentsandMedia/EPRI-NRDC_PHEV_GHG_report.pdf (accessed on 25 July 2017).
12. Heinen, S.; Elzinga, D.; Kim, S.K.; Ikeda, Y. Impact of Smart Grid Technologies on Peak Load to 2050. Available online: https://www.iea.org/publications/freepublications/publication/smart_grid_peak_load.pdf (accessed on 2 August 2011).
13. Technology Roadmap. Available online: https://www.iea.org/publications/freepublications/publication/smartgrids_roadmap.pdf (accessed on 27 April 2011).
14. Lewis, M.F.; Amr, S.; Francisco, B.; Alessandra, S.; Deger, S. *Electric Vehicles: Technology Brief*; International Renewable Energy Agency: Abu Dhabi, United Arab Emirates, 2017.
15. Maria, L.T.; Michael, L.A. A review of the development of smart grid technologies. *Renew. Sustain. Energy Rev.* **2016**, *59*, 710–725.
16. Mathiesen, B.; Lund, H.; Connolly, D.; Wenzel, H.; Østergaard, P.; Möller, B.; Nielsen, S.; Ridjan, I.; Karnøe, P.; Sperling, K.; Hvelplun, F. Smart Energy Systems for coherent 100% renewable energy and transport solutions. *Appl. Energy* **2015**, *145*, 139–154. [CrossRef]
17. Hooman, F.; Christopher, N.-H.D.; Jose, A.P.-O. An integrated supply-demand model for the optimization of energy. *J. Clean. Prod.* **2016**, *114*, 268–285.
18. Michael, S.; Ontje, L.; Jörg, B.; Martin, T. Decentralized control of units in smart grids for the support of renewable energy supply. *Environ. Impact Assess. Rev.* **2015**, *52*, 40–52.

19. John, B.; Margaret, O.-M. Modelling charging profiles of electric vehicles based on real-world electric vehicle charging data. *Sustain. Cities Soc.* **2016**, *26*, 203–216.
20. Alexander, S.; Christoph, M.F.; Sebastian, G. Quantifying load flexibility of electric vehicles for renewable energy. *Appl. Energy* **2015**, *151*, 335–344.
21. Pero, P.; Goran, G.; Goran, K.; Neven, D. Long-term energy planning of Croatian power system using multi-objective optimization with focus on renewable energy and integration of electric vehicles. *Appl. Energy* **2016**, *184*, 1493–1507.
22. Guilherme, A.D.; Nivalde, J.-C.; Roberto, B.; Rubens, R.; Alexandre, L. Prospects for the Brazilian electricity sector in the 2030s: Scenarios and guidelines for its transformation. *Renew. Sustain. Energy Rev.* **2017**, *68*, 997–1007.
23. Shang, D.; Sun, G. Electricity-price arbitrage with plug-in hybrid electric vehicle: Gain or loss? *Energy Policy* **2016**, *95*, 402–410. [CrossRef]
24. Thomas, K.; Patrick, J.; Wolf, F. Solar energy storage in germany households: profitability, load changes and flexibility. *Energy Policy* **2016**, *98*, 520–532.
25. Arif, A.I.; Babar, M.; Ahamed, T.P.I.; Al-Ammar, E.A.; Nguyen, P.H.; Kamphuis, I.G.R.; Malik, N.H. Online scheduling of plug-in vehicle in dynamic pricing schemes. *Sustain. Energy Grid Netw.* **2016**, *7*, 25–36. [CrossRef]
26. João, S.; Mohammad, A.F.G.; Nuno, B.; Zita, V. A stochastic model for energy resources management considering demand response in smart grids. *Electr. Power Syst. Res.* **2017**, *143*, 599–610.
27. Lance, N.; Benjamin, K.S. Why did better placeFail?: Rangeanxiety, interpretive flexibility, and electric vehicle promotion in Denmarkand Israel. *Energy Policy* **2016**, *94*, 377–386.
28. Joy, C.M.; Saurabh, S.; Arobinda, G. Mobility aware scheduling for imbalance reduction through charging coordination of electric vehicles in smart grid. *Pervasive Mob. Comput.* **2015**, *21*, 104–118.
29. Eva, N.; Floortje, A. How is value created and captured in smart grids? A review of the literature and ananalysis of pilot projects. *Renew. Sustain. Energy Rev.* **2016**, *53*, 629–638.
30. Sousa, T.; Soares, T.; Morais, H.; Castro, R.; Vale, Z. Simulated annealing to handle energy and ancillary services joint management considering electric vehicles. *Electr. Power Syst. Res.* **2016**, *136*, 383–397. [CrossRef]
31. Li, B.; Shen, J.; Wang, X.; Jiang, C. From controllabe loads to generalized demand-side resources: A review on developments of demand-side resources. *Renew. Sustain. Energy Rev.* **2016**, *53*, 936–944.
32. Ayodele, T.R.; Ogunjuyigbe, A.S.O. Mitigation of wind power intermittency: Storage technology approach. *Renew. Sustain. Energy Rev.* **2015**, *44*, 447–456. [CrossRef]
33. Lee, J.; Park, G.-L. Dual battery managment for renewable energy integration in EV charging station. *Neurocomputing* **2015**, *148*, 181–186. [CrossRef]
34. Zhao, H.R.; Wu, Q.W.; Hu, S.J.; Xu, H.H.; Claus, N.R. Review of energy storage system for wind power integration support. *Appl. Energy* **2015**, *137*, 545–553. [CrossRef]
35. Opitz, A.; Badami, P.; Shen, L.; Vignarooban, K.; Kannan, A.M. Can Li-ion batteries be the panacea for automotive application. *Renew. Sustain. Energy Rev.* **2017**, *68*, 685–692. [CrossRef]
36. United Nations. *Transport of Dangerous Goods*, 5th ed.; United Nations Publication: New York, NY, USA, 2009.
37. Doughty, D.H. *Vehicle Battery Safety Roadmap Guidance*; National Renewable Energy Laboratory: Golden, CO, USA, 2012.
38. Hu, X.S.; Clara, M.M.; Yang, Y.L. Charging, power management, and battery degradation mitigation in plug-in hybrid electric vehicles: A unified cost-optimal approach. *Mech. Syst. Signal. Process.* **2017**, *87*, 4–16. [CrossRef]
39. Branimir, S.; Joško, D. A novel model of electric vehicle fleet aggregate battery for energy planning studies. *Energy* **2015**, *92*, 444–455.
40. Kate, E.F.; Brian, T.; Li, Z.; Brendan, S.; Scott, S. Charging a renewable future: The impact of electric vehicle charging intelligence on energy storage requirements to meet renewable portfolio standards. *J. Power Sources* **2016**, *336*, 63–74.
41. Alexander, F.; Wladislaw, W.; Andrea, M.; Dirk, U. Critical review of on-board capacity estimation techniques for lithiumion batteries in electric and hybrid electric vehicles. *J. Power Sources* **2015**, *281*, 114–130.
42. Le, D.; Cheng, C.-C. Energy savings by energy managment system: A review. *Renew. Sustain. Syst.* **2016**, *56*, 760–777.

43. Hsin, W.; Edgar, L.-C.; Evan, T.R.; Clinton, S.W. Mechanical abuse simulation and thermal runaway risks of large format Li-ion batteries. *J. Power Sources* **2017**, *342*, 913–920.

44. Cheng, L.; Aihua, T. Simplification and efficient simulation of electrochemical model for Li-ion battery in EVs. *Energy Procedia* **2016**, *104*, 68–73.

45. National Transportation Safety Board. Available online: https://www.ntsb.gov/investigations/AccidentReports/Reports/AIR1401.pdf (accessed on 8 January 2013).

46. Jyri, S.; Topi, R.; Juuso, L.; Peter, D.L. Flexibility of electric vehicles and space heating in net zero energy houses: an optimal control model with thermal dynamics and battery degradation. *Appl. Energy* **2017**, *190*, 800–812.

47. Wang, Q.; Jiang, B.; Xue, Q.; Sun, H.; Li, B.; Zou, H.; Yan, Y. Experimental investigation on EV battery cooling and heating by heat pipes. *Appl. Therm. Eng.* **2015**, *88*, 54–60. [CrossRef]

48. Saxena, S.; Floch, C.L.; MacDonald, J.; Moura, S. Quantifying EV battery end-of-life through analysis of travel needs with vehicle powertrain models. *J. Power Source* **2015**, *282*, 265–276. [CrossRef]

49. Jan, B.; Benjamin, N.; Sebastian, T.; Christoph, H. Integrating on-site renewable electricity generation into a manufacturing system with intermittent battery storage from electric vehicles. *Procedia CIRP* **2016**, *48*, 483–488.

50. Darcovich, K.; Kenney, B.; MacNeil, D.; Armstrong, M. Control strategies and cycling demands for Li-ion storage batteries in residential micro-cogeneration systems. *Appl. Energy* **2015**, *141*, 32–41. [CrossRef]

51. Girish, S.; Simona, O. A control-oriented cycle-life model for hybrid electric vehicle lithiumion batteries. *Energy* **2016**, *96*, 644–653.

52. Siwar, K.; Mouna, R.; Lotfi, K. A flexible control strategy of plug-in electric vehicles operating in seven modes for smoothing load power curves in smart grid. *Energy* **2017**, *118*, 197–208.

53. Calvillo, C.F.; Sanchez-Miralles, A.; Villar, J. Energy Managment and planing in smart cities. *Renew. Sustain. Syst.* **2016**, *55*, 273–287. [CrossRef]

54. Jakobus, G.; Thomas, G.; James, M. Accelerated energy capacity measurement of lithium-ion cells to support future circular economy strategies for electric vehicles. *Renew. Sustain. Syst.* **2017**, *69*, 98–111.

55. Li, M.H.; Chang, D.-S. Allocative efficiency of high-power Li-ion batteries from automotive mode (AM) to storage mode (SM). *Renew. Sustain. Syst.* **2016**, *64*, 60–67.

56. Stuart, S.; Thomas, B. Leaving the grid—The effect of combining home energy storage with renewable energy generation. *Renew. Sustain. Syst.* **2016**, *60*, 1213–1224.

57. Faessler, B.; Kepplinger, P.; Petrasch, J. Decentralized price-driven grid balancing via repurposed electric vehicle batteries. *Energy* **2017**, *118*, 446–455. [CrossRef]

58. Sun, X.H.; Toshiyuki, Y.; Takayuki, M. Charge timing choice behavior of battery electric vehicle users. *Transp. Res. Part. D: Transp. Environ.* **2015**, *37*, 97–107. [CrossRef]

59. Magnor, D.; Lunz, B.; Sauer, D.U. 'Double Use' of Storage Systems. In *Electrochemical Energy Storage for Renewable Sources and Grid Balancing*; Patrick, T., Jurgen, G., Eds.; Elsevier: Amsterdam, Holland, 2015; pp. 453–463.

60. Franziska, S.; Claudia, M.; Susen, D.; Bettina, K.; Ramona, W.; Josef, F.K.; Andreas, K. User responses to a smart charging system in Germany: Battery electric vehicle driver motivation, attitudes and acceptance. *Energy Res. Soc. Sci.* **2015**, *9*, 60–71.

61. Atmaga, T.D.; Amin. Energy storage system using battery and ultracapacitor on mobile charging station for electric vehicle. *Energy Procedia* **2015**, *68*, 429–437. [CrossRef]

62. Muhammad, A.; Takuya, O.; Takao, K. Extended utilization of electric vehicles and their re-used batteries to support the building energy management system. *Energy Procedia* **2015**, *75*, 1938–1943.

63. Martin, P.; Eric, P.; Valérie, S.-M. Development of an empirical aging model for Li-ion batteries and application to assess the impact of Vehicle-to-Grid strategies on battery lifetime. *Appl. Energy* **2016**, *172*, 398–407.

64. Harighi, T.; Bayindir, R.; Hossain, E. Overview of Quality of Service Evaluation of a Charging Station for Electric Vehicle. In Proceeding of the 2017 IEEE 6th International Conference on Renewable Energy Research and Applications (ICRERA), San Diego, CA, USA, 5–8 November 2017; pp. 1180–1185.

65. Hussain, S.; Mainul, I.M.; Mohamed, A. A review of the stage-of-the-art charging technologies, placement methodologies, and impacts of electric vehicle. *Renew. Sustain. Syst.* **2016**, *64*, 403–420.

66. Yong, J.Y.; Ramachandaramurthy, V.K.; Tan, K.M.; Mithulananthan, N. A review on the state-of-the-art technologies of electric vehicle, its impacts and prospects. *Renew. Sustain. Syst.* **2015**, *49*, 365–385. [CrossRef]

67. Kafeel, A.K.; Muhammad, A.; Saad, M. Inductively coupled power transfer (ICPT) for electric vehicle charging—A review. *Renew. Sustain. Syst.* **2015**, *47*, 462–475.

68. Radu, G.; Eduardo, R.; João, M.; João, P.S.C. Smart electric vehicle charging scheduler for overloading prevention of an industry client power distribution transformer. *Appl. Energy* **2016**, *178*, 29–42.

69. Yang, H.M.; Xiong, T.L.; Qiu, J.; Qiu, D.; Dong, Z.Y. Optimal operation of DES/CCHP based regional multi-energy prosumer with demand response. *Appl. Energy* **2016**, *167*, 353–365. [CrossRef]

70. Zhou, B.; Feng, Y.; Tim, L.; Zhang, H.G. An electric vehicle dispatch module for demand-side energy participation. *Appl. Energy* **2016**, *177*, 464–474. [CrossRef]

71. Pedro, N.; Figueiredo, R.; Brito, M.C. The use of parking lots to solar-charge electric vehicles. *Renew. Sustain. Syst.* **2016**, *66*, 679–693.

72. Christian, W.; Alexander, S. Understanding user acceptance factors of electric vehicle smart charging. *Transp. Res. Part C Emerg. Technol.* **2016**, *71*, 198–214.

73. Bi, Z.C.; Kan, T.Z.; Chunting, C.M.; Zhang, Y.M.; Zhao, Z.M.; Gregory, A.K. A review of wireless power transfer for electric vehicles: Prospects to enhance sustainable mobility. *Appl. Energy* **2016**, *179*, 413–425. [CrossRef]

74. Pan, Z.J.; Zhang, Y. A novel centralized charging station planning strategy considering urban power network struture strength. *Electr. Power Syst. Res.* **2016**, *136*, 100–109. [CrossRef]

75. Saeid, M.; Farshid, K.; Mohammad, R.R.; Akbar, M. Generation expansion planning by considering energy-efficiency programs in a competitive environment. *Electr. Power Energy Syst.* **2016**, *80*, 109–118.

76. Mehdi, A.; Mehdi, N.; Omer, T. Getting to net zero energy building: investigating the role of vehicle to home technology. *Energy Build.* **2016**, *130*, 465–476.

77. Zamani, A.G.; Zakariazadehand, A.; Jadid, S. Day-ahead resource scheduling of a renewable energy based virtual power plant. *Appl. Energy* **2016**, *169*, 324–340. [CrossRef]

78. Barisa, A.; Rosa, M.; Laicane, I.; Sarmins, R. Application of low-carbon technologies for cutting household GHG emissions. *Energy Procedia* **2015**, *72*, 230–237. [CrossRef]

79. Hu, J.J.; Hugo, M.; Tiago, S.; Morten, L. Electric vehicle fleet management in smart grids: A review of services, optimization and control aspects. *Renew. Sustain. Syst. Energy Rev.* **2016**, *56*, 1207–1226. [CrossRef]

80. Zhao, Y.; Nuri, C.; Murat, K.; Omer, T. Carbon and energy footprints of electric delivery trucks: A hybrid multi-regional input-output life cycle assessment. *Transp. Res. Part. D Transp. Environ.* **2016**, *47*, 195–207. [CrossRef]

81. Lee, A.H.I.; Chen, H.H.; Chen, J. Building smart grid to power the next century in Taiwan. *Renew. Sustain. Energy Rev.* **2017**, *68*, 126–135. [CrossRef]

82. Ying, L.; Chris, D.; Zofia, L.; Margot, W. Electric vehicle charging in China's power system: Energy, economic and environmental trade-offs and policy implications. *Appl. Energy* **2016**, *173*, 535–554.

83. Farid, A.; Esther, P.L.; Nathan, W.; Bart, D.S.; Zofia, L. Fuel cell cars in a microgrid for synergies between hydrogen and electricity networks. *Appl. Energy* **2017**, *192*, 296–304.

84. Seyed, M.M.T.; Hassan, R.; Mehdi, J.; Hamid, K. A probabilistic unit commitment model for optimal operation of plug-in electric vehicle in microgrid. *Renew. Sustain. Energy Rev.* **2016**, *66*, 934–947.

85. Liu, L.C.; Zhu, T.; Pan, Y.; Wang, H. Multiple energy complementation based on distributed energy systems—Case study of chongming county, China. *Aoolied Energy* **2017**, *192*, 329–336. [CrossRef]

86. Dominković, D.F.; Bacekovic, I.; Cosic, B.; Krajacic, G.; Puksec, T.; Duic, N.; Markovska, N. Zero carbon energy system of South East Europe in 2050. *Appl. Energy* **2016**, *184*, 1517–1528. [CrossRef]

87. Katja, L.; Arne, V.; Daniel, J. Business Model for Electric Mobility. *Procedia CIRP* **2016**, *47*, 483–488.

88. Hu, Z.C.; Zhan, K.Q.; Zhang, H.C.; Song, Y.H. Pricing mechanisms design for guiding electric vehicle charging to fill load valley. *Appl. Energy* **2016**, *178*, 155–163. [CrossRef]

89. Andrenacci, N.; Ragona, R.; Valenti, G. A demand-side approach to the optimal deployment of electric vehicle charging station in metroplitan areas. *Appl. Energy* **2016**, *182*, 36–46. [CrossRef]

90. Morais, H.; Sousa, T.; Soares, J.; Faria, P.; Vale, Z. Distributed energy resource management using plug-in hybrid electric vehicle as a fuel-shifting demand. *Energy Convers. Manag.* **2015**, *97*, 78–93. [CrossRef]

91. Guo, Z.M.; Julio, D.; Fan, Y.Y. Infrastructure planning for fast charching stations in a compactitive market. *Transp. Res. Part. C Emerg. Technol.* **2016**, *68*, 215–227. [CrossRef]

92. Florian, S.; Jens, P.I.; Christoph, M.F.; Hauke, B.; Clemens, D. Impact of electric vehicles on distribution substations: A Swiss case stady. *Appl. Energy* **2015**, *137*, 88–96.

93. Nikolaos, G.P.; Madeleine, G. A methodology to generate power profiles of electric vehicle parking lots under different operational strategies. *Appl. Energy* **2016**, *173*, 111–123.

94. World Population Data: Focus on Youth. Available online: http://www.worldpopdata.org/ (accessed on 15 August 2017).

95. Anu, G.K.; Anmol, M.; Akhil, V.S. A strategy to Enhance Electric Vehicle Penetration Level in india. *Procedia Technol.* **2015**, *21*, 552–559.

96. Tarroja, B.; Zhang, L.; Wifvat, V.; Shaffer, B.; Samuelsen, S. Assessing the stationary energy storage equivalency of vehicle-to-grid charging battery electric vehicle. *Energy* **2016**, *106*, 673–690. [CrossRef]

97. Pankaj, M.; Pushkin, K.; Alexander, P.; Romesh, K. Development of control models for the planning of sustainable transportation systems. *Transp. Res. Part. C Emerg. Technol.* **2015**, *55*, 474–485.

98. Islam, S.B.; Taha, S.U. A survey on behind the meter energy managment system in smart grid. *Renew. Sustain. Energy Rev.* **2017**, *72*, 1208–1232.

99. Yasin, K. A survey on smart metering and smart grid communication. *Renew. Sustain. Energy Rev.* **2016**, *57*, 302–318.

100. Ali, S.M.; Jawad, M.; Khan, B.; Mehmood, C.A.; Zeb, N.; Tanoli, A.; Farid, U.; Glower, J.; Khan, S.U. Wide area smart grid architectural model and control: A survey. *Renew. Sustain. Energy Rev* **2016**, *64*, 311–328. [CrossRef]

101. Luo, Y.G.; Zhu, T.; Wan, S.; Zhang, S.W.; Li, K.Q. Optimal charging schoulding for large-scale EV (electric vehicle) deployment base on the interaction of the smart-grid and intelligent-transport systems. *Energy* **2016**, *97*, 359–368. [CrossRef]

102. Fang, X.L.; Yang, Q.; Wang, J.H.; Yan, W.J. Coordinated dispatch in multiple cooperative autonomous islanded microgrids. *Appl. Energy* **2016**, *162*, 40–48. [CrossRef]

103. Nasim, N.; Yong, W. Risk management and participation planning of electric vehicle in smart grid for demand response. *Energy* **2016**, *116*, 836–850.

104. Masoud, H.; Alireza, Z.; Shahram, J. Self-scheduling of electric vehicle in an intelligent parking lot using stochastic optimization. *J. Frankl. Inst.* **2015**, *352*, 449–467.

105. Bharati, G.R.; Paudyal, S. Coordinated control of distribution grid and electric vehicle loads. *Electr. Power Syst. Res.* **2016**, *140*, 761–768. [CrossRef]

106. Michael, E.; Ramesh, R. Communication technologies for smart grid applications: A survey. *J. Netw. Comput. Appl.* **2016**, *74*, 133–148.

107. Martin-Martínez, F.; Sanchez-Miralles, A.; Rivier, M. A literature review of microgrids: A functional layer based classifiction. *Renew. Sustain. Energy Rev.* **2016**, *62*, 1133–1153. [CrossRef]

108. Moein, M.; Hassan, F.; Ali, P.; Siamak, A. Smart grid adaptive volt-VAR optimization: Challenges for sustainable future grids. *Sustain. Cities Soc.* **2017**, *28*, 242–255.

109. Nazmus, S.N.; Khandakar, A.; Mark, A.G.; Manoj, D. A survey of smart grid architectures, applications, benefits and standardization. *J. Netw. Comput. Appl.* **2016**, *76*, 23–36.

110. López, G.; Moreno, J.; Amarís, H.; Salazar, F. Paving the road toward Smart Grids through large-scale advanced metering infrastructures. *Electr. Power Syst. Res.* **2015**, *120*, 194–205. [CrossRef]

111. Sudip, M.; Samaresh, B.; Tamoghna, O.; Hussein, T.M.; Alagan, A. ENTRUST: Energy trading under uncertainty in smart grid systems. *Comput. Netw.* **2016**, *110*, 232–242.

112. Adriano, F.; Ângela, F.; Olivier, C.; Paulo, L. Extension of holonic paradigm to smart grids. *IFAC* **2015**, *48*, 1099–1104.

113. Ozan, E.; Nikolaos, G.P.; Iliana, N.P.; Anastasios, G.B.; João, P.S.C. A new perspective for sizing of distributed generation and energy storage for smart households under demand response. *Appl. Energy* **2015**, *143*, 26–37.

114. Dušan, B.; Miloš, P. Impact of electric-drive vehicles on power system reliability. *Energy* **2015**, *83*, 511–520.

115. Ozan, E.; Nikolaos, G.P.; João, P.S.C. Overview of insular power systems under increasing penetration of renewable energy sources: Opportunities and challenges. *Renew. Sustain. Energy Rev* **2015**, *52*, 333–346.

116. Andreas, P. Sustainable options for electric vehicle technologies. *Renew. Sustain. Energy Rev.* **2015**, *41*, 1277–1287.

117. Antonio, C.-S.; Cipriano, R.-R.; David, B.-D.; Eduardo, C.-F. Distributed generation: A review of factors that can contribute most to achieve a scenario of DG units embedded in the new distribution networks. *Renew. Sustain. Energy Rev.* **2016**, *59*, 1130–1148.

118. Bhatti, A.R.; Salam, Z.; Aziz, M.J.B.; Yee, K.P.; Ashique, R.H. Electric vehicles charging using photovoltaic: Status and technological review. *Renew. Sustain. Energy Rev.* **2016**, *54*, 34–47. [CrossRef]

119. Olivier, B.; Eric, L.; Ghislain, R.; Eric, B. Efficiency-optimal power partitioning for improved partial load efficiency of electric drives. *Electr. Power Syst. Res.* **2017**, *142*, 176–189.

120. Ahmed, T.E.; Ahmed, A.M.; Osama, A.M. DC microgrids and distribution systems: An overview. *Electr. Power Syst. Res.* **2015**, *119*, 407–417.

121. Amany, E.-Z. Application of smart grid specifications to overcome excessive load shedding in Alexandria, Egypt. *Electr. Power Syst. Res.* **2015**, *124*, 18–32.

122. Jagruti, T.; Basab, C. Intelli-grid: Moving towards automation of electric grid in India. *Renew. Sustain. Energy Rev.* **2015**, *42*, 16–25.

123. Meng, J.; Mu, Y.; Jia, H.; Wu, J.; Yu, X.; Qu, B. Dynamic frequency response from electric vehicles considering travelling behavior in the Great Britain power system. *Appl. Energy* **2016**, *162*, 966–979. [CrossRef]

124. Saber, F.; Seyed, A.T.; Mohammad, S. A new smart charging method for EVs for frequency control of smart grid. *Electr. Power Energy Syst.* **2016**, *83*, 458–469.

125. Sun, Y.; Li, N.; Zhao, X.; Wei, Z.; Sun, G.; Huang, C. Robust H∞ load frequency control of delayed multi-area power system with stochastic disturbances. *Neurocomputing* **2016**, *193*, 58–67. [CrossRef]

126. Chinthaka, S.; Ozansoy, C. Frequency response due to a large generator loss with the increasing penetration of wind/PV generation—A literatu rereview. *Renew. Sustain. Energy Rev.* **2016**, *57*, 659–668.

127. Lakshmanan, V.; Marinelli, M.; Hu, J.; Bindner, H.W. Provision of secondary frequency control via demand response activation on thermostatically controlled loads: Solutions and experiences from Denmark. *Appl. Energy* **2016**, *173*, 470–480. [CrossRef]

128. Miao, K.; Ramachandaramurthy, V.K.; Yong, J.Y. Integration of electric vehicles in smart grid: A review on vehicle to grid technologies and optimization techniques. *Renew. Sustain. Energy Rev.* **2016**, *53*, 720–732.

129. Heydarian-Forushani, E.; Golshan, M.; Shafie-khah, M. Flexible interaction of plug-in electric vehicle parking lots for efficient wind integration. *Appl. Energy* **2016**, *179*, 338–349. [CrossRef]

130. Zhou, B.; Littler, T.; Meegahapola, L.; Zhang, H. Power System steady-state analysis with large-scale electric vehicle integration. *Energy* **2016**, *115*, 289–302.

131. Tsado, Y.; Lund, D.; Gamage, K.A.A. Resiliet communication for smart grid ubiquitous sensor network: State of the art and prospects for next generation. *Comput. Commun.* **2015**, *71*, 34–49. [CrossRef]

132. Faeze, B.; Masoud, H.; Shahram, J. Optimal electrical and thermal energy managment of a residential energy hub, integrating demand response and energy storage system. *Energy Build.* **2015**, *90*, 65–75.

133. Howlader, H.O.R.; Matayoshi, H.; Senjyu, T. Distributed generation integrated with thermal unit commitment considering demand response for energy storage optimization of smart grid. *Renew. Energy* **2016**, *99*, 107–117. [CrossRef]

Article

Maximum Power Point Tracking for Brushless DC Motor-Driven Photovoltaic Pumping Systems Using a Hybrid ANFIS-FLOWER Pollination Optimization Algorithm

Neeraj Priyadarshi [1], Sanjeevikumar Padmanaban [2,*], Lucian Mihet-Popa [3,*], Frede Blaabjerg [4] and Farooque Azam [1]

[1]   Department of Electrical and Electronics Engineering, Millia Institute of Technology, Purnea 854301, India; neerajrjd@gmail.com (N.P.); farooque53786@gmail.com (F.A.)
[2]   Department of Energy Technology, Aalborg University, 6700 Esbjerg, Denmark
[3]   Faculty of Engineering, Østfold University College, Kobberslagerstredet 5, 1671 Kråkeroy-Fredrikstad, Norway
[4]   Center for Reliable Power Electronics (CORPE), Department of Energy Technology, Aalborg University, Aalborg 9220, Denmark; fbl@et.aau.dk
*   Correspondence: san@et.aau.dk (S.P.); lucian.mihet@hiof.no (L.M.-P.); Tel.: +47-922-713-53 (L.M.-P.)

Received: 18 March 2018; Accepted: 24 April 2018; Published: 26 April 2018

Abstract: In this research paper, a hybrid Artificial Neural Network (ANN)-Fuzzy Logic Control (FLC) tuned Flower Pollination Algorithm (FPA) as a Maximum Power Point Tracker (MPPT) is employed to amend root mean square error (RMSE) of photovoltaic (PV) modeling. Moreover, Gaussian membership functions have been considered for fuzzy controller design. This paper interprets the Luo converter occupied brushless DC motor (BLDC)-directed PV water pump application. Experimental responses certify the effectiveness of the suggested motor-pump system supporting diverse operating states. The Luo converter, a newly developed DC-DC converter, has high power density, better voltage gain transfer and superior output waveform and can track optimal power from PV modules. For BLDC speed control there is no extra circuitry, and phase current sensors are enforced for this scheme. The most recent attempt using adaptive neuro-fuzzy inference system (ANFIS)-FPA-operated BLDC directed PV pump with advanced Luo converter, has not been formerly conferred.

Keywords: ANFIS; artificial neural network; brushless DC motor; FPA; maximum power point tracking; photovoltaic system; root mean square error

---

1. Introduction

As conventional energy sources are depleting day by day, the demand forrenewable energy sources is raising [1–3]. Solar energy sources are promising renewable energy sources for developed and developing nations due to being free, abundant, and environmentally friendly. Standalone photovoltaic (PV) systems for water-pumping applications are employed in remote areas [4,5]. Because of grid absence in remote places, standalone PV water pumping is installed for agricultural and household applications. Various electric motors have been used to drive the pumping system [6,7]. The DC motor-based pumping system requires maintenance because of commutator and brush presence. Therefore, DC motors are not frequently used for PV pumping applications. Single-phase induction motors have also been used for driving low-inertia torque load. Due to a complex control strategy, the induction motors are not efficient for pumping applications. Therefore, in this research work, a brushless DC (BLDC) motor has been considered as it has simple design control, low power range and requires maintenance-free operation compared to AC motors [8]. Distinct DC-DC converters were contenders for optimizing PV module generated power with a soft-starting and controlling motor pump system [9–11]. The contemporary PV system has insubstantial converse competency. Therefore, Maximum Power Point Trackers (MPPT) is the indispensable constituents required for optimal power tracking from PV modules. Numerous MPPT methods have been occupied viz. Perturb and Observe (P&O), Increment Conductance (INC), Fraction Short/Open circuit etc. [12–14]. Under steady-state operating conditions, particular algorithms provide high outturn. However, these algorithms are found lacking under adverse weather conditions showing slow convergence velocity and being unable to achieve a global power point (GPP) for partial shading situations with high power oscillations around this point. Recently, different intelligent techniques viz. Fuzzy Logic Control (FLC) and Artificial Neural Network (ANN) have been employed for PV tracking [15]. However, because of complex fuzzy inference rules and individual sensor requirements, meta-heuristic algorithms have been employed recently. Genetic algorithms and artificial immune systems (AIS) are meta-heuristic algorithms used for non-linear stochastic problem solutions [16,17]. These algorithms are capable to resolve non-linear complication. However, due to a large population size and adaptive immune cell mechanism, the GA and AIS algorithms, respectively, have low velocity of convergence with large computational period. However, the implementation of selection, mutation and crossover process is complex with reduced convergence computational period. Currently, bio-inspired and swarm optimization have been derived as MPPT techniques. The particle swarm optimization is an evolutionary methodology based onthe nature of a swarm that can reduce oscillations around GPP [18]. The classical particle swarm optimization PSO technique has randomness in acceleration value with high regulation parameters as major problems. Nevertheless, variance of this algorithm is capitulated when randomness is miniaturized. Surrogating to swarm techniques, current bio-inspired algorithms viz. Firefly Algorithms (FA), Artificial Bee Colony (ABC), Cuckoo Search etc. are considered as bio-inspired MPPT and have the advantages of high convergence speed, and less transience with fast tracked performance [19–21]. Nevertheless, because of a lower number of bees, ABC technique has a slow velocity of convergence under fluctuating weather situations. Because of the deviating movement of a large number of fireflies, the response of the system becomes slow with the prerequisite high computational period. The Cuckoo search algorithm provides an efficient solution to non-linear problems. However, because of complex nest population and inadequate contingency, the Cuckoo technique has slow convergence velocity under varying environmental conditions. However, the implementation complexities with the tuning of parameters are amajor hindrance of this finding. The above-mentioned algorithms' drawbacks can be handled by applying Flower Pollination (FPA) as an MPPT technique. Ram et al. [22] has discussed the FPA algorithm for PV MPPT under dynamic operating conditions. This algorithm provides single-stage global searching, simpler coding, and lower tuned specification requirements with low cost implementation and has fast response compared to P&O and PSO techniques under dynamic weather conditions. Included in this work, a novel flower pollination algorithm is contemplated and associated with the hybrid ANFIS MPPT [23] algorithm. Compared to the FPA algorithm, the merits of the hybrid ANFIS-Flower Pollination Algorithm (FPA) are simple implementation, high convergence speed with tune parameters and easier code compilation. Due to presence of parasitic components, the voltage outcomes and power transform adequacy are restrained in classical switched-power converters. However, re-lift/triple lift methodology is employed by Luo converter to enhance the voltage balance and run-over the limiter issue. Equated with Zeta, single-ended primary-inductor converter (SEPIC) and Cuk converters, the Luo converter has accurate steady and dynamic system behavior. However, the Cuk converter comprises maximum transients and more settled periods with average system performance of SEPIC and Zeta compared to the Luo converter, which is applicable for upraised power utility and electrical drive operation [24,25]. In contrast with different employed power converters, modern Luo convertershave been considered for this research approach as they deliver better power/density ratio with economical implementation. The most recent attempt, using adaptive neuro-fuzzy inference system (ANFIS)-FPA-operated BLDC directed PV pump with advanced Luo converter, has not been formerly conferred and examined using dSPACE (DS1104) platform under changing weather conditions.

2. Complete System Formation

Figure 1 illustrates the Luo converter-employed BLDC-driven PV pumping for a remote location. A hybrid ANFIS-FPA MPPT controller is operated to produce required pulse for power switched of Luo converter. This converter delivers better power/density ratio with economical implementation with interface between the inverter power circuit and solar system. Moreover, electronic commutation methodology controls voltage source inverter (VSI) employed BLDC motor in which winding current is adjusted with the help of a decoder in a proper sequence.

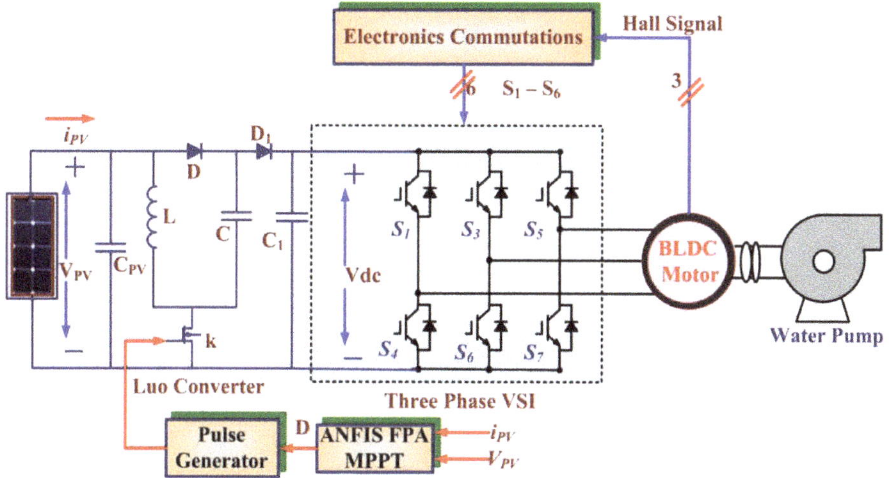

Figure 1. BLDC-driven Photovoltaic Complete System Formation.

2.1. PV Generator

In this research work, a two-diode PV cell model is considered (Figure 2) because it is a simple and accurate model compared to the single-diode PV cell. By means of photoelectric effect, the conversion of solar energy to electricity takes place and output power can be enhanced by connecting numerous solar cells in shunt or series as required. Both diodes are employed to represent polarization occurrence with current source exhibiting sun insolation, followed by power loss delivered by resistances (series/Parallel) used. The prognosis of the overall system is calculated based on accurate equivalent modeling. The output of the PV current is expressed mathematically as [26]:

$$I_{PVo} = I_{PVG} - I_{RSC}(I' + 2) - \left(\frac{V_{PVo} + I_{PVo} \times R_{series}}{R_{Parallel}} \right) \tag{1}$$

where,

$$I' = exp\left(\frac{V_{PVo} + I_{PVo} \times R_{series}}{V_{Thermal}} \right) + exp\left(\frac{V_{PVo} + I_{PVo} \times R_{series}}{A \times V_{Thermal}} \right) \tag{2}$$

I_{PVG} = Photo Current

I_{RSC} = Diode reverse saturation current

I_{PVo} = Output PV current

V_{PVo} = PV output voltage

R_{series} = Resistance in series

$R_{Parallel}$ = Resistance in parallel
$V_{Thermal}$ = PV module thermal voltage
A = Ideality constant of diode

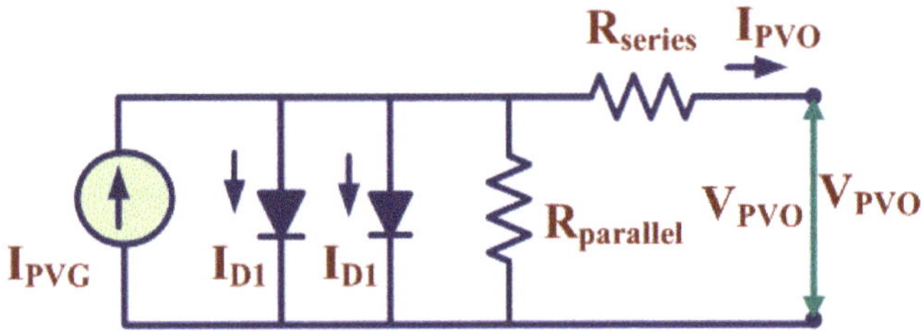

Figure 2. Two diode PV cell model.

2.2. Luo Converter Mathematical Modeling

Renewable technology comprises DC-DC topologies for yield of energy harvest with admissible proficiency. With respect to other DC-DC converters, modern Luo topology depicted in Figure 3 delivers reasonable cost, better power/density ratio and enhanced transformation efficiency. It comprises the least ripple content with geometric output voltage and surpasses the parasitic element action. The auxiliary benefit of this topology is switched components, which take ground as a reference. In addition to that, the input inductor smoothes the ripple present to input source. Employed capacitors get charged to stated value to accomplish high voltage leveled. Table 1 presents the designed parameters of Luo Converter used during practical implementation.

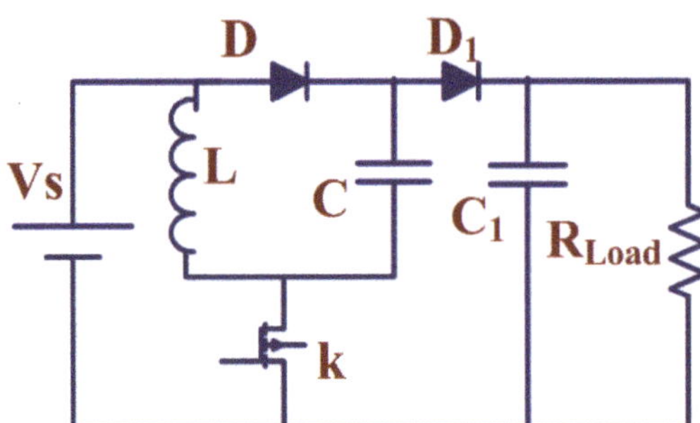

Figure 3. Power Circuit Luo converter.

Table 1. Luo converter parameter.

S.N	Parameters	Values
1.	Inductor (L)	0.02 mH
2.	Capacitor (C and C_1)	20 µF, 15 µF
3.	Switching Frequency (f_{pulse})	10 KHz
4.	Duty Ratio (d_{duty})	0.58

Transfer gain voltage is evaluated as [24,25]:

$$\frac{V_0}{V_S} = \frac{2 - d_{duty}}{1 - d_{duty}} \tag{3}$$

Relation between inductor ripple current and duty cycle is expressed as:

$$\Delta I_{L_{Ripple}} = \frac{V_S \times d_{duty}}{f_{Pulse} \times L} \tag{4}$$

Capacitors ($C=C_1$) values are determined mathematically as:

$$C = C_1 = \frac{\left(1 - d_{duty}\right) \times V_0}{f_{Pulse} \times R_{Load} \times \Delta V_0} \tag{5}$$

where,

d_{duty} = Duty ratio

f_{Pulse} = Frequency of Switched pulse

V_0 = Output Voltage of Luo Converter

2.3. A Hybrid Proposed FLC-ANN Tuned FPA MPPT

In this proposed scheme, the hybrid ANFIS-FPA MPPT algorithm is realized for maximizing PV outturn and accurate motion control with PV-pump interface. The FLC data is trained by ANN which is finally optimized by FPA method, leading to minimum RMSE of FLC and ANN. It comprises the dominance of both FLC and ANN. The threshold and weight of NN models are optimized by FPA algorithm to produce minimum RMSE. Figure 4 depicts the complete structure of hybrid learning in which learning data has been achieved from FLC architecture. The FLC architecture comprises fuzzification, Inference Rule base and defuzzification as elemental constituents. Real variables are converted to linguistic parameters using fuzzification. The requisite output is introduced by the Mamdani fuzzy inference rule deployed by max-min composition. With the help of centroid method, the defuzzification process converts the linguistic parameters to real values. Parameters used in FLC and ANN are presented using Table 2. Employed membership values are illustrated in Figure 5. The FPA method of MPPT is predicted by reproduction of flower of transferring pollen. This convection is possible through biotic/cross and abiotic/self-pollination. In cross-pollination the pollens are translated between two unlike flowers. On the other hand, abiotic pollination takes place between distant species. It is noted that in flower pollination 90% possibility of cross-pollination and only 10% possibility of self-pollination happen, which is limited in the probability range Rε[1,0]. Table 3 describes ANFIS-FPA parameters used for practical validation of BLDC-driven PV pumping. The complete process is based on the following 4 rules [22]. The min-max composition (Mamdani's rule) is employed to calculate the fuzzy error (E) and change in error (dE/CE) input as [23]:

$$E(r) = \frac{dP_{PV}(r)}{dV_{PV}(r)} \tag{6}$$

$$dE = E(r) - E(r-1) \tag{7}$$

$$\mu_{P \to Q}(xy) = min\left[\mu_P(x), \mu_Q(y)\right], \qquad \forall P \in X, \forall Q \in Y \tag{8}$$

where,

$\mu_P(x)$ = Membership function of P fuzzy set in X Universe of discourse

$\mu_Q(y)$ = Membership function of Q fuzzy set in Y Universe of discourse

$X, Y = x, y$ variables defined in Universe X and Y, respectively

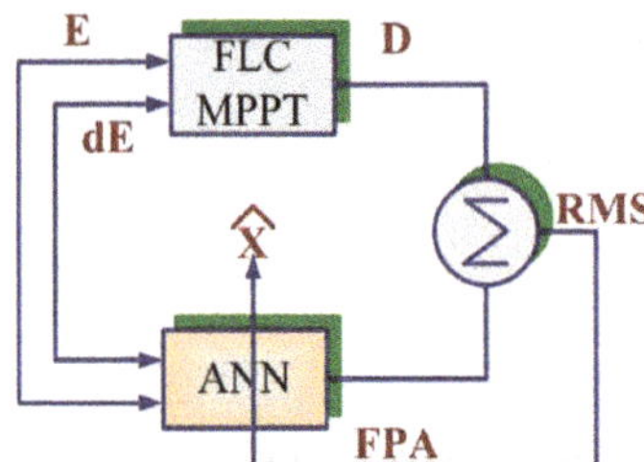

Figure 4. Complete structure of hybrid ANFIS-FPA.

Table 2. Parameters used in FLC and ANN.

S.N	Parameters	Value
1	Total fuzzy rule base fired	25
2	Total number of Epoch	740
3	Types of membership function	Gaussian type
4	Total layer (neural network)	5
5	Total neural network training data sets	200

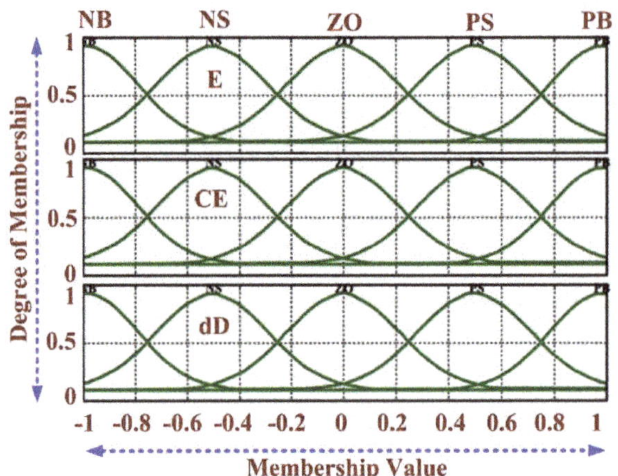

Figure 5. Employed membership values.

Table 3. ABC-FPA Parameters.

S.N	Parameters	Values
1.	Switched Probability (P_f)	0.7
2.	Scaling Factor	1.25
3.	No of Epoch	740
4.	RMSE (Obtained)	106×10^{-6}
5.	Total Rule Based Fired	25
6.	ANFIS Obtained (Training Error)	0.6255×10^{-6}

Output D is calculated as:

$$\hat{D} = \frac{D \in S \int \mu_D(D)dD}{D \in S \int \mu_D(D)} \tag{9}$$

where,

$\hat{D}$ = Crisp output
$\mu_D(D)$ = Membership function (Aggregated)
D = Fuzzy output
S = Subarea/Universe of discourse

Also, neural-fuzzy network output ($\overline{D}$) expressed mathematically as:

$$\overline{D} = \mu_A(E) \times \mu_B(dE) \times W_{Li} \tag{10}$$

where,

$\mu_A(E)$ = Membership function of fuzzy set A in E universe of discourse
$\mu_B(dE)$ = Membership function of fuzzy set B in dE universe of discourse
W_{Li} = Weight of consequent ith layer

The ANFIS objective function is expressed mathematically as:

$$RMSE = \left[\frac{1}{P} \sum_{i=1}^{P} (D - \overline{D})^2 \right]^{1/2} \tag{11}$$

where,

P = Total sample
D = Fuzzy output
$\overline{D}$ = Neural network output

Rule I: Biotic pollination uses levy flight for transferring pollens and is called global pollination in which the ith pollen solution vector is expressed mathematically using Equation (12). The Levy flight factor is accountable for pollens transport which improves the methodology of pollination where scaling factor is responsible to limit step size.

$$X_i^{T+1} = X_i^T + L_f \times \gamma_{scaling} \times \left(X_i^T - G_{best} \right) \tag{12}$$

where,

X_i^T = Vector representing solution
T = No. of iteration
L_f = Levy flight factor
$\gamma_{scaling}$ = Scaling factor
G_{best} = Global best solution

Rule II: Self-pollination is termed as local pollination and characterized mathematically as:

$$X_i^{T+1} = X_i^T + P_f \times \left(X_m^T - X_n^T \right) \tag{13}$$

X_m^T and X_n^T = two unlike pollen in the species
P_f = Switched probability

Rule III: The performance of the flower is assumed to be identical to the probability of reproduction which is equivalent to resemblance of two concerned flowers.

Rule IV: Pollination is interchanged between global and local, which depends on switching probability which lies between 0 and 1.

The proposed nature-inspired FPA algorithms are responsible for providing proper learning of the neural network to reduce root mean square error between outcomes of D(Fuzzy output) and $\overline{D}$ (Neural network output). Pollen position (duty ratio) is updated using biotic/abiotic pollination for the next iteration. Under step variation in solar insolation, the corresponding variance in voltage/current threshold is expressed mathematically as [22]:

$$\frac{dP_{PV}(n)}{dV_{PV}(n)} \geq 0.2 \tag{14}$$

$$\frac{dI_{PV}(n)}{I_{PV}(n)} \geq 0.1 \tag{15}$$

where,

$V_{PV}(n)$ =nth iteration PV voltage

$I_{PV}(n)$ =nth iteration PV current

$dV_{PV}(n)$ = change in PV voltage (nth and $(n-1)$th iteration)

2.4. Electronic BLDC Commutator and VSI Switching

Commutation in Permanent Magnet DC Motor (PMDC) is obtained by a commutator and brushes. Nevertheless, hall sensors are important components employed in BLDC motors which sense the position of a rotor as the commutation wave. Coils and permanent magnets are employed as stator and rotor respectively, in which stator's magnetic field rotates the rotor. Armature of a BLDC motor consists of a permanent magnet as a substitute of the coil which does not require brushes. Figure 6 demonstrates BLDC-driven structure with induced EMF and reference current. The electronic commutation process is used to control the VSI-employed BLDC motor in which winding current is adjusted with the help of decoder in proper sequence. In this method, symmetrical DC currents are situated in the phase voltage at $120°$. Based on the motor alignment, the hall sensors produce signals of $60°$ phase difference. The gating signal for 3-phase VSI generated by transforming hall signals using the decoder is illustrated in Figure 7. The pulse width modulated pulses are generated by comparing triangular signal with duty cycle produced through MPPT. Table 4 portrays Hall signals and Switching states of BLDC used with electronic commutation. The high-frequency PWM pulses and six fundamental signals are operated with an AND gate, which produces 6 gating pulses for VSI inverter. As the atmospheric conditions change, the duty cycle is also regulated using MPPT methods which control the VSI and finally the BLDC motor is adjusted accordingly.

The BLDC motor is analyses mathematically as [27]:

$$\begin{bmatrix} V_{ap} \\ V_{bp} \\ V_{cp} \end{bmatrix} = \begin{bmatrix} R_T & 0 & 0 \\ 0 & R_T & 0 \\ 0 & 0 & R_T \end{bmatrix} \begin{bmatrix} I_{ap} \\ I_{bp} \\ I_{cp} \end{bmatrix} + \begin{bmatrix} L_1 - M_1 & 0 & 0 \\ 0 & L_1 - M_1 & 0 \\ 0 & 0 & L_1 - M_1 \end{bmatrix} \frac{d}{dx} \begin{bmatrix} I_{ap} \\ I_{bp} \\ I_{cp} \end{bmatrix} + \begin{bmatrix} E_{ba} \\ E_{bb} \\ E_{bc} \end{bmatrix} \tag{16}$$

Developed electromagnetic torque (T_{EM}) by BLDC motor can be expressed mathematically as:

$$T_{EM} = \frac{E_{ba} \times I_{ap} + E_{bb} \times I_{bp} + E_{bc} \times I_{cp}}{\omega_{rotor}} \tag{17}$$

where,

V_{ap}, V_{bp}, V_{cp} = Phase voltage of a 3-Phase BLDC motor

I_{ap}, I_{bp}, I_{cp} = Phase Currents

E_{ba}, E_{bb}, E_{bc} = Phase Back EMF of BLDC motor

L_1 = Each Phase self-inductance

M_1 = Two phase's mutual inductance

T_{EM} = Developed Electromagnetic torque of BLDC motor

ω_{Rotor} = Rotor Speed

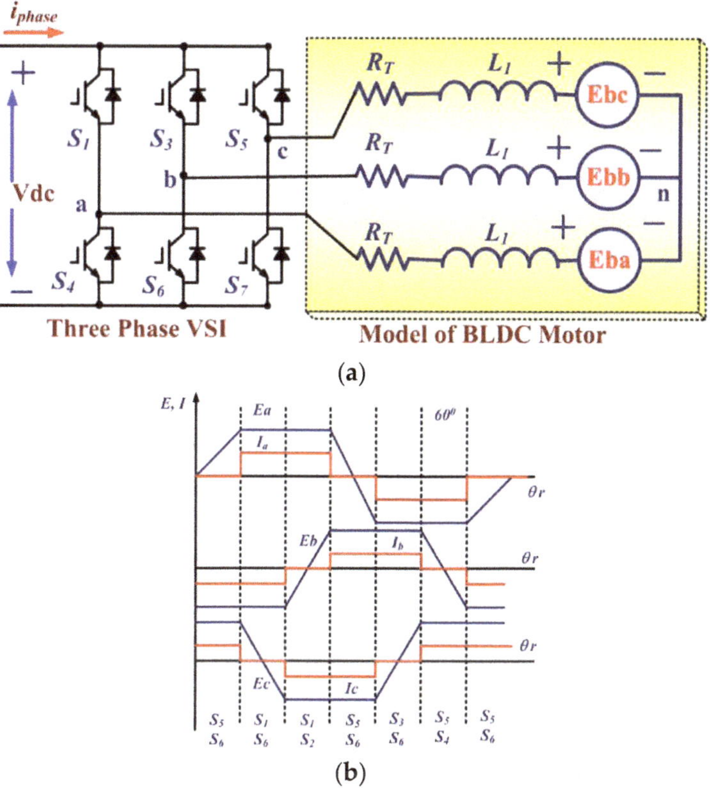

Figure 6. (**a**) BLDC driven structure (**b**) induced EMF and reference current.

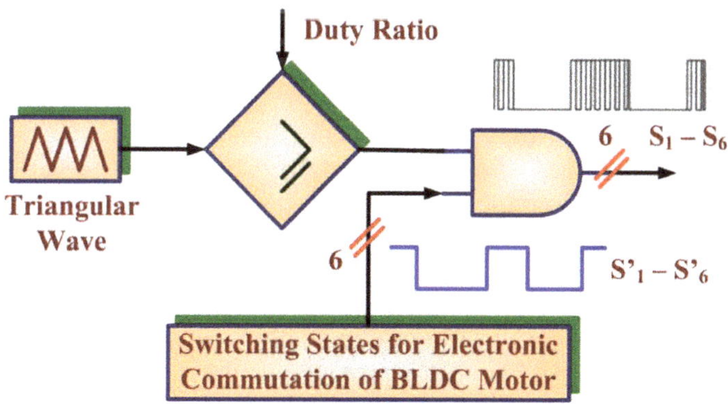

Figure 7. Gating signal for 3-phase VSI.

Table 4. Hall signals and Switching states.

Angle	Hall Signals			Switching States					
	H_1	H_2	H_3	S_1	S_2	S_3	S_4	S_5	S_6
$0-\pi/3$	1	0	1	0	1	1	0	0	0
$\pi/3-2\pi/3$	0	0	1	0	1	0	0	1	0
$2\pi/3-\pi$	0	1	1	0	0	0	1	1	0
$\pi-4\pi/3$	0	1	0	1	0	0	1	0	0
$4\pi/3-5\pi/3$	1	1	0	1	0	0	0	0	1
$5\pi/3-2\pi$	1	0	0	0	0	1	0	0	1

3. Experimental Results

Performance justification of the BLDC-driven PV pumping-employed Luo converter has been done through the dSPACE controller. For the purposes of MPPT operation, LA-55/LV-25 as current/voltage sensors are employed during practical implementation. Figure 8 portrays the BLDC-driven Luo converter-employed PV-pumping hardware developed in the laboratory. With the help of an A/D converter, analog pulses are transformed to digital and fed to the dSPACE interface. Electronic commutation/controlling BLDC has been executed by obtaining hall pulses from the input/output terminal and then generated pulses are outturned to the inverter.

Figure 8. BLDC-driven Luo converter-employed PV-pumping hardware.

3.1. Steady-State Performance

The experimental behaviors of the PV module and motor pumping system have been tested under steady-state condition of irradiance level 1000 W/m^2. The proposed MPPT design technique is working effectively and tracks optimal power from PV module with unity duty cycle at 1000 W/m^2 solar insolation level depicted in Figure 9. Practical results obtained for the BLDC-driven Luo converter-employed PV pumping are described in Figure 9a PVG at 1000 W/m^2. (Figure 9b) BLDC performance at 1000 W/m^2. (Figure 9c) generated hall sensor pulses at 1000 W/m^2 (Figure 9d) switched and hall pulses at 1000 W/m^2 (Figure 9e) BLDC performance at 300 W/m^2 (Figure 9f) switched and hall pulses at 300 W/m^2.The corresponding BLDC motor and torque (1500 rpm) has been demonstrated in Figure 9d presents the obtained hall sensor pulses with motor torque. The performance of the BLDC motor-pumping system has been evaluated with 300 W/m^2 solar irradiance. The motor torque is experimentally obtained, which is sufficient to operate PV water pumping. Based on duty cycle generation using the MPPT algorithm, the corresponding hall signals have been generated to trigger six switches of the inverter.

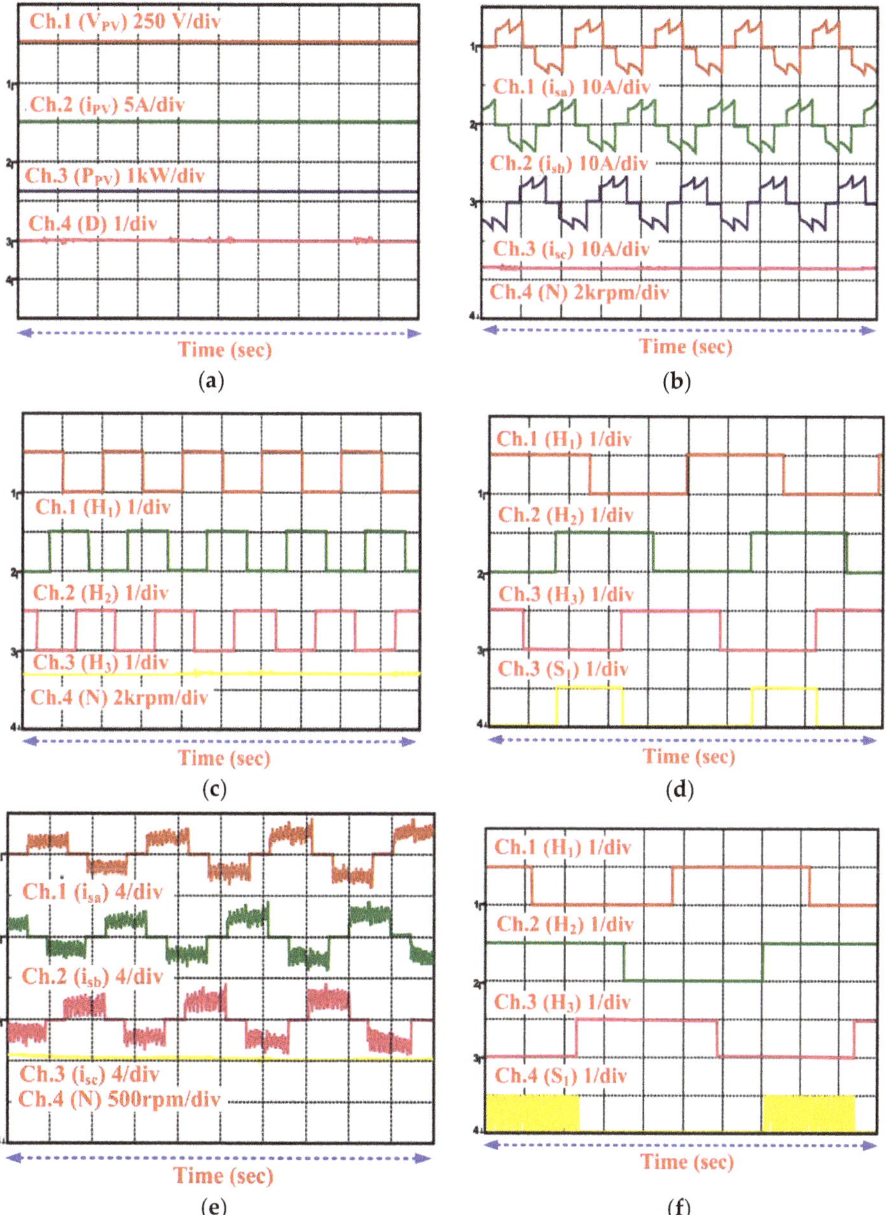

Figure 9. BLDC-driven Luo converter-employed PV pumping (**a**) PVG at 1000 W/m^2;(**b**) BLDC performance at 1000 W/m^2; (**c**) generated hall sensor pulses at 1000 W/m^2; (**d**) switched and hall pulses at 1000 W/m^2; (**e**) BLDC performance at 300 W/m^2; (**f**) switched and hall pulses at 300 W/m^2.

3.2. Dynamic Behavior of PV System

The effective practice of recommended PV pumping system was proved under varying sun insolation levels. In this experiment, solar irradiance level is varied from 300 W/m^2 to 1000 W/m^2. According to variation in sun irradiance level, corresponding changes in PV current, DC link voltage, BLDC stator current and motor torque have been verified (Figure 10) and PV pumping is running without any interruption. The duty cycle for BLDC-PV pump control is generated with variation in sun insolation accordingly and outstanding motion control has been comprehended.

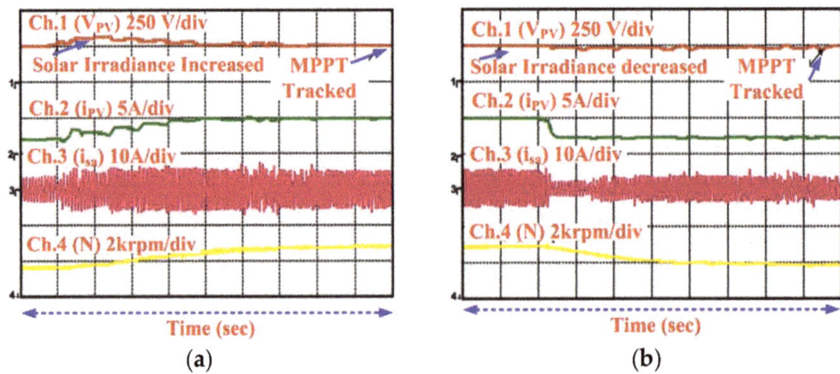

Figure 10. BLDC-driven Luo converter (**a**) increased solar irradiance (**b**) decreased solar irradiance.

3.3. Behavior at Starting

Practical results found in Figure 11 interpret the safe starting of the BLDC motor under irradiance level 1000 W/m^2 and 300 W/m^2. Initially, the duty cycle is kept at 0.5 to run the motor. The sufficient motor speed is obtained by controlling the starting current, which runs the motor-pump system successfully. Figure 11 portrays the successful action of BLDC-PV pump at the start by limiting starting current, which reveals the progression with safe and soft start. The obtained results prove the more relevant performance conducted for the EMI reduction and soft starting for the experimental test conducted in [28].

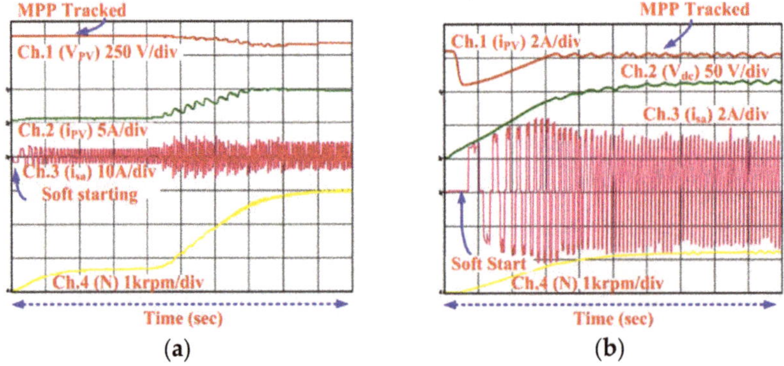

Figure 11. BLDC driven Luo converter employed PV pumping under soft starting (**a**) 1000 W/m^2; (**b**) 400 W/m^2

Table 5 portrays laboratory-adopted BLDC specification for a motion-controlled PV pump. Figure 12 interprets the existing global nature of the PV system under divergent sun radiation, which is demonstrated by the dark line. The operation begins with open-circuit voltage ($V_{OPENCkt}$ state) and reaches a global power point with variable solar irradiance. With application of hybrid ANFIS-FPA MPPT, steady GPP is attained over a complete day. The performance of the MPPT controllers for two algorithms ANFIS-FPA and FPA are tested with stepped irradiance input. Figure 13a illustrates that the proposed ANFIS-FPA imparts accurate and precise PV system outcomes with zero variation around GPP with fluctuating sun insolation. However, the FPA employed algorithm provides inconsistent and more oscillation nearby GPP that equates to the ANFIS-FPA algorithm described using Figure 13b. Under these situations, ANFIS-FPA has high tracked PV power with proportionately less GPP time. Practical results demonstrate that ANFIS-FPA algorithm contributes rapid and insignificant swinging differentiated with FPA MPPT illustrated in Figure 13a,b. Figure 14 demonstrates the behavior of numerous MPPT Viz. FPA, PSO, FLC and P and O control under standard test conditions. Under standard test conditions, ANFIS-FPA has better PV tracking efficiency compared to ANFIS-PSO, FLC and P and O methods, as illustrated with Figure 14. A hybrid ANFIS-FPA algorithm has global power point trajectory with the most tracked power and has zero oscillation throughout, equated with different controllers. The PV tracked trajectories are also examined under fluctuating weather situations (Figure 15). Under dynamic weather conditions, the PV tracking trajectory is found to be more accurate compared to conventional algorithms and has a zero GPP oscillation around this point, which is explained by Figure 15. Practical results reveal that ANFIS-FPA-optimized MPPT provides optimal tuning with high performance index.

Table 5. Laboratory adopted BLDC specification.

S.N	Parameters	Value
1	Resistance of stator	4.16 Ω
2	Inductance value of stator	2.2 mH
3	Speed rating	1500 rpm
4	Number of Pole pair	2
5	Constants(Voltage & torque)	86 V_{LL}/KRPM & 0.85 Nm/Ampere

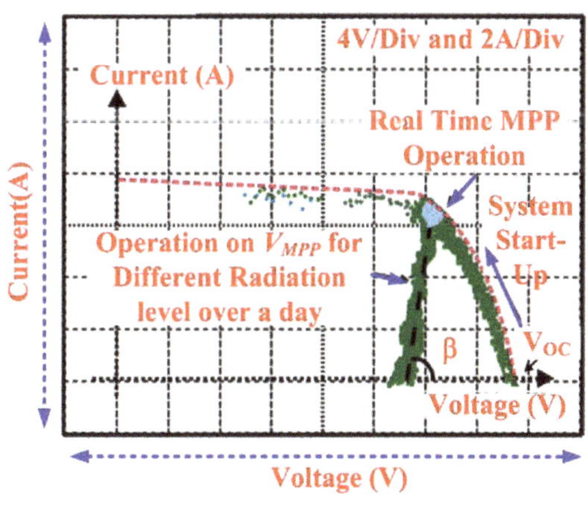

Figure 12. Existent global nature of PV system under divergent sun radiation.

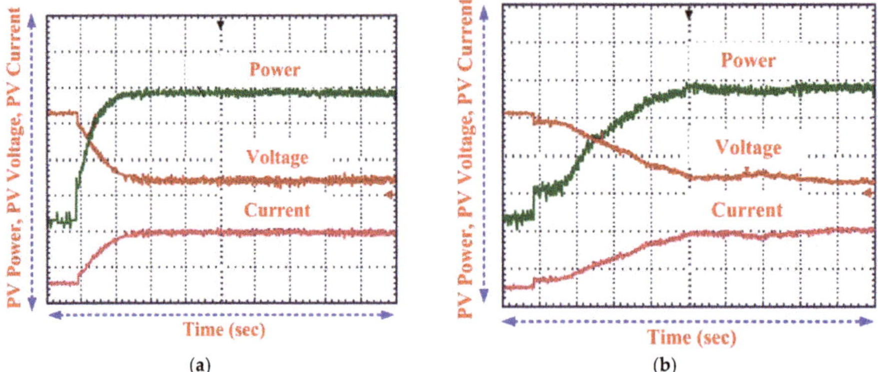

Figure 13. Behavior of MPPT under stepped irradiance (**a**) Hybrid ANFIS-FPA; (**b**) FPA.

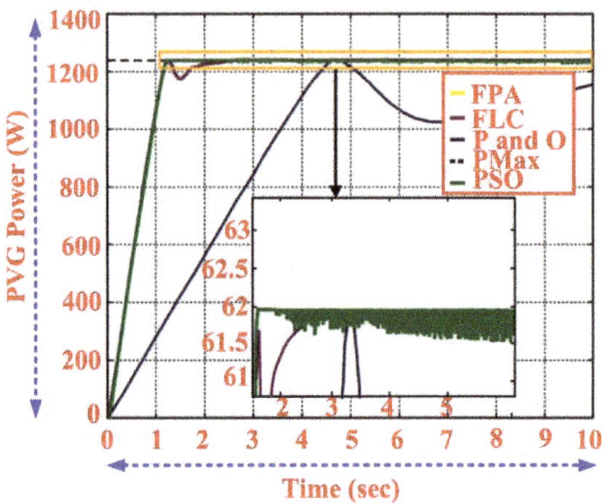

Figure 14. Behavior of numerous MPPT control under standard test conditions.

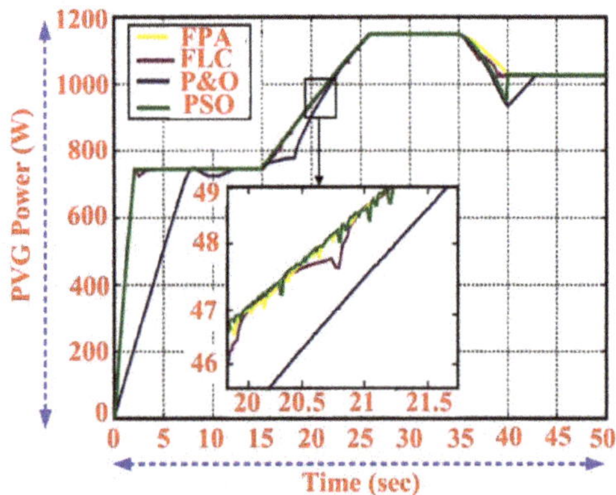

Figure 15. PV tracked trajectories examined under fluctuating weather situations.

4. Conclusions

The Luo converter-based BLDC-driven PV pumping with ANFIS-FPA MPPT has been demonstrated under varying weather conditions using the dSPACE platform. The Luo converter has been proposed for desired GPP functions and is responsible for updating the duty ratio in each iteration using biotic/abiotic pollination. The PV-fed BLDC motor drive pumping system operates effectively under steady, dynamic states and soft-starting operating conditions, which have been validated through experimentally obtained responses. The enforcement of the ANFIS-FPA MPPT controller has been equated with the general P&O and ANFIS-PSO methods, which gives high tracking efficiency, fast design, and rapid convergence time under varying solar irradiance level. Performed experimental responses reveal that, compared to different bio-inspired, swarm-intelligence and classical MPPT techniques reviewed in literature, the ANFIS-FPA has superior power tracking ability, fast convergence velocity and accurate system response. The designed PV-based BLDC-driven pumping system provides the following functions viz. Luo converter-based MPPT tracking reducing switching losses based on electronic commutation/VSI switching, maintenance of the DC-link voltage, and BLDC motor speed regulation under low sun insolation, validated through practical responses using the dSPACE board.

Author Contributions: All authors contributed equally for the decimation of the research article in current form.

Conflicts of Interest: The authors declare no conflict of interest.

References

1. Priyadarshi, N.; Sanjeevikumar, P.; Maroti, P.K.; Sharma, A. An Extensive Practical Investigation of FPSO based MPPT for Grid Integrated PV System under Variable Operating Conditions with Anti Islanding Protection. *IEEE Syst. J.* **2018**. [CrossRef]
2. Lakshman, N.P.; Palanisamy, K.; Babu, N.M.; Sanjeevikumar, P. Photovoltaic- STATCOM with Low Voltage Ride-Through Strategy and Power Quality Enhancement in Grid Integrated Wind-PV System. *Electronics* **2018**, *7*, 51.
3. Jain, S.; Ramulu, C.; Sanjeevikumar, P.; Ojo, O.; Ertas, A.H. Dual MPPT Algorithm for Dual PV Source Fed Open-End Winding Induction Motor Drive for Pumping Application. *Eng. Sci. Technol.* **2016**, *19*, 1771–1780. [CrossRef]
4. Kumar, R.; Singh, B. BLDC Motor Driven Solar PV Array Fed Water Pumping System Employing Zeta Converter. *IEEE Trans. Ind. Appl.* **2016**, *52*, 2315–2322. [CrossRef]
5. Montorfano, M.; Sbarbaro, D.; Mor'an, L. Economic and technical evaluation of solar assisted water pump stations for mining applications: A case of study. *IEEE Trans. Ind. Appl.* **2016**, *52*, 4454–4459. [CrossRef]
6. Alghuwainem, S.M. Speed Control of a PV Powered DC Motor Driving a Self- Excited 3-Phase Induction Generator for Maximum Utilization Efficiency. *IEEE Trans. Energy Convers.* **1996**, *11*, 768–773. [CrossRef]
7. Jain, S.; Thopukara, A.K.; Karampuri, R.; Somasekhar, V.T. A Single-Stage Photo Voltaic System for a Dual-Inverter fed Open-End Winding Induction Motor Drive for Pumping Applications. *IEEE Trans. Power Electron.* **2015**, *30*, 4809–4818. [CrossRef]
8. Sashidhar, S.; Fernandes, B.G. A Novel Ferrite SMDS Spoke-Type BLDC Motor for PV Bore-Well Submersible Water Pumps. *IEEE Trans. Power Electron.* **2017**, *64*, 104–114. [CrossRef]
9. Killi, M.; Samanta, S. An Adaptive Voltage Sensor Based MPPT for Photovoltaic Systems with SEPIC Converter including Steady State and Drift Analysis. *IEEE Trans. Power Electron.* **2015**, *62*, 7609–7619. [CrossRef]
10. Priyadarshi, N.; Kumar, V.; Yadav, K.; Vardia, M. An Experimental Study on Zeta buck-boost converter for Application in PV system. In *Handbook of Distributed Generation*; Springer: Cham, Switzerland, 2017; pp. 393–406.
11. Priyadarshi, N.; Anand, A.; Sharma, A.K.; Azam, F.; Singh, V.K.; Sinha, R.K. An Experimental Implementation and Testing of GA based Maximum Power Point Tracking for PV System under Varying Ambient Conditions Using dSPACE DS 1104 Controller. *Int. J. Renew. Energy Res.* **2017**, *7*, 255–265.

12. Kumar, N.; Hussain, I.; Singh, B.; Panigrahi, B.K. Framework of Maximum Power Extraction from Solar PV Panel using Self Predictive Perturb and Observe Algorithm. *IEEE Trans. Sustain. Energy* **2017**, *9*, 895–903. [CrossRef]

13. Elgendy, M.A.; Zahawi, B.; Atkinson, D.J. Assessment of the Incremental Conductance Maximum Power Point Tracking Algorithm. *IEEE Trans. Sustain. Energy* **2013**, *4*, 108–117. [CrossRef]

14. Zamora, A.C.; Vazquez, G.; Sosa, J.M.; Rodriguez, P.R.M.; Juarez, M.A. Efficiency Based Comparative Analysis of Selected Classical MPPT Methods. In Proceedings of the IEEE International Autumn Meeting on Power, Electronics and Computing, Ixtapa, Mexico, 8–10 November 2017; pp. 1–6.

15. Abu-Rub, H.; Iqbal, A.; Ahmed, SK.M.; Peng, F.Z.; Li, Y.; Baoming, G. Quasi-Z-Source Inverter-Based Photovoltaic Generation System With Maximum Power Tracking Control Using ANFIS. *IEEE Trans. Sustain. Energy* **2013**, *4*, 11–20. [CrossRef]

16. Mohamed, A.A.S.; Berzoy, A.; Mohammed, O. Design and Hardware Implementation of FL-MPPT Control of PV Systems Based on GA and Small-Signal Analysis. *IEEE Trans. Sustain. Energy* **2017**, *8*, 279–290. [CrossRef]

17. Wang, L.; Singh, C. Population-Based Intelligent Search in Reliability Evaluation of Generation Systems with Wind Power Penetration. *IEEE Trans. Power Syst.* **2008**, *23*, 1336–1345. [CrossRef]

18. Koad, R.B.A.; Zobaa, A.F.; El-Shahat, A. A Novel MPPT Algorithm Based on Particle Swarm Optimisation for Photovoltaic Systems. *IEEE Trans. Sustain. Energy* **2017**, *8*, 468–476. [CrossRef]

19. Priyadarshi, N.; Sharma, A.K.; Azam, F. A Hybrid Firefly-Asymmetrical Fuzzy Logic Controller based MPPT for PV-Wind-Fuel Grid Integration. *Int. J. Renew. Energy Res.* **2017**, *7*, 1546–1560.

20. Sundareswaran, K.; Sankar, P.; Nayak, P.S.R.; Simon, S.P.; Palani, S. Enhanced Energy Output From a PV System Under Partial Shaded Conditions Through Artificial Bee Colony. *IEEE Trans. Sustain. Energy* **2015**, *6*, 198–209. [CrossRef]

21. Kalaam, R.N.; Muyeen, S.M.; Al-Durra, A.; Hasanien, H.N.; Al-Wahedi, K. Optimisation of controller parameters for grid tied photovoltaic system at faulty network using artificial neural network-based cuckoo search algorithm. *IET Renew. Power Gen.* **2017**, *11*, 1517–1526. [CrossRef]

22. Ram, J.P.; Rajasekar, N. A novel Flower Pollination based Global Maximum Power Point method for Solar Maximum Power Point Tracking. *IEEE Trans. Power Electron.* **2017**, *32*, 8486–8499.

23. Abouzeid, M.; Sood, V.; Youssef, M. A Comparative Study of a PV-MPPT grid-integrated system under Different Control Techniques. In Proceedings of the IEEEElectrical Power and Energy Conference (EPEC), London, ON, Canada, 26–28 October 2015; pp. 256–261.

24. Pansare, C.; Sharma, S.K.; Jain, C.; Saxena, R. Analysis of a Modified Positive Output Luo Converter and its application to Solar PV system. In Proceedings of the IEEE Industry Applications Society Annual Meeting, Cincinnati, OH, USA, 1–5 October 2017; pp. 1–6.

25. Pachauri, R.K.; Chauhan, Y.K. Modeling and Simulation Analysis of PV Fed Cuk, Sepie, Zeta and Luo DC-DC Converter. In Proceedings of the IEEE1st International Conference on Power Electronics. Intelligent Control and Energy Systems (ICPEICES-2016), Delhi, India, 4–6 July 2016; pp. 1–6.

26. Kaced, K.; Larbes, C.; Ait-Chikha, S.M.; Bounabi, M.; Dahmane, Z.E. FPGA implementation of PSO based MPPT for PV systems under partial shading conditions. In Proceedings of the IEEE6th International Conference on Systems and Control, Batna, Algeria, 7–9 May 2017; pp. 150–155.

27. Niapoor, S.A.KH.M.; Danyali, S.; Sharifian, M.B.B. PV Power System Based MPPT Z-Source Inverter to Supply a Sensorless BLDC Motor. In Proceedings of the 1st Power Electronic & Drive Systems & Technologies Conference, Tehran, Iran, 17–18 February 2010; pp. 111–116.

28. Saponara, S.; Gabriele, C.; Groza, V.Z. Design and Experimental Measurement of EMI Reduction Techniques for Integrated Switching DC/DC Converters. *IEEE Can. J. Electr. Comput. Eng.* **2017**, *40*, 116–127.

Article

Boundary Detection and Enhancement Strategy for Power System Bus Bar Stabilization—Investigation under Fault Conditions for Islanding Operation

Aref Pouryekta [1], Vigna K. Ramachandaramurthy [1], Sanjeevikumar Padmanaban [2,*], Frede Blaabjerg [3] and Josep M. Guerrero [4]

[1] Institute of Power Engineering, Department of Electrical Power Engineering, Universiti Tenaga Nasional, Kajang 43000, Malaysia; aref.pouryekta@gmail.com (A.P.); vigna@uniten.edu.my (V.K.R.)
[2] Department of Energy Technology, Aalborg University, 6700 Esbjerg, Denmark
[3] Centre for Reliable Power Electronics, Department of Energy Technology, Aalborg University, 9000 Aalborg, Denmark; fbl@et.aau.dk
[4] Department of Energy Technology, Aalborg University, 9000 Aalborg, Denmark; joz@et.aau.dk
* Correspondence: sanjeevi_12@yahoo.co.in; Tel.: +91-984-310-8228

Received: 9 December 2017; Accepted: 4 January 2018; Published: 11 April 2018

Abstract: Distribution systems can form islands when faults occur. Each island represents a subsection with variable boundaries subject to the location of fault(s) in the system. A subsection with variable boundaries is referred to as an island in this paper. For operation in autonomous mode, it is imperative to detect the island configurations and stabilize these subsections. This paper presents a novel scheme for the detection of island boundaries and stabilizing the system during autonomous operation. In the first stage, a boundary detection method is proposed to detect the configuration of the island. In the second stage, a dynamic voltage sensitivity factor (DVSF) is proposed to assess the dynamic performance of the system. In the third stage, a wide area load shedding program is adopted based on DVSF to shed the load in weak bus-bars and stabilize the system. The proposed scheme is validated and tested on a generic 18-bus system using a combination of EMTDC/PSCAD and MATLAB software's.

Keywords: islanding; load management; power distribution; voltage fluctuations

---

1. Introduction

In recent decades, environmental concerns about greenhouse gases and global warming caused by fossil fuels have attracted many countries to adopt renewable energy sources (RES). RES-based micro grids are new structures in modern electrical distribution networks. Predefined clusters of dispersed generator (DG) units associated with their vicinity loads provide the opportunity to form a micro grid (MG), which functions as a subsystem in the distribution network. Two operation modes can be defined for micro grids: grid-connected mode and autonomous mode. In a grid-connected mode, MGs are able to exchange the energy with the upstream network. For the autonomous mode, during a grid failure or islanding, MG assures continues power supply to local loads and increases system reliability. A schematic diagram of a radial distribution network with MG and islands is shown in Figure 1.

Electrical connections between unspecified number of bus-bars and main grid can be lost due to faults in the electrical distribution system. Hence, provided that at least one DG unit remains in the separated section, an active island can be formed. Based on IEEE-Standard 1547, islanding should be detected and DG units must cease to supply within 2 s.

Due to the high penetration of RES in distribution systems, islanded operation mode for DG units are considered by many utility providers as a suitable approach to maintain continuity and reliability of supply. However, as faults can occur at any point in a distribution system, the configuration of the active subsections formed is unpredictable. As shown in Figure 1, there are several possibilities to form islands via opening breakers BK1-BK37. The identification of the island configuration is one of the important requirements for the power system analysis. Moreover, in order to stabilize the system and to conduct a voltage stability study, it is important to detect new system configuration(s) immediately after removing the faulty section. Furthermore, the new topology of the system is vital for triggering the self-healing scenarios including load and generator tripping, generation controls, intentional islanding and system reconfiguration.

Once islanding takes place, if the generation does not meet the load demand, load shedding is unavoidable to prevent voltage collapse. In order to enhance the voltage stability of the island, less stable bus-bars can be chosen for load shedding. Thus, different voltage stability indices (VSI) can be used based on the system characteristics to determine weak bus-bars in the system. VSIs indicate the stability status of the bus-bars and line sections. When islanding takes place in the system, an online voltage stability index can be utilized based on the new system structure results to determine the weakest bus-bars for load shedding. In order to detect islanding, three islanding detection methods (IDMs) [1–9] have been developed; Active Detection Method (ADM) [1,8–20], Passive Detection Method (PDM) and Remote Detection Method (RDM) [8,13,14,21–26].

The philosophy of the PDM is to measure the variables at PCC and compare with the reference values to detect the islanding. The advantages of these methods include simplicity and low cost to implement as they use conventional metering and protection devices to detect islanding. There are also several passive methods such as over/under voltage/frequency ones [1–7,27,28]. The over/under voltage and frequency detection methods are easy to implement but their speed is unpredictable, moreover, they don't have a detrimental impact on power quality [3–6]. The ROCOF detection method is discussed in [8,9]. The ROCOF detection method is able to detect islanding in balanced conditions and it does not have negative impact on power quality, however, any disturbance can result in misdetection of islanding as well as it is hard to choose thresholds. The ROCOF Over Power detection method is discussed in [9]. The ROCOF over Power method can detect islanding effectively when the power imbalance is small and has a small non-detection zone. The characteristics of Phase Jump Detection (PJD) are presented in [3,9]. PJD is difficult to implement and the threshold is hard to adjust. The main drawback is that the PDMs have a large non-detection zone (NDZ), which leads to failure in islanding detection when the load and power generation are balanced [29]. The performance of ADM is based on the perturbation and observation concept. A DG parameter is chosen to be distorted by injecting a perturbation. In a stiff grid, the amplitude of the variations at the DG unit terminal is negligible since the grid parameters are dominant. However, injecting a disturbance into the terminal results in a significant variation of the DG parameters during islanding phenomena.

The Impedance Measurement detection method [8,10–13] has a low performance in the presence of multiple DG units, introduces harmonics and it is hard to adjust the thresholds. For Active Frequency Drift (AFD), NDZ is decreased as well as the error ratio and it is sensitive to the load [1,8,14–16]. The Frequency Jump (FJ) detection method also is unable to detect the islanding in the presence of multiple DG units [8]. The Negative Sequence Current Injection Method is not sensitive to load changes and has a better performance in the presence of multiple DG units [8,9,18–20]. There are several drawbacks for ADMs such as reduction in power quality via the injection of distortion into the system and the possibility of interfering distortion signals in the presence of multiple DG units.

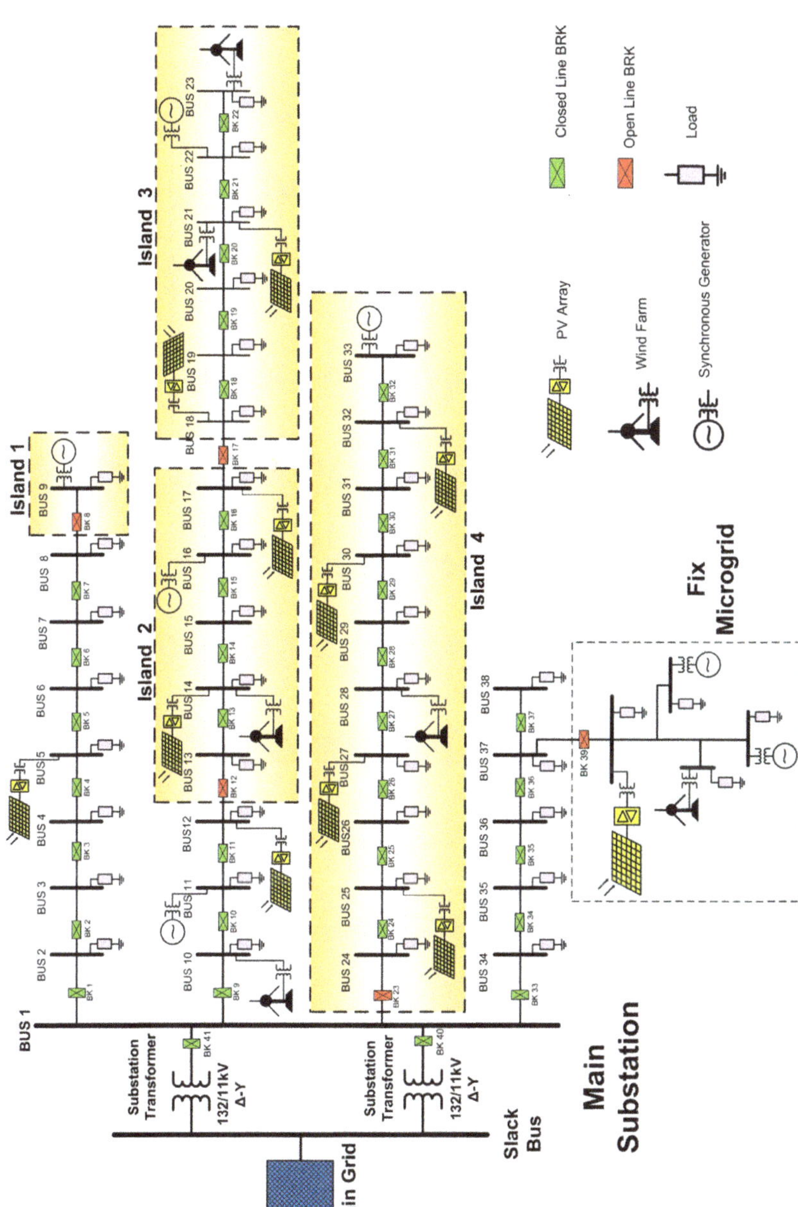

Figure 1. Single line diagram of a RES base integrated distribution network include MG and islands.

Remote detection methods (RDMs) were developed to overcome the problems faced by existing methods. The RDMs employ a communication pattern between DG units and the main grid to alert and disconnect the DG units once unintentional islanding occurs. The Supervisory Control and Data Acquisition (SCADA) detection method provides more control signals to control DG units, as well as detection speed dependent on the system characteristics [8,14,21,22]. Power Line Carrier Signaling (PLCS) is applicable for both DG types, and uses the existing electrical network as a path for signaling. The PLCS method is suitable for large networks with high penetration of DG units since it is costly [8,23–25]. RDMs are applicable to the systems with multiple DG units which include both inverter-based sources and synchronous generators.

In [30–36], different load shedding methods have been developed for micro grids. Measurement of the system parameters such as voltages, frequency and rate of change of frequency reveals the imbalance between generation and load demand. Hence, the load shedding method would assure the stable operation. However, not all these methods are able to do the load shedding for the islands since the bus-bars for load shedding cannot be predefined.

Since these IDMs monitor voltage, current, etc. at the point of common coupling (PCC), they can detect the islanding of DG units. However, these methods are not capable of identifying the boundary of the island. The method proposed here is able to detect the boundary of the island and determine the available generation and load in each island. Moreover, the proposed method can be applied on meshed networks. The proposed method is also applicable to systems with non-consecutive bus-bar numbering.

In order to determine the system voltage stability status, various indices have been developed [37–39]. Voltage stability analysis can be categorized as dynamic analysis and static analysis and uses static voltage stability indices (SVSIs). Pre-islanding information or micro grid information is used to assess the voltage stability of the system using SVSIs. Hence, SVSIs do not reflect the dynamic operation of the power system during contingencies. However, dynamic analysis identifies the system parameters during transient and steady state conditions. Thus, an index is required to identify dynamic performance and system stability.

Modified graph theory is utilized to detect the boundary of each island. In the second step, in order to perform a wide area load shedding, a dynamic voltage sensitivity factor (DVSF) is proposed to specify the weak bus-bars in the system. The proposed DVSF shows the relevancy of reactive power and voltage at each bus-bar. Thus, the most sensitive bus-bars based on new configuration of the system are chosen for load shedding.

This paper is organized as follows: Section 2 explains the methodology of the proposed algorithm. Section 3 presents the boundary detection method for islands. Section 4 elaborates the calculation of dynamic sensitivity factor for the islands. In Section 5, a practical test network is utilized to validate the proposed scheme. Simulation results are investigated in Section 6 and the conclusion is presented in Section 7.

2. Proposed Scheme to Stabilize the System

The proposed scheme assumes that voltage and current measurements are available at each bus-bar. When islanding occurs in the system, the number of islands is identified using the proposed detection method. The proposed algorithm also specifies the bus-bars belonging to each island. A power system with many bus-bars can be divided into several active and reactive islands depending on the availability of generation in islanded sections. The islands topology is not predictable due to the nature of the fault. In Figure 2, an islanded network is shown. The main idea of this method is to determine the configuration of the islands and maintain system stability. When island configuration is determined, the amount of available generation and load demand are disclosed. Load shedding in the weakest bus-bar prevents voltage collapse in the system.

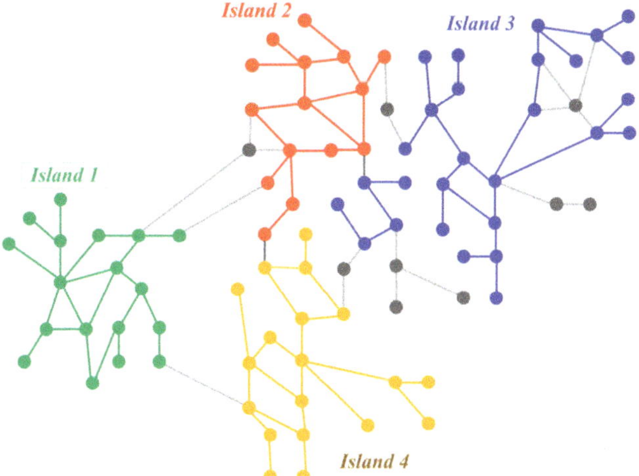

Figure 2. A Power system with four islands.

Weak bus-bars are chosen based on the dynamic voltage sensitivity factor. In order to conform to IEEE-1547 standard, after the first round of load shedding, if the system frequency division exceeds the ±1% of its rating, further load shedding will be done at the next weak bus-bar based on the new DVSF. This procedure is repeated until a stable operation point for all islands is gained. A flowchart of the proposed algorithm is depicted in Figure 3.

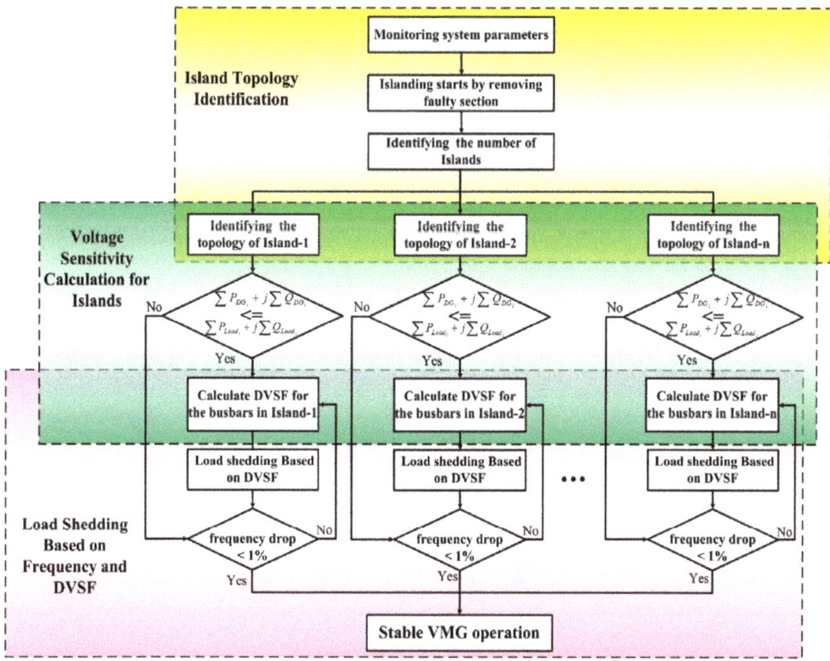

Figure 3. Flowchart of the proposed scheme to stabilize a system after islanding.

3. Boundary Detection Using Graph Model

The focus of existing detection methods is to detect the islanding at PCC. However, in this paper, a boundary detection algorithm based on modified graph theory is proposed to identify the islands topology once they are formed. A simplified graph theory is utilized for different purposes such as micro grid reconfiguration [40] and load flow calculation [41].

3.1. Node Integration Matrix

In this section, mathematical model of the proposed boundary detection method is presented. For this purpose, a graph consisting of 10 edges and 8 vertices with three subsections is considered. The graph is shown in Figure 4. As a first step, all the vertices should be numbered. Numbering of the vertices is random and each vertex should get a number. In the second step, connections between each vertex are investigated. In order to determine all connections in this graph, a node integration matrix $[NIM]_{n,n}$ where n represents the number of vertices is defined as follows:

$$\begin{cases} NIM_{i,j} = 1, & \text{If there is a connection between } i \text{ and } j. \\ NIM_{i,j} = 0, & \text{If there is no connection between } i \text{ and } j. \\ NIM_{i,i} = 1. \end{cases}$$

If there is a connection between two vertices, the related elements will be unity. The value 0 indicates no connection between two vertices. Moreover, in this matrix, the diagonal elements are assumed to be 1, since the proposed search algorithm will use these particular numbers and copy them in $[TMC]$ matrix. Equation (1) represents the $[NIM]$ for the graph of Figure 4.

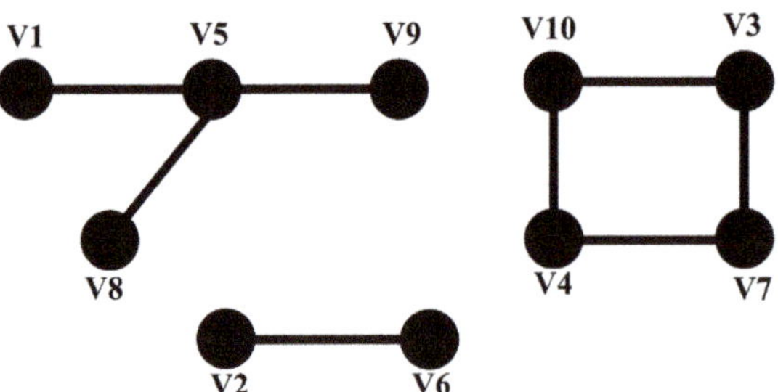

Figure 4. A graph consisting of 10 edges and 8 vertices with non-consecutive numbering.

This matrix represents the existence of edges between vertices of the graph. However, the topology of the subsections still should be extracted from this matrix:

$$[NIM] = \begin{bmatrix} 1 & 0 & 0 & 0 & 1 & 0 & 0 & 0 & 0 & 0 \\ 0 & 1 & 0 & 0 & 0 & 1 & 0 & 0 & 0 & 0 \\ 0 & 0 & 1 & 0 & 0 & 0 & 1 & 0 & 0 & 1 \\ 0 & 0 & 0 & 1 & 0 & 0 & 1 & 0 & 0 & 1 \\ 1 & 0 & 0 & 0 & 1 & 0 & 0 & 1 & 1 & 0 \\ 0 & 1 & 0 & 0 & 0 & 1 & 0 & 0 & 0 & 0 \\ 0 & 0 & 1 & 1 & 0 & 0 & 1 & 0 & 0 & 0 \\ 0 & 0 & 0 & 0 & 1 & 0 & 0 & 1 & 0 & 0 \\ 0 & 0 & 0 & 0 & 1 & 0 & 0 & 0 & 1 & 0 \\ 0 & 0 & 1 & 1 & 0 & 0 & 0 & 0 & 0 & 1 \end{bmatrix} \tag{1}$$

3.2. Topology Connection Matrix

Once the node integration matrix is calculated, the topology Connection Matrix $[TCM]_{n,n}$ is carried out. The topology connection matrix represents the subsections and the vertices in each subsection. The flowchart to calculate the $[TCM]$ is depicted in Figure 5. Like the node integration matrix, $[TCM]$ is a square matrix and the number of rows and columns depends on the number of vertices. The $[NIM]$ is utilized to determine all routes in the system. The steps of determining the subsection is as follows:

(a) As a first step, vertex V1 is chosen and the route starting from this vertex to other vertices is determined. For this purpose, existence of any edge between V2 and V1 is investigated. If any connection is found between V2 and V1, then connection between V3 and V2 will be checked and this will continue for the next vertex. Using this method, the route starts from V1 can be determined.

(b) In the second step, since there is a possibility of more than one edge connected to V1, the connection between other unchecked vertices and V1 should be investigated. This step is repeated to find a link between all other vertices and V1.

The same steps are applied to the rest of the vertices to determine all the subsections. In Figure 5, the flowchart for calculating the $[TCM]$ is depicted. According to this flowchart, the following are the steps to determine the island's boundary:

(1) For a graph with n vertices, an identity square matrix is formed.
(2) Start to build the ith row.
(3) If $NIM_{j,k} = 1$, there is a connection between jth vertex and upstream kth vertices. Then, set the $TCM_{i,j} = TCM_{i,k}$.
(4) Repeat the procedure from step (2) to build the next row of $[TCM]$ for all the vertices.

The proposed algorithm in Figure 5 is used to determine the routes connecting two or more vertices. As explained above, the connectivity of each vertex with all upstream vertices is examined. Using this method, all the routes connecting the examined vertices can be detected.

Since the connection between one vertex to other vertices is checked just once by this algorithm, time will be saved. Moreover, since all the routes are determined using the above procedures, this algorithm can be applied on graphs with loops as well. Equation (2) represents the $[TCM]$ matrix for the graph of Figure 4. When a consecutive numbering for vertices is used, The $[TMC]$ can be calculated using Equation (2):

$$[TCM] = \begin{bmatrix} 1 & 0 & 0 & 0 & 1 & 0 & 0 & 1 & 1 & 0 \\ 0 & 1 & 0 & 0 & 0 & 1 & 0 & 0 & 0 & 0 \\ 0 & 0 & 1 & 1 & 0 & 0 & 1 & 0 & 0 & 1 \\ 0 & 0 & 1 & 1 & 0 & 0 & 1 & 0 & 0 & 1 \\ 1 & 0 & 0 & 0 & 1 & 0 & 0 & 1 & 1 & 0 \\ 0 & 1 & 0 & 0 & 0 & 1 & 0 & 0 & 0 & 0 \\ 0 & 0 & 1 & 1 & 0 & 0 & 1 & 0 & 0 & 1 \\ 1 & 0 & 0 & 0 & 1 & 0 & 0 & 1 & 1 & 0 \\ 1 & 0 & 0 & 0 & 1 & 0 & 0 & 1 & 1 & 0 \\ 0 & 0 & 1 & 1 & 0 & 0 & 1 & 0 & 0 & 1 \end{bmatrix} \tag{2}$$

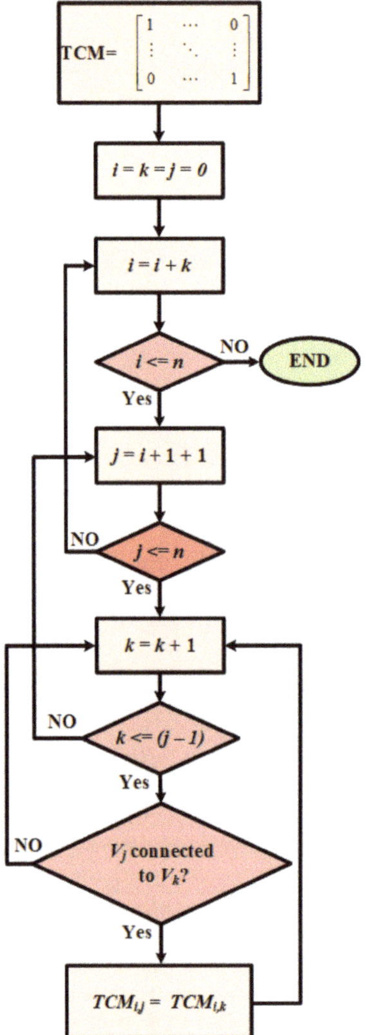

Figure 5. Flowchart for [*TCM*] calculation.

3.3. Boundary Detection Matrix

In the previous section, the topology connection matrix is formed and the subsections are marked. In this step, vertices in each subsection are extracted and Boundary Detection Matrix $[BDM]_{n_s,n_v}$ is formed. In this matrix, n_s represents the number of subsections and n_v represents the number of vertices in the largest subsection. The maximum n_s are equal to n when all the vertices are separated and no edge in the graph. In contrast, the maximum value for n_v is obtained for integrated graphs without any subsection. As shown in Equation (2), each row indicates one subsection. The steps for extracting the vertices and determining the boundary of each subsection are as follows:

(a) First, in the first row of [*TCM*], the column numbers of those elements that are equal to 1 are taken to the [*BDM*].

(b) Second, when column numbers (for example: column 1 and column 2) has been taken to [*BDM*], all the elements in the in the [*TCM*] rows with same number (row 1 and row2) have been changed to 0.

(c) Third, the next row with non-zero elements is checked with the same procedure that explained in (a).

When the entire rows of the [*TCM*] are checked, the [*BDM*] matrix will be completed. The flowchart to identify the subsections and vertices in each section is depicted in Figure 6. The following are the steps to calculate [*BDM*]:

(1) Start with *m*th row of [*TMC*] to find if there is 1 in this row.
(2) Set the $BDM_{i,j}$ to *m* (ith is based on number of subsections).
(3) Set entire *m*th row of [*TMC*] to zero and go to step 1.

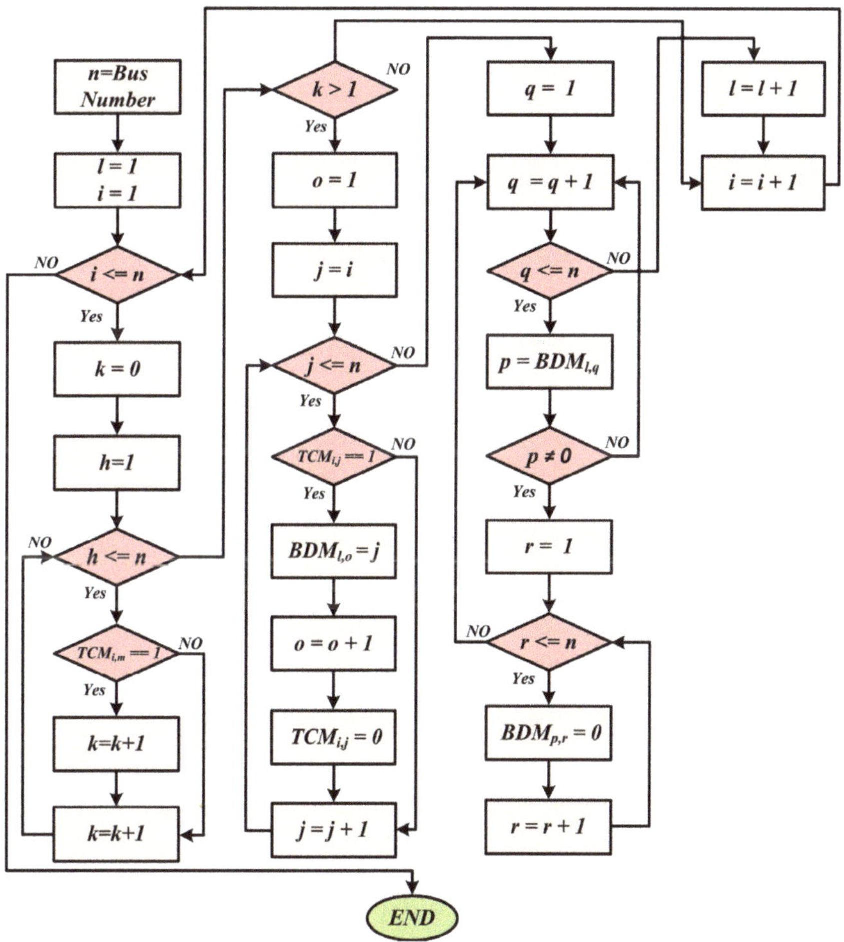

Figure 6. Procedure to extract subsections and bus-bars.

Equation (3) shows the [*BDM*] matrix calculated using the procedures mentioned previously. Three different subsections including the number of vertices in each section are shown using Equation (3). One important advantage of the proposed method is the ability to detect the boundaries in meshed networks with various loops and non-consecutive bus-bar numbering.

$$[BDM] = \begin{bmatrix} 1 & 2 & 3 & 4 & 5 \\ 6 & 7 & 8 & 9 & 0 \\ 10 & 11 & 12 & 0 & 0 \end{bmatrix} \tag{3}$$

4. Dynamic Voltage Sensitivity

In this paper, a novel dynamic voltage sensitivity factor is proposed to provide useful information about the system's dynamic performance during islanding. Online calculation of instantaneous static voltage stability provides a pattern of points, which can be used as a dynamic index to determine the system voltage stability for each bus-bar. The DVSF reflects the influence of changes in system parameters on each bus-bar's stability status and determines the weak bus-bar before and after islanding. A conventional load flow Jacobian matrix is used to determine the DVSF. In order to calculate the DVSF, bus-bar admittance matrix [Y_{bus}] and the Jacobian matrix is re-calculated based on the system configuration. A three-dimensional [Y_{bus}] is defined according to the new configuration of the system. Figure 7 demonstrates the [Y_{bus}] for a system with n_s islands. The size of the new [Y_{bus}] is (n_v, n_v, n_s). In this matrix, mth layer represents the calculated elements for mth island.

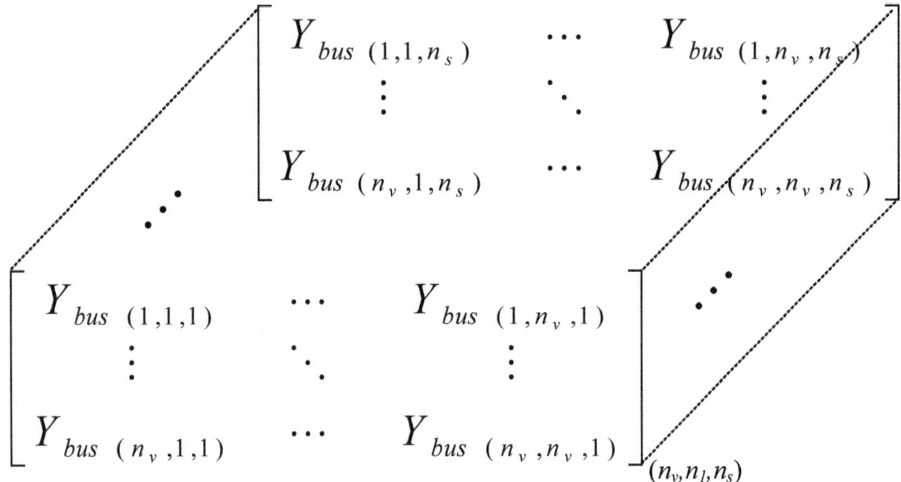

Figure 7. Three-dimensional admittance matrix.

Similarly, a new Jacobian matrix according to the new system configuration is calculated. The dimensions of the new Jacobian matrix are ($2n_v \times 2n_v \times 2n_s$). Equations (4)–(8) represents the mth layer of the Jacobian matrix calculated for the mth island in the system:

$$[J]^{(m)} = \left[\begin{array}{c|c} J_1^m & J_2^m \\ \hline J_3^m & J_4^m \end{array} \right] \tag{4}$$

$$J_1^m = \begin{cases} \dfrac{\partial P_{i,m}}{\partial \delta_{i,m}} = \sum_{j \neq i} |V_{i,m}||V_{j,m}||Y_{i,j,m}| \sin(\theta_{i,j,m} - \delta_{i,m} + \delta_{j,m}) \\[2mm] \dfrac{\partial P_{i,m}}{\partial \delta_{j,m}} = -|V_{i,m}||V_{j,m}||Y_{i,j,m}| \sin(\theta_{i,j,m} - \delta_{i,m} + \delta_{j,m}) \forall j \neq i \end{cases} \tag{5}$$

$$J_2^m = \begin{cases} \dfrac{\partial P_{i,m}}{\partial |V_{i,m}|} = 2|V_{i,m}||Y_{i,i,m}| \cos(\theta_{i,j,m} - \delta_{i,m} + \delta_{j,m}) + \sum_{j \neq i} |V_{j,m}||Y_{i,j,m}| \cos(\theta_{i,j,m} - \delta_{i,m} + \delta_{j,m}) \\[2mm] \dfrac{\partial P_{i,m}}{\partial |V_{j,m}|} = |V_{i,m}||Y_{i,j,m}| \cos(\theta_{i,j,m} - \delta_{i,m} + \delta_{j,m}) \forall j \neq i \end{cases} \tag{6}$$

$$J_3^m = \begin{cases} \dfrac{\partial Q_{i,m}}{\partial \delta_{i,m}} = \sum_{j \neq i} |V_{i,m}||V_{j,m}||Y_{i,j,m}| \cos(\theta_{i,j,m} - \delta_{i,m} + \delta_{j,m}) \\[2mm] \dfrac{\partial Q_{i,m}}{\partial \delta_{j,m}} = -|V_{i,m}||V_{j,m}||Y_{i,j,m}| \cos(\theta_{i,j,m} - \delta_{i,m} + \delta_{j,m}) \forall j \neq i \end{cases} \tag{7}$$

$$J_4^m = \begin{cases} \dfrac{\partial Q_{i,m}}{\partial |V_{i,m}|} = -2|V_{i,m}||Y_{i,i,m}| \sin(\theta_{i,j,m} - \delta_{i,m} + \delta_{j,m}) - \sum_{j \neq i} |V_{j,m}||Y_{i,j,m}| \cos(\theta_{i,j,m} - \delta_{i,m} + \delta_{j,m}) \\[2mm] \dfrac{\partial Q_{i,m}}{\partial |V_{j,m}|} = -|V_{i,m}||Y_{i,j,m}| \sin(\theta_{i,j,m} - \delta_{i,m} + \delta_{j,m}) \forall j \neq i \end{cases} \tag{8}$$

The new Jacobian matrix for the the system with m islands is shown in Equation (9):

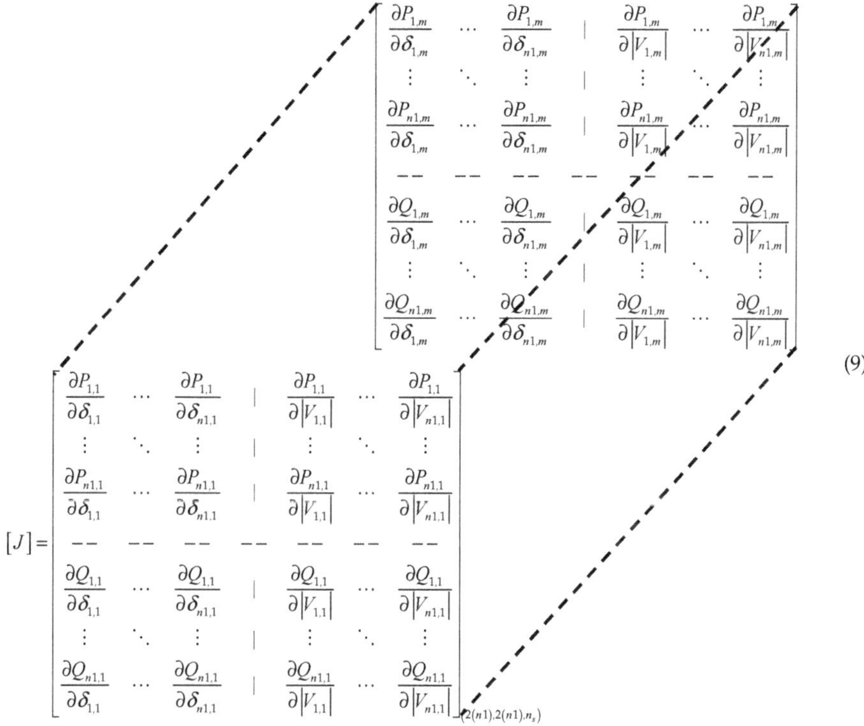

$$\tag{9}$$

Accordingly, Equation (10) represents the relation between reactive power and the voltage for all the bus-bars in islands when active power exchange is assumed to be zero ($\Delta P = 0$):

$$[\Delta Q]_{n_v,1,n_s} = [J_R]_{n_v,n_v,n_s}[\Delta V]_{n_v,1,n_s} \tag{10}$$

Equation (11) represents the reduced matrix calculated for the mth island in the system. In this equation, $[J_R]^{(m)}$ is known as a reduced Jacobian matrix of the mth island in the system:

$$[J_R]^{(m)} = \left[J_4{}^{(m)} - J_3{}^{(m)} J_1^{-1(m)} J_2{}^{(m)} \right] \tag{11}$$

Change in voltage with regards to the injected reactive power for the *m*th island can be defined by Equation (12):

$$[\Delta V]^{(m)} = \left[J_R^{-1} \right]^{(m)} [\Delta Q]^{(m)} \tag{12}$$

Diagonal values of the $\left[J_R^{-1} \right]^{(m)}$, represents the voltage sensitivity for each bus-bar of the *m*th island in the system. DVSF is defined individually for each bus-bar. Positive and small values for DVSF indicate stable operation. On the contrary, high values for sensitivity factor are obtained when the system is near its stability margins. Negative values represent unstable operation and very small negative values represent very unstable operation. Since the relation between voltage and reactive power is nonlinear, the magnitude of the sensitivity factor in different operating conditions does not provide the degree of the stability [42].

5. Generic Test Distribution System

A modified 11 kV, 50 Hz generic network from a Malaysian electrical distribution utility was modeled using EMTDC/PSCAD in order to investigate the proposed boundary detection method and dynamic voltage sensitivity factor. The single line diagram of this network is presented in Figure 8. The network has 18 bus-bars and 17 overhead lines. Both transformers in the main substation are 132/11 kV, Δ-Y_g, rated at 30 MVA and impedance 10%. Fault level at 132 kV bus-bar is 14 kA and at 11 kV is 17.8 kA. Three 2 MVA synchronous generators were used as DG units as shown in Figure 8. The DG units are equipped with droop control to adjust the voltage and frequency after islanding. Each generator is connected to the distribution network via 0.4/11 kV, 3 MVA Y_g-Y_g transformer with 5% impedance. Simulation results are discussed in detail in the next section.

6. Results and Discussion

A case study with three islands was investigated to verify the performance of the proposed methods. In order to model the system, PSCAD/EMTDC and MATLAB were utilized. The network was modeled in PSCAD and the proposed algorithms were analyzed in MATLAB. A user defined block was used as an interface to exchange the data between the software's and to send the wide area load shedding commands from MATLAB to PSCAD. Synchronous generator 'start event' was simulated via releasing the rotor at 2 s. Intentional islanding was initiated in the system after 15 s via opening the line sections L1, L9 and L13 as in Figure 8. The opening of three line sections split the network to four subsystems. In the first step, results for the island boundary detection method were investigated. In the second step, DVSFs were shown for one island. In the third step, the wide area load shedding results based on dynamic voltage sensitivity factor were presented.

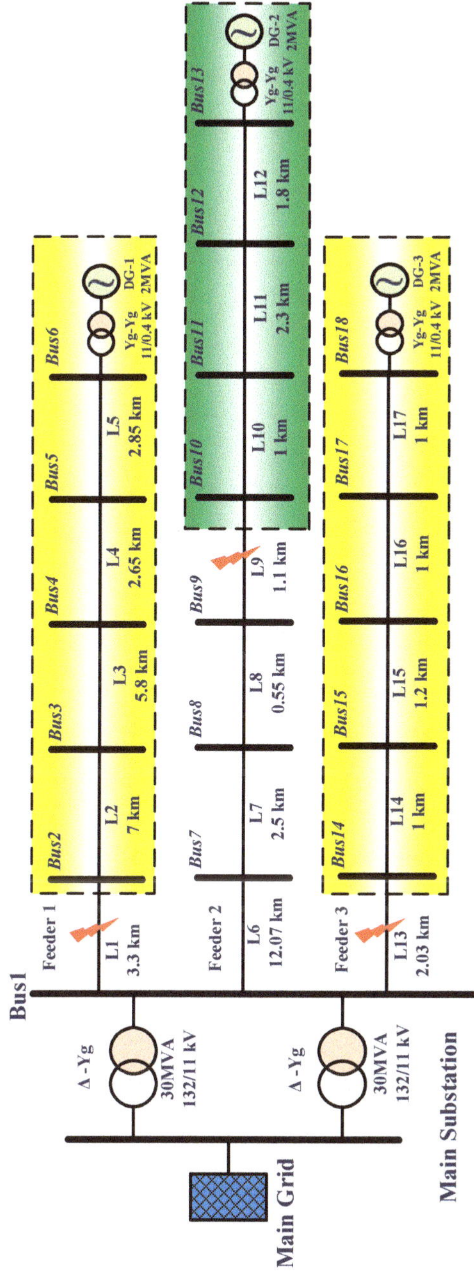

Figure 8. A single line diagram for a practical 18-bus-bar distribution network.

6.1. Island Boundary Detection Results

The equivalent graph for the given test system is shown in Figure 9. It was assumed that the voltage and current measurements were available in each bus-bar. After the tie line breakers opened, current would be reduced to zero in that line section. Hence, all downstream bus-bars were considered as an island, provided that a source existed. As explained in section (II), the node integration matrix $[NIM]_{18 \times 18}$ was calculated as follows:

$$[NIM] = \begin{bmatrix} 1 & 0 & 0 & 0 & 0 & 0 & 1 & 0 & 0 & 0 & 0 & 0 & 0 & 0 & 0 & 0 & 0 & 0 \\ 0 & 1 & 1 & 0 & 0 & 0 & 0 & 0 & 0 & 0 & 0 & 0 & 0 & 0 & 0 & 0 & 0 & 0 \\ 0 & 1 & 1 & 1 & 0 & 0 & 0 & 0 & 0 & 0 & 0 & 0 & 0 & 0 & 0 & 0 & 0 & 0 \\ 0 & 0 & 1 & 1 & 1 & 0 & 0 & 0 & 0 & 0 & 0 & 0 & 0 & 0 & 0 & 0 & 0 & 0 \\ 0 & 0 & 0 & 1 & 1 & 1 & 0 & 0 & 0 & 0 & 0 & 0 & 0 & 0 & 0 & 0 & 0 & 0 \\ 0 & 0 & 0 & 0 & 1 & 1 & 0 & 0 & 0 & 0 & 0 & 0 & 0 & 0 & 0 & 0 & 0 & 0 \\ 1 & 0 & 0 & 0 & 0 & 0 & 1 & 1 & 0 & 0 & 0 & 0 & 0 & 0 & 0 & 0 & 0 & 0 \\ 0 & 0 & 0 & 0 & 0 & 0 & 0 & 1 & 1 & 0 & 0 & 0 & 0 & 0 & 0 & 0 & 0 & 0 \\ 0 & 0 & 0 & 0 & 0 & 0 & 0 & 1 & 1 & 0 & 0 & 0 & 0 & 0 & 0 & 0 & 0 & 0 \\ 0 & 0 & 0 & 0 & 0 & 0 & 0 & 0 & 0 & 1 & 1 & 0 & 0 & 0 & 0 & 0 & 0 & 0 \\ 0 & 0 & 0 & 0 & 0 & 0 & 0 & 0 & 0 & 1 & 1 & 1 & 0 & 0 & 0 & 0 & 0 & 0 \\ 0 & 0 & 0 & 0 & 0 & 0 & 0 & 0 & 0 & 0 & 1 & 1 & 1 & 0 & 0 & 0 & 0 & 0 \\ 0 & 0 & 0 & 0 & 0 & 0 & 0 & 0 & 0 & 0 & 0 & 1 & 1 & 0 & 0 & 0 & 0 & 0 \\ 0 & 0 & 0 & 0 & 0 & 0 & 0 & 0 & 0 & 0 & 0 & 0 & 0 & 1 & 1 & 0 & 0 & 0 \\ 0 & 0 & 0 & 0 & 0 & 0 & 0 & 0 & 0 & 0 & 0 & 0 & 0 & 1 & 1 & 1 & 0 & 0 \\ 0 & 0 & 0 & 0 & 0 & 0 & 0 & 0 & 0 & 0 & 0 & 0 & 0 & 0 & 1 & 1 & 1 & 0 \\ 0 & 0 & 0 & 0 & 0 & 0 & 0 & 0 & 0 & 0 & 0 & 0 & 0 & 0 & 0 & 1 & 1 & 1 \\ 0 & 0 & 0 & 0 & 0 & 0 & 0 & 0 & 0 & 0 & 0 & 0 & 0 & 0 & 0 & 0 & 1 & 1 \end{bmatrix} \tag{13}$$

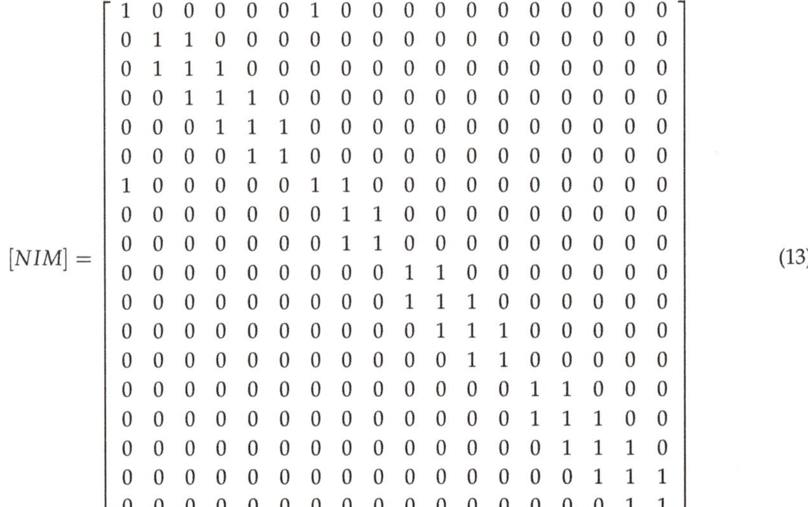
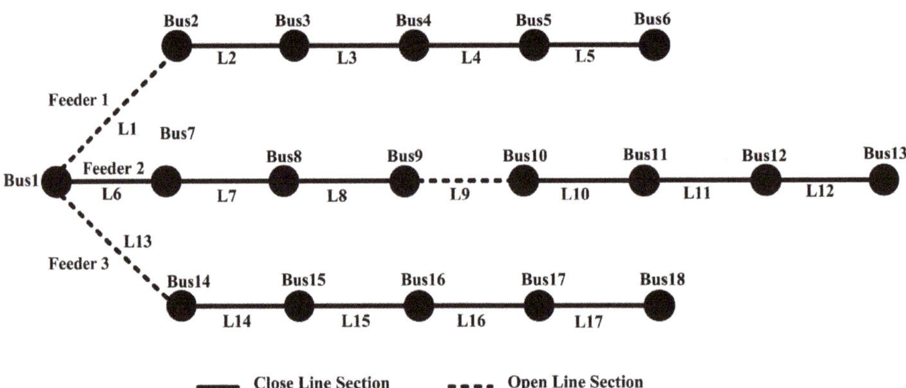

Figure 9. Equivalent graph for the test system with four subsections.

In order to form the node integration matrix $[NIM]$, the system topology was monitored using current measurements. The line section with none zero current was considered as a closed line. Using the procedure explained in Section 2, the calculated $[TCM]_{18 \times 18}$ is shown in Equation (14) for the test distribution network. To extract the number of bus-bars and corresponding Islands, the procedure, which is depicted in Figure 6, was applied on the calculated topology connection matrix. The calculated topology matrix $[TCM]$ for the test network is shown in Equation (14):

$$[TCM] = \begin{bmatrix} 1 & 0 & 0 & 0 & 0 & 0 & 1 & 1 & 1 & 0 & 0 & 0 & 0 & 0 & 0 & 0 & 0 & 0 \\ 0 & 1 & 1 & 1 & 1 & 1 & 0 & 0 & 0 & 0 & 0 & 0 & 0 & 0 & 0 & 0 & 0 & 0 \\ 0 & 0 & 1 & 1 & 1 & 1 & 0 & 0 & 0 & 0 & 0 & 0 & 0 & 0 & 0 & 0 & 0 & 0 \\ 0 & 0 & 0 & 1 & 1 & 1 & 0 & 0 & 0 & 0 & 0 & 0 & 0 & 0 & 0 & 0 & 0 & 0 \\ 0 & 0 & 0 & 0 & 1 & 1 & 0 & 0 & 0 & 0 & 0 & 0 & 0 & 0 & 0 & 0 & 0 & 0 \\ 0 & 0 & 0 & 0 & 0 & 1 & 0 & 0 & 0 & 0 & 0 & 0 & 0 & 0 & 0 & 0 & 0 & 0 \\ 0 & 0 & 0 & 0 & 0 & 0 & 1 & 1 & 1 & 0 & 0 & 0 & 0 & 0 & 0 & 0 & 0 & 0 \\ 0 & 0 & 0 & 0 & 0 & 0 & 0 & 1 & 1 & 0 & 0 & 0 & 0 & 0 & 0 & 0 & 0 & 0 \\ 0 & 0 & 0 & 0 & 0 & 0 & 0 & 0 & 1 & 0 & 0 & 0 & 0 & 0 & 0 & 0 & 0 & 0 \\ 0 & 0 & 0 & 0 & 0 & 0 & 0 & 0 & 0 & 1 & 1 & 1 & 1 & 0 & 0 & 0 & 0 & 0 \\ 0 & 0 & 0 & 0 & 0 & 0 & 0 & 0 & 0 & 0 & 1 & 1 & 1 & 0 & 0 & 0 & 0 & 0 \\ 0 & 0 & 0 & 0 & 0 & 0 & 0 & 0 & 0 & 0 & 0 & 1 & 1 & 0 & 0 & 0 & 0 & 0 \\ 0 & 0 & 0 & 0 & 0 & 0 & 0 & 0 & 0 & 0 & 0 & 0 & 1 & 0 & 0 & 0 & 0 & 0 \\ 0 & 0 & 0 & 0 & 0 & 0 & 0 & 0 & 0 & 0 & 0 & 0 & 0 & 1 & 1 & 1 & 1 & 1 \\ 0 & 0 & 0 & 0 & 0 & 0 & 0 & 0 & 0 & 0 & 0 & 0 & 0 & 0 & 1 & 1 & 1 & 1 \\ 0 & 0 & 0 & 0 & 0 & 0 & 0 & 0 & 0 & 0 & 0 & 0 & 0 & 0 & 0 & 1 & 1 & 1 \\ 0 & 0 & 0 & 0 & 0 & 0 & 0 & 0 & 0 & 0 & 0 & 0 & 0 & 0 & 0 & 0 & 1 & 1 \\ 0 & 0 & 0 & 0 & 0 & 0 & 0 & 0 & 0 & 0 & 0 & 0 & 0 & 0 & 0 & 0 & 0 & 1 \end{bmatrix} \tag{14}$$

$$[BDM] = \begin{bmatrix} 1 & 7 & 8 & 9 & 0 \\ 2 & 3 & 4 & 5 & 6 \\ 10 & 11 & 12 & 13 & 0 \\ 14 & 15 & 16 & 17 & 18 \end{bmatrix} \tag{15}$$

The boundary detection matrix $[BDM]$ shows that the system is divided into four subsystems. The system with bus-bar (1) is not considered as an island as it is connected to the main grid. As shown in the results, the proposed method is capable of detecting the configuration of Islands in the distribution network. Moreover, the proposed method is capable of tracking any changes in the island configuration. If the island's configuration changed due to any reason such as faults in the system or self-healing strategies, the proposed method can detect the new configuration without applying any further modification.

6.2. Dynamic Voltage Sensitivity Factor

When intentional islanding event is initiated via opening the line sections L1, L9 and L13, the stable operation of each island should be assured. The remaining generation and load in each island should be balanced to avoid voltage collapse. A proposed dynamic voltage sensitivity factor is used to identify weak bus-bars in each island. In Island-2, total available generation is 2 MVA while Table 1 shows the load demand in Island-2 bus-bars. In Table 1, the load consumption in each bus-bar is presented. The total active load in island-2 is equal to 3.5 MW and total reactive load is equal to 1 MVAr. Hence, 1.5 MW of additional load should be shedded to maintain the stability of Island-2.

Table 1. Load demand in island-2.

Bus-bar	Active Power (MW)	Reactive Power (MVar)
Bus10	1.055	0.2842
bus11	0.5916	0.2615
Bus12	0.9267	0.2255
Bus13	0.9268	0.2278

Since the Island-2 is not stable due to lack of generation, the Island-2 was used to validate the proposed DVSF. The bus-bar voltages for Island-2 are shown in Figure 10. As shown in Figure 10, the bus-bar voltages collapsed after islanding because the generation is not enough to cover the loads in the vicinity. Similarly, generator angular velocity and system frequency as shown in Figures 10 and 11a also collapses. As seen in Figures 10 and 11, generator stalls due to heavy load in the system. Hence, it is vital to identify the unstable bus-bars to perform the wide area load shedding.

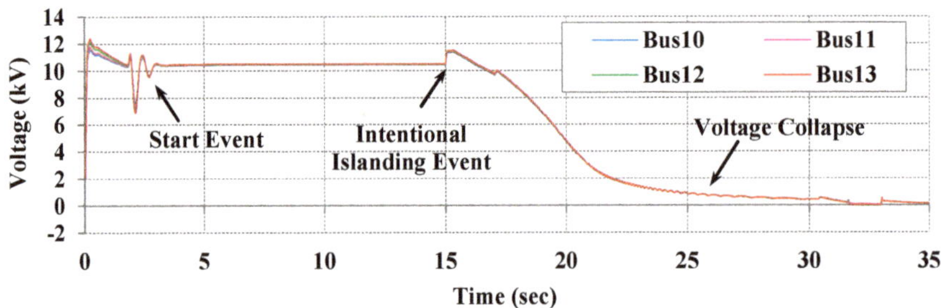

Figure 10. Voltage collapse in the island-2 after intentional islanding event.

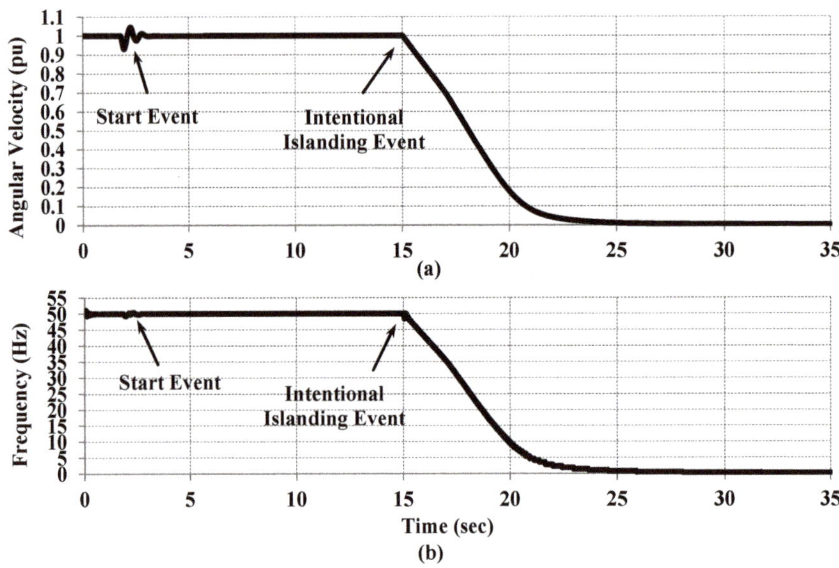

Figure 11. System parameters (a) synchronous generator angular velocity, (b) island-2 frequency.

The full window of the calculated conventional voltage sensitivity index using pre-islanding data is shown in Figure 12a and the expanded window during the islanding is demonstrated in Figure 12b. With reference to the results, the most stable bus-bar is Bus11, since it has the lowest sensitivity index and the most sensitive bus-bar is Bus13. Hence, according to results calculated using conventional method, load-shedding program should start with Bus13 by virtue of being the weakest bus-bar and proceed to Bus10, 12 and 11. However, in the proposed scheme, once islanding occurs, the new system topology is used to calculate the DVSF. The calculated DVSF for Island-2 is illustrated in Figure 13. According to these results, once islanding takes place, the sensitivity of Bus13 is higher compared to the other bus-bars. Hence, it can be concluded that Bus13 is still the weakest bus-bar followed by Bus10, Bus11 and Bus12.

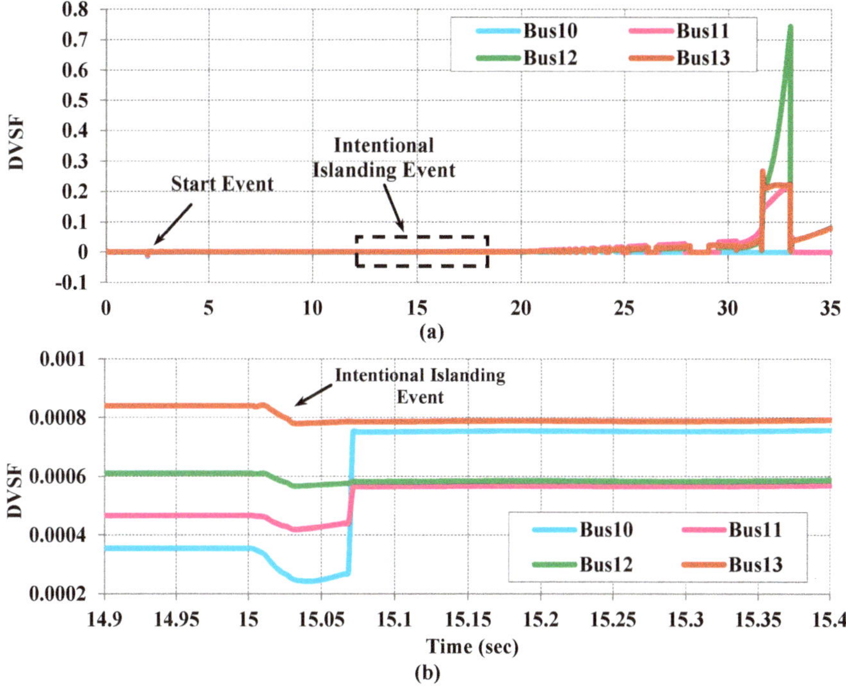

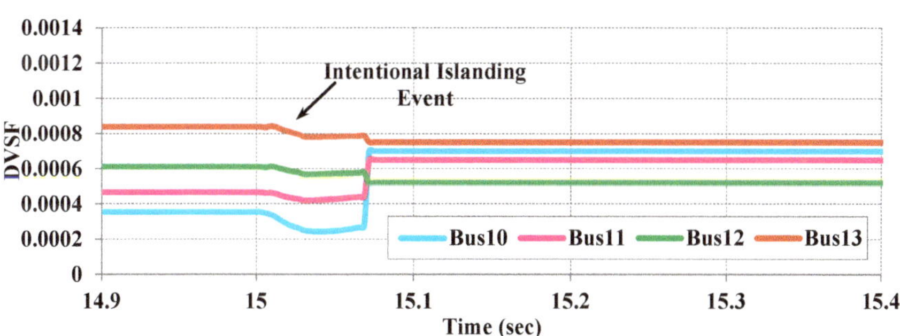

Figure 12. Dynamic Voltage Sensitivity Factor (**a**) full window of DVSF for bus-bars in the island-2 (**b**) expanded window for DVSF during 'Intentional Islanding Event'.

Figure 13. Voltage sensitivity indices using proposed boundary detection method.

6.3. Wide Area Load Shedding

In this section, Island-2 was used to verify the robustness of the proposed wide area load shedding to improve the voltage profile of the micro grid. Once the weak bus-bars were identified, the load shedding was performed to prevent voltage collapse. The amount of load shedding is related to the system frequency. Moreover, the DG unit should be able to work ±10% of its rated value for 20–30 min [23].

In order to control the voltage of the system, the power factor of the DG unit was varied between 0.85 lagging to 0.9 leading. According to the IEEE-Standard 1547, the system should be capable of operating for 300 s when frequency deviation was less than ±5% of the rated value.

For the new topology, there is 1.82 MW of excess load in island-2. Thus, according to the previous section, Bus13 as the weakest bus-bar and Bus10 as the second weakest bus-bar were chosen to perform the load shedding. The results for the bus-bar voltages are shown in Figure 14. As seen in this figure, the bus-bar voltages decreases after islanding. However, the load shedding prevents the voltage collapse in the island and all the bus-bar voltages recover to the acceptable range.

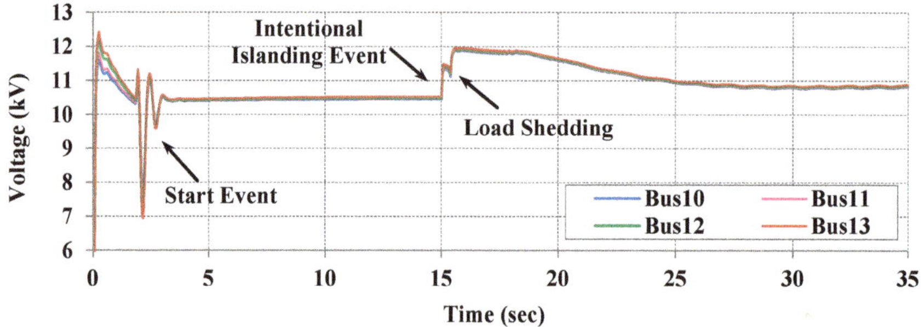

Figure 14. Voltage profile in island-2 recovers after load shedding.

Synchronous generator speed and frequency of the system are shown in Figure 15a,b.

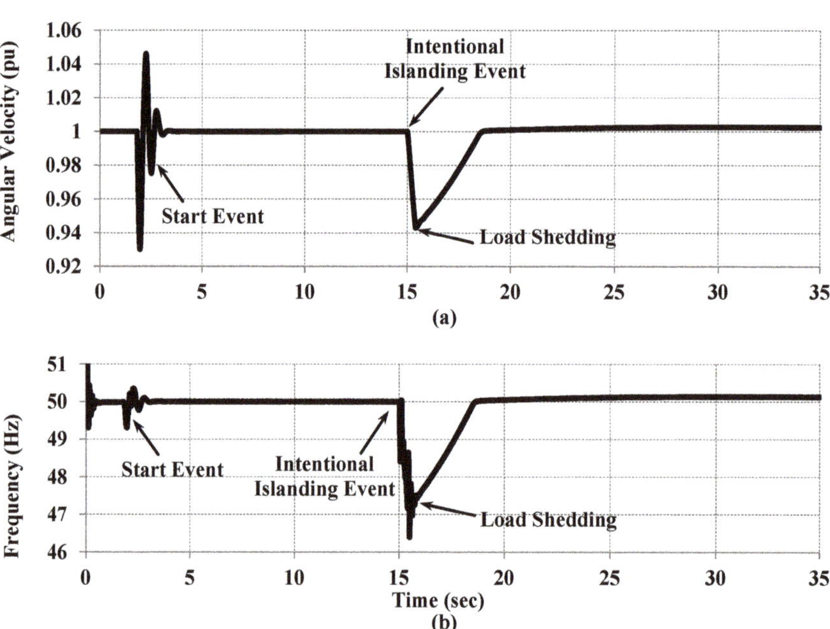

Figure 15. System parameters (**a**) synchronous generator angular velocity, (**b**) island-2 frequency.

Figure 15a illustrates the angular velocity of the generator during contingency. The speed of the generator recovers to 1 p.u. after load shedding. Similarly, the system frequency recovers to 50 Hz after the wide area load shedding. By comparing Figures 10 and 14, it confirms that the proposed load shedding prevented voltage collapse in the system. A full window of the dynamic voltage sensitivity factor is shown in Figure 16a. An expanded window during 'intentional islanding event' and 'load shedding' is shown in Figure 16b. As seen in Figure 16, the voltage sensitivity index of the bus-bars change due to islanding and load shedding. Moreover, the DVSF for all bus-bars remained near zero, which confirms the voltage stability in the Island-2.

Measurement of the system parameters such as voltages, frequency and rate of change of frequency reveals the imbalance between generation and load demand. Since existing detection methods monitor the point of common coupling, they can detect the islanding of DG units. However, these methods are not capable of identifying the boundary of the island. The proposed method here is able to detect the boundary of the island and determine the available generation and load in each island. Moreover, the proposed method can be applied on a meshed network and for systems with non-consecutive bus-bar numbering.

It can be concluded from the results of this section, that the proposed load shedding represented a vast improvement to voltage stability by considering the dynamic performance of the system during the islanding operation for Island-2. Hence, the proposed boundary detection method and the corresponding load shedding have the ability to reconfigure the islands and improve the profile of the voltages in all bus-bars.

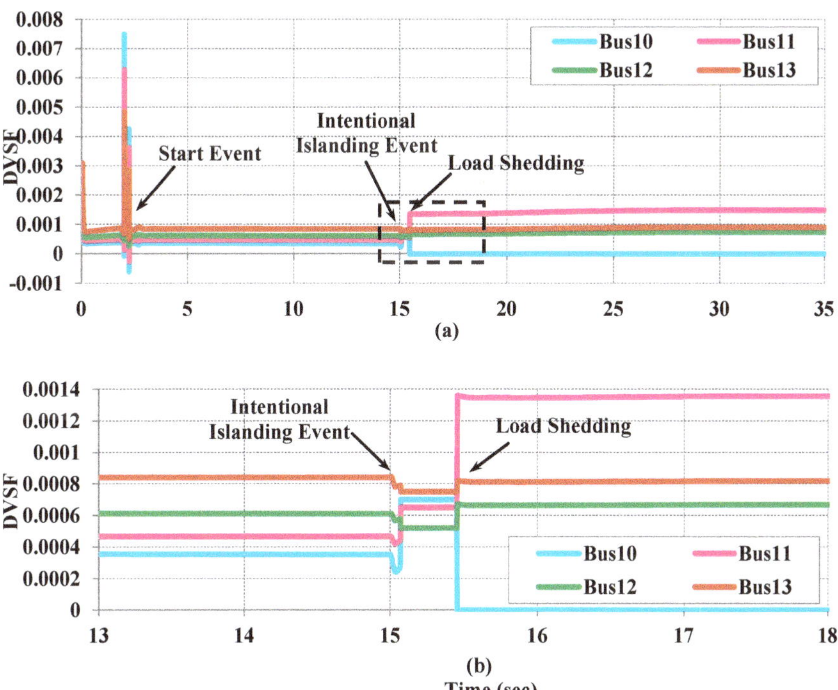

Figure 16. Dynamic voltage sensitivity factor (**a**) full window of DVSF for bus-bars in the island-2, (**b**) Expanded window for DVSF during 'intentional islanding event' and 'load shedding'.

7. Conclusions

This paper presents a novel scheme to detect the configuration of islands in a power system. The results show that the proposed boundary detection method is capable of accurately identifying the system topology. Thus, the available generation and loads in each island can be calculated to ensure the available generation meets the load demand in the islanded subsections. Furthermore, the proposed dynamic sensitivity factor is calculated for all bus-bars in every island to determine the weak bus-bars. The obtained results prove that the DVSF reflects the system's dynamic performance and can be used to detect the weakest bus-bar in the system. Wide area load shedding was performed on weak bus-bars and the results conform to the IEEE-1547 standard in terms of voltage and frequency in the islands. Accordingly, the proposed scheme assures stable operation of the islands using boundary detection along with wide area load shedding during autonomous mode. The proposed method can be used to reconfigure the islands to get an optimized operating condition.

Acknowledgments: The authors would like to thank the Power Quality Research Group, Universiti Tenaga Nasional, Malaysia for providing the funding resources and laboratory facilities.

Author Contributions: Aref Pouryekta, Vigna K. Ramachandaramurthy, Sanjeevikumar Padmanaban, Frede Blaabjerg and Josep M. Guerrero developed the original research work and contributed their experience in boundary detection and power systems stability.

Conflicts of Interest: The authors declare no conflict of interest.

References

1. Yu, B.; Matsui, M.; Yu, G. A review of current anti-islanding methods for photovoltaic power system. *Sol. Energy* **2010**, *84*, 745–754. [CrossRef]
2. Zeineldin, H.; Kirtley, J.L., Jr. A simple technique for islanding detection with negligible nondetection zone. *IEEE Trans. Power Deliv.* **2009**, *24*, 779–786. [CrossRef]
3. Khamis, A.; Shareef, H.; Bizkevelci, E.; Khatib, T. A review of islanding detection techniques for renewable distributed generation systems. *Renew. Sustain. Energy Rev.* **2013**, *28*, 483–493. [CrossRef]
4. Syamsuddin, S.; Rahim, N.; Selvaraj, J. Implementation of TMS320F2812 in islanding detection for Photovoltaic Grid Connected Inverter. In Proceedings of the 2009 International Conference for Technical Postgraduates (TECHPOS), Kuala Lumpur, Malaysia, 14–15 December 2009.
5. Lidula, N.; Rajapakse, A. A pattern recognition approach for detecting power islands using transient signals—Part I: Design and implementation. *IEEE Trans. Power Deliv.* **2010**, *25*, 3070–3077. [CrossRef]
6. Laghari, J.; Mokhlis, H.; Karimi, M.; Bakar, A.; Mohamad, H. Computational Intelligence based techniques for islanding detection of distributed generation in distribution network: A review. *Energy Convers. Manag.* **2014**, *88*, 139–152. [CrossRef]
7. Zeineldin, H.; El-Saadany, E.F.; Salama, M. Impact of DG interface control on islanding detection and nondetection zones. *IEEE Trans. Power Deliv.* **2006**, *21*, 1515–1523. [CrossRef]
8. Li, C.; Cao, C.; Cao, Y.; Kuang, Y.; Zeng, L.; Fang, B. A review of islanding detection methods for microgrid. *Renew. Sustain. Energy Rev.* **2014**, *35*, 211–220. [CrossRef]
9. Raza, S.; Mokhlis, H.; Arof, H.; Laghari, J.; Wang, L. Application of signal processing techniques for islanding detection of distributed generation in distribution network: A review. *Energy Convers. Manag.* **2015**, *96*, 613–624. [CrossRef]
10. Ropp, M.; Ginn, J.; Stevens, J.; Bower, W.; Gonzalez, S. Simulation and Experimental Study of the Impedance Detection Anti-Islanding Method in the Single-Inverter Case. In Proceedings of the Conference Record of the 2006 IEEE 4th World Conference on Photovoltaic Energy Conversion, Waikoloa, HI, USA, 7–12 May 2006; Volume 2, pp. 2379–2382.
11. O'Kane, P.; Fox, B. Loss of mains detection for embedded generation by system impedance monitoring. In Proceedings of the Sixth International Conference on (Conf. Publ. No. 434) Developments in Power System Protection, Nottingham, UK, 25–27 March 1997; pp. 95–98.
12. Mohamad, H.; Mokhlis, H.; Ping, H.W. A review on islanding operation and control for distribution network connected with small hydro power plant. *Renew. Sustain. Energy Rev.* **2011**, *15*, 3952–3962. [CrossRef]

13. Ahmad, K.N.E.K.; Selvaraj, J.; Rahim, N.A. A review of the islanding detection methods in grid-connected PV inverters. *Renew. Sustain. Energy Rev.* **2013**, *21*, 756–766. [CrossRef]

14. Balaguer-Alvarez, I.J.; Ortiz-Rivera, E.I. Survey of Distributed Generation Islanding Detection Methods. *IEEE Lat. Am. Trans.* **2010**, *8*, 565–570. [CrossRef]

15. Hanif, M.; Basu, M.; Gaughan, K. A discussion of anti-islanding protection schemes incorporated in a inverter based DG. In Proceedings of the 10th International Conference on Environment and Electrical Engineering (EEEIC), Rome, Italy, 8–11 May 2011; pp. 1–5.

16. Ropp, M.; Begovic, M.; Rohatgi, A. Analysis and performance assessment of the active frequency drift method of islanding prevention. *IEEE Trans. Energy Convers.* **1999**, *14*, 810–816. [CrossRef]

17. Kim, B.; Sul, S.; Lim, C. Anti-islanding detection method using Negative Sequence Voltage. In Proceedings of the 7th IEEE International Power Electronics and Motion Control Conference (IPEMC), Harbin, China, 2–5 June 2012; Volume 1, pp. 604–608.

18. Karimi, H.; Yazdani, A.; Iravani, R. Negative-Sequence Current Injection for Fast Islanding Detection of a Distributed Resource Unit. *IEEE Trans. Power Electron.* **2008**, *23*, 298–307. [CrossRef]

19. Bahrani, B.; Karimi, H.; Iravani, R. Nondetection Zone Assessment of an Active Islanding Detection Method and its Experimental Evaluation. *IEEE Trans. Power Deliv.* **2011**, *26*, 517–525. [CrossRef]

20. Tuyen, N.D.; Fujita, G. Negative-sequence Current Injection of Dispersed Generation for Islanding Detection and Unbalanced Fault Ride-through. In Proceedings of the 46th International Universities' Power Engineering Conference (UPEC), Berlin, Germany, 5–8 September 2011; pp. 1–6.

21. Yin, J.; Chang, L.; Diduch, C. Recent developments in islanding detection for distributed power generation. In Proceedings of the IEEE Large Engineering systems Conference on Power Engineering, LESCOPE-04, Halifax, NS, Canada, 28–30 July 2004; pp. 124–128.

22. Gao, W. Comparison and review of islanding detection techniques for distributed energy resources. In Proceedings of the 40th North American Power Symposium, Calgary, AB, Canada, 28–30 September 2008; pp. 1–8.

23. Xu, W.; Zhang, G.; Li, C.; Wang, W.; Wang, G.; Kliber, J. A power line signaling based technique for anti-islanding protection of distributed generators—Part I: Scheme and analysis. *IEEE Trans. Power Deliv.* **2007**, *22*, 1758–1766. [CrossRef]

24. Wang, W.; Kliber, J.; Zhang, G.; Xu, W.; Howell, B.; Palladino, T. A power line signaling based scheme for anti-islanding protection of distributed generators—Part II: Field test results. *IEEE Trans. Power Deliv.* **2007**, *22*, 1767–1772. [CrossRef]

25. De Mango, F.; Liserre, M.; Aquila, A.D. Overview of anti-islanding algorithms for pv systems. part ii: Activemethods. In Proceedings of the 12th IEEE International Power Electronics and Motion Control Conference, EPE-PEMC 2006, Portoroz, Slovenia, 30 August–1 September 2006; pp. 1884–1889.

26. Velasco, D.; Trujillo, C.; Garcerá, G.; Figueres, E. Review of anti-islanding techniques in distributed generators. *Renew. Sustain. Energy Rev.* **2010**, *14*, 1608–1614. [CrossRef]

27. Zhihong, Y.; Kolwalkar, A.; Zhang, Y.; Pengwei, D.; Walling, R. Evaluation of anti-islanding schemes based on nondetection zone concept. In Proceedings of the 34th IEEE Annual Power Electronics Specialist Conference, Acapulco, Mexico, 15–19 June 2003; Volume 4, pp. 1735–1741.

28. Wang, F.; Mi, Z. Notice of Retraction Passive Islanding Detection Method for Grid Connected PV System. In Proceedings of the IEEE International Conference on Industrial and Information Systems, Haikou, China, 24–25 April 2009; pp. 409–412.

29. Raipala, O.; Makinen, A.; Repo, S.; Jarventausta, P. The effect of different control modes and mixed types of DG on the non-detection zones of islanding detection. In Proceedings of the Integration of Renewables into the Distribution Grid, CIRED 2012 Workshop, Lisbon, Portuga, 29–30 May 2012; pp. 1–4.

30. Lin, Z.; Xia, T.; Ye, Y.; Zhang, Y.; Chen, L.; Liu, Y.; Tomsovic, K.; Bilke, T.; Wen, F. Application of wide area measurement systems to islanding detection of bulk power systems. *IEEE Trans. Power Syst.* **2013**, *28*, 2006–2015. [CrossRef]

31. Lei, D.; Zhencun, P.; Wei, C.; Jianye, P. An Integrated Automatic Control System for Distributed Generation Hierarchical Islanding. In Proceedings of the PowerCon 2006, International Conference on Power System Technology, Chongqing, China, 22–26 October 2006; pp. 1–6.

32. Qin, L.; Peng, F.Z.; Balaguer, I.J. Islanding control of DG in microgrids. In Proceedings of the IEEE 6th International Power Electronics and Motion Control Conference, Wuhan, China, 17–20 May 2009; pp. 450–455.
33. You, H.; Vittal, V.; Zhong, Y. Self-healing in power systems: An approach using islanding and rate of frequency decline-based load shedding. *IEEE Trans. Power Syst.* **2003**, *18*, 174–181. [CrossRef]
34. Zadeh, S.G.; Madani, R.; Seyedi, H.; Mokari, A.; Zadeh, M.B. New approaches to load shedding problem in islanding situation in distribution networks with distributed generation. In Proceedings of the CIRED 2012 Workshop: Integration of Renewables into the Distribution Grid, Lisbon, Portugal, 29–30 May 2012. [CrossRef]
35. Zahidi, R.A.; Abidin, I.Z.; Hashim, H.; Omar, Y.R.; Ahmad, N.; Ali, A.M. Study of static voltage stability index as an indicator for Under Voltage Load Shedding schemes. In Proceedings of the 3rd International Conference on Energy and Environment, Malacca, Malaysia, 7–8 December 2009; pp. 256–261.
36. Gu, W.; Liu, W.; Zhu, J.; Zhao, B.; Wu, Z.; Luo, Z.; Yu, J. Adaptive Decentralized Under-Frequency Load Shedding for Islanded Smart Distribution Networks. *IEEE Trans. Sustain. Energy* **2014**, *5*, 886–895. [CrossRef]
37. Arya, L.; Choube, S.; Shrivastava, M. Technique for voltage stability assessment using newly developed line voltage stability index. *Energy Convers. Manag.* **2008**, *49*, 267–275. [CrossRef]
38. Wang, Y.; Li, W.; Lu, J. A new node voltage stability index based on local voltage phasors. *Electr. Power Syst. Res.* **2009**, *79*, 265–271. [CrossRef]
39. Sinha, A.; Hazarika, D. A comparative study of voltage stability indices in a power system. *Int. J. Electr. Power Energy Syst.* **2000**, *22*, 589–596. [CrossRef]
40. Shariatzadeh, F.; Vellaithurai, C.B.; Biswas, S.S.; Zamora, R.; Srivastava, A.K. Real-Time Implementation of Intelligent Reconfiguration Algorithm for Microgrid. *IEEE Trans. Sustain. Energy* **2014**, *5*, 598–607. [CrossRef]
41. Alwash, S.F.; Ramachandaramurthy, V.K.; Mithulananthan, N. Fault-Location Scheme for Power Distribution System with Distributed Generation. *IEEE Trans. Power Deliv.* **2015**, *30*, 1187–1195. [CrossRef]
42. Kundur, P.; Balu, N.J.; Lauby, M.G. *Power System Stability and Control*; McGraw-Hill: New York, NY, USA, 1994.

Article

An Original Transformer and Switched-Capacitor (T & SC)-Based Extension for DC-DC Boost Converter for High-Voltage/Low-Current Renewable Energy Applications: Hardware Implementation of a New T & SC Boost Converter

Sanjeevikumar Padmanaban [1,*], Mahajan Sagar Bhaskar [2], Pandav Kiran Maroti [3], Frede Blaabjerg [4] and Viliam Fedák [5]

1   Department of Energy Technology, Aalborg University, 6700 Esbjerg, Denmark
2   Department of Electrical Engineering, Qatar University, P.O. Box 2713 Doha, Qatar; sagar25.mahajan@gmail.com
3   Department of Electrical and Electronics Engineering, Marathwada Institute of Technology, Aurangabad, Maharashtra 431028, India; kiranpandav88@yahoo.co.in
4   Centre for Reliable Power Electronics (CORPE), Department of Energy Technology, Aalborg University, 9100 Aalborg, Denmark; fbl@et.aau.dk
5   Department of Electrical Engineering and Mechatronics, FEI TU of Košice, Letná 9, 04200 Košice, Slovakia; viliam.fedak@tuke.sk
*   Correspondence: san@et.aau.dk; Tel.: +45-71-682-084

Received: 25 September 2017; Accepted: 6 November 2017; Published: 29 March 2018

Abstract: In this article a new Transformer and Switched Capacitor-based Boost Converter (T & SC-BC) is proposed for high-voltage/low-current renewable energy applications. The proposed T & SC-BC is an original extension for DC-DC boost converter which is designed by utilizing a transformer and switched capacitor (T & SC). Photovoltaic (PV) energy is a fast emergent segment among the renewable energy systems. The proposed T & SC-BC combines the features of the conventional boost converter and T & SC to achieve a high voltage conversion ratio. A Maximum Power Point Tracking (MPPT) controller is compulsory and necessary in a PV system to extract maximum power. Thus, a photovoltaic MPPT control mechanism also articulated for the proposed T & SC-BC. The voltage conversion ratio (V_o/V_{in}) of proposed converter is $(1 + k)/(1 - D)$ where, k is the turns ratio of the transformer and D is the duty cycle (thus, the converter provides 9.26, 13.88, 50/3 voltage conversion ratios at 78.4 duty cycle with k = 1, 2, 2.6, respectively). The conspicuous features of proposed T & SC-BC are: (i) a high voltage conversion ratio (V_o/V_{in}); (ii) continuous input current (I_{in}); (iii) single switch topology; (iv) single input source; (v) low drain to source voltage (V_{DS}) rating of control switch; (vi) a single inductor and a single untapped transformer are used. Moreover, the proposed T & SC-BC topology was compared with recently addressed DC-DC converters in terms of number of components, cost, voltage conversion ratio, ripples, efficiency and power range. Simulation and experimental results are provided which validate the functionality, design and concept of the proposed approach.

Keywords: DC-DC boost converter; transformer; switched capacitor; maximum power point tracking; renewable energy; high-voltage; low current

---

1. Introduction

In recent years the importance of using renewable energies has grown significantly due to the fact that the usage of fossil fuels such as oil, coal, and gas results in environmental pollution and serious greenhouse effects which have a huge influence on the world [1,2]. The demands for energy and power converters have been increasing for the last several decades, due to the greater industrialization, rising population and increased living standards of society [3–9]. The International Energy Agency (IEA) has anticipated that the developing nations are raising their energy utilization at a quicker pace than developed ones and will need to nearly the double their present installed generation facility by the year 2020 for fulfill their energy requirements. The total energy consumption of the world from 1980 to 2025 is depicted in Figure 1a [10–12]. The IEA has also details that more than 1.3 billion people in the developing nations are living with insufficient or without any access to electricity because of the unavailability of electric grid in these regions and other constraints [10–12]. Thus, there is a need for new sources of energy that are cheaper and sustainable with less carbon emissions [13,14]. Photovoltaic energy is considered as a reliable, promising and favorable source and it has various advantages such as being pollution free, long life, low maintenance, etc. [15–18]. In Figure 1b the global cumulated PV capacity (in Gigawatts) from 1996 to 2012, and an estimation (E) by 2020 is shown. It is observed that PV is a fast emergent segment among renewable energy systems [19]. To increase the effectiveness and efficiency of power conditioning to tracking Maximum Power Point (MPP) plays an important role in increasing the conversion efficiency. A PV system has non-linear P-V and I-V characteristics and the generated power depends on the environmental conditions such as solar irradiation and temperature [20–22]. The power of a PV system is higher at the knee point of the P-V characteristic curve, as the MPP keeps on changing according to the varying irradiation levels, a MPPT method is used to track the MPP of the system. The PV grid-connected power system in domestic applications is becoming a fast-rising segment in the PV market [23–26]. Unfortunately, the output voltage of the PV arrays or panels is relatively low. Thus, PV series-connected configurations are generally used in order to satisfy the demand and the high bus voltage requirements of the half and full-bridge, multilevel inverters (MLI) needed to transfer energy to electric grid. This type of system suffers from partial shading, reduced PV module efficiency, and mismatched MPPT control [27,28]. Thus boost converters with high voltage conversion ratio and inverters are required to feed energy to the electric grid. In practice DC-DC converters with high efficiency with low input voltage, high input current, high output voltage and high voltage conversion ratio provide a practicable solution for photovoltaic systems to transfer photovoltaic energy to the electric grid via inverters [29,30].

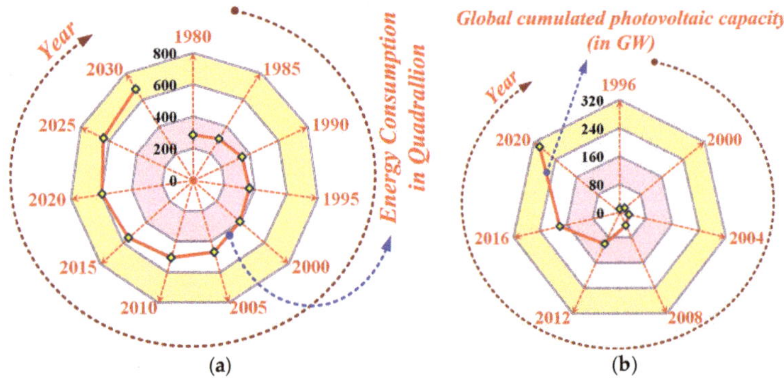

Figure 1. (a) Radar plot of energy consumption in quadrillion from 1980 to 2030; (b) Radar plot of global cumulated photovoltaic capacity (in GW) from 1996 to 2020.

Additionally, traditional boost converters are not a practicable solution to achieve high conversion ratios due to the leakage resistance of the inductor, high switch stress and the performance of a boost converter is deteriorated by the high duty cycle of the power switch and thus, not able to achieve conversion ratios of more than four. To operate the DC-DC converters for getting high output voltage without using high duty cycles for power semiconductor controlled switches, isolated converters can be employed which contain transformers, coupled inductors, etc. [29–37], but the usage of such a large number of magnetic components increases the size of the circuit as well as the leakage reactance of the converter and also produces electromagnetic interference which reduces the converter function ability and efficiency. Recently many boost converter topologies are addressed by extending the power circuit of the traditional boost converter. In [38], a single switch n-stage Cascaded Boost Converter (n-stage CBC) is discussed to achieve high voltage but it requires a large number of inductors, diodes and capacitors. Figure 2a depicts the power circuit of a single switch n-stage Cascaded Boost Converter (single switch n-stage CBC). In [38], a single switch boost converter with a voltage multiplier extension to achieve high voltage is discussed, but it requires a large number of diodes and capacitor circuitry. Figure 2b depicts the power circuit of a single switch boost converter with voltage multiplier. In [39], a new coupled inductor-based step-up converter is discussed with a large pump. The voltage can be easily achieved by modifying the turns-ratio of coupled inductors but the leakage energy induces high voltage stress and switching losses. The power circuit of coupled inductor based step-up converter is shown in Figure 2c. In [40], a new Quadratic Boost Converter (QBC) with coupled inductor in second boost converter is proposed and shown in Figure 2d. This QBC achieves high step-up voltage gain with an appropriate duty ratio and low voltage stress on the power switch.

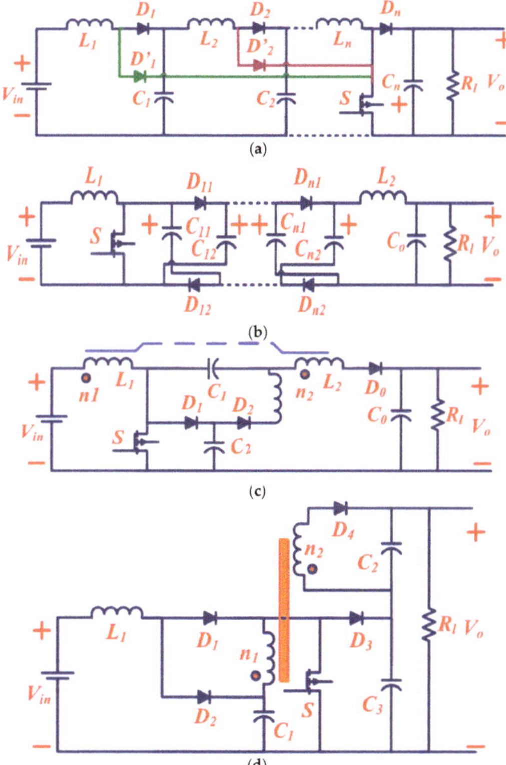

Figure 2. Recently addressed DC-DC power converter circuit (**a**) single switch n-stage Cascaded Boost Converter (n-stage CBC) (**b**) single switch boost converter with voltage multiplier (**c**) coupled inductor based step-up converter and (**d**) Quadratic Boost Converter (QBC) with coupled inductor.

In [41], a DC-DC non-inverting Nx Interleaved Multilevel Boost Converter (Nx IMBC) is proposed to achieve a high voltage conversion ratio (V_o/V_{in}) and to reduce voltage/current ripple. Nx-IMBC provides N times more voltage conversion ratio compared to traditional boost converters but requires large number of diodes and capacitors. In [42,43], DC-DC non-inverting 2Nx and 4Nx Interleaved Boost converters (2Nx IMBC and 4Nx IMBC) are proposed to achieve high voltage conversion ratio (V_o/V_{in}) and to reduce voltage/current ripple. 2Nx IMBC and 4Nx IMBC provide 2N and 4N times more voltage conversion ratio compared to traditional boost converters, but also require large numbers of diodes and capacitors. To achieve a high inverting voltage conversion ratio, a new inverting Nx and 2Nx Multilevel Boost Converter (MBC) was proposed for renewable energy applications [44]. In [45,46], a new family of DC-DC converters called "X-Y Converter Family" was proposed to achieve high conversion ratios for renewable applications. These converters are well suited to achieve high voltage, but need more number of devices, thus increasing the size and cost of the converter.

In this article, a new T & SC-BC is proposed for high-voltage/low-current renewable applications to overcome the drawback of recently addressed converters. The power circuit and block diagram of proposed converter scheme system is shown in Figure 3. The proposed T & SC-BC is an original extension for DC-DC boost converters which is based on T & SC. The proposed T & SC-BC converter combines the features of a conventional boost converter and T & SC [47] to achieve high voltage conversion ratios. In a PV cell model the current source is basically associated in parallel with the reversed diode and also has series and parallel resistance as shown in Figure 3. Resistance in series (R_S) is due to barrier in the pathway of flow of electrons from n to p junction and resistance in shunt (R_{Sh}) is due to the leakage current [48]. The relation between the currents of a PV cell is shown in Equation (1):

$$\left. \begin{array}{l} I = I_{ph} - I_d - I_{Sh}, I_d = I_0(e^{\frac{V+IR_S}{nV_T}} - 1), I_{Sh} = \frac{V+IR_S}{R_{Sh}} \\ V_T = \frac{kT}{q}, V_T = 0.0259V \; at \; T = 25 \, ^{\circ}C \end{array} \right\} \tag{1}$$

where I_{ph} is the photocurrent generated by the cell, I_d is the current flowing through the diode of the solar cell, I_{Sh} is the shunt current flowing through the shunt resistance (R_{Sh}), I is the output current or current flowing through the series resistance (R_S), V is the voltage at the output terminal of the cell, I_0 is the reverse saturation current, n is the diode ideal factor, V_T is the thermal voltage, k is the Boltzmann constant and T is the absolute temperature. When irradiance strikes the flat surface of a PV module or cell, an electrical field is produced within the cell. In the presence of an electric field, these charges can create a current that can be used in a peripheral circuit. This current depends on the concentration and intensity of the incident solar radiation. The higher the level of the light intensity, the more electrons can be allowed to run free from the flat surface, and the more current is created. All the time it is necessary to track MPP due to deviation of hotness and irradiation of the array [49,50]. MPPT techniques to track MPP have been addressed and published over years of research [50–53]. Every MPPT method has its own merits and demerits like required sensors, complexity, cost, range of effectiveness, convergence speed, correct tracking when the irradiation and temperature change. The Perturb & Observe (P & O) algorithm and Incremental Conductance algorithm is the most popular and simple methods to track MPP. The P & O algorithm is depends on hill climbing concept hence also called "hill-climbing P & O Method". This is one of most used algorithms due to its simplicity and ease of implementation and low cost [49,53]. It operates with a cyclic perturbation (increase or decrease) of the array terminal voltage and by comparing the PV power of last perturbation. If the power increases the perturbation goes with the same direction otherwise it will goes with the reverse direction. In this method, the sign of the last perturbation and the sign of the last increment in the power are used to decide what the next perturbation should be [53]. The concept of the P & O MPPT algorithm is shown in Figure 4a–c with the P-V and I-V characteristics. To extract the maximum power, a P & O MPPT control mechanism is used to locate the MPP for the proposed T & SC-BC.

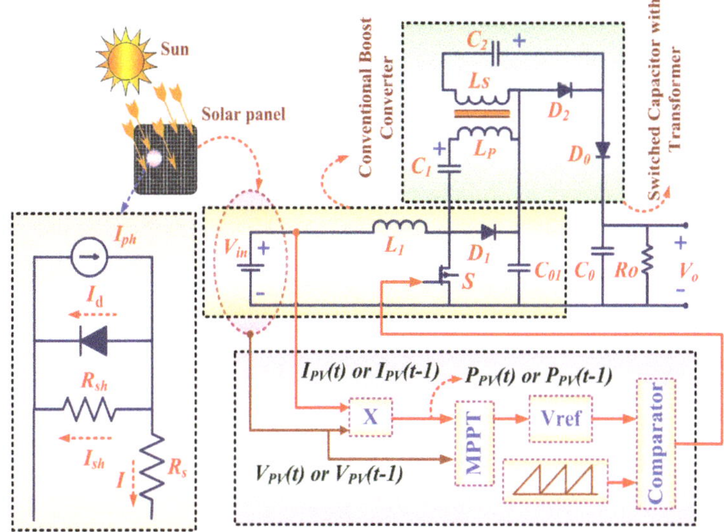

Figure 3. Power circuit and block diagram of proposed converter system: Transformer and Switch Capacitor Based Boost Converter (T & SC-BC) with Maximum Power Point Tracking (MPPT) for high-voltage/low-current renewable applications.

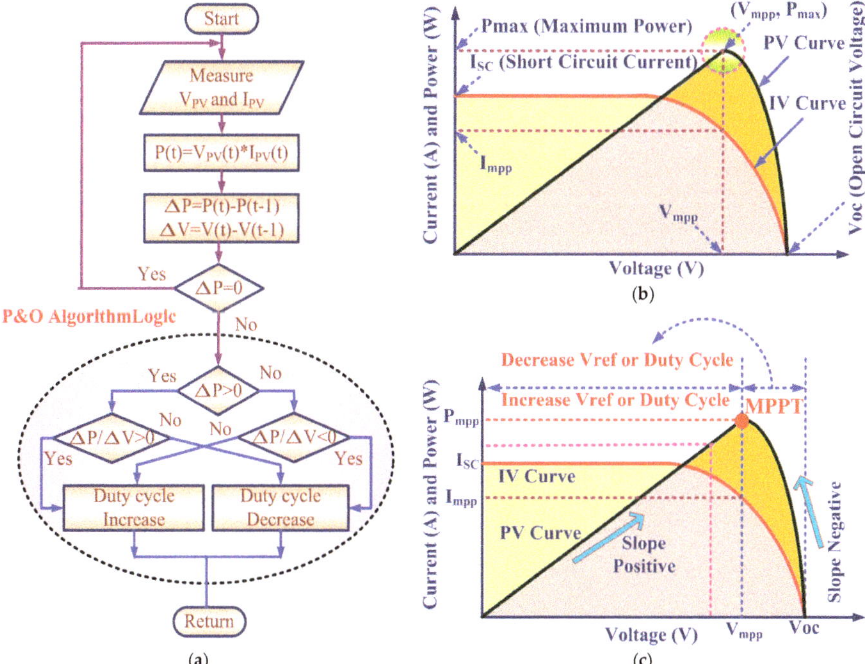

Figure 4. Concept of Maximum Power Point Tracking for proposed T & SC-BC (**a**) Perturb and Observed algorithm (**b**) PV and IV characteristics of photovoltaic cell and (**c**) Concept to track Maximum Power Point (MPP) to extract maximum power.

The conspicuous features of proposed T & SC-BC are:

(1)   High voltage conversion ratio (V_o/V_{in}),
(2)   Continuous input current (I_{in}),
(3)   Single switch topology,
(4)   Single input source,
(5)   Low Drain to Source voltage (V_{DS}) rating of switch,
(6)   Single inductor and single untapped transformer.

Simulation and experimental result are provided which validate the functionality, design and concept of the proposed approach.

2. Power Circuit and Operation Modes of Transformer and Switch Capacitor Based Boost Converter (T & SC-BC)

A conventional boost converter with SC or Voltage Doubler (VD) is shown in Figure 5a. A circuit connection of capacitor C_1, C_0 and diode D_2, D_0 forms a SC stage or VD stage. A transformer is employed in the SC stage to provide a T & SC stage for high voltage conversion. The power circuit of the proposed T & SC-BC is shown in Figure 5b. The proposed T & SC-BC is designed by using a conventional/traditional boost converter by employing a T & SC stage at the output side. In Figure 5, inductor L_1, switch S, diode D_1 and capacitor C_{01} form a conventional boost converter. The input supply is directly connected to the conventional boost converter. Capacitor C_1, C_2, diode D_0, D_2 and transformer form the newly designed T & SC stage. L_P is the primary winding of the transformer whose one terminal is connected via capacitor C_1 at the inductor of conventional boost converter or at the drain terminal of a switch and the other terminal is directly connected at the cathode of diode D_1 or at output capacitor C_{01} of the conventional boost converter. L_S is the secondary winding of the transformer whose one terminal is directly connected to the anode of diode D_2 and another terminal connected to capacitor C_2.

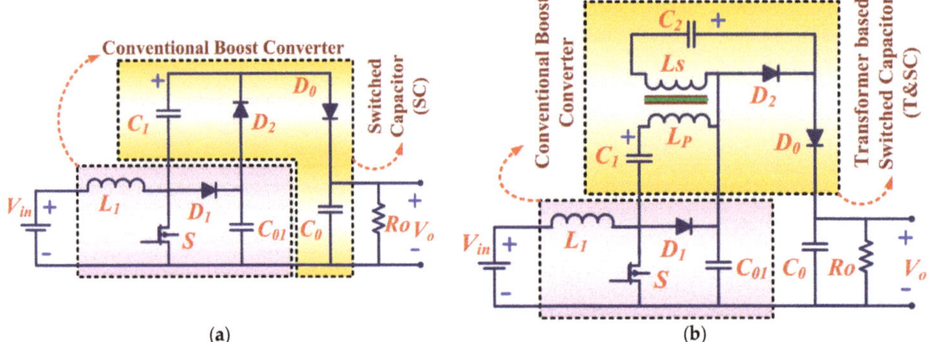

Figure 5. Power circuit of converter (**a**) Conventional boost converter with Switched Capacitor or Voltage Doubler stage and (**b**) Proposed Transformer and Switched Capacitor Based Converter.

The operation of the proposed T & SC-BC is divided into two main modes; one when switch S is turned ON and another when switch S is turned OFF. These two modes are divided into five sub-modes to explain the CCM operation of the proposed T & SC-BC in detail. The CCM characteristic waveforms with the five sub-modes are shown in Figure 6.

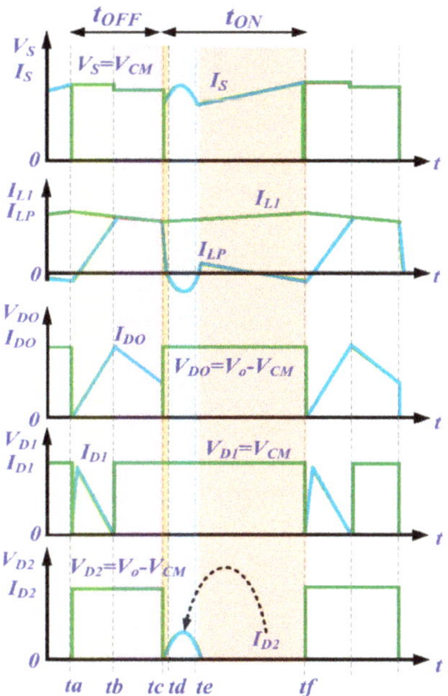

Figure 6. Characteristic waveforms of the proposed T & SC-BC in CCM Mode. (Mode-1: time *ta* to *tb*, Mode-2: time *tb* to *tc*, Mode-3: time *tc* to *td*, Mode-4: time *td* to *te* and Mode-5: time *te* to *tf*).

2.1. Mode-1 (Time ta–tb)

In Mode-1, switch S is turned OFF and inductor L_1 is demagnetized. Thus, the energy stored in the inductor L_1 is transferred to capacitor C_{01} through diode D_1. The energy of capacitor C_1, C_2, transformer windings is also transferred through diode D_0 to the output capacitor C_0. The equivalent circuit of the proposed converter for this mode is shown in Figure 7a. In this mode diode D_1, D_0 are operated in forward biased condition whereas diode D_2 are operated in reverse biased condition.

$$\left.\begin{array}{c} Loop-1 \rightarrow V_{in} - V_{L1} - VC_{01} = 0 \\ Loop-2 \rightarrow V_{in} - V_{L1} + VC_1 - V_{LP} - VC_{01} = 0 \\ Loop-3 \rightarrow V_{LS} + VC_2 - V_{CO} + VC_{01} = 0 \end{array}\right\} \tag{2}$$

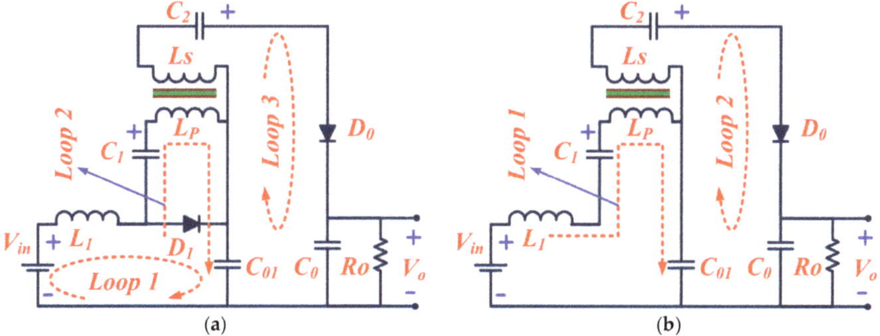

Figure 7. Equivalent circuit of proposed converter when switch S is in OFF state (**a**) Mode-1 (Time *ta–tb*); and (**b**) Mode-2 (Time *tb–tc*).

2.2. Mode-2 (Time tb–tc)

In Mode-2, switch S is turned OFF and inductor L_1 is still demagnetized but the energy stored in the inductor L_1 is not directly transferred to capacitor C_{01} through diode D_1 (diode D_1 is acting as an open circuit) but rather transferred through the windings. The energy of capacitor C_1, C_2, transformer windings is also transferred through diode D_0 to output capacitor C_0. The equivalent circuit of proposed converter for this mode is shown in Figure 7b. In this mode diode D_0 are operated in forward biased condition whereas diode D_1, D_2 are operated in reverse biased condition.

$$\left.\begin{array}{l} Loop-1 \rightarrow V_{in} - V_{L1} + VC_1 = V_{LP} + VC_{01} \\ Loop-2 \rightarrow V_{LS} + VC_2 = V_{CO} - VC_{01} \end{array}\right\} \tag{3}$$

2.3. Mode-3 (Time tc–td)

In Mode-3, switch S is turned ON and inductor L_1 is magnetized through switch S. The energy of capacitor C_1, C_2, transformer windings is transferred through diode D_0 to output capacitor C_0 but the diode D_0 current start decreasing and current slope (di/dt) is limited by the transformer. Hence, the diode reverse current recovery problem is decreased. The equivalent circuit of the proposed converter in this mode is shown in Figure 8a. In this mode diode D_0 are operated in forward biased condition whereas diodes D_1, D_2 are reverse biased.

$$\left.\begin{array}{l} Loop-1 \rightarrow V_{in} - V_{L1} = 0 \\ Loop-2 \rightarrow VC_1 - V_{LP} - VC_{01} = 0 \\ Loop-2 \rightarrow V_{LS} + VC_2 + VC_{01} = V_{CO} \end{array}\right\} \tag{4}$$

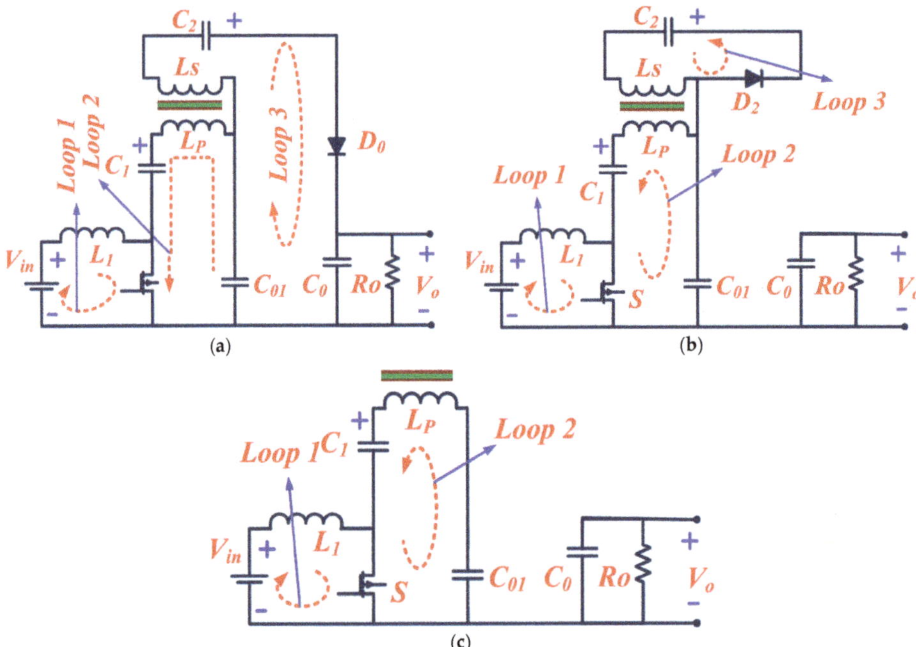

Figure 8. Equivalent circuit of proposed converter when switch S is in ON state (**a**) Mode-3 (Time *tc–td*); (**b**) Mode-4 (Time *td–te*); and (**c**) Mode-5 (Time *te–tf*).

2.4. Mode-4 (Time td–te)

In Mode-4, switch S is turned ON and inductor L_1 is continuously magnetized through switch S. The energy of the transformer windings Ls is transferred to capacitor C_2 through diode D_2. The energy conversion takes place in a resonant approach and the leakage inductance confines the current. The output capacitor C_0 provides energy to load. At the end of the mode capacitor C_2 is fully charged and diode D_2 is blocked. The equivalent circuit of proposed converter for this mode is shown in Figure 8b. In this mode diode D_2 are operated in forward biased condition whereas diode D_1, D_0 are operated in reverse biased.

$$\left. \begin{array}{l} Loop - 1 \rightarrow V_{in} - V_{L1} = 0 \\ Loop - 2 \rightarrow VC_1 - V_{LP} - VC_{01} = 0 \\ Loop - 3 \rightarrow VC_2 = -V_{LS} \end{array} \right\} \tag{5}$$

2.5. Mode-5 (Time te–tf)

In Mode-5, switch S is turned ON and inductor L_1 is continuously magnetized through switch S. The energy of transformer windings Ls is not transferred to capacitor C_2 due to diode D_2 is in reverse biased. The output capacitor C_0 provides energy to load. The equivalent circuit of proposed converter for this mode is shown in Figure 8c. In this mode no diodes are operated in forward biased condition whereas diodes D_0, D_1, D_2 are in reverse biased condition.

$$\left. \begin{array}{l} Loop - 1 \rightarrow V_{in} - V_{L1} = 0 \\ Loop - 2 \rightarrow VC_1 = V_{LP} + VC_{01} \end{array} \right\} \tag{6}$$

The voltage conversion ratio (V_o/V_{in}) and Drain to Source voltage of switch (V_{DS}) of conventional boost converter with VD (power circuit of converter shown in Figure 5a) is calculated by Equation (7), where T is the total time of one switching cycle. The voltage conversion ratio (V_o/V_{in}) and (V_{DS}/V_{in}) of conventional boost converter with VD is shown graphically in Figure 9a. From graph it is observed that drain to source switch voltage is exactly half of the output voltage (for example at D = 0.75, V_o/V_{in} = 8 and V_{DS}/V_{in} = 4). Hence the drain to source voltage rating of the switch must be above four times of input to operate converter at 75% duty cycle.

$$\left. \begin{array}{l} \frac{V_o}{V_{in}} = \frac{2}{1-D} = \frac{2T}{T-t_{on}} \\ V_{DS} = \frac{1}{1-D} V_{in} = \frac{V_o}{2} = \frac{T}{T-t_{on}} V_{in} \end{array} \right\} \tag{7}$$

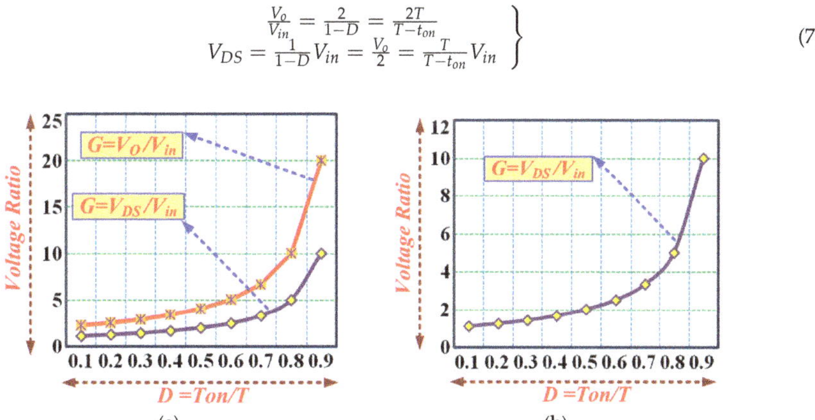

Figure 9. *Cont.*

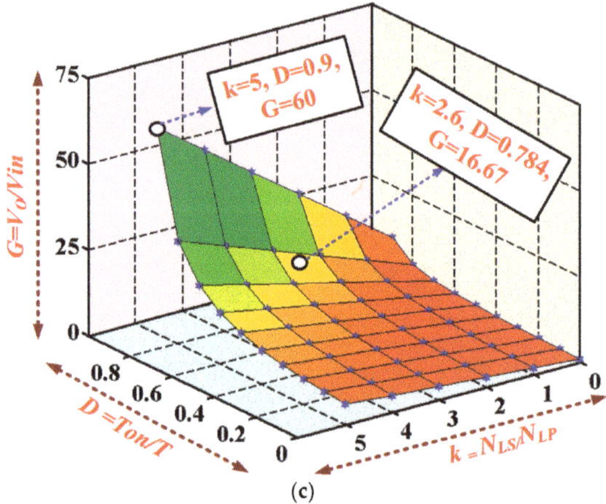

(c)

Figure 9. Graphical plot (**a**) voltage conversion ratio (V_o/V_{in}) and Drain to Source switch voltage (V_{DS}) of conventional Boost converter with Voltage Doubler (V_D) versus duty cycle (D); (**b**) Drain to Source switch voltage (V_{DS}) of proposed T & SC-BC versus duty cycle (D); and (**c**) Relation between voltage conversion ratio (V_o/V_{in}), turns ratio, and duty cycle of proposed T & SC-BC.

The voltage conversion ratio (V_o/V_{in}) and Drain to Source voltage of switch (V_{DS}) of the proposed T & SC-BC are calculated by Equation (8), where T is the total time of one switching cycle. The relation of V_{DS}/V_{in} with duty cycle (D) is graphically shown in Figure 9b. The relation of voltage conversion ratio (V_o/V_{in}), turns ratio (k), and duty cycle (D) of the proposed T & SC-BC is shown graphically in Figure 9c. From the graphs it is observed that voltage conversion ratio is linearly increased with duty cycle and turns ratio. It is observed that the voltage conversion ratio is 16.67 at $D = 0.784$ when $k = 2.6$. The voltage conversion ratio is 60 at $D = 0.9$ when $k = 5$. It is also investigated that the drain to source switch voltage is exactly equal to the drain to source of a traditional boost converter with VD.

$$\left. \begin{array}{c} \dfrac{V_o}{V_{in}} = \dfrac{1+k}{1-D} = \dfrac{(1+k)T}{T-t_{on}} \\[2mm] V_{DS} = \dfrac{1}{1-D}V_{in} = \dfrac{V_o}{1+k} = \dfrac{T}{T-t_{on}}V_{in} \\[2mm] k = \dfrac{\text{Turns of Secondary winding of transformer } (N_{LS})}{\text{Turns of Primary winding of transformer } (N_{LP})} \end{array} \right\} \qquad (8)$$

3. Steady State Analysis of Transformer and Switch Capacitor Based Boost Converter (T & SC-BC)

In this section a steady state analysis of the proposed T & SC-BC is explained and the conversion losses, efficiency and voltage conversion ratio are calculated. In order to analyze the T & SC-BC, we consider the T & SC is working in steady state and the following assumptions are considered throughout the switching: (i) ripple free DC Source (V_{in}) (ii) diode D_1 semiconductor loss is V_{D1}; (iii) forward conduction loss of diode D_1 is modelled by ON-state resistance R_{D1} (efficiency of diode is 100% if V_{D1} and $R_{D1} = 0$); (iv) R_S is ON-state resistance of controlled switch. R_{L1} is internal resistance of winding of inductor L_1; (v) f_s is switching frequency (vi) ripple across capacitor is very small (vii) T & SC cell losses include copper loss, iron loss and diode conduction loss (viii) $R_{T\&SC}$ is the equivalent resistance of the T & SC cell. The steady state equivalent circuits of the T & SC-BC ON and OFF state are shown in Figure 10a,b respectively.

Consider i_{L1}, *is* and i_{c01} current is flowing through inductor L_1, switch S and capacitor C_{01} respectively. I_{L1}, *Is* and I_{C01} average current is flowing through inductor L_1, switch S and capacitor C_{01} respectively.

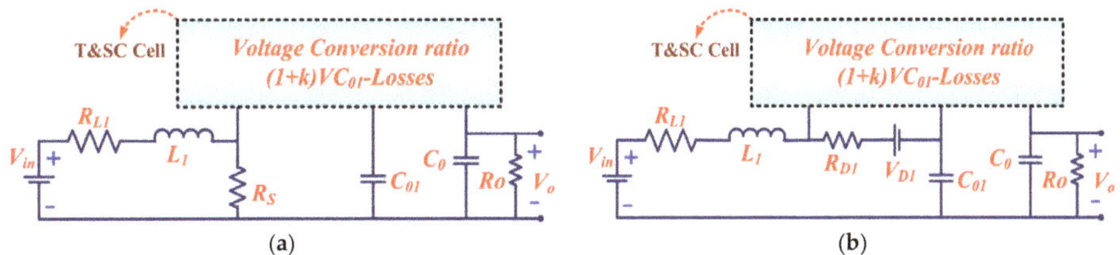

Figure 10. Equivalent circuit of T & SC-BC (**a**) ON state and (**b**) OFF state.

When switch is in ON state, diode D_1 is in reversed biased and inductor is charged by input supply V_{in}:

$$\left.\begin{array}{c} V_{in} - v_{L1} - i_{in}R_{L1} - i_s R_s = 0, i_{in} = i_s \\ v_{L1} = V_{in} - i_{in}(R_{L1} + R_s) \approx V_{in} - I_{in}(R_{L1} + R_s) \\ i_c(t) = -V_{c01}/R_{T\&SC} \end{array}\right\} \quad (9)$$

when switch is in OFF state, diode D_1 is conduct and inductor is discharged to charge capacitor C_{01}:

$$\left.\begin{array}{c} V_{in} - v_{L1} - i_{in}R_{L1} - i_{D1}R_{D1} - V_{D1} - V_{c01} = 0, i_{in} = i_{D1} \\ v_{L1} = V_{in} - i_{in}(R_{L1} + R_{D1}) - V_{D1} - V_{c01} \approx V_{in} - I_{in}(R_{L1} + R_{D1}) - V_{D1} - V_{c01} \\ i_c(t) = i_{in} - V_{c01}/R_{T\&SC} \end{array}\right\} \quad (10)$$

Inductor volt second balance method and capacitor charge method is used to calculate the voltage conversion ratio equation:

$$\left.\begin{array}{c} V_{c01} = (\frac{1}{1-D})(V_{in} - (1-D)V_{D1})(\frac{(1-D)^2 R_{T\&SC}}{(1-D)^2 R_{T\&SC} + R_{L1} + DR_s + (1-D)R_{D1}}) \\ \frac{V_{c01}}{V_{in}} = (\frac{1}{1-D})(1 - \frac{(1-D)V_{D1}}{V_{in}})(\frac{1}{1 + \frac{R_{L1} + DR_s + (1-D)R_{D1}}{(1-D)^2 R_{T\&SC}}}) \end{array}\right\} \quad (11)$$

$$V_o = V_{c01}(1+k) - losses \text{ of transformer} \quad (12)$$

$$\left.\begin{array}{c} V_o = (1+k)(\frac{1}{1-D})(V_{in} - (1-D)V_{D1})(\frac{(1-D)^2 R_{T\&SC}}{(1-D)^2 R_{T\&SC} + R_{L1} + DR_s + (1-D)R_{D1}}) \\ \frac{V_o}{V_{in}} = (1+k)(\frac{1}{1-D})(1 - \frac{(1-D)V_{D1}}{V_{in}})(\frac{1}{1 + \frac{R_{L1} + DR_s + (1-D)R_{D1}}{(1-D)^2 R_{T\&SC}}}) \end{array}\right\} \quad (13)$$

$$Conversion\ Losses = (\frac{1+k}{1-D})\frac{(1-D)V_{D1}}{V_{in}}(\frac{1}{1 + \frac{R_{L1} + DR_s + (1-D)R_{D1}}{(1-D)^2 R_{T\&SC}}}) \quad (14)$$

$$\left.\eta, efficiency = \frac{(1 - \frac{(1-D)V_{D1}}{V_{in}})}{1 + \frac{R_{L1} + DR_s + (1-D)R_{D1}}{(1-D)^2 R_{T\&SC}}}\right\} \quad (15)$$

If all the components efficiency is 100% (ideal components) then:

$$Losses = 0,\ \eta = 100\%\ and \frac{V_o}{V_{in}} = (1+k)(\frac{1}{1-D}) \quad (16)$$

The ideal voltage conversion ratio of T & SC-BC is given in Equation (16).

4. Comparison of Proposed Transformer and Switch Capacitor Based Boost Converter (T & SC-BC) with Newly Addressed DC-DC Converters

In this section the proposed T & SC-BC is compared with recently addressed DC-DC converters. Recently addressed converters are discussed in Section 1 of the article. In Table 1, the converter comparison is summarized in terms of voltage conversion ratio, number of inductors, number of switches, and number of diodes. First, it is observed that voltage conversion ratio of single switch

n-stage CBC is depends on the number of cascaded stages (n) and duty cycle. To design single switch n-stage CBC [30], n-number of inductors, n-number of capacitors and $2n - 1$ number of diodes along with single switch is required. Second, it is observed that voltage conversion ratio of single switch boost converter with voltage multiplier is depends on the odd and even number of voltage multiplier stage and duty cycle. To design single switch boost converter with voltage multiplier, 2 inductors, $2n + 1$ number of capacitors and $2n$ number of diodes along with single switch is required. Third, it is observed that voltage conversion ratio of coupled inductor based step-up converter is depends on the coupling coefficient of coupled inductor ($k = n_2/n_1$) and duty cycle. To design coupled inductor based step-up converter, 3 inductors (1 without coupling and 2 with coupled), 3 capacitors and 3 diodes along with single switch is required. Fourth, it is observed that voltage conversion ratio of Quadratic Boost Converter (QBC) with coupled inductor is depends on the coupling coefficient of coupled inductor ($k = n_2/n_1$) and duty cycle. To design Quadratic Boost Converter (QBC) with coupled inductor, 3 inductors (1 without coupling and 2 with coupled), 3 capacitors and 4 diodes along with single switch is required. Fifth, it is observed that to design conventional boost converter with VD 1 inductor, 3 capacitors, 1 switch and 3 diodes are required. Sixth, it is observed that voltage conversion ratio of proposed converter (T & SC-BC) is depends on the transformer turns ratio (k) and duty cycle. To design proposed converter (T & SC-BC), 1 inductor, 1 transformer, 4 capacitors and 3 diodes along with single switch is required.

Table 1. Comparison summary of the proposed T & SC-BC with recently addressed DC-DC converters.

DC-DC Converter	Voltage Conversion Ratio (V_o/V_{in})	Number of Switches	Number of Inductors	Number of Capacitors	Number of Diodes
Single switch n-stage CBC [38]	$\frac{1}{(1-D)^n}$	1	n	n	$2n - 1$
Single switch boost converter with voltage multiplier [38]	$\frac{n+D}{1-D}, n = 1,3\ldots$ $\frac{n+1+D}{1-D}, n = 2,4\ldots$	1	2	$2n + 1$	$2n$
Coupled inductor based step-up converter [39]	$\frac{1+(1+k)D}{1-D}, k = \frac{n_2}{n_1}$	1	3 (2 Inductor are Coupled)	3	3
Quadratic Boost Converter (QBC) with coupled inductor [40]	$\frac{1+kD}{(1-D)^2}, k = \frac{n_2}{n_1}$	1	3 (2 Inductor are Coupled)	3	4
Boost converter with voltage doubler (Figure 5a)	$\frac{2}{1-D}$	1	1	3	3
Proposed converter	$\frac{1+k}{1-D}, k = \frac{n_{LS}}{n_{LP}}$	1	1 Inductor with 1 transformer	4	3

n = Number of stages and k = Turns ratio.

Thus from Table 1 it is seen that the proposed converter required less components and has a higher voltage conversion ratio compared to the other discussed converters. In Table 2 the cost of the power circuit of the proposed T & SC-BC and recently addressed converters (discussed in Section 1) is tabulated and it is observed that the proposed converter requires less cost.

Table 2. Comparison of the proposed T & SC-BC with recent DC-DC converters in terms of cost.

DC-DC Converter	Cost of Converter
Single switch n-stage CBC [38]	$C_S + [n \times C_L] + [n \times C_C] + [(2n - 1) \times C_D]$
Single switch boost converter with voltage multiplier [38]	$C_S + [2 \times C_L] + [(2n + 1) \times C_C] + [(2n) \times C_D]$
Coupled inductor based step-up converter [39]	$C_S + [(1 \times C_{cL}) + (1 \times C_L)] + [3 \times C_C] + [3 \times C_D]$
QBC with coupled inductor [40]	$C_S + [(1 \times C_{cL}) + (1 \times C_L)] + [3 \times C_C] + [4 \times C_D]$
Boost converter with voltage doubler (Figure 5a)	$C_s + C_L + [3 \times C_C] + [3 \times C_D]$
Proposed T & SC-BC	$C_S + [(1 \times C_{cL}) + (1 \times C_T)] + [4 \times C_C] + [3 \times C_D]$

C_s = Cost of single switch, C_L = Cost of single inductor, C_C = Cost of single capacitor, C_D = Cost of single diode, C_T = Cost of Transformer, C_{CL} = Cost of two inductor with couple effect.

The plots of the voltage conversion ratio of a single switch CBC and single switch boost converter with voltage multiplier versus duty cycle considering the number of stages, n = 1 to 5 is depicted in

Figure 11a,b respectively. It is observed that the voltage conversion ratio increases with the increase in the number of cascaded stages, but the quasi-linear region of the converter decreases. The plot of the voltage conversion ratio of the coupled inductor based step-up converter and QBC with coupled inductor versus duty cycle considering coupling coefficients, k = 1 to 5 is depicted in Figure 11c,d respectively. It is observed that the voltage conversion ratio is greatly increased with the increase in coupling coefficient (k), but the quasi-linear region of the converter decreases.

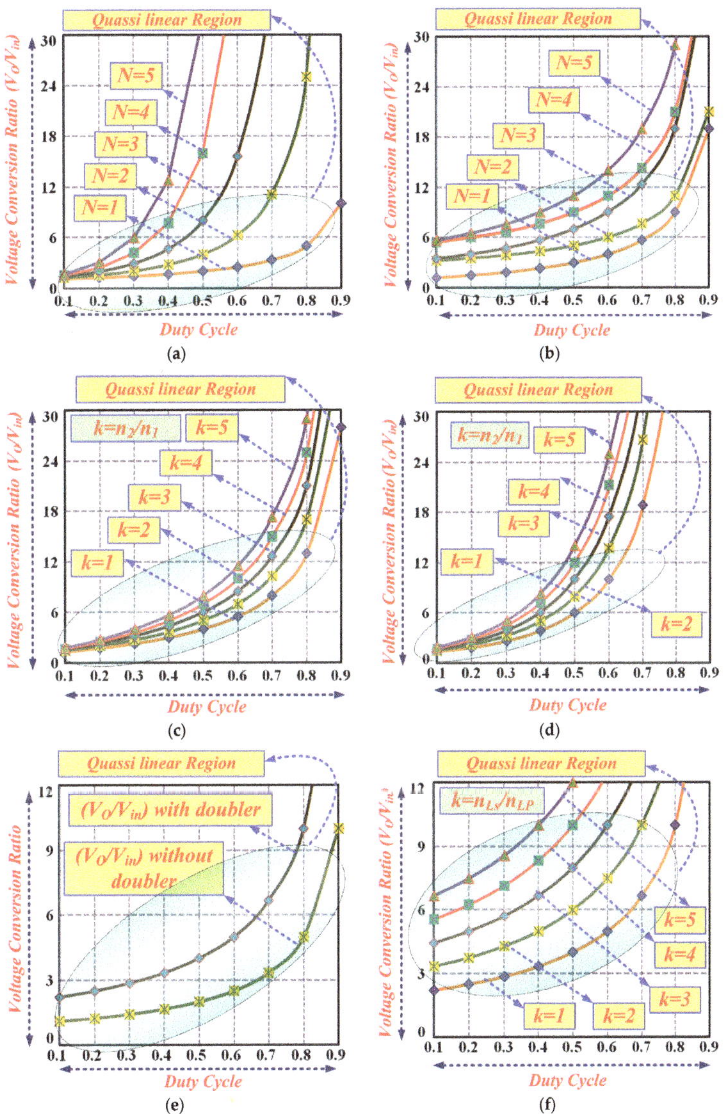

Figure 11. Plot of Voltage conversion ration versus Duty cycle (**a**) single switch *n*-stage CBC; (**b**) Single switch boost converter with voltage multiplier; (**c**) Coupled inductor based step-up converter; (**d**) Quadratic Boost Converter (QBC) with coupled inductor; (**e**) Conventional boost converter with Switched Capacitor (SC) or Voltage Doubler (VD) stage; and (**f**) Proposed Transformer and Switched Capacitor Based Converter (T & SC-BC).

The voltage conversion ratio and capacitor voltage plot of a conventional boost converter with or without SC or VD stage is depicted in Figure 11e. It is observed that the conversion ratio is double with the voltage doubler compared to without the voltage doubler but the quasi-linear region is slightly reduced. The voltage conversion ratio plot of the proposed T & SC-BC versus duty cycle considering transformer ratio, $k = 1$ to 5 is depicted in Figure 11f. It is observed that the voltage conversion ratio is greatly increased with the increase in number of the transformer ratio (k) but the quasi-linear region of the converter is decreased.

In Figure 12a, the plot of VC_n/V_{in} (ratio of capacitor voltage to input voltage) versus duty cycle for a single switch n-stage CBC is shown considering stages $n = 1$ to 5. It is observed that the capacitor rating increases with the increase in stages and duty cycle. Thus, a single switch n-stage CBC requires more rated capacitors compared to a single switch $n - 1$ stage CBC. In Figure 12b the plot of VC_{n1}/V_{in} and VC_{n2}/V_{in} (ratio of capacitor voltage to input voltage) versus duty cycle for a single switch boost converter with voltage multiplier is shown considering stage $n = 1$ to 5. It is observed that the rating of the capacitor increases with the increase in voltage multiplier stages and duty cycle. In Figure 12c the plot of VC_2/V_{in} (ratio of capacitor voltage to input voltage) versus duty cycle for a coupled inductor-based step-up converter is shown with coupling coefficients $k = 1$ to 5. It is observed that the capacitor voltage is independent of the coupling coefficient but depends upon the duty cycle.

In Figure 12d the plot of VC_{02}/VC_{01} (ratio of capacitor C_{02} voltage to capacitor C_{01} voltage) versus duty cycle for the QBC with coupled inductor is shown with coupled coefficients $k = 1$ to 5. It is observed that the capacitor voltage depends on the coupling coefficient and also on the duty cycle. Thus, the rating of the capacitor is increased with an increase in coupling coefficient (k) and duty cycle (D). In Figure 12e the plot of VC_{01}/V_{in} (ratio of capacitor C_{01} voltage to input voltage (V_{in})) versus duty cycle for the proposed T & SC-BC with transformer turn ratio $k = 1$ to 5. It is observed that the capacitor voltage is independent of the transformer ratio but depends upon the duty cycle. T & SC-BC is compared with recently addressed DC-DC converters (discussed in Section 1) in terms of the voltage conversion ratio (V_o/V_{in}) and the comparison plot is shown in Figure 12f. It is observed that the proposed T & SC-BC provides higher voltage conversion ratios compared to other DC-DC converters. T & SC-BC is compared with recently addressed DC-DC converters in term of V_{DS}/V_{in} (ratio of Drain to source voltage and input voltage) and the comparison plot is shown in Figure 12g and also tabulated in Table 3. It is observed that the proposed converter has a smaller V_{DS}/V_{in} ratio (ratio of Drain to Source voltage with respect input voltage), hence low rating components are suitable to design the T & SC-BC compared to recently proposed DC-DC converters (discussed in Section 1). Also in Table 3, T & SC-BC is compared in terms of efficiency; power range and output ripple with recent DC-DC converters.

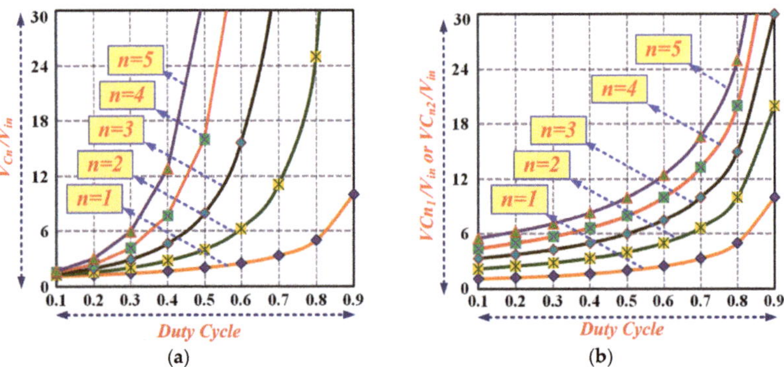

Figure 12. *Cont.*

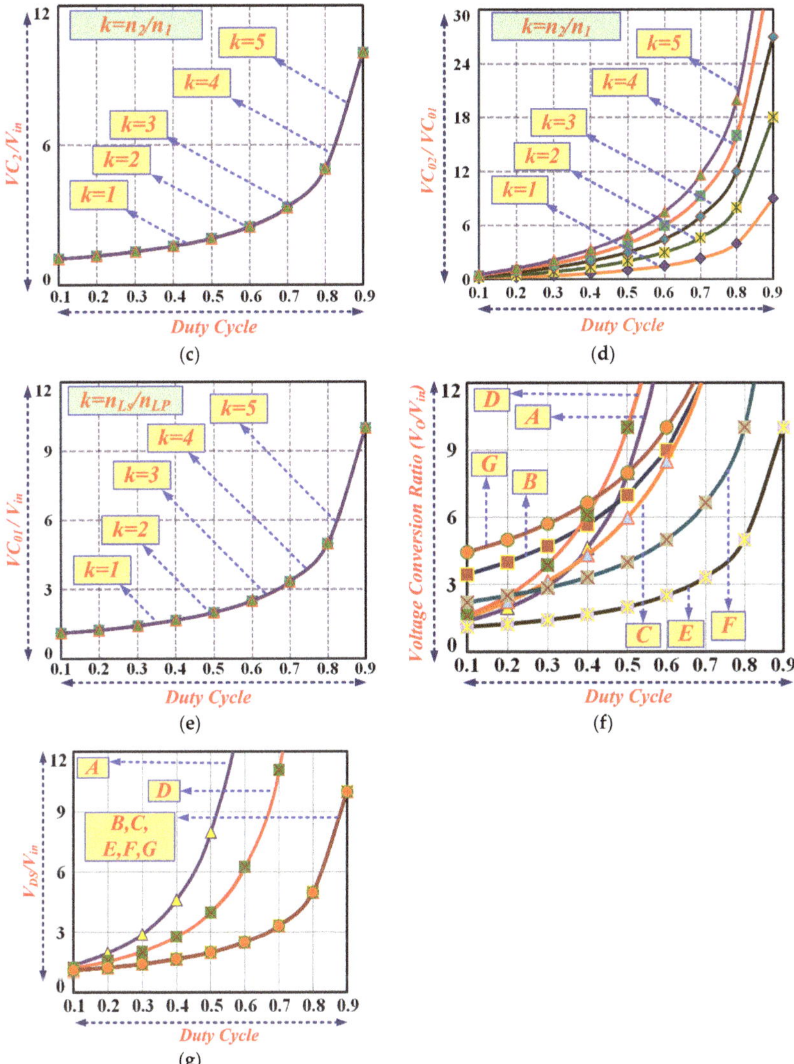

Figure 12. (**a**) Plot of (VC_n/V_{in}) versus duty cycle for single switch n-stage CBC; (**b**) Plot of VC_{n1}/V_{in} and VC_{n2}/V_{in} versus duty cycle for single switch boost converter with voltage multiplier; (**c**) Plot of VC_2/V_{in} versus duty cycle for coupled inductor based step-up converter; (**d**) Plot of VC_{02}/VC_{01} versus duty cycle for QBC with coupled inductor; (**e**) Plot of VC_{01}/V_{in} versus duty cycle for proposed T & SC-BC; (**f**) Comparison of T & SC-BC and recently addressed DC-DC converter (discussed in Section 1) in terms of voltage conversion ratio (V_o/V_{in}); and (**g**) Comparison of T & SC-BC and recently addressed DC-DC converter (discussed in Section 1) in term of V_{DS}/V_{in} (ratio of Drain to source voltage and input voltage) with considering n = 3 and k = 3. (*A*: single switch n-stage CBC, *B*: single switch boost converter with voltage multiplier, *C*: coupled inductor based step-up converter, *D*: QBC with coupled inductor, *E*: Traditional Boost Converter, *F*: Traditional Boost Converter with Voltage Doubler (VD), *G*: Proposed T & SC-BC).

Table 3. Comparison of the proposed T & SC-BC with recently addressed DC-DC converters in terms of switch drain to source voltage.

DC-DC Converter	Switch Drain to Source Voltage (V_{DS})	Efficiency	Applications	Output Ripple
Single switch n-stage CBC [38] (Figure 2a)	$\frac{V_{in}}{(1-D)^n}, n = 1,2,3\ldots$	moderate (85–90%)	low/medium power	high
Single switch boost converter with voltage multiplier [38], (Figure 2b)	$\frac{V_{in}}{1-D}$	moderate (85–90%)	low/medium power	high
Coupled inductor based step-up converter [39], (Figure 2c)	$\frac{V_{in}}{1-D}$	high (90–95%)	medium/high power	low
Quadratic Boost Converter (QBC) with coupled inductor [40], (Figure 2d)	$\frac{V_O}{1+kD}, k = \frac{n_2}{n_1} = 1,2,3\ldots$	high (90–95%)	medium/high power	low
Conventional boost converter with voltage doubler (Figure 5a)	$\frac{V_{in}}{1-D}$	high (90–95%)	low/medium power	high
Proposed converter (Figure 5b)	$\frac{V_{in}}{1-D}$	high (90–95%)	medium/high power	low

5. Simulation and Experimental Results

To verify the proposed converter functionality, the proposed converter is simulated in Matrix Laboratory (MATLAB) and the parameters are tabulated in Table 4. The components are chosen and designed according to Equations (17)–(24).

$$Duty\ cycle, D = \frac{V_O - V_{in}(1+k)}{V_O} = \frac{250 - 15(1+2.6)}{250} = 0.784\ or\ 78.4\% \tag{17}$$

$$Switch\ Voltage, V_{DS} = \frac{V_{in}}{1 - \frac{T_{ON}}{T}} = \frac{V_{in}}{1 - fT_{ON}} = \frac{V_{in}}{1 - D} = \frac{15}{1 - 0.784} = 69.44V \approx 70V \tag{18}$$

$$\left. \begin{array}{c} Diode\ D_0\ Voltage, V_{D0} = \frac{n_{LS}}{n_{LP}} \times \frac{V_{in}}{1 - \frac{T_{ON}}{T}} \\ = \frac{kV_{in}}{1 - fT_{ON}} = \frac{kV_{in}}{1 - D} = \frac{2.6 \times 15}{1 - 0.784} = 180.5V \approx 181V \end{array} \right\} \tag{19}$$

$$Diode\ D_1\ Voltage, V_{D1} = \frac{V_{in}}{1 - \frac{T_{ON}}{T}} = \frac{V_{in}}{1 - fT_{ON}} = \frac{V_{in}}{1 - D} = \frac{15}{1 - 0.784} = 69.44V \approx 70V \tag{20}$$

$$\left. \begin{array}{c} Diode\ D_2\ Voltage, V_{D2} = \frac{n_{LS}}{n_{LP}} \times \frac{V_{in}}{1 - \frac{T_{ON}}{T}} \\ = \frac{kV_{in}}{1 - fT_{ON}} = \frac{kV_{in}}{1 - D} = \frac{2.6 \times 15}{1 - 0.784} = 180.5V \approx 181V \end{array} \right\} \tag{21}$$

$$Inductor, L_1 = \frac{V_{in} \times \frac{T_{ON}}{T}}{\Delta iL_1 \times f} = \frac{V_{in} \times DT}{\Delta iL_1} = \frac{15 \times 0.784}{5 \times 20000} = 117.6\mu H \tag{22}$$

$$Transformer\ Primary\ Winding, L_P = \frac{V_{in} \times \frac{T_{ON}}{T}}{\Delta iL_1 \times \frac{1}{T}} = \frac{V_{in} \times DT}{\Delta iL_1} = \frac{15 \times 0.784}{5 \times 20000} = 117.6\mu H \tag{23}$$

$$Transformer\ Secondary\ Winding, L_S = k^2 \times \frac{V_{in} \times \frac{T_{ON}}{T}}{\Delta iL_1 \times \frac{1}{T}} = k^2 \times \frac{V_{in} \times DT}{\Delta iL_1} = 794.97\mu H \tag{24}$$

The capacitors value of the proposed converter is calculated with the help of parameters voltage ripple through capacitors, switching frequency (f_s), current through capacitor and turn ratio of transformer (k). In Figure 13a the output voltage and input voltage waveform is depicted. It is observed that a constant 250 V is achieved with an input voltage of 15 V, hence the conversion ratio is 16.67.

The transient period analysis is investigated and it is observed that the time constant (τ) of output and input voltage waveforms is 0.006 s. The output and input voltage at 0.006 s is 157.5 V (63% of 250 V) and 9.45 V (63% of 15 V), respectively. It is observed that a constant 250 V is achieved at 0.035 s (approximately). In Figure 13b the output voltage and current waveform are depicted. It is observed that the time constant (τ) of output voltage and current waveform is 0.006 s.

Table 4. Simulation parameters of the proposed T & SC-BC.

Parameter	Value
Input voltage (V_{in}), Output voltage (V_o)	15 V, 250 V
Input current (I_{in}), Output current (I_o)	3.34 A, 0.2 A
Power (P)	50 W
Duty cycle (D)	0.784
Switching frequency (f_s)	20 kHz
Switch Drain to Source voltage (V_{DS})	70 V (Minimum Voltage)
Inductor (L_1)	117.6 µH
Transformer windings	L_P = 117.6 µH, L_S = 794.97 µHTurns ratio, k = 2.6
Capacitors (C_1, C_2)	2.5 µF, 250 V
Capacitor C_0 and C_{01}	100 µF, 400 V

The output voltage and current at 0.006 s is 157.5 V and 0.126 A (63% of 0.2 A), respectively. It is observed that a constant 0.2 A is achieved at 0.035 s (approximately). In Figure 13c the output power waveform is depicted and it is observed that the output power at 0.006 s is 19.84 W. It is observed that a constant 50 W power is achieved at 0.035 s (approximately). In Figure 13d, the inductor (L_1) current waveform is depicted with gate pulse of the switch and slope of the waveforms is calculated by Equation (17). It is observed that the inductor (L_1) current slope is positive when switch S is in ON state and slope is negative when switch S is in OFF state. In Figure 13e, the switch drain to source voltage (V_{DS}) waveform is shown and nearly 70 V appears across the switch when it is in OFF state. A fluctuation is observed in the switch voltage due to the voltage across diode D_1 in the sub-modes. The voltage across capacitor C_1, C_2 and C_{01} is depicted in Figure 13f and it is observed that the voltage across capacitor C_{01} is 70 V (approximately). The voltage across diode D_0, D_1 and D_2 is depicted in Figure 13g and it is observed that diode D_0 is in forward biased when switch S is in OFF state. It is also that the voltage across D_1 is less than the voltage across diode D_0 and D_2. The proposed T & SC boost converter is implemented and the details of parameters or components are provided in Table 5. The experimental setup of the proposed converter is shown in Figure 14. The output voltage and input voltage waveforms are shown in Figure 15a,b. It is observed that 249.6 V is achieved from 15.1 V input supply.

$$\left. \begin{array}{l} \dfrac{di_{L1}}{dt} = \dfrac{VL_{1ON}}{L_1} = Positive\ Slope \\[2mm] \dfrac{di_{L1}}{dt} = \dfrac{VL_{1OFF}}{L_1} = Negative\ Slope \end{array} \right\} \tag{25}$$

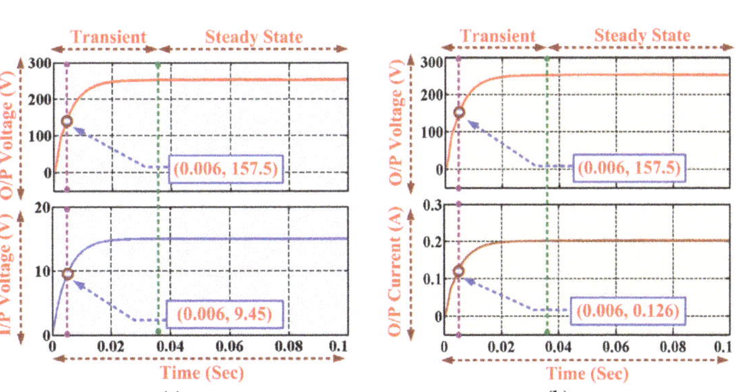

Figure 13. *Cont.*

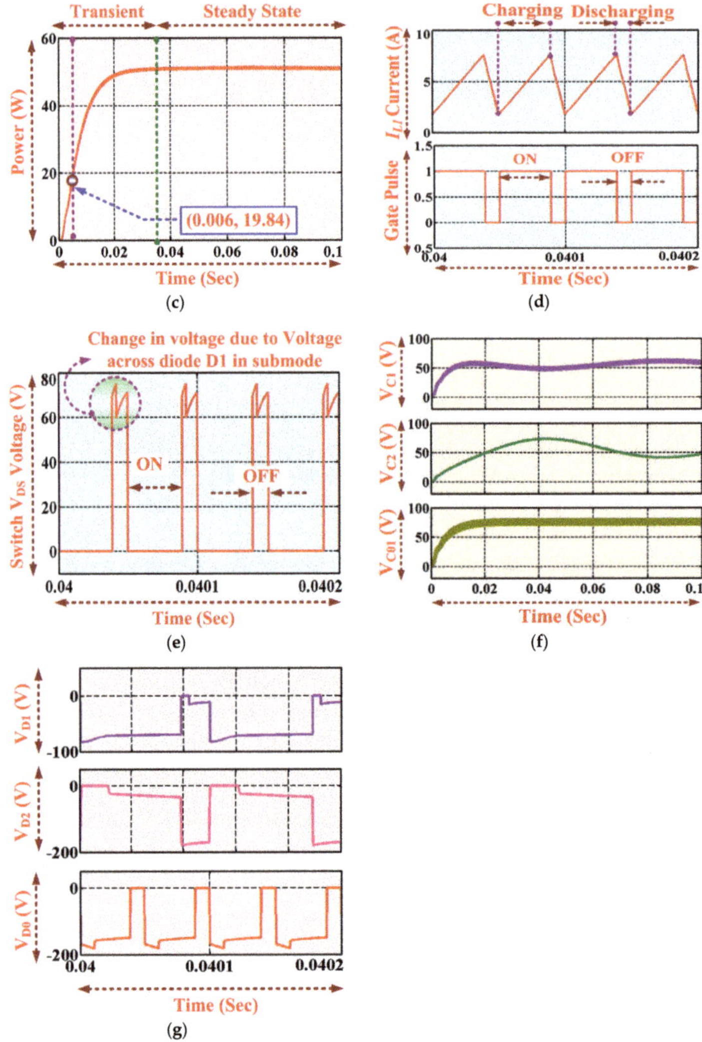

Figure 13. Simulation result of the proposed T & SC-BC (**a**) Output and input voltage waveform; (**b**) Output voltage and current waveform; (**c**) Output power waveform; (**d**) Inductor L_1 current waveform and gate pulse (V_{GS}); (**e**) Drain to Source switch voltage (V_{DS}) waveform; (**f**) Voltage waveform across capacitor C_1, C_2 and C_{01}; and (**g**) Voltage waveform across diode D_0, D_1 and D_2.

Table 5. Parameters and component details of the prototype of the proposed T & SC-BC.

Parameter	Value
PV Panel	3 PV Panels (each of 18 V)
Input Voltage (V_{in}), Output Voltage (V_o)	15 V, 250 V
Input Current (I_{in}), Output Current (I_o)	3.34 A, 0.2 A
Power (P)	50 W
Duty Cycle (D)	0.784
Switching frequency (f_s)	20 kHz
Switch Drain to Source Voltage (V_{DS})	70 V (Minimum Voltage)
Inductor (L_1)	120 µH
Transformer Windings	L_P = 120 µH, L_S = 800 µH Turns ratio = 2.6
Capacitors (C_1 and C_2)	3.3 µF, 250 V
Capacitor C_0 and C_{01}	100 µF, 400 V
Diode (D_1, D_2, D_0)	MUR 860
Control Switch	IRF 540

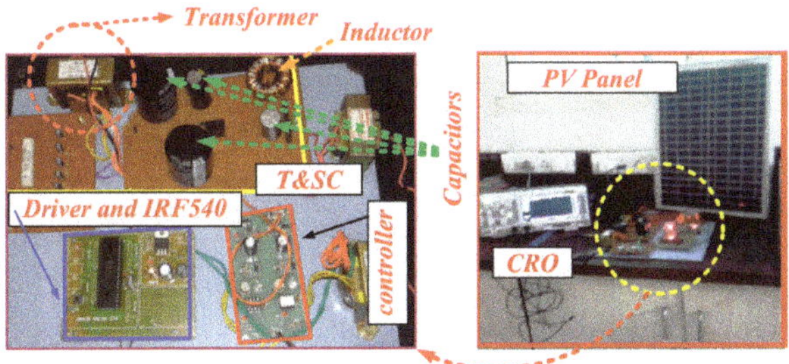

Figure 14. Experimental setup of the proposed T & SC Boost Converter.

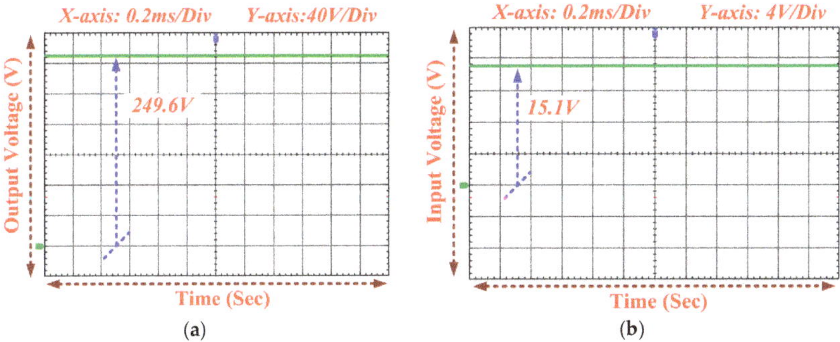

(a)   (b)

Figure 15. Experimental result (**a**) Output voltage waveform and (**b**) Input voltage waveform.

6. Future Extension and Combination of the Proposed T & SC-BC Topology with a Voltage Multiplier

To obtain a high voltage, the proposed T & SC-BC also provides a suitable solution by combining the features of a voltage multiplier. The power circuit of the proposed T & SC-BC with voltage multiplier is depicted in Figure 16. The voltage conversion ratio is analyzed and provided in Equation (26). The voltage across a switch is not affected by the voltage multiplier. The voltage multiplier level can be increased without disturbing the main power circuit of the T & SC to raise the voltage conversion ratio more.

$$
\left.
\begin{aligned}
&\frac{V_{C01}}{V_{in}} = \frac{1+k}{1-D} = \frac{(1+k)T}{T-t_{on}},\ \frac{V_{C03}}{V_{in}} = 2 \times \frac{1+k}{1-D} = 2 \times \frac{(1+k)T}{T-t_{on}} \\
&\quad\ \frac{V_{C05}}{V_{in}} = 3 \times \frac{1+k}{1-D} = 3 \times \frac{(1+k)T}{T-t_{on}}, \\
&\quad\ \frac{V_{C02N-1}}{V_{in}} = \frac{V_o}{V_{in}} = N\left(\frac{1+k}{1-D}\right) = N\frac{(1+k)T}{T-t_{on}} \\
&V_{DS} = \frac{1}{1-D}V_{in} = \frac{V_{C01}}{1+k} = \frac{V_o}{N(1+k)} = \frac{T}{T-t_{on}}V_{in} \\
&\quad k = \frac{Turns\ of\ Secondary\ winding\ of\ transformer\ (N_{LS})}{Turns\ of\ primary\ winding\ of\ transformer\ (N_{LP})} \\
&\quad N = number\ of\ voltage\ multiplier\ stage
\end{aligned}
\right\}
\tag{26}
$$

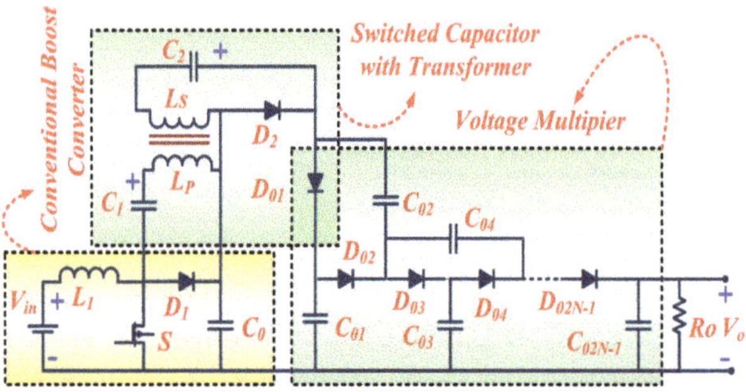

Figure 16. Power circuit of T & SC-BC with voltage multiplier (VM).

7. Conclusions

A new Transformer and Switch Capacitor-Based Boost Converter (T & SC-BC) is articulated for high-voltage/low-current renewable energy source applications. Conventional boost converter, transformer and switched capacitors functions are combined to design the proposed T & SC-BC for high voltage conversion ratio. A $(1+k)/(1-D)$ voltage conversion ratio (V_o/V_{in}) is achieved from the proposed converter where, k is the turns ratio of the transformer and D is the duty cycle. Conspicuous features of the proposed T & SC-BC are: (i) high voltage conversion ratio (V_o/V_{in}); (ii) continuous input current (I_{in}); (iii) single switch topology; (iv) single input source; (v) low Drain to Source voltage (V_{DS}) rating of the switch; and (vi) single inductor and single untapped transformer. The proposed T & SC-BC topology has low drain to source switch voltage, low stress on output diodes and requires less components to achieve high voltage compared to recently addressed DC-DC converters. Moreover the cost of the proposed T & SC-BC topology is less and it is also suitable for combination with a voltage multiplier to achieve more voltage at the output. The proposed converter is designed for 50/3 voltage conversion ratio at 78.4% duty cycle and the turns ratio is 2.6. Simulation and experimental results are provided which validate the functionality, design and concept of the proposed approach.

Author Contributions: All authors contributed equally for the final decimation of the research article in its current form.

Conflicts of Interest: The authors declare no conflict of interest.

References

1. Hussain, S.; Al-ammari, R.; Iqbal, A.; Jafar, M.; Padmanaban, S. Optimization of Hybrid Renewable Energy System Using Iterative Filter Selection Approach. *IET Renew. Power Gener.* **2017**, *11*, 1440–1445. [CrossRef]
2. Chandramohan, K.; Padmanaban, S.; Kalyanasundaram, R.; Bhaskar, M.S.; Mihet-Popa, L. Grid Synchronization of Seven-phase Wind Electric Generator using d-q PLL. *Energies* **2017**, *10*, 926. [CrossRef]

3. REN21. Renewable 2017: Global Status Report. Available online: http://www.ren21.net/ (accessed on 14 June 2017).

4. REN21. Global Futures Report: Scenario Profiles Report. Available online: http://www.ren21.net (accessed on 16 January 2013).

5. Coster, E.J.; Myrzik, J.M.A.; Kruimer, B.; Kling, W.L. Integration issues of distributed generation in distribution grids. *Proc. IEEE* **2011**, *99*, 28–39. [CrossRef]

6. Teodorescu, R.; Liserre, M.; Rodriguez, P. *Grid Converters for Photovoltaic and Wind Power Systems*; Wiley: Hoboken, NJ, USA, 2011.

7. Padmanaban, S.; Pecht, M. An Isolated/Non-Isolated Novel Multilevel Inverter Configuration for Dual Three-Phase Symmetrical/Asymmetrical Converter. *Eng. Sci. Technol. Int. J.* **2016**, *19*, 1763–1770. [CrossRef]

8. Gunabalan, R.; Sanjeevikumar, P.; Blaabjerg, F.; Olorunfemi, O.; Subbiah, V. Analysis and Implementation of Parallel Connected Two Induction Motor Single Inverter Drive by Direct Vector Control for Industrial Application. *IEEE Trans. Power Electron.* **2015**, *30*, 6472–6475. [CrossRef]

9. Dragonas, F.A.; Nerrati, G.; Sanjeevikumar, P.; Grandi, G. High-Voltage High-Frequency Arbitrary Waveform Multilevel Generator for DBD Plasma Actuators. *IEEE Trans. Ind. Appl.* **2015**, *51*, 3334–3342. [CrossRef]

10. Türkay, B.; Telli, A.Y. Economic analysis of standalone and grid connected hybrid energy systems. *Renew. Energy* **2011**, *36*, 1931–1943. [CrossRef]

11. Energy Information Administration (EIA). International Energy Outlook 2009. United States Department of Energy; 2009. Available online: www.eia.doe.gov (accessed on 27 May 2009).

12. International Energy Agency (IEA). World Energy Outlook 2012. Available online: www.worldenergyoutlook.org (accessed on 12 November 2012).

13. Das, V.; Padmanaban, S.; Venkitusamy, K.; Selvamuthukumaran, R.; Blaabjerg, F.; Siano, P. Recent Advances and Challenges of Fuel Cell Based Power System Architectures and Control—A Review. *Renew. Sustain. Energy* **2017**, *73*, 10–18. [CrossRef]

14. Vavilapalli, S.; Sanjeevikumar, P.; Umashankar, S.; Mihet-Popa, L. Power Balancing Control for Grid Energy Storage System in PV Applications—Real Time Digital Simulation Implementation. *Energies* **2017**, *10*, 928. [CrossRef]

15. European Photovoltaic Industry Association (EPIA). *Global Market Outlook for Photo-Voltaic until 2013*; EPIA: Brussels, Belgium, 2009.

16. Timilsina, G.R.; Kurdgelashvili, L.; Narbel, P.A. Solar energy: Markets, economics and policies. *Renew. Sustain. Energy* **2012**, *16*, 449–465. [CrossRef]

17. Blaabjerg, F.; Yang, Y.; Ma, K.; Wang, X. Power Electronics—The key Technology for renewable energy system Integration. In Proceedings of the 4th International Conference on Renewable Energy Research and Application (ICRERA), Palermo, Italy, 22–25 November 2015.

18. Joshi, A.S.; Dincer, I.; Reddy, B.V. Performance analysis of photovoltaic systems: A review. *Renew. Sustain. Energy* **2009**, *13*, 1884–1897. [CrossRef]

19. Blaabjerg, F.; Ma, K.; Yang, Y. Power Electronics for Renewable Energy Systems—Status and Trends. In Proceedings of the 8th International Conference on Integrated Power Systems (CIPS), Nuremberg, Germany, 25–27 February 2014.

20. Liu, C.; Wu, B.; Cheung, R. Advanced Algorithm for Control of Photovoltaic Systems. In Proceedings of the Canadian Solar Buildings Conference, Montreal, QC, Canada, 20–24 August 2004.

21. Gupta, A.; Chauhan, Y.; Pachauri, R. A comparative investigation of maximum power point tracking methods for solar PV system. *Sol. Energy* **2016**, *136*, 236–253. [CrossRef]

22. Sanjeevikumar, P.; Grandi, G.; Wheeler, P.; Blaabjerg, F.; Loncarski, J. A Simple MPPT Algorithm for Novel PV Power Generation system by High Output Voltage DC-DC Boost Converter. In Proceedings of the 24th IEEE International Symposium on Industrial Electronics, Rio de Janeiro, Brazil, 3–5 June 2015; pp. 214–220.

23. Sanjeevikumar, P.; Blaabjerg, F.; Wheeler, P.; Ojo, J.O.; Ertas, A. High-Voltage DC-DC Converter Topology for PV Energy Utilization—Investigation and Implementation. *J. Electr. Power Compon. Syst.* **2016**, 1–12. [CrossRef]

24. Mahajan, S.B.; Sanjeevikumar, P.; Blaabjerg, F. A Multistage DC-DC Step-up Self Balanced and Magnetic Component–Free Converter for photovoltaic Application: Hardware Implementation. *Energies* **2017**, *10*, 719. [CrossRef]

25. Tofoli, F.L.; Pereira, D.C.; Paula, W.J. Comparative Study of Maximum Power Point Tracking Techniques for Photovoltaic Systems. *Int. J. Photoenergy* **2015**, *2015*, 812582. [CrossRef]

26. Jain, S.; Ramulu, C.; Padmanaban, S.; Ojo, J.O.; Ertas, A.H. Dual MPPT Algorithm for Dual PV Source Fed Open-End Winding Induction Motor Drive for Pumping Application. *Eng. Sci. Technol. Int. J.* **2016**, *19*, 1771–1780. [CrossRef]

27. Subudhi, B.; Pradhan, R. A comparative study on maximum power point tracking techniques for photovoltaic power systems. *IEEE Trans. Sustain. Energy* **2013**, *4*, 89–98. [CrossRef]

28. Gules, R.; dos Santos, W.M.; dos Reis, F.A.; Romaneli, E.F.R.; Badin, A.A. A Modified Sepic Converter with High Static gain for Renewable Applications. *IEEE Trans. Power Electron.* **2014**, *29*, 5860–5871. [CrossRef]

29. Meneses, D.; Blaabjerg, F.; Garcia, O.; Cobos, J. Review and comparison of step-up transformerless topologies for photovoltaic AC-Module application. *IEEE Trans. Power Electron.* **2013**, *28*, 2649–2663. [CrossRef]

30. Forouzesh, M.; Siwakoti, Y.; Gorji, S.; Blaabjerg, F.; Lehman, B. Step-up DC–DC Converters: A Comprehensive Review of Voltage Boosting Techniques, Topologies, and Applications. *IEEE Trans. Power Electron.* **2017**. [CrossRef]

31. Tofoli, F.L.; Dênis de Castro, P.; Josias de Paula, W.; de Sousa Oliveira Júnior, D. Survey on non-isolated high-voltage step-up dc–dc topologies based on the boost converter. *IET Power Electron.* **2015**. [CrossRef]

32. Prudente, M.; Pfitscher, L.; Emmendoerfer, G.; Romaneli, E.; Gules, R. Voltage multiplier cells applied to non-isolated DC–DC converters. *IEEE Trans. Power Electron.* **2008**, *23*, 871–887. [CrossRef]

33. Zhou, D.; Pietkiewicz, A.; Cuk, S. A Three-Switch high-voltage converter. *IEEE Trans. Power Electron.* **1999**, *14*, 177–183. [CrossRef]

34. Henn, G.; Silva, R.; Praca, P.; Barreto, L.; Oliveira, D. Interleaved boost converter with high voltage gain. *IEEE Trans. Power Electron.* **2010**, *25*, 2753–2761. [CrossRef]

35. Wai, R.; Lin, C.; Duan, R.; Chang, Y. High-Efficiency DC-DC converter with high voltage gain and reduced switch stress. *IEEE Trans. Ind. Electron.* **2007**, *54*, 1354–1364. [CrossRef]

36. Wai, R.; Duan, R. High-efficiency power conversion for low power fuel cell generation system. *IEEE Trans. Power Electron.* **2005**, *20*, 847–856. [CrossRef]

37. Li, W.; He, X. A family of interleaved DC–DC converters deduced from a basic cell with winding-cross-coupled inductors (WCCIs) for high step-up or step-down conversions. *IEEE Trans. Power Electron.* **2008**, *23*, 1791–1801. [CrossRef]

38. Young, C.; Chen, M.; Chang, T.; Ko, C.; Jen, K. Cascade Cockcroft–Walton Voltage Multiplier Applied to Transformer-less High Step-up DC–DC Converter. *IEEE Trans. Ind. Electron.* **2013**, *60*, 523–537. [CrossRef]

39. Tomaszuk, A.; krupa, A. High efficiency high step-up DC/DC converters—A review. *Bull. Pol. Acad. Sci. Tech. Sci.* **2011**, *59*. [CrossRef]

40. Chen, S.; Liang, T.; Yang, L.; Chen, J. A Cascaded High Step-up DC–DC Converter with Single Switch for Microsource Applications. *IEEE Trans. Power Electron.* **2011**, *26*. [CrossRef]

41. Mahajan, S.B.; Kulkarni, R.; Sanjeevikumar, P.; Siano, P.; Blaabjerg, F. Hybrid Non-Isolated and Non Inverting Nx Interleaved DC-DC Multilevel Boost Converter for Renewable Energy Applications. In Proceedings of the 16th IEEE International Conference on Environment and Electrical Engineering, Florence, Italy, 7–10 June 2016; pp. 1–6.

42. Mahajan, S.B.; Kulkarni, R.; Sanjeevikumar, P.; Blaabjerg, F.; Fedák, V.; Cernat, M. Non Isolated and Non-Inverting Cockcroft Walton Multiplier Based Hybrid 2Nx Interleaved Boost Converter for Renewable Energy Applications. In Proceedings of the 17th IEEE Conference on the Power Electronics and Motion Control, Varna, Bulgaria, 25–30 September 2016; pp. 146–151.

43. Mahajan, S.B.; Sanjeevikumar, P.; Blaabjerg, F.; Norum, L.; Ertas, A. 4Nx Non-Isolated and Non-Inverting Hybrid Interleaved Boost Converter Based on VLSI Cell and Cockroft Walton Voltage Multiplier for Renewable Energy Applications. In Proceedings of the IEEE International Conference on Power Electronics, Drives and Energy Systems, Trivandrum, India, 14–17 December 2016; pp. 1–6.

44. Mahajan, S.B.; Sanjeevikumar, P.; Ojo, O.; Marco, R.; Kulkarani, R. Non-Isolated and Inverting Nx Multilevel Boost Converter For Photovoltaic DC Link Applications. In Proceedings of the IEEE International Conference on Automatica, XXII Congress of the Chilean Association of Automatic Control, Talca, Chile, 19–21 October 2016; pp. 1–8.

45. Mahajan, S.B.; Sanjeevikumar, P.; Blaabjerg, F.; Kulkarni, R.; Seshagiri, S.; Hajizadeh, A. Novel LY Converter Topologies for High Gain Transfer Ratio-A New Breed of XY Family. In Proceedings of the 4th IET International Conference on Clean Energy and Technology (IET_CEAT16), Kuala Lumpur, Malaysia, 14–15 November 2016.

46. Mahajan, S.B.; Sanjeevikumar, P.; Wheeler, P.; Blaabjerg, F.; Rivera, M.; Kulkarni, R. XY Converter Family: A New Breed of Buck Boost Converter for High Step-up Renewable Energy Applications. In Proceedings of the Proceedings of the IEEE International Conference on Automatica, XXII Congress of the Chilean Association of Automatic Control (IEEE-ICA/ACCA'16), Talca, Chile, 19–21 October 2016; pp. 1–8.

47. Axelrod, B.; Berkovich, Y.; Ioinovici, A. Switched-Capacitor/Switched-Inductor structures for getting transformerless hybrid DC–DC PWM converters. *IEEE Trans. Circuits Syst. I Regul. Pap.* **2008**, *55*, 687–696. [CrossRef]

48. Villalva, M.; Gazoli, J.R.; Filho, E. Comprehensive Approach to Modeling and Simulation of Photovoltaic Arrays. *IEEE Trans. Power Electron.* **2009**, *24*, 1198–1208. [CrossRef]

49. Bendib, B.; Belmili, H.; Krim, F. A survey of the most used MPPT methods: Conventional and advanced algorithms applied for photovoltaic systems. *Renew. Sustain. Energy Rev.* **2015**, *45*, 637–648. [CrossRef]

50. Safari, A.; Mekhilef, S. Simulation and Hardware Implementation of Incremental Conductance MPPT with Direct Control Method Using Cuk Converter. *IEEE Trans. Ind. Electron.* **2011**, *58*. [CrossRef]

51. Saad Saoud, M.; Abbassi, H.; Kermiche, S.; Nada, D. Improved incremental conductance method for maximum power point tracking using cuk converter. *Mediterr. J. Model. Simul.* **2014**, *57–65*. Available online: http://www.webreview.dz/IMG/pdf/mjms_01_2014_057-065.pdf (accessed on 1 January 2014).

52. Hartmann, L.V.; Vitorino, M.A.; de Rossiter Correa, M.B.; Lima, A.M.N. Combining model-based and heuristic techniques for fast tracking the maximum power point of photovoltaic systems. *IEEE Trans. Power Electron.* **2013**, *28*, 2875–2885. [CrossRef]

53. Esram, T.; Chapman, P. Comparison of Photovoltaic Array Maximum Power Point Tracking Techniques. *IEEE Trans. Energy Convers.* **2007**, *22*. [CrossRef]

Article

Sliding Mode Controller and Lyapunov Redesign Controller to Improve Microgrid Stability: A Comparative Analysis with CPL Power Variation

Eklas Hossain [1,*], Ron Perez [2], Sanjeevikumar Padmanaban [3,*], Lucian Mihet-Popa [4], Frede Blaabjerg [5] and Vigna K. Ramachandaramurthy [6]

[1] Department of Electrical Engineering & Renewable Energy, Oregon Tech, Klamath Falls, OR 97601, USA
[2] Department of Mechanical Engineering, University of Wisconsin-Milwaukee, Milwaukee, WI 53211, USA; perez@uwm.edu
[3] Department of Electrical and Electronics Engineering, University of Johannesburg, Auckland Park 2006, South Africa
[4] Faculty of Engineering, Østfold University College, Kobberslagerstredet 5, 1671 Kråkeroy-Fredrikstad, Norway; lucian.mihet@hiof.no
[5] Centre for Reliable Power Electronics (CORPE), Department of Energy Technology, Aalborg University, 9000 Aalborg, Denmark; fbl@et.aau.dk
[6] Institute of Power Engineering, Department of Electrical Power Engineering, Universiti Tenaga Nasional, Kajang 43000, Selangor, Malaysia; Vigna@uniten.edu.my
*   Correspondence: eklas.hossain@oit.edu (E.H.); sanjeevi_12@yahoo.co.in (S.P.); Tel.: +1-541-885-1516 (E.H.); +27-79-219-9845 (S.P.)

Received: 11 September 2017; Accepted: 25 October 2017; Published: 24 November 2017

Abstract: To mitigate the microgrid instability despite the presence of dense Constant Power Load (CPL) loads in the system, a number of compensation techniques have already been gone through extensive research, proposed, and implemented around the world. In this paper, a storage based load side compensation technique is used to enhance stability of microgrids. Besides adopting this technique here, Sliding Mode Controller (SMC) and Lyapunov Redesign Controller (LRC), two of the most prominent nonlinear control techniques, are individually implemented to control microgrid system stability with desired robustness. CPL power is then varied to compare robustness of these two control techniques. This investigation revealed the better performance of the LRC system compared to SMC to retain stability in microgrid with dense CPL load. All the necessary results are simulated in Matlab/Simulink platform for authentic verification. Reasons behind inferior SMC performance and ways to mitigate that are also discussed. Finally, the effectiveness of SMC and LRC systems to attain stability in real microgrids is verified by numerical analysis.

Keywords: sliding mode control; Lyapunov redesign control; constant power load; robustness analysis; variation of CPL power; microgrid stability

---

1. Introduction

Since the beginning of the 21st century, the conventional utility grid system has started to be replaced by the newly adopted microgrid system due to several reasons. Microgrid systems offer environment-friendly distributed generation by local renewable energy resources [1–10]. From an economic aspect, it reduces the overall cost (combining the generation, transmission, and distribution) considerably. Apart from that, it is a great tool to distribute electricity to those areas where the utility grid-based electricity cannot be reached. However, though a microgrid is easy to construct and implement, the stability maintenance of the microgrid system is a matter of concern to system engineers, professionals, and researchers globally. The stability of the microgrid system is basically hampered due to the CPL (constant power load) based load in the system. The CPL exhibits negative incremental load characteristics (shown in Figure 1) and easily creates exponential and random oscillation in the system, thus instability is forming in the system [2,11–13]. For compensating the instabilities caused by CPL, a lot of research has been conducted. Research regarding instabilities in microgrids started during 1998–1999, but as the electrification industry and microgrid technology grew gradually, this issue drew attention of researchers all over the world. Research timeline on CPL compensation is shown in Figure 2. The increase in research on microgrid is easily noticeable from this figure. Figure 3 shows the research work done on CPL compensations techniques in different countries. The United States of America is currently in the lead, but China, Norway, France, as well as India are churning up significant contributions.

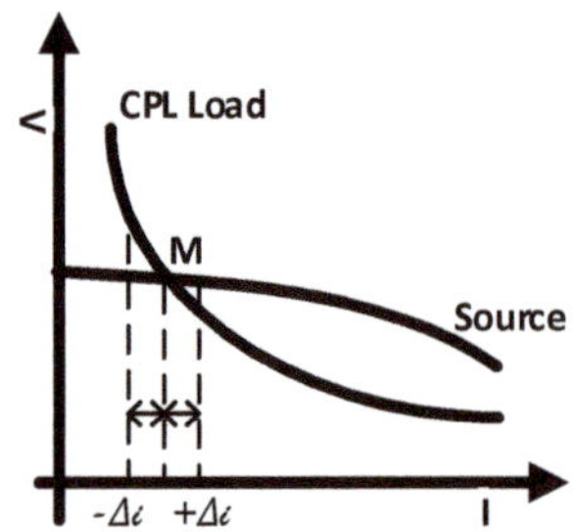

Figure 1. Negative impedance characteristic of constant power load.

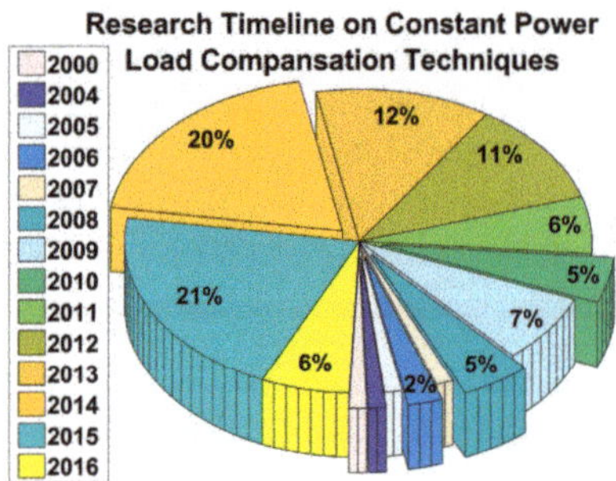

Figure 2. Research timeline on Constant Power Load Compensation Techniques, considering the published research works.

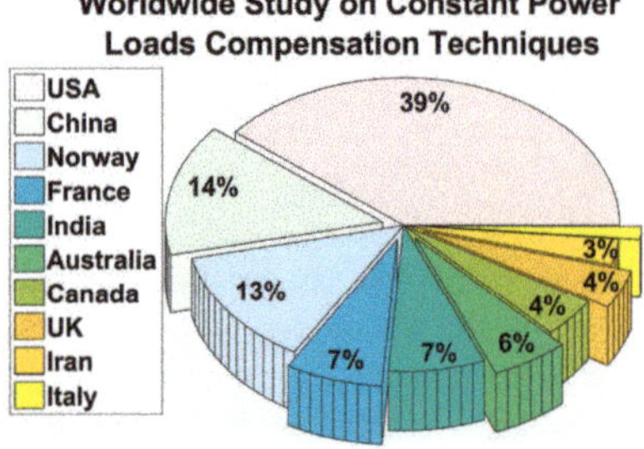

Figure 3. Contributions of different countries on Constant Power Load Instability Compensation research.

Several investigations have been conducted by researchers and system engineers all over the globe to ameliorate the stability scenario of microgrids. For direct current (DC) microgrid, several researches are reviewed at [14–17]. Sliding Mode Control (SMC) and Lyapunov Redesign Control (LRC) techniques are two of the most prominent nonlinear control techniques used to improve microgrid stability [18,19]. Prior to this, several studies have been carried out on the SMC technique. The stability characteristics become harder to establish in large systems. Sliding mode control has been applied in direct current (DC) microgrids to use the actual nonlinear models [20,21]. It has been accomplished by discovering a sliding surface and employing a sliding mode controller, which is discontinuous, for making the system voltage more stable. Later on, in [22], Vinicius Stramosk and Daniel J. Pagano presented a novel Sliding Mode Controller for precise governing of DC bus voltage. In like manner, a non-linear sliding surface is put forward by the two Indian Institute of Technology Jodhpur researchers: Suresh Singh and Deepak Fulwani in [23–25] to moderate CPL instability. The non-linear surface that they had proposed confirmed maintaining the constant power by the converter in practice. In this way, the proposed controller succeeded in mitigating the oscillating effect of the CPL of Point of Loads (POL) which are tightly regulated, and assured that the DC microgrids will operate stably under several disturbance conditions. Researchers Aditya R. Gautam et al. demonstrated, in [23], a robust sliding mode control technique to examine CPL instability. In like manner, in the case of alternating current (AC) microgrid, several researches have been reviewed in [12,26–31].

To achieve better controlled performance for polynomial nonlinear systems, the Lyapunov redesign of adaptive controller has been implemented by Qian Zheng and Fen Wu in [32]. Apart from the microgrid system, Wen-Ching Chung et al has implemented the Lyapunov redesign technique in vehicle dynamics to experience better steering control [33]. Then, Attaullah Y. Memon et al, in [34], used conditional servomotor to experiment with output control of a nonlinear system. In this course, they have implemented the Lyapunov redesign control technique. There are three basic compensation techniques to handle the microgrid instability: (i) feeder side compensation technique, (ii) intermediate circuitry based compensation technique, and (iii) load side compensation technique. In this paper, the storage-based load side compensation technique is adopted due to superior robustness and cost effectiveness among these techniques [35–42]. Adopting storage-based load side compensation in this paper, a comparative performance analysis will be presented for SMC and LRC techniques with the variation of the CPL power. The following are the contributions of this paper: besides modeling of the storage-based load side compensation technique (Section 2), SMC and LRC theories will be presented (Section 3), the robustness of the SMC and the LRC technique will be presented with the variation of CPL power load (Section 4). Then, the comparative performance analysis will be presented between SMC and LRC technique (Section 5) that will justify why the Lyapunov Redesign Control technique shows better robustness than the former one in microgrid application with dense CPL loaded condition. Reasons behind inferior SMC performance and ways to mitigate them will be discussed in Section 6. Section 7 will present numerical analysis of the control systems in real microgrid situations which verifies their effectiveness. Finally, the conclusion will be drawn in Section 8.

2. Modeling Microgrid with CPL

To mitigate purturbation caused by CPL loads, a compensation technique at the load side is the rational choice rather than compensating at the feeder side or using the intermediate circuitry approach. The load side compensation technique does required manipulation at the load side of the system to shield it from experiencing the effects caused by constant power loads. To elucidate this method, schematic models of storage-based real power compensation and reactive power compensation techniques (load side) are presented below in Figures 4 and 5 [18].

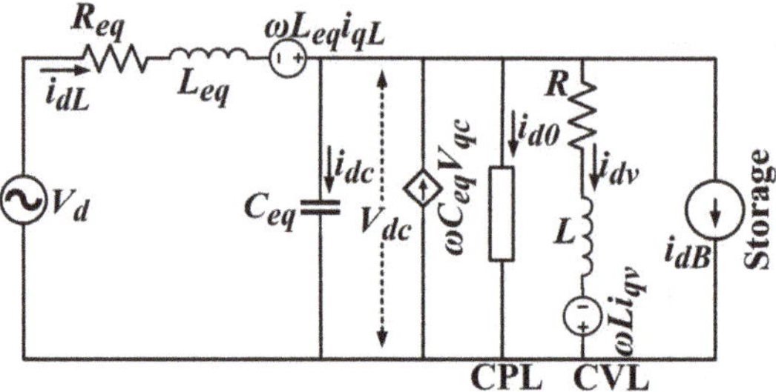

Figure 4. Real power compensation method at the load side, modeled for *d*-axis.

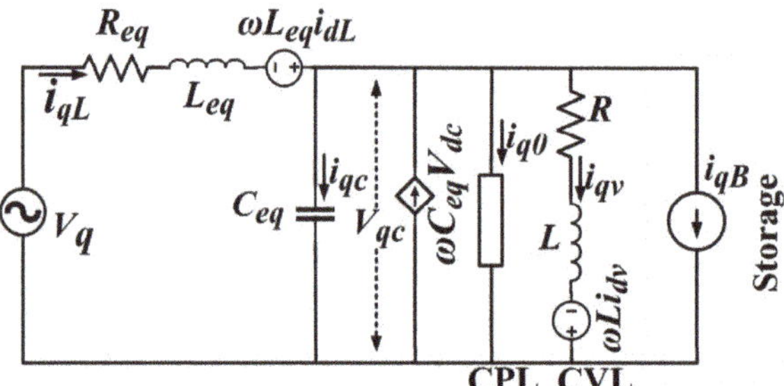

Figure 5. Reactive power compensation method at the load side, modeled for *q*-axis.

From the dq-axis models demonstrated above, the combined state space equation of the espoused load side compensation technique is shown in Equation (1) [19].

$$
\begin{bmatrix}
\frac{di_{dL}}{dt} \\[4pt]
\frac{di_{qL}}{dt} \\[4pt]
\frac{dV_{dC}}{dt} \\[4pt]
\frac{dV_{qC}}{dt} \\[4pt]
\frac{di_{dV}}{dt} \\[4pt]
\frac{di_{qV}}{dt}
\end{bmatrix}
=
\begin{bmatrix}
\omega i_{qL} - \frac{R_1}{L_1} i_{dL} - \frac{V_{dc}}{L_1} + \frac{V_d}{L_1} \\[4pt]
-\omega i_{dL} - \frac{R_1}{L_1} i_{qL} - \frac{V_{qc}}{L_1} + \frac{V_q}{L_1} \\[4pt]
\omega V_{qC} + \frac{1}{C} i_{dL} - \frac{1}{C} \frac{P_O}{V_{dC}} - \frac{1}{C} i_{dV} - \frac{1}{C} i_{dB} \\[4pt]
-\omega V_{dC} + \frac{1}{C} i_{qL} - \frac{1}{C} \frac{Q_O}{V_{qc}} - \frac{1}{C} i_{qV} - \frac{1}{C} i_{qB} \\[4pt]
\omega i_{qV} + \frac{1}{L} V_{dC} - \frac{R}{L} i_{dV} \\[4pt]
-\omega i_{dV} + \frac{1}{L} V_{qC} - \frac{R}{L} i_{qV}
\end{bmatrix} ,
\tag{1}
$$

3. Introduction to SMC and LRC

Sliding Mode Control (SMC) is a type of Variable Structure Control (VSC) in control theory. It gets switched from one continuous structure to a different one, based on the current state-space location. That makes SMC a variable structure control method. Its various control structures are configured to move the trajectories to a switching condition all the time, and therefore, the final trajectory will not be wholly within a single control structure. Instead of that, the final trajectory will slide along the control structure boundaries. The system's motion while sliding along such boundaries is known as a Sliding Mode. The geometrical locus involving the boundaries is known as the sliding (hyper) surface. The sliding surface is defined by $\sigma = 0$, and after the limited time when the trajectories of the system have reached the surface, the sliding mode along the surface begins.

3.1. Sliding Mode Controller (SMC)

3.1.1. Control Statement of Sliding Mode

Considering a nonlinear dynamic system affine in control:

$$\dot{\bar{x}}(t) = f(\bar{x}, t) + B(\bar{x})\,\bar{u}(t), \tag{2}$$

$$\bar{x}(t) \in \Re^n, \bar{u}(t) \in \Re^m, f(\bar{x}, t) \in \Re^n, B(\bar{x}) \in \Re^{n \times m} \tag{3}$$

The components of the discontinuous feedback are given by:

$$u_i(t) = \left\{ \begin{array}{l} u_i^+(\bar{x}, t)\, if\, \sigma_i(x) > 0 \\ u_i^-(\bar{x}, t)\, if\, \sigma_i(x) < 0 \end{array} \right. \quad i = 1, 2, \cdots, m, \tag{4}$$

where $\sigma_i(x) = 0$ is the i-th component of the sliding surface, and $\sigma(x) = [\sigma_1(x), \sigma_2(x), \cdots, \sigma_m(x)]^T = 0$ is the $(n - m)$ dimensional sliding manifold. The sliding mode control structure includes selecting a manifold or a hypersurface (i.e., the sliding surface) so that the system trajectory demonstrates desired performance when restricted within this manifold, and finding discontinuous feedback gains to make the trajectory of the system intersect and stay on the manifold. Vicinity of the switching surface can be viewed from Figure 6.

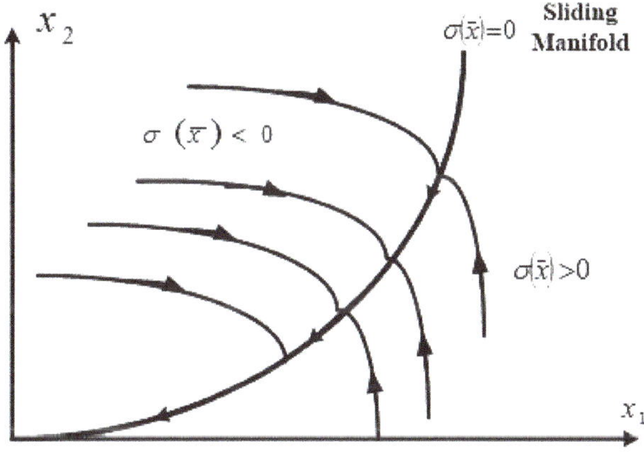

Figure 6. Vicinity of the switching surface.

A sliding mode exists, given that in the environs of the switching surface, $\sigma(x) = 0$, the state trajectory's velocity vector, $\dot{\bar{x}}(t)$, is always directed toward the switching surface. The control laws of the sliding mode not being continuous, it is able of driving trajectories to the sliding mode in finite time (i.e., the sliding surface's stability is superior to asymptotic). Nevertheless, the character of the sliding mode is taken on by the system (e.g., on this surface, the origin $x = 0$ can only possess asymptotic stability) once the trajectories reach the sliding surface.

3.1.2. Chattering

Due to the presence of external disturbance—noise and inertia of the sensors and actuators—the switching around the sliding surface occurs at a very high (but finite) frequency. The main consequence is that the sliding mode occurs in a small vicinity of the sliding manifold, which is called boundary layer, and which has a dimension that is inversely proportional to the control switching frequency. The effect of high frequency switching is known as chattering (shown in Figure 7).

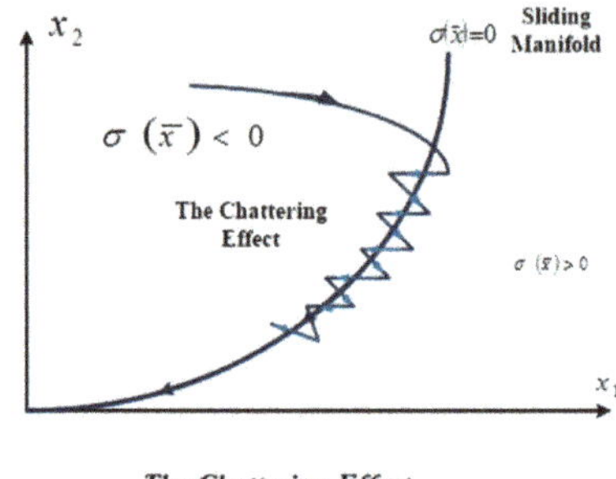

Figure 7. The chattering effect of Sliding Mode Controller (SMC).

The high-frequency switching propagate through the system exciting the fast dynamics and undesired oscillations that affect the system output. To prevent the chattering effect different techniques are used. One of the techniques is the use of continuous approximations of *sign*(.) using *sat*(.) or *tanh*(.) function in the implementation of the control law. A consequence of this method is that the invariance property is lost.

3.1.3. Chattering Reduction

Nowadays, typical approaches have been developed to reduce the amount of chattering. Slotine [43–45] based their original proposal on the generalized event of the nth-order single input variant of nonlinear system: $x^{(n)} = f\left(x,t\right) + B\left(x,t\right)u$; here x is the state variable; $x = [x, \dot{x}, \ddot{x}, \ldots, x^{(n-1)}]$; $x^{(n)}$ is the x's nth-order derivative; B is the gain; f is a nonlinear function and u is the control input. Furthermore, a formula for the switching manifold of the above system and the distance between the state trajectory: s, is stated as: $s(t) = \left(\frac{d}{dt} + \lambda\right)^{(n-1)}\tilde{x}$; while $\lambda > 0$ is a design constant, and $\tilde{x}$ is the tracking error defined as: $\tilde{x} = x - x_d$; whereas x_d is the state variable for the desired trajectory. Henceforth the corresponding switching manifold is: $s(t) = 0$. Meanwhile, Slotine also proposed to smooth the previously mentioned discontinuity via a thin boundary layer closely surrounding the switching manifold. In such case continuous control within this boundary layer was attained by changing the switching term in the control law to a saturation function. Although the system would be driven to the boundary layer, yet the trajectory would not be staying on the switching manifold and thus the sliding mode would not exist [46]. Later Hung and Gao [47] offered the technique of reaching mode and reaching law, which was based upon nth-order m-input systems. To guarantee the state trajectory's attraction towards the switching manifold within the reaching mode, their suggestion was to control the reaching speed by applying certain reaching law. They put forward three certain kinds of reaching laws besides the general form. Among these types they claimed that the power rate reaching law would eliminate chattering and provide fast reaching as well: $\dot{s}_i = -k_i|s_i|^{\alpha}sgn(s_i)$. The reaching time T_i was deduced to: $T_i = \dfrac{|s_i(0)|^{1-\alpha}}{(1-\alpha)k_i}, i = 1, 2, \ldots, m$; where $\dot{s}_i$ was the reaching speed; $\dot{s}_i$ was defined as according to Equations (6) and (7); $\dot{s}_i(0)$ was the initial value of $\dot{s}_i$; $k_i > 0$ was the switching gain (in the i-th dimension), and $0 < \alpha < 1$. Yet typically it has been found that chattering cannot be totally eliminated by such method. The above approaches are bounded by defects. Besides, Luo & Feng's switching zone [48] appears mainly theoretical, whereas the Ground Validation System (GVS) of Hamerlan et al will have minimal effect on speed and position of the controlled subject [49].

3.2. Lyapunov Redesign Controller (LRC)

Unlike sliding mode controller (SMC), Lyapunov redesign controller, or LRC, is based only on Lyapunov function [50,51]. Consider a nonlinear system that is described by:

$$\dot{x} = f(x) + G(x)u, \tag{5}$$

where $x \in \Re^n$ is the state and $u \in \Re^m$ is the controlled input. Assuming the matrix $G(x)$ and the vector field $f(x)$ each has two components: an unknown part and a known nominal part. Therefore,

$$f(x) = f_0(x) + f^*(x), \tag{6}$$

$$G(x) = G_0(x) + G^*(x), \tag{7}$$

where f_0 and G_0 represent the known nominal plant, and f^*, G^* characterize the uncertainty. Later let us assume the unknown portion to conform to a certain bounding condition. Additionally, it is assumed that the uncertainty fulfills a so-called matching condition:

$$f^*(x) = G_0(x)\Delta_f(x), \tag{8}$$

$$G^*(x) = G_0(x)\Delta_G(x), \tag{9}$$

The matching condition suggests that terms of uncertainty are present in the same equations with the control inputs u, and consequently, it will be possible to control them by controller. By replacing (6)−(9) in (5) we obtain:

$$\dot{x} = f_0(x) + G_0(x)(u + \eta(x, u)), \tag{10}$$

which includes all of the uncertainty terms, and is defined by:

$$\eta(x, u) = \Delta_f + \Delta_G u, \tag{11}$$

The Lyapunov redesign method works on the ensuing problem: supposing the equilibrium of the nominal model $\dot{x} = f(x) + G(x)u$ been made asymptotically stable uniformly by employing a feedback control law $u = p_0(x)$, the goal is to devise a control function $p^*(x)$, which is corrective in nature, so that the enhanced control law $u = p_0(x) + p^*(x)$ can stabilize the system (defined by Equation (10)) faced by the uncertainty (x, u) getting constrained by a known function.

Then, let us think about the specifics of the Lyapunov redesign technique, that is comprehensively offered for a more common case. Let us assume a control law: $u = p_0(x)$ to exist so that $x = 0$ becomes a stable equilibrium point which is uniformly asymptotically of the closed-loop nominal system $\dot{x} = f(x) + G_0(x)p_0(x)$. We also assume to know a Lyapunov function $V_0(x)$ that fulfills:

$$\alpha_1(||x||) \leq V_0(x) \leq \alpha_2(||x||), \tag{12}$$

$$\frac{\partial V_0}{\partial x}[f(x) + G_0(x)p_0(x)] \leq -\alpha_3(||x||), \tag{13}$$

where $\alpha_1, \alpha_2, \alpha_3 : \Re^+ \rightarrow \Re^1$ are stringently increasing functions that satisfy $\alpha_i(0) = 0$ and $\alpha_i(r) \rightarrow \infty$ as $r \rightarrow \infty$. These types of functions are sometimes called as class K_∞ functions. The term of uncertainty is presumed to satisfy the bound

$$||\eta(x, u)||_\infty \leq \bar{\eta}(t, x), \tag{14}$$

where the bounding function $\bar{\eta}$ is presumed to be known 'a priori', or accessible for measurement. At this point, let us proceed to designing the corrective "control component" $p^*(x)$ so that the system classes described by (10) and conforming to (14) are stabilized by $u = p_0 + p^*$. An approach adhering to the nominal Lyapunov function V_0 is used as the base to design the corrective control term, thus the name 'Lyapunov redesign method' is justified. Considering the exact same Lyapunov function V_0 guaranteeing the nominal closed-loop system's asymptotic stability, let us think about the time derivative of V_0 which is alongside the solutions of the full system (10). We have:

$$\dot{V_0} = \frac{\partial V_0}{\partial x}[f_0(x) + G_0(x)(u + \eta(x,u))]$$
$$= \frac{\partial V_0}{\partial x}[f_0(x) + G_0(x)p_0(x)] + \frac{\partial V_0}{\partial x}G_0(x)p^*(x) + \eta(x,u)) \leq -\alpha_3(\|x\|) + \omega(x)^T p^*(x) + \omega(x)^T \eta(x,u) \tag{15}$$

where,

$$\omega(x) = [\frac{\partial V_0}{\partial x}G_0(x)]^T \in \Re^m, \tag{16}$$

which is a recognized function. We obtain by taking limits:

$$\dot{V_0} \leq -\alpha_3(\|x\|) + \sum_{i=1}^{m} \omega_i(x)p_i^*(x) + \|\omega(x)\|1\|\eta(x,u)\|_\infty$$
$$= -\alpha_3(\|x\|) + \sum_{i=1}^{m} \omega_i(x)p_i^*(x) + \bar{\eta}(x,t)|\omega_i(x)|) \tag{17}$$

The second term at the right-hand side of (17) can be made equal to zero if $p_i^*(x)$ is taken as:

$$p_i^*(x) = -\bar{\eta}(x,t)\text{sgn}(\omega_i(x)), \tag{18}$$

Every term of the corrective control vector $p^*(x)$ is chosen to be of the form $p^*(x) = \pm\bar{\eta}(x,t)$, where the sign of $p^*(x)$ is contingent on the sign of $_i(x)$ and changes as $_i(x)$ changes its sign. Substituting Equation (18) in Equation (17), the desired "stability" property is obtained.

$\dot{V_0} \leq -\alpha_3(\|x\|)$; which infers that the closed-loop system is stable asymptotically. The augmented control law $u = p_0(x) + p^*(x)$ is discontinuous since each element $p_i^*(x)$ is discontinuous at $_i(x) = 0$. Moreover, the discontinuity jump $\bar{\eta}(x,t) \to -\bar{\eta}(x,t)$ can have great magnitude if the bound of uncertainty $\bar{\eta}$ is large. As demonstrated earlier, chattering can be caused by discontinuities in the control law; hence smoothing the discontinuity is desirable and is expected to retain some degree the nice stability properties at the same time from the original discontinuous control law. It is achievable by replacing Equation (18) with

$$p_i^*(x) = -\bar{\eta}(x,t)\tanh(\frac{\omega_i(x)}{\varepsilon}), \tag{19}$$

where $\varepsilon > 0$ is a small design constant. It can be noted with ε approaching zero, the function $\tanh(\frac{\omega_i}{\varepsilon})$ gets converged to the $\text{sgn}(_i)$ function, which is discontinuous. By substituting Equation (19) in Equation (17) we obtain:

$$\dot{V_0} \leq -\alpha_3(\|x\|) + \bar{\eta}(x,t) \sum_{i=1}^{m}(|\omega_i(x)| - \omega_i(x)\tanh(\frac{\omega_i(x)}{\varepsilon})), \tag{20}$$

Using Lemma:

$$\dot{V_0} \leq -\alpha_3(\|x\|) + \varepsilon m k \bar{\eta}(x,t), \tag{21}$$

α_3 being a strictly increasing class k_∞ function, for all $r > 0$ and any uniformly bounded function $\bar{\eta}$, there can exist a sufficiently small ε, so that $\dot{V_0} \leq 0$ for x outside of a region $D_\varepsilon = \{x \; V(x) \leq r\}$. Consequently, the trajectory becomes convergent to the invariant set D_ε. A Lyapunov function's level surfaces are shown in Figure 8. It demonstrates the Lyapunov surfaces for increasing values of k. The condition $\dot{V_0} \leq 0$ suggests that the a trajectory moves within the set $\Omega_k = \{x \in \Re^n | V(x) \leq k\}$ when it crosses the Lyapunov surface $V(x) = k$, and it cannot ever come out. The trajectory moves to an inner Lyapunov surface with smaller values of k when $V < 0$. The Lyapunov surface $V(x) = k$ reduces back to the origin as k decreases, which shows that the approach of the trajectory to the origin with progressing time.

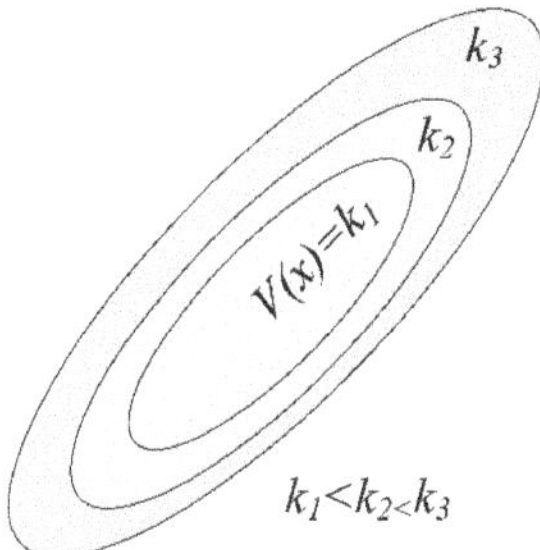

Figure 8. A Lyapunov function's level surfaces.

4. Implementation and Robustness Analysis of SMC and LRC

The intended outputs, or control objectives of the proposed controllers (each of SMC and LRC controller) is:

Y1 = $V_{dC} \approx V_d \approx$ 480 Volt

Y2 = $V_{qC} \approx V_q \approx$ (the lowest possible) Volt

Equation (22) gives the general system form affined within the control(s):

$$\dot{x} = f(x) + g(x)u, \tag{22}$$

4.1. Implementation and Robustness Analysis of Sliding Mode Controller against Parametric Uncertainties Including Uncertainties in Power of CPL

Sliding mode control, or SMC, is an advanced non-linear control technique featuring prominent characteristics of accuracy, robustness, and ease of tuning. By using the discontinuous control signal that forces the output of the system to 'slide' along with sliding surface or a distinct cross-section of the minimal behavior of the system, it can adjust the dynamics of the system in a way [47]. The state feedback control law is a discontinuous time function here, and can shift from one structure to the next depending on the prevailing location in space in a continuous manner. Hence, sliding mode control can be described as a control technique with variable structures. As the system's certain operation mode slides along the predetermined control structure boundaries, it is called the sliding mode. The geometrical locus, which consists of the boundaries, is called the system's sliding surface. To implement the sliding mode controller, the state space model equation below can be rewritten as Equation (23). In this section, the robustness will be enhanced by considering the uncertainties in active power of CPL (P_0) and reactive power of CPL (Q_0). When P_0 is unknown in case of designing u_1, we will also consider x_3 as unknown to avoid any complexity. Similarly, in case of u_2, we will also consider x_4 as unknown.

$$
\begin{bmatrix} \dot{x}_1 \\ \dot{x}_2 \\ \dot{x}_3 \\ \dot{x}_4 \\ \dot{x}_5 \\ \dot{x}_6 \end{bmatrix}
=
\begin{bmatrix}
\omega x_2 - \frac{R_1}{L_1}x_1 - \frac{x_3}{L_1} \\
-\omega x_1 - \frac{R_1}{L_1}x_2 - \frac{x_4}{L_1} \\
\omega x_4 + \frac{1}{C}x_1 - \frac{1}{C}\frac{P_0}{x_3} - \frac{1}{C}x_5 \\
-\omega x_3 + \frac{1}{C}x_2 - \frac{1}{C}\frac{Q_0}{x_4} - \frac{1}{C}x_6 \\
\omega x_6 + \frac{1}{L}x_3 - \frac{R}{L}x_5 \\
-\omega x_5 + \frac{1}{L}x_4 - \frac{R}{L}x_6
\end{bmatrix}
+
\begin{bmatrix} 0 \\ 0 \\ -\frac{1}{C}u_1 \\ -\frac{1}{C}u_2 \\ 0 \\ 0 \end{bmatrix}
+
\begin{bmatrix} \frac{r_1}{L_1} \\ \frac{r_2}{L_1} \\ 0 \\ 0 \\ 0 \\ 0 \end{bmatrix},
\tag{23}
$$

Although P_0 and Q_0 are unknown, they satisfy $P_0 \le \delta_P$ and $Q_0 \le \delta_Q$ for some known bounds δ_P and δ_Q. The variation on CPL power can be summarized as:

$$d_P = \Delta_P / \Delta x_3, \tag{24}$$

$$d_Q = \Delta_Q / \Delta x_4, \tag{25}$$

where d_P represents the uncertainties of P_0, d_Q represents the uncertainties of Q_0, Δx_3 is the uncertainties in x_3, and Δx_4 is the uncertainties in x_4. As x_3 and x_4 are in the denominator, we need lower bounds of these parameters. Power uncertainty is expressed in term of current. We know that x_3 is the voltage of "d-axis" and it satisfies $\Delta x_3 \le \delta_{x3}$ for some known, stringently positive bound δ_{x3}. Similarly, x_4 is the "q-axis" voltage. It satisfies $\Delta x_4 \le \delta_{x4}$ for some known, stringently positive bound δ_{x4}. Overall, there are six unknowns with known bounds. The Sliding Mode Control input, u_1 will be designed first, with the similar method adopted to design u_2, the other control input. Using the similar method as discussed in the previous section, let

$$e_1 = \int (x_3 - x_{3d})dt, \tag{26}$$

$$e_2 = \dot{e}_1 = x_3 - x_{3d}, \tag{27}$$

$$\dot{e}_2 = \dot{x}_3 - \dot{x}_{3d} = f_3(x) + g_3(x)u_1 - \dot{x}_{3d}, \tag{28}$$

Expanding $f_3(x)$ and $g_3(x)$

$$\dot{e}_2 = \omega x_4 + \frac{1}{c}x_1 - \frac{1}{c}\frac{P_0}{x_3} - \frac{1}{c}x_5 - \frac{1}{c}u_1 - \dot{x}_{3d}, \tag{29}$$

Let, the sliding surface be

$$s = e_1 + e_2, \tag{30}$$

After differentiating and considering the uncertainties:

$$\dot{s} = \dot{e}_1 + \dot{e}_2, \tag{31}$$

$$\dot{s} = e_2 + (\omega(\hat{x}_4 + \Delta x_4) + \frac{1}{c}(\hat{x}_1 + \Delta x_1) - \frac{1}{c}(\frac{P_0}{x_3} + d_P) - \frac{1}{c}x_5 - \frac{1}{c}u_1 - \dot{x}_{3d}), \tag{32}$$

where $x_4 = \hat{x}_4 + \Delta x_4$. Then the total parametric uncertainty including uncertainty of CPL power can be represented as:

$$d = \frac{1}{c}\Delta x_1 + \omega \Delta x_4 - \frac{1}{c}d_P; \ \|d\| \le dmax, \tag{33}$$

here $dmax$ is the limit of the total disturbance d.

$$dmax = \frac{1}{c}\delta_{x1} + \omega \delta_{x4} - \frac{1}{c}\delta_P / \delta_{x3}, \tag{34}$$

Then,

$$\dot{s} = e_2 - \frac{1}{c}x_5 - \dot{x}_{3d} + \omega \hat{x}_4 + \frac{1}{c}\hat{x}_1 - \frac{1}{c}\frac{P_0}{x_3} - \frac{1}{c}u_1 + d, \tag{35}$$

Let it be considered as the Lyapunov candidate function.

$$V = \frac{1}{2}s^2, \tag{36}$$

$$\dot{V} = s\dot{s} = s(e_2 - \frac{1}{c}x_5 - \dot{x}_{3d} + \omega \hat{x}_4 + \frac{1}{c}\hat{x}_1 - \frac{1}{c}\frac{P_0}{x_3} - \frac{1}{c}u_1 + d), \tag{37}$$

We use u_1.

$$u_1 = -c\left[-e_2 + \frac{1}{c}x_5 + \dot{x}_{3d} - \omega\hat{x}_4 - \frac{1}{c}\hat{x}_1 + \frac{1}{c}\frac{P_0}{x_3} + v\right],$$ (38)

Now, we can obtain:

$$\dot{V} = s(d + v),$$ (39)

$\|d\| \leq dmax$, put into consideration, $\dot{V}$ will be made negative by the subsequent discontinuous control, v. Consequently, it will guarantee stability.

$$v = -dmax * sat\left(\frac{s}{\varepsilon}\right); \ \varepsilon > 0,$$ (40)

In total, the control input is:

$$u_1 = -c\left[-e_2 + \frac{1}{c}x_5 + \dot{x}_{3d} + \frac{1}{c}\frac{P_0}{x_3} - \omega\hat{x}_4 - \frac{1}{c}\hat{x}_1 - dmax * sat\left(\frac{s}{\varepsilon}\right)\right],$$ (41)

Such an analysis is also presented here for u_2, let,

$$e_3 = \int(x_4 - x_{4d})dt,$$ (42)

$$e_4 = \dot{e}_3 = x_4 - x_{4d},$$ (43)

$$\dot{e}_4 = \dot{x}_4 - \dot{x}_{4d} = f_4(x) + g_4(x)u_2 - \dot{x}_{4d},$$ (44)

Taking the sliding surface as:

$$s = e_3 + e_4,$$ (45)

After differentiation and considering the uncertainties:

$$\dot{s} = e_4 + \left(-\omega(\hat{x}_3 + \Delta x_3) + \frac{1}{c}(\hat{x}_2 + \Delta x_2) - \frac{1}{c}(\frac{Q_0}{x_4} + d_Q) - \frac{1}{c}x_6 - \frac{1}{c}u_2 - \dot{x}_{4d}\right),$$ (46)

where $x_3 = \hat{x}_3 + \Delta x_3$. Then the total parametric uncertainty including uncertainty of CPL power can be represented as:

$$d = \frac{1}{c}\Delta x_2 - \omega\Delta x_3 - \frac{1}{c}d_Q; \ \|d\| \leq dmax,$$ (47)

where *dmax* is the limit for d, the total disturbance.

$$dmax = \frac{1}{c}\delta_{x2} - \omega\delta_{x3} - \delta_Q/\delta_{x4},$$ (48)

Then,

$$\dot{s} = e_3 - \frac{1}{c}x_6 - \dot{x}_{4d} + \omega\hat{x}_3 + \frac{1}{c}\hat{x}_2 - \frac{1}{c}\frac{Q_0}{x_4} - \frac{1}{c}u_2 + d,$$ (49)

Considering this as the Lyapunov candidate function:

$$V = \frac{1}{2}s^2,$$ (50)

$$\dot{V} = s\dot{s} = s(e_3 - \frac{1}{c}x_6 - \dot{x}_{4d} + \omega\hat{x}_3 + \frac{1}{c}\hat{x}_2 - \frac{1}{c}\frac{Q_0}{x_4} - \frac{1}{c}u_2 + d),$$ (51)

We then use u_2.

$$u_2 = -c\left[-e_3 + \frac{1}{c}x_6 + \dot{x}_{4d} - \omega\hat{x}_3 - \frac{1}{c}\hat{x}_2 + \frac{1}{c}\frac{Q_0}{x_4} + v\right],$$ (52)

Then, we can obtain:

$$\dot{V} = s(d + v), \tag{53}$$

Considering $\|d\| \leq dmax$, $\dot{V}$ will be made negative by the subsequent discontinuous control, v. Consequently, it will guarantee stability.

$$v = -dmax * sat\left(\frac{s}{\varepsilon}\right); \ \varepsilon > 0, \tag{54}$$

In total, the control input is:

$$u_2 = -c\left[-e_3 + \frac{1}{c}x_6 + \dot{x}_{4d} - \omega\hat{x}_3 - \frac{1}{c}\hat{x}_2 + \frac{1}{c}\frac{Q_0}{x_4} - dmax * sat\left(\frac{s}{\varepsilon}\right)\right], \tag{55}$$

4.2. Implementation and Robustness Analysis of Lyapunov Redesign Controller against Parametric Uncertainties Including Uncertainties in Power of CPL

The LRC is based only on Lyapunov function. Its nominal controller is designed to ensure the nominal system or disturbance-free system to be stable by forcing the Lyapunov function derivative of the nominal system to be negative. If there is disturbance in the system, the discontinuous control is used alone to handle the disturbance. The discontinuous controller is formulated by redesigning the Lyapunov function of the nominal system. In the redesigning process, the disturbance is introduced to the Lyapunov function of the nominal system and then solved for the discontinuous control to overcome that disturbance and force the new derivative Lyapunov function or be negative and consequently, the system to be globally stable. It has some chattering issues because of the discontinuous controller. The chattering magnitude is dependent on the magnitude of dmax. Having large dmax makes the system stable against large disturbance but it can cause larger chattering if it is set as a very large value. If the disturbance happens to be greater than the set dmax, the system can become unstable. But, the LRC has greater margin for stability because its nominal system is also ensured to be stable, thus provides better performance for large disturbance.

First of all, the Lyapunov Redesign Control input, u_1 will be designed, with the same approach followed next to design the other control input, u_2. Using the similar method as discussed in last section, we introduce new state variables:

$$e_1 = \int (x_3 - x_{3d})dt, \tag{56}$$

$$e_2 = \dot{e}_1 = x_3 - x_{3d}, \tag{57}$$

$$\dot{e}_2 = \dot{x}_3 - \dot{x}_{3d} = f_3(x) + g_3(x)u_1 - \dot{x}_{3d}, \tag{58}$$

Expanding $f_3(x)$ and $g_3(x)$:

$$\dot{e}_2 = \omega x_4 + \frac{1}{c}x_1 - \frac{1}{c}\frac{P_0}{x_3} - \frac{1}{c}x_5 - \frac{1}{c}u_1 - \dot{x}_{3d}, \tag{59}$$

Considering the uncertainties:

$$\dot{e}_2 = \omega(\hat{x}_4 + \Delta x_4) + \frac{1}{c}(\hat{x}_1 + \Delta x_1) - \frac{1}{c}\left(\frac{P_0}{x_3} + d_P\right) - \frac{1}{c}x_5 - \frac{1}{c}u_1 - \dot{x}_{3d}, \tag{60}$$

Then the total parametric uncertainty including uncertainty of CPL power can be represented as:

$$d = \frac{1}{c}\Delta x_1 + \omega\Delta x_4 - \frac{1}{c}d_P; \ \|d\| \leq dmax, \tag{61}$$

here $dmax$ is the limit of d, the total disturbance.

$$dmax = \frac{1}{c}\delta_{x1} + \omega\delta_{x4} - \frac{1}{c}\delta_P / \delta_{x3}, \tag{62}$$

Following the methodology of Lyapunov redesign, the over-all input is $u_1 = u_0 + v$; where u_0 is the nominal stabilizing controller and v is to handle the disturbances. We get the linear state space of error as in Equation (63):

$$\dot{e} = \begin{bmatrix} 0 & 1 \\ -k_1 & -k_2 \end{bmatrix} e, \tag{63}$$

Now, we define the desired Eigen values for the linearized system. Desired Eigen values would be -10.

Let, Equation (63) be written as $\dot{e} = Ae$ and $A = \begin{bmatrix} 0 & 1 \\ -k_1 & -k_2 \end{bmatrix}$

Generalized Eigen values of matrix "A":

$$sI - A = \begin{bmatrix} s & -1 \\ k_1 & s + k_2 \end{bmatrix}, \tag{64}$$

$$|sI - A| = s^2 + k_2 s + k_1, \tag{65}$$

Characteristic polynomial (desired):

$$(s + 10)(s + 10) = s^2 + 20s + 100, \tag{66}$$

Comparing Equations (65) and (66):

$$k_2 = 20, \; k_1 = 100$$

So, the values of k_1 and k_2 will become $+100$ and $+20$ respectively.

$$\dot{e} = \begin{bmatrix} 0 & 1 \\ -100 & -20 \end{bmatrix} e, \tag{67}$$

$$A = \begin{bmatrix} 0 & 1 \\ -100 & -20 \end{bmatrix}, \tag{68}$$

$$PA + A^T P = -I, \tag{69}$$

$$P = \begin{bmatrix} \frac{21}{8} & \frac{1}{200} \\ \frac{1}{200} & \frac{101}{4000} \end{bmatrix}, \tag{70}$$

$$V(e) = e^T P e, \tag{71}$$

$$w = 2e^T PG = 2 \begin{bmatrix} e_1 & e_2 \end{bmatrix} \begin{bmatrix} \frac{21}{8} & \frac{1}{200} \\ \frac{1}{200} & \frac{101}{4000} \end{bmatrix} \begin{bmatrix} 0 \\ 1 \end{bmatrix}, \tag{72}$$

$$w = \frac{1}{100}e_1 + \frac{101}{2000}e_2, \tag{73}$$

Then, we can choose the Lyapunov function for the nominal system or disturbance-free system to be:

$$V = \frac{1}{2}e_2^2, \tag{74}$$

$$\dot{V} = e_2\dot{e}_2 = e_2\left(\omega\hat{x}_4 + \frac{1}{c}x_1 - \frac{1}{c}\frac{P_0}{x_3} - \frac{1}{c}x_5 - \frac{1}{c}u_0 - \dot{x}_{3d}\right), \tag{75}$$

If we choose,

$$u_0 = -c\left[\frac{1}{c}\frac{P_0}{x_3} - \omega\hat{x}_4 + \frac{1}{c}x_5 + \dot{x}_{3d} - k_1e_1 - k_2e_2\right],\tag{76}$$

then, $\dot{V} < 0$. The terms $[-k_1e_1 - k_2e_2]$ guarantee the global stability of the nominal system which is absent in SMC method. The overall system is stabilized using the discontinuous control in the presence of disturbances. Redesigning the Lyapunov function considering disturbances,

$$V = \frac{1}{2}e_2^2,\tag{77}$$

$$\dot{V} = e_2\dot{e}_2 = e_2\left(\left(\omega\hat{x}_4 + \frac{1}{c}x_1 - \frac{1}{c}\frac{P_0}{x_3} - \frac{1}{c}x_5 - \frac{1}{c}u_0 - \dot{x}_{3d}\right) + \left(\frac{1}{c}v + d\right)\right),\tag{78}$$

$\left[\omega\hat{x}_4 + \frac{1}{c}x_1 - \frac{1}{c}\frac{P_0}{x_3} - \frac{1}{c}x_5 - \frac{1}{c}u_0 - \dot{x}_{3d}\right]$ is assured to be negative, then the discontinuous control can be designed as:

$$v = -c * dmax * sat\left(\frac{dmax * \omega}{\mu}\right),\tag{79}$$

Then, the overall input is:

$$u_1 = -c\left[\frac{1}{C}\frac{P_0}{x_3} - \omega\hat{x}_4 + \frac{1}{C}x_5 + \dot{x}_{3d} - 100e_1 - 20e_2 - dmax * sat\left(\frac{dmax\left(\frac{1}{100}e_1 + \frac{101}{2000}e_2\right)}{\mu}\right)\right],\tag{80}$$

Therefore, there is $\mu > 0$ so that for $\mu < \mu^*$, the closed-loop system's origin is asymptotically stable globally according to absolute stability theorem. Similarly, we have Equation (81), when we design a controller for u_2 with same desired points.

$$u_2 = -c\left[\frac{1}{C}\frac{Q_0}{x_4} + \omega\hat{x}_3 + \frac{1}{C}x_6 + \dot{x}_{4d} - 100e_3 - 20e_4 - dmax * sat\left(\frac{dmax\left(\frac{1}{100}e_3 + \frac{101}{2000}e_4\right)}{\mu}\right)\right],\tag{81}$$

where,

$$e_3 = \int(x_4 - x_{4d})dt,\tag{82}$$

$$e_4 = \dot{e}_3 = x_4 - x_{4d},\tag{83}$$

$$d = \frac{1}{c}\Delta x_2 - \omega\Delta x_3 - \frac{1}{c}d_Q; \ \|d\| \le dmax = \frac{1}{c}\delta_{x2} - \omega\delta_{x3} - \delta_Q/\delta_{x4},\tag{84}$$

5. Results

Here, the parameters and the parametric values regarding the simulation done for comparative analysis by varying the CPL power have been shown in Table 1.

Table 1. Table of Parameters.

Parameter	Value	Parameter	Value
Ω	60 Hz	δ_{x4}	100 V
X_3	600 V	δ_P	30 kW
X_4	50 V	δ_Q	2 kVar
d_P	50 A	ρ_{x3}	200 V
d_Q	20 A	ε	100
δ_{x3}	1000 A	R_{eq}	0.25 Ohm
L_{eq}	0.5×10^{-3} H	C_{eq}	10×10^{-6} F
R_{CVL}	15 Ohm	L_{CVL}	5×10^{-3} H
R_B	10 Ohm	C_B	1×10^{-6} F
L_B	1×10^{-3} H		

In Figure 9a,b, performance comparisons have been illustrated between SMC (colored in blue) and LRC (colored in green) for *d*-axis output voltage and *q*-axis output voltage respectively. The control objective for d-axis output voltage has been considered as 480 Volt, for normal conditions—where the variance value has been set as 10% to simulate noise. From Figure 9a, it is evident that the LRC controller shows considerably superior performance than that of the SMC controller, as its output stayed closer to the control objective. For *q*-axis output voltage, the control objective has been considered as low as possible and negligible in practice. In Figure 9b, the *q*-axis output voltage fluctuates more in the case of the SMC controller than that of the LRC controller. To determine the controller behaviors in a very noisy environment, more noise is added by setting the variance value as 100%, and noise rejection capabilities of SMC and LRC are tested. The results are shown in Figure 10a,b, which demonstrate LRC's superior capability to stick to the reference value with close proximity, whereas SMC has fluctuations of great magnitudes. For nonlinearity, LRC is capable of attaining the reference d-axis value with negligible time-delay, but SMC needs some time to reach that (Figure 11a). However, for *q*-axis, both controllers exhibit a similar performance (Figure 11b). In case of parametric uncertainties, LRC again proves to be the better suited one, displaying less fluctuations than SMC to maintain the control objective (Figure 12a,b). Hence, LRC offers appreciable stability considering CPL power variation and parametric uncertainties. In Figure 13a,b, performance comparisons have been presented between SMC (blue colored) and LRC (green colored) in the case of *d*-axis control input current, I_d (u_1) and *q*-axis control input current, I_q (u_2) respectably considering CPL power variation and parametric uncertainties. Here, the more the fluctuation in control input current, the more the stress will be imposed on the storage system to compensate, consequently degrading the storage performance and overall life time. This situation, in practice, makes it harder to retain microgrid stability. The mean squared errors (MSE) obtained from these analyses for both SMC and LRC controllers are presented in Table 2. It is obvious from the presented values that LRC is the better controller, and in a noisy environment, SMC is no match for LRC, as the former displays error significantly greater than LRC. This observation also leads to believe that, for practical applications, LRC can provide better performance than LRC. Therefore, from the comparative analysis presented here, the Lyapunov Redesign Controller shows better performance to retain system stability in face of CPL power variation. Hence, the LRC controller is preferred to be adopted for storage-based load side compensation technique for microgrid stability improvement with dense CPL loads present.

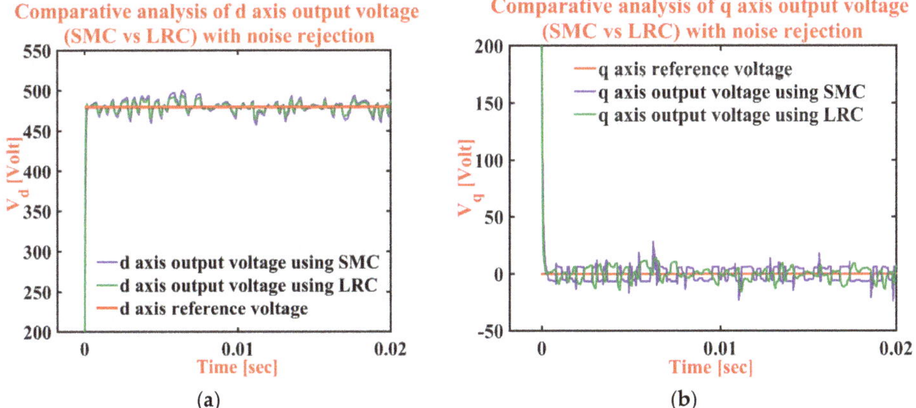

Figure 9. Comparison of performance between SMC (blue colored) and LRC (green colored) for normal condition in case of (**a**) *d*-axis output voltage; (**b**) *q*-axis output voltage considering CPL power variation and parametric uncertainties. LRC controller shows considerably better performance than SMC by staying closer to the reference voltage.

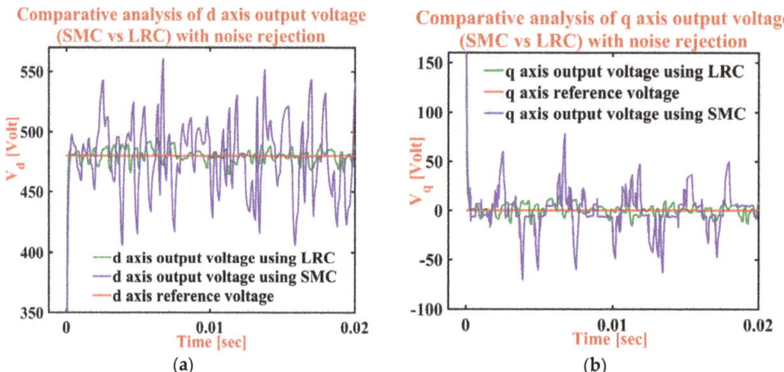

Figure 10. Comparison of performance between SMC (blue colored) and LRC (green colored) for very noisy environment in case of (**a**) *d*-axis output voltage; (**b**) *q*-axis output voltage considering CPL power variation and parametric uncertainties. LRC controller shows far better performance than SMC, as the latter shows high fluctuations from the reference voltage, while LRC stays close to it.

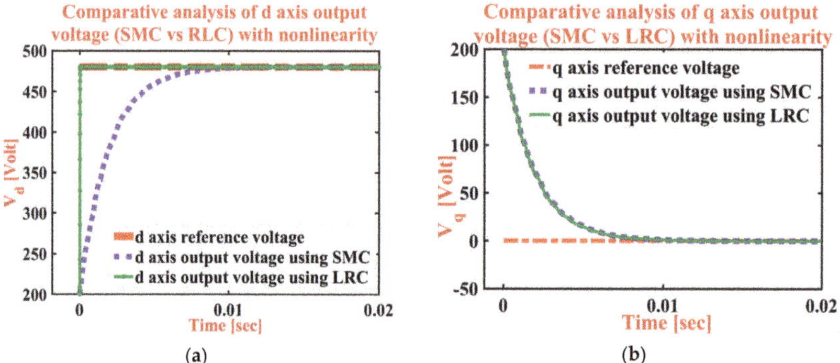

Figure 11. Comparison of performance between SMC (colored in blue) and LRC (colored in green) for nonlinearity in case of (**a**) *d*-axis output voltage, (**b**) *q*-axis output voltage. Unlike SMC, LRC is capable of attaining the reference d-axis value with negligible time-delay.

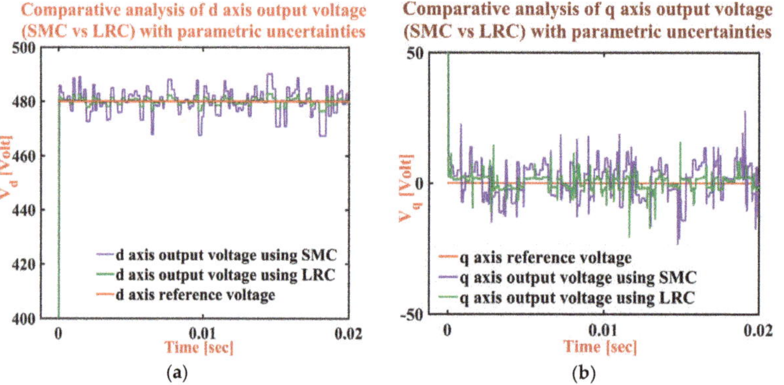

Figure 12. Comparison of performance between SMC (colored in blue) and LRC (colored in green) in case of (**a**) *d*-axis output voltage, (**b**) *q*-axis output voltage considering parametric uncertainties. LRC controller shows considerably better performance than SMC by staying closer to the reference voltage.

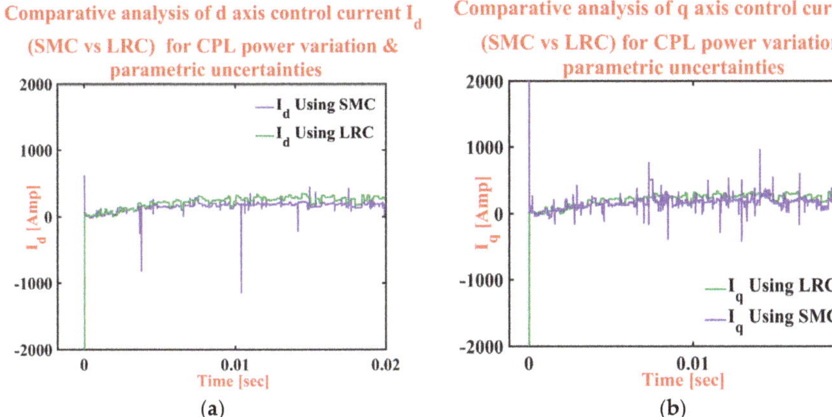

Figure 13. Comparison of performance among SMC (colored in blue) and LRC (colored in green) in case of (**a**) *d*-axis control input current, I_d (u_1); (**b**) *q*-axis control input current, I_q (u_2) considering CPL power variation and parametric uncertainties. LRC fluctuated less than SMC, causing less stress on the system and thus providing a longer lifetime.

Table 2. Mean squared error (MSE) values of SMC and LRC controllers for the different conditions.

Conditions	Parameter	Mean Squared Error		Relative Error (SMC-LRC)
		SMC	LRC	
Noise Rejection (normal condition)	X_3, *d*-axis voltage	0.00270320	0.00139074	1.31246×10^{-3}
	X_4, *q*-axis voltage	0.00138790	0.00113038	2.5752×10^{-4}
Noise Rejection (very noisy condition)	X_3, *d*-axis voltage	0.03661991	0.00139411	3.52258×10^{-2}
	X_4, *q*-axis voltage	0.00818637	0.00112001	7.06636×10^{-3}
Nonlinearity	X_3, *d*-axis voltage	0.00321589	0.00002730	3.18859×10^{-3}
	X_4, *q*-axis voltage	0.00942499	0.00924679	1.782×10^{-4}
Parameter Uncertainty	X_3, *d*-axis voltage	0.00116666	0.00011259	1.04076×10^{-3}
	X_4, *q*-axis voltage	0.00126333	0.00017554	1.08779×10^{-3}

6. Reason behind Inferior SMC Performance and Solutions

Sliding Mode Control presents many fascinating challenges to the mathematicians. It is also extensively used in engineering applications because of the comparatively easy implementation which does not require a deep understanding of the complex mathematical background. These two reasons put it in a unique position among control theories. There are three main stages of designing a Sliding Mode Controller: designing the sliding surface, selecting the control law that will hold the system trajectory on the sliding surface, and implementing in a chatter-free setup—which is the most important one of these three. Although in theory, Sliding Mode Control is a robust one, experiments show otherwise—SMC has some serious shortcomings. The most prominent one of them is chattering—the high frequency oscillation around the sliding surface. It reduces the control performance significantly. As an example, the following second-order system can be considered:

$$\dot{x}_1 = x_2, \tag{85}$$

$$\dot{x}_2 = ax_1 + bx_2 + csinx_1 + du, \tag{86}$$

a and b are negative constant values here whereas c and d are positive constants. For $c > |a|$, the system is known to be unstable. For the actuator, existence of fast dynamics is posited, and it is stable. These are not considered in the ideal model. The equations governing them are:

$$w_1 = w, \tag{87}$$

$$\dot{w}_1 = w_2, \tag{88}$$

$$\dot{w}_2 = -\frac{1}{\mu^2}w_1 - \frac{2}{\mu}w_2 + \frac{1}{\mu^2}u, \tag{89}$$

μ is a constant, considered to have a positive, sufficiently small value. As demonstrated in Figure 14, with actuator unmodeled dynamics present, $w(t)$ is the actual input of the system, not $u(t)$ directly from the sliding mode controller. The sliding mode surface and the control input is chosen as:

$$u = -Msign(\sigma), \tag{90}$$

$$\sigma = \lambda x_1 + x_2, \tag{91}$$

where λ and M are positive constants, with M is required to be large enough to enforce sliding mode into the ideal model ($\dot{\sigma}\sigma > 0$). $\dot{x}$ becomes a continuous time function in real system, thus making the expectation of sliding mode to occur invalid; and causes chatter.

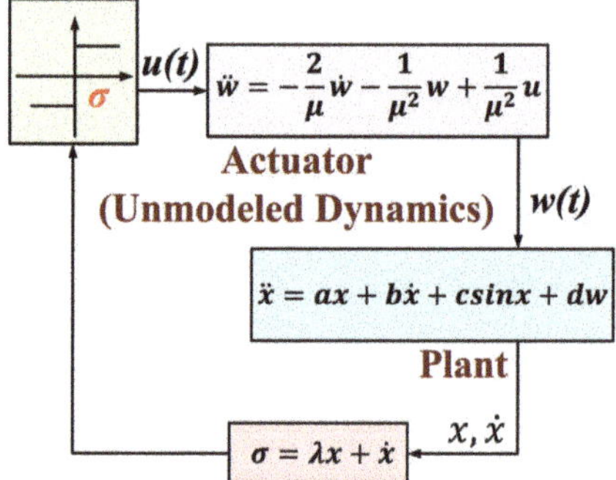

Figure 14. An example system demonstrating sliding mode control for systems as described in Equations (85) and (86). There are actuator dynamics that are not included in the ideal system. Chattering is caused from the excitation of these unmodeled dynamics by the high frequency switching action.

According to theory, the unmodeled dynamics present in the system causes the chattering effect. A sliding mode control, which is "chattering free", is not attainable as the model used in designing the controller can never capture all the system dynamics. But, the chattering can be curtailed. The sliding mode is normally implemented with a relay—which represents the sign function. It creates a common problem with relative degree equal to one. An alternative to this approach is using approximations of the sign function, which is widely used. Sigmoids, saturation, and hysteresis functions are used often too, providing a continuous or smooth control signal, but also losing the invariance property of the sliding mode control along the way. Table 3 shows some methods to improve the effectiveness of SMC. Fuzzy Sliding Mode Control (FSMC)—which uses a low pass filter, and estimates the sliding variable through a disturbance estimator—is the one with the least effectiveness. Integral Sliding Mode Control (ISMC), High Order Sliding Mode (HOSM), and Sliding Mode Extended State Observer (SMESO) offers better effectiveness. However, Type-2 Fuzzy-Neural Network Indirect Adaptive Sliding Mode

Control (T2FNNAS) is the way to achieve the best performance, which is based on the synthesis approach of Lyapunov [52,53].

Table 3. Methods to improve SMC technique.

Technique Name	Base of the Technique	Working Principle	Effectiveness
ISMC	Has a equal dimension to the state space	Control signal composed by a linear term with a continuous low excitation of the unmodeled dynamics	**
HOSM	High-gain control with saturations used for overcoming the effect of chattering by approximation of the sign function within a boundary layer around the switching manifold	The order of the mode is determined by the smoothness of tangency of the sliding manifold	**
T2FNNAS	Type-2 Fuzzy Neural Network	Based on the synthesis method Lyapunov, the adaptive FNN's free parameters are tuned on-line	***
SMESO	Extended state observer with active disturbance rejection control	Dramatically reduced chattering phenomenon on the control input channel with respect to Linear Extended State Observer	**
FSMC	Low pass filter	Estimation of the sliding variable via a disturbance estimator	*

*** = Excellent, ** = Satisfactory, * = Acceptable.

7. Numerical Verification of Results for Microgrid Application

The results obtained so far demonstrate the capabilities of both SMC and LRC to maintain microgrid stability. To ascertain the effectiveness of these methods in real-life conditions, both of them are simulated numerically with data obtained from physical microgrids. These simulations confirm the efficacy of these control systems to sustain stability in real microgrids.

7.1. SMC Technique

To verify the global stability, we have to calculate the equation below:

$$\dot{V} = s(d + v), \tag{92}$$

where,

$$d = \frac{1}{c}\Delta x_1 + \omega\Delta x_4 - \frac{1}{c}d_P, \tag{93}$$

$$v = -dmax * sat\left(\frac{s}{\varepsilon}\right); \ \varepsilon > 0, \tag{94}$$

So,

$$\dot{V} = s\left(\frac{1}{c}\Delta x_1 + \omega\Delta x_4 - \frac{1}{c}d_P - dmax * sat\left(\frac{s}{\varepsilon}\right)\right), \tag{95}$$

where,

$$dmax = \frac{1}{c}\delta_{x1} + \omega\delta_{x4} - \frac{1}{c}\delta_P / \delta_{x3}, \tag{96}$$

Putting these all together,

$$\dot{V} = s\left(\frac{1}{c}\Delta x_1 + \omega\Delta x_4 - \frac{1}{c}d_P - \left(\frac{1}{c}\delta_{x1} + \omega\delta_{x4} - \frac{1}{c}\delta_P / \delta_{x3}\right) * sat\left(\frac{s}{\varepsilon}\right)\right), \tag{97}$$

Now let,

$\omega = 60\,\text{Hz}, \ \Delta x_1 = 200\,\text{A}, \ \Delta x_4 = 50\,\text{V}, \ d_P = 50\,\text{A}, \ \delta_{x1} = 4000\,\text{A}, \ \delta_{x3} = 100\,\text{A}, \ \delta_{x4} = 100\,\text{V},$

$$\delta_P = 30 \text{ kW}, \ \varepsilon = 100, \ c = 10 \ \mu\text{F}$$

Putting these values, we get:

$$\dot{V} = s\left(\tfrac{1}{10\mu}(200) + (60)(50) - \tfrac{1}{10\mu}(50) - \left(\tfrac{1}{10\mu}(4000) + (60)(100) - \tfrac{1}{10\mu}\left(\tfrac{30k}{100}\right)\right) * sat\left(\tfrac{s}{100}\right)\right), \tag{98}$$

$$\dot{V} = s\left[15.003 \times 10^6 - [370.006 \times 10^6] \, sat\left(\frac{s}{100}\right)\right], \tag{99}$$

Now, if s is either positive or negative, we will obtain $\dot{V} \leq 0$, which guarantees global stability.

7.2. LRC Technique

We have,

$$\dot{V} = e_2 \dot{e}_2, \tag{100}$$

where,

$$\dot{e}_2 = \left(\omega x_4 + \frac{1}{c}x_1 - \frac{1}{c}\frac{P_0}{x_3} - \frac{1}{c}x_5 - \frac{1}{c}u_0 - \dot{x}_{3d}\right) + \left(\frac{1}{c}v + d\right), \tag{101}$$

And here,

$$v = -c * dmax * sat\left(\frac{dmax * w}{\mu}\right), \tag{102}$$

$$d = \Delta\omega n_4 + \Delta\omega x_4 + \omega n_4 + \frac{1}{c}\Delta x_1 - \frac{1}{c}(n_5 + \Delta x_5) - \frac{1}{c}d_P, \tag{103}$$

And,

$$dmax = \frac{1}{c}\delta_{x1} + \delta_\omega \delta_{n4} + \delta_\omega \delta_{x4} + \omega \delta_{n4} - \frac{1}{c}\delta_5 - \frac{1}{c}\delta_P / \delta_{x3}, \tag{104}$$

If we choose

$$u_0 = -c\left[\frac{1}{c}\frac{P_0}{x_3} - \omega \hat{x}_4 + \frac{1}{c}x_5 + \dot{x}_{3d} - k_1 e_1 - k_2 e_2\right], \tag{105}$$

Putting it all together,

$$\dot{V} = e_2\left(\left(\omega x_4 + \tfrac{1}{c}x_1 - \tfrac{1}{c}\tfrac{P_0}{x_3} - \tfrac{1}{c}x_5 - \tfrac{1}{c}u_0 - \dot{x}_{3d}\right) + \left(\tfrac{1}{c}\left(-c * dmax * sat\left(\tfrac{dmax*w}{\mu}\right)\right)\right.\right.$$
$$\left.\left. + \left(\Delta\omega n_4 + \Delta\omega x_4 + \omega n_4 + \tfrac{1}{c}\Delta x_1 - \tfrac{1}{c}(n_5 + \Delta x_5) - \tfrac{1}{c}d_P\right)\right)\right), \tag{106}$$

$$\dot{V} = e_2\left(\left(\omega x_4 + \frac{1}{c}x_1 - \frac{1}{c}\frac{P_0}{x_3} - \frac{1}{c}x_5 - \frac{1}{c}\left(-c\left[\frac{1}{c}\frac{P_0}{x_3} - \omega \hat{x}_4 + \frac{1}{c}x_5 + \dot{x}_{3d} - k_1 e_1 - k_2 e_2\right]\right) - \dot{x}_{3d}\right)\right.$$
$$+ \left(\frac{1}{c}\left(-c * \left(\frac{1}{c}\delta_{x1} + \delta_\omega \delta_{n4} + \delta_\omega \delta_{x4} + \omega \delta_{n4} - \frac{1}{c}\delta_5 - \frac{1}{c}\delta_P / \delta_{x3}\right)\right.\right.$$
$$\left. * sat\left(\frac{\left(\frac{1}{c}\delta_{x1} + \delta_\omega \delta_{n4} + \delta_\omega \delta_{x4} + \omega \delta_{n4} - \frac{1}{c}\delta_5 - \frac{1}{c}\delta_P / \delta_{x3}\right) * w}{\mu}\right)\right) \tag{107}$$
$$\left.\left. + \left(\Delta\omega n_4 + \Delta\omega x_4 + \omega n_4 + \frac{1}{c}\Delta x_1 - \frac{1}{c}(n_5 + \Delta x_5) - \frac{1}{c}d_P\right)\right)\right)$$

Now let,

$\omega = 60 \text{ Hz}, \ x_3 = 600 \text{ V}, \ x_4 = 10 \text{ V}, \ \Delta x_1 = 200 \text{ A}, \ \Delta x_2 = 200 \text{ A}, \ \Delta x_5 = 10 \text{ A}, \ n_3 = 50 \text{ V}, \ n_4 = 50 \text{ V},$
$n_5 = 30 \text{ A}, \ n_6 = 30 \text{ A}, \ \Delta\omega = 10 \text{ Hz}, \ d_P = 50 \text{ A}, \ d_Q = 20 \text{ A}, \ \delta_{x1} = 4000 \text{ A}, \ \delta_{x3} = 200 \text{ A}, \ \delta_{x4} = 100 \text{ V},$
$\delta_\omega = 70 \text{ Hz}, \ \delta_P = 30 \text{ kW}, \ \delta_Q = 20 \text{ Var}, \ \delta_{n3} = \delta_{n4} = \delta_{n5} = \delta_{n6} = 100 \text{ A}, \ \rho_{x3} = 200 \text{ V}, \ \text{and} \ \mu = 100,$
$\delta_{x5} = 50 \text{ A}, \ \delta_{x6} = 3 \text{ A}, \ \delta_5 = 150 \text{ A} \ \text{and} \ \delta_6 = 13 \text{ A}, \ c = 10 \ \mu\text{F}.$

Putting the values, we get:

$$\dot{V} = e_2\Big(\big(-(100)(600-480)-(20)(0)\big) + \Big(\tfrac{1}{10\mu}\big(-10\mu * \big(\tfrac{1}{10\mu}(4000)+(70)(100)$$
$$+(70)(100)+(60)(100) - \tfrac{1}{10\mu}(150) - \tfrac{1}{10\mu}\big(\tfrac{30k}{200}\big)\big)$$
$$*sat\Big(\frac{\big(\tfrac{1}{10\mu}(4000)+(70)(100)+(70)(100)+(60)(100) - \tfrac{1}{10\mu}(150) - \tfrac{1}{10\mu}\big(\tfrac{30k}{200}\big)\big)*(60)}{100}\Big)\Big)$$
$$+\big((10)(50)+(10)(10)+(60)(50) + \tfrac{1}{10\mu}(100) - \tfrac{1}{10\mu}(30+10) - \tfrac{1}{10\mu}(50)\big)\Big)\Big), \tag{108}$$

$$\dot{V} = e_2\big[988900 - [370.02 \times 10^6] * sat(222.012 \times 10^6)\big] \tag{109}$$

As we are getting $\dot{V} \leq 0$ from this equation, the system is globally stable.

8. Conclusions

A microgrid system has several advantages over the conventional utility grid system, such as unlimited renewable fuel resources, environment-friendly power generation, easy implementation, cost effectiveness, and so on. However, the maintenance of the microgrid electrification has been confronted by the challenge of continually increasing instability issues due to the growth of modern electronic devices. For improving the stability scenario of the microgrid system despite the presence of dense CPL loads, a storage-based load side compensation technique has been adopted in this paper. Besides that, Sliding Mode Controller (SMC) and Lyapunov Redesign Controller (LRC), two of the most prominent nonlinear control techniques, have been implemented individually to retain microgrid system stability. After that, SMC and LRC controller robustness analysis have been presented with the variation of CPL power. Next, the comparative analysis between the SMC controller and the LRC controller robustness has been illustrated which ascertains that Lyapunov Redesign Controller has a superior performance than the former one to retain microgrid stability in dense CPL-loaded conditions. Reasons for inferior SMC performance and ways to overcome them have been discussed afterwards, followed by numerical analysis of both of the control techniques to verify their performance in real microgrids. All the necessary results have been simulated in Matlab/Simulink platform with appreciable aftermath.

Acknowledgments: No funding has been received for this research project.

Author Contributions: All the authors contributed equally for the research article to be decimated in its current version.

Conflicts of Interest: The authors declare no conflict of interest.

References

1. Bayindir, R.; Hossain, E.; Kabalci, E.; Perez, R. A comprehensive study on microgrid technology. *Int. J. Renew. Energy Res.* **2014**, *4*, 1094–1107.
2. Hossain, E.; Kabalci, E.; Bayindir, R.; Perez, R. Microgrid testbeds around the world: State of art. *Energy Convers. Manag.* **2014**, *86*, 132–153. [CrossRef]
3. Reddy, K.R.; Babu, N.R.; Sanjeevikumar, P. A review on grid codes and reactive power management in power grids with wecs. In *Advances in Smart Grid and Renewable Energy*; Springer: Berlin, Germany, 2018; pp. 525–539.
4. Un-Noor, F.; Padmanaban, S.; Mihet-Popa, L.; Mollah, M.N.; Hossain, E. A comprehensive study of key electric vehicle (ev) components, technologies, challenges, impacts, and future direction of development. *Energies* **2017**, *10*, 1217. [CrossRef]
5. Ganesan, S.; Padmanaban, S.; Varadarajan, R.; Subramaniam, U.; Mihet-Popa, L. Study and analysis of an intelligent microgrid energy management solution with distributed energy sources. *Energies* **2017**, *10*, 1419. [CrossRef]

6. AL-Nussairi, M.K.; Bayindir, R.; Padmanaban, S.; Mihet-Popa, L.; Siano, P. Constant power loads (cpl) with microgrids: Problem definition, stability analysis and compensation techniques. *Energies* **2017**, *10*, 1656. [CrossRef]

7. Mihet-Popa, L.; Isleifsson, F.; Groza, V. Experimental Testing for Stability Analysis of Distributed Energy Resorces Components with Storage Devices and Loads. In Proceedings of the 2012 IEEE International Instrumentation and Measurement Technology Conference, Graz, Austria, 13–16 May 2012; pp. 588–593.

8. Mihet-Popa, L.; Han, X.; Bindner, H.; Pihl-Andersen, J.; Mehmedalic, J. Development and Modeling of different scenarios for a Smart Distribution Grid. In Proceedings of the 2013 IEEE 8th International Symposium on Applied Computational Intelligence and Informatics, Timisoara, Romania, 23–25 May 2013; pp. 257–261.

9. Mihet-Popa, L.; Zong, Y.; You, S.; Groza, V. Simulation Platform Developed to Study and Identify Critical Cases in a Future Smart Grid. In Proceedings of the 2016 IEEE Electrical Power and Energy Conference (EPEC), Ottawa, ON, Canada, 12–14 October 2016.

10. Camacho, O.M.F.; Nørgård, P.B.; Rao, N.; Mihet-Popa, L. Electrical Vehicle Batteries Testing in a Distribution Network using Sustainable Energy. *IEEE Trans. Smart Grid* **2014**, *5*, 1033–1042. [CrossRef]

11. Emadi, A.; Khaligh, A.; Rivetta, C.H.; Williamson, G.A. Constant power loads and negative impedance instability in automotive systems: Definition, modeling, stability, and control of power electronic converters and motor drives. *IEEE Trans. Veh. Technol.* **2006**, *55*, 1112–1125. [CrossRef]

12. Jelani, N.; Molinas, M.; Bolognani, S. Reactive power ancillary service by constant power loads in distributed ac systems. *IEEE Trans. Power Deliv.* **2013**, *28*, 920–927. [CrossRef]

13. Rahimi, A.M.; Williamson, G.A.; Emadi, A. Loop-cancellation technique: A novel nonlinear feedback to overcome the destabilizing effect of constant-power loads. *IEEE Trans. Veh. Technol.* **2010**, *59*, 650–661. [CrossRef]

14. Huddy, S.R.; Skufca, J.D. Amplitude death solutions for stabilization of dc microgrids with instantaneous constant-power loads. *IEEE Trans. Power Electron.* **2013**, *28*, 247–253. [CrossRef]

15. Kwasinski, A.; Onwuchekwa, C.N. Dynamic behavior and stabilization of dc microgrids with instantaneous constant-power loads. *IEEE Trans. Power Electron.* **2011**, *26*, 822–834. [CrossRef]

16. Sanchez, S.; Molinas, M. Large signal stability analysis at the common coupling point of a dc microgrid: A grid impedance estimation approach based on a recursive method. *IEEE Trans. Energy Convers.* **2015**, *30*, 122–131. [CrossRef]

17. Wu, M.; Lu, D.D.-C. A novel stabilization method of lc input filter with constant power loads without load performance compromise in dc microgrids. *IEEE Trans. Ind. Electron.* **2015**, *62*, 4552–4562. [CrossRef]

18. Hossain, E.; Perez, R.; Nasiri, A.; Bayindir, R. Development of lyapunov redesign controller for microgrids with constant power loads. *Renew. Energy Focus* **2017**, *19*, 49–62. [CrossRef]

19. Hossain, E.; Perez, R.; Padmanaban, S.; Siano, P. Investigation on the development of a sliding mode controller for constant power loads in microgrids. *Energies* **2017**, *10*, 1086. [CrossRef]

20. Singh, S.; Fulwani, D.; Kumar, V. Robust sliding-mode control of dc/dc boost converter feeding a constant power load. *IET Power Electron.* **2015**, *8*, 1230–1237. [CrossRef]

21. Padmanaban, S.; Ozsoy, E.; Fedák, V.; Blaabjerg, F. Development of sliding mode controller for a modified boost ćuk converter configuration. *Energies* **2017**, *10*, 1513. [CrossRef]

22. Stramosk, V.; Pagano, D.J. Nonlinear control of a bidirectional dc-dc converter operating with boost-type constant-power loads. In Proceedings of the 2013 Brazilian Power Electronics Conference (COBEP), Gramado, Brazil, 27–31 October 2013; IEEE: Piscataway, NJ, USA, 2013; pp. 305–310.

23. Gautam, A.R.; Singh, S.; Fulwani, D. DC bus voltage regulation in the presence of constant power load using sliding mode controlled DC-DC bi-directional converter interfaced storage unit. In Proceedings of the 2015 IEEE First International Conference on DC Microgrids (ICDCM), Atlanta, GA, USA, 7–10 June 2015; IEEE: Piscataway, NJ, USA, 2015; pp. 257–262.

24. Singh, S.; Fulwani, D. On design of a robust controller to mitigate CPL effect—A DC micro-grid application. In Proceedings of the 2014 IEEE International Conference on Industrial Technology (ICIT), Busan, Korea, 26 February–1 March 2014; IEEE: Piscataway, NJ, USA, 2014; pp. 448–454.

25. Padmanaban, S.; Blaabjerg, F.; Wheeler, P.; Ojo, J.O.; Ertas, A.H. High-voltage dc-dc converter topology for pv energy utilization—Investigation and implementation. *Electr. Power Compon. Syst.* **2017**, *45*, 221–232. [CrossRef]

26. Liu, Z.; Liu, J.; Bao, W.; Zhao, Y. Infinity-norm of impedance-based stability criterion for three-phase ac distributed power systems with constant power loads. *IEEE Trans. Power Electron.* **2015**, *30*, 3030–3043. [CrossRef]

27. Emadi, A. Modeling of power electronic loads in ac distribution systems using the generalized state-space averaging method. *IEEE Trans. Ind. Electron.* **2004**, *51*, 992–1000. [CrossRef]

28. Sun, J. Small-signal methods for ac distributed power systems—A review. *IEEE Trans. Power Electron.* **2009**, *24*, 2545–2554.

29. Karimipour, D.; Salmasi, F.R. Stability analysis of ac microgrids with constant power loads based on popov's absolute stability criterion. *IEEE Trans. Circuits Syst. II Express Briefs* **2015**, *62*, 696–700. [CrossRef]

30. Vavilapalli, S.; Padmanaban, S.; Subramaniam, U.; Mihet-Popa, L. Power balancing control for grid energy storage system in photovoltaic applications—Real time digital simulation implementation. *Energies* **2017**, *10*, 928. [CrossRef]

31. Swaminathan, G.; Ramesh, V.; Umashankar, S.; Sanjeevikumar, P. Investigations of microgrid stability and optimum power sharing using robust control of grid tie pv inverter. In *Advances in Smart Grid and Renewable Energy*; Springer: Singapore, 2018; pp. 379–387.

32. Zheng, Q.; Wu, F. Lyapunov redesign of adaptive controllers for polynomial nonlinear systems. In Proceedings of the American Control Conference (ACC '09), St. Louis, MO, USA, 10–12 June 2009; IEEE: Piscataway, NJ, USA, 2009; pp. 5144–5149.

33. Chung, W.-C.; Liaw, D.-C.; Chang, S.-T. The steering control of vehicle dynamics via a lyapunov redesign approach. In Proceedings of the 48th IEEE Conference on Decision and Control, 2009 Held Jointly with the 2009 28th Chinese Control Conference (CDC/CCC 2009), Shanghai, China, 15–18 December 2009; IEEE: Piscataway, NJ, USA, 2009; pp. 6321–6326.

34. Memon, A.Y.; Khalil, H.K. Lyapunov redesign approach to output regulation of nonlinear systems using conditional servocompensators. In Proceedings of the American Control Conference, Seattle, WA, USA, 11–13 June 2008; IEEE: Piscataway, NJ, USA, 2008; pp. 395–400.

35. Farrell, J.A.; Polycarpou, M.M. *Adaptive Approximation Based Control: Unifying Neural, Fuzzy and Traditional Adaptive Approximation Approaches*; John Wiley & Sons: Chichester, UK, 2006; Volume 48.

36. Khalil, H.K. *Noninear Systems*; Prentice-Hall: Upper Saddle River, NJ, USA, 1996; Volume 2.

37. Vilathgamuwa, D.; Zhang, X.; Jayasinghe, S.; Bhangu, B.; Gajanayake, C.; Tseng, K.J. Virtual resistance based active damping solution for constant power instability in ac microgrids. In Proceedings of the IECON 2011—37th Annual Conference on IEEE Industrial Electronics Society, Melbourne, Australia, 7–10 November 2011; IEEE: Piscataway, NJ, USA, 2011; pp. 3646–3651.

38. Tiwari, R.; Babu, N.R.; Arunkrishna, R.; Sanjeevikumar, P. Comparison between pi controller and fuzzy logic-based control strategies for harmonic reduction in grid-integrated wind energy conversion system. In *Advances in Smart Grid and Renewable Energy*; Springer: Berlin, Germany, 2018; pp. 297–306.

39. Padmanaban, S.; Grandi, G.; Blaabjerg, F.; Wheeler, P.W.; Siano, P.; Hammami, M. A comprehensive analysis and hardware implementation of control strategies for high output voltage dc-dc boost power converter. *Int. J. Comput. Int. Syst.* **2017**, *10*, 140–152. [CrossRef]

40. Padmanaban, S.; Daya, F.J.; Blaabjerg, F.; Wheeler, P.W.; Szcześniak, P.; Oleschuk, V.; Ertas, A.H. Wavelet-fuzzy speed indirect field oriented controller for three-phase ac motor drive–investigation and implementation. *Eng. Sci. Technol. Int. J.* **2016**, *19*, 1099–1107. [CrossRef]

41. Febin Daya, J.L.; Subbiah, V.; Iqbal, A.; Padmanaban, S. Novel wavelet-fuzzy based indirect field oriented control of induction motor drives. *J. Power Electron.* **2013**, *13*, 656–668. [CrossRef]

42. Febin Daya, J.; Subbiah, V.; Sanjeevikumar, P. Robust speed control of an induction motor drive using wavelet-fuzzy based self-tuning multiresolution controller. *Int. J. Comput. Int. Syst.* **2013**, *6*, 724–738. [CrossRef]

43. Slotine, J.-J.E.; Li, W. *Applied Nonlinear Control*; Prentice Hall: Englewood Cliffs, NJ, USA, 1991; Volume 199.

44. Slotine, J.-J.E. Sliding controller design for non-linear systems. *Int. J. Control* **1984**, *40*, 421–434. [CrossRef]

45. Utkin, V.I. *Sliding Modes in Control and Optimization*; Springer Science & Business Media: Berlin/Heidelberg, Germany, 2013.

46. Sastry, S.S. *Nonlinear Systems: Analysis, Stability, and Control*; Springer Science & Business Media: New York, NY, USA, 2013; Volume 10.

47. Hung, J.Y.; Gao, W.; Hung, J.C. Variable structure control: A survey. *IEEE Trans. Ind. Electron.* **1993**, *40*, 2–22. [CrossRef]

48. Ning-Su, L.; Chun-Bo, F. A new method for suppressing chattering in variable structure feedback control systems. *IFAC Proc. Vol.* **1989**, *22*, 279–284. [CrossRef]

49. Loh, A.M.; Yeung, L. Chattering reduction in sliding mode control: An improvement for nonlinear systems. *WSEAS Trans. Circuits Syst.* **2004**, *3*, 2090–2097.

50. Freeman, R.; Kokotovic, P.V. *Robust Nonlinear Control Design: State-Space and Lyapunov Techniques*; Springer Science & Business Media: Boston, MA, USA, 2008.

51. Krishna S, M.; Daya JL, F.; Padmanaban, S.; Mihet-Popa, L. Real-time analysis of a modified state observer for sensorless induction motor drive used in electric vehicle applications. *Energies* **2017**, *10*, 1077. [CrossRef]

52. Sanjeevikumar, P.; Febin Daya, J.L.; Wheeler, P.; Blaabjerg, F.; Fedák, V.; Ojo, J.O. Wavelet Transform with Fuzzy Tuning Based Indirect Field Oriented Speed Control of Three-Phase Induction Motor Drive. In Proceedings of the 2015 International Conference on Electrical Drives and Power Electronics (EDPE), Tatranska Lomnica, Slovakia, 21–23 September 2015; pp. 111–116.

53. Hosseyni, A.; Trabelsi, R.; Iqbal, A.; Sanjeevikumar, P.; Mimouni, M.F. An Improved Sensorless Sliding Mode Control/Adaptive Observer of a Five-Phase Permanent Magnet Synchronous Motor Drive. *Int. J. Adv. Manuf. Technol.* **2017**, *93*, 1–11. [CrossRef]

Article

Minimization of Load Variance in Power Grids—Investigation on Optimal Vehicle-to-Grid Scheduling

Kang Miao Tan [1], Vigna K. Ramachandaramurthy [1], Jia Ying Yong [1], Sanjeevikumar Padmanaban [2,*], Lucian Mihet-Popa [3] and Frede Blaabjerg [4]

[1] Power Quality Research Group, Institute of Power Engineering, Department of Electrical Power Engineering, Universiti Tenaga Nasional, Jalan IKRAM-UNITEN, Kajang, Selangor 43000, Malaysia; tankangmiao@gmail.com (K.M.T.); vigna@uniten.edu.my (V.K.R.); yongjiaying89@gmail.com (J.Y.Y.)

[2] Department of Electrical and Electronics Engineering, University of Johannesburg, 2006 Auckland Park, South Africa

[3] Faculty of Engineering, Østfold University College, Kobberslagerstredet 5, 1671 Kråkeroy-Fredrikstad, Norway; lucian.mihet@hiof.no

[4] Centre for Reliable Power Electronics (CORPE), Department of Energy Technology, Aalborg University, 9000 Aalborg, Denmark; fbl@et.aau.dk

* Correspondence: sanjeevi_12@yahoo.co.in; Tel.: +27-79-219-9845

Received: 16 September 2017; Accepted: 3 November 2017; Published: 16 November 2017

Abstract: The introduction of electric vehicles into the transportation sector helps reduce global warming and carbon emissions. The interaction between electric vehicles and the power grid has spurred the emergence of a smart grid technology, denoted as vehicle-to grid-technology. Vehicle-to-grid technology manages the energy exchange between a large fleet of electric vehicles and the power grid to accomplish shared advantages for the vehicle owners and the power utility. This paper presents an optimal scheduling of vehicle-to-grid using the genetic algorithm to minimize the power grid load variance. This is achieved by allowing electric vehicles charging (grid-to-vehicle) whenever the actual power grid loading is lower than the target loading, while conducting electric vehicle discharging (vehicle-to-grid) whenever the actual power grid loading is higher than the target loading. The vehicle-to-grid optimization algorithm is implemented and tested in MATLAB software (R2013a, MathWorks, Natick, MA, USA). The performance of the optimization algorithm depends heavily on the setting of the target load, power grid load and capability of the grid-connected electric vehicles. Hence, the performance of the proposed algorithm under various target load and electric vehicles' state of charge selections were analysed. The effectiveness of the vehicle-to-grid scheduling to implement the appropriate peak load shaving and load levelling services for the grid load variance minimization is verified under various simulation investigations. This research proposal also recommends an appropriate setting for the power utility in terms of the selection of the target load based on the electric vehicle historical data.

Keywords: electric vehicles; energy management; load variance; optimal scheduling; optimization; vehicle-to-grid

---

1. Introduction

Electric vehicles (EVs) have gained popularity due to their emission free and fuel independence characteristics. Governments across the nations have established several schemes and organizations to facilitate the electrification process of roadway vehicles in terms of the technology, market, policy and finance. The efforts from numerous parties have successfully stimulated the deployment of the EV market. The recent Global EV Outlook report showed improved signs of EV adoption. In the near future, the global EV stock is predicted to reach more than one hundred million by the year 2050 [1–5]. As a result, the anticipated large scale of EV charging will bring technical challenges to the power grid. Extensive studies on the harmful impacts of large scale EV charging on the power grid had been performed. The potential impacts include the overloading of power equipment, harmonics, voltage drop, voltage instability and power losses [6–16]. A review in [17] concluded that the main impact of EV integration into the power grid was the overloading issue, which further led to equipment overheating and rapid aging issues. Meanwhile, the grid power quality was assessed under various EV penetration and charging level scenarios in [18,19]. The results revealed that a large number of fast charging EVs would introduce serious voltage deviations and voltage instability problems.

EVs utilize relatively large capacity batteries as their energy source for the vehicle propulsion. Hence, the large scale of the grid-connected EVs can be considered as an enormous distributed energy storage in the power grid. Moreover, the potential of EVs to share energy with the power grid has created a new opportunity to improve the system reliability and sustainability. This concept, known as Vehicle-to-Grid (V2G) technology, was firstly introduced by Kempton and his research team in [20–24]. The authors proposed the adoption of V2G technology to improve the reliability of the power grid, as well as to facilitate the large scale integration of renewable energy [20]. The economic potential of EVs in providing energy support to the power grid was also studied by Kempton in [25]. The findings assured a significant revenue stream to the EV fleet for providing the V2G support. For a more accurate revenue estimation, the authors had developed a calculation model in [26] to evaluate the cost and profit in providing V2G support, mainly in the form of peak load shaving, spinning reserves and energy regulation services. Kempton and his research team also paid attention to the willingness of people to pay for EVs and their attributes in [27]. The research results suggested that EV cost needs to be reduced significantly in order to have higher EV adoption in the marketplace. Additionally, V2G management strategies and implementation steps were proposed in [28] to assist the transition of V2G technology.

The potential of the V2G concept has urged other research and development in the supporting technology for V2G application. The literature has presented many potential benefits for the power grid by adopting V2G technology. These benefits include peak load shaving, load levelling, grid voltage regulation, improvement in energy efficiency and mitigation of renewable energy intermittency [29–32]. For instance, a Vehicle-to-Home (V2H) concept was presented in [33], where EVs collaborated with the renewable energy in a smart home system to minimize the pollution, fossil fuel depletion and investment cost. In addition, a study in [34] utilized EVs to support the autonomous operation mode of microgrids. During the transition to separate from the power grid, the available EVs were used by the system frequency controller as a buffering feature to stabilize the microgrid operation.

Nevertheless, the V2G implementation requires frequent charging and discharging processes which can cause extra deterioration to the EV batteries [35]. This downside has created a strong social barrier that can prevent public support for the V2G concept. Hence, optimal V2G charging and discharging scheduling with compensated incentives given to the participating EV owners will be especially crucial for the realization of this technology. Research has been carried out to develop the optimal planning and scheduling for the proper V2G execution. The complexity in managing the large amount of energy exchange between the power grid and EVs represents a challenge for a power utility. The first logical step to implement the V2G technology was proposed in [36]. The proposed unidirectional spinning reserve V2G algorithm was able to modulate the EV charging rate according to a Preference Operating Point (POP), where the minimal preliminary investment and EV batteries degradation were achieved.

On the other hand, the bidirectional V2G technology has tremendous flexibility to accomplish the EV scheduling according to the preferences of the EV owners and power grid requirements. Many researchers have proposed EV scheduling that utilizes the Grid-to-Vehicle (G2V) and Vehicle-to-Grid (V2G) operations to help the power grid in achieving various benefits [37–39]. From the economic standpoint, the authors in [40,41] emphasized the maximization of power utility profit while developing the V2G algorithm. Meanwhile, the minimization of power system losses is also a popular topic in the V2G application. In [42], EVs were used as reactive power resources to reduce the power grid energy losses. Furthermore, V2G technology was also utilized to maximize the renewable energy generation by solving the renewable energy intermittency issue [43,44]. Intelligent optimization algorithms were proposed by the authors in [45–47] to solve the renewable energy intermittency problem while maximizing the power grid energy efficiency.

Various optimal V2G scheduling strategies have been introduced in the literature to minimize the power grid load variance [48,49]. This concept is well known for its effectiveness in reducing power grid operation losses [50]. In general, this is achieved by enabling the grid-connected EVs to absorb the extra electricity during the power grid valley load period and feed the energy back to the power grid during the peak load period [51–54]. For instance, a V2G scheduling algorithm was proposed in [55] to perform power grid load shaving and valley filling operations using grid-connected EVs. The operation was achieved by regulating the power grid load at the target load pre-determined in the scheduling algorithm. A similar demand side management concept was adopted in [56], where considerable savings and emissions reduction were demonstrated by the results. Meanwhile, the authors in [57] proposed a peak shaving algorithm to reduce the peak demand of the power grid. The application of this algorithm has shown good potential in extending the lifespan of transformers, as well as maintaining a healthy grid operation voltage profile. Another peak shaving strategy was introduced in [58,59], which provided dynamic adjustments in EVs discharging rate without affecting the battery usage for EV traveling purpose. This strategy limited the maximum load demand by utilizing the extra energy in the grid-connected EVs for power grid support. In these literatures, many of the proposed V2G scheduling for power grid load variance minimization utilized a preset target loading in the scheduling algorithm. Nevertheless, most of the related studies have not discussed the methodology used in determining these targets. Inaccurate setting of this target loading will reduce the effectiveness of the V2G algorithm to achieve the minimization of load variance. The selection of this target loading can be a challenging task, as the power grid and grid-connected EVs are dynamic in nature. Thus, this paper will focus on the development of a V2G optimization algorithm, as well as analyzing the influence of different target load curves on the performance of the proposed algorithm.

This paper presents a G2V/V2G optimization algorithm, which is focused on the minimization of grid load variance by performing load levelling and peak load shaving using the available grid-connected EVs. During the event where the actual power grid loading is lower than the target loading, EVs are allowed to receive charging power from the power grid. This G2V operation falls under the load leveling scenario. Meanwhile, when the actual power grid loading is larger than the target loading, EVs are encouraged to discharge for power grid support. This V2G operation is called the peak load shaving scenario. In general, this paper will present the proposed algorithm, which consists of G2V and V2G operations as the V2G optimization algorithm. The V2G algorithm instructs EV charging and discharging operations according to the proposed objective function with the aim of minimizing the difference between the actual power grid loading and target loading. The V2G optimization algorithm includes thorough consideration of both power grid and EV constraints. The algorithm can also deliberate the V2G scheduling by considering the uncertain mobility characteristics of EVs, which have random initial State of Charge (SOC) and grid connection duration, as well as the dynamic grid connection probability. In addition, extensive analyses are conducted to investigate the effect of different target load curves and various average initial SOC of EV batteries on the performance of the proposed V2G optimization algorithm. A performance index was introduced to compare and evaluate the success rate of each optimized scenario. Then, a proper selection of the V2G target load curve based on the average initial SOC of the available grid-connected EV batteries is recommended. In summary, the contributions of this paper include: (i) the development of a V2G optimization algorithm to minimize grid load variance via peak load shaving and load levelling, and (ii) a performance analysis of the proposed V2G optimization under various target load and average initial SOC of EV batteries.

The rest of the paper is organized into sections. Section 2 describes the power grid model for the implementation of V2G optimization algorithm. In Section 3, the detailed formulation of the objective function, constraints and optimization algorithm are presented. The simulation results to validate the effectiveness of the proposed V2G optimization algorithm are discussed in Section 4. Section 5 concludes the paper.

2. Power Grid Model for V2G Implementation

The implementation of V2G technology requires extensive planning and careful management in order to ensure the reliability of the power grid. In the initial stage of the V2G implementation, the V2G location shall be well planned for the proper V2G application. For instance, the quantity of EV mobility within the V2G location is crucial for the practicality of V2G technology. In this paper, an adequate quantity of the grid-connected EVs was required to ensure the proper amount of EV batteries' storage capacities was available to absorb or deliver the power during the interaction with the power grid. This is to ensure an average distributed amount of EVs shall be accessible within the V2G location throughout the day. Thus, the selected V2G location shall be a relatively large township, which consisted of both residential and commercial areas to ensure a sufficient and consistent EV mobility throughout the day. Figure 1 presents an intelligent township for the V2G implementation consisting of the commercial offices, residential areas and smart V2G car parks. The components to realize the V2G technology are also shown in Figure 1. For instance, the township was equipped with the dual communication system across all the commercial loads, residential loads and smart V2G car parks with the local aggregator. Furthermore, the bidirectional V2G chargers were installed in the smart V2G car parks, which allowed each EV to charge from and discharge to the power grid.

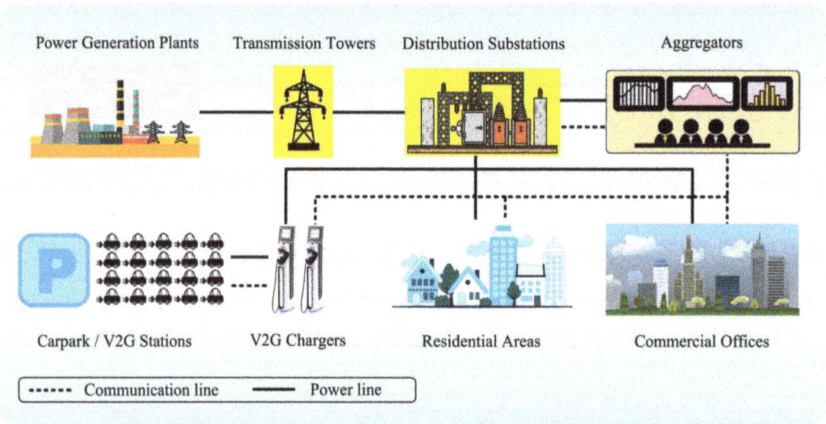

Figure 1. Intelligent Vehicle-to-Grid (V2G) township.

Figure 2 shows the single-line diagram of a generic township used in this paper for the implementation of V2G technology. The generic township was a typical radial configured power distribution grid. This township had a total of 200 units of the commercial offices and 800 units of the residential condominiums. The maximum demand for each commercial office was 15 kW, whilst 10 kW for each residential condominium. All the loads within the generic township were supplied via the 22 substations, where each substation had a step down transformer rated at 1 MW. Several assumptions were made for the V2G optimization algorithm and are listed as follows:

- All the residential, commercial and EV loads were evenly distributed to each substation in the proposed generic township.

- The town had a total of 1800 EV mobility daily, which includes EVs from the generic and nearby townships.
- The smart car parks had sufficient parking spaces equipped with the bidirectional V2G chargers.
- The aggregator had the appropriate bidirectional communication system to monitor the load profile of each substation and access the information of each EV, such as the EV availability and SOC level of EV battery.
- The aggregator was given the full authority to manage and control the amount of power sharing between the EVs and power grid.

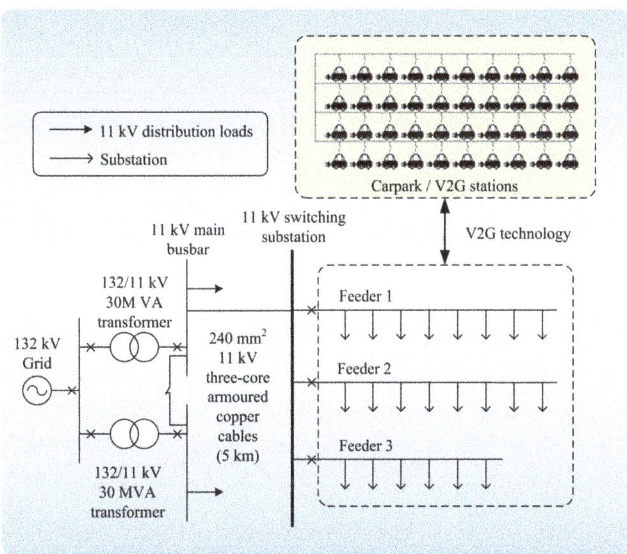

Figure 2. Single-line diagram of a generic township.

3. Problem Formulation

The implementation of V2G technology in the power grid usually involves multiple conflicting objectives, which are plagued with numerous uncertainties and nonlinearities. Therefore, the utilization of the optimization technique is crucial to execute the V2G concept in the power grid. In this paper, the Genetic Algorithm (GA) optimization technique is employed to solve the V2G optimization problem. This section discusses the objective function and constraints for the optimization problem.

3.1. Objective Function

In this paper, the proposed V2G optimization algorithm has the objective to optimize the grid-connected EV charging and discharging power in order to minimize the power grid load variance. A generalized daily power load curve is depicted in Figure 3. Besides, the desired grid loading is achieved by performing peak load shaving and load levelling to minimize the variance between the power grid loading and target loading as illustrated in Figure 3. The proposed V2G optimization algorithm is performed by enabling the EV charging during the period where the power grid loading is less than the target loading (G2V operation). On the contrary, EVs are required to discharge the energy from the batteries when the power grid loading is larger than the target loading (V2G operation). No power flows between the EVs and power grid when the target loading is equal to the power grid loading. Equations (1) and (2) express the objective function in terms of grid load variance, target loading, charging and discharging rate:

$$\min \Delta P = P_{\text{target}}(t) - P_{\text{load}}(t) - P_{\text{EV}}(t) \tag{1}$$

$$P_{\text{EV}}(t) = \begin{cases} \left(\sum_{n=1}^{N} A_n K_n \right) \times P_{\text{EV,charging}} & \text{, when } P_{\text{load}} < P_{\text{target}} \\ \left(\sum_{n=1}^{N} A_n K_n \right) \times P_{\text{EV,discharging}} & \text{, when } P_{\text{load}} > P_{\text{target}} \\ 0 & \text{, when } P_{\text{load}} = P_{\text{target}} \end{cases} \tag{2}$$

where ΔP is the grid load variance, t is time, P_{target} is the target loading, P_{load} is the existing load of the power grid, P_{EV} is the total power of EV loads/sources, n is the EV index, N is the maximum EV index, A_n is the availability of nth EV for the V2G application, K_n is the indicator of nth EV for V2G application, $P_{\text{EV,charging}}$ is the EV charging rate, and $P_{\text{EV,discharging}}$ is the EV discharging rate. In Equation (1), the grid load variance (ΔP) was minimized by optimizing the number of grid-connected EV to charge or discharge during valley load and peak load period, respectively. The determined optimal grid-connected EV numbers are then reflected in the indicator of nth EV for V2G application (K_n). Meanwhile, A_n refers to the availability of nth EV to the power grid prepared for the V2G application. This variable depends on the dynamic EV mobility characteristics, where the detailed EV grid-connection probability is discussed in Section 3.2.3. In the condition where K_n and A_n are available, the nth EV will be instructed to charge or discharge the battery.

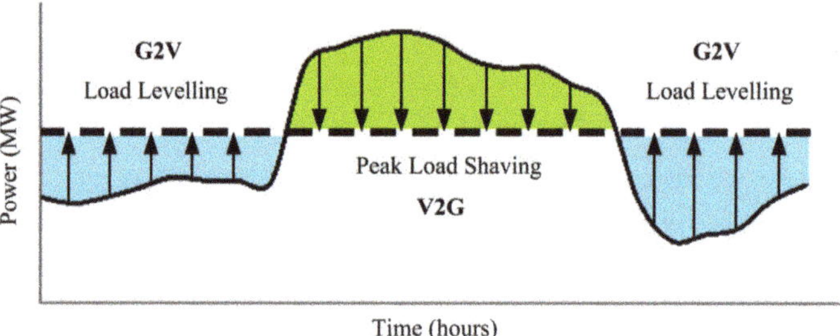

Figure 3. The concept of peak load shaving and load levelling.

3.2. Optimization Constraints

The operation of V2G system is restricted under plenty of system constraints and uncertainties. Therefore, the proposed V2G optimization algorithm required the compliance with these power grid and EV constraints.

3.2.1. Power Balance

The important power grid constraint to be considered in the proposed V2G optimization algorithm is the power balance between the grid generation and demand. The supplied power from the generation plants and distributed EV battery sources must satisfy the power grid load and EV charging demands:

$$P_{\text{grid}}(t) + \sum_{n=1}^{N} A_n K_n P_{\text{EV,discharging}}(t) = P_{\text{load}}(t) + \sum_{n=1}^{N} A_n K_n P_{\text{EV,charging}}(t) \tag{3}$$

where P_{grid} is the active power from the generation plant.

3.2.2. SOC of EV Battery

The SOC level of EV battery is one of the EV constraints which must be kept within certain limits during the V2G operation. It is important to establish these limits for two reasons. The first reason is to protect the health of the battery. Studies show that the SOC level of a lithium ion battery should best to be limited to a 60% swing (between 30% and 90% or 25–85%) to minimize the battery health degradation [60,61]. The second reason to keep the SOC level of the EV battery within certain limits is to reserve a certain amount of energy for the EV travel usage. Therefore, by taking both reasons into consideration, each EV is prevented from discharging if the battery SOC is lower than the SOC_{min} which is set at 55%. Meanwhile, the EV charging process is only allowed if the SOC level of each EV battery is below the SOC_{max}, which is set at 90% to prevent the battery overcharging issue. Both of the EV charging and discharging processes are allowed if the battery SOC is within the SOC_{min} and SOC_{max}:

$$SOC_n \leq SOC_{min}, \text{ allow for charging only} \tag{4}$$

$$SOC_{min} \leq SOC_n \leq SOC_{max}, \text{ allow for charging and discharging} \tag{5}$$

$$SOC_n \geq SOC_{max}, \text{ allow for discharging only} \tag{6}$$

where SOC_n is the battery SOC for nth EV.

3.2.3. EV Grid Connection Probability

The feasibility of V2G technology requires EVs to be connected to the power grid. Nevertheless, each EV can be connected to the power grid at different arrival and departure times in the car park, which will lead to the dynamic EV mobility characteristics [62–69]. Hence in this paper, the EV grid connection probability is estimated based on the driving behaviour of the township's residents and commercial workers. In Figure 4, the EV grid connection probability of the residential car park showed that more EVs are available in the car park, since the beginning of the day until 06:00 o'clock in the morning.

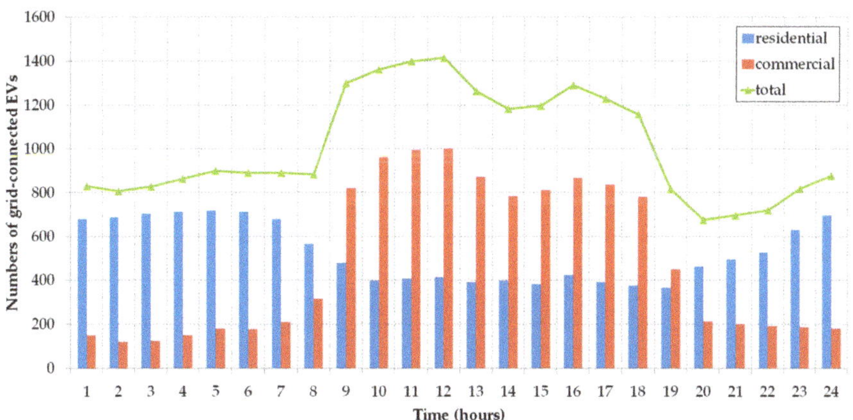

Figure 4. Electric Vehicle (EV) grid connection probability of the township.

Later on, most of the EV owners are away to their workplaces and schools. From 18:00 o'clock onwards, the residents started to return home and occupied the residential car park. On the other hand, the EV grid connection probability of the commercial car parks depicted that most of the parking spaces are occupied during the daytime office hours. Other than these periods, the commercial car park is almost empty. The combination of the EV grid connection probability of the residential and commercial car parks gave the total EV availability in the proposed generic township, as shown in Figure 4.

$$A_n = 1, \text{ when } n\text{th EV was connected to the power grid} \tag{7}$$

$$A_n = 0, \text{ when } n\text{th EV was not connected to the power grid} \tag{8}$$

3.2.4. EV Power Exchange Rate

The power exchange rate between the EV batteries and power grid shall be restricted within a safe margin to protect the safety and health of EV batteries throughout the power exchange periods. A common EV battery available in the market is considered in this research. The EV battery used in this paper a lithium-ion battery, which has the rated capacity of 50 Ah. The power exchange rate of each EV battery is limited below 3.3 kW, which is the typical power rating for slow charging:

$$P_{\text{EV,charging}} \leq P_{\text{EV,max}} \tag{9}$$

$$P_{\text{EV,discharging}} \leq -P_{\text{EV,max}} \tag{10}$$

where $P_{\text{EV,max}}$ is the maximum EV exchange rate.

3.3. Optimization Algorithm

The proposed V2G optimization algorithm is implemented in MATLAB software tool (R2013a, MathWorks, Natick, MA, USA) as shown in Figure 5.

In order to handle a large number of parameters, the GA optimization technique is adopted. The GA is an iteration method that is capable of searching for the global optimal solution within an execution time limit. Furthermore, the GA is inspired by the living organism evolutionary process, which requires the representation of a potential solution as the genetic chromosome. A proper fitness function is utilized to compute and evaluate the score of this genetic chromosome. After the evaluation, the GA principle is repeated again to reproduce a new generation of chromosome until this iteration converges to an optimal solution.

In the initial stage of the proposed V2G algorithm, the system parameters are obtained and updated into the database. With this information, the GA algorithm evaluates the fitness function of the grid load variance minimization, which later produce the next generation of solution. The evaluation repeated itself until the iteration converged to an optimal EV charging or discharging power. This optimization process is bound by the power grid and EV constraints. Meanwhile, the V2G optimization algorithm is executed for every hour.

With respect to the constraint limits, the proposed V2G optimization algorithm performed the peak load shaving and load levelling services by utilizing the EV battery storages in order to minimize the load variance, as it is shown in Figure 5.

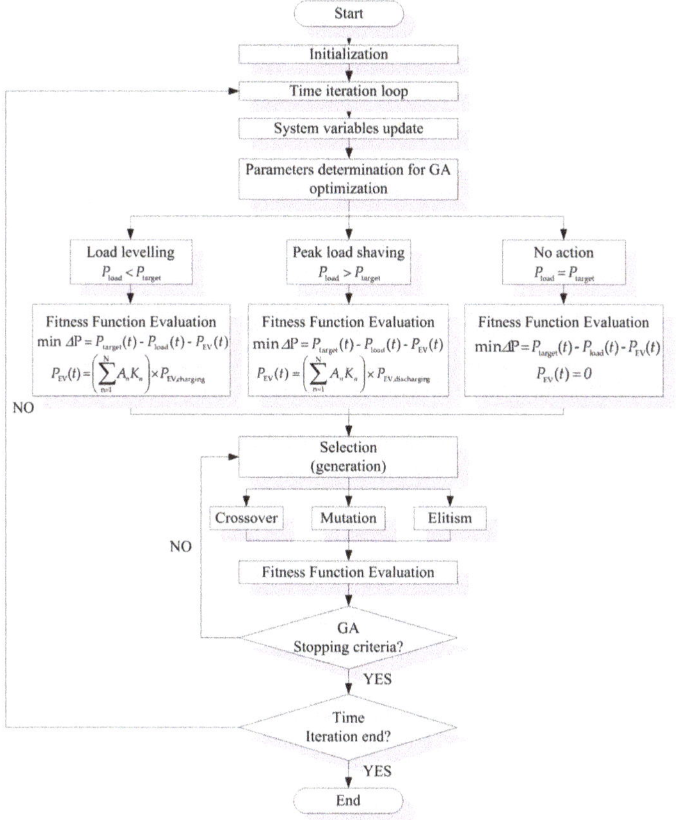

Figure 5. Flowchart for the proposed V2G optimization algorithm.

4. Results and Discussion

In this section, the feasibility of the proposed V2G optimization algorithm that was investigated under different scenarios was elaborated. Various average initial SOC of EV batteries ($SOC_{i,ave}$) and target load curve in percentage (TLC_{pct}) were considered to examine the performance of the proposed V2G optimization algorithm. The definitions of the $SOC_{i,ave}$ and TLC_{pct} are shown in Equations (11) and (12), correspondingly:

$$\text{Average initial SOC of EV Batteries } (SOC_{i,ave}) = \frac{\sum\limits_{n=1}^{N} K_n A_n SOC_n}{N} \tag{11}$$

$$\text{Percentage of Target Load Curve } (TLC_{pct}) = \frac{\text{target load curve}}{\text{peak load}} \times 100\% \tag{12}$$

All the scenarios were conducted in the generic commercial-residential township as depicted in Figure 2. Figures 6–8 present the comparison between the optimized power load curves and original power load curves, where the TLC_{pct} were set at 50%, 55% and 60%, respectively. In each scenario, different $SOC_{i,ave}$ of 40%, 50%, 60%, 70% and 80% were applied. Based on the preset values of the $SOC_{i,ave}$ and TLC_{pct}, different optimized power load curves were acquired with the implementation of the proposed V2G optimization algorithm. For instance, the scenario in Figure 6a shows that the

$SOC_{i,ave}$ was set at 40% and the TLC_{pct} was kept at 50% of the peak loading of the power load curve. Since the $SOC_{i,ave}$ was low, the proposed V2G optimization algorithm can perform the load levelling service by the EV charging operation, but not the EV discharging operation for the peak load shaving service. In contrast, reversed outcomes can be observed for scenario in Figure 8e due to higher $SOC_{i,ave}$ and TLC_{pct}.

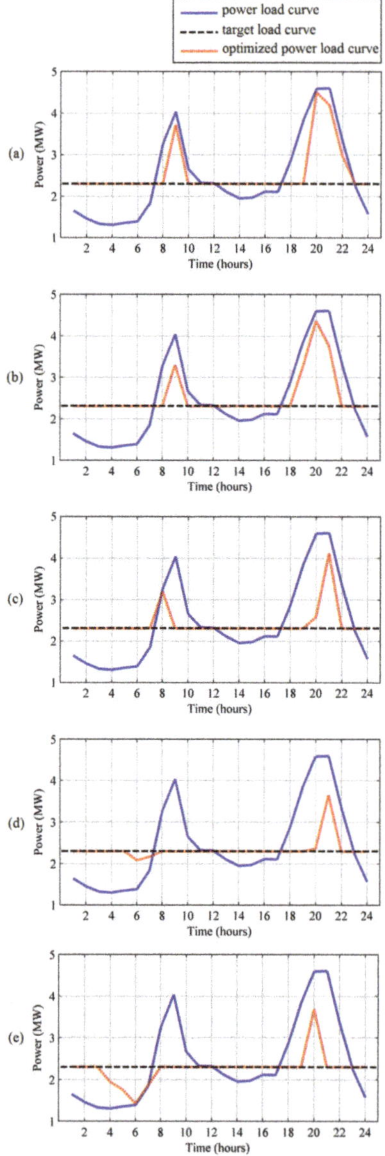

Figure 6. Optimized power load curves with 50% TLC_{pct} and $SOC_{i,ave}$ of (**a**) 40%, (**b**) 50%, (**c**) 60%, (**d**) 70% and (**e**) 80%.

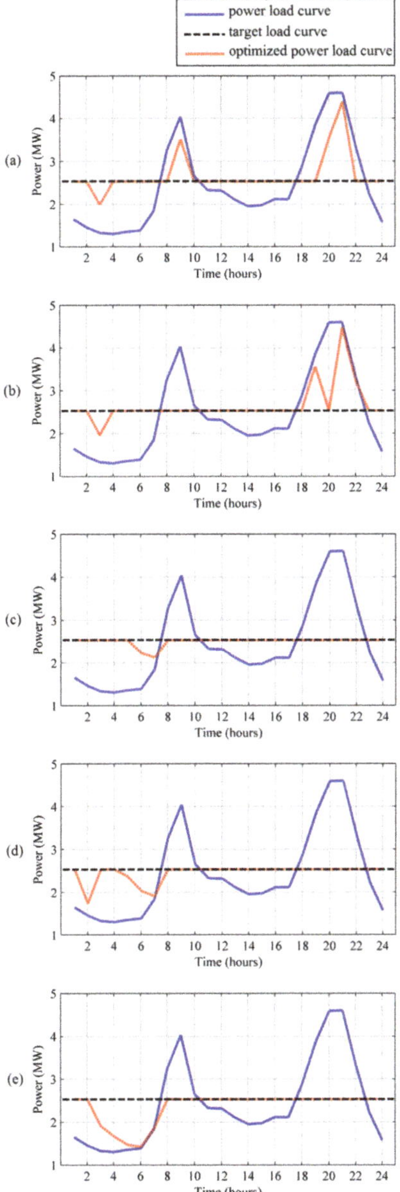

Figure 7. Optimized power load curves with 55% TLC_{pct} and $SOC_{i,ave}$ of (**a**) 40%, (**b**) 50%, (**c**) 60%, (**d**) 70% and (**e**) 80%.

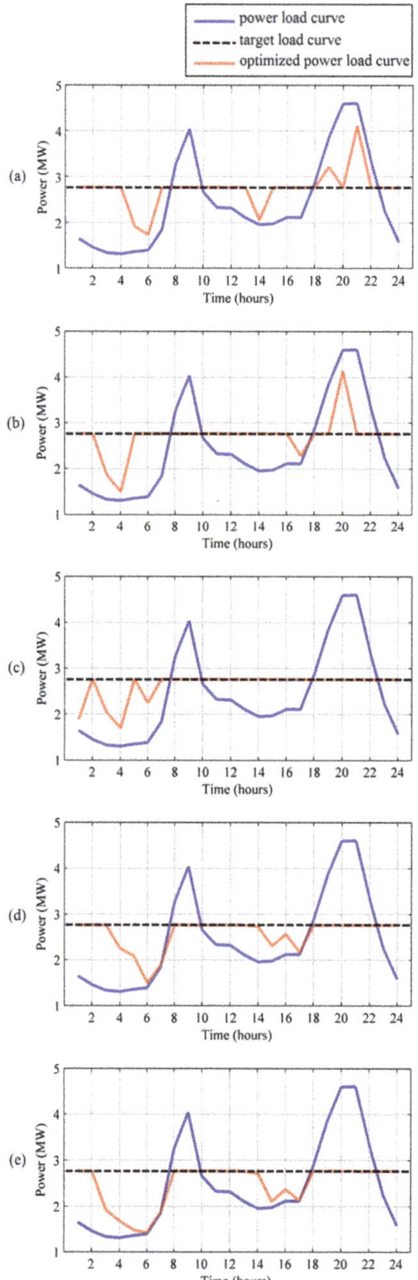

Figure 8. Optimized power load curves with 60% TLC_{pct} and $SOC_{i,ave}$ of (**a**) 40%, (**b**) 50%, (**c**) 60%, (**d**) 70% and (**e**) 80%.

The consideration of various values of $SOC_{i,ave}$ and TLC_{pct} had led to different optimization results. In certain scenarios, the optimized load curves were not completely flat due to the fact the required energy demand for peak load shaving and load levelling services exceeded the available energy capacity of the EV batteries. In other words, EV batteries cannot supply or absorb the required energy for the minimization of grid load variance. Therefore, an index denoted as Performance Index was introduced to allow better comparison among the optimized power load curves. The Performance Index indicated the percentage of successful operations of the peak load shaving, load levelling or combination of both to achieve the preset target load curve over a day. The Performance Index can range from zero to one, where greater value indicates higher successful rate. Table 1 presents the Performance Index of the proposed V2G optimization algorithm for all the scenarios depicted in Figures 6–8. Three sets of the Performance Index were calculated for peak load shaving, load levelling and both services together.

With reference to the Performance Index in Table 1, the capability of peak load shaving was enhanced while the capability of load levelling was reduced with the increase of $SOC_{i,ave}$. These situations were due to the increase in the available EV energy to be discharged for the peak load shaving, as well as the reduced need of EV batteries to receive charging from the power grid and thus limited the load levelling. Likewise, the occurrence of similar trends can be determined with the increase of the TLC_{pct}, where more peak load shaving service can be accomplished while fewer load levelling can be achieved using the proposed optimization algorithm. The reason was due to the required EV discharging energy for the peak load shaving was greatly reduced when the set point of TLC_{pct} was increased. However, the load levelling became more difficult to be achieved due to the significant increase of energy required to be charged into the EV batteries.

Table 1. Performance Index of the V2G optimization algorithm under various scenarios.

Region	Percentage of Target Load Curves (TLC_{pct})	Average Initial SOC of EV Batteries ($SOC_{i,ave}$)				
		40%	50%	60%	70%	80%
Peak load shaving	50%	0.429	0.496	0.724	0.870	0.872
	55%	0.571	0.590	1.000	1.000	1.000
	60%	0.751	0.809	1.000	1.000	1.000
Load levelling	50%	1.000	1.000	1.000	0.956	0.713
	55%	0.953	0.950	0.937	0.818	0.623
	60%	0.830	0.826	0.791	0.694	0.526
Overall	50%	0.667	0.706	0.839	0.905	0.806
	55%	0.785	0.792	0.965	0.898	0.789
	60%	0.805	0.821	0.858	0.793	0.679

Other than the two individual set of Performance Index values for peak load shaving and load levelling, the overall Performance Index shows the success percentage for both peak load shaving and load levelling. The shaded region of the overall Performance Index in Table 1 indicates the best TLC_{pct} to be selected with respect to the $SOC_{i,ave}$. Figures 9 and 10 illustrate the Performance Index under various scenarios and gives a clearer perception on the overall optimized scenarios. The peak point in Figure 10 shows the best optimized scenario with the overall Performance Index of 0.965, which dropped under the 60% of $SOC_{i,ave}$ and 55% of TLC_{pct} category. This has verified that the selection of target loading for the V2G optimization algorithm played a significant role in ensuring the performance of the algorithm.

A detailed analysis on the best scenario is further presented in Figure 11. With the implementation of the proposed V2G optimization algorithm, the achieved optimized power load curve for the category of 60% of $SOC_{i,ave}$ and 55% of TLC_{pct} is shown in Figure 11a. The optimized power load curve has almost met all the target loadings throughout the day, except during the periods from 05:00 to 08:00 o'clock.

Figure 11b presents a clearer insight to explain the operations during these periods. The bar graphs in the positive and negative regions in Figure 11b depict the maximum available capacity of the EV batteries and stored energy for the load levelling and peak load shaving, correspondingly. These bar graphs were obtained by considering all the constraint limits as specified in Section 3.2 and therefore, served as the limits for the optimization process. The curve shown in Figure 11b presents the optimized EV charging and discharging power required to achieve the optimized power load curve as shown in Figure 11a. During the periods from 05:00 to 08:00 o'clock, the optimized EV charging power had reached the maximum limit of the available capacity of EV batteries for the load levelling service. Consequently, the load levelling service was not completely achieved in the optimization process.

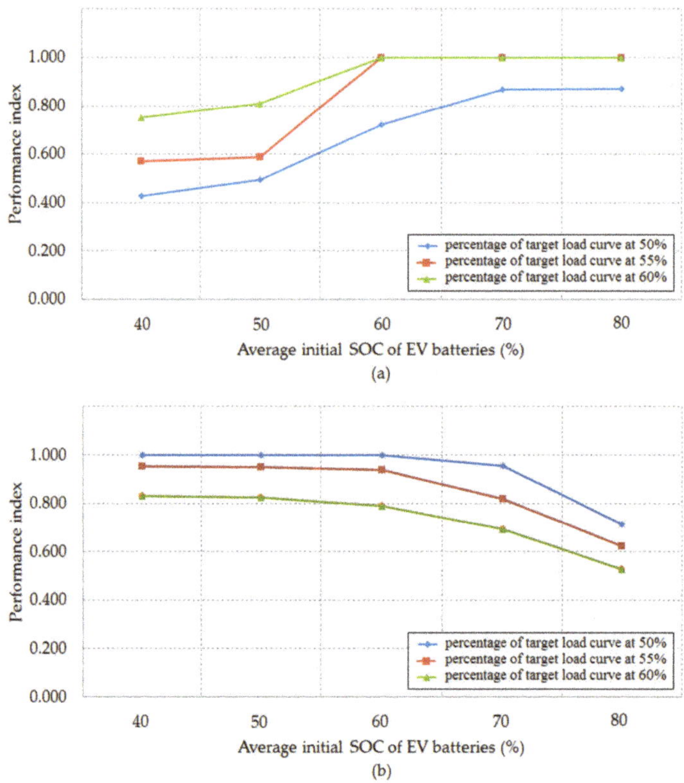

Figure 9. Performance Index under various scenarios: (a) peak load shaving and (b) load levelling.

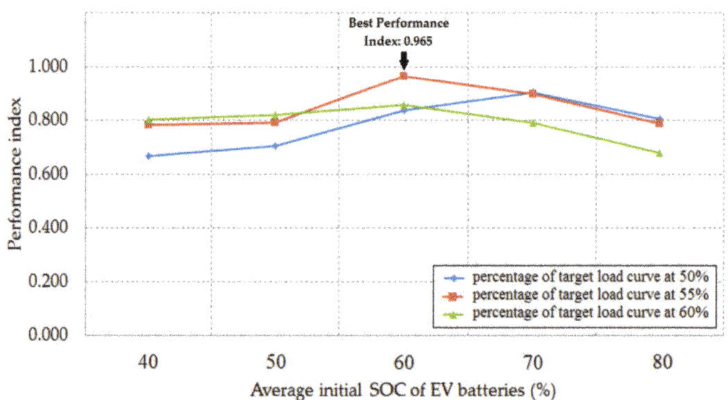

Figure 10. Overall Performance Index under various scenarios.

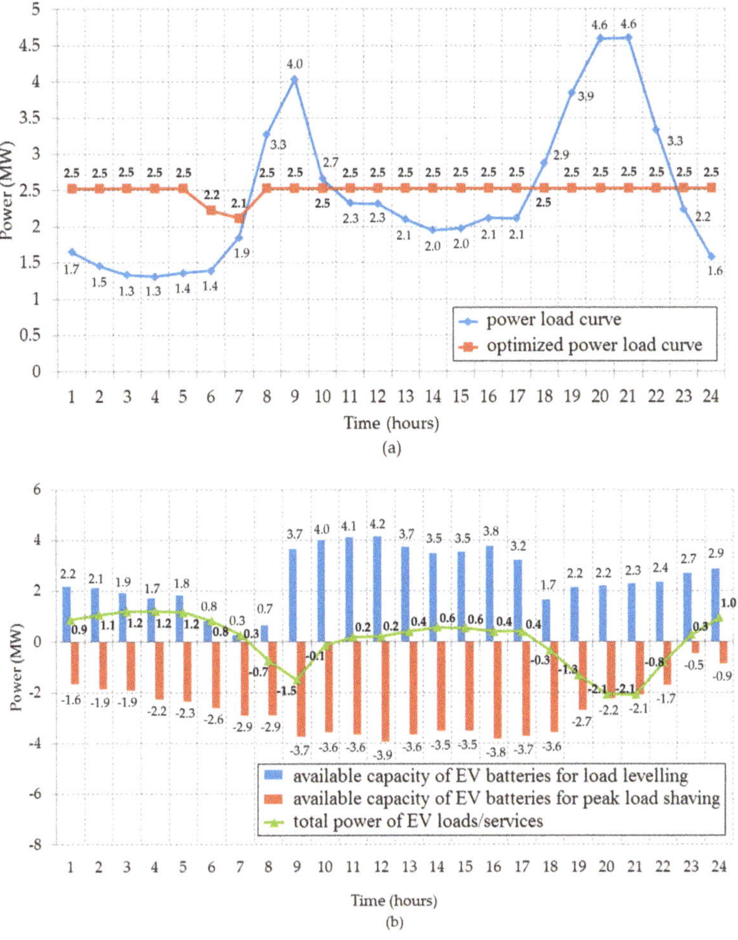

Figure 11. Detailed description of the best optimized scenario: (**a**) optimized power load curve (**b**) optimized EV charging and discharging power within constraint limits.

Figure 12 shows the SOC status of some random selected EVs under the best optimized scenario. During the load levelling periods, all the EVs experienced the charging process, except for the EVs with the SOC level higher than the SOC_{max} of 90%. Meanwhile, all the EVs discharged their battery energy during the peak shaving periods, due to having all the SOC levels of EV batteries higher than the SOC_{min} of 55%. Another observation from Figure 12 is that the EVs participating in the V2G optimization program had the tendency to reach to a similar SOC level at the end of the optimization process. These findings indicated that the proposed V2G optimization algorithm can flatten the power load curve with respect to the pre-determined SOC constraints.

The performance of the proposed V2G optimization algorithm was compared with other algorithms in the latest literature. A comparative parameter defined as the percentage improvement of peak and valley load difference, P_i was introduced to assess the algorithm performances that were conducted under different scopes and conditions. The formulation of P_i is expressed in (13):

$$P_i = \left| \frac{P_{d,after} - P_{d,before}}{P_{d,before}} \right| \times 100\% \tag{13}$$

where $P_{d,after}$ is the peak and valley load difference after V2G optimization and $P_{d,before}$ is the peak and valley load difference before V2G optimization.

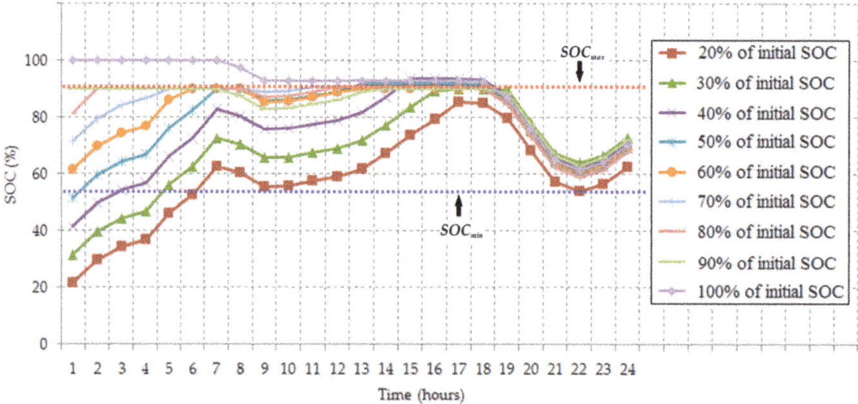

Figure 12. The SOC status of random selected EVs under the best optimized scenario.

Both $P_{d,after}$ and $P_{d,before}$ can be calculated using Equations (14) and (15), respectively:

$$P_{d,after} = P_{p,after} - P_{v,after} \tag{14}$$

$$P_{d,before} = P_{p,before} - P_{v,before} \tag{15}$$

where $P_{p,after}$ is the peak load value after V2G optimization, $P_{v,after}$ is the valley load value after V2G optimization, $P_{p,before}$ is the peak load value before V2G optimization, and $P_{v,before}$ is the valley load value before V2G optimization.

The V2G algorithm proposed in [55] was capable of performing peak shaving and valley filling control. The power grid loading before the implementation of V2G algorithm had $P_{p,before}$ of 1090 MW and $P_{v,before}$ of 855 MW. With the execution of V2G algorithm, the power grid loading presented $P_{p,after}$ of 1080 MW and $P_{v,after}$ of 950 MW. Hence, the computed P_i using (13) was 44.68%. A similar investigation was conducted for the researches in [56,57], where the computed P_i acquired was 46.89% and 36.84%, respectively. The analysis was also investigated on the best scenario (V2G optimization with 60% of $SOC_{i,ave}$ and 55% of TLC_{pct}) in this paper. The power grid loading before the implementation of the proposed V2G algorithm presented $P_{p,before}$ of 4.6 MW and $P_{v,before}$ of 1.3 MW. Meanwhile, the power grid loading after the employment of the V2G algorithm achieved $P_{p,after}$ of 2.5 MW and $P_{v,after}$ of 2.1 MW. Thus, P_i was acquired to be 87.88%. By comparing with the other literatures, the proposed algorithm in this paper shows a better performance in terms of the percentage improvement of peak and valley load difference. The details of the comparative analysis are presented in Table 2.

Table 2. Comparative analysis of the proposed algorithm with other approaches for grid load variance minimization.

Parameter	Ref. [55]	Ref. [56]	Ref. [57]	Proposed Algorithm
$P_{p,before}$ (MW)	1090	188	0.56	4.6
$P_{v,before}$ (MW)	855	98	0.37	1.3
$P_{d,before}$ (MW)	235	90	0.19	3.3
$P_{p,after}$ (MW)	1080	159.8	0.49	2.5
$P_{v,after}$ (MW)	950	112	0.37	2.1
$P_{d,after}$ (MW)	130	47.8	0.12	0.4
P_i (%)	44.68	46.89	36.84	87.88

5. Conclusions

This paper has presented the development of a V2G optimization algorithm with the objective of minimizing the grid load variance by utilizing the grid-connected EVs to provide the peak load shaving (V2G) and load levelling services (G2V). The proposed algorithm was examined under various scenarios of varying TLC_{pct} and $SOC_{i,ave}$ while complying with the crucial constraints, such as the grid power balance, initial SOC of EV battery, EV grid connection probability and EV grid connection duration. The simulation results had verified the effectiveness of the proposed algorithm in achieving the objective of the grid load variance minimization in all scenarios. A Performance Index was introduced in this paper to provide an excellent indication on the overall performance of the proposed V2G optimization algorithm. The best optimized scenario was achieved at 55% of TLC_{pct} and 60% of $SOC_{i,ave}$, with a Performance Index of 0.965. Moreover, all the participated EVs had complied with the preset SOC limits and tended to reach to a similar SOC level at the end of the V2G scheduling. This analysis aims to serve as a recommendation for the selection of TLC_{pct} based on the available historical data of $SOC_{i,ave}$. As a future work, the optimal energy scheduling of an integrated system of EVs and renewable energy resources will be investigated.

Author Contributions: All authors contributed equally for the decimation of the research article in current form.

Conflicts of Interest: The authors declare no conflict of interest.

References

1. Global EV Outlook 2016: Beyond One Million Electric Cars. Available online: https://www.iea.org/publications/freepublications/publication/Global_EV_Outlook_2016.pdf/ (accessed on 3 May 2017).
2. Hussain, S., Al-ammari, R.; Iqbal, A.; Jafar, M.; Padmanaban, S. Optimization Of Hybrid Renewable Energy System Using Iterative Filter Selection Approach. *IET Renew. Power Gener.* **2017**, *11*, 1440–1445. [CrossRef]
3. Tiwaria, R.; Babu, N.R.; Sanjeevikumar, P. A Review on GRID CODES—Reactive power management in power grids for Doubly-Fed Induction Generator in Wind Power application. In *Lecture Notes in Electrical Engineering*; Springer: Berlin, Germany, 2017.
4. Hajizadeh, A.; Norum, L.E.; Hassanzadehc, F.; Sanjeevikumar, P. An Intelligent Power Controller for Hybrid DC Micro Grid Power System. In *Lecture Notes in Electrical Engineering*; Springer: Berlin, Germany, 2018; in press.
5. Saraswathi, A.; Sanjeevikumar, P.; Sutha, S.; Blaabjerg, F.; Ertas, A.H.; Fedák, V. Analysis of Enhancement in Available Power Transfer Capacity by STATCOM Integrated SMES by Numerical Simulation Studies. *Eng. Sci.Technol. Int. J.* **2016**, *19*, 671–675.
6. Un-Noor, F.; Padmanaban, S.; Mihet-Popa, L.; Mollah, M.N.; Hossain, E. A comprehensive study of key electric vehicle (EV) components, technologies, challenges, impacts, and future direction of development. *Energies* **2017**, *10*, 1217. [CrossRef]
7. Vavilapalli, S.; Umashankar, S.; Sanjeevikumar, P.; Ramachandramurthy, V.K. Design and Real-Time Simulation of an AC Voltage Regulator based Battery Charger for Large-Scale PV-Grid Energy Storage Systems. *IEEE Access J.* **2017**. [CrossRef]

8. Sridhar, V.; Sanjeevikumar, P.; Ramesh, V.; Mihet-Popa, L. Study and Analysis of Intelligent Microgrid Energy Management Solution with Distributed Energy Sources. *Energies* **2017**, *10*, 1419.

9. Bharatiraja, C.; Sanjeevikumar, P.; Siano, P.; Ramesh, K.; Raghu, S. Real Time Foresting of EV Charging Station Scheduling for Smart Energy System. *Energies* **2017**, *10*, 337.

10. Al-Nussairif, M.; Bayindir, R.; Sanjeevikumar, P.; Mihet-Popa, L.; Siano, P. Constant Power Loads (CPL) with Microgrids: Problem Definition, Stability Analysis and Compensation Techniques. *Energies* **2017**, *10*, 1656.

11. Hossain, E.; Perez, R.; Sanjeevikumar, P.; Siano, P. Investigation on Development of Sliding Mode Controller for Constant Power Loads in Microgrids. *Energies* **2017**, *10*, 1086. [CrossRef]

12. Ali, A.; Sanjeevikumar, P.; Twala, B.; Marwala, T. Electric Power Grids Distribution Generation System For Optimal Location and Sizing—An Case Study Investigation by Various Optimization Algorithms. *Energies* **2017**, *10*, 960.

13. Vavilapalli, S.; Sanjeevikumar, P.; Umashankar, S.; Mihet-Popa, L. Power Balancing Control for Grid Energy Storage System in PV Applications—Real Time Digital Simulation Implementation. *Energies* **2017**, *10*, 928.

14. Swaminathan, G.; Ramesh, V.; Umashankar, S.; Sanjeevikumar, P. Investigations of Microgrid Stability and Optimum Power sharing using Robust Control of grid tie PV Inverter. In *Lecture Notes in Electrical Engineering*; Springer: Berlin, Germany, 2018; in press.

15. Tamvada, K.; Umashankar, S.; Sanjeevikumar, P. Impact of Power Quality Disturbances on Grid Connected Double Fed Induction Generator. In *Lecture Notes in Electrical Engineering*; Springer: Berlin, Germany, 2018; in press.

16. Yong, J.Y.; Ramachandaramurthy, V.K.; Tan, K.M. A review on the state-of-the-art technologies of electric vehicle, its impacts and prospects. *Renew. Sust. Energy Rev.* **2015**, *49*, 365–385. [CrossRef]

17. Mwasilu, F.; Justo, J.J.; Kim, E.K.; Do, T.D.; Jung, J.W. Electric vehicles and smart grid interaction: A review on vehicle to grid and renewable energy sources integration. *Renew. Sustain. Energy Rev.* **2014**, *34*, 501–516. [CrossRef]

18. Gary, M.K.; Morsi, W.G. Power quality assessment in distribution systems embedded with plug-in hybrid and battery electric vehicles. *IEEE Trans. Power Syst.* **2015**, *30*, 663–671. [CrossRef]

19. Dharmakeerthi, C.H.; Mithulananthan, N.; Saha, T.K. A comprehensive planning framework for electric vehicle charging infrastructure deployment in the power grid with enhanced voltage stability. *Int. Trans. Electr. Energy* **2015**, *25*, 1022–1040. [CrossRef]

20. Kempton, W.; Letendre, S.E. Electric vehicles as a new power source for electric utilities. *Transp. Res. Part D Transp. Environ.* **1997**, *2*, 157–175. [CrossRef]

21. Krishna, M.; Daya, F.J.L.; Sanjeevikumar, P.; Mihet-Popa, L. Real-time Analysis of a Modified State Observer for Sensorless Induction Motor Drive used in Electric Vehicle Applications. *Energies* **2017**, *10*, 1077. [CrossRef]

22. Lund, H.; Kempton, W. Integration of renewable energy into the transport and electricity. *Energy Policy* **2008**, *36*, 3578–3587. [CrossRef]

23. Apostolaki-Iosifidou, E.; Codani, P.; Kempton, W. Measurement of power loss during electric vehicle charging and discharging. *Energy* **2017**, *127*, 730–742. [CrossRef]

24. Shinzaki, S.; Sadano, H.; Maruyama, Y.; Kempton, W. Deployment of vehicle-to-grid technology and related issues. *SAE Tech. Pap.* **2015**. [CrossRef]

25. Tomić, J.; Kempton, W. Using fleets of electric-drive vehicles for grid support. *J. Power Sources* **2007**, *168*, 459–468. [CrossRef]

26. Kempton, W.; Tomić, J. Vehicle-to-grid power fundamentals: Calculating capacity and net revenue. *J. Power Sources* **2005**, *144*, 268–279. [CrossRef]

27. Hidrue, M.K.; Parsons, G.R.; Kempton, W.; Gardner, M.P. Willingness to pay for electric vehicles and their attributes. *Resour. Energy Econ.* **2011**, *33*, 686–705. [CrossRef]

28. Kempton, W.; Tomić, J. Vehicle-to-grid power implementation: From stabilizing the grid to supporting large-scale renewable energy. *J. Power Sources* **2005**, *144*, 280–294. [CrossRef]

29. Liu, C.; Chau, K.T.; Wu, D.; Gao, S. Opportunities and challenges of vehicle-to-home, vehicle-to-vehicle, and vehicle-to-grid technologies. *Proc. IEEE* **2013**, *101*, 2409–2427. [CrossRef]

30. Ghofrani, M.; Arabali, A.; Etezadi-Amoli, M.; Fadali, M.S. Smart scheduling and cost-benefit analysis of grid-enabled electric vehicles for wind power integration. *IEEE Trans. Smart Grid* **2014**, *5*, 2306–2313. [CrossRef]

31. Manbachi, M.; Farhangi, H.; Palizban, A.; Arzanpour, S. A novel volt-VAR optimization engine for smart distribution networks utilizing vehicle to grid dispatch. *Int. J. Electr. Power Energy Syst.* **2016**, *74*, 238–251. [CrossRef]

32. Tan, K.M.; Ramachandaramurthy, V.K.; Yong, J.Y. Integration of electric vehicles in smart grid: A review on vehicle to grid technologies and optimization techniques. *Renew. Sust. Energy Rev.* **2016**, *53*, 720–732. [CrossRef]

33. Rao, S.; Berthold, F.; Pandurangavittal, K.; Blunier, B.; Bouquain, D.; Williamson, S.; Miraoui, A. Plug-in hybrid electric vehicle energy system using home-to-vehicle and vehicle-to-home: Optimization of power converter operation. In Proceedings of the IEEE Transportation Electrification Conference and Expo, Detroit, MI, USA, 16–19 June 2013; pp. 1–6.

34. Khederzadeh, M.; Maleki, H. Coordinating storage devices, distributed energy sources, responsive loads and electric vehicles for microgrid autonomous operation. *Int. Trans. Electr. Energy* **2015**, *25*, 2482–2498. [CrossRef]

35. Mejdoubi, A.E.; Oukaour, A.; Chaoui, H.; Gualous, H.; Sabor, J.; Slamani, Y. State-of-charge and state-of-health lithium-ion batteries' diagnosis according to surface temperature variation. *IEEE Trans. Ind. Electron.* **2015**, *63*, 2391–2402. [CrossRef]

36. Sortomme, E.; El-Sharkawi, M.A. Optimal combined bidding of vehicle-to-grid ancillary services. *IEEE Trans. Smart Grid* **2012**, *3*, 70–79. [CrossRef]

37. Shafie-Khah, M.; Heydarian-Forushani, E.; Osório, G.J.; Jamshid Aghaei, F.A.S.G.; Barani, M.; Catalão, J.P.S. Optimal behavior of electric vehicle parking lots as demand response aggregation agents. *IEEE Trans. Smart Grid* **2016**, *7*, 2654–2665. [CrossRef]

38. Nguyen, H.N.T.; Zhang, C.; Mahmud, M.A. Optimal coordination of G2V and V2G to support power grids with high penetration of renewable energy. *IEEE Trans. Transp. Electrification* **2015**, *1*, 188–195. [CrossRef]

39. Xie, S.; Zhong, W.; Xie, K.; Yu, R.; Zhang, Y. Fair energy scheduling for vehicle-to-grid networks using adaptive dynamic programming. *IEEE Trans. Neural Netw. Learn. Syst.* **2016**, *27*, 1697–1707. [CrossRef] [PubMed]

40. Amirioun, M.H.; Kazemi, A. A new model based on optimal scheduling of combined energy exchange modes for aggregation of electric vehicles in a residential complex. *Energy* **2014**, *69*, 186–198. [CrossRef]

41. Saber, A.Y.; Venayagamoorthy, G.K. Intelligent unit commitment with vehicle-to-grid—A cost-emission optimization. *J. Power Sources* **2010**, *9*, 898–911. [CrossRef]

42. Nafisi, H.; Abyaneh, H.A.; Abedi, M. Energy loss minimization using PHEVs as distributed active and reactive power resources: A convex quadratic local optimal solution. *Int. Trans. Electr. Energy* **2016**, *26*, 1287–1302. [CrossRef]

43. Nunes, P.; Farias, T.; Brito, M.C. Enabling solar electricity with electric vehicles smart charging. *Energy* **2015**, *87*, 10–20. [CrossRef]

44. Lee, W.; Xiang, L.; Schober, R.; Wong, V.W.S. Electric vehicle charging stations with renewable power generators: A game theoretical analysis. *IEEE Trans. Smart Grid* **2015**, *6*, 608–617. [CrossRef]

45. Jin, C.; Sheng, X.; Ghosh, P. Energy efficient algorithms for electric vehicle charging with intermittent renewable energy sources. In Proceedings of the IEEE Power and Energy Society General Meeting, Vancouver, BC, Canada, 21–25 July 2013; pp. 1–5.

46. Fazelpour, F.; Vafaeipour, M.; Rahbari, O.; Rosen, M.A. Intelligent optimization to integrate a plug-in hybrid electric vehicle smart parking lot with renewable energy resources and enhance grid characteristics. *Energy Convers. Manag.* **2014**, *77*, 250–261. [CrossRef]

47. Nan, Z.; Nian, L.; Jianhua, Z.; Jinyong, L. Multi-objective optimal sizing for battery storage of PV-based microgrid with demand response. *Energies* **2016**, *9*, 591. [CrossRef]

48. Gerards, M.E.T.; Hunrink, J.L. Robust peak-shaving for a neighborhood with electric vehicles. *Energies* **2016**, *9*, 594. [CrossRef]

49. López, M.A.; de la Torre, S.; Martín, S.; Aguado, J.A. Demand-side management in smart grid operation considering electric vehicles load shifting and vehicle-to-grid support. *Electr. Power Energy Syst.* **2015**, *64*, 689–698. [CrossRef]

50. Sortomme, E.; Hindi, M.M.; MacPherson, S.D.J.; Venkata, S.S. Coordinated charging of plug-in hybrid electric vehicles to minimize distribution system losses. *IEEE Trans. Smart Grid* **2011**, *2*, 198–205. [CrossRef]

51. Sheikhi, A.; Bahrami, S.; Ranjbar, A.M.; Oraee, H. Strategic charging method for plugged in hybrid electric vehicles in smart grids; a game theoretic approach. *Int. J. Electr. Power Energy Syst.* **2013**, *53*, 499–506. [CrossRef]

52. Jian, L.; Zheng, Y.; Xiao, X.; Chan, C.C. Optimal scheduling for vehicle-to-grid operation with stochastic connection of plug-in electric vehicles to smart grid. *Appl. Energy* **2015**, *146*, 150–161. [CrossRef]

53. Jian, L.; Zhu, X.; Shao, Z.; Niu, S.; Chan, C.C. A scenario of vehicle-to-grid implementation and its double-layer optimal charging strategy for minimizing load variance within regional smart grids. *Energy Convers. Manag.* **2014**, *78*, 508–517. [CrossRef]

54. Kordkheili, R.A.; Pourmousavi, S.A.; Savaghebi, M.; Guerrero, J.M.; Nehrir, M.H. Assessing the potential of plug-in electric vehicles in active distribution networks. *Energies* **2016**, *9*, 34. [CrossRef]

55. Wang, Z.; Wang, S. Grid power peak shaving and valley filling using vehicle-to-grid systems. *IEEE Trans. Power Deliv.* **2013**, *28*, 1822–1829. [CrossRef]

56. Shinde, P.; Swarup, K.S. Optimal electric vehicle charging schedule for demand side management. In Proceedings of the International Conference on Sustainable Green Building and Communities (SGBC), Chennai, India, 18–20 December 2016; pp. 1–6.

57. Turker, H.; Hably, A.; Bacha, S. Housing peak shaving algorithm (HPSA) with plug-in hybrid electric vehicles (PHEVs): Vehicle-to-home (V2H) and vehicle-to-grid (V2G) concepts. In Proceedings of the Fourth International Conference on Power Engineering, Energy and Electrical Drives (POWERENG), Istanbul, Turkey, 13–17 May 2013; pp. 1–7.

58. Alam, M.J.E.; Muttaqi, K.M.; Sutanto, D. A controllable local peak-shaving strategy for effective utilization of PEV battery capacity for distribution network support. *IEEE Trans. Ind. Appl.* **2015**, *51*, 2030–2037. [CrossRef]

59. Sridhar, V.; Umashankar, S.; Sanjeevikumar, P. Decoupled Active and Reactive Power Control of Cascaded H-Bridge PV-Inverter for Grid Connected Applications. In *Lecture Notes in Electrical Engineering*; Springer: Berlin, Germany, 2018; in press.

60. Millner, A. Modeling lithium ion battery degradation in electric vehicles. In Proceedings of the IEEE Conference on Innovative Technologies for an Efficient and Reliable Electricity Supply, Waltham, MA, USA, 27–29 September 2010; pp. 349–356.

61. Camacho, O.M.F.; Nørgård, P.B.; Rao, N.; Mihet-Popa, L. Electrical vehicle batteries testing in a distribution network using sustainable energy. *IEEE Trans. Smart Grid* **2014**, *5*, 1033–1042. [CrossRef]

62. Jerin, A.R.A.; Palanisamy, K.; Sanjeevikumar, P.; Umashankar, S.; Ramachandramurthy, V.K. Improved Fault Ride Through Capability in DFIG based Wind Turbines using Dynamic Voltage Restorer with Combined Feed-Forward and Feed-Back Control. *IEEE Access J.* **2017**. [CrossRef]

63. Tiwari, R.; Babu, N.R.; Arunkrishna, R.; Sanjeevikumar, P. Comparison between PI controller and Fuzzy logic based control strategies for harmonic reduction in Grid integrated Wind energy conversion system. In *Lecture Notes in Electrical Engineering*; Springer: Berlin, Germany, 2018; in press.

64. Tiwari, R.; Ramesh Babu, N.; Sanjeevikumar, P.; Martirano, L. Coordinated DTC and VOC Control for PMSG based Grid Connected Wind Energy Conversion System. In Proceedings of the 2017 IEEE International Conference on Environment and Electrical Engineering and 2017 IEEE Industrial and Commercial Power Systems Europe (EEEIC/I&CPS Europe), Milan, Italy, 6–9 June 2017.

65. Chinthamalla, R.; Sanjeevikumar, P.; Karampuria, R.; Jain, S.; Ertas, A.H.; Fedak, V. A Solar PV Water Pumping Solution Using a Three-Level Cascaded Inverter Connected Induction Motor Drive. *Eng. Sci. Technol. Int. J.* **2016**, *19*, 1731–1741.

66. Jain, S.; Ramulu, C.; Sanjeevikumar, P.; Ojo, O.; Ertas, A.H. Dual MPPT Algorithm for Dual PV Source Fed Open-End Winding Induction Motor Drive for Pumping Application. *Eng. Sci. Technol. Int. J.* **2016**, *19*, 1771–1780. [CrossRef]

67. Swaminathan, G.; Ramesh, V.; Umashankar, S.; Sanjeevikumar, P. Fuzzy Based Micro Grid Energy Management System using Interleaved Boost Converter and Three Level NPC Inverter with Improved Grid Voltage Quality. In *Lecture Notes in Electrical Engineering*; Springer: Berlin, Germany, 2018; in press.

68. Awasthi, A.; Karthikeyan, V.; Rajasekar, S.; Sanjeevikumar, P.; Siano, P.; Ertas, A.H. Dual Mode Control of Inverter to Integrate Solar-Wind Hybrid fed DC-Grid with Distributed AC grid. In Proceedings of the 16th IEEE International Conference on Environment and Electrical Engineering (IEEE-EEEIC'16), Florence, Italy, 7–10 June 2016.
69. Camacho, O.M.F.; Mihet-Popa, L. Fast Charging and Smart Charging Tests for Electric Vehicles Batteries using Renewable Energy. *Oil Gas Sci. Technol.* **2016**, *71*, 3.

Article

Hardware Implementation and a New Adaptation in the Winding Scheme of Standard Three Phase Induction Machine to Utilize for Multifunctional Operation: A New Multifunctional Induction Machine

Mahajan Sagar Bhaskar [1], Sanjeevikumar Padmanaban [1,*], Sonali A. Sabnis [2], Lucian Mihet-Popa [3], Frede Blaabjerg [4] and Vigna K. Ramachandaramurthy [5]

[1] Department of Electrical and Electronics Engineering, University of Johannesburg, P.O. Box 524, Auckland Park, Johannesburg 2006, South Africa; sagar25.mahajan@gmail.com
[2] Department of Electrical and Electronics Engineering, Marathwada Institute of Technology, Aurangabad 431001, India; sabnis.sonali@gmail.com
[3] Faculty of Engineering, Østfold University College, Kobberslagerstredet 5, 1671 Kråkeroy-Fredrikstad, Norway; lucian.mihet@hiof.no
[4] Centre for Reliable Power Electronics (CORPE), Department of Energy Technology, Aalborg University, Aalborg 9000, Denmark; fbl@et.aau.dk
[5] Institute of Power Engineering, Department of Electrical Power Engineering, Universiti Tenaga Nasional, 43000 Kajang, Selangor, Malaysia; vigna@uniten.edu.my
* Correspondence: sanjeevi_12@yahoo.co.in; Tel.: +27-79-219-9845

Received: 23 July 2017; Accepted: 23 October 2017; Published: 1 November 2017

Abstract: In this article a new distinct winding scheme is articulated to utilize three phase induction machines for multifunctional operation. Because of their rugged construction and reduced maintenance induction machines are very popular and well-accepted for agricultural as well as industrial purposes. The proposed winding scheme is used in a three phase induction machine to utilize the machine for multifunctional operation. It can be used as a three-phase induction motor, welding transformer and phase converter. The proposed machine design also works as a single phase induction motor at the same time it works as a three-phase to single phase converter. This new design does not need any kind of special arrangement and can be constructed with small modifications to any standard three-phase induction motor. This modified induction machine is thoroughly tested to determine its efficiency and other parameters and also hardware implementation results are provided in the article, which validate the design and construction.

Keywords: three phase; induction machine; winding scheme; multifunctional; welding transformer; phase converter

---

1. Introduction

Alternating current (AC) systems are employed in most countries due to their benefits like ease in high voltage (HV) transmission and distribution (T and D) of electricity. Because of the AC at the input side, power modulators are required to convert it into direct current (DC) or leave it as alternating current (AC) to operate DC and AC machine-based drives, respectively [1–4]. The overwhelming disadvantage of DC brushed machines (both series wound and other types) are the brushes themselves. The mechanical nature of this rotary switching of high power produces considerable electrical arcing and also mechanical wear of the carbon brushes and the copper segments of the commutator. Brushes bear mechanical force in the form of friction which causes wear of the brushes and increases the maintenance of the machine and additionally the carbon dust produced from wear is conductive and may cause unintended contacts between high voltage terminals. Induction AC motors are employed to avoid these types of problems [5–8]. They are generally more efficient, offering virtually maintenance-free operation, they are sealed or splash resistant, and with modern power electronic (PE) devices [9–11] it is possible to operate the machine at desired varying speeds. Also, these motors are available for three-phase or single-phase supplies and also available in multiphase Nowadays open winding induction machines are more popular for drive applications [2,9–11]. Even though the induction machine is less costly, most of its practical use is for limited types of load at constant speeds although multiphase concepts and advanced control schemes have recently also been proposed [12–26]. In [27], an integrated converter for a 48 V micro/mid hybrid electric vehicle (HVE) was proposed. The integrated converter is a compact DC-DC converter specially designed for the belt driven starter generator in a 48 V HVE and with less electromagnetic interference (EMI) problems. In [28], a multifunctional induction motor (IM) using double winding motor concepts and modified stator windings for motoring as well as for welding operations was proposed, but this IM has a drawback of high winding stress. In [29], high power transformer faults were diagnosed by using the networking vibration mode based on integrated signal processing. The diagnostic system performance is observed by vibration measurements on the high power three phase transformer used in railways. This paper also deals with railway interlocking signaling installations with Uninterrupted Power Supply UPS capability for safety applications and predictive diagnostics. In [30], fault diagnostics of electric motors by non-invasive methods utilizing acoustic signals generated by the motor are proposed. The proposed approached reduces the cost and faults in motors.

In this paper, a new adaptation of the winding scheme of a standard three-phase induction machine (IM) is proposed to utilize the motor for multifunctional operation. The proposed multifunctional induction machine is able to run on single-phase supply at the same time it can also be used as a phase converter. When it runs as a three-phase induction motor it also can be cast to work as a welding transformer. Both single-phase and three-phase windings are placed in the same slots to allow the rotor to move on both types of supply. While performing the single phase operation, the capacitor can be used to produce a starting torque then after acceleration, the starting winding can be disconnected by a simple arrangement.

When three-phase supply is given to the motor, an electromotive force (EMF) is induced in the single phase winding as it is also placed in the same slots. Start and end connections of these coils are taken out which when connected in parallel step downs the voltage value, thereby producing high currents ideal for welding purposes. On the other hand, if the motor is operated on a single-phase supply, the voltage across the open circuit phases of the three phase winding terminals is stepped up. The operation mode of proposed induction motor is discussed in the next section.

2. Multifunctional Operation of the New Induction Machine Based on a New Winding Scheme

This proposed induction machine (IM) has all the advantages of AC or induction machines and with some modifications, in it can be made multi-functional. The proposed design can work as:

1. A three-phase induction motor,
2. A single phase induction motor (capacitor start),
3. A rotary phase converter and
4. A welding transformer.

2.1. Working as a Three-Phase Induction Motor

The proposed induction machine (IM) is derived from the standard three-phase induction motor which functions on the principle of rotating magnetic fields. When a three-phase supply is applied to the stator windings separated electrically by 120° in space, a rotating magnetic field is established in the stator winding. This voltage is induced in the rotor conductors to cause motion. The proposed induction machine (IM) has three phase winding in the stator and a cage rotor which makes the machine operate as a three-phase induction motor. The winding connection is similar to that of a conventional three-phase induction machine.

2.2. Working as Single-Phase Induction Motor

In operation of a single phase induction motor starting and running windings are placed in the same slot as the three-phase winding slots. The starting or auxiliary winding and running winding are separated electrically by 90° in space. An external capacitor is provided for starting the motor. The single-phase motor connection diagram is shown in Figure 1.

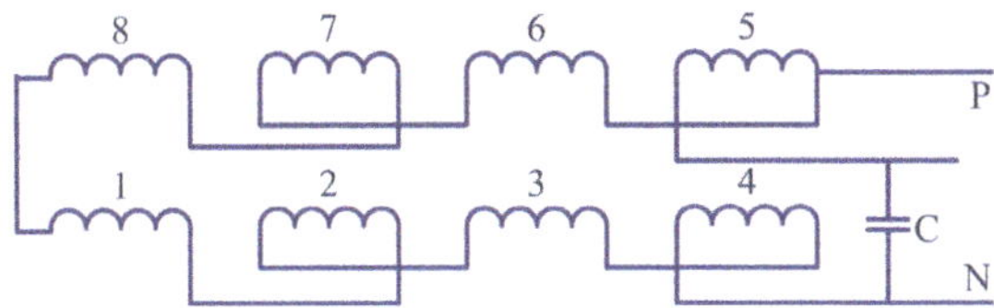

Figure 1. Proposed induction motor winding arrangement for working as a single-phase motor.

2.3. Working as a Welding Transformer

Figure 2 describes the working diagram of the machine configured to run as a welding transformer. In welding transformers high currents are produced at the secondary side. In the welding transformer operation, the running windings of the single-phase motor are taken into consideration to get a voltage step down. Four running windings are then connected in parallel to meet the high current requirements sufficient for welding.

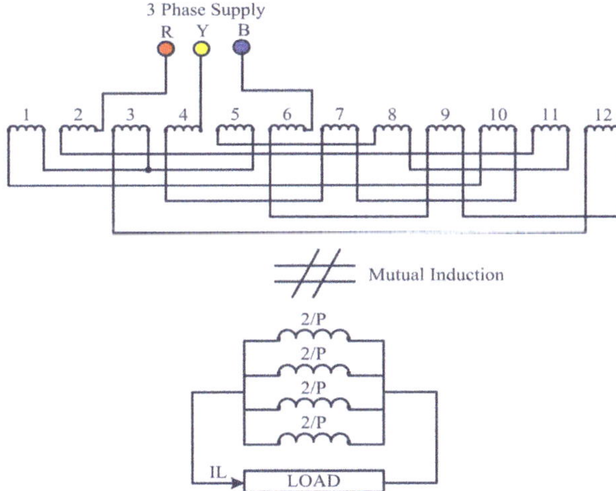

Figure 2. Proposed induction motor winding arrangement working as a welding transformer.

2.4. Working as a Phase Converter

If there is only single phase supply then the motor cannot operate as a three-phase machine or if there is only three-phase supply then the motor cannot operate a sinlge-phase machine. Then phase conversion is possible from 1-phase to 3-phase and 3-phase to 1-phase and vice-versa. Phase converters are applied where it is difficult to erect a three-phase loading system because of arrangement requirements and cost. The proposed induction motor consists of three-phase as well as single-phase windings in its stator. When the machine is supplied at the stator EMF also gets induced in single-phase windings across which a single-phase voltage occurs. Figure 3 shows an equivalent diagram of the proposed IM winding design operating as a phase converter.

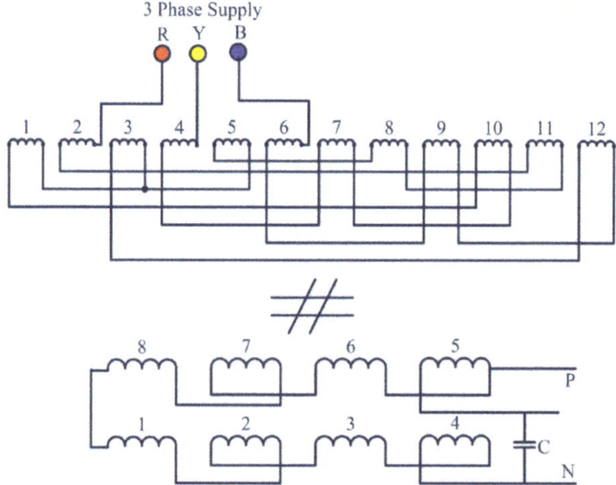

Figure 3. Proposed induction motor winding arrangement working as a phase converter.

3. Design of the Proposed Multifunctional Induction Machine Based on a New Winding Scheme

To design the proposed multifunctional induction machine, the original windings of the conventional induction motor (IM) are divided into three distinct windings with four groups on the basis of the number of turns. The first winding is with the same gauge wire and half of the original number of turns. Hence, this is a winding of a 3-phase induction motor and as the number of turns is half the motor has only half the capacity. The other windings are used for the purpose of welding and act as a tap of the welding transformer. As the welding application requires high current rating, a triple layer winding is used to improve the current rating. The same motor is used for a 1-phase induction motor. Hence, these windings are also used as starting and running windings of the 1-phase induction motor.

The winding factor initially assumed is 0.955, which is the value of the winding factor for infinitely distributed winding with full pitched coils. Flux per pole and stator winding per phase are calculated by Equations (1) and (2) respectively, where θ_m is flux per pole, Vs is stator voltage per phase, B_{av} is the average value of fundamental flux density, τ is pole pitch, d is the inner diameter of stator, l is length of the induction motor, P is number of poles, V_S is stator voltage per phase, f is supply frequency in Hz, T_S is the number of turns per phase in stator and K_{WS} is stator winding factor.

The current density in the stator winding is usually between 3 to 5 A/mm^2. For a lower current value round conductors would be more convenient to use while for higher current bar or stripped conductor with 1 cm^2 in cross sectional area should be adopted [22]. The area of each stator winding (A) is calculated by Equation (3) where, I_S is stator current per phase and ∂_S is the current density in the stator. The approximate area of each slot is equal to ratio of copper area per slot to space factor.

The space factor ordinarily varies from 0.25 to 0.4. High voltage machines have a lower space factor owing to the large thickness of insulation. After obtaining the area of the slot, the dimensions of the slot should be adjusted. The width of the slot should be within limits to maintain the required width of the teeth. The width of the slot should be so adjusted such that the mean flux density (Wb/m^2)

in the teeth lies in between 1.3 to 1.7. Table 1 shows the ratings and the parameters of the IM designed as a multifunctional motor:

$$\theta_m = B_{av}\tau l = B_{av}\frac{\pi dl}{P}$$ (1)

$$V_S = 4.44 f \theta_m T_S K_{WS}$$ (2)

$$A = I_S \partial_S$$ (3)

Table 1. Designed ratings and parameters of the proposed multifunctional induction machine (IM).

Parameter	Value
Rated voltage	400 V (L-L)
Power Supply	Three Phase
Frequency	50 Hz
Phase Voltage	230 V
Winding	Delta Connected
Power rating	2.5 HP = 1865 W
Power Factor	0.8 (lagging)
Rated current	3.05 A

3.1. Three-Phase Winding Design

With the help of measuring instruments the slot pitch is calculated and the type of winding is decided from it. It will be easy to do coil groups of both single and double layer windings. The film papers with the proper size are first inserted in order to protect the conductor insulation, which may be damaged by rubbing stator. Again the coils are also covered on the upper portion.

Then the free space will be added in order to do the single-phase connection. In this way the three- phase winding can be designed. For three-phase winding there are 17 turns per coil and hence for 36 slots, 12 slots per phase, hence 12 coils per phase and total number of 3-phase winding turns = 17 turns × 36 slots = 612 turns. For good conductivity super enameled wire is used ("super" is marketing name given by the enameled wire manufacturer). It is of class B type insulation. For designing the motor, there will be double layer winding with consideration of the stator winding parameters given in Table 2. Figure 4a,b depict the three-phase winding arrangement of the motor in slots and the actual placement of the three-phase windings on the stator, respectively. A three-phase 2.5 HP IM current is calculated by Equation (4).

$$\left.\begin{array}{c} 2.5hp = 1865 Watt \\ P = \sqrt{3}VI_{ph}\cos\phi \\ 1865 = \sqrt{3} \times 440 \times I_{ph} \times 0.8 \\ I_{ph} = 3.05 A \end{array}\right\}$$ (4)

Table 2. Design values of the stator winding of the proposed multifunctional induction machine (IM).

Parameter	Value
No. of poles	4
No. of slots	36
No. of slot/pole	9
No. of slots/pole/phase	3
Slot pitch	1 to 8
Coil pitch	1 to 8
Turns/slot	17
Total turns	612 (17 × 36)

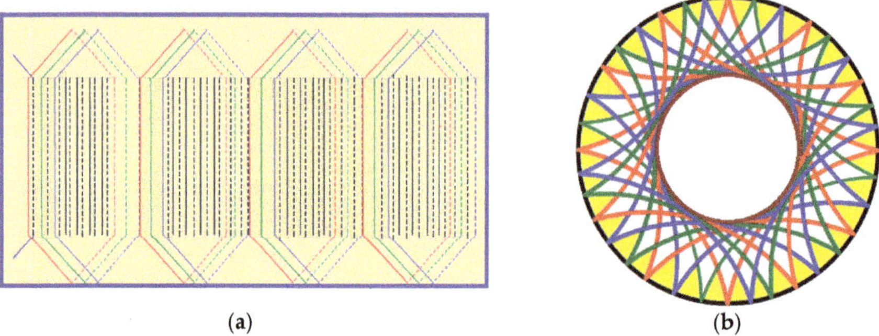

Figure 4. (**a**) Arrangement of the three-phase windings of the motor in slots. (**b**) Placement of the three-phase windings on the stator.

3.2. Single-Phase Winding Design

The rated current and voltage for the single-phase motor is 10.15 A, 230 V while the power and frequency remain same. Here, design of the single-phase winding is very important as the same single-phase winding is used for welding purposes, so during the welding process the current carrying capability of the winding should withstand the current demands. At that time the machine should withstand high currents without any problems like over-heating, over-current/voltage or insulation damage. For this reason conductors having a current capacity double than the rated current are employed. The winding arrangement of the proposed motor is done with the placement of 90° electrical degree separation between the starting and running windings. Four coil groups of starting windings and four coil groups of running windings are designed. There are two starting and running windings placed electrically at 90° with the capacitor star arrangement in the single-phase induction motor. There are 17 turns per coil, meaning that the double layer turns per coil is equal to 34. Running winding having three layers is selected, so the total number of running winding turns is equal to 408 turns. Figure 5 shows the actual placement of the single phase windings on the stator.

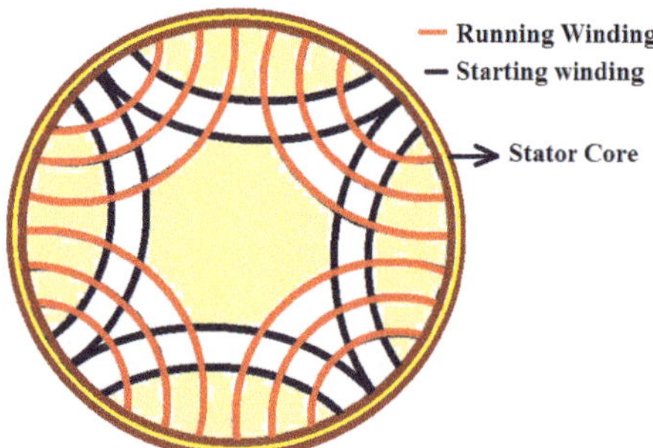

Figure 5. Placement of single phase windings on the stator.

A single-phase 2.5HP IM current is calculated by Equation (5):

$$\left.\begin{array}{c} P = VI_{ph}\cos\phi \\ 1865 = 230 \times I_{ph} \times 0.8 \\ I_{ph} = 10.15A \end{array}\right\} \tag{5}$$

3.3. Welding Transformer Design

For this application the three-phase winding functions as the primary of the transformer and the running winding of the single-phase motor works as the secondary of the welding transformer. Hence, the welding transformer application is only related to running windings and it is also possible to use the starting winding for higher current rating welding applications. The transformation ratio of the welding transformer is given in Equation (6) where N_1, N_2 are the turns of the primary and secondary windings and I_1, I_2 are primary and secondary winding currents of the welding transformer:

$$\text{Transformation Ratio,}\ \frac{N_2}{N_1} = \frac{I_1}{I_2} \tag{6}$$

$$\left.\begin{array}{l} \frac{N_2}{N_1} = \frac{E_2}{E_1} = \frac{I_1}{I_2} \\[2mm] \frac{102}{612} = \frac{E_2}{400} = \frac{3.05}{I_2},\ I_2 = 18.30A \end{array}\right\} \tag{7}$$

From the above Equation (4), I_2 can be calculated by using Equation (7) and is equal to the current in each winding of single phase which is 18.30 A, as three-phase winding turns (welding transformer primary winding turn) N_1 is equal to 612 turns, single-phase winding turns (welding transformer secondary turn) N_2 are equal to 102 turns and the rated three phase current I_1 = 3.05 A. From this calculation the welding transformer current is 18.31 A per coil, hence by connecting these four coils in parallel this current is added and finally the welding transformer gets nearly 70 to 75 A current for welding. This current is able to provide a smooth output for welding with 2.5 mm welding rods.

3.4. Phase Converter Design

Phase converter calculations can be done with the same procedure as the welding transformer. Here the complete single-phase winding is considered in a circuit including auxiliary winding which states that the total number of single-phase windings is equal to the number of three-phase windings, i.e., primary turns is equal to secondary turns and the transformation ratio is 1. However, the input is three-phase line voltage of 400 Volts, so the output phase voltage, V_{ph} = 400/1.7321 = 230 Volts are achieved as single-phase output. Hence single-phase supply can be obtained from a proposed design of the motor. The phase converter calculation is similar to transformer calculation because here the energy transformation is achieved by mutual induction; hence the transformation ratio (Equation (7)) and the EMF equation of the transformer are used to calculate voltages.

4. Hardware Assembly and Testing Results of the Proposed Induction Machine

An old scrap inductor motor (IM) was used to verify the proposed concept. In Figure 6a the old IM is depicted and its windings are redesigned as shown in Figure 6b.

(a)   (b)   (c)

Figure 6. (**a**) Old scrap motor. (**b**) Redesigned winding arrangement for the proposed multifunctional motor. (**c**) Squirrel cage rotor (rotor of the old machine).

In Figure 6c the rotor of old machine which is a squirrel cage type is shown. In Table 3 the specifications of the old IM are given. Figure 7a depicts the hardware implementation of the stator winding of the proposed machine design. Figure 7b depicts the complete stator winding arrangement which consists of both windings. Figure 7c depicts the end terminals of the proposed designed machine. A conventional two wattmeter (W_1 and W_2) open circuit test (OCT) and blocked rotor test (BRT) are performed on the three-phase and single-phase induction motor to calculate its efficiency. The multiplying factor is calculated for three-phase and single-phase operation by using Equation (8). Tables 4 and 5 describes the OCT and BRT results of the proposed IM as a three-phase induction motor and single-phase induction motor, respectively.

$$\text{Multiplying factor} = \frac{\text{Voltage rang} \times \text{Current Range} \times \text{Power factor}}{\text{Maximum Scale Deflection}} \tag{8}$$

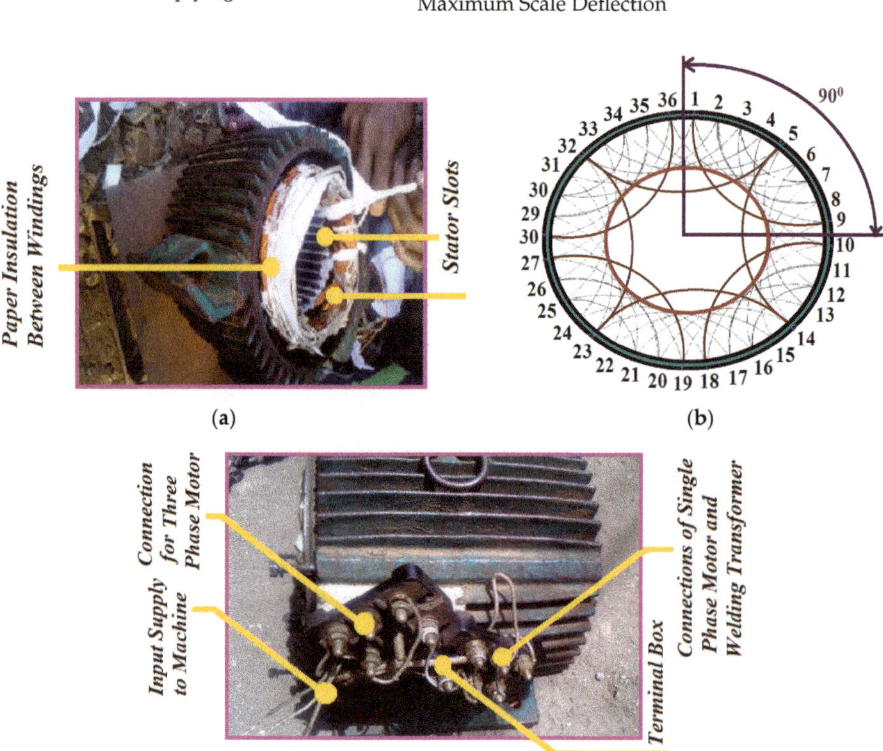

Figure 7. Design of the proposed induction machine: (**a**) hardware implementation of the stator winding; (**b**) complete stator winding arrangement; (**c**) proposed induction machine end terminal connection.

Table 3. Parameters of the old induction machine (IM).

Parameter	Value
Speed	1430 rpm
Connection	delta
Insulation	Class-B
No of slots	36
No of pole-4	4
Slot pitch	1 to 8
No of conductor per slot	72
Old wire guige	18
Total no of coils	$12 \times 3 = 36.$
Name plate rating	3.7 kW (5 HP).
Voltage, Current and Frequency	415 V, 3.5 A, 50 Hz

Table 4. Testing of the three-phase proposed induction machine (IM).

TEST	V_O for OCT / V_{BR} for BRT in (V)	I_O for OCT / I_{BR} for BRT in (V)	W_1 (W)	W_2 (W)	Wo ($W_1 + W_2$) / Multiplying Factor = 8
OCT	390	3.1	-60	105	$45 \times 8 = 360$
BRT	61	3.0	0	20	$20 \times 8 = 160$

Table 5. Testing of the single-phase induction machine (IM).

TEST	V_O for OCT / V_{BR} for BRT in (V)	I_O for OCT / I_{BR} for BRT in (V)	Wo (W) Multiplying Factor = 4
OCT	220	7.3	$170 \times 4 = 680$
BRT	60	7	$130 \times 4 = 520$

In Tables 4 and 5, V_O is the line to line voltage, I_O is the stator phase current, V_{BR} is the stator line to line voltage in blocked rotor mode, I_{BR} is the stator phase current in a blocked rotor mode, W_1 and W_2 are the readings of two wattmeters used in the OCT and BRT test. Table 6 shows the output voltage and current of the welding transformer which is sufficient to perform arc welding and Table 7 shows the output voltages of the phase converter at an input of 230 V. In Table 8 the welding transformer winding specifications are given.

Table 6. Output of the welding transformer.

Connection	Secondary Winding Voltage (Volts)	Secondary Current (Amperes)
Running winding	61.3	79.8

Table 7. Output voltages of the phase converter.

Serial Number	Terminals	Voltage (L-L) in Volts
1	R-Y	402
2	Y-B	395
3	B-R	400

Table 8. Welding transformer winding specifications.

Winding	Layer	No. of Turns	Guage
3 Ph. I.M.	Single	17	21
Running Winding 1	Triple	$17 \times 2 = 34$	21
Starting Winding 2	Triple	17	21

For three phase operation:

Input power (P_{in}) = 1.7321 × 440 × 3.05 × 0.8 = 1860 W

Output power (P_{out}) = 1860 − losses (no load + block rotor) = 1860 − (360 + 160) = 1340

Hence, Efficiency (ή) = P_{out}/P_{in} × 100 = (1340/1860) × 100 = Efficiency (ή) = 72.64%

For Single phase operation:

Input power (P_{in}) = 220 × 10.5 × 0.8 = 1860 W

Output power (P_{out}) = 1860 − losses (no load + block rotor) = 1860 − (680 + 520) = 660 W

Hence, Efficiency (ή) = P_{out}/P_{in} × 100 = (660/1860) × 100

Efficiency (ή) = 35.48%

5. Merits and Applications of the Proposed Induction Machine

The following are the noticeable merits of the proposed induction machine:

- The multifunctional motor is simple and robust in construction.
- The motor is able to do at a time two operations, i.e., motoring and welding.
- The motor requires less space.
- Another advantage is that it has less weight compared to separate combinations of welding transformer and induction motor as well.
- Hence, the cost required for two different machines is reduced, with high efficiency.
- The cost is less.

The following are a few applications of the proposed induction machine.

- The use of multipurpose motors is very convenient for use in mega workshops where both welding and motoring applications are required.
- These types of motors are also useful in an electric traction systems where the space requirements is small.
- Metal cutting workshops are another example where multifunctional motors would be useful.
- It can also be used in heavy fabrication industry.
- Multifunctional motors are very useful in the steel industry.

6. Conclusions

A new winding scheme is articulated to utilize a three-phase induction machine for multifunctional operation. The proposed motor provides an operative solution for agricultural as well as industrial purposes because of rugged construction and less maintenance needs. The proposed winding scheme is adapted to utilize a three-phase induction machine as a three-phase induction motor, single-phase induction machine, and welding transformer and phase converter. This new design does not need any kind of special arrangement and can be constructed with small modifications of any standard three-phase induction motor. The proposed concept is verified by designing a motor by modifying the windings of an old induction motor (IM) and the proposed motor is well tested to find its efficiencies and the experimental results are provided in the article to validate the design and construction.

Author Contributions: Mahajan Sagar Bhaskar, Sanjeevikumar Padmanaban, Sonali A. Sabnis has developed the concept of the original proposal and designed, performed the real time implementation. Frede Blaabjerg provided the guidelines for theoretical validation in line with the developed theoretical background. Sanjeevikumar Padmanaban, Lucian Mihet-Popa, Vigna K. Ramachandramurthy provided the support for technical validation for the mathematical background and validation with results. All authors contributed equally for articulate the research work in its current form as full research manuscript.

Conflicts of Interest: The authors declare no conflict of interest.

References

1. Sanjeevikumar, P.; Grandi, G.; Blaabjerg, F.; Wheeler, P.; Ojo, J.O. Analysis and Implementation of Power Management and Control Strategy for Six-Phase Multilevel AC Drive System in Fault Condition. *Eng. Sci. Technol. Int. J.* **2016**, *19*, 31–39. [CrossRef]
2. Sanjeevikumar, P.; Bhaskar, M.S.; Blaabjerg, F.; Norum, L.E.; Seshagiri, S.; Hajizadeh, A. Nine-Phase Hex-tuple Inverter for Five-Level Output Based on Double Carrier PWM Technique. In Proceedings of the 4th IET Clean Energy and Technology Conference (CEAT 2016), Kuala Lumpur, Malaysia, 14–15 November 2016.
3. Oleschuk, V.; Bojoi, R.; Profumo, F.; Tenconi, A.; Stankovic, A.M. Multifunctional Six-Phase Motor Drives with Algorithms of Synchronized PWM. In Proceedings of the IEEE 32nd Annual Conference on Industrial Electronics (IECON 2006), Paris, France, 7–10 November 2006.
4. Bhaskar, M.S.; Sanjeevikumar, P.; Blaabjerg, F. Multistage DC-DC Step-Up Self Balanced and Magnetic Component Free Converter for Photovoltaic Applications—Hardware Implementation. *Energies* **2017**, *10*, 719. [CrossRef]
5. Duan, Y.; Harley, R.G. A Novel Method for Multiobjective Design and Optimization of Three Phase Induction Machines. *IEEE Trans. Ind. Appl.* **2011**, *17*, 1707–1715. [CrossRef]
6. Misir, O.; Raziee, S.M.; Hammouche, N.; Klaus, C.; Kluge, R.; Ponick, B. Calculation method of three-phase induction machines equipped with combined star-delta windings. In Proceedings of the XXII International Conference on Electrical Machines (ICEM), Lausanne, Switzerland, 4–7 September 2016.
7. Schreier, L.; Bendl, J.; Chomat, M. Comparison of five-phase induction machine operation with various stator-winding arrangements. In Proceedings of the 39th Annual Conference of the IEEE Industrial Electronics Society (IECON 2013), Vienna, Austria, 10–13 November 2013.
8. Misir, O.; Ponick, B. Analysis of three-phase induction machines with combined Star-Delta windings. In Proceedings of the IEEE 23rd International Symposium on Industrial Electronics (ISIE), Istanbul, Turkey, 1–4 June 2014.
9. Sanjeevikumar, P.; Bhaskar, M.S.; Pandav, K.M.; Siano, P.; Oleschuk, V. Hexuple-Inverter Configuration for Multilevel Nine-Phase Symmetrical Open-Winding Converter. In Proceedings of the IEEE International Conference on Power Electronics, Intelligent Control and Energy Systems (ICPEICES), Delhi, India, 4–6 July 2016; pp. 1837–1844.
10. Sanjeevikumar, P.; Blaabjerg, F.; Wheeler, P.; Khanna, S.; Mahajan, B.; Dwivedi, S. Optimized Carrier Based Five-Level Generated Modified Dual Three-Phase Open-Winding Inverter For Medium Power Application. In Proceedings of the IEEE International Transportation Electrification Conference and Expo, (ITEC Asia-Pacific), Busan, Korea, 1–4 June 2016; pp. 40–45.
11. Sanjeevikumar, P.; Blaabjerg, F.; Wheeler, P.; Lee, K.; Mahajan, S.B.; Dwivedi, S. Five-Phase Five-Level Open-Winding/Star-Winding Inverter Drive For Low-Voltage/High-Current Applications. In Proceedings of the IEEE International Transportation Electrification Conference and Expo (ITEC Asia-Pacific), Busan, Korea, 1–4 June 2016; pp. 66–71.
12. Chatterjee, D. Impact of core losses on parameter identification of three-phase induction machines. *IET Power Electron.* **2014**, *7*, 3126–3136. [CrossRef]
13. Mengoni, M.; Sala, G.; Zarri, L.; Tani, A.; Serra, G.; Gritli, Y.; Duran, M. Control of a fault-tolerant quadruple three-phase induction machine for More Electric Aircrafts. In Proceedings of the 42nd Annual Conference of the IEEE Industrial Electronics Society, Florence, Italy, 23–26 October 2016.
14. Subtirelu, G.; Dobriceanu, M.; Linca, M. Virtual instrumentation for no-load testing of induction motor. In Proceedings of the IEEE International Power Electronics and Motion Control Conference (PEMC), Varna, Bulgaria, 25–30 September 2016.

15. Hackner, T.; Pforr, J. Comparison of different winding schemes of an asynchronous machine driven by a multi-functional converter system. In Proceedings of the 2010 IEEE Energy Conversion Congress and Exposition (ECCE), Atlanta, GA, USA, 12–16 September 2010.

16. Ferreira, F.; Cistelecan, M.; Almeida, A. Comparison of Different Tapped Windings for Flux Adjustment in Induction Motors. *IEEE Trans. Energy Convers.* **2014**, *29*, 375–391. [CrossRef]

17. Gunabalan, R.; Sanjeevikumar, P.; Blaabjerg, F.; Ojo, O.; Subbiah, V. Analysis and Implementation of Parallel Connected Two Induction Motor Single Inverter Drive by Direct Vector Control for Industrial Application. *IEEE Trans. Power Electron.* **2015**, *30*, 6472–6475. [CrossRef]

18. Muteba, M.; Jimoh, A. Performance analysis of a three-phase induction motor with double-triple winding layout. In Proceedings of the International Future Energy Electronics Conference (IFEEC), Tainan, Taiwan, 3–6 November 2013; pp. 131–136. [CrossRef]

19. Dragonas, F.A.; Nerrati, G.; Sanjeevikumar, P.; Grandi, G. High-Voltage High-Frequency Arbitrary Waveform Multilevel Generator for DBD Plasma Actuators. *IEEE Trans. Ind. Appl.* **2015**, *51*, 3334–3342. [CrossRef]

20. Alberti, L.; Bianchi, N. Impact of winding arrangement in dual 3-phase induction motor for fault tolerant applications. In Proceedings of the XIX International Conference on Electrical Machines (ICEM), Rome, Italy, 6–8 September 2010.

21. Kocabas, D.A. Space harmonic effect comparison between a standard induction motor and a motor with a novel winding arrangement. In Proceedings of the IEEE International Electric Machines and Drives Conference, Miami, FL, USA, 3–6 May 2009; pp. 65–69.

22. Sanjeevikumar, P.; Grandi, G. *Multiphase-Multilevel Inverter for Open-Winding Loads*; LAP LAMBERT Academic Publishing: Saarbrücken, Germany, 2012; ISBN 978-3-659-30278-7.

23. Kalaivani, C.; Sanjeevikumar, P.; Rajambal, K.; Bhaskar, M.S.; Mihet-Popa, L. Grid Synchronization of Seven-phase Wind Electric Generator using d-q PLL. *Energies* **2017**, *10*, 926. [CrossRef]

24. Cistelecan, M.V.; Cosan, H.B.; Popescu, M. Part-winding starting improvement of three-phase squirrel-cage induction motor. In Proceedings of the 8th IEEE International Symposium on Advanced Electromechanical Motion Systems & Electric Drives Joint Symposium, Lille, France, 1–3 July 2009.

25. Kocabas, D.A. Novel Winding and Core Design for Maximum Reduction of Harmonic Magnetomotive Force in AC Motors. *IEEE Trans. Magn.* **2009**, *45*, 735–746. [CrossRef]

26. United States Army. Electrical Conductors. In *Introduction to Marine Electricity*; United States Army: Arlington County, VA, USA, 1994; pp. 1–22.

27. Saponara, S.; Tisserand, P.; Chassard, P.; Ton, D. Design and Measurement of Integrated Converters for Belt-Driven Starter-Generator in 48 V Micro/Mild Hybrid Vehicles. *IEEE Trans. Ind. Appl.* **2017**, *53*, 3936–3949. [CrossRef]

28. Dattu, N.; Deepthi, G.; Pundlik, G. Multifunctional Induction Machine. In Proceedings of the IEEE International WIE Conference on Electrical and Computer Engineering (WIECON-ECE), Pune, India, 19–21 December 2016.

29. Saponara, S.; Fanucci, L.; Bernardo, F.; Falciani, A. Predictive Diagnosis of High-Power Transformer Faults by Networking Vibration Measuring Nodes with Integrated Signal Processing. *IEEE Trans. Instrum. Meas.* **2016**, *65*, 1749–1759. [CrossRef]

30. Glowacz, A. Diagnostics of DC and Induction Motors Based on the Analysis of Acoustic Signals. *Meas. Sci. Rev.* **2014**, *14*, 257–262. [CrossRef]

Article

Development of Sliding Mode Controller for a Modified Boost Ćuk Converter Configuration

Sanjeevikumar Padmanaban [1,*], Emre Ozsoy [1], Viliam Fedák [2] and Frede Blaabjerg [3]

[1]   Department of Electrical and Electronics Engineering, University of Johannesburg, Auckland Park 2092, South Africa; eemreozsoy@yahoo.co.uk
[2]   Department of Electrical Engineering and Mechatronics, FEI TU of Košice, Letná 9, 04200 Košice, Slovakia; viliam.fedak@tuke.sk
[3]   Centre for Reliable Power Electronics (CORPE), Department of Energy Technology, Aalborg University, 9000 Aalborg, Denmark; fbl@et.aau.dk
*   Correspondence sanjeevi_12@yahoo.co.in; Tel.: +27-79-219-9845

Received: 18 June 2017; Accepted: 13 July 2017; Published: 29 September 2017

Abstract: This paper introduces a sliding mode control (SMC)-based equivalent control method to a novel high output gain Ćuk converter. An additional inductor and capacitor improves the efficiency and output gain of the classical Ćuk converter. Classical proportional integral (PI) controllers are widely used in direct current to direct current (DC-DC) converters. However, it is a very challenging task to design a single PI controller operating in different loads and disturbances. An SMC-based equivalent control method which achieves a robust operation in a wide operation range is also proposed. Switching frequency is kept constant in appropriate intervals at different loading and disturbance conditions by implementing a dynamic hysteresis control method. Numerical simulations conducted in MATLAB/Simulink confirm the accuracy of analysis of high output gain modified Ćuk converter. In addition, the proposed equivalent control method is validated in different perturbations to demonstrate robust operation in wide operation range.

Keywords: closed loop control; Ćuk converter; sliding mode control; robustness; active hysteresis control

---

1. Introduction

Direct current to direct current (DC-DC) converters play a vital role in electrical systems due to the increasing penetration of renewable sources in electrical networks. In addition to high efficiency and reliability requirements, robust performance of the converter in a wide operating range is of great importance, since DC-DC converters are also used in diverse special-purpose applications, such as electrical vehicles, DC motor drives, and telecommunication systems.

Different DC-DC converter topologies can be encountered in the literature. Classical converter topologies suffer from the lack of voltage gain ratio. Higher output voltage gain ratio with improved efficiency increases the performance of the converter, which is especially crucial for solar applications [1]. Diverse DC-DC converter topologies are proposed in [2–9] to improve the voltage gain ratio and efficiency. Important voltage lift methods are also reviewed and compared in [10].

A DC-DC converter circuit topology must be upgraded for higher voltage output gain and improved efficiency, lowering the conduction losses, designing a smaller size converter, and minimizing voltage and current stress on the semiconductor switch. In addition to circuit modification for achieving the above goals, controller structure is also of great importance to improve the performance, robustness, and reliability in a wide operation range. Unfortunately, these converters are still bottlenecked in terms of system reliability and performance [1]. In addition to circuit and controller design requirements, the availability and reliability of a complete system in harsh environments is also an important task to be considered. The studies given in [11–14] outline the harsh environment requirements of electronic circuits and implement different types of electronic circuit applications for automotive systems.

High performance control of a DC-DC converter is a challenge for both control engineering and power electronics practitioners due to the highly nonlinear nature of DC-DC converters. Furthermore, fast response in terms of rejection of load variations, input voltage disturbances, and parameter uncertainties is mandatory for robust operation.

A Ćuk converter is a kind of buck/boost converter topology; the inverted output is either lower or higher than the input voltage. Different modifications are applied to classical Ćuk circuit [15,16] to enhance the performance. Modeling and control of Ćuk converters has been investigated with different approaches. Linear methods [17,18] and proportional integral (PI) controllers [19] are well-known design procedures with ease of implementation. However, these classical methods do not guarantee the stability and high performance in different perturbations due to highly nonlinear behavior of Ćuk converters. Thus, different nonlinear control algorithms are also implemented in Ćuk converters to overcome this drawback, such as passivity-based control [20], neural networks [21], direct control methods [22], fuzzy logic [23], and sliding mode control (SMC) [24].

SMC for variable structure systems [25] is a robust control method of nonlinear systems due to its insensitivity to parameter variations, fast dynamic response, and ease of implementation. SMC was first applied to DC-DC converters in [26,27], and many diverse implementation examples are available in [27]. Design criteria for SMC application to DC-DC converters is outlined in [28]. SMC-based equivalent controllers are applied to buck/boost and Ćuk converter topologies in [29,30]. However, SMC is not popularly implemented in DC-DC converters due to its unavailability of integrated circuit forms for power electronic applications. Moreover, its variable switching frequency (SF) behavior depending on the converter parameters and operation regions complicates electromagnetic interference filter design and practical implementation. A scheme given in [31] outlines the SF fixing and reduction methods in SMC applications. In addition, it is known that DC-DC converters are unwanted noise generators, and this problem can be overcome with fixed frequency operation [28].

Different control techniques have been proposed to achieve constant SF operation to DC-DC converters. An equivalent controller is designed and the output of the controller is compared with a saw-tooth signal to fix the SF in [32]. Frequency locking techniques are applied in [33] to achieve constant SF operation of SMC for buck converter. An analog circuit design perspective for fixed frequency operation of SMC is given in [34]. Dynamic hysteresis control [35,36] is another contribution which is commonly used for fixed SF operation.

This study aims to improve the output voltage gain of a Ćuk converter circuit by inclusion of a single inductor and capacitor. The efficiency of the overall system is increased, and it is verified that voltage transformation ratio (V_o/V_i) is increased to $1/(1 - \delta)$, just as in classical boost converters, where δ is the duty ratio of the converter. The proposed model is mathematically analyzed, and numerical simulations conducted on MATLAB/Simulink validate the accuracy of the analysis.

Moreover, an SMC-based cascaded equivalent controller is implemented for robust operation of the proposed converter. The general structure of SMC for DC-DC converters consists of external voltage controllers to achieve the desired output voltage requirements, and inner SMC performs the control of input current [26]. In general, a PI controller is sufficient for voltage requirements. Therefore, cascaded PI+SMC structure achieves robust operation of a novel high output gain Ćuk converter in a wide operation range. Constant SF operation is achieved at different loading and disturbance conditions by using a simple dynamic hysteresis controller. The control algorithm is implemented in the MATLAB/Simulink environment in different scenarios: (1) A high value of output reference voltage step; (2) Output resistance variation; (3) Input voltage drop; (4) Input inductor parameter variation. The proposed method effectively achieves performance goals for all aforementioned perturbations.

2. High Output Gain Modified Ćuk Converter

The developed Ćuk converter is depicted in Figure 1. A classical Ćuk converter was modified with an additional inductor (L_3) and capacitor (C_2). Figure 2a and b provides the equivalent circuit representation of the modified Ćuk converter with the semiconductor switch S turned ON and OFF, respectively.

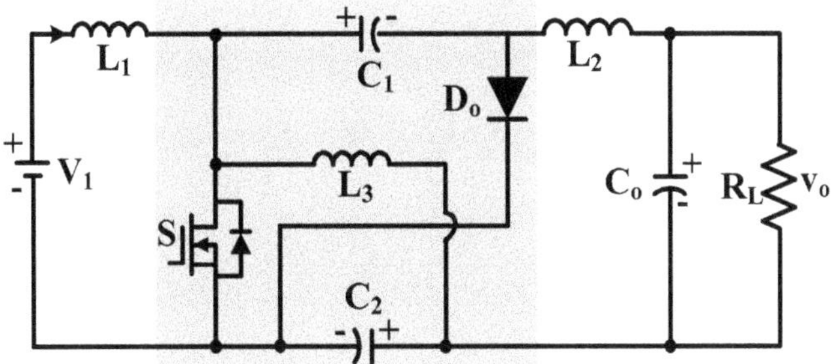

Figure 1. Topology of proposed novel Ćuk converter.

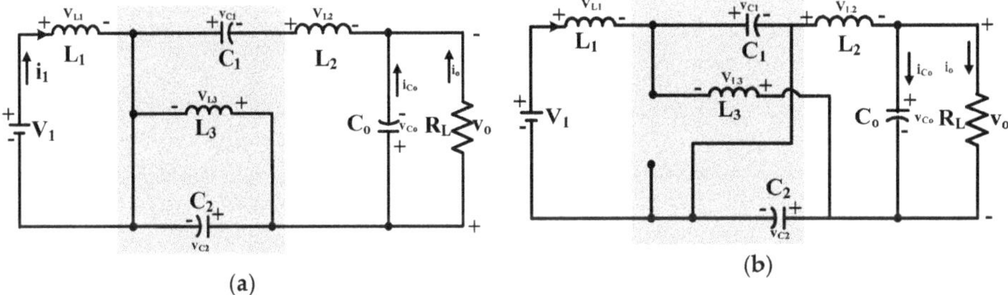

(a)

(b)

Figure 2. Circuit configuration (**a**) ON state; (**b**) OFF state.

When the switch S is turned ON and OFF, the following inductance voltage equations can be written to the circuit over one period for steady-state conditions. When the S is ON:

$$V_{L1} = V_1, \; V_{L2} = V_o - V_{C2} - V_{C1}, \; V_{L3} = - V_{C2} \tag{1}$$

When the S is OFF:

$$V_{L1} = V_1 - V_{C1}, \; V_{L2} = - V_o + V_{C2}, \; V_{L3} = V_{C1} + V_{C2} \tag{2}$$

According to Faraday's Law, the average voltage across an inductor is zero at steady-state. Hence, the voltage gain ratio of the converter can be obtained by starting the commonly used equation given below.

$$\delta \, V_{L(ON)} = (1 - \delta)V_{L(OFF)} \tag{3}$$

The term δ means the duty ratio of the switch S. Equation (3) will be written for L_1, L_2, and L_3, and required voltage gain ratio equation will be obtained.

First, (3) is written for L_1, and the equation given below is obtained;

$$\delta \, V_1 = (1 - \delta)(V_1 - V_C) \tag{4}$$

The above Equation (4) can be simplified as:

$$(2\delta - 1)V_1 = (\delta - 1)V_{C1} \tag{5}$$

Second, (3) can be written for L_2 as given below:

$$\delta(V_o - V_{C2} - V_{C1}) = (1 - \delta)(V_o + V_{C2}) \tag{6}$$

Equation (6) can be arranged as given below:

$$V_o(2\delta - 1) = V_{C2} + \delta V_{C1} \tag{7}$$

Finally, (3) can be written for L_3:

$$\delta V_{C2} = (1 - \delta)(V_{C1} + V_{C2}) \tag{8}$$

This simplifies to:

$$V_{C2} = (1 - \delta)V_{C1} \tag{9}$$

If (5), (7), and (9) are combined, the duty ratio of the converter can be obtained. If (9) is inserted into (7),

$$V_o(2\delta - 1) = V_{C1} \tag{10}$$

If (10) is inserted into (5), the duty ratio of the system can be finalized.

$$\frac{V_o}{V_i} = -\frac{1}{1 - \delta} \tag{11}$$

A sample design circuit can be conducted by using the circuit parameters given in Table 1 in MATLAB/Simulink. Different δ values are applied in the simulation, as shown in Figure 3c. Output voltage (V_o) and input current (i_i) curves change accordingly, as shown in Figure 3a,b, respectively. Output voltage and input currents are zoomed; it is observed in simulations that the frequency of the ripples is equal to the SF (150 kHz).

Table 1. Design parameters of Modified Ćuk Converter.

Symbol	Quantity	Unit
Input Voltage	15	V
Duty Ratio (Δ)	0.1–0.9	–
Switching Frequency	150	kHz
Inductances (L_1, L_2, L_3)	100	μH
Capacitors (C_0, C_1, C_2)	5	μF
Load Resistance	100	Ω

The performance of the modified Ćuk converter was compared to classical Ćuk and buck/boost converter circuits. Figure 4a shows δ comparison of converters. A simulation platform is constructed in MATLAB/Simulink with the same parameters given in Table 1. Theoretical and simulation values of the modified Ćuk converter validate the results. Efficiency comparison of simulated buck/boost, Ćuk, and modified Ćuk converter is depicted in Figure 4b. Modified Ćuk converter efficiency is higher than classical Ćuk and buck/boost converter. It can be stated that the proposed modified Ćuk converter produces higher efficiency due to the inclusion of additional passive elements. This reduces several parasitic effects and switching/conduction losses and increases voltage gain ratio, as emphasized in [37].

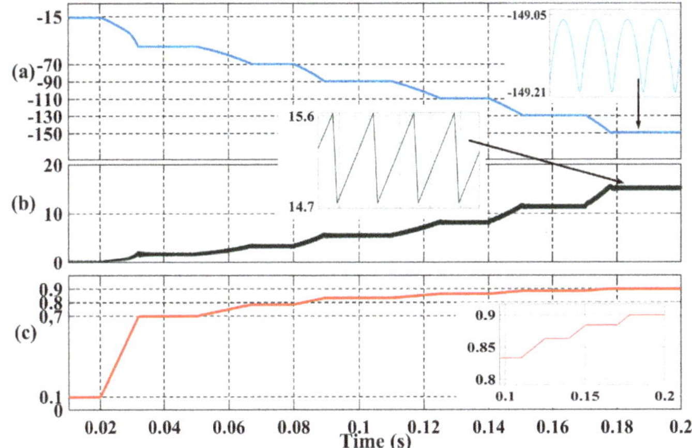

Figure 3. Simulation of modified Ćuk converter. (**a**) Output Voltage (V); (**b**) Input Current (A); (**c**) δ.

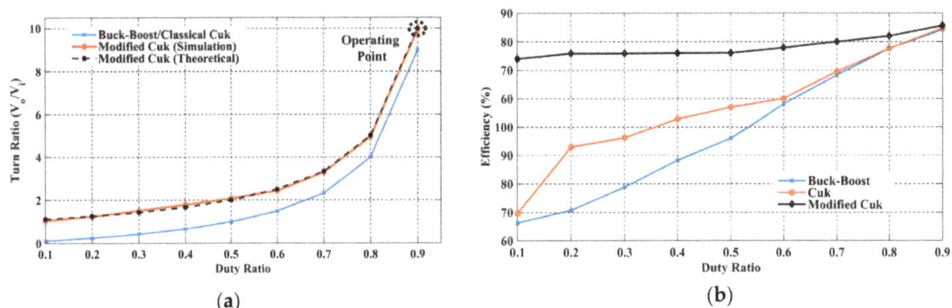

(**a**)   (**b**)

Figure 4. Comparison of buck/boost, Ćuk, and modified Ćuk converter. (**a**) Duty Ratio; (**b**) Efficiency.

3. Equivalent Control of Modified Ćuk Converter

A cascaded PI+SMC controller structure could be used for ease of implementation to modified Ćuk converter as depicted in Figure 5. A simple external voltage controller can generate input current reference, while equivalent controller controls the input current [29,30].

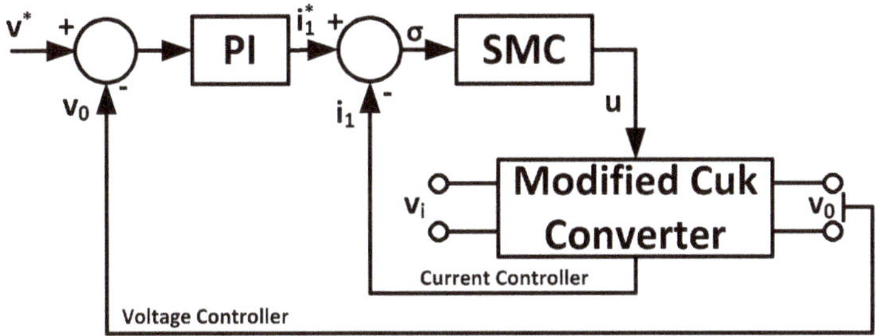

Figure 5. Cascaded control of modified Ćuk converter. PI: proportional integral; SMC: sliding mode controller.

Although the modified Ćuk converter is a third-order nonlinear model, only an input current equation is required to construct an equivalent controller. This is the main advantage of SMC-based equivalent controllers, since the performance is independent of all system dynamics and parameter variations. Input current of the modified Ćuk converter in terms of Kirchhoff's voltage law can be written in the following form:

$$\frac{di_{L1}}{dt} = \frac{1}{L_1}(V_i - (1-u)V_{C1}) \tag{12}$$

The terms V_i and V_{c1} are input C_1 voltages, and u is the switching signal of semiconductor switch as explained below.

$$u(t)\begin{bmatrix} 1 & Switch \rightarrow ON & D_o \rightarrow OFF \\ 0 & Switch \rightarrow OFF & D_o \rightarrow ON \end{bmatrix} \tag{13}$$

The external PI controller aims to achieve the reference voltage target, and the output of the PI controller acts as reference current (i^*). The internal SMC-based equivalent current controller aims to track current trajectory, and the analytical design procedure is detailed below. The switching surface can be given as [29,30]:

$$\sigma = i_{L1} - i^* \tag{14}$$

The time derivative of the switching surface is:

$$\dot{\sigma} = \dot{i_{L1}} - \dot{i^*} \tag{15}$$

If the input current equation of the modified Ćuk converter in (12) is written to derivative of the switching surface in Equation (15), the following equation can be obtained:

$$\dot{\sigma} = \frac{V_i}{L_1} - \frac{1}{L_1}(1-u)\dot{V}_{c1} - \dot{i^*} \tag{16}$$

If $\dot{\sigma}$ is assumed to be zero at steady-state, the equivalent control signal can be generated as given below.

$$u_{eq} = 1 + \frac{L_1}{V_{c1}}\dot{i^*} - \frac{V_i}{V_{c1}} \tag{17}$$

The switching surface can be simplified as given below, considering $\dot{\sigma}$ is zero at steady-state. The continuous function u_{eq} will be converted into discontinuous form as follows:

$$\dot{\sigma} = u - u_{eq} \tag{18}$$

Closed loop control signal from switching surface can be given as:

$$u = \hat{u}_{eq} - K\sigma \tag{19}$$

The term K is positive definite control gain, and if (19) is inserted in (18), the switching surface can be written as follows:

$$\dot{\sigma} = \hat{u}_{eq} - K\sigma - u_{eq} \tag{20}$$

where $\hat{u}_{eq}$ is the estimated equivalent control input. It can be written in steady-state that $\hat{u}_{eq} = u_{eq}$.

$$\dot{\sigma} = -K\sigma \tag{21}$$

Finally, stability and existing conditions for sliding mode control must be clarified [18]. The definition $\dot{\sigma}\sigma < 0$ must be satisfied, and it can be derived from (21) that;

$$\begin{cases} \sigma > 0 & \dot{\sigma} < 0 \\ \sigma < 0 & \dot{\sigma} > 0 \end{cases} \tag{22}$$

Thus, the stability of the sliding surface is satisfied. Controller structure can be constructed by estimating $\hat{u}_{eq}$. Estimation of the equivalent control can be formed as:

$$v = \hat{u}_{eq} + l\sigma \tag{23}$$

Where the term l is the filter gain of the estimator. It is assumed that $\hat{u}_{eq}$ is constant, and the time derivative of (23) can be written as given below:

$$\dot{v} = l\dot{\sigma} \tag{24}$$

The time derivative of $\dot{\sigma}$ in (18) can be inserted into (23) as given below:

$$\dot{v} = l\left(u - u_{eq}\right) \tag{25}$$

It can be stated that $\hat{u}_{eq} = u_{eq}$ in steady state:

$$\dot{v} = l\left(u - \hat{u}_{eq}\right) \tag{26}$$

If (26) is written in the form of $\hat{u}_{eq} = v - l\sigma$, the following equation can be obtained [29,30]:

$$\dot{v} = l(u - v + l\sigma) \tag{27}$$

Finally, the simple equivalent controller structure in Figure 6 can be obtained from the definitions written above.

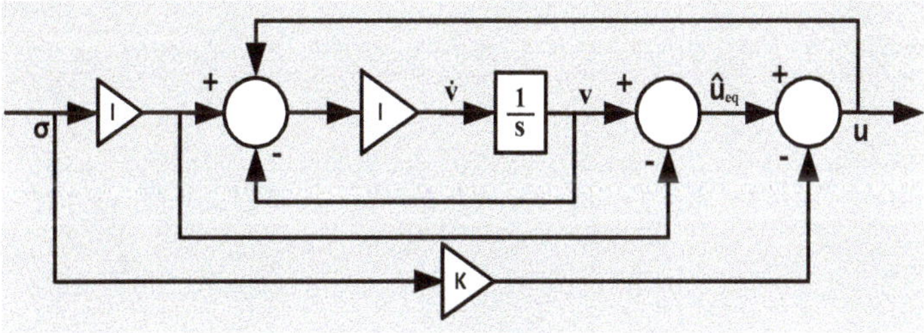

Figure 6. SMC-based equivalent controller.

As detailed in [25,29,30], Figure 6 shows that a dynamic system can be formed as a series of integrators, and it can be assumed that the output of this system can be estimated by an upper bound of the integral. Finally, a dynamic relay with hysteresis function as given in Figure 7 can be applied to control a signal to generate a sliding surface which oscillates with the magnitude of M.

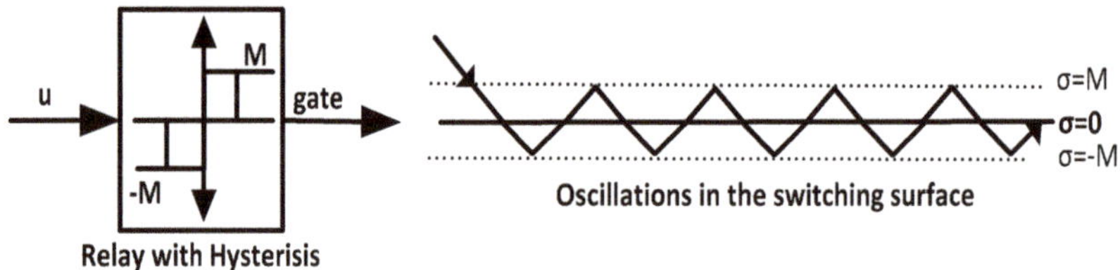

Figure 7. Dynamic relay with hysteresis function.

The main disadvantage of the SMC-based equivalent controller is variable switching frequency (SF), because the magnitude of oscillations in sliding surface is highly dependent on circuit parameters and operating conditions due to the nonlinear behavior of the converter. One of the methods that can constrain the sliding surface to constant SF operation can be a dynamic hysteresis controller [35] which dynamically changes the magnitude of sliding surface σ according to the desired switching frequency value. An additional PI controller which intermittently operates to bring back the SF to the desired value can be a simple and practical solution. The output of the PI controller dynamically changes the hysteresis of dynamic relay (M). Thus, SF can settle to a desired interval accordingly.

Another drawback of the method is the requirement of the SF measurement. A SF measurement algorithm could be implemented by counting the rising edge of the gate signals at certain instants. If the number of rising signals is divided into a predefined time interval, SF can be easily calculated. As a result, an intermittent PI controller structure can keep the SF constant at a specified interval as shown in Figure 8. The output of the intermittent PI controller is the resultant M value of the dynamic relay with hysteresis function.

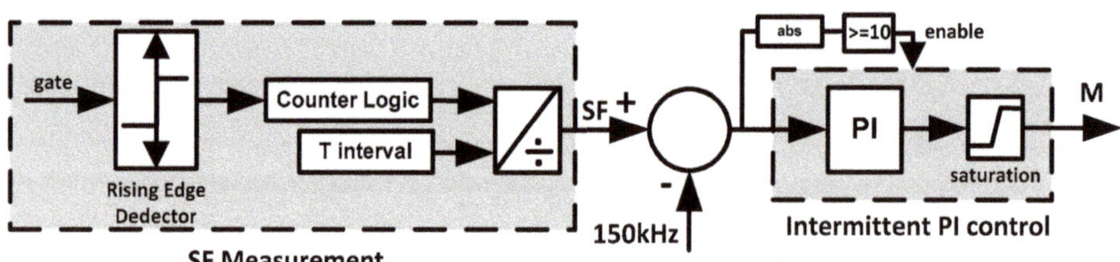

Figure 8. Switching frequency (SF) measurement and intermittent PI controller.

4. Simulation Results

Four different scenarios are implemented in a single simulation in MATLAB/Simulink SimPowerSystem platform at different time instants. Variable Step Ode23tb (stiff/TR-BDF2) solver is used in simulations. Circuit parameters given in Table 1 are used in simulation. Stable gains for external PI controller and equivalent controller are given in Table 2.

Table 2. Controller parameters of modified Ćuk converter.

Symbol	Quantity
K_p of voltage PI controller	0.001
K_I of PI voltage controller	90
Target switching frequency	140–160 kHz
K of equivalent controller	0.01
L of equivalent controller	1
K_p of hysteresis controller	0.00005
K_I of hysteresis controller	0.6

Standard Routh–Hurwitz criterion for determination of PI gain values is omitted in this paper, and all gain values are determined using trial–error methods. For further details of Routh–Hurwitz criterion for external PI controllers, one can refer to [24]. The target SF value was selected as 150 kHz, and intermittent PI hysteresis controller was only enabled when the absolute value of error exceeded 10, as shown in Figure 8. It is not possible to realize a precise controller for SF control in all perturbations due to unexpected oscillations in M value.

Applied step and disturbance instants are given below.

0.03–0.06 s: Output voltage reference is changed from −150 V to −50 V.

0.08–0.11 s: Load resistance is decreased from 100 Ω to 50 Ω (100% load increase)

0.13–0.16 s: Input voltage is decreased from 15 V to 12 V (20% input voltage dip)

0.18–0.21 s: Input inductor (L_1) is decreased from 100 µH to 85 µH (15% L_1 reduction)

Figure 9 shows the output voltage and input current response at different perturbations. Required trajectories are successfully tracked at all disturbances. Figure 9a shows the output voltage trajectory at all peak values at the instants of perturbations. The controller successfully passed all perturbations. Figure 9b,c shows the input and output currents of the converter. All ripple values are zoomed, and matches the design circuit results of Figure 3.

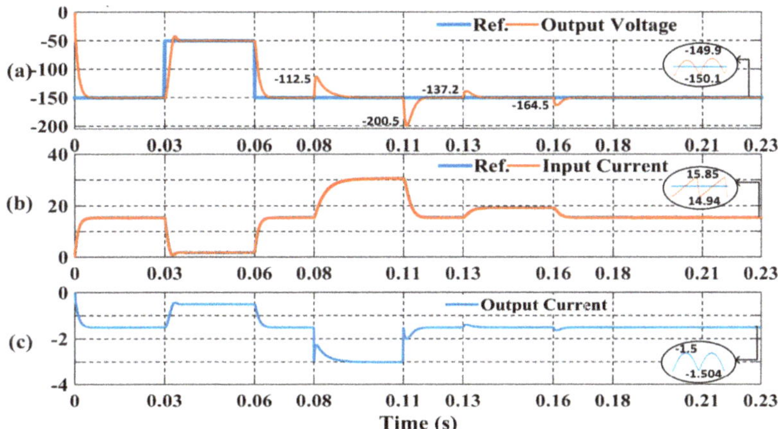

Figure 9. Performance of proposed equivalent controller. (**a**) Output Voltage (V); (**b**) Input Current (A); (**c**) Output Current (A).

Figure 10 shows the performance of the intermittent hysteresis controller. Figure 10a shows the load resistance variation to validate the applied load variation. Figure 10b shows the SF variations at all perturbations. Target SF is achieved at all different perturbations by changing M value as shown in Figure 10c. SF exceeds the target value at transient conditions, but settles to target interval at all steady-state instants of perturbations. Figure 11 shows the output voltage and input current ripples at the instants of semiconductor ON and OFF states in simulation. Figure 11a shows the instants of the gate signal, and Figure 11b shows the control signal u generated by the SMC-based equivalent controller. Figure 11c,d show the ripple contents of output voltage and input current. The ripple values verify the ripple values at design circuit simulation in Figure 3.

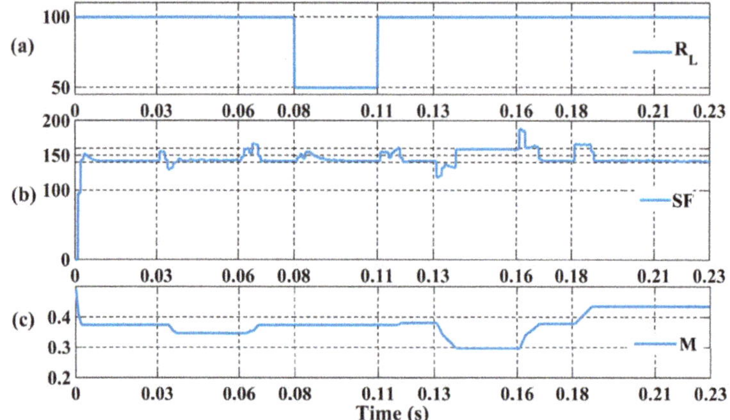

Figure 10. Performance of the proposed equivalent controller. (**a**) R_L; (**b**) SF; (**c**) M.

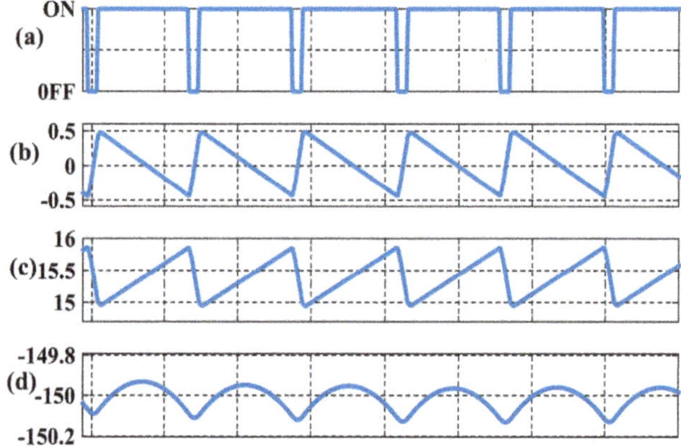

Figure 11. Output voltage and current ripple transients. (**a**) Gate Signal; (**b**) Control Signal (u); (**c**) Input current; (**d**) Output voltage (V_o).

A comparison between classical linear control methods and the proposed equivalent controller was also attempted to show the effectiveness of proposed method. A cascaded controller structure which consists of voltage and current PI controllers is depicted in Figure 12. In particular, some studies use single voltage PI controllers to control a DC-DC converter. However, this type of controller has a very limited operational range, and comparison of a single PI controller would be inconsistent due to the cascaded controller structure of the proposed equivalent controller. An additional current controller introduces additional left half plane zeros to the controller and increases the performance of the controller structure. It is a difficult task to design a linear controller for a third-order modified Ćuk converter, and all states must be observed or measured. Perturbations applied to the controller are given below.

0.03–0.06 s: Output voltage reference is changed from -145 V to -45 V.
0.08–0.11 s: Load resistance is decreased from 100 Ω to 85 Ω (15% load increase)
0.13–0.16 s: Input voltage is decreased from 15 V to 12 V (20% input voltage dip)
0.18–0.21 s: Input inductor (L_1) is decreased from 100 µH to 85 µH (15% L_1 reduction)

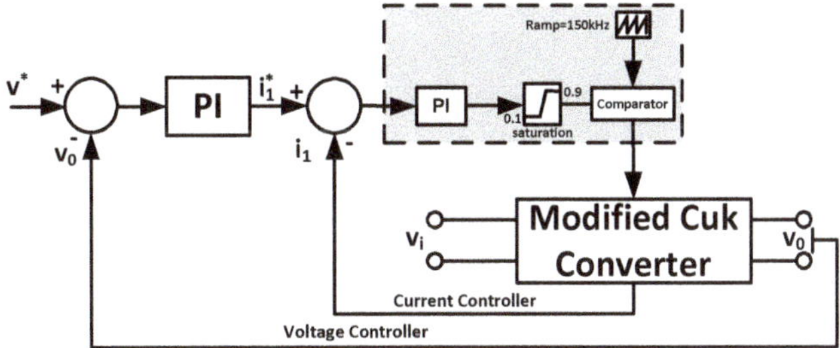

Figure 12. Comparison controller structure.

Higher perturbations as applied to proposed SMC equivalent controller could not be accomplished due to stability problems. Maximum allowable δ is 0.9, and higher δ values could not be achieved. Therefore, -150 V output voltage reference could not be accomplished.

PI controller gain values are optimized with trial and error methods, which are given in Table 3. Controllers are tuned at maximum allowable proportional and integral coefficients to achieve the highest dynamic performance. Higher proportional and integral gains could not be achieved due to higher oscillations in voltage output.

Table 3. Controller parameters of comparison PI controller.

Symbol	Quantity
K_p of PI voltage controller	0.001
K_I of PI voltage controller	50
K_p of PI current controller	0.001
K_I of PI current controller	500
Switching Frequency	150 kHz

Figure 12 shows the performance comparison of PI controller. Figure 12a shows that output voltage performance could not be achieved for all perturbations. Dynamic response is more sluggish compared to proposed SMC based equivalent controller, and load resistance change at 0.11th second causes steady state error and oscillations. Input voltage perturbation cannot be responded due to unavailability of higher δ than 0.9. Steady state error exists at the instant of input voltage perturbation due to sluggish dynamic performance and unavailability of higher δ. Figure 12b shows the responded input currents, and Figure 12c depicts the resultant δ.

Finally, performance indices of the proposed controller structure at PI controller. Figure 13a shows that output voltage performance could not be achieved for all perturbations. Dynamic response is more sluggish compared to the proposed SMC-based equivalent controller, and load resistance change at the 0.11th second causes steady-state error and oscillations. Input voltage perturbation cannot be responded due to the unavailability of higher δ than 0.9. Steady-state error exists at the instant of input voltage perturbation due to sluggish dynamic performance and unavailability of higher δ. Figure 13b shows the responded input currents, and Figure 13c depicts the resultant δ. All perturbations are summarized in Table 4. Maximum peak overshoots are outlined and it is shown that the controller passed high load impact values and other disturbances. The value of M is dynamically changed according to SF requirements at different conditions. Input current and output voltage ripples changed according to varying M value.

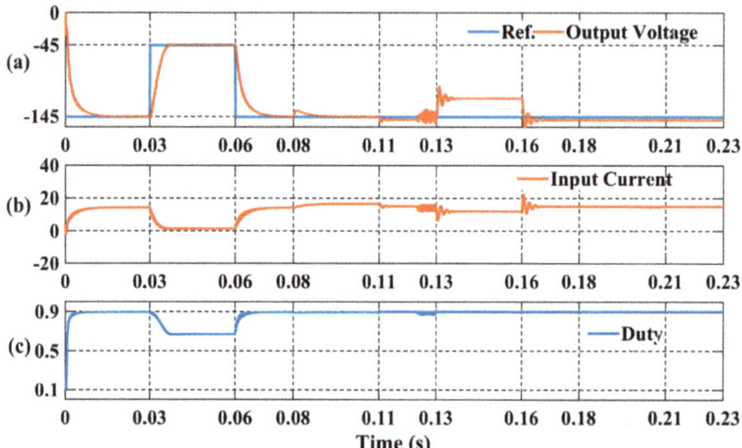

Figure 13. Comparison controller performance. (**a**) Output voltage; (**b**) Input current (u); (**c**) δ.

Table 4. Performance indices of modified Ćuk converter.

Perturbation	V_o Peak Overshoot	M	i_i Ripple (A)	V_o Ripple (V)
Steady State (No Perturbation)	N/A	0.3750	0.9	0.15
0.03 s: V_o −150 to −50 V 0.06 s: V_o −150 to −50 V	5.2 V 0	0.3476	0.7	0.14
0.08 s: R_L 100 to 50 Ω 0.11 s: R_L 50 to 100 Ω	37.5 V 50.5 V	0.3748	0.9	0.17
0.13 s: V_i 15 to 12 V 0.16 s: V_i 12 to 15 V	12.8 V 14.5 V	0.297	0.6	0.12
0.18 s: L_1 100 to 85 μH 0.21 s: L_1 85 to 100 μH	0 0	0.435	1.8	0.16

5. Conclusions

This paper proposed a modified high output gain Ćuk converter with an SMC-based equivalent controller. The efficiency and performance of a classical Ćuk converter was improved by the simple inclusion of a single inductor and capacitor. Moreover, a constant switching frequency cascaded equivalent controller structure is proposed. Simulation results show the effectiveness and robustness of the proposed method, and the constant switching frequency approach to the SMC-based controller provides the opportunity of simple application to real systems.

Acknowledgments: No source of funding for this project.

Author Contributions: All authors contributed and involved equally in framing the full version of the research article in its current form of decimation.

Conflicts of Interest: The authors declare no conflict of interest.

References

1. Sanjeevikumar, P.; Grandi, G.; Wheeler, P.; Blaabjerg, F.; Loncarski, J. A simple mPPT algorithm for novel PV power generation system by high output voltage DC-DC boost Converter. In Proceedings of the 24th International Symposium on Industrial Electronics, Rio de Janeiro, Brazil, 3–5 June 2015; pp. 214–220.
2. Mahajan, S.B.; Sanjeevikumar, P.; Blaabjerg, F. A Multistage DC-DC step-up self-balanced and magnetic component-free converter for photovoltaic application: Hardware implementation. *Energies* **2017**, *10*, 719.

3.   Mahajan, S.B.; Sanjeevikumar, P.; Blaabjerg, F.; Ojo, S.; Seshagiri, S.; Kulkarni, R. Inverting Nx and 2Nx non isolated multilevel boost converter for renewable energy application. In Proceedings of the 4th IET International Conference on Clean Energy and Technology, Kuala Lumpur, Malaysia, 14–15 November 2016; pp. 1–8.

4.   Sanjeevikumar, P.; Grandi, G.; Blaabjerg, F.; Wheeler, P.; Siano, P.; Hammami, M. A Comprehensive Analysis and Hardware Implementation of Control Strategies for High Output Voltage DC-DC Boost Power Converter. *Int. J. Comput. Intell. Syst.* **2017**, *10*, 140–152.

5.   Mahajan, S.B.; Sanjeevikumar, P.; Wheeler, P.; Blaabjerg, F.; Rivera, M.; Kulkarni, R. XY converter family: A new breed of buck boost converter for high step-up renewable energy applications. In Proceedings of the IEEE International Conference on Automatica, Curico, Chile, 19–21 October 2016; pp. 1–8.

6.   Bhaskar, M.S.; Sanjeevikumar, P.; Kulkarni, R.; Blaabjerg, F.; Seshagiri, S.; Hajizadeh, A. Novel LY converter topologies for high gain transfer ratio—A new breed of XY family. In Proceedings of the 4th IET International Conference on Clean Energy and Technology, Kuala Lumpur, Malaysia, 14–15 November 2016; pp. 4–8.

7.   Mahajan, S.B.; Kulkarni, R.; Sanjeevikumar, P.; Blaabjerg, F.; Fedák, V.; Cernat, M. Non-isolated and non-inverting Cockcroft Walton multiplier based hybrid 2Nx interleaved boost converter for renewable energy applications. In Proceedings of the IEEE Conference on 17th The Power Electronics and Motion Control, Varna, Bulgaria, 25–28 September 2016; pp. 146–151.

8.   Mahajan, S.B.; Kulkarni, R.; Sanjeevikumar, P.; Siano, P.; Blaabjerg, F. Hybrid Non-isolated and non-inverting Nx interleaved DC-DC multilevel boost converter for renewable energy applications. In Proceedings of the 16th IEEE International Conference on Environment and Electrical Engineering, Florence, Italy, 7–10 June 2016; pp. 1–6.

9.   Sanjeevikumar, P.; Kabalci, E.; Iqbal, A.; Abu-Rub, H.; Ojo, O. Control strategy and hardware implementation for DC-DC boost power conversion based on proportional-integral compensator for high voltage application. *Eng. Sci. Tech. Int. J.* **2014**, *18*, 163–170.

10.   Li, C.W.; He, X. Review of non-isolated high step-up DC/DC converters in photovoltaic grid-connected applications. *IEEE Trans. Ind. Electron.* **2011**, *58*, 1239–1250. [CrossRef]

11.   Costantino, N.; Serventi, R.; Tinfena, F.; D'Abramo, P.; Chassard, P.; Tisserand, P.; Fanucci, L. Design and test of an HV-CMOS intelligent power switch with integrated protections and self-diagnostic for harsh automotive applications. *IEEE Trans. Ind. Electron.* **2011**, *58*, 2715–2727. [CrossRef]

12.   Saponara, S.; Pasetti, G.; Tinfena, F.; Fanucci, L.; D'Abramo, P. HV-CMOS design and characterization of a smart rotor coil driver for automotive alternators. *IEEE Trans. Ind. Electron.* **2013**, *60*, 2309–2317. [CrossRef]

13.   Baronti, F.; Lazzeri, A.; Roncella, R.; Saletti, R.; Saponara, S. Design and characterization of a robotized gearbox system based on voice coil actuators for a Formula SAE Race Car. *IEEE/ASME Trans. Mechatron.* **2013**, *18*, 53–61. [CrossRef]

14.   Saponara, S.; Fanucci, L.; Bernardo, F.; Falciani, A. Predictive diagnosis of high-power transformer faults by networking vibration measuring nodes with integrated signal processing. *IEEE Trans. Inst. Meas.* **2016**, *65*, 1749–1760. [CrossRef]

15.   Zhu, M.; Luo, F.L. Enhanced self-lift Ćuk converter for negative to positive voltage conversion. *IEEE Trans. Power Electron.* **2010**, *25*, 2227–2233. [CrossRef]

16.   Sabzali, A.J.; Ismail, E.H.; Al-Saffar, M.A.; Fardoun, A.A. New bridgeless DCM Sepic and Ćuk PFC rectifiers with low conduction and switching losses. *IEEE Trans. Ind. Appl.* **2011**, *47*, 873–881. [CrossRef]

17.   Ćuk, S.; Middlebrook, R.D. Advances in switched-mode power conversion part I. *IEEE Trans. Ind. Electron.* **1983**, *1*, 10–19. [CrossRef]

18.   Ćuk, S.; Middlebrook, R.D. Advances in switched-mode power conversion part II. *IEEE Trans. Ind. Electron.* **1983**, *1*, 19–29. [CrossRef]

19.   Rashid, M.H. *Power Electronics Handbook: Circuits, Devices and Applications*; Elsevier: New York, NY, USA, 2010; pp. 249–265.

20.   Flores, J.L.; Avalos, J.; Espinoza, C.A.B. Passivity-based controller and online algebraic estimation of the load parameter of the DC-to-DC power converter Ćuk type. *IEEE Lat. Am. Trans.* **2011**, *9*, 50–57. [CrossRef]

21.   Mahdavi, J.; Nasiri, M.R.; Agah, A.; Emadi, A. Application of neural networks and state-space averaging to DC/DC PWM converters in sliding mode operation. *IEEE/ASME Trans. Mechatron.* **2005**, *10*, 60–67. [CrossRef]

22. Safari, A.; Mekhilef, S. Simulation and hardware implementation of incremental conductance MPPT with direct control method using Ćuk converter. *IEEE Trans. Ind. Electron.* **2011**, *58*, 1154–1161. [CrossRef]
23. Balestrino, A.; Landi, A.; Sani, L. Ćuk converter global control via fuzzy logic and scaling factors. *IEEE Trans. Ind. Appl.* **2002**, *38*, 406–413. [CrossRef]
24. Chen, Z. PI and sliding mode control of a Ćuk converter. *IEEE Trans. Power Electron.* **2012**, *27*, 3695–3703. [CrossRef]
25. Utkin, V.; Guldner, J.; Jingxin, S. *Sliding Mode Control in Electro-Mechanical Systems*; CRC Press: Boca Raton, FL, USA, 2009; pp. 325–355.
26. Venkataramanan, R.; Sabanovic, A.; Ćuk, S. Sliding mode control of DC–DC converters. In Proceedings of the 2012 IEEE 51st Annual Conference on Decision and Control, Maui, HI, USA, 10–13 December 2012.
27. Venkataramanan, R. Sliding Mode Control of Power Converters. Ph.D. Thesis, California Institute of Technology, Pasadena, CA, USA, 1986.
28. Tan, S.C.; Lai, Y.M.; Chi, K.T. General design issues of sliding-mode controllers in DC-DC converters. *IEEE Trans. Power Electron.* **2008**, *55*, 1160–1174.
29. Ahmad, F.; Rasool, A.; Ozsoy, E.E.; Sabanovic, A.; Elitas, M. A robust cascaded controller for DC-DC boost and Ćuk converters. *World J. Eng.* **2017**, *14*. [CrossRef]
30. Ahmad, F.; Rasool, A.; Ozsoy, E.E.; Sabanovic, A.; Elitas, M. Design of a robust cascaded controller for Cuk converter. In Proceedings of the IEEE International Power Electronics and Motion Control Conference, Varna, Bulgaria, 25–28 September 2016; pp. 80–85.
31. Cardoso, B.J.; Moreira, A.F.; Menezes, B.R.; Cortizo, P.C. Analysis of switching frequency reduction methods applied to sliding mode controlled DC-DC converters. In Proceedings of the 7th Annual Applied Power Electronics Conference and Exposition, Boston, MA, USA, 23–27 February 1992; pp. 403–410.
32. He, Y.; Luo, F.L. Sliding-mode control for dc–dc converters with constant switching frequency. *IEEE Proc. Control Theory Appl.* **2006**, *153*, 37–45. [CrossRef]
33. Agostinelli, M.; Priewasser, R.; Marsili, S.; Huemer, M. Constant switching frequency techniques for sliding mode control in DC-DC converters. In Proceedings of the 2011 Joint 3rd Int'l Workshop on Nonlinear Dynamics and Synchronization (INDS) & 16th Int'l Symposium on Theoretical Electrical Engineering (ISTET), Klagenfurt, Austria, 25–27 July 2011; pp. 1–5.
34. Tan, S.C.; Lai, Y.M.; Tse, C.K.; Cheung, M.K. A fixed-frequency pulse width modulation based quasi-sliding-mode controller for buck converters. *IEEE Trans. Power Electron.* **2005**, *20*, 1379–1392. [CrossRef]
35. Leung, K.K.S.; Chung, H.S.H. Dynamic hysteresis band control of the buck converter with fast transient response. *IEEE Trans. Circuits Syst. II Expr. Briefs* **2015**, *52*, 398–402. [CrossRef]
36. Ho, C.N.M.; Cheung, V.S.; Chung, H.S.H. Constant-frequency hysteresis current control of grid-connected VSI without bandwidth control. *IEEE Trans. Power Electron.* **2009**, *24*, 2484–2495. [CrossRef]
37. Park, S.H.; Park, S.R.; Yu, J.S.; Jung, Y.C.; Won, C.Y. Analysis and design of a soft-switching boost converter with an HI-Bridge auxiliary resonant circuit. *IEEE Trans. Power Electron.* **2010**, *25*, 2142–2149. [CrossRef]

Article

Coordinated Control Strategies for a Permanent Magnet Synchronous Generator Based Wind Energy Conversion System

Ramji Tiwari [1], Sanjeevikumar Padmanaban [2,*] and Ramesh Babu Neelakandan [1]

[1]   School of Electrical Engineering, VIT University, Vellore 632014, India; ramji.tiwari2015@vit.ac.in (R.T.); nrameshbabu@vit.ac.in (R.B.N.)
[2]   Department of Electrical and Electronics Engineering, University of Johannesburg, Auckland Park 2006, South Africa
*   Correspondence: sanjeevi_12@yahoo.co.in; Tel.: +27-79-219-9845

Academic Editor: Lucian Mihet
Received: 19 June 2017; Accepted: 14 August 2017; Published: 28 September 2017

Abstract: In this paper, a novel co-ordinated hybrid maximum power point tracking (MPPT)-pitch angle based on a radial basis function network (RBFN) is proposed for a variable speed variable pitch wind turbine. The proposed controller is used to maximise output power when the wind speed is low and optimise the power when the wind speed is high. The proposed controller provides robustness to the nonlinear characteristic of wind speed. It uses wind speed, generator speed, and generator power as input variables and utilises the duty cycle and the reference pitch angle as the output control variables. The duty cycle is used to control the converter so as to maximise the power output and the reference pitch angle is used to control the generator speed in order to control the generator output power in the above rated wind speed region. The effectiveness of the proposed controller was verified using MATLAB/Simulink software.

Keywords: wind energy conversion system; permanent magnet synchronous generator; maximum power point tracking; pitch angle; radial basis function network

---

1. Introduction

Renewable energy systems (RES) have drastically increased their contribution in the production of electrical energy, which is environmentally friendly, and thus minimising the dependency of the fossil fuel based power generation technique. Among the renewable based energy production systems, wind energy conversion systems (WECS) are mostly preferred for their wide availability and enhanced technology in implementation of the large capacity wind turbine. Due to steady growth in the increase of the power rating of wind turbines (WT) and integration to the power grid, additional advanced control strategies are required to make the wind energy systems more reliable and feasible for grid integration [1].

Currently, variable speed wind turbine (VSWT) systems are mostly preferred over fixed speed wind turbines (FSWT) because VSWT can harness the electrical power from all wind speed regions by controlling their shaft speed based on the wind velocity. Various variable speed turbines are commercially available on the market such as the doubly fed induction generator (DFIG) and the permanent magnet synchronous generator (PMSG). Among the VSWT, PMSG based WECS has received much attention because of its several advantages such as gear-less operation, high efficiency, low maintenance, less noise, and high robustness. Direct driven PMSG provides higher power output and lower mechanical stress [2]. The typical diagram of direct driven PMSG with basic topology and control strategies is shown in Figure 1.

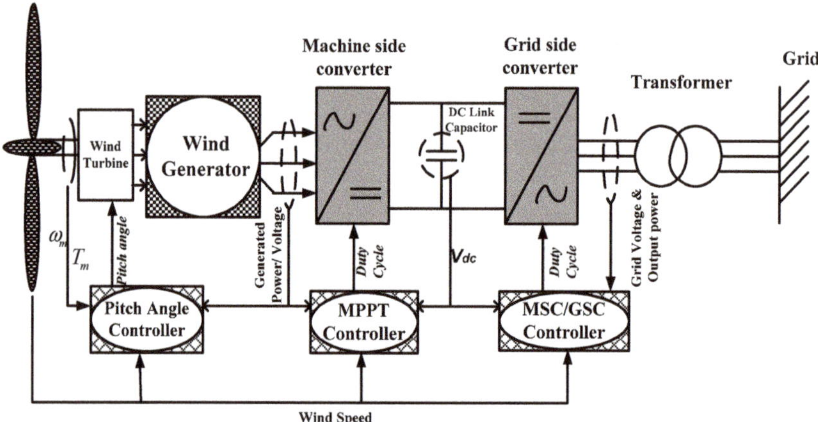

Figure 1. Basic topology of wind energy conversion systems (WECS) with control strategies.

The trend of installing large scale wind turbines will increase in the near future. Regardless of this, there are numerous challenges for the effective generation of wind power from the available wind speed as wind velocity is highly non-linear. Thus to optimise the WECS, various control strategies have been presented in the literature such as maximum power point tracking (MPPT), pitch angle and grid and machine side controller as shown in Figure 1 [1]. The machine and grid side controller are mainly used to penetrate the stabilised power in the grid following the grid code. The MPPT control strategy is employed to extract the maximum available power from the available wind whereas the pitch angle controller is used to regulate the power within the rated power when the power exceeds the rated power of the wind generator, which may lead to serious damage to the system [3]. The operating regions of MPPT and the pitch angle controller can be seen in Figure 2.

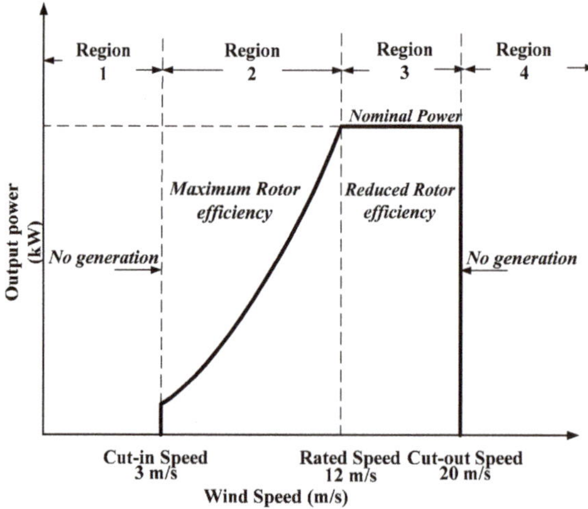

Figure 2. WECS operating regions based on wind speed.

The rapid variation and frequent discontinuity of wind pattern is one of the major operational problems which exist in WECS. The older wind turbines are equipped with basic control strategies which are not suitable for the current scenario [4]. The repowering of an existing wind turbine with modern technology is the most preferred solution to upgrade wind farms. The new innovation in wind power technology should be installed in each wind turbine in order to achieve maximum efficiency and to be capable enough for grid integration [5]. The repowering of a wind turbine with the new technology is very costly which is an additional burden to the country or power producers.

In this paper, a novel hybrid MPPT and pitch angle control strategy is employed which can be equipped with the existing wind energy conversion system. The older wind turbines are usually of low rating. Thus they are usually not operated when the wind speed exceeds the rated wind speed. Thus to prevent this, pitch angle control strategy is employed. Pitch angle control strategy regulates the power when the wind velocity exceeds the rated speed and hence WECS is operated in that region. The MPPT strategy is used to increase the power yield when the wind velocity is below the rated wind speed. The MPPT control technique extracts the maximum available power which is present in the available wind speed. Thus by employing this intelligent hybrid control, the performance of WECS is improved as the system becomes robust and further the cost of repowering an existing wind turbine is also eliminated. This controller works on all the operating regions of the wind system.

2. Modelling of the Wind Energy Conversion System

The configuration of the proposed WECS is shown in Figure 3. The system consists of a direct driven PMSG, diode rectifier, boost converter and a proposed hybrid control strategy for a standalone system. The boost converter is incorporated with the MPPT control strategy which generates duty cycle according to the available wind speed [6]. The pitch angle is initiated when the wind velocity exceeds the rated wind speed. The pitch angle command controls the angle of the blade in such a way that the rotational speed of the shaft is kept at the optimum value.

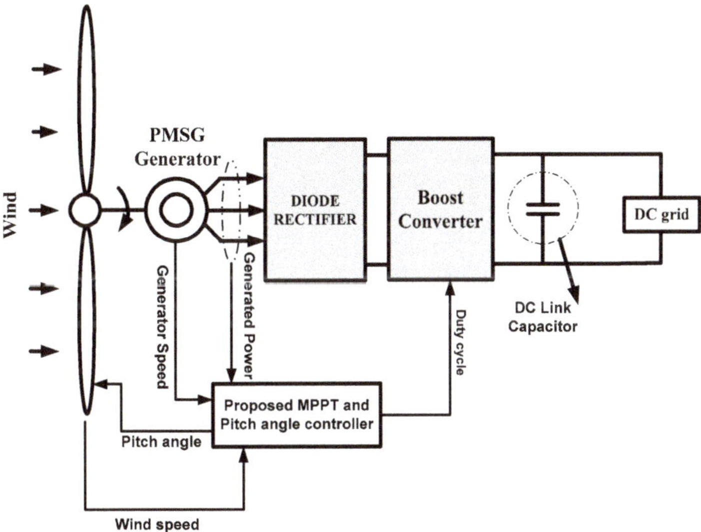

Figure 3. Proposed configuration for the permanent magnet synchronous generator (PMSG) based wind energy conversion system.

2.1. Wind Turbine Aerodynamic Model

The mechanical power (P_m) that can be extracted from the available wind by the wind turbine is given by [7],

$$P_m = \frac{1}{2}\rho A C_p(\lambda, \beta)v^3 \tag{1}$$

where ρ is the air density, A is the total area swept by the blades, v is the wind velocity and the power coefficient (C_p) of the wind system is determined using the tip speed ratio (λ) and the blade pitch angle (β). The pitch angle is always kept constant when the MPPT control strategy is operational. The optimal value at which the wind turbine extracts the maximum power (P_{max}) for the available wind speed is given by [8],

$$\lambda = \frac{\omega_r R}{v} \tag{2}$$

$$P_{max} = \frac{1}{2}\rho A \frac{C_{pmax} R^3}{\lambda_{opt}^3} * \omega_r^3 \tag{3}$$

where ω_r is the rotor rotational speed, R is the rotor radius and C_{pmax} is the maximum power coefficient at the optimal tip speed ratio (λ_{opt}). Figure 3 represents that the power generated (P_m) by the turbine purely depends upon the power coefficient (C_p) at a particular wind speed. The tendency of the wind turbine to extract the maximum power from the available wind speed declines by 9% after a life span of five years of operation [8]. Thus an efficient and intelligent control strategy should able to cope with the deficiency and extract the maximum power for an entire operational period.

2.2. PMSG Modelling

PMSG based WECS is the mostly preferred wind turbine for its flexibility and efficiency. The PMSG can deliver power at the desired power factor based on the requirement [9]. The dynamic equations of the three-phase salient pole PMSG in the d–q reference frame are described as [10],

$$\left.\begin{aligned} v_q &= Ri_q + p\lambda_q + \omega_s\lambda_d \\ v_d &= Ri_d + p\lambda_d - \omega_s\lambda_q \end{aligned}\right\} \tag{4}$$

where v_d, i_d represents the stator voltage and current in the d axis and v_q, i_q in the q axis respectively. R denotes the stator resistance of PMSG. λ_q, λ_d are the stator flux linkages of the d, q axis respectively given as,

$$\left.\begin{aligned} \lambda_q &= L_q i_q \\ \lambda_d &= L_d i_d + L_{md}I_{fd} \end{aligned}\right\} \tag{5}$$

where L_{md}, I_{fd} are the mutual inductance and magnetising current in the d axis. L_d, L_q are the self inductances of the d axis and q axis respectively. ω_s is the stator frequency, which is represented as,

$$\omega_s = n_p\omega_r \tag{6}$$

where n_p is the number of poles and ω_r is the rotation rotor speed.

3. Control Strategy of WECS

Wind energy conversion systems demonstrate the challenges of rapid variation in wind speed, nonlinearity and uncertainty. Thus an advanced controller is required to solve them efficiently. Integrating an advanced controller into WECS is done in order to increase efficiency in terms of power conversion and blade control design. Many researches have been carried out to develop a control strategy of WECS which can be integrated into the grid. The controllers must be simple, reliable, and cost effective. Moreover, they must able to withstand the fluctuations caused during their operation.

The most preferred control strategy for producing optimal and quality power from WECS are the MPPT and pitch angle controller [1].

3.1. Maximum Point Tracking Controller

To improve the energy capture efficiency in a modern WECS, an efficient and advanced MPPT technique is required. To operate the WECS at that specific point, various MPPT algorithms have been proposed in the literature [9,11–13]. The foremost controllers which are widely used are power signal feedback (PSF), hill climb search (HCS) or perturb and observe (P&O), tip speed ratio (TSR), optimal torque control (OTC), and soft-computing based techniques like fuzzy logic control (FLC) and artificial neural network (ANN) [1].

The power signal feedback (PSF) based MPPT controller technique tends to reduce the error between actual power and reference power. The PSF controller requires pre knowledge of the wind turbine. The value is recorded in a lookup table and the optimal power is obtained based on the available wind speed. The most advanced PSF controller uses DC voltage and DC current as an input rather than a power and speed shaft which reduces the use of a speed sensor. The major drawback of this system is the complexity in implementation. The HCS or P&O method is the widely preferred MPPT algorithm for its simplicity and low cost. This method is based on comparing the obtained power from the previous power and generating the appropriate duty cycle based on the comparison. The major drawback of this methodology in tracking the maximum power in the wind energy system is that it fails to follow the rapid variation in wind velocity. The convergence speed and efficiency is reduced when the P&O algorithm is subjected to the highly non-linear system. OTC control strategy adjusts the generated torque of the wind turbine based on the reference torque where the maximum power can be extracted at the particular wind speed. The reference torque is compared with the actual torque and an error signal is generated which is fed into the controller to maintain the optimal torque of the generator. OTC control strategy is simple, fast and efficient but the major drawback in this control strategy is that it does not measure the wind speed directly and hence the change in wind speed is not observed in the reference directly.

Thus to overcome the above issues, a soft computing based MPPT strategy like FLC and ANN controllers is implemented. Soft computing controllers do not require the mathematical knowledge of the system. Soft computing control strategy has a faster convergence speed and is highly reliable. Fuzzy logic based MPPT controller is suitable for a region where there is rapid and continuous variation in wind velocity. FLC technique has a fast response towards a change in system dynamics without the knowledge of system parameters. The FLC can overcome the high non-linearity of the system which is an important parameter for wind power system. The efficiency of FLC is based purely based on the selection of the input parameters and rule implementation. Thus an error caused by the control designer can have a severe effect on the efficiency. The major setback of FLC is the flexibility towards the changing of system parameters after implementation. To overcome this issue an ANN based controller is used. The ANN controllers use reference values to determine the maximum power of the system. The ANN method has a faster convergence speed than that of previous MPPT controllers. To realise an MPPT operation a power electronic converter (PEC) is necessary. In this paper, the boost converter is used as the PEC to interface the wind generator with the load.

3.2. Pitch Angle Controller

The pitch angle controller is implemented to limit the aerodynamic power captured by the wind turbine when the wind velocity is above the rated value [14]. Several pitch angle controllers have been suggested in past literature [1,15–18]. The most common pitch angle controllers are the proportional-integral (PI) controllers. They are simple and cost effective. However, the major disadvantage of the system in the performance of the controller is minimised due to frequent changing of the operating point because of the rapid variation in the wind speed. Another method to vary blade angle is the H-infinity. This controller gives enhanced performance of output power and provides robustness to the variation of the wind. However, the major problem is that it is a complex system to design and when a constraint changes, the redesigning process is time consuming and difficult. Linear quadratic Gaussian (LQG) based pitch angle controller is robust in nature towards wind turbine parameter but it lacks in adaptation to the non-linearity of the wind system. The other technique which is employed in controlling the speed of the shaft is the sliding mode control. It is suitable for a highly non-linear system. The major disadvantage of the system is the requirement of a mathematical model of the system as well as sudden large changes in the wind turbine which increase the mechanical stress of the WECS thus damaging the mechanical parts associated with it.

To vary the pitch angle of the blade a more feasible and accurate soft computing technique is implemented. The fuzzy logic controller and artificial neural network are some of the soft computing based control strategies which are implemented to control the blade angle. These methods are reliable and robust with regard to the non-linear characteristics of pitch angle with wind speed. These controllers require wind speed information which uses an anemometer thus the cost of the overall system is increased. The reliance of the system on the sensor degrades the performance of the overall efficiency. In addition, installation of the anemometer is not feasible for old wind turbines when repowering is considered a primary agenda.

3.3. Proposed Coordinated Hybrid MPPT-Pitch Angle Control Strategy

The operating regions of the wind turbine are generally classified into two regions based on the wind speed as shown in Figure 2. In the region where the wind speed is lower than that of the rated wind speed, the turbine speed is controlled at the optimal value using MPPT strategy so that the maximum power can be extracted from the wind system based on the available wind speed (Region 2). In the region where the wind speed exceeds the rated value, the pitch angle control strategy optimises the output power by controlling the blade angle which limits the turbine speed (Region 3). The block diagram of the proposed control strategy using the radial basis function network (RBFN) is shown in Figure 4. The proposed topology consists of the boost converter and PMSG generator. The DC link capacitor (C_L) and output capacitor (C_O) are also present in the topology. The inputs for the RBFN control techniques are wind speed, generated speed, and generated power. The outputs of the proposed control are the duty cycle which is activated in the low wind speed region and the pitch angle which is triggered in the high wind speed region. In a region where the wind speed is below the rated wind velocity, the reference power (P_{ref}) is determined using Equation (3),

$$P_{ref} = K_{opt} * \omega_r^3 \tag{7}$$

where

$$K_{opt} = \frac{1}{2}\rho A \frac{C_{pmax} R^3}{\lambda_{opt}^3} \tag{8}$$

The pitch angle during Region 2 is kept near zero so as to increase the power co-efficient (C_p).

In the high wind speed region, the reference power (P_{ref}) is considered the same as the rated power of the wind turbine. The pitch angle should increase so as to decrease the power coefficient (C_p) thus limiting the power to the nominal value. In Region 3, the output power is maintained constant at the rated power of the wind turbine. Equation (3) is used to determine the relation between the power coefficient (C_p) and the pitch angle which is deduced in Equation (9)

$$C_p(\lambda, \beta) = \frac{2P_{max}}{\rho A v^3} \tag{9}$$

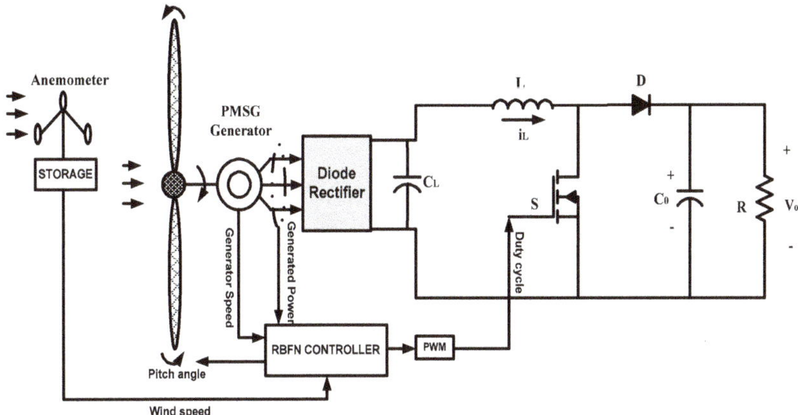

Figure 4. Proposed coordinated maximum power point tracking (MPPT) control strategy.

By replacing the nominal value of power, the optimal value of the power coefficient $C_{p\,opt}$ which maintains the power at the nominal (P_{nom}) value for variable wind speed is obtained from Equation (10).

$$C_p(\lambda, \beta) = C_{p\,opt}(\lambda, \beta) = \frac{2P_{nom}}{\rho A v^3} \tag{10}$$

Thus by optimising the power coefficient value and obtaining the new $C_p(\lambda, \beta)$ value the corresponding value of the pitch angle can be obtained using Equation (11),

$$C_p(\lambda, \beta) = (0.5 - 0.0167(\beta - 2)) \sin\left[\frac{\pi(\lambda + 0.1)}{18 - 0.3(\beta - 2)}\right] - 0.00184(\lambda - 3)(\beta - 2) \tag{11}$$

From Equation (11), the corresponding pitch angle value is obtained and fed to the system when the wind speed is higher than that of the rated wind speed.

Thus by utilising the Coordinated hybrid MPPT-Pitch angle control strategy, the output power can be maximised in the low wind speed region and optimised in the high wind speed region. Thus improving the overall efficiency of WECS and eliminating the need for repowering.

The WECS is highly non-linear in nature, thus an efficient and complex problem solving controller is required to enhance the overall performance. In this paper, RBFN based ANN control strategy was used to predict the precise duty cycle for the MPPT controller when the wind speed was below the rated speed as well as the specific angle for the blade for the pitch angle control strategy when the wind velocity exceeded the rated value.

Radial Basis Function Network

RBFN is a type of feed forward neural network which uses radial basis network as an activation function. The radial basis network is determined by the distance between the input and the prototype vector [19]. The RBFN is similar to the multi-layer perceptron (MLP) network. RBFN has three layers of input layer, hidden layer, and output layer as shown in Figure 5. The network neurons are connected to each other. From Figure 5, it can be seen that there is no weight coefficient between the input layer and the hidden layer. Hence the neuron in the hidden layer receives the same variables as in input layer. The training process of the RBFN network is carried out in two stages [20]. In the first stage, the unsupervised method is implemented where the parameter is governed by the radial basis function. In the second stage, the supervised training method is employed to train the weights which are the same as the back propagation algorithm [21].

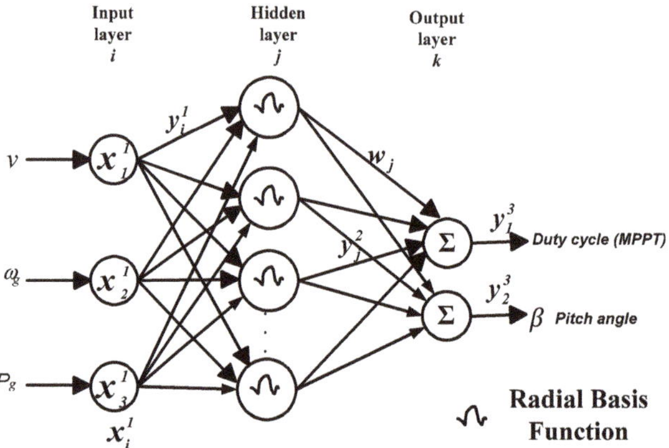

Figure 5. Proposed architecture of radial basis function network (RBFN) based hybrid control strategy.

In this paper, the RBFN controller is implemented to control the WECS based on the wind speed regions. In Region 2 where the wind speed is lower than that of the rated wind speed, the RBFN network generates the duty cycle for the PEC based on the corresponding wind speed to maximise the output power of the system. In Region 3 where the wind speed is greater than that of the rated wind velocity, the RBFN gives the appropriate pitch angle to the blades of the wind turbine to limit the output power. The wind speed, generated output power, and generator speed are fed to the input neurons of the RBFN which are used to compute the duty cycle and pitch angle as the output neuron based on the wind speed. The key considerations of the selection of input variables are the relevant data, training network, computational effort, dimensionality, and comprehensibility. The important parameters which are suggested for input selection in the neural network are the availability of the variables, correlation between the selected input variables, and inputs with minimum or zero prediction [22].

The basic nodes of operation are characterized into three layers [23],

The inputs of the three neurons in this layer are transmitted directly to the next layer. The net input and output are represented as,

$$\left.\begin{array}{l} net_i^1 = x_i^1(N) \\ y_i^1(N) = f_i^1\left(net_i^1(N)\right) = net_i^1(N) \end{array}\right\}_{i=1,2,3} \tag{12}$$

where x_i^1 is the input layer which consists of x_1^1 as the wind speed, x_2^1 as the generator power, and x_3^1 as the generator speed. The net_i^1 represents the net sum of nodes of the input layer and y_i^1 is the output of the input layer which is fetched to the hidden layer with respect to the node i.

The neurons in the hidden layer perform using the Gaussian membership function in RBFN. The net input and output of the hidden layer are represented as,

$$\left.\begin{array}{l} net_j^2(N) = -\left(X - M_j\right)^T \sum_j \left(X - M_j\right) \\ y_j^2(N) = f_j^2\left(net_j^2(N)\right) = \exp\left(net_j^2(N)\right) \end{array}\right\}_{j=1,2,\ldots,800} \tag{13}$$

where $M_j = \left[m_{1j}, m_{2j}, \ldots, m_{ij}\right]^T$ is the mean of the Gaussian function and the standard deviation of the Gaussian function is denoted as $\sum_j = diag\left[1/\sigma_{1j}^2, 1/\sigma_{2j}^2, \ldots, 1/\sigma_{ij}^2\right]^T$.

The output layer computes two neurons which are determined by node *k*. The MPPT control signal and pitch angle control signal are generated in this layer by summing all the incoming signals with the linear activation function. The node *k*-1 represents the duty cycle based on the wind speed and node *k*-2 represents the pitch angle.

$$\left. \begin{array}{l} net_k^3 = \sum_j w_j y_j^2(N) \\ y_k^3(N) = f_{ki}^3\left(net_k^3(N)\right) = net_k^3(N)_{k=1,2} \end{array} \right\} \tag{14}$$

where w_j is the weight which connects the hidden layer and output layer.

Supervised learning is implemented once the RBFN is initialised to train the system. The training method is the same as the back propagation algorithm which is used to adjust the RBFN parameters using the training patterns. The error of each layer is calculated and updated by the supervised learning algorithm in order to track the performance of the wind system and act appropriately. The error of each output neuron is calculated using the squared error function. The total error of the system is given as [24,25]

$$E_{total} = \sum \frac{1}{2}(target - output)^2 \tag{15}$$

where the target refers to the pre-defined data and the output is the data obtained.

To evaluate the effectiveness of the proposed controller, the RBFN based controller is utilised for the system when only the MPPT controller and pitch angle control strategy are employed as shown in Figures 6 and 7. The wind speed and generated power are taken as the input parameters of the controller technique whereas the duty cycle is generated as the output of the power electronic converter (PEC) when only MPPT control strategy is used. The RBFN based MPPT control strategy implemented in this topology is shown in Figure 6 [25].

The wind speed and generated speed are considered as the input for the pitch angle control strategy. The ripple components of the generator speed and the output power are also eliminated and the parameters are subjected within the rated value. Figure 7 shows the pitch angle control strategy [26–33].

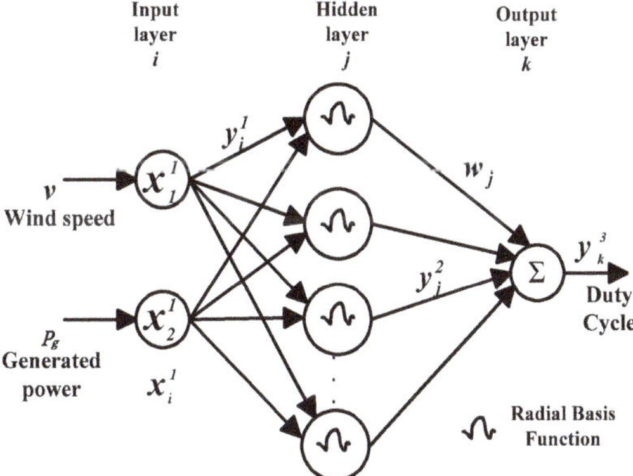

Figure 6. RBFN based MPPT control strategy.

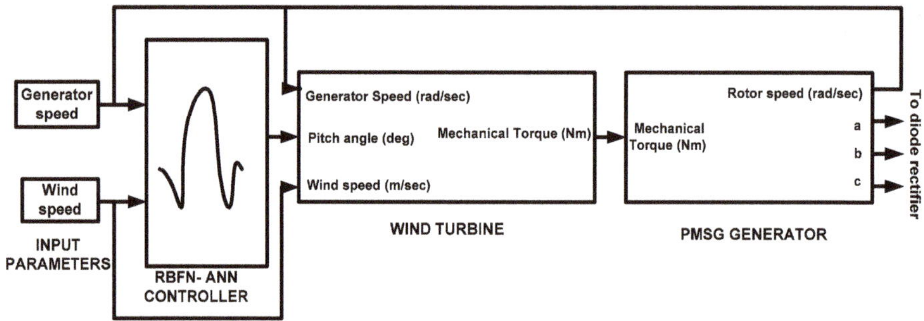

Figure 7. RBFN based pitch angle control strategy.

4. Results and Discussion

To validate the proposed system, the simulation is performed in MATLAB/Simulink for the AEOLOS 3kW wind turbine system. The parameters of AEOLOS 3kW wind turbine and PMSG are given in Table 1. To verify the performance of the coordinated hybrid MPPT-pitch angle control strategy, the simulation is carried out and compared with the control technique employing only the MPPT technique and only the pitch angle control technique when subjected to rapid variations in wind velocity as shown in Figure 8. The rated wind speed considered here is 12 m/s. The performance of the proposed system was validated and compared using the system employing only MPPT and pitch angle control technique as shown in following sections.

Table 1. Parameters of the Aeolos 3 kW system.

Parameters	Rating
Rated power	3 kW
Rated wind speed	12 m/s
Cut-in win speed	3.0 m/s
Cut-out wind speed	25 m/s
Frequency	50 Hz
Voltage	220–240 V
Rotor diameter	5.0 m
Generator type	Three phase PMSG

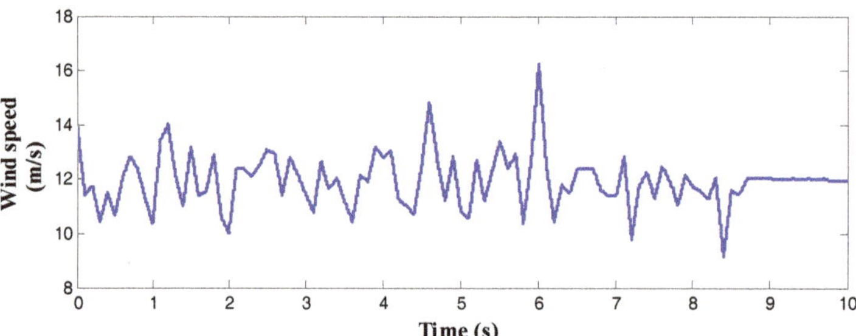

Figure 8. Wind speed—Input pattern.

4.1. Only MPPT

The performance of WECS when subjected to wide wind speed region employing RBFN based MPPT control strategy is discussed in this section [26]. The input parameters are the wind speed and generator power to bring uniformity in the selection of parameters. The WECS comprises the same wind turbine parameters and wind speed as stated earlier. The pitch angle at this operation is kept fixed at $-5°$. The MPPT technique is utilised in both the operating regions (Region 2 and Region 3). In Region 2 where the wind velocity is below the rated value, the MPPT tends to extract the maximum available power at the present wind speed and when the speed surpasses the rated value the MPPT technique optimises the voltage and power to the rated value by adjusting the duty cycle. The performance of the MPPT controller in Region 2 is far better than when compared to Region 3.

Figure 9 shows the DC voltage obtained when the MPPT control strategy is used as the control strategy. As the wind variation is between 9 m/s to 16 m/s the DC voltage obtained is constant. The rated value of the DC voltage is kept as 380 V.

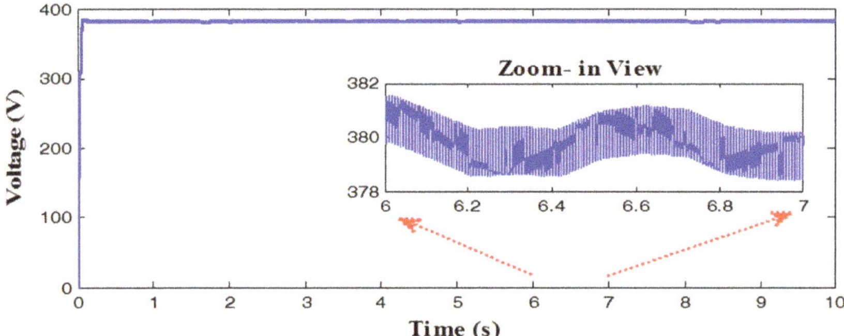

Figure 9. DC voltage—MPPT control strategy.

The maximum power extracted from the available wind speed is shown in Figure 10. The MPPT control strategy can also optimise the power when the wind speed is above the rated speed.

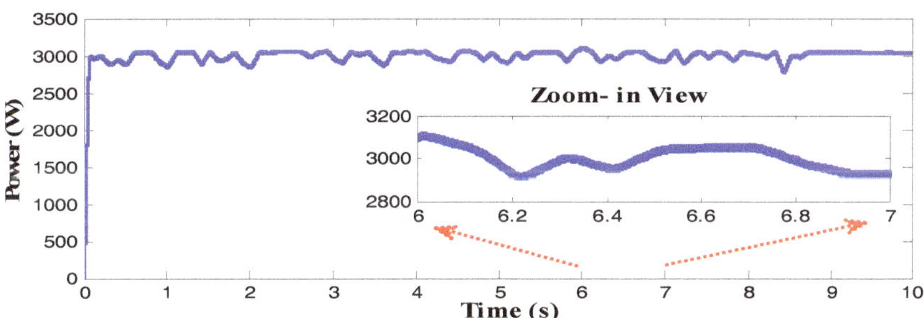

Figure 10. Output power—MPPT control strategy.

The generated power when the wind velocity is above the rated speed is optimised to some extent but still, there is a peak which may increase the turbine stress and damage the system. The PEC associated with the WECS also is designed to withstand up to 5% of the rated value as when there is a sudden hike in the power the whole power electronic components will also be damaged. Thus the wind turbine is subjected to freewheel once the threshold wind speed is obtained thus generating no or very low power during high wind speed conditions. The zoom-in view of the results are also presented between the time range 6 s to 7 s, where the wind input is subjected to sudden gusts in order to validate the performance of the control strategy.

4.2. Only Pitch Angle

The performance of pitch angle control strategy using RBFN control technique is discussed in this section. The input parameters are chosen as wind speed and generator speed. The duty cycle during this strategy is kept constant at an optimum value of 42.1%. The pitch angle control strategy is employed for both the operating regions. In Region 2 the pitch angle control strategy tends to extract the maximum available power from the wind turbine by controlling the power coefficient of the wind turbine. The main application of the pitch angle controller is to optimise the output power beyond the rated power so as to keep the generator producing power in high wind speed conditions. Since the pitch angle is mostly preferred for Region 3, their performance suffers in Region 2. Thus operation of the pitch angle frequently at the all wind speed regions also increases the stress in the system hence the occurrence of break downs is also increased.

Figure 11 shows the output voltage obtained when the pitch angle control strategy is implemented. A constant rated DC voltage of 380 V is obtained at the load despite the high non-linearity.

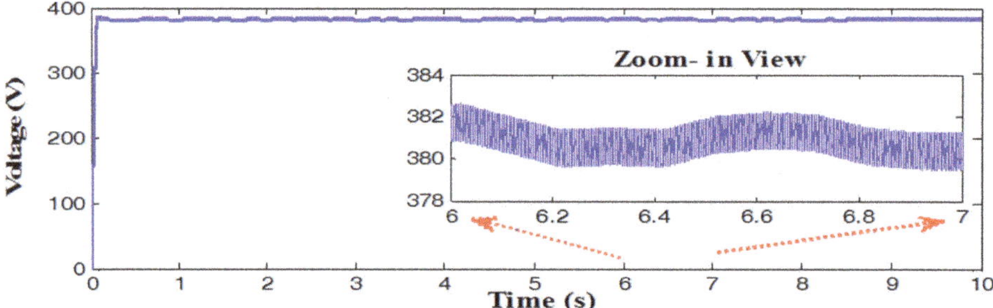

Figure 11. DC voltage—Pitch angle control strategy.

The generated power obtained when the pitch angle control strategy is implemented is shown in Figure 12.The comparison of maximum power obtained when the system is operating in Region 2 is low compared to when MPPT control strategy is used. However, in Region 3, where the system is operating at high wind speed the pitch angle control strategy successfully optimises the output power without any fluctuations or peaking.

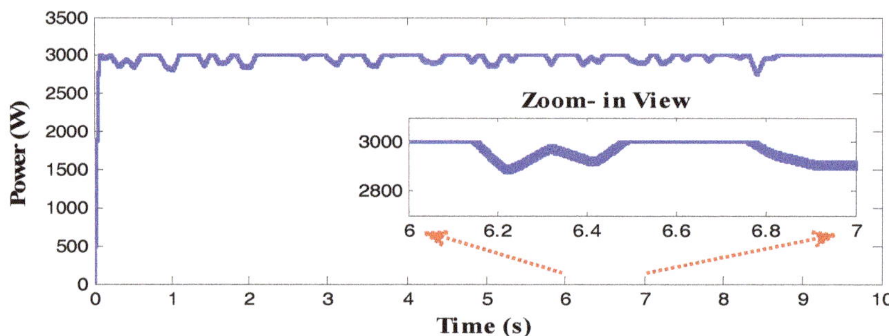

Figure 12. Output power—Pitch angle control strategy.

The pitch angle control strategy when operated to obtain maximum power is subjected to the power coefficient of the turbine. Thus the power coefficient obtained when the pitch angle control strategy operates in the stated system is shown in Figure 13.

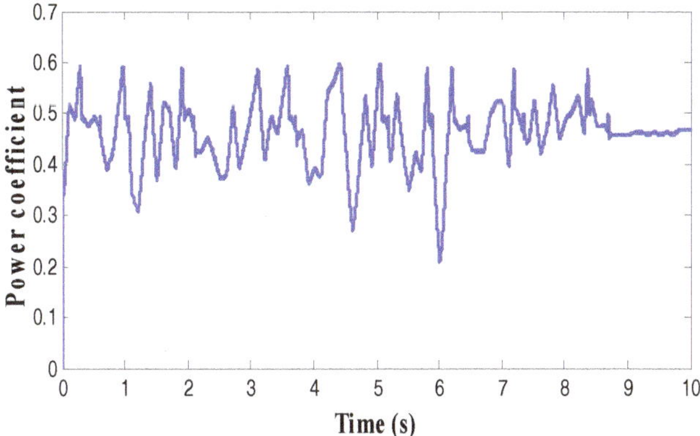

Figure 13. Power coefficient—Pitch angle control strategy.

The performance analysis of the individual controller in the wide wind speed operating region is shown in Table 2. Thus from the analysis, it is observed that to operate the WECS in all wind speed operating regions efficiently it is necessary to develop a new intelligent control strategy.

Table 2. Performance analysis of individual maximum power point tracking (MPPT) and pitch angle control strategy.

Control Strategy	Average Output Power		Average Output Voltage	
	Below Rated Wind Speed	Above Rated Wind Speed	Below Rated Wind Speed	Above Rated Wind Speed
MPPT	2921 W	3097 W	376 V	382 V
Pitch angle	2883 W	3002 W	377 V	380 V

4.3. Proposed Coordinated Hybrid MPPT-Pitch Angle Control Strategy

The proposed control strategy utilises the MPPT control technique when the wind speed is lower than that of the rated wind speed and the pitch angle control strategy when the wind velocity surpasses the rated value. Thus by combining both the control strategies the drawbacks of each individual strategy in maximising and optimising the power can be eliminated. Taking high non-linearity into account, RBFN based ANN control technique can be considered for this technique. The input parameters considered for the proposed technique are wind speed based on which the control technique is decided, generator power to determine the maximum power obtained, and generator speed to obtain the speed of the system in order to optimise it to the rated value. As stated earlier, to fairly evaluate the performance of hybrid control strategy and compare it with the individual control technique the same input parameters are considered. The output of the RBFN based control strategy is the duty cycle for the below rated wind speed condition and the pitch angle for the above rated wind speed condition.

The DC voltage of WECS obtained at the load when the proposed controller is used is shown in Figure 14. The DC voltage which is kept constant at the rated value of 380 V is achieved by using this control strategy. The wind speed variation considered here varies only from 9 m/s to 16 m/s. Thus the rated voltage is obtained and kept constant throughout the operation. In standalone wind systems when the wind speed drops below the transition region (7 m/s in this topology), the rated voltage value also dips. Thus the wind speed should be above the transition value in order to generate a constant DC voltage.

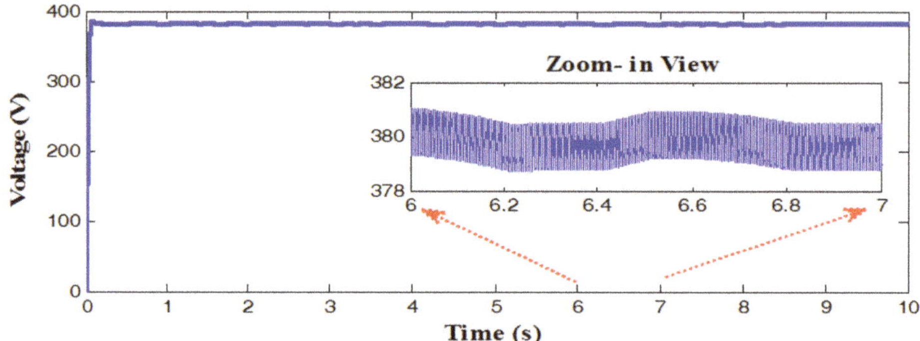

Figure 14. DC voltage—Proposed controller.

The output power which is obtained by using the proposed control strategy is shown in Figure 15. The output power which is achieved in Region 2 is similar to that of the system which uses only MPPT control strategy. Additionally, the output power obtained when operating the WECS in Region 3 is similar to the system which is employed for only pitch angle control strategy. Thus by using the proposed coordinated controller, the performance and efficiency can be enhanced over a wide speed range and prevent the future damage caused by over speeding of the turbine due to high wind speed.

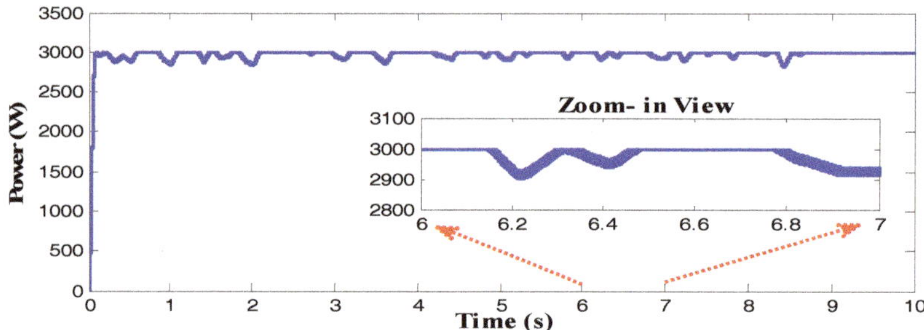

Figure 15. Output power—Proposed controller.

The important parameter which is used to maximise the output power and determine the optimisation of the power in Region 3 is the power coefficient (C_p) which is shown in Figure 16. When the wind speed is lower than the rated wind speed, the generated power is lower than the rated power thus MPPT control strategy is applied in the range of 42.1% to 46.5% for wind speed ranging from 9 m/s to 16 m/s with the pitch angle tuned to $-5°$. Thus the power co-efficient is fixed at 0.456. When the wind speed is higher than that of the rated wind speed (Region 3) the duty cycle of the system is fixed at 42.1% and the optimal pitch angle in the range $-5°$ to 18.4° is generated based on the wind speed in order to limit the output power. Region 4 is a no generation region. The WECS produces no power in that region. The optimal power coefficient is obtained using Equation (10) where the corresponding pitch angle is generated using Equation (11).

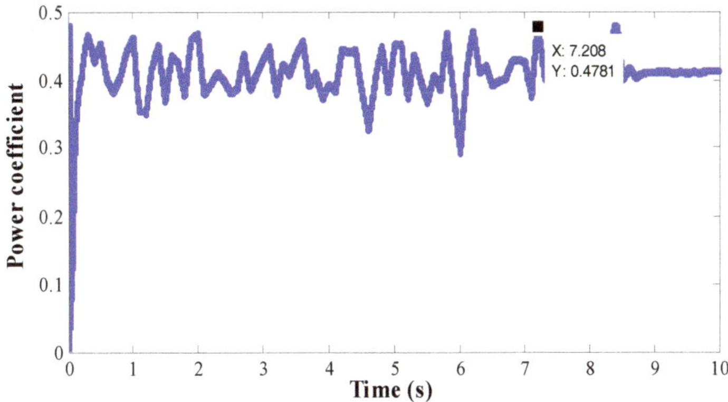

Figure 16. Power coefficient—Proposed controller.

The overall performance of the proposed control strategy is shown in Table 3. The optimum voltage of 380 V is kept as the rated voltage for WECS which is termed as the standard value for the DC microgrid. A rated power of 3 kW is used which is suitable for both standalone and grid connected application.

Table 3. Performance analysis of individual MPPT and pitch angle control strategy.

Control Strategy	Average Output Power		Average Output Voltage	
	Below Rated Wind Speed	**Above Rated Wind Speed**	**Below Rated Wind Speed**	**Above Rated Wind Speed**
Proposed coordinated MPPT and Pitch angle	2943 W	3008 W	376.5 V	381 V

5. Conclusions

In this research, a novel coordinated hybrid MPPT-Pitch angle control strategy employing RBFN based ANN technique for PMSG based WECS was proposed to enhance the overall efficiency of wind power generation in all operating regions. The proposed controller tracks the maximum power when the wind speed is below the rated wind speed and limits the output power in the high wind speed region. To develop the controller, a highly non-linear wind speed input was considered. The wind speed, generator speed, and generated power were selected as the control input variables. The duty cycle and pitch command to the blade were considered as the controlled output variables. The selection of which output variable to be activated is based on the wind speed. In the low wind speed region, the duty cycle is generated to control the power electronics converter thus obtaining maximum available power from the wind speed. When the wind speed exceeds the rated wind velocity, the blade tends to change its angle in order to limit the output power. Thus by using the proposed hybrid control strategy, the output power is maximised with an efficiency of 98.1% compared to 97.3% achieved using an individual MPPT control strategy and 96% using an individual control strategy in below rated wind speed; generator power can be optimised and regulated it the rated value of 3 kW in high wind speed regions. The proposed controller is a suitable alternative for the repowering of small scale WECS, many of which have been installed in developing countries.

Author Contributions: All the authors contributed equally to this work.

Conflicts of Interest: The authors declare no conflict of interest.

References

1. Tiwari, R.; Babu, N.R. Recent developments of control strategies for wind energy conversion system. *Renew. Sustain. Energy Rev.* **2016**, *66*, 268–285. [CrossRef]
2. Babu, N.R.; Arulmozhivarman, P. Wind energy conversion systems-a technical review. *J. Eng. Sci. Technol.* **2013**, *8*, 493–507.
3. Van, T.L.; Nguyen, T.H.; Lee, D.C. Advanced pitch angle control based on fuzzy logic for variable-speed wind turbine systems. *IEEE Trans. Energy Convers.* **2015**, *30*, 578–587. [CrossRef]
4. Vidal, Y.; Acho, L.; Cifre, I.; Garcia, A.; Pozo, F.; Rodellar, J. Wind Turbine Synchronous Reset Pitch Control. *Energies* **2017**, *10*, 770. [CrossRef]
5. Del, R.P.; Silvosa, A.C.; Gómez, G.I. Policies and design elements for the repowering of wind farms: A qualitative analysis of different options. *Energy Policy* **2011**, *39*, 1897–1908. [CrossRef]
6. Daili, Y.; Gaubert, J.P.; Rahmani, L. Implementation of a new maximum power point tracking control strategy for small wind energy conversion systems without mechanical sensors. *Energy Convers. Manag.* **2015**, *97*, 298–306. [CrossRef]
7. Tiwari, R.; Babu, N.R. Fuzzy Logic Based MPPT for Permanent Magnet Synchronous Generator in wind Energy Conversion System. *IFAC-PapersOnLine* **2016**, *49*, 462–467. [CrossRef]
8. Wei, C.; Zhang, Z.; Qiao, W.; Qu, L. An adaptive network-based reinforcement learning method for MPPT control of PMSG wind energy conversion systems. *IEEE Trans. Power Electron.* **2016**, *31*, 7837–7848. [CrossRef]
9. Tripathi, S.M.; Tiwari, A.N.; Singh, D. Grid-integrated permanent magnet synchronous generator based wind energy conversion systems: A technology review. *Renew. Sustain. Energy Rev.* **2015**, *51*, 1288–1305. [CrossRef]
10. Rahimi, M. Modeling, control and stability analysis of grid connected PMSG based wind turbine assisted with diode rectifier and boost converter. *Int. J. Electr. Power Energy Syst.* **2017**, *93*, 84–96. [CrossRef]
11. Heo, S.Y.; Kim, M.K.; Choi, J.W. Hybrid intelligent control method to improve the frequency support capability of wind energy conversion systems. *Energies* **2015**, *8*, 11430–11451. [CrossRef]
12. Kumar, D.; Chatterjee, K. A review of conventional and advanced MPPT algorithms for wind energy systems. *Renew. Sustain. Energy Rev.* **2016**, *55*, 957–970. [CrossRef]
13. Kumar, K.; Babu, N.R.; Prabhu, K.R. Design and Analysis of an Integrated Cuk-SEPIC Converter with MPPT for Standalone Wind/PV Hybrid System. *Int. J. Renew. Energy Res.* **2017**, *7*, 96–106.
14. Dahbi, A.; Nait-Said, N.; Nait-Said, M.S. A novel combined MPPT-pitch angle control for wide range variable speed wind turbine based on neural network. *Int. J. Hydrog. Energy* **2016**, *41*, 9427–9442. [CrossRef]
15. Whei-Min, L.; Chih-Ming, H.; Ou, T.C.; Tai-Ming, C. Hybrid intelligent control of PMSG wind generation system using pitch angle control withRBFN. *Energy Convers. Manag.* **2011**, *52*, 1244–1251.
16. Tiwari, R.; Babu, N.R. Comparative Analysis of Pitch Angle Controller Strategies for PMSG Based Wind Energy Conversion System. *Int. J. Intell. Syst. Appl.* **2017**, *9*, 62–73.
17. Moradi, H.; Vossoughi, G. Robust control of the variable speed wind turbines in the presence of uncertainties: A comparison between H∞ and PID controllers. *Energy* **2015**, *90*, 1508–1521. [CrossRef]
18. Duong, M.Q.; Grimaccia, F.; Leva, S.; Mussetta, M.; Ogliari, E. Pitch angle control using hybrid controller for all operating regions of SCIG wind turbine system. *Renew. Energy* **2014**, *70*, 197–203. [CrossRef]
19. Assareh, E.; Biglari, M. A novel approach to capture the maximum power from variable speed wind turbines using PI controller, RBF neural network and GSA evolutionary algorithm. *Renew. Sustain. Energy Rev.* **2015**, *51*, 1023–1037. [CrossRef]
20. Babu, N.R.; Arulmozhivarman, P. Forecasting of wind speed using artificial neural networks. *Int. Rev. Model. Simul.* **2012**, *5*, 2276–2280.
21. Poultangari, I.; Shahnazi, R.; Sheikhan, M. RBF neural network based PI pitch controller for a class of 5-MW wind turbines using particle swarm optimization algorithm. *ISA Trans.* **2012**, *51*, 641–648. [CrossRef] [PubMed]
22. May, R.; Dandy, G.; Maier, H. Review of input variable selection methods for artificial neural networks. In *Artificial Neural Netwoks—Methological Advances in Biomedical Applications*; InTech: Croatia, Europe, 2011; Volume 2, pp. 19–44.
23. Saravanan, S.; Babu, N.R. RBFN based MPPT algorithm for PV system with high step up converter. *Energy Convers. Manag.* **2016**, *122*, 239–251. [CrossRef]

24. Park, D.C.; El-Sharkawi, M.A.; Marks, R.J.; Atlas, L.E.; Damborg, M.J. Electric load forecasting system using an artificial neural network. *IEEE Trans. Power Syst.* **1991**, *6*, 442–449. [CrossRef]

25. Sanjeevikumar, P.; Paily, B.; Basu, M.; Conlon, M. Classification of Fault Analysis of HVDC Systems using Artificial Neural Network. In Proceedings of the Conference on 49th IEEE Universities' Power Engineering (IEEE–UPEC'14), Cluj-Napoca, Romania, 2–5 September 2014; pp. 1–5.

26. Ramulu, C.; Sanjeevikumar, P.; Karampuria, R.; Jain, S.; Ertas, A.H.; Fedak, V. A Solar PV Water Pumping Solution Using a Three-Level Cascaded Inverter Connected Induction Motor Drive. *Eng. Sci. Technol. Int. J.* **2016**, *19*, 1731–1741. [CrossRef]

27. Chokkalingam, B.; Padmanaban, S.; Siano, P.; Krishnamoorthy, R.; Selvaraj, R. Real Time Foresting of EV Charging Station Scheduling for Smart Energy System. *Energies* **2017**, *10*, 377. [CrossRef]

28. Ali, A.; Sanjeevikumar, P.; Twala, B.; Marwala, T. Electric Power Grids Distribution Generation System For Optimal Location and Sizing—An Case Study Investigation by Various Optimization Algorithms. *Energies* **2017**, *10*, 1–13.

29. Chandramohan, K.; Padmanaban, S.; Kalyanasundaram, R.; Bhaskar, M.S.; Mihet-Popa, L. Grid Synchronization of Seven-phase Wind Electric Generator using d-q PLL. *Energies* **2017**, *10*, 926. [CrossRef]

30. Vavilapalli, S.; Padmanaban, S.; Subramaniam, U.; Mihet-Popa, L. Power Balancing Control for Grid Energy Storage System in PV Applications - Real Time Digital Simulation Implementation. *Energies* **2017**, *10*, 928. [CrossRef]

31. Hossain, E.; Perez, R.; Padmanaban, S.; Siano, P. Investigation on Development of Sliding Mode Controller for Constant Power Loads in Microgrids. *Energies* **2017**, *10*, 1086. [CrossRef]

32. Un-Noor, F.; Padmanaban, S.; Mihet-Popa, L.; Mollah, M.N.; Hossain, E. A Comprehensive Study of Key Electric Vehicle (EV) Components, Technologies, Challenges, Impacts, and Future Direction of Development. *Energies* **2017**, *10*, 1217. [CrossRef]

33. Ozsoy, E.; Padmanaban, S.; Mihet-Popa, L.; Fedák, V.; Ahmad, F.; Akhtar, R.; Sabanovic, A. Control Strategy for Grid-Connected Inverters Under Unbalanced Network Conditions-A DOb Based Decoupled Current Approach. *Energies* **2017**, *10*, 1067. [CrossRef]

Article

Study and Analysis of an Intelligent Microgrid Energy Management Solution with Distributed Energy Sources

Swaminathan Ganesan [1], Sanjeevikumar Padmanaban [2,*], Ramesh Varadarajan [1], Umashankar Subramaniam [1] and Lucian Mihet-Popa [3]

1   School of Electrical Engineering, Vellore Institute of Technology (VIT) University, Vellore, Tamilnadu 632014, India; gswami@yahoo.co.in (S.G.); vramesh@vit.ac.in (R.V.); umashankar.s@vit.ac.in (U.S.)
2   Department of Electrical and Electronics Engineering, University of Johannesburg, Auckland Park, Johannesburg 2006, South Africa
3   Faculty of Engineering, Østfold University College, Kobberslagerstredet 5, 1671 Kråkeroy-Fredrikstad, Norway; lucian.mihet@hiof.no
*   Correspondence: sanjeevi_12@yahoo.co.in; Tel.: +27-79-219-9845

Received: 11 July 2017; Accepted: 12 September 2017; Published: 16 September 2017

Abstract: In this paper, a robust energy management solution which will facilitate the optimum and economic control of energy flows throughout a microgrid network is proposed. The increased penetration of renewable energy sources is highly intermittent in nature; the proposed solution demonstrates highly efficient energy management. This study enables precise management of power flows by forecasting of renewable energy generation, estimating the availability of energy at storage batteries, and invoking the appropriate mode of operation, based on the load demand to achieve efficient and economic operation. The predefined mode of operation is derived out of an expert rule set and schedules the load and distributed energy sources along with utility grid.

Keywords: energy management system; microgrid; distributed energy sources; energy storage system

1. Introduction

The traditional bulk power generation, transmission and distribution system is facing a lot of technological challenges to fulfil the growing demand and increased penetration of distributed energy resources. The existing infrastructures are also outdated, which hinders the integration of newer technology for capacity enhancement and sophisticated monitoring and control. Hence the need has arisen for distributed generation which can co-exist with existing bulk power networks [1]. In recent years, there has been significant growth in renewable energy generation through wind and solar resources. A microgrid is a miniature version of the bulk power system with distributed energy resources capable of serving as an independent electrical island separated from the bulk power system [2]. Microgrids employ environmentally benign energy sources like solar, wind, and fuel cells [3]. The higher the penetration of sustainable energy sources the more the socio-economic benefits will be. The recent advances in control and communication technology facilitate robust and intelligent control of microgrids [3–5]. In emerging economies, to encourage independent sustainable energy generation, there is a strong regulatory framework which in turn will constitute the microgrid building blocks.

The Figure 1 depicts the microgrid architecture under consideration for an energy management system (EMS). The proposed microgrid system comprises sources like the utility grid, a diesel generator, photovoltaic (PV) generator, and a battery energy storage system (BESS) [3,6]. The loads are classified into secure and non-secure loads [7]. All secure loads are supplied from an uninterruptible power supply (UPS), while the rest of the loads are supplied directly either from the utility grid or from distributed energy sources (DES) [8,9]. All the sources and loads are connected through appropriate circuit breakers. The current and voltage feedback signals from the loads and local feeder lines are fed to the EMS controller. The control signals to circuit breakers are sent from the EMS controller. The input and output data of the EMS is shown in Figure 2. Figure 3 depicts the typical data flow between sources, load and controller. The main controller receives active power, reactive power, voltage, and current data from the local/embedded controller from the DES. Table 1 lists the specification of loads and sources used in this analysis. The cost of energy data from the grid is fed from the utility side. The cost of energy for local generation using DES within the microgrid are fed manually into the EMS for decision-making purposes to achieve economic operation. The user interface of the EMS will allow users to manually enter the specific parameters based on which the power flow decisions to be made. The central database which stores historical load demand, and the actual forecast data will be processed in the EMS for effective load management and power delivery.

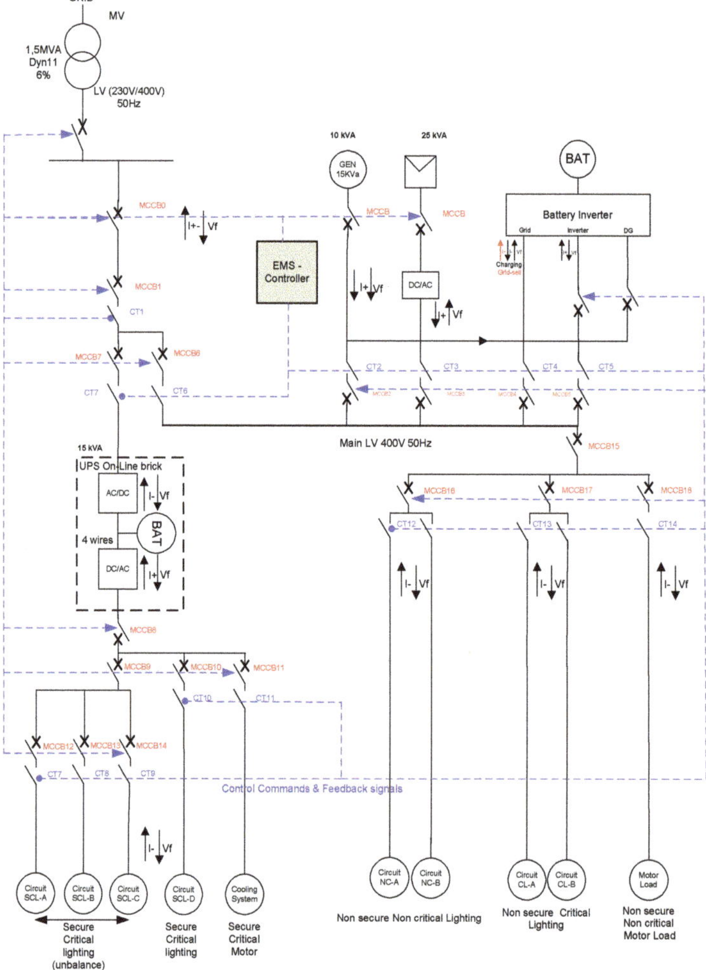

Figure 1. Microgrid schematic diagram.

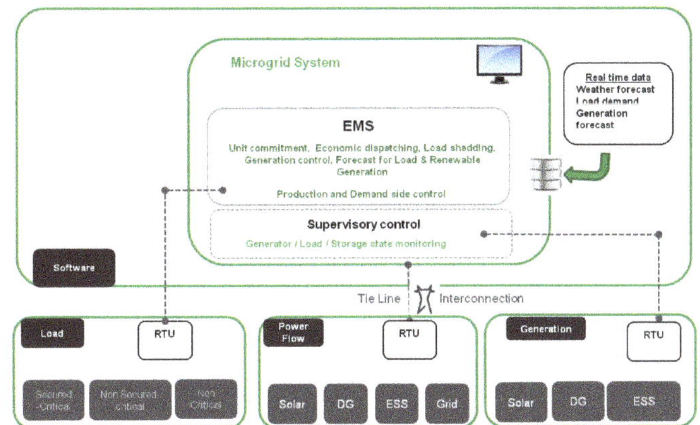

Figure 2. EMS controller Input and Output data.

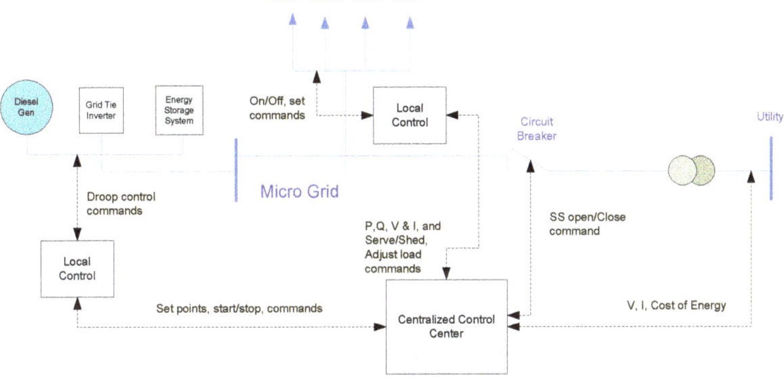

Figure 3. EMS controller data flow between controller, load and sources.

Table 1. System specification considered for analysis.

Serial No.	Type of Source/Load	Specification
1	Total Network capacity	100 kVA, 400 V, 3 PH, TT grounding system
2	PV Generator	25 kW
3	Diesel Generator	50 kW
4	BESS	25 kW, 50 kWh
5	UPS	15 kVA, 400 V, 3 PH
6	Managed Loads	400 kVA, Air conditioner, Heater, & Standard 16 A Loads, 10 kVA
7	Priority unmanaged loads (Single phase)	PH 1-N 230 V, Lighting: 13 kVA, PF 0.7 & Loads: 12 kVA, PF 0.8 PH 2-N 230 V, Lighting: 8 kVA, PF 0.55 & Loads: 7 kVA, PF 0.6 PH 3-N 230 V, Lighting: 16 kVA, PF 0.8 & Loads: 3.5 kVA, PF 0.67
8	Priority unmanaged loads (Three phase)	400 V, 3 PH + N: 20 kVA, PF 0.85 (Motor Loads)
9	Critical unmanaged loads (Three phase)	400 V, 3 PH + N: 6.45 kVA, PF 0.85 (Miscellaneous Loads)

The communication network will carry the control and feedback signals over the network. This will facilitate having proper co-ordination and control among the loads, sources and utility. All the measured critical parameters of the respective devices connected to the network will be transmitted to the central/local controller over the specified communication protocol for processing and take appropriate decisions and actions based on the control algorithm. The parameters to be measured are defined in the EMS data flow diagram in Figure 2. The Modbus RTU protocol has been deployed to acquire the data from various sources and loads. The EMS controller gets the weather forecast and cost of energy from the utility and then computes the energy forecast based on the historical consumption patterns. The forecast of renewable energy generation is estimated by the EMS controller using the weather data input. The decisions for controlling loads and DES are sent to the respective devices through RS485 or the TCP/IP protocol based on the device compatibility. Table 2 lists various parameters that are acquired from the sources and loads connected in the microgrid system to EMS controller and the respective output command. Figure 3 presents the single line representation of the data flow from all the connected devices in the microgrid network to the EMS controller. Figure 4 presents the communication architecture used in the microgrid system. The communication is divided into three parts: (i) device level; (ii) unit level and (iii) system level. Device level communication is point to point data transfer, unit level communication is controller to controller data exchange, and system level communication is like unit level, but over a long distance and bulk data exchange between microgrid networks. For system level communication, the IEC 61850 protocol has been considered, whereby the IEC 61850 9-2 process bus protocol facilitates Generic Object Oriented System Event (GOOSE) messages for data exchange with the EMS controller. For device level, since it is shorter distance the RS485 Modbus protocol has been considered. Modbus TCP has been considered for unit level communication. Further, this proposed communication architecture has a provision to be expanded for ZigBee and Wi-Fi protocols as per IEEE 802.15.4.

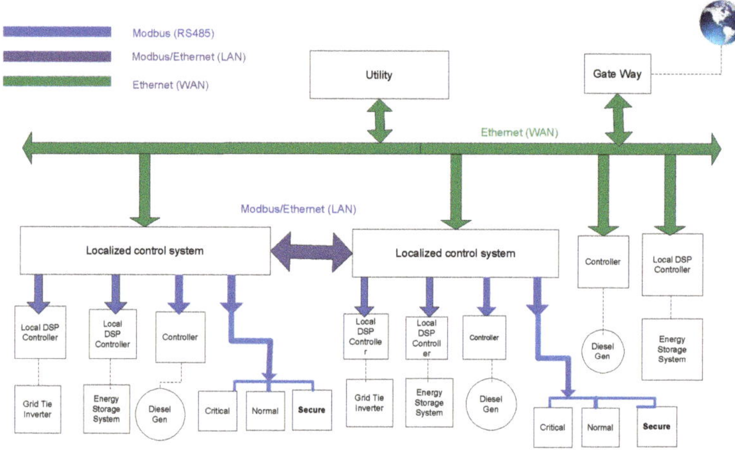

Figure 4. EMS controller data flow between controller, load and sources.

Table 2. List of parameters acquired from loads and sources to EMS controller and corresponding output control from EMS.

Serial No.	Type	Description	Acquired Data to EMS	Control Command from EMS
1	Source	PV Generator	P, Q, I, V, F	P, Q
2	Source	BESS	V, I, SOC	Charge/Discharge
3	Source	DG	P, Q, I, V and Fuel level	P, Q
4	Load	Cooling	T, C, Occupancy	On/Off
5	Load	Lighting	L, Occupancy	On/Off
6	Load	Pump	Water level	On/Off

Presently there are many microgrid architectures under research, and the focus is predominantly on developing energy management solutions through sophisticated artificial intelligence technologies [5] for achieving superior economic benefits, but the same amount of focus is not present in developing coordinated control of DER, grid and loads with centralized controllers [10]. Having precise control at the individual device or source level and at the network controller level will facilitate the faster response, seamless transition of load sharing between sources, and more reliable operation of microgrids [4]. Keeping this in mind, the authors proposed a microgrid energy management system (EMS) to establish control at the device level and overall system level with the help of state of art communication technology [11,12]. In load level control, the proposed EMS enables precise management of power flows by forecasting renewable energy generation, estimating the availability of energy at storage batteries, and invoking the appropriate mode of operation, based on the load demand to achieve efficient and economic operation. The predefined mode of operation is derived out of an expert rule set and schedules the load and distributed energy sources along with the utility grid. In system level control, the focus is mainly on system stability and power sharing. The proposed new controller ensures the stability of the system during transition modes and steady state operating conditions which are validated with different load and source dynamics within the microgrid system. The connection and disconnection of PV generator from the grid in islanded mode and corresponding power sharing of diesel generator (DG) and battery energy storage system (BESS) are recorded and validated for conformance to the intended operation to ensure optimum power flow from different sources to loads.

2. Load Management

2.1. Classification of Loads

The connected loads in the microgrid are classified into multiple clusters to have efficient load management. The major classification is in three clusters: (i) secured critical loads; (ii) non-secured-critical loads and (iii) non-secured, non-critical loads [13]. Typically, all critical loads are a non-shedable loads, which must be served at all the time irrespective of the source of generation and cost of energy [14]. Loads that are classified as non-critical can be scheduled to achieve economic operation. Under these two major classifications, there are subsets that are distinguished as forecastable and non-forecastable loads [15]. This piece of information will help in realizing economic demand and response management (DRM) [16].

2.2. Control of Air Conditioning Loads

The cooling system is one of the major and critical energy consuming loads in any premise. Efficient management of this critical load is important to achieve economic operation. The control algorithm of a cooling system is based on an analogy between the caloric behaviour and electro kinetics and steady state operating conditions have been considered for modelling. The generation of cold source is equivalent to the cold energy stored in the walls and in the atmospheric air, during this process, there will be a loss of energy. The cooling production can be derived using Equation (1), where Tout is the outdoor temperature, $T(t)$ is the indoor temperature, C is the total thermal building Capacity, R_{th} is the total thermal building resistance, ϕ_s is the building cooling production. In Equation (3), $\tau = R_{th}C$ is the building time constant.

$$\phi_s = C\frac{dT(t)}{dt} + \frac{1}{R_{th}}(T(t) - Tout) \tag{1}$$

During load shedding operation, the cooling production is stopped: $\phi_s = 0$.

$$Tout = R_{th}C\frac{dT(t)}{dt} + T(t) \tag{2}$$

$$T(t) = Tout + (T_0 - Tout) \times e^{-t/\tau} \tag{3}$$

Figure 5 shows the cycle diagram for cooling system load management, where based on the occupancy and temperature sensor input the load will be operated as per the cycle diagram. During time interval T1, the shedding command for the cooling system is activated, which will allow the room temperature to rise, but this will be maintained so as to not to reach the discomfort zone. The entire cooling system will remain shut off during the time interval T2, this is the maximum time interval for shedding affecting the comfort of occupants. During time interval T3, the shedding command will be withdrawn and temperature will start reducing till the cool set limit is reached. T4 time interval is minimum time duration required to restore the comfort temperature level. Equation (4) helps to calculate T2, while the production is stopped: $\phi_s = 0$ and Equation (5) helps to calculate T4. The main objective of this algorithm is to determine the time interval to attain T_{cool} after T2 duration, where temperature is equal to T_{warm}. The flow chart for the cooling system control based on the mode of operation and cost of energy is shown in Figure 6. When the grid is available (Is Grid ok = 1), then the controller will look for the cost of energy, and based on the user set point for high, low and medium cost values through the user interface, the controller will activate the respective mode of operation.

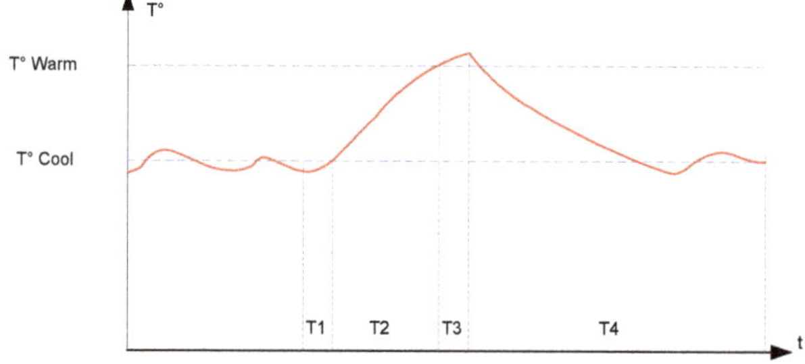

Figure 5. Cooling load management cycle diagram.

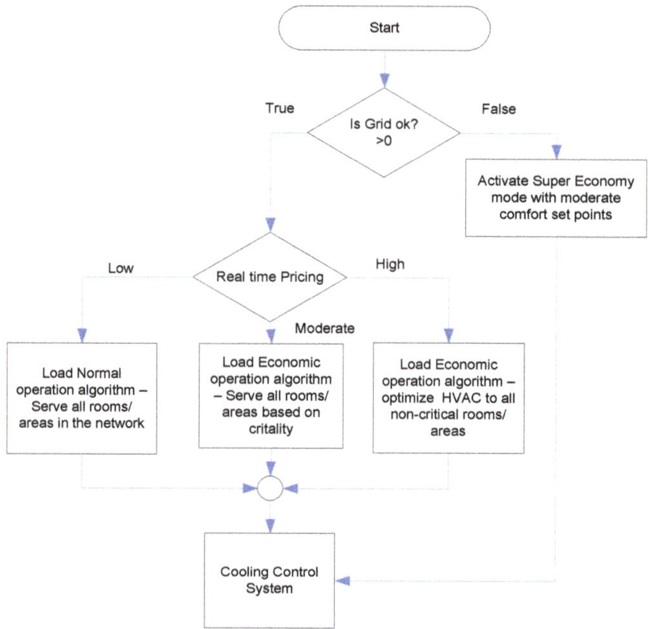

Figure 6. Cooling system control flow chart.

$$\frac{dT(t)}{dt} = -\frac{1}{R_{th}C}(T(t) - Tout(t))$$

(4)

$$\frac{dT(t)}{dt} = \frac{\phi_s}{C} - \frac{1}{R_{th}C}(T(t) - Tout(t))$$

(5)

Figure 7 shows the typical cooling load control during time interval T2, where the load will be in off condition. Here the T_{cool} limit is set to 25 °C, T_{warm} limit is set to 30 °C, Total thermal resistance R_{th} = 0.00018 K/W, building time constant τ = 30, with these parameters, the experimental results are confirmed to be 38 min of T2, which means the cooling load was in off state with the occupancy of two people.

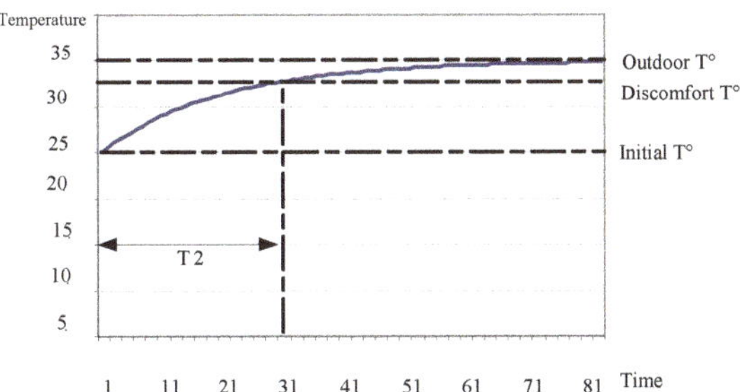

Figure 7. Typical cooling load off cycle during T2 time till discomfort level.

2.3. Control of Lighting Loads

Figure 8 shows the lighting system controller. And the Figure 9 shows the control flow chart for lighting load control. The lighting load control algorithm is based on the input from a photo sensor, occupancy sensor and the energy tariff. The EMS also has the provision to override the control by selecting manual mode of operation. If daylight is partially available, then dim control mode will be invoked to reduce the energy consumption. If occupancy is not sensed, then all the lighting loads will be turned off. The lighting controller is designed using the natural light availability from the photo sensor. The optimum required illumination is derived for the total area of surface using the standard lumen method. The availability of natural light is derived from Equation (6). The lighting controller actuator command for illuminating artificial light is calculated from the difference between the required illumination and natural light availability.

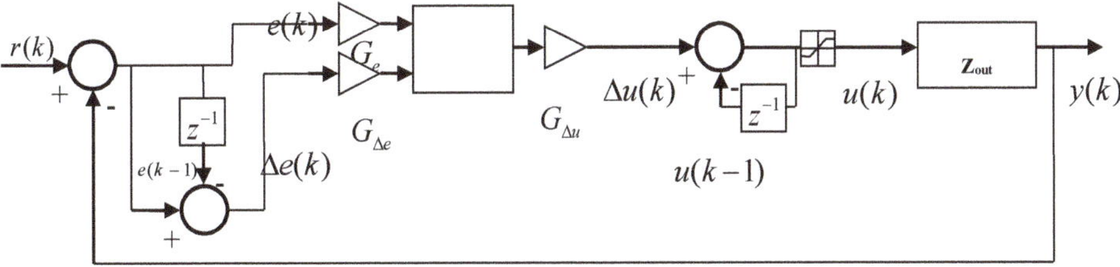

Figure 8. Lighting system controller.

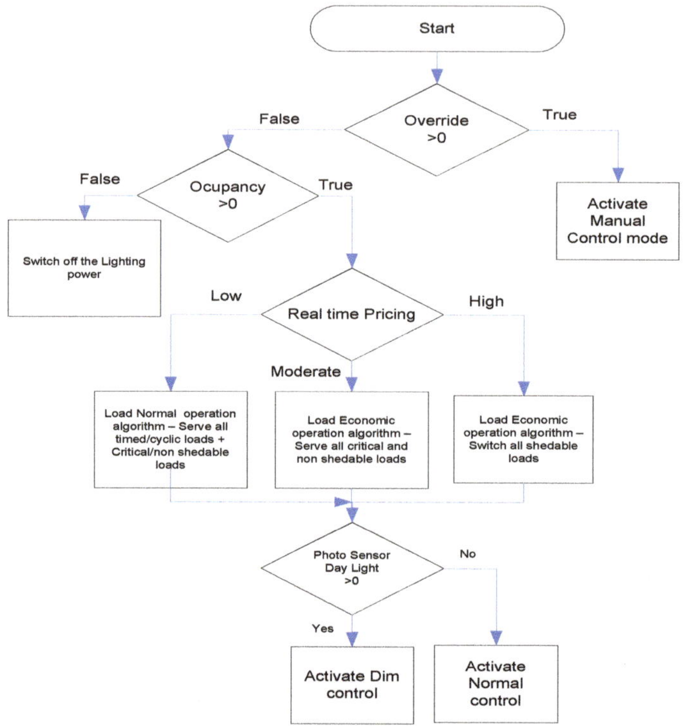

Figure 9. Lighting load control flow chart.

$$Lx_{in} = \frac{A_w \tau E_v}{A_{in}(1 - \rho)} \qquad (6)$$

where A_w is surface area of window in square meter, τ is the light transmittance of the window, E_v is the luminance available on the window in lux, A_{in} is the total indoor area of surfaces in square meter, and ρ is the mean reflectance of the weighted area of all indoor surfaces. For design purpose, the following parameters have been considered in the analysis: building with a south-facing glass window of area (2.75 m^2), and total room area of 48 m^2, volume 138 m^3 with reasonable thermal inertia, good light transmittance of the window glazing with τ = 0.817, the reflectance of all room inner surfaces considered as ρ = 0.4. Total electric lights of 13 lamps, 0–1000 lux, 950 W total, and a shading beam. The controller's reference set point for indoor Illuminance = {500–800} lux. With these values the controller was validated to maintain the luminous intensity to the preferred set value based on the other input parameters. Figure 10 shows the integration of lighting load control system into the EMS. If the user decides to disable automatic lighting control through EMS, override option can be used.

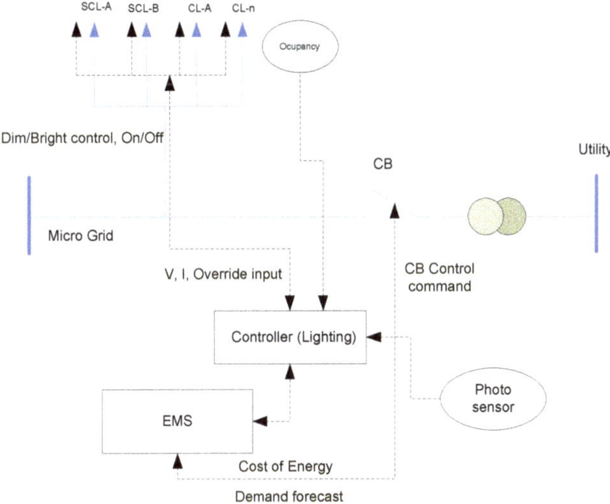

Figure 10. Lighting load control integration with EMS.

2.4. Control of Water Pump Loads

Water pumps constitute a considerable amount of load in a microgrid system. Hence the efficient control and scheduling of water pump control is critical for EMS. The level of water from the storage will be detected using a water level sensor, and this input is compared with the set point of the required level and the error signal is supplied to PID controller, the output of this controller is fed to servo motor and in turn operates the gave valve to increase or decrease the water flow to maintain the required level of water. Normally the set point is derived from the upper level sensor, and there will be a provision in the user interface to enter a manual value to supersede the sensor input limit. Figure 11 shows the control system block diagram for the pump controller. The input to the controller comes from the EMS, and Figure 12 shows the flow chart for the control algorithm based on real time pricing.

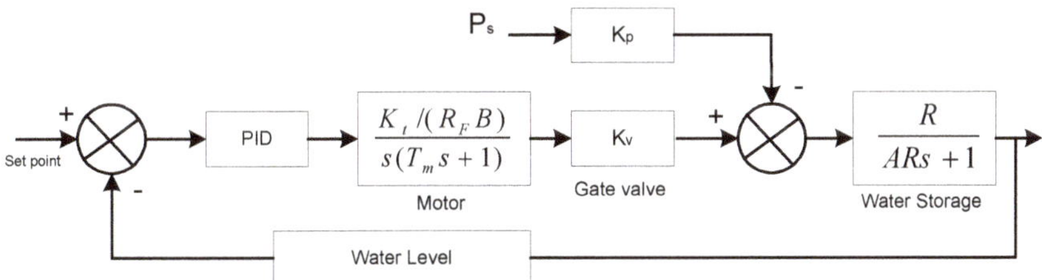

Figure 11. Water pump control system block diagram.

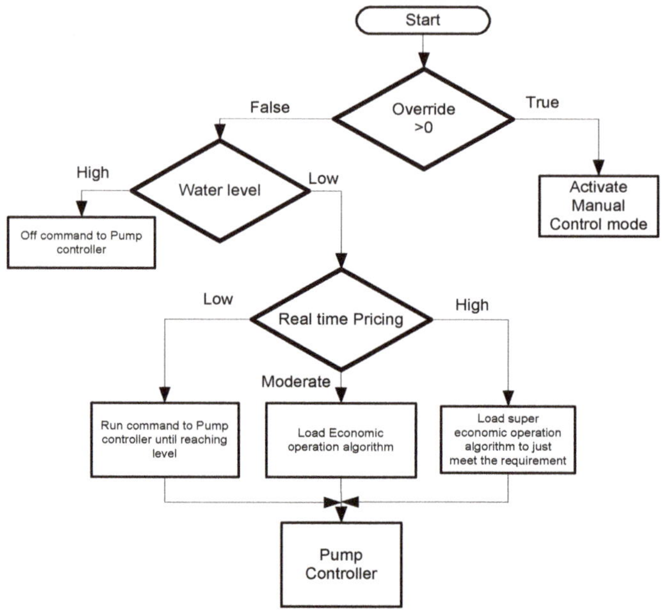

Figure 12. Water pump motor control flow chart.

3. Modes of Operation

The microgrid operation has been classified into two major categories: On grid mode and Off grid mode. During On grid mode of operation, the entire system is powered by the utility grid as well as sustainable energy sources. The sharing of loads between DES is controlled by the EMS as per the defined control algorithm. In Off grid mode of operation, the entire microgrid will be in islanded mode from the utility grid, all the connected loads will be served from the local energy sources and storage system connected in the network. In Transition mode, all the critical loads are served by the UPS and this mode is a state in between On grid mode and Off grid mode [17].

3.1. On Grid Mode of Operation

Table 3 shows the power flow control between DES and loads based on the cost of energy. The EMS is designed for three main tariff classifications. When the cost of energy is low and PV generation is available, then all the loads are shared between the utility grid and PV source, and any surplus power is used for charging the UPS and BESS based on their SOC [4,18]. When the cost of energy is medium, the available power from the PV source is completely utilized to serve the load and only for any power requirement deficit, the utility grid is used partially. BESS will also share the loads from the stored energy. During this tariff mode, no power is being used for charging UPS and BESS, assuming there are no surplus power available from DES. When the cost of energy is high, the load demand is shared by the DES as a priority and then partially from the utility grid for the deficit. Since the cost of energy is high, non-critical loads will be removed from the network and will be scheduled to operate later during off-peak time. Also, the critical loads will be operated at optimum power consumption mode to reduce the energy bill, like the cooling system will be operated to exploit thermal inertia without compromising comfort levels along with using natural cooling to the possible extent. During this mode, all forecastable loads will be served as the energy will be preserved in storage devices for the loads based on the demand pattern.

Table 3. Power sharing between DES, Grid and Loads during On Grid mode of operation.

Cost of Energy	Grid Power	PV	State of UPS	UPS SOC	Battery Storage (BESS)	DG	Critical Secure Loads	Non-Secure & Critical Loads	Non-Secure & Non-Critical Loads
Low	Full	Share load & charge BESS	Online	Charge	Charge	Off	Grid	Grid	Grid
Medium	Partial	Share load & charge BESS	Online	Off	Supply	Off	Grid	Grid + BESS	Grid + BESS
High	Partial	Share load only	Online	Off	Supply	Off	UPS	BESS	Shed All loads

3.2. Off Grid Mode of Operation

During Off Grid mode of operation, the utility gird will be completely shut off from the microgrid network, and only DES will be present. There are three different combinations of DES operation in Off Grid mode [3,19]. Considering sufficient PV generation is available, during this time, UPS will serve all secured critical loads and BESS will be charged to serve during intermittency periods. All non-critical loads will be erased from the network as it will be scheduled during surplus power flow. When the BESS is drained, then the DG will be turned on and during this period, PV and DG will share the load. UPS will serve all the secured critical loads and simultaneously will charge its back up. Non-critical loads will be curtailed as they will only be served partially due to higher cost of DG power generation. Table 4 shows the power sharing among DES during Off Grid mode [20].

Table 4. Power sharing between DES, Grid and Loads during Off Grid mode of operation.

DES Availability	Grid Power	PV	State of UPS	UPS SOC	Battery Storage (BESS)	DG	Critical Secure Loads	Non-Secure, Critical Loads	Non-Secure, Non-Critical Loads
PV	Off	Share load & charge BESS	Serve Secure load	Dis charge	Charge	Off	UPS	PV	Shed all loads
PV and BESS	Off	Share load	Serve Secure load	Dis charge	Supply	Off	UPS	PV + BESS	Curtail
PV and DG	Off	Share load & charge BESS	Serve Secure load	Charge	Charge	ON	UPS	PV + DG	Curtail

Stability and Power Sharing

In Off Grid mode it is essential that all the connected DES synchronize properly and form a grid. For grid tied inverters, a reference voltage source is required for pumping PV power into the microgrid [21]. In the proposed network, the UPS will provide the reference voltage and help the PV inverter to build power. Once the grid is formed, the stability of the grid is to be maintained by proper control of voltage (V), frequency (F), active power (P) and reactive power (Q). The grid impedance will be checked by PV inverters to sense the grid presence. The PV inverter's active power (P) is controlled as a function of frequency (F) to ensure sustained operation in islanded mode. Figure 13 shows the graph for power versus frequency control to maintain the microgrid stability. When the microgrid frequency reaches a F_{start} (50.2 Hz) then the PV inverter starts to reduce the active power generation. Further, if the microgrid frequency increases and reaches a maximum allowable limit F_{stop} (54 Hz), then the PV inverter disconnects from the microgrid network. The slope will define the derating as % power reduction with regards to change in frequency. The slope can be configured by the user through the HMI.

$$P_e = \frac{E \cdot V}{X} \sin \delta \cong \frac{E \cdot V}{X} \delta \tag{7}$$

$$E - V \cong \frac{XQ}{E} \tag{8}$$

$$\frac{d\delta}{dt} = \omega(t) - \omega_{MG} \tag{9}$$

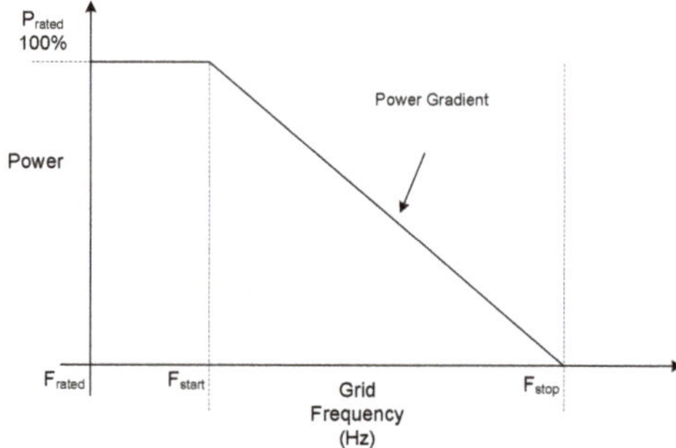

Figure 13. Power gradient diagram for active power (P) versus frequency (F) control function in the PV inverter.

The generic power equation for a generator and corresponding output is given in Equation (7), where P_e is electric power delivered to Load, E and V are generated voltage and terminal voltage at load respectively. δ is the angular displacement between E and V. From Equation (8), it is understood that, the reactive power (Q) controls the voltage (V), when reactive power increases, the output voltage decreases. Figure 14 shows the relation between frequency versus active power and voltage versus reactive power.

Here, ω is the angular velocity of the generator (DG or PV generator output), ω_{MG} is the angular velocity of the microgrid at the point of common coupling (PCC) [22]. The microgrid output power (P) is controlled by changing the frequency of the PV inverter or prime mover input to the DG, and similarly the reactive power (Q) is controlled by changing the output voltage or excitation to the DG. In Off grid mode, the secondary control is vital to ensure the stability as the sources are not stiff and will easily fall out of sync [23]:

$$f - f_o = -k_P(P - P_o) \tag{10}$$

$$V - V_o = -k_Q(Q - Q_o) \tag{11}$$

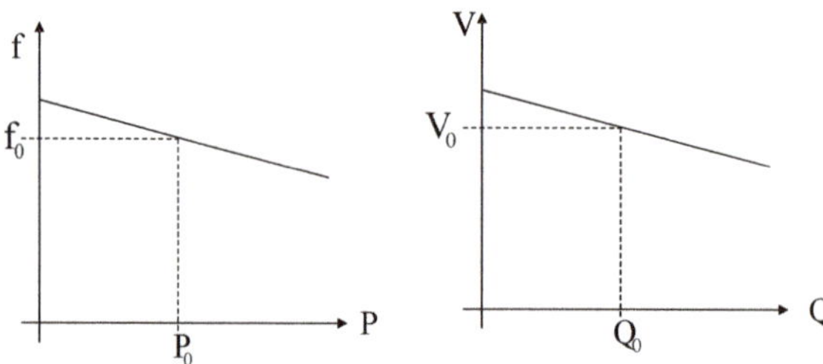

Figure 14. Active power (P) vs Frequency (F) relation and Reactive power (Q) vs. Voltage (V) relation at PCC.

Table 5 lists the voltage and frequency trip limits and corresponding disconnection and reconnection time for an Indian grid; these limits are critical for both on grid and Off grid mode of operation [24]. When the line to neutral voltage limit falls below 195.5 V, then within two seconds the system should disconnect, similarly when the line to neutral voltage goes above 310 V, then within 2 s the system should cease operation to avoid any damage to the connected equipment. The same conditions have been tested and validated for reliable operation and control.

Table 5. Power sharing between DES, grid and loads during Off Gr.

Parameter	Limit	Value
	Under voltage LV1 Tripped value (V)	195.5 V
	Under voltage LV1 Tripping Time (s)	$\leq$2 s
	Over voltage LV2 Tripped value (V)	310.5 V
Voltage limits and disconnection time	Over voltageLV2 Tripping Time (s)	$\leq$50 ms
	Under voltage LV2 Tripped value (V)	195.5 V
	Under voltage LV2 Tripping Time (s)	$\leq$100 ms
	Over voltage LV1 Tripped value (V)	253 V
	Over voltage LV1 Tripping Time (s)	$\leq$2 s
	Under frequency LV1 Tripped value (Hz)	49 Hz
	Under frequency LV1 Tripping Time (s)	$\leq$200 ms
Grid frequency limits and disconnection time	Over frequency LV1 Tripped value (Hz)	51 Hz
	Over frequency LV1 Tripping Time (s)	$\leq$200 ms
	Reconnection Time (s)	20–300 s

4. Simulation Results

The proposed microgrid network shown in Figure 1 has been modelled in MATLAB/Simulink brick by brick and the simulation results are presented in Figures 15 and 16. In the simulation, all the individual sources and loads are modelled using the Simulink libraries. The sources and loads are integrated and simulated for the different operating scenarios. Figure 15 shows the simulation results of On grid mode. From 0 to 0.6 s, only the utility grid and PV generation sources are supplying the entire load and charging the storage units, at t = 0.6 s, the storage unit started sharing loads by discharging its stored energy, hence reducing power drawn from the utility grid. Initially BESS charge was 77 percentage of its full capacity and UPS was at 80 percentage of full charge. From 0 to 0.6 s, the cost of energy from utility is low, at 0.6 s; the system goes to islanded mode of operation, hence the utilization is optimized to ensure reliable operation. During this entire period, DG is in off condition, and the results are captured for active and reactive power supply from all the sources.

Figure 16 shows the simulation results for change in mode of operation from Off grid to On grid mode of operation. From 0 to 0.6 s, the system operates in islanded mode, at t = 0.6 s, the utility grid is restored and the entire network is supported by PV and utility. In the absence of the utility grid, the UPS will continue to provide the reference voltage to the PV inverter and that will ensure continuity of PV generation. All the secure loads are served from the UPS and non-critical loads are curtailed to preserve energy. DG is in off condition during this period. Initially the SOC of the BESS is considered as 77 % and that of the UPS is 80 %. From time 0 to 0.6 s, the BESS discharges and load consumption is optimized to serve all critical loads for a longer duration. The results are confirmed to match the intended results and the transition from one mode to other mode is achieved smoothly without disturbing the system stability and dynamics. Figure 17 shows the results for a change in utility energy tariff, the moment the tariff increases, immediately the EMS controller changes the load sharing pattern and Figure 18 presents the results for load sharing by DES.

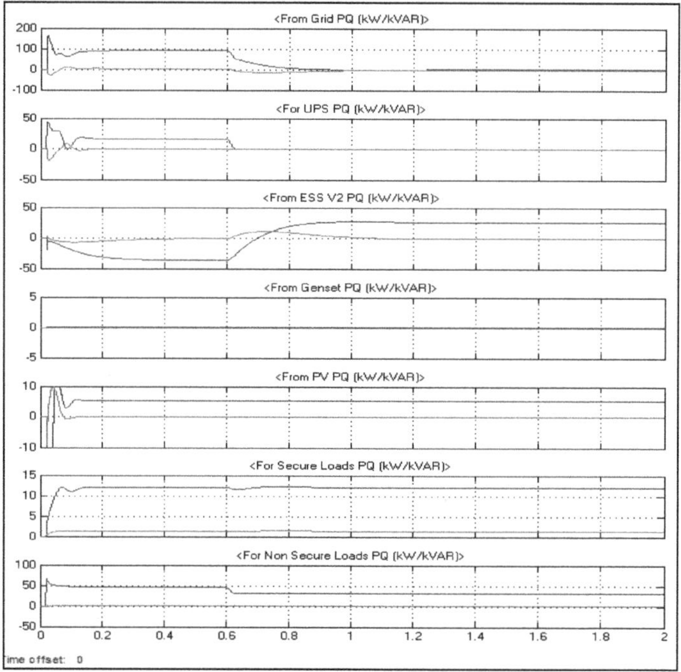

Figure 15. Simulation results in On Grid mode during low cost of energy tariff.

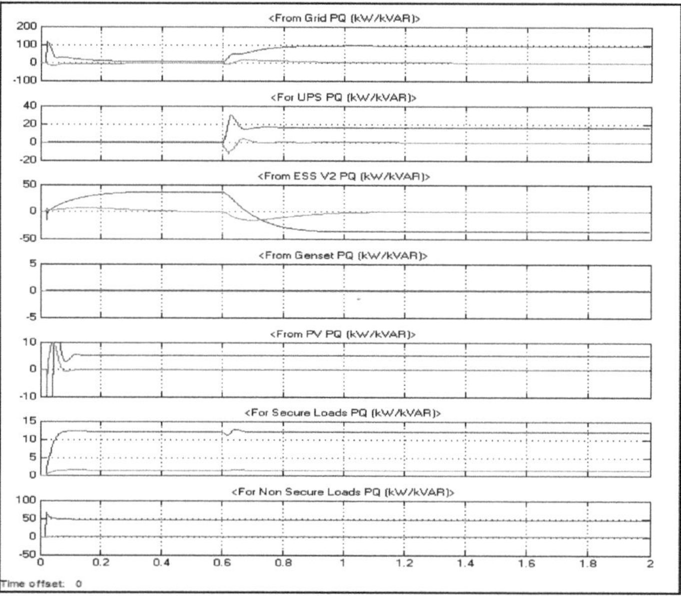

Figure 16. Simulation results in Off Grid mode with PV and BESS as source.

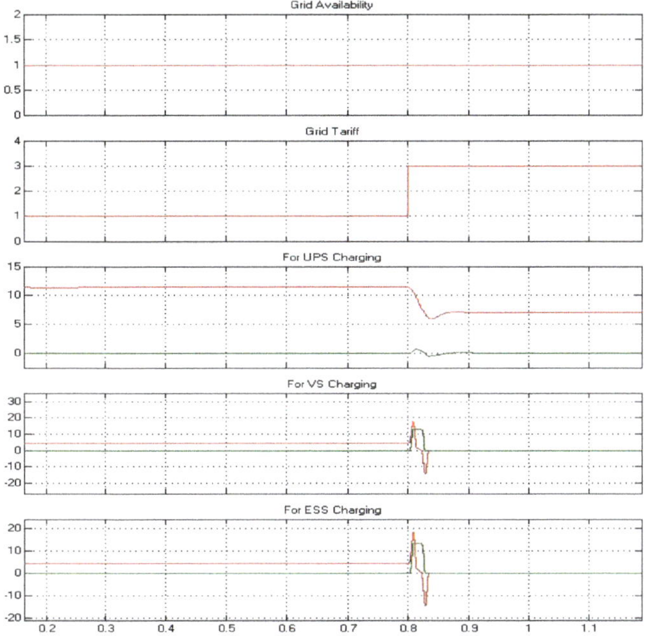

Figure 17. Simulation results for power transition in on grid mode during utility tariff change from low to high.

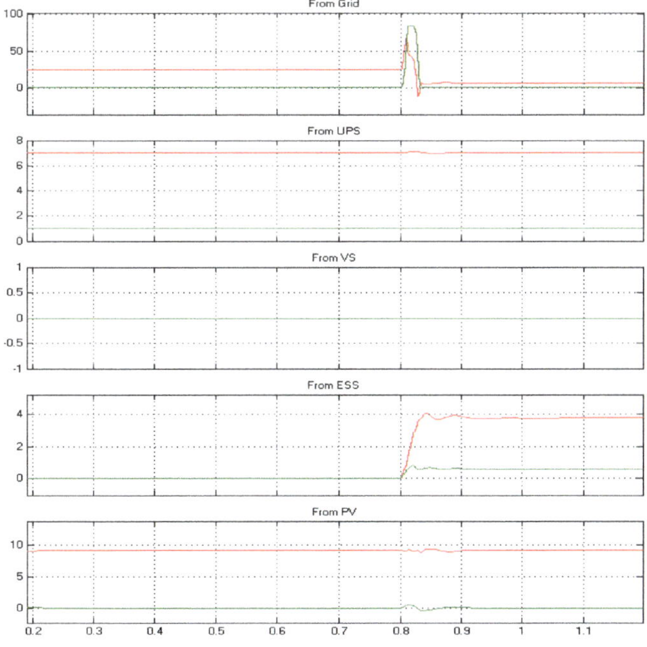

Figure 18. Simulation results for load sharing by DES in on grid mode during tariff change from low to high.

Figures 17 and 18 present the response of various sources and the control action by the EMS for the change in tariff for the cost of energy from low to high. Figure 17 shows the grid availability; it is available throughout the simulation period. From 0 to 0.6 s the cost of energy is low and during this period, the BESS stores energy and PV source and grid share the entire load and changes the storage units connected in the network. At 0.6 s, the tariff changes and cost of energy becomes high, the change in graph 1 to 3 represents per unit value, as the increase denotes the change is three times the cost that was present till 0.6 s. Immediately the EMS controller activates economic mode of operation to optimize the consumption and save cost of operation. The charging of BESS and UPS is stopped and the load sharing is done by the BESS, PV and grid. This enables minimum consumption from the utility and facilitates economic operation. However, the consumption from the PV source is unchanged. Figure 18 shows the active and reactive power supplied by PV, BESS and utility grid for the change in tariff condition at 0.6 s.

5. Experimental Results

To test and validate the proposed microgrid energy management system, an experimental prototype microgrid setup has been developed comprising a PV generator (25 kW), BESS (10 kW), UPS (8 kW) and utility grid [25]. The control of individual bricks is achieved by a TMS320F28335 programmable digital signal processor chip from Texas Instruments (Dallas, TX, USA). The overall centralized controller was developed using a model M258LF42DT Programmable Logic Controller (PLC, Modicon, Schneider Electric, Reuil-Malmaison, France). For communication between the subsystems, a TSXETG100 (Schneider Electric, Reuil-Malmaison, France) serial to TCP/IP converter communication module is used.

To set the parameters, a FDM 121 Human Machine Interface (HMI) has been used (Schneider Electric, Reuil-Malmaison, France). The energy management control algorithm has been programmed in the PLC. To emulate the load conditions, lamp load and resistive load banks were used. Figure 19 shows the experimental set up used for this analysis. Detailed testing has been done to validate the performance of EMS under different load and source operating conditions. Figure 20 shows the active power output from the PV generator in islanded mode of operation. In islanded mode of operation, with the reduction in load, the network voltage will increase. To ensure stability of operation, the active power to be derated for the increase in voltage, if the voltage increases to the cut-off limit, the PV generator to be shut off. Figure 21 shows the control of active power as a function of grid frequency in islanded mode of operation [26]. Under this condition, if the generated power from DES is more than the load demand, the surplus power generated from PV source will be diverted to charge the BESS. Immediately after the battery reaches full charge level, the BESS unit will increase the frequency of its output. On detecting this increase in frequency, PV inverter active power to be curtailed as a function of frequency to follow the predefined gradient. The PV generation will ramp up if the BESS voltage level goes down. The results are in conformance to the design. Figure 22 presents the oscilloscope waveform for under frequency fault condition in islanded mode of operation, at instance 'a' the frequency has been reduced to 48.9 Hz, immediately all the loads got shut down command from EMS controller within 120 ms. Figure 23 shows a snapshot of the EMS control screen where the user can activate and deactivate the sources, and loads and change the modes of operation and set the operating parameters.

Figure 19. Experimental test set up with EMS controller and DES with load banks.

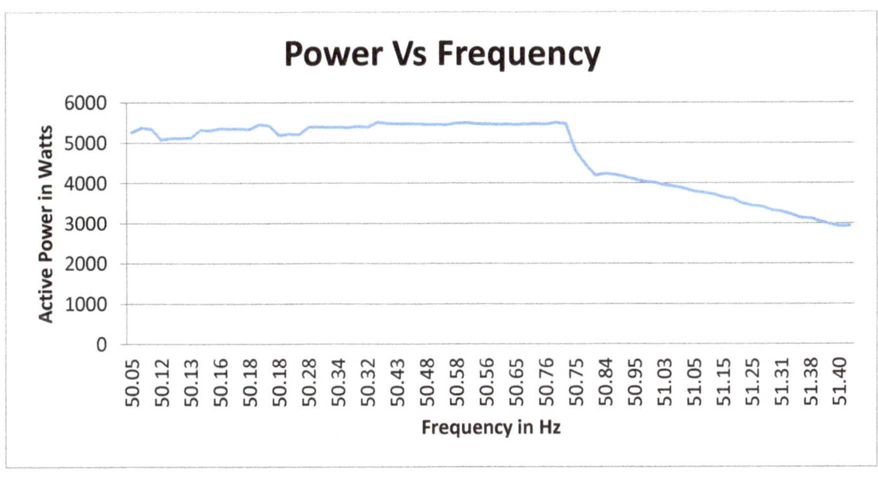

Figure 20. Active power control for the variation in frequency in islanded mode.

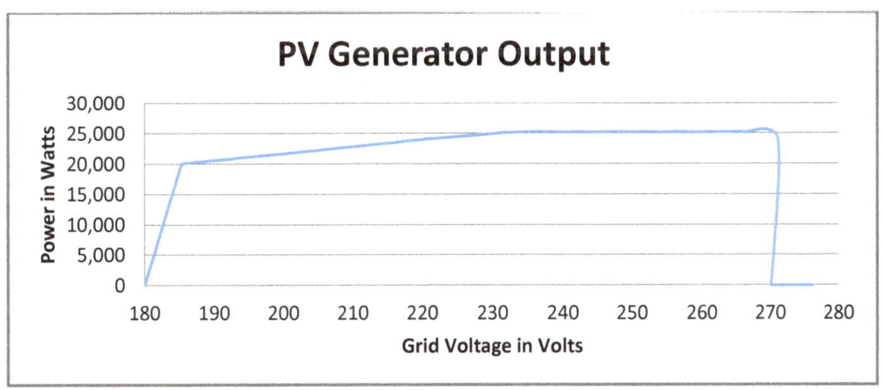

Figure 21. Active power derating with increase in grid voltage in islanded mode of operation.

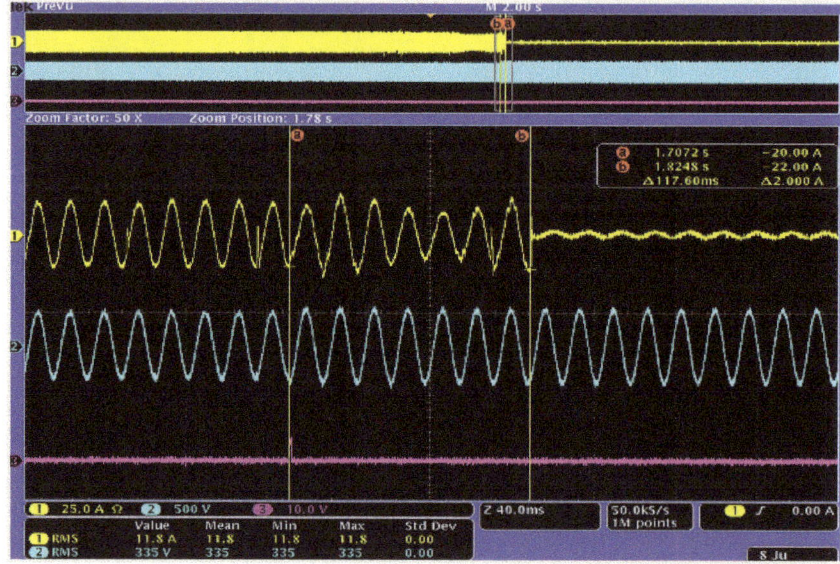

Figure 22. Microgrid system disconnection during under frequency fault in Off-grid mode of operation.

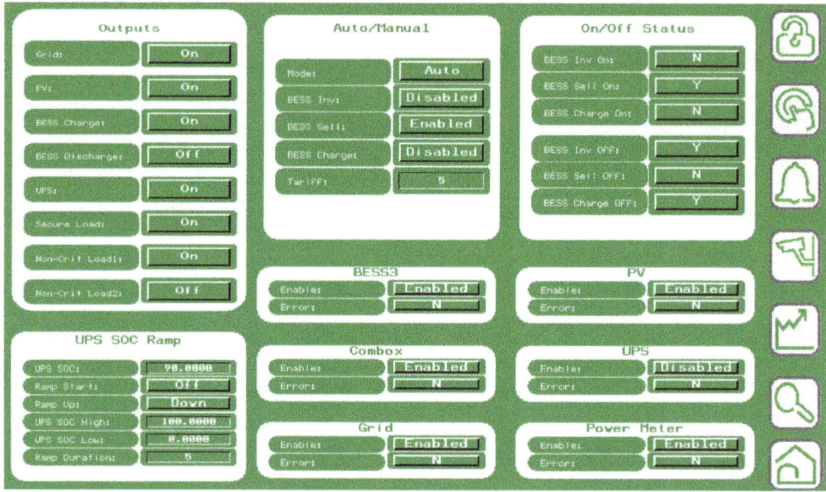

Figure 23. EMS control screen snapshot from HMI.

6. Conclusions

In this paper, a detailed study of a microgrid system has been carried out with the objective to formulate an efficient control algorithm to integrate, manage and control the various power generation sources like PV, DG, BESS, UPS and utility grid with the connected loads. The complexity lies in the smooth transition of load sharing from the PV source to a BESS or the utility grid during the clouding effect on the PV source. Since the penetration of PV sources are on the rise, the intermittency will pose a bigger concern to ensure the stability of microgrid operation during those conditions [27–30]. This problem was tackled by having coordinated control of PV generation and BESS management for sharing of loads in the network. The authors have proposed a solution to effectively manage the DES and load to achieve stable operation and economic load dispatch in the entire microgrid network [15]. Faster communication topologies were deployed to achieve better response time for the control commands at local as well as centralized controllers. This work can be enhanced by interlinking multiple microgrid networks with a more complex source and load system. The load management and control can be further improvised by using artificial intelligence and optimization techniques as future work. A robust control methodology has been developed and demonstrated in a deterministic way to operate the microgrid network in a sustainable mode. The proposed methodology largely depends on the historical data of load consumption pattern, power generation forecast, and demand forecast.

Acknowledgments: There were no funding sources for the proposed investigation, research and its results decimation.

Author Contributions: All authors involved and contributed for the proposed research work, and articulated the manuscript for its current decimation format.

Conflicts of Interest: The authors declare no conflict of interest.

List of Acronyms

EMS	Energy Management System
PV	Photovoltaic
UPS	Uninterrupted Power Supply
DES	Distributed Energy Sources
BESS	Battery Energy Storage System
DER	Distributed Energy Resources
DRM	Demand and Response Management
SOC	State of Charge
DG	Diesel Generator
PCC	Point of Common Coupling
ESS	Energy Storage System
HMI	Human Machine Interface
TCP	Transmission Control Protocol
IP	Internet Protocol
PLC	Programmable Logic Controller

References

1. Gaurav, S.; Chirag, B.; Aman, L.; Umashankar, S.; Swaminathan, G. Energy Management of PV–Battery Based Microgrid System. *Procedia Technol.* **2015**, *21*, 103–111. [CrossRef]
2. Swaminathan, G.; Ramesh, V.; Umashankar, S. Performance Improvement of Micro Grid Energy Management System using Interleaved Boost Converter and P&O MPPT Technique. *Int. J. Renew. Energy Res.* **2016**, *6*, 2.
3. Mihet-Popa, L.; Isleifsson, F.; Groza, V. Experimental Testing for Stability Analysis of Distributed Energy Resorces Components with Storage Devices and Loads. In Proceedings of the IEEE I2MTC-International Instrumentation & Measurement Technology Conference, Gratz, Austria, 12–15 May 2012; pp. 588–593.
4. Mihet-Popa, L.; Bindner, H. Simulation models developed for voltage control in a distribution network using energy storage systems for PV penetration. In Proceedings of the 39th Annual Conference of the IEEE Industrial Electronics Society-IECON'13, Vienna, Austria, 10–13 November 2013; pp. 7487–7492.
5. Zong, Y.; Mihet-Popa, L.; Kullman, D.; Thavlov, A.; Gehrke, O.; Bindner, H. Model Predictive Controller for Active Demand Side Management with PV Self-Consumption in an Intelligent Building. In Proceedings of the IEEE PES Innovative Smart Grid Technologies Europe, Berlin, Germany, 14–17 October 2012.
6. Wu, B.; Kouro, S.; Malinowski, M.; Pou, J.; Franquelo, L.G.; Gopakumar, K.; Rodriguez, J. Recent advanncement in industrial application of multilevel converter. *IEEE Trans. Ind. Electron.* **2010**, *57*, 2553–2580.
7. Kim, S.-K.; Jeon, J.-H.; Cho, C.-H.; Kon, S.-H. Dynamic modeling and controls of a grid-connected hybrid genration system for versatile power transfers. *IEEE Trans. Ind. Electron.* **2008**, *55*, 1677–1688. [CrossRef]

8. Sanjeevikumar, P.; Grandi, G.; Blaabjerg, F.; Wheeler, P.; Hammami, M.; Siano, P. A Comprehensive Analysis and Hardware Implementation of Control Strategies for High Output Voltage DC-DC Boost Power Converter. *Int. J. Comput. Intell. Syst. (IJCIS)* **2017**, *10*, 140–152.

9. Hemanshu, R.; Hossain, M.J.; Mahmud, M.A.; Gadh, R. Control for Microgrids with Inverter Connected Renewable energy Resources. In Proceedings of the IEEE PES General Meeting, Washington, DC, USA, 27–31 July 2014.

10. Adhikari, S.; Li, F.X. Coordinated V-f and P-Q Control of Solar Photovoltaic Generators With MPPT and Battery Storage in Microgrids. *IEEE Trans. Smart Grid* **2014**, *5*, 1270–1281. [CrossRef]

11. Yang, M.; Li, H.P. Analysis of Parallel Photovoltaic Inverters with Improved Droop Control Method. In Proceedings of the International Conference on Modeling and Applied Mathematics (MSAM 2015), Phuket, Thailand, 23–24 August 2015; Atlantis Press: Amsterdam, The Netherlands.

12. Ganesan, S.; Ramesh, V.; Umashankar, S.; Sanjeevikumar, P. Fuzzy Based Micro Grid Energy Management System using Interleaved Boost Converter and Three Level NPC Inverter with Improved Grid Voltage Quality. *LNEE Springer J.* **2016**. Accepted for Publication.

13. Hosseinzadeh, M.; Salmasi, F.R. Power management of an isolated hybrid AC/DC micro-grid with fuzzy control of battery banks. *IET Renew. Power Gener.* **2015**, *9*, 484–493. [CrossRef]

14. Guo, Z.Q.; Sha, D.S.; Liao, X.Z. Energy management by using point of common coupling frequency as an agent for islanded microgrids. *IET Power Electron.* **2014**, *7*, 2111–2122. [CrossRef]

15. Chen, C.; Duan, S.; Cai, T.; Liu, B.; Hu, G. Smart energy management system for optimal microgrid economic operation. *IET Renew. Power Gener.* **2011**, *5*, 258–267. [CrossRef]

16. Karavas, C.S.; Kyriakarakos, G.; Arvanitis, K.G.; Papadakis, G. A multi-agent decentralized energy management system based on distributed intelligence for the design and control of autonomous polygeneration microgrids. *Energy Convers. Manag.* **2015**, *103*, 166–179. [CrossRef]

17. Singh, S.; Singh, M.; Kaushik, S.C. Optimal power scheduling of renewable energy systems in microgrids using distributed energy storage system. *IET Renew. Power Gener.* **2016**, *10*, 1328–1339. [CrossRef]

18. Yang, H.-T.; Liao, J.-T. Hierarchical energy management mechanisms for an electricity market with microgrids. *J. Eng.* **2014**. [CrossRef]

19. Kanchev, H.; Lu, D.; Colas, F.; Lazarov, V.; Francois, B. Energy Management and Operational Planning of a Microgrid With a PV-Based Active Generator for Smart Grid Application. *IEEE Trans. Ind. Electron.* **2011**, *58*. [CrossRef]

20. Ishigaki, Y.; Kimura, Y.; Matsusue, I.; Miyoshi, H.; Yamagishi, K. Optimal Energy Management System for Isolated Micro Grids. *SEI Tech. Rev.* **2014**, *78*, 73–78.

21. Asghari, B.; Guo, F.; Hooshmand, A.; Patil, R.; Pourmousavi, S.A.; Shi, D.; Ye, Y.Z.; Sharma, R. Resilient Microgrid Management Solution. *NEC Tech. J.* **2016**, *10*, 103–106.

22. Zhang, Y.; Gatsis, N.; Giannakis, G.B. Robust Energy Management for Microgrids With High-Penetration Renewables. *IEEE Trans. Sustain. Energy* **2013**, *4*, 944–953. [CrossRef]

23. Shi, W.B.; Lee, E.-K.; Yao, D.Y.; Huang, R.; Chu, C.-C.; Gadh, R. Evaluating Microgrid Management and Control with an Implementable Energy Management System. In Proceedings of the International Conference on Smart Grid Communications, Venice, Italy, 3–6 November 2014.

24. Tantimaporn, T.; Jiyajan, S.; Payakkarueng, S. Microgrid Islanding Operation Experience. In Proceedings of the 22nd International Conference on Electricity Distribution (CRIED 2013), Stockholm, Sweden, 10–13 June 2013.

25. Lidula, N.W.A.; Rajapakse, A.D. Microgrids research: A review of experimental microgrids and test systems. *Renew. Sustain. Energy Rev.* **2011**, *15*, 186–202. [CrossRef]

26. Kyriakarakos, G.; Piromalis, D.; Dounis, A.I.; Arvanitis, K.G.; Papadakis, G. Intelligent Demand Side Energy Management System for Autonomous Polygeneration Smart Microgrids. *Appl. Energy* **2013**, *103*, 39–451. [CrossRef]

27. Nikos, H.; Asano, H.; Iravani, R.; Marnay, C. Microgrids. *IEEE Power Energy Mag.* **2007**, *5*, 78–94. [CrossRef]

28. Hossain, E.; Perez, R.; Padmanaban, S.; Siano, P. Investigation on Development of Sliding Mode Controller for Constant Power Loads in Microgrids. *Energies* **2017**, *10*, 1086. [CrossRef]

29. Swaminathan, G.; Ramesh, V.; Umashankar, S.; Sanjeevikumar, P. *Investigations of Microgrid Stability and Optimum Power Sharing Using Robust Control of Grid Tie PV Inverter*; Lecture Notes in Electrical Engineering; Springer: Berlin/Heidelberg, Germany, 2017.

30. Tamvada, K.; Umashankar, S.; Sanjeevikumar, P. *Impact of Power Quality Disturbances on Grid Connected Double Fed Induction Generator*; Lecture Notes in Electrical Engineering; Springer: Berlin/Heidelberg, Germany, 2017.

Article

A Modular AC-DC Power Converter with Zero Voltage Transition for Electric Vehicles

Jazmin Ramirez-Hernandez [1], Ismael Araujo-Vargas [1,*] and Marco Rivera [2]

[1] Escuela Superior de Ingenieria Mecanica y Electrica, Unidad Culhuacan, Instituto Politecnico Nacional of Mexico, ESIME Cul. Av. Santa Ana No. 1000, Col. San Francisco Culhuacan, C.P. 04430 Mexico City, Mexico; iaraujo@ipn.mx or jazzrh@hotmail.com

[2] Electrical Engineering Department, Universidad de Talca, Merced 437, C.P. 3341717 Curicó, Chile; marcoesteban@gmail.com

* Correspondence: pinguix76@hotmail.com; Tel.: +52-55-5729-6000 (ext. 73208 or 73058)

Academic Editor: Sergio Saponara
Received: 9 August 2017; Accepted: 8 September 2017; Published: 12 September 2017

Abstract: A study of the fundamental of operation of a three-phase AC-DC power converter that uses Zero-Voltage Transition (ZVT) together with Space Vector Pulse Width Modulation (SVPWM) is presented. The converter is basically an active rectifier divided into two converters: a matrix converter and an H bridge, which transfer energy through a high-frequency transformer, resulting in a modular AC-DC wireless converter appropriate for Plug-in Electric Vehicles (PEVs). The principle of operation of this converter considers high power quality, output regulation and low semiconductor power loss. The circuit operation, idealized waveforms and modulation strategy are explained together with simulation results of a 5 kW design.

Keywords: PEV's; AC-DC converter; ZVT; SVPWM; matrix converter; H bridge

---

1. Introduction

AC-DC power converters are typically used in Plug-in Electric Vehicles (PEVs) that require high reliability, reduced total harmonic distortion (THD) of the drawn currents and output regulation to charge batteries or supercapacitors with high performance and satisfy stringent power quality and density standards [1–3]. Electric Vehicle (EV) systems need to consider power quality regulations that typically include the harmonic emission during the charging and transient states, which have been analyzed with different standards and methods as the works described in [4,5]. Moreover, high-frequency switching techniques can increase the power quality, density and energy transfer reliability of AC-DC power converters without altering the basic topology configuration, keeping output controllability and reliability [6]; however, high AC line currents, drawn by the converter to rapidly charge the storage devices distributed along the traction DC link of the EV, may endanger the EV charger operator or impair the charger connector and, therefore, the supply drawn currents are limited to gain reliability causing slow energy transfer [7].

An available method of ensuring operator reliability with high currents, obtaining a simple way of connecting the charge uses wireless power transfer [8–10]. This is a twofold method, since the technique commonly uses a transformer to effectively isolate the power transfer from the user, and provides possible circuit splitting since the transformer is operated as an intermediary between two modules. For example, the converter presented in [11] uses a Medium-Frequency-Transformer for wind power conversion applications; however, the system uses an indirect matrix converter with three output phases that increases the circuit complexity and the magnetic components. EVs typically us two modules, where the first module is located on a charging station and the other on board the vehicle, improving the power density of the converter and increasing the EV distance range since the on-board section may be free of heavy devices [12]. The use of a single-phase high-frequency transformer was proposed in [13] as a wireless device to convert power from the utility to a DC-link capacitor bank; however, the circuit incurs in the use of several power stages required to generate high-frequency power signals from the utility supply to the transformer, limiting the efficiency of the converter and causing high complexity.

This paper presents the principle of operation of a modular AC-DC converter with Zero-Voltage Transition (ZVT), whose aim is to produce high-quality current waveforms, output regulation and soft switching of the semiconductor devices through the use of an off-board current-feed matrix converter and an on-board high-frequency H bridge, in such a way that the size of the converter located inside the vehicle becomes reduced in contrast with other topologies, [14–16]. The method potentially exhibits the following advantages: first, the current-feed matrix converter is used for straightforward generation of low-ripple, AC line currents without the need of several power stages; second a high-frequency transformer is used to isolate the matrix converter from the H bridge and regulate the output voltage; and third, the circuit uses a ZVT technique together with Space Vector Pulse Width Modulation (SVPWM) to achieve soft commutation of the power semiconductors and generate high-quality sinusoidal currents. Simulation results obtained in Saber with a 5 kW model are provided in the article, demonstrating the correspondence with the developed theoretical background and idealized waveforms to verify its feasibility.

2. ZVT AC-DC Converter

2.1. Circuit Description

The topology arises from the idea of splitting a conventional active rectifier in two modules as is shown in Figure 1. The off-board module of the converter is a three-phase current fed matrix converter that generates a single phase high-frequency current. This converter is intended to be located outside the electric vehicle in the charging station. The matrix converter has six bi-directional switches (Q_1–Q_6) and is fed by the line current of inductors L. The primary side of a high-frequency transformer is connected at the output of the matrix converter to allow wireless power transfer between the matrix converter and the second module.

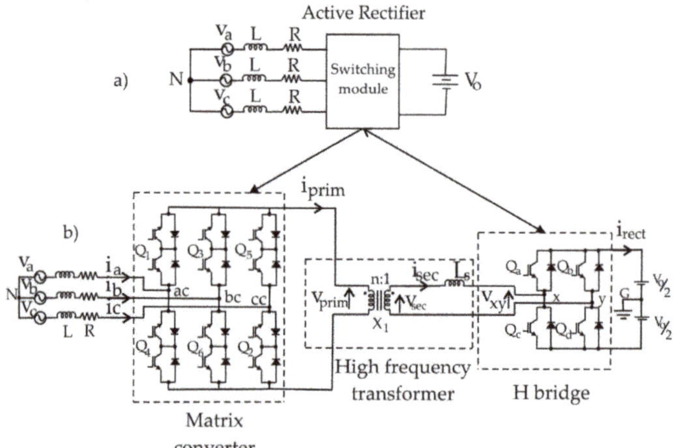

Figure 1. (a) Conventional active rectifier; (b) Circuit diagram of the proposed modular AC-DC converter.

The on-board module uses the secondary side of the high-frequency transformer and a series inductor, L_s, partially formed by the transformer leakage inductance and an additional inductor, together with an H bridge to regulate the output voltage, V_o. The H bridge inverts the output voltage with a phase-shifted Pulse Width Modulation (PWM) technique, such that a quasi-square with a $\pm nV_o$ amplitude is generated on the primary side of the transformer, v_{prim}. In this way, the conventional SVPWM technique is modified to obtain three-level PWM voltage waveforms which are used to control the line currents together with the supply voltage.

2.2. Principle of Operation

The control strategy used in the AC-DC converter is based in a SVPWM technique allowing DC output voltage regulation using an index modulation, being easy to implement with a fast response compared to other methods of PWM [17–19]. The conventional SVPWM technique is modified with a ZVT strategy, in such a way that the semiconductor devices are soft switched. The fundamental of operation of the SVPWM technique with ZVT is described below assuming negligible output voltage ripple and lossless components. Figure 2a–c show the active rectifier switching states, S_a, S_b and S_c, of a conventional SVPWM scheme for one switching period in the first sector, which ranges from $0°$ to $60°$, where 1 and 0 indicate on and off states respectively. These switching states are splitted to define the switching states of the H bridge and those of the matrix converter.

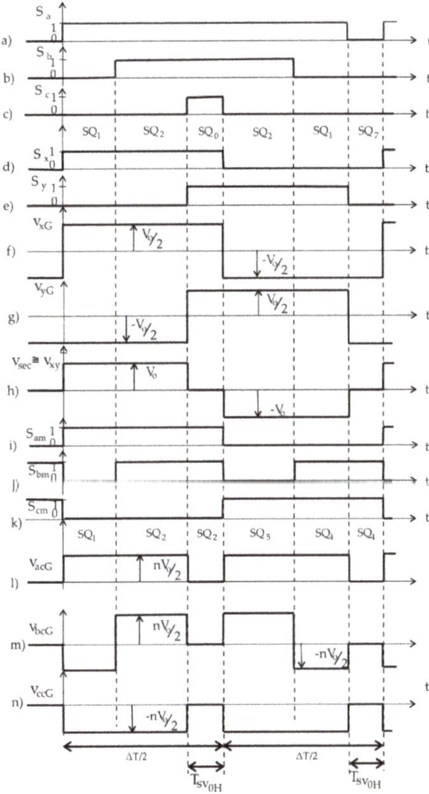

Figure 2. Derivation of the transistor switching states of the modular active rectifier of Figure 1b together with their voltage converter waveforms: (**a**) switching state S_a; (**b**) switching state S_b; (**c**) switching state S_c; (**d**) switching state S_x; (**e**) switching state S_y; (**f**) voltage v_{xG}; (**g**) voltage v_{yG}; (**h**) voltage v_{sec}; (**i**) switching state S_{am}; (**j**) switching state S_{bm}; (**k**) switching state S_{cm}; (**l**) voltage v_{acG}; (**m**) voltage v_{bcG}; (**n**) voltage v_{ccG}.

The H-bridge switching states S_x and S_y are presented in Figure 2d,e, with a duty cycle of 50% and a short-circuit period, T_{SV0H}, that generates the voltages v_{xG} (Figure 2f), and v_{yG} (Figure 2g) referred to the G node of the DC rail. The H-bridge voltage (Figure 2h), $v_{xy} = v_{xG} - v_{yG}$, clamps the secondary side of the transformer to zero Volts when the H bridge is in the short-circuit state. During the first semi cycle, $\Delta T/2$, the voltage in the secondary side of the transformer, v_{sec}, is clamped to V_0 and the switching states of the matrix converter legs, S_{am}, S_{bm} and S_{cm} (Figure 2i–k) are equal to S_a, S_b and S_c respectively. When S_a, S_b and S_c are equalized in the three inverters legs, a short-circuit occurs; which is caused in the modular version through the H bridge during T_{SV0H}, such that the matrix converter retains the last switching state combination. During the second semi cycle, a mirrored state sequence occurs, since the output voltage in the H bridge becomes $v_{sec} = -V_0$ and the switching states in the matrix converter are the complement of S_a, S_b and S_c. The H-bridge short-circuit is caused again and the matrix converter retains the last switching states when S_a, S_b and S_c are equal. In this way, the neutral combination is again caused by the H bridge instead of the matrix converter.

The matrix converter voltages referred to the DC rail node G, v_{acG}, v_{bcG} and v_{ccG} are shown in Figure 2l–n, which are the product of the primary voltage v_{prim} ($v_{prim} = nv_{sec}$) and the individual switching state of each leg. When the short-circuit is caused by the H bridge, the matrix converter voltages are clamped to zero Volts.

2.3. SVPWM Technique

Six active voltage space vectors, sv_1 to sv_6, are obtained at the central nodes of the matrix converter shown in Figure 1b by using six switching states combinations, SQ_1 to SQ_6, which are listed in Table 1, with respect to the states of S_{am}, S_{bm} and S_{cm}. These space vectors are plotted in the bi-dimensional α-β plane of Figure 3 using the Clarke transform [20]. sv_1 to sv_6 are generated using the switching states of the matrix converter together with its output voltage $\pm nV_o$. A neutral space vector, sv_0, is produced by using any state combination of the matrix converter together with the H-bridge short-circuit. sv_0 is located at the origin of the α-β plane.

Table 1. Switching states vectors of matrix converter.

Switching State Combination	S_{am}, S_{bm} and S_{cm} States
SQ_1	$(1, 0, 0)$
SQ_2	$(1, 1, 0)$
SQ_3	$(0, 1, 0)$
SQ_4	$(0, 1, 1)$
SQ_5	$(0, 0, 1)$
SQ_6	$(1, 0, 1)$

An arbitrary averaged voltage vector, $v_{cav} = V_{cpk}\angle\theta_c$, can be generated at the converter input using a Volts-seconds balance to control the input line currents together with the supply voltages. Since the matrix converter output voltage reverses its biasing during half of the switching cycle, the Volts-seconds balance utilises the operating sector and its opposite sector of the α-β plane. For example, during the first semi cycle of a switching period in sector S_1, v_{cav} is determined by:

$$v_{cav} = a_1 sv_1 + a_2 sv_2 + a_0 sv_0 \tag{1}$$

where $\alpha_1 = \dfrac{T_{sv1}}{\frac{\Delta T}{2}}$, $\alpha_2 = \dfrac{T_{sv2}}{\frac{\Delta T}{2}}$ and $\alpha_0 = \dfrac{T_{sv0H}}{\frac{\Delta T}{2}}$.

Whereas, during the second semi cycle, v_{cav} is defined by:

$$v_{cav} = a_4 sv_4 + a_5 sv_5 + a_0 sv_0 \tag{2}$$

where $\alpha_4 = \dfrac{T_{sv4}}{\frac{\Delta T}{2}}$, $\alpha_5 = \dfrac{T_{sv5}}{\frac{\Delta T}{2}}$, since $v_{prim} = -nV_o$.

The same procedure applies for the rest of the sectors. Table 2 summarizes the switching states according to the sector location of θ_c together with the biasing of v_{prim}.

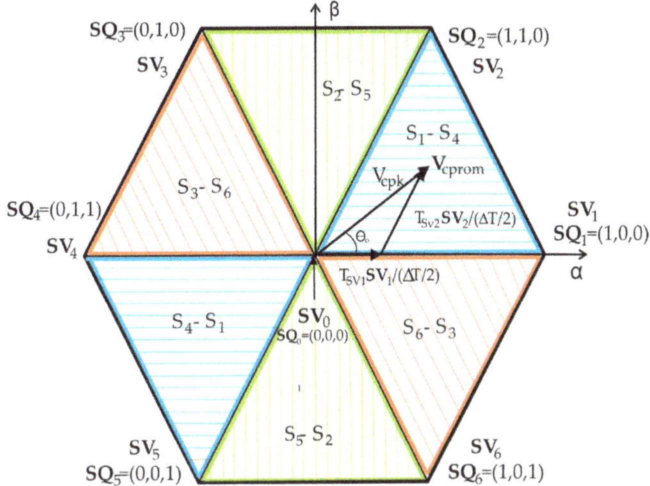

Figure 3. Space Vector Bi-dimensional plane.

Table 2. Switching states vectors.

Angle θ_c	Sector	v_{prim} (+)	v_{prim} (−)
		Switching State	Switching State
1–60°	S_1	SQ_1, SQ_2	SQ_4, SQ_5
61–120°	S_2	SQ_2, SQ_3	SQ_5, SQ_6
121–180°	S_3	SQ_3, SQ_4	SQ_6, SQ_1
181–240°	S_4	SQ_4, SQ_5	SQ_1, SQ_2
241–300°	S_5	SQ_5, SQ_6	SQ_2, SQ_3
301–360°	S_6	SQ_6, SQ_1	SQ_3, SQ_4

v_{xy} and v_{sec} are shown in Figure 4a,b respectively for straight comparison to describe the generation of the secondary transformer current, i_{LS}, shown in Figure 4c. During the first semicycle, i_{LS} is assumed positive together with v_{sec} clamped to V_0. The slope of i_{LS} is negative since the converter voltage is greater than the AC supply voltage even when the matrix converter switches from SQ_1 to SQ_2; whilst during the H-bridge short circuit the matrix switching state SQ_2 is retained, such that the slope of i_{LS} is inverted. A current reversal occurs in i_{LS} since v_{xy} is clamped to $-V_0$ and v_{sec} to zero during an overlap period T_{ovl} that is calculated with Equation (3):

$$T_{ovL} = \frac{n I_{primpk} L_s}{V_0} \tag{3}$$

where I_{primpk} is the peak magnitude of the current in the primary side of the transformer, i_{prim}.

In this way, the ZVT is performed in the semiconductor devices during the current reversal to obtain soft commutation. The overlap period finishes when i_{LS} reaches the same magnitude with an opposite sign, such that the first semi cycle becomes to an end. The second semi cycle is a mirror of the first since the complementary switching states of SQ_1 and SQ_2, SQ_4 and SQ_5 respectively, are used to operate the matrix converter and because the H bridge has clamped v_{xy} to $-V_0$.

i_{Ls} is equal to ni_{prim}, where i_{prim} depends on the switching of the line currents. Table 3 lists the equivalence of i_{prim} respective to the line currents and the matrix switching states.

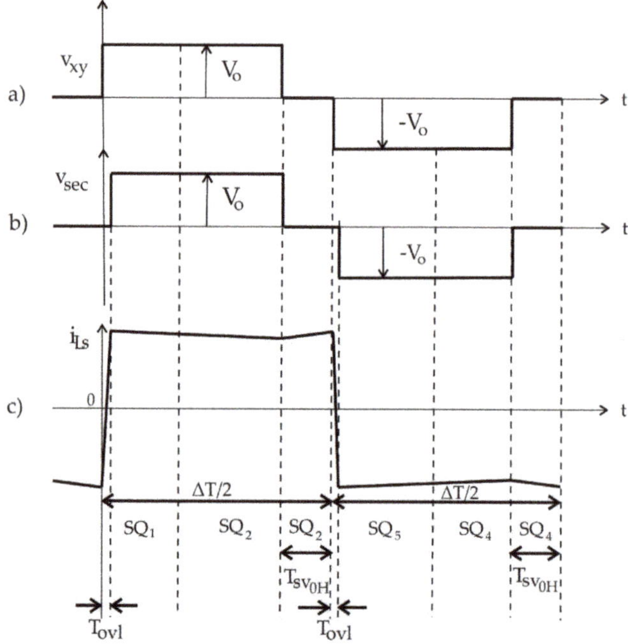

Figure 4. Ideal wavfeorms: (**a**) Voltage v_{xy}; (**b**) voltage v_{sec}; and (**c**) current through the transformer, i_{LS}.

Table 3. i_{LS} magnitude for each switching state.

Switching State	i_{prim}
SQ_1	$i_{prim} = i_a - i_b - i_c$
SQ_2	$i_{prim} = i_a + i_b - i_c$
SQ_3	$i_{prim} = i_b - i_a - i_c$
SQ_4	$i_{prim} = i_b + i_c - i_a$
SQ_5	$i_{prim} = i_c - i_b - i_a$
SQ_6	$i_{prim} = i_a - i_b + i_c$

3. ZVT AC-DC Converter

A block diagram for the AC-DC operation of the circuit of Figure 1b is shown in Figure 5. In this figure θ_c and the index modulation, M_a, are the inputs required for the SVPWM scheme. Since the converter of Figure 1b is divided into two parts, a sequenced operation is used and described below:

(a) The conventional active rectifier switching states are generated using M_a and θ_c. These are equivalent to the control signals of Figure 2a–c.

(b) S_{am}, S_{bm} and S_{cm}, shown in Figure 2i–k, are generated and multiplexed to assign the control signals for each bi-direccional switch.

(c) S_x and S_y, shown in Figure 2d,e, are generated from the active periods of the conventional active rectifier.

(d) An overlap is required to turn on and off the bidirectional switches during the reversal of current i_{LS}.

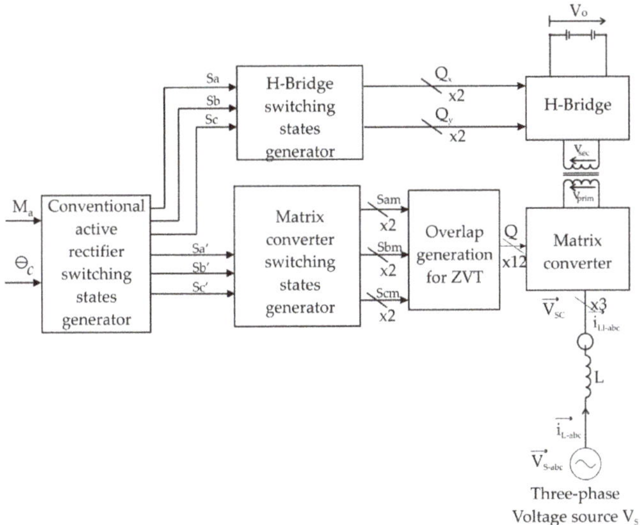

Figure 5. Block diagram for AC-DC operation of the modular converter of Figure 1b.

3.1. H-Bridge and Matrix Converter Switching States

The derivation of the H-bridge and matrix converter control switching states is described using the block diagram of Figure 6, which are obtained from the switching states of the conventional active rectifier, as described in Section 2.2. Firstly, S_a, S_b and S_c are generated using θ_c and M_a. The conventional SVPWM active periods, T_a, T_b and T_c, are compared with a high-frequency carrier triangular waveform to produce three digital signals, PWM_a, PWM_b and PWM_c, which are shown in Figure 7. These signals are multiplexed with respect to the sector location to derive the switching states S_a, S_b and S_c, (Figure 7e–g). The H-bridge switching states, Q_a, Q_c, Q_b an Q_d, are obtained with a sequential circuit using S_a, S_b and S_c as inputs.

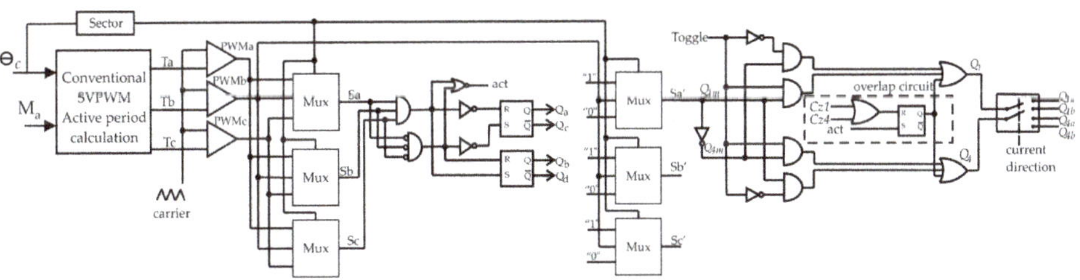

Figure 6. Block diagram to generate the switching states for the matrix converter and the H bridge.

Other set of multiplexers are utilized to derive three digital signals, $S_a{}'$, $S_b{}'$ and $S_c{}'$, that produce the active switching states by using off and on states instead of the short and long pulse trains of PWM_a and PWM_c respectively, as is shown in Figure 7h–j. $S_a{}'$, $S_b{}'$ and $S_c{}'$ are also multiplexed with respect to the sector location and are processed to derive the switching state of each matrix converter leg. The digital circuit to generate the matrix converter control signals for the first leg, Q_{1a}, Q_{1b}, Q_{4a} and Q_{4b}, is shown at the right-hand of Figure 6. This circuit commutates the switching states to the opposite side of the α-β plane, sending the complement states when v_{sec} is clamped to negative voltage. The same circuit is used to generate the control signals for the second and third matrix converter legs.

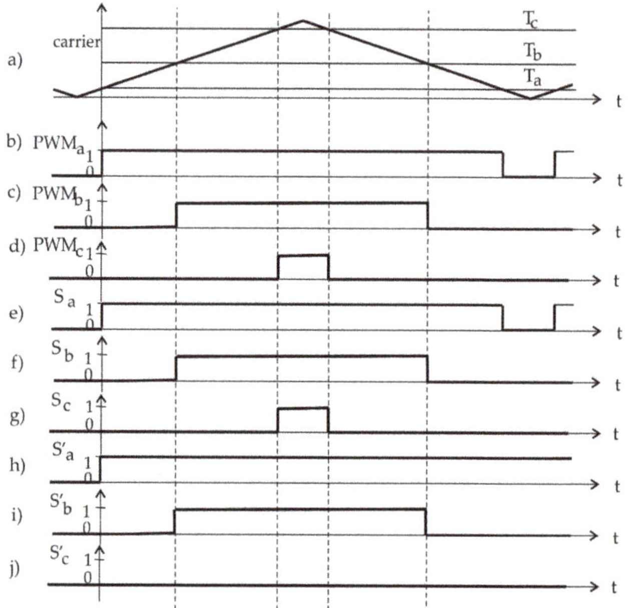

Figure 7. Generation of PWM signals, and conventional active rectifier switching states. (**a**) High-frequncy triangular carrier signal; (**b**) digital signal PWM_a; (**c**) digital signal PWM_b; (**d**) digital signal PWM_c; (**e**) switching state S_a; (**f**) switching state S_b; (**g**) switching state S_c; (**h**) digital signal S'_a; (**i**) digital signal S'_b; (**j**) digital signal S'_c.

3.2. Neutral-to-Active Switching Transition in the Matrix Converter

Since each bi-directional switch in the matrix converter is built with two semiconductor devices with a common collector configuration that allows the current flow in both directions, an overlap in all the matrix converter legs is required to commutate the flowing of current in one direction to the opposite direction during the negative biasing of v_{sec}. Figure 8 shows the switching sequence of Q_{1a}, Q_{1b} and Q_{4a}, Q_{4b} to turn off Q_1 and turn on Q_4.

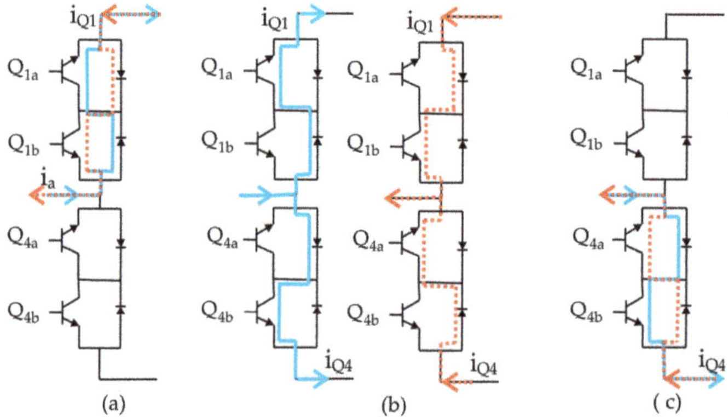

Figure 8. Switching sequence in the on-to-off transition from neutral-to-active switching state. (**a**) initial current flow; (**b**) overlap time; (**c**) current flow in the opposite direction.

In Figure 8a, Q_{1a} and Q_{1b} are conducting the current in the blue or red arrow direction, i_{Q1}. When Q_1 and Q_4 are commutated to the on and off states respectively, Q_{1b} is turned off and Q_{4b} is turned on when the current is flowing in the blue arrow direction; and Q_{1a} is turned off and Q_{4a} is turned on when the current is flowing in the red arrow direction (Figure 8b). Finally, when the current flow stops due to the biasing inversion of v_{sec} through the corresponding arrow, the transistors of Q_4 are turned on, (Figure 8c), and, therefore, the current in Q_4, i_{Q4}, flows in the opposite direction through the matrix converter leg.

The logic circuit that generates the overlap in the first matrix converter leg is included in the right-hand of Figure 6. This diagram shows the zero crossing detector signals C_{z1} and C_{z2} that are used to indicate the end of the overlap.

3.3. Active-to-Active Switching Transition in the Matrix Converter

The transition between two active vectors takes place twice in a switching period. In Figure 2a the transition occurs from SQ_1 to SQ_2 during the first semi cycle; whereas in the second semi cycle, the transition occurs from SQ_5 to SQ_4. The transition between adjacent active vectors implies a switching state change in one matrix converter leg. By instance, the state of the second matrix leg, S_{bm} in Figure 2, switches from the off to the on state producing an active-to-active transition. The turn on-to-off sequence of Q_3 and Q_6 is shown in Figure 9.

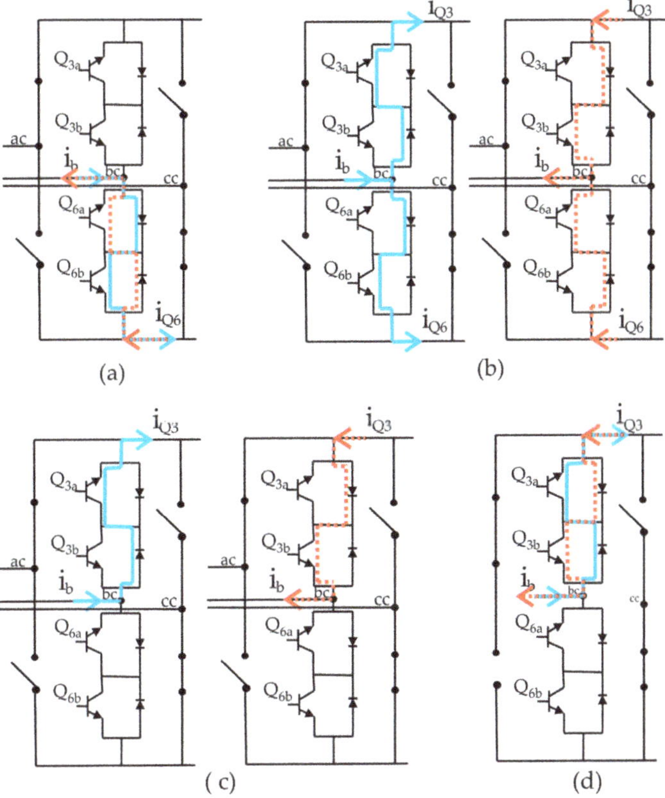

Figure 9. Switching sequence in the on-to-off transition from active-to-active switching state. (**a**) initial current flow; (**b**) overlap time; (**c**) current flow in the opposite direction; (**d**) bidirectional switch in on state.

In Figure 9a, Q_{6a} and Q_{6b} are firstly turned on and the current flows in the red or blue arrow direction, i_{Q6}; then, in Figure 9b Q_{6a} and Q_{3a} are turned off and on respectively to allow the current reversal in the matrix converter leg when the current is flowing in the blue arrow direction; or Q_{6b} and Q_{3b} are turned off and on respectively to allow the current reversal when the current is flowing in the red arrow direction. In this figure, the transistors of Q_6 are turned on for a period of time t_D longer than the turning off time of the semiconductor device, t_r, to guarantee that the current stops flowing through it. This bidirectional switch configuration comes to an end when the transistor of Q_6 is turned off as shown in Figure 9c. Lastly, the sequence finishes by turning on transistors of Q_3, as shown in Figure 9d, with i_{Q3} flowing in the opposite direction.

4. Steady-State Analysis and Parameters Selection

A steady-state analysis of the AC-DC modular converter is derived using the generalized diagram of Figure 10, where v_s is the three-phase source voltage vector, L is the line inductor, v_L is the line inductor voltage vector, v_{cav} is the converter voltage vector, $n{:}1$ is the transformer turns ratio that links the off-board AC-AC module with the AC-DC module; V_o is the DC output voltage and R is the output load.

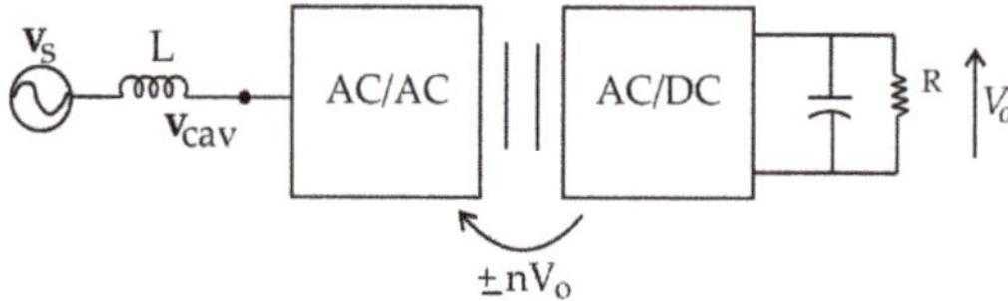

Figure 10. Generalized diagram for steady-state analysis

Assuming a vectorial current control to drive the phase and magnitude of the line current i_L and produce high power factor, a phasorial diagram is shown in Figure 11 which describes the behavior of the AC-DC converter of Figure 1. In this diagram, v_s, v_L and v_{cav} are defined by:

$$v_s = V_{spk}\angle 0°, \tag{4}$$

$$v_L = V_{Lpk}\angle 90°, \tag{5}$$

$$v_{cav} = V_{cpk}\angle \varphi° \tag{6}$$

In Figure 11 the amplitudes of v_{cav} and v_L are defined by:

$$V_{cpk} = \frac{V_{spk}}{\cos \varphi} \tag{7}$$

$$V_{Lpk} = V_{cpk} \sin \varphi \tag{8}$$

Whereas the current i_L is obtained with Equation (9):

$$i_L = \frac{v_L}{\omega L} \tag{9}$$

Considering the phasorial diagram of Figure 11, v_L is calculated using Equation (10):

$$v_L = v_s - v_c \tag{10}$$

Equations (9) and (10) are used to derive the amplitude of i_L, I_{Lpk}, which is defined by:

$$I_{Lpk} = \frac{V_{spk} tg\varphi}{\omega L} \tag{11}$$

where $\omega = 2\pi f$. Assuming ideal conditions, a power balance is derived:

$$P_{in} = P_{out} \tag{12}$$

where P_{in} and P_{out} are the input and output power converter respectively and are equivalent to Equation (13):

$$\frac{3V_{spk}I_{Lpk}}{2} = \frac{V_o^2}{R} \tag{13}$$

Substituting Equation (11) in (13), the power balance becomes:

$$\frac{3V_{spk}^2 tg\varphi}{2\omega L} = P_o \tag{14}$$

According to Figure 11, the AC-DC converter can operate under two extreme conditions; a minimum supply voltage, V_{spkmin}, obtaining a maximum phase, φ_{max}, with a maximum power demand at the output, P_{omax}; and a maximum supply voltage, V_{spkmax} obtaining a minimum phase, φ_{min}, with a minimum power demand at the output, P_{omin}. Therefore, φ_{max} and φ_{min} are obtained using Equations (15) and (16) respectively:

$$\varphi_{max} = tg^{-1}\frac{2P_{omax}\omega L}{3V_{spkmin}^2} \tag{15}$$

$$\varphi_{min} = tg^{-1}\frac{2P_{omin}\omega L}{3V_{spkmax}^2} \tag{16}$$

Considering $P_{omax} = 5$ kW, $P_{omin} = 500$ W, $V_{spkmin} = 144$ V and $V_{spkmax} = 216$ V, φ_{max} and φ_{min} are obtained for different values of L. Figure 12 is used to select the optimal L that allows an appropriate range of phase control. L was judged to be 3 mH, in such a way that φ can be ranged from 0.46° to 10.3°.

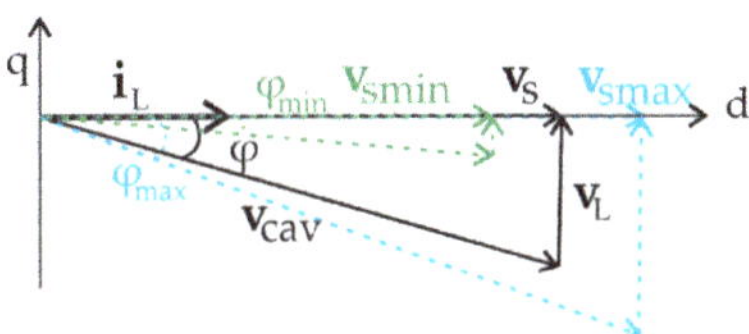

Figure 11. Phasorial diagram for the AC-DC converter operation.

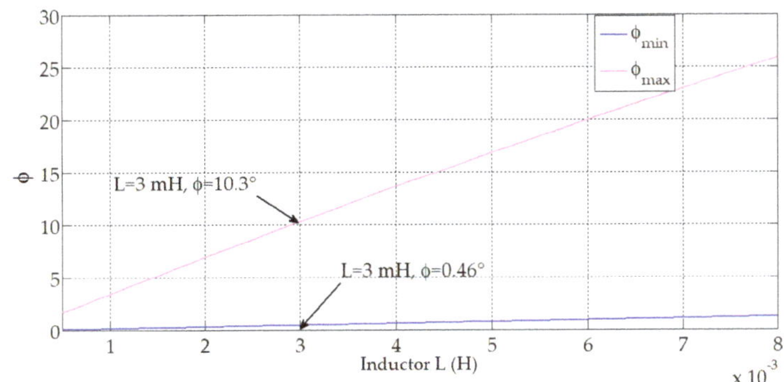

Figure 12. φ_{min} and φ_{max} obtained for different values of L.

The voltage V_{cpk} is expressed in function of the modulation index, M_a, when the converter operates with a SVPWM scheme [14], as follows:

$$V_{cpk} = M_a n V_o \frac{\sqrt{3}}{4} \tag{17}$$

The minimum and maximum converter voltages, V_{cpkmin} and V_{cpkmax}, are utilized to obtain the minimum and maximum modulation indexes, M_{amin} and M_{amax} respectively, which are shown in (18) and (19):

$$M_{amin} = \frac{4V_{cpkmin}}{\sqrt{3}nV_o} \tag{18}$$

$$M_{amax} = \frac{4V_{cpkmax}}{\sqrt{3}nV_o} \tag{19}$$

where V_{cpkmin} = 144.004 Volts and V_{cpkmax} = 219.53 Volts were calculated using Equation (7) for V_{spkmin} and V_{spkmax} respectively. An optimal value for n was determined by using Equations (18) and (19) and ranging n from 2 to 10, such that the results are plotted in Figure 13. n was selected to be 5 since M_a can be set between 0.66 and 1 to control the amplitude of the converter input voltage. It was judged in this work that n should be 5 to allow the converter an appropriate SVPWM operation and leave a small upper range of M_a for the ZVT effects since the maximum value of M_a is 1.16 for SVPWM [20].

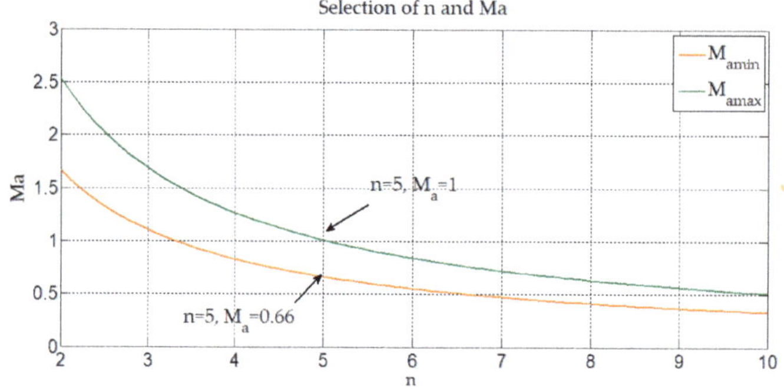

Figure 13. M_{amin} and M_{amax} obtained for different n.

Equations (14) and (17) show that there is a compromise between the selection of the parameters n, φ and L to reach a wide range of M_a within the conventional active rectifier operation range.

5. Numerical Verification

To verify the principle of operation of the modular AC-DC converter and the SVPWM with ZVT control strategy, a simulation in Saber was performed using ideal components and the parameters listed in Table 4.

Table 4. Simulation Parameters.

Parameter	Value
Source voltage v_a, v_b and v_c	180 V peak
Source frequency	60 Hz
Switching fequency	7.2 kHz
Input inductor L	3 mH
Index modulation M_a	0.628
Input resistor R	0.1 Ω
Leakage Inductance L_s	50 μH
Turns ratio n	5:1
Output voltage V_o	100 V
Output Power P_o	5 kW

5.1. Verification of the Modified SVPWM

A Saber simulation was performed synchronizing the operation of the divided rectifier control signals with the fundamental frequency of the supply, using the scheme of Figure 14. I_{Lpk} is the current reference used to define θ_c and M_a for the SVPWM operation of the rectifier and cause high power factor as shown in Figure 11. The converter voltage was phase shifted to align the line currents with the supply using the vectorial control system of Figure 14. A reference current of I_{Lpk} = 19.54 A was used to obtain a 100 V, 5 kW output with high power factor supply. The component of V_{cav} in the q axis, V_{cavq}, was determined using Equation (7) and the phasorial diagram of Figure 11, V_{cavq} = 22.1 V, such that V_{cavd} = V_{spk} = 180 V, therefore, the phase between v_s and v_{cav} was φ = 7°.

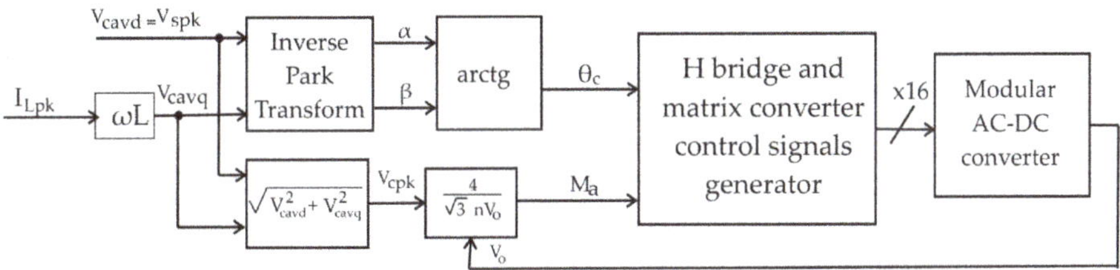

Figure 14. General scheme used in simulation.

To confirm the correct operation of the scheme shown in Figure 14, Figure 15 shows a Saber result plot of the supply and converter phases, θ_S and θ_C, for current references of 9.4 A and 19.54 A and cause 2.5 kW and 5 kW respectively. In Figure 15a a phase shift of 3° is obtained, whereas in Figure 15b the phase shift becomes of 7° since the current reference was increased. The effectiveness of the SVPWM rectifier operation of the circuit of Figure 1b was verified analyzing the Saber simulation results of the line currents and input voltages of the converter. Figure 16a shows Saber results of the source voltage v_{aN} and the line current i_a. The resultant line current i_a is sinusoidal with a 3.3% ripple. Figure 16b shows the five-level converter voltage, v_{acN}, to verify the correct operation of the space vector strategy. The current and voltage in the primary side of the transformer, i_{prim} and v_{prim} are shown in Figure 16c, where i_{prim} is the rectified version of the line currents, but, inverted in one quadrant of the switching cycle producing a high-frequency AC square current.

The operation of the H bridge was verified contrasting the conventional SVPWM states with the phase-shifted control signals of the H bridge. Figure 17 plots the Saber results of S_a, S_b and S_c (Figure 17a–c, respectively) in contrast to S_x and S_y (Figure 17d,e). The H-bridge input voltages v_{xG} and v_{yG} are shown in Figure 17f,g, respectively, and its difference v_{xy} is shown in Figure 17h. In the latter figure, v_{xy} is a $\pm V_o$ quasi-square waveform, whose zero level is effectively produced by the H-bridge overlap, which can be used to produce the neutral vectors required by the conventional rectifier states in a switching cycle.

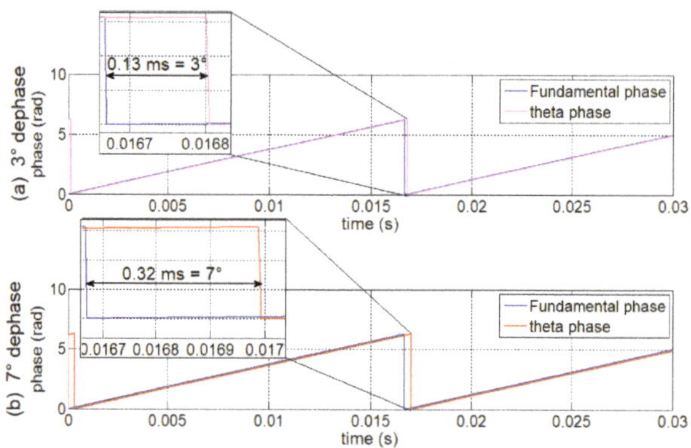

Figure 15. Variation of φ: (**a**) $\varphi = 3°$; (**b**) $\varphi = 7°$.

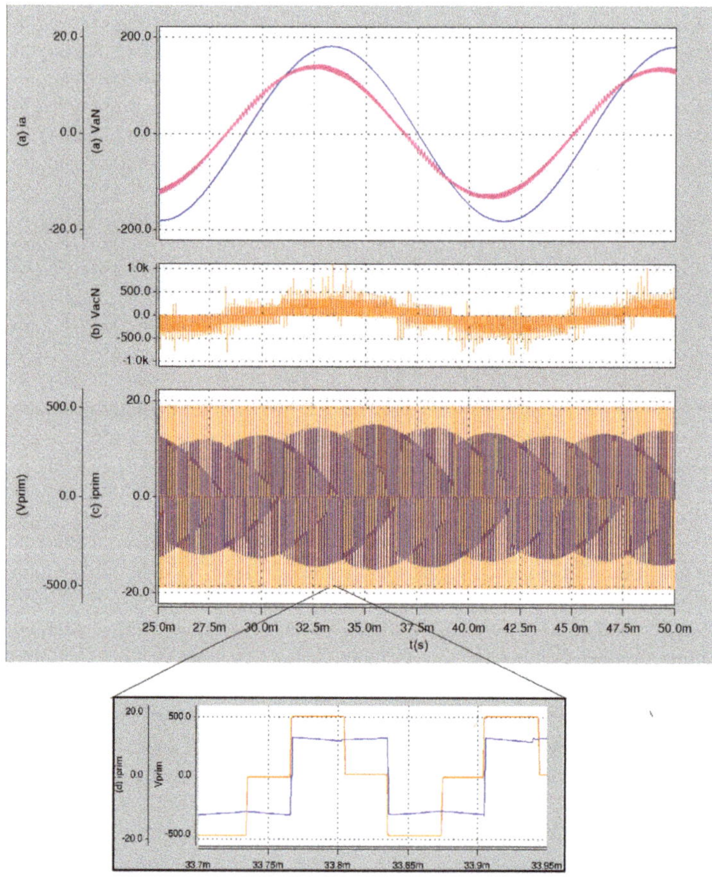

Figure 16. Simulation results (**a**) Supply voltage v_{aN} and line current i_a; (**b**) converter voltage in phase a; and (**c**) High-frequency current and voltage in the primary side of the transformer. Supply: 127 V, 60 Hz, Output: 100 V, 5 kW.

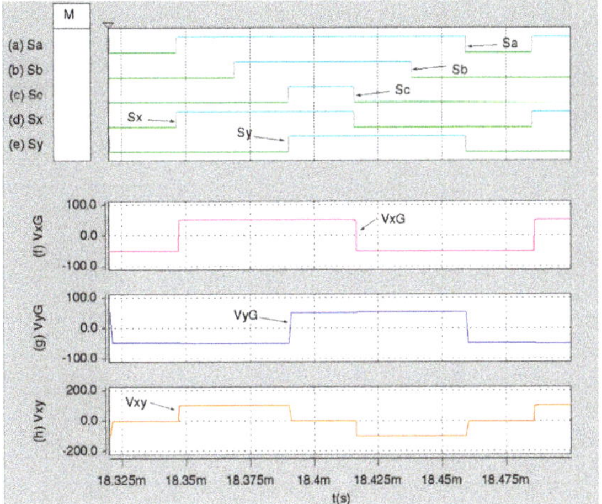

Figure 17. Simulations results: conventional switching states (**a**) S_a, (**b**) S_b and (**c**) S_c; H-bridge switching states (**d**) S_x and (**e**) S_y and voltages (**f**) v_{xG} (**g**) v_{yG} and (**h**) v_{xy}. Supply: 127 V, 60 Hz, Output: 100 V, 5 kW.

The SVPWM strategy for the divided AC-DC converter of Figure 1b was verified by contrasting a switching cycle of the states S_{am}, S_{bm} and S_{cm} with S_a, S_b and S_c. The upper plot of Figure 18 shows S_a, S_b and S_c together with S_{am}, S_{bm} and S_{cm}; whilst the three-phase input converter voltages with respect to the G node of the DC link, v_{acG}, v_{bcG} and v_{ccG}, are shown in the lower plots of Figure 18 (Figure 18g–i) for straightforward comparison with those plotted in Figure 2. In these plots v_{acG} and v_{ccG} are positive and negative rectified waveforms of v_{xy}; furthermore, v_{bcG} becomes a fully AC waveform due to the state reversals of the matrix converter leg since this switching cell utilizes the active-to-active switching transition.

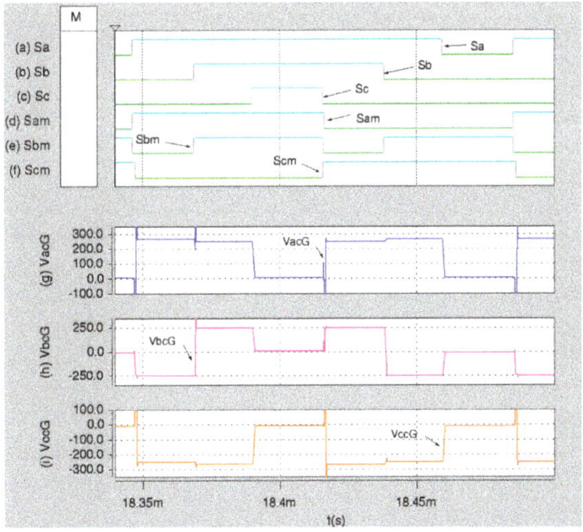

Figure 18. Simulations results: (**a**) conventional switching states S_a; (**b**) S_b; and (**c**) and S_c; (**d**) matrix converter switching states S_{am}; (**e**) S_{bm}; and (**f**) S_{cm}; (**g**) Voltage v_{acG}; (**h**) v_{bcG}; and (**i**) v_{ccG}. Supply: 127 V, 60 Hz, Output: 100 V, 5 kW.

5.2. ZVT Verification

The effects of the ZVT in the transistors of the matrix converter were initially verified analyzing the quasi-square voltages v_{xy} and v_{sec} together with the transformer secondary current, $i_{sec} = i_{Ls}$, as shown in Figure 19 for direct comparison with Figure 4. In Figure 19a,b, during the first semi cycle, v_{xy} and v_{sec} are clamped to V_o such that i_{Ls} (Figure 19c) has a smooth negative slope; however, this slope becomes positive when the H bridge is in its overlap state to produce a neutral voltage vector at the converter AC input. A expanded portion of the first semi cycle of i_{Ls} is shown in Figure 19d.

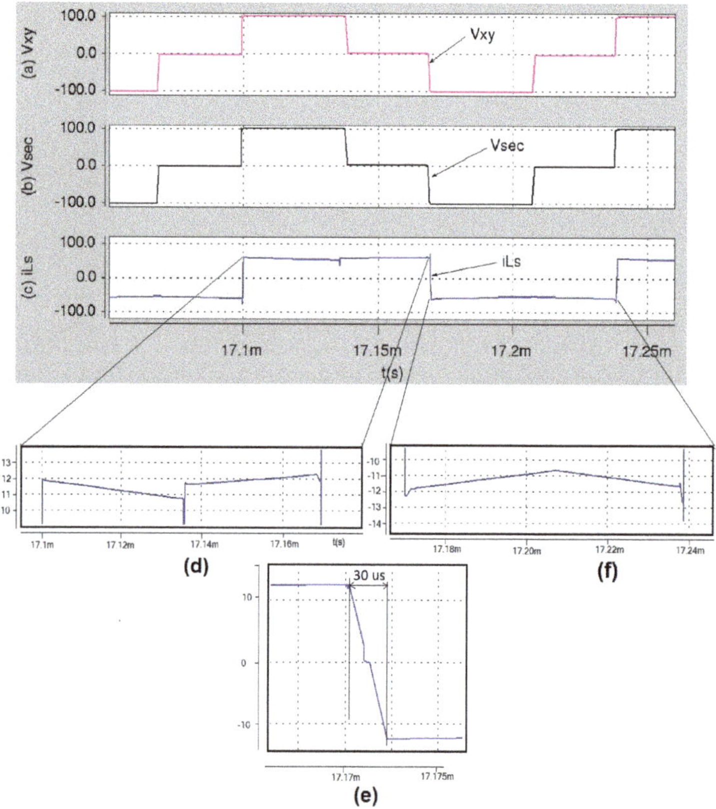

Figure 19. Simulation results: (**a**) voltage v_{xy}; (**b**) v_{sec}; (**c**) current i_{Ls}; (**d**) Expanded portion of the first semicycle; (**e**) expanded portion of T_{ovL}; and (**f**) expanded portion of the second semicycle. Supply: 127 V, 60 Hz, Output: 100 V, 5 kW.

At the beginning of the second semicycle, i_{Ls} is reversed in an overlap period of $T_{ovL} = 3$ μs, as shown in the expanded portion of Figure 19e, which was confirmed using Equation (3), since v_{xy} and v_{sec} are now clamped to $-V_o$ and zero respectively; whereas the rest of the second semicycle i_{Ls} becomes a mirrored wave of the first, which is shown in the expanded portion of Figure 19f. i_{Ls} is an AC trapezoidal waveform that is shaped by the switched operation of the H bridge and the matrix converter, which needs to be controlled to ensure stability and prevent saturation by the aid of a DC blocking capacitor, or a peak current control [21], since a high-frequency transformer is used in the proposed converter.

A zero-voltage switching transition is achieved in the matrix converter legs due to the zero-voltage biasing of its bidirectional switches whilst i_{Ls} is being reversed. This was verified analyzing the voltage and current waveforms in the first matrix leg switches Q_1 and Q_4 when an i_{LS} reversal occurs. Figure 20 depicts the simulated switching transition result described in Figure 8 to turn on Q_4 and turn off Q_1.Initially in Figure 20, the switches of Q_1, Q_{1a} and Q_{1b}, and those of Q_4, Q_{4a} and Q_{4b}, are in the on and off states respectively. Later in Figure 20a, Q_{1a} and Q_{4a} have an overlap period that cause to the voltage of Q_4, v_{Q4}, decrease to zero, and then i_{Q4} increases as shown in Figure 20b, achieving a ZVT turn on in Q_4 during the current reversal. Whereas i_{Q4} increases, i_{Q1} falls to set Q_1 to the off state, as shown in Figure 20c, which depicts a hard switch off transition. During the overlap, v_{sec} is clamped to zero and then biased to $-V_o$ at the end of the overlap, whilst the current reversal in i_{Ls} occurs (Figure 20d).

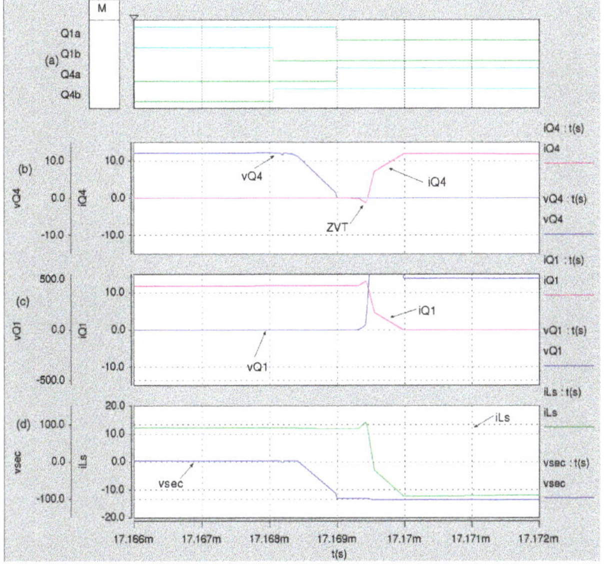

Figure 20. Simulation results to verify ZVT during the switching transition to turn on Q_4, (**a**) control signals Q_{1a}, Q_{1b}, Q_{4a}, Q_{4b}; (**b**) i_{Q4} and v_{Q4}; (**c**) i_{Q1} and v_{Q1}; and (**d**) v_{sec} and i_{Ls}. Supply: 127 V, 60 Hz, Output: 100 V, 5 kW.

Figure 21a shows the simulated results of the opposite switching transition to turn on Q_1 and turn off Q_4. Mirrored waveforms are obtained for i_{Q1}, i_{Q4}, v_{Q1} and v_{Q4} in contrast to Figure 20. The ZVT is shown in Figure 21b when i_{Q1} becomes positive during the zero voltage in Q_1. v_{sec} is clamped to $+V_o$ whilst i_{Ls} is reversed.

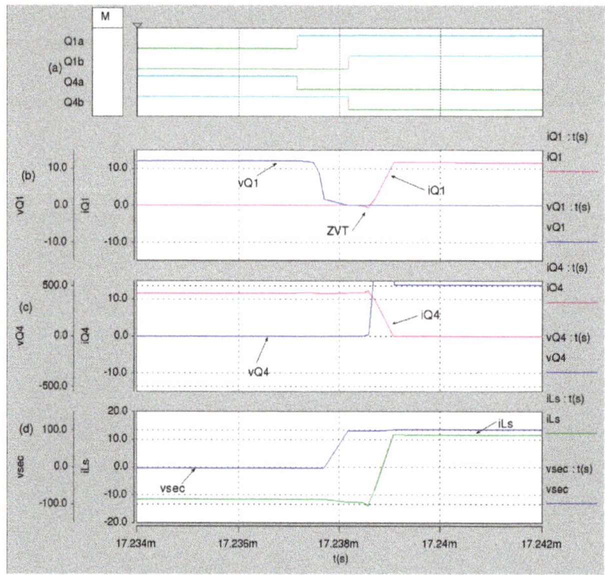

Figure 21. Simulation results to verify ZVT during the switching transition to turn on Q_1, (**a**) control signals Q_{1a}, Q_{1b}, Q_{4a}, Q_{4b}; (**b**) i_{Q1} and v_{Q1}; (**c**) i_{Q4} and v_{Q4}; and (**d**) v_{sec} and i_{Ls}. Supply: 127 V, 60 Hz, Output: 100 V, 5 kW.

5.3. Steady-State Power Balance Verification

To verify the input-to-output active power balance, the DC output power was calculated by the aid of the H-bridge output current i_{rect}. The current i_{Ls} and i_{rect} are shown in Figure 22a,b, respectively. In these figures i_{Ls} is seen to be rectified by the H bridge since i_{rect} may become either $\pm i_{Ls}$, during the $\pm V_o$ clamping of v_{xy} respectively or zero when the H bridge is in its short-circuit state. The average or i_{rect}, I_{rect}, was calculated using the simulation results shown in Figure 22, and it was found that $I_{rect} = 50$ A and, therefore, the output power is 5 kW, since the Saber simulation was performed assuming a constant DC output voltage of $V_o = 100$ Volts. In this fashion, the output power equalizes the active supply power, demonstrating the principle of operation of the AC-DC divided converter of Figure 1b.

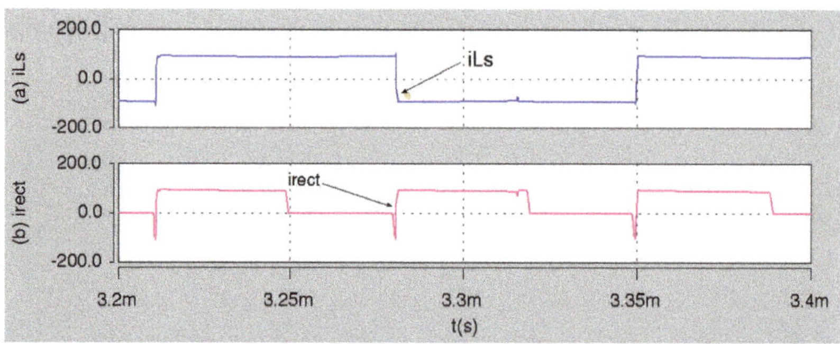

Figure 22. Simulations results for (**a**) i_{Ls}; (**b**) i_{rect}. Supply: 127 V, 60 Hz, Output: 100 V, 5 kW.

To verify the high-quality supply currents the harmonic content in the line current i_a was calculated, as shown in Figure 23; the measured current THD was 4.43%, being the power rated at 5 kW. Two main current harmonic clusters were found at the harmonic order of $n = 242$ and $n = 488$, corresponding to the frequencies of 14.5 kHz and 29.28 kHz, which were increased in amplitude since four switching transitions take place between space vector combinations during a switching period.

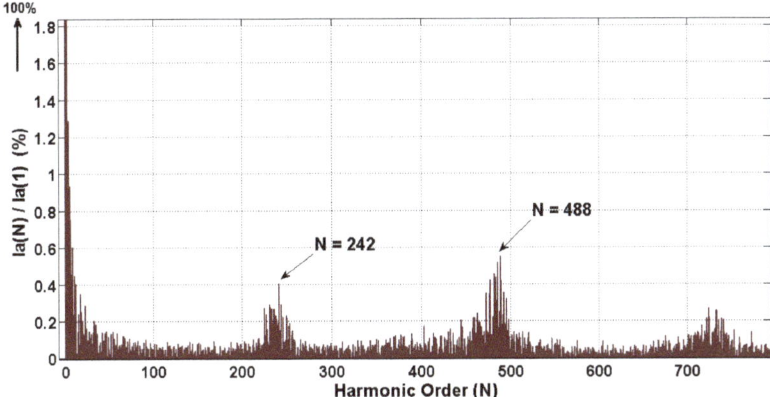

Figure 23. Harmonic content for line current i_a.

The low order harmonic content of the input current was compared with the Standard EN61000-3-2 [22], for the limits of Class A converters. Figure 24 shows that the components are within the standard limits; for example, for $n = 3$, the amplitude is 0.39 A, 2% of the fundamental, that is less than limit, 2.3 A.

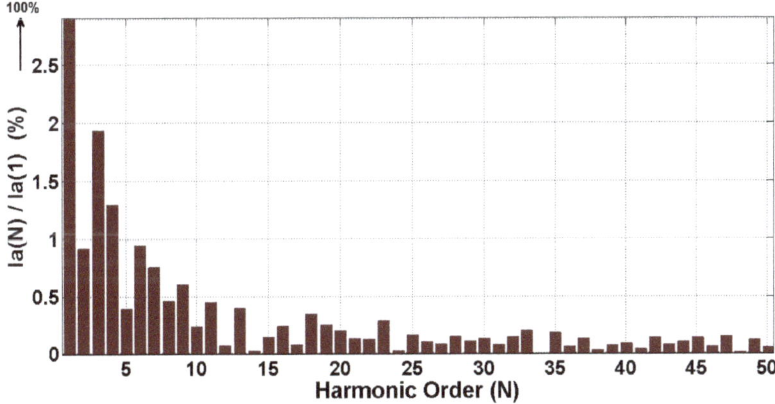

Figure 24. Low Harmonic content for line current i_a for comparison with Standard EN 61000-3-2.

6. Comparison of the Proposed Converter with Other AC-DC Topologies

Table 5 presents a detailed comparison of the proposed topology with four other AC-DC converters of different levels. The first row of this table is referred to the charging power levels for EVs, which are currently three according to the International Electrotechnical Commission standard IEC61851 [23]. Level 1 is for small on-board battery chargers with typical use in home or office; Level 2 is for medium power battery chargers that can be used in private or public outlets, and Level 3 is generally designed for a recharging station for commercial and public transportation. The proposed topology is intended for Level 3 applications, which justifies the use of a three-phase voltage supply.

Table 5. Comparison with three other AC-DC converters.

Factor \ Topology	Proposed AC-DC Modular Converter	Three-Phase PFC Rectifier with DC-DC Converter, [14]	AC-DC Matrix Converter, [14]	Isolated On-Board Vehicle Battery Charger Utilizing SiC Power Devices, [15]	Inductively Coupled Multi-Phase Resonant Wireless Converter, [16]
Level	3	2	2–3	1–2	1
Supply voltage phases	3	3	3	1	1
Switching Devices	16 (4 on board)	12	12	6	6
THD	4.40%	<5%	<1%	4.20%	<5%
Switching losses	Virtual 0 W (ZVT)	241.1 W	165.2 W	0 W (using ZVT)	0 W (using ZVT)
Switching Frequency	7.2 kHz	10 kHz	10 kHz	250 kHz	83–88 kHz
Capability to reverse power flow	Yes	No	Yes	No	No
Possibility to split the converter	Yes	No	No	No	Yes
Output Power	5–20 kW	22.6 kW	20.4 kW	6.1 kW	1 kW
Efficiency	95.8% (estimated)	97.72%	96.80%	94%	93.34%
Total Volume On-Board Converter	1700 cm^3 (Estimated)	8430 cm^3	6668.5 cm^3	1742 cm^3	5250 cm^3 (Estimated)
Power Density	10 kW/dm^3 (Estimated)	3.8 kW/dm^3	4.3 kW/dm^3	5 kW/dm^3	192 W/dm^3 (Estimated)
Advantages over others	Reduces the size of the converter located on-boar the vehicle. The SVPWM together with ZVT generate high-quality sinusoidal currents with null switching losses	Eliminates harmonics, improves the power factor, great simplicity, stable and reliable operation	The volume of the reactive components is reduced. Passive components are not needed in intermediate steps	The switching frequency is increased, the size and weight is reduced	Full-range regulation from zero to full power without switching losses
Major Drawbacks	The efficiency can be reduced by using the transformer	Need to be followed by a step-down DC-DC converter. Passive components are required in intermediate steps	The converter is on-board the vehicle. When the switching devices reach the temperature of 145°, the maximum output power decreases	The conversion is made by three steps with intermediate passive components. Not suitable for high power applications	Not suitable for high power applications. More than transformers are used, increasing the losses

Table 5 shows that the number of switching devices used in the proposed converter is slightly higher in contrast to the topologies described in [14–16]; however, only four of them are located on-board the vehicle. The use of the bidirectional switches in the matrix converter allows the possibility power flow reversal through the converter, being more suitable to future smart grids. The implementation of the proposed AC-DC modular converter is considering the use of high-frequency, nanocrystal magnetic materials for the transformer core. In this fashion, higher power capability may be obtained with distributed gapped cores, such as the used in [15], increasing the efficiency limit by the actual wireless coupling techniques used in actual AC-DC chargers for wireless electric vehicles [22].

7. Conclusions

The splitting of a conventional active rectifier into a matrix converter and an H bridge linked through a high-frequency transformer resulted in an AC-DC modular converter topology ideal for high power density applications. The converter portion on board makes the topology particularly attractive for PEV's; nevertheless, the technique is suitable for other applications. A SVPWM technique with ZVT was proposed for the described converter which allows symmetric generation of virtually square current waves that are attractive for high-frequency wireless transmission of high power.

The topology was verified using a numerical prediction performed in Saber which resulted in high-quality supply currents with a current THD of 4.43%. An input-to-output power balance was verified ensuring reliable power transmission. The high-frequency switching of 7.2 kHz allowed ZVT of the semiconductor devices, since a short overlap period was caused by a simple sequential logic circuitry which aids to the reduction of the switching power losses of typical SVPWM schemes; nevertheless, semiconductor current monitoring is required to obtain correct switching behavior of the matrix converter.

Future aims of research of this topology consider its practical development at higher switching frequencies, allowing further reduction of size and weight, possibility of wireless power transmission, which would be particularly attractive for reliable and risk free charging of electric vehicles. Furthermore, the application of the proposed topology could be examined for other power electronic topologies.

Acknowledgments: The authors are grateful to the Fronteras de la Ciencia Research Project No. 101 of the Consejo Nacional de Ciencia y Tecnología of Mexico (CONACyT), the Instituto Politécnico Nacional of Mexico (IPN) for their encouragement and kind economic support to realize the research project. In addition, we are grateful to the Fondecyt Regular 1160690 Research Project.

Author Contributions: Jazmin Ramirez-Hernandez and Ismael Araujo-Vargas developed the principle of operation of the proposed AC-DC converter. The topology of the AC-DC converter was proposed by Marco Rivera. The numerical simulation results were obtained by Jazmin Ramirez-Hernandez using the Synopsis Saber Software, which was validated by Ismael Araujo-Vargas. Furthermore, the paper writing was done by Jazmin Ramirez-Hernandez and revised by Ismael Araujo-Vargas with a technical revision scope of Marco Rivera. All authors were involved and contributed in each part of the article for its final depiction as a research paper.

Nomenclature

C_{z1}, C_{z2}	Zero crossing detector signals
i_a, i_b, i_c	Line current in phase a, b and c respectively
i_L	Three-phase line current vector
i_{LS}	Current through the leakage inductance
i_{Q1}, i_{Q4}, i_{Q3}, i_{Q6}	Current through switches Q_1, Q_4, Q_3 and Q_6 respectively
i_{prim}	Current in the primary side of the transformer
i_{sec}	Current in the secondary side of the transformer
I_{primpk}	Peak magnitude of the current in the primary side of the transformer
i_{rect}	H-bridge output current
I_{rect}	Peak magnitude of H-bridge output current

L	Line inductor
L_s	Transformer leakage inductance
M_a	Modulation index
M_{amin}, M_{amax}	Minimum and maximum modulation index variation
n	Transformer turns ratio
P_{in}, P_{out}	Input and output power
P_{omin}, P_{omax}	Minimum and maximum output power variation
PWM_a, PWM_b, PWM_c	Digital signals obtained from the comparison between conventional active periods and a high-frequency carrier triangular waveform
Q_1-Q_6	Bi-directional switches in the matrix converter
Q_a, Q_b, Q_c, Q_d	H-bridge control signals
$Q_{1a}, Q_{1b}, Q_{4a}, Q_{4b}$	Control signals in the first matrix converter leg
$Q_{3a}, Q_{3b}, Q_{6a}, Q_{6b}$	Control signals in the second matrix converter leg
R	Output load
S_a, S_b, S_c	Conventional active rectifier switching states
S_a', S_b', S_c'	Intermediate signals to obtain the matrix converter switching states
S_{am}, S_{bm}, S_{cm}	Matrix converter switching states
S_x, S_y	H-bridge switching states
sv_1 to sv_6	Active voltage space vectors
sv_0	Neutral voltage space vector
SQ_1 to SQ_6	Vector of matrix converter switching states combinations
S_1 to S_6	Sectors in the α-β plane
T	Switching period
T_a, T_b, T_c	Conventional SVPWM active periods
t_D	Overlap period required in the active-to-active switching transition
T_{ovL}	Matrix converter legs overlap period
t_r	Turning off time in the semiconductor devices
T_{SV0H}	Short-circuit period in the H-bridge
$T_{SV1}, T_{SV2}, T_{SV4}, T_{SV5}$	Active times of the space vectors sv_1, sv_2, sv_4 and sv_5 respectively
v_a, v_b, v_c	Phase a, b and c voltages
$v_{acG}, v_{bcG}, v_{ccG}$	Matrix converter voltages referred to the G node
v_{cav}	Averaged converter voltage vector
V_{cpk}	Peak magnitude of the averaged converter voltage vector
v_{cavd}, v_{cavq}	Converter d-q-axis voltage
V_{cpkmin}, V_{cpkmax}	Minimum and maximum peak converter voltage variation
v_L	Line inductor voltage vector
V_{Lpk}	Peak magnitude of line inductor voltage vector
V_o	Output voltage
v_{prim}	Voltage in the primary side of the transformer
v_{Q1}, v_{Q4}	Voltages in switches Q_1 and Q_4 respectively
v_{sec}	Voltage in the secondary side of the transformer
v_s	Three-phase source voltage vector
v_{smin}, v_{smax}	Minimum and maximum three-phase source voltage vector variation
V_{spk}	Peak magnitude of the three-phase source voltage vector
V_{spkmin}, V_{spkmax}	Minimum and maximum voltage supply variation
v_{xy}	Voltage generated by the H bridge
v_{xG}, v_{yG}	H-bridge legs voltages referred to the G node
θ_c	Converter operation phase
θ_s	Source phase
φ	Phase between v_s ad v_{cav} vectors
$\varphi_{min}, \varphi_{max}$	Minimum and maximum phase φ variation

References

1. Fariborz, M.; Wilson, E.; Wiliam, G.D. Efficiency evaluation of single-phase solutions for AC-DC PFC boost converters for plug-in-hybrid electric vehicle battery chargers. In Proceedings of the IEEE Vehicle Power and Propulsion Conference, Lille, France, 1–3 September 2010.

2. Lingxiao, X.; Zhiyu, S.; Dushan, B.; Paolo, M.; Daniel, D. Dual Active Bridge-Based Battery Charger for Plug-in Hybrid Electric Vehicle with Charging Current Containing Low Frequency Ripple. *IEEE Trans. Power Electron.* **2015**, *30*, 7299–7307. [CrossRef]

3. Muntasir, U.; Wilson, E.; Fariborz, M. A hybrid resonant bridgeless AC-DC power factor correction converter for off-road and neighborhood electric vehicle battery charging. In Proceedings of the Applied Power Electronics Conference and Exposition, Fort Worth, TX, USA, 16–20 March 2014.

4. Hernandez, J.; Ortega, M.; Medina, A. Statistical characterization of harmonic current emission for large photovoltaic plants. *Int. Trans. Electr. Energy Syst.* **2014**, *24*, 1134–1150. [CrossRef]

5. European Committee for Electrotechnical Standardization. *Limits for Harmonic Current Emissions*; CENELEC: Brussels, Belgium, 1995.

6. Bhim, S.; Sanjeev, S.; Ambrish, C. Comprehensive Study of Single-Phase AC-DC Power Factor Corrected Converters with High-Frequency Isolation. *IEEE Trans. Ind. Inform.* **2011**, *7*, 540–556. [CrossRef]

7. Rao, S.; Berthold, F.; Pandurangavittal, K.; Benjamin, B.; David, B.; Sheldon, W.; Abdellatif, M. Plug-in Hybrid Electric Vehicle energy system using home-to-vehicle and vehicle-to-home: Optimizaton of power converter operation. In Proceedings of the IEEE Transportation Electrification Conference and Expo, Dearborn, MI, USA, 16–19 June 2013.

8. Fariborz, M.; Wilson, E. Overview of wireless power transfer technologies for electric vehicle battery charging. *IET Power Electr.* **2014**, *7*, 60–66. [CrossRef]

9. Deepak, R.; Vamsi, K.; Akash, R.; Najath, A.A.; Sheldon, S.W. Modified resonant converters for contactless capacitive power transfer systems used in EV charging applications. In Proceedings of the 42nd Annual Conference of the IEEE Industrial Electronics Society, Florence, Italy, 24–27 October 2016.

10. Chia-Ho, O.; Hao, L.; Weihua, Z. Investigating Wireless Charging and Mobility of Electric Vehicles on Electricity Market. *IEEE Trans. Ind. Electron.* **2015**, *5*, 3123–3133. [CrossRef]

11. Chunyang, G.; Krishnamoorthy, H.; Prasad, N.; Yongdong, L. A novel medium-frequency-transformer isolated matrix converter for wind power conversion applications. In Proceedings of the IEEE Energy Conversion Congress and Exposition, Pittsburgh, PA, USA, 14–18 September 2014.

12. Peschiera, B.; Williamson, S. Review and comparison of inductive charging power electronic converter topologies for electric and plug-in hybrid electric vehicles. In Proceedings of the IEEE Transportation Electrification Conference and Expo, Dearborn, MI, USA, 16–19 June 2013.

13. Zhao, J.; Jiang, J.; Yang, X. AC-DC-DC isolated converter with bidirectional power flow capability. *IET Power Electron.* **2010**, *4*, 472–479. [CrossRef]

14. Rizzoli, G.; Zarri, M.; Mengoni, A.; Tani, A.; Attilio, L.; Serra, G.; Casadei, D. Comparison between an AC-DC matrix converter and an interleaved DC-DC converter with power factor corrector for plug-in electric vehicles. In Proceedings of the IEEE International Electric Vehicle Conference, Florence, Italy, 17–19 December 2014.

15. Whitaker, B.; Barkley, A.; Cole, Z. A High-Density, High-Efficiency, Isolated On-Board Vehicle Battery Charger Utilizing Silicon Carbide Power Devices. *IEEE Trans. Power Electron.* **2014**, *29*, 2606–2617. [CrossRef]

16. Bojarski, M.; Asa, E.; Colak, K.; Dariusz, C. Analysis and Control of Multi-Phase Inductively Coupled Resonant Converter for Wireless Electric Vehicle Charger Applications. *IEEE Trans. Transp. Electr.* **2017**, *3*, 312–320. [CrossRef]

17. Il-Oun, L. Hybrid PWM-Resonant Converter for Electric Vehicle On-Board Battery Chargers. *IEEE Trans. Power Electron.* **2016**, *31*, 3639–3649. [CrossRef]

18. Dong-Gyun, W.; Yun-Sung, K.; Byoung-Kuk, L. Effect of PWM schemes on integrated battery charger for plug-in hybrid electric vehicles: Performance, power factor, and efficiency. In Proceedings of the IEEE Applied Power Electronics Conference and Exposition, Fort Worth, TX, USA, 16–20 March 2014.

19. Onur, S.; Erkan, M. Investigating DC link current ripple and PWM modulation methods in Electric Vehicles. In Proceedings of the 3rd International Conference on Electric Power and Energy Conversion Systems, Istanbul, Turkey, 2–4 October 2013.

20. Grahame, D.; Lipo, T. Modulation of Three-Phase Voltage Source Inverters. In *Pulse Width Modulation for Power Converter Principles and Practice*; Wiley-IEEE Press: Piscataway, NJ, USA, 2003; pp. 215–258, ISBN 9780470546284.

21. Forsyth, A.J.; Mollov, S.V. Modelling and control of DC-DC converters. *Power Eng. J.* **1998**, *12*, 229–236. [CrossRef]

22. Jianyong, L.; Yitong, C.; Deliang, L.; Lin, G.; Fei, D.; Fangjun, J. Novel network model for dynamic stray capacitance analysis of planar inductor with nanocrystal magnetic core in high frequency. In Proceedings of the IEEE Conference on Electromagnetic Field Computation, Chicago, IL, USA, 9–12 May 2010.

23. Pedersen, A.; Martinenas, S.; Andersen, P.; Thomas, M.S.; Henning, S.H. A method for remote control of EV charging by modifying IEC61851 compliant EVSE based PWM signal. In Proceedings of the IEEE International Conference on Smart Grid Communications, Miami, FL, USA, 2–5 November 2015.

 *energies*

Article

Interconnecting Microgrids via the Energy Router with Smart Energy Management

Yingshu Liu *, Yue Fang and Jun Li

School of Electrical and Information Engineering, Tianjin University, Tianjin 300072, China;
fangyue@tju.edu.cn (Y.F.); lijunxaut@163.com (J.L.)
* Correspondence: liu_ysh@tju.edu.cn; Tel.: +86-159-2209-7591

Received: 14 July 2017; Accepted: 28 August 2017; Published: 30 August 2017

Abstract: A novel and flexible interconnecting framework for microgrids and corresponding energy management strategies are presented, in response to the situation of increasing renewable-energy penetration and the need to alleviate dependency on energy storage equipment. The key idea is to establish complementary energy exchange between adjacent microgrids through a multiport electrical energy router, according to the consideration that adjacent microgrids may differ substantially in terms of their patterns of energy production and consumption, which can be utilized to compensate for each other's instant energy deficit. Based on multiport bidirectional voltage source converters (VSCs) and a shared direct current (DC) power line, the energy router serves as an energy hub, and enables flexible energy flow among the adjacent microgrids and the main grid. The analytical model is established for the whole system, including the energy router, the interconnected microgrids and the main grid. Various operational modes of the interconnected microgrids, facilitated by the energy router, are analyzed, and the corresponding control strategies are developed. Simulations are carried out on the Matlab/Simulink platform, and the results have demonstrated the validity and reliability of the idea for microgrid interconnection as well as the corresponding control strategies for flexible energy flow.

Keywords: interconnection of microgrids; energy storage equipment; energy router; voltage source converter

---

1. Introduction

Increasing amounts of distributed renewable energy sources (RESs), such as solar and wind, are expected to be integrated into the power system to meet the ever-growing energy demand, and more importantly, to meet the need to preserve fossil fuel resources. As a result, the existing electrical power grid will be experiencing a shift from the traditional single-sourced, radial framework network to the multi-sourced, meshed network of the next generation—the Internet of Energy (IOE) [1–4]. However, as the demand keeps rising for higher penetration of RESs, there are many technical challenges yet to be overcome [5]. One of them is that existing technologies rely heavily on energy storage (ES) systems to maintain instant power balance and minimize the impact to the power grid due to the intermittent nature of the RES outputs [6–9]. Unfortunately, the costs of ES units (such as batteries) still remain high and are not yet economically practical for large scale deployment [10]. For example, the capital cost of lithium-ion batteries for microgrid and residential implementations are up to $1000 and $1500 per kWh, respectively [11]. Besides batteries, there are other types of energy storage technologies such as the flywheel, pumped hydro power, compressed air [12], vehicle-to-grid [13] and power-to-gas [14]. Each technology has unique features, but none of them can meet all the requirements of being robust, reliable and economically competitive [15–17]. Therefore, there is a need to develop flexible and sophisticated energy management strategies as a way of maintaining the energy balance and operational stability of the power system, and to alleviate the dependency on large-scale energy storage units for the time being.

Owing to technical advances in the fields of power electronics and communication, in recent years, research on intelligent and multi-functional energy converter—the energy router (ER)—has been booming rapidly [18–21]. With the functionalities of flexible power flow control and wide area information interaction, the energy router has demonstrated great potential to serve as the "energy hub" in the Internet of Energy. More encouragingly, it is now technically and economically feasible for the energy router to serve as the "energy and information sink node" at the levels of the microgrid [22] and home-area network [23,24], by providing multiple functionalities: (1) working as a smart electrical interface, by enabling flexible, adjustable and bidirectional energy flow between the microgrid and the power grid [25,26]; (2) facilitating optimal energy management within the microgrid and improving the efficiency, reliability and economy of the system [22]; and (3) enabling wide-area data collection from various devices (including RESs, electric equipment and loads) in real time, and providing these data to the control center of the microgrid or to the main grid for load forecasting [27], operational status monitoring [28] and fault diagnosis [29,30]. With all these functionalities, the energy router is naturally considered to be a perfect candidate to deploy flexible and sophisticated energy management strategies.

This paper elaborates on a novel and flexible interconnecting framework for microgrids, and corresponding energy management strategies are presented, in response to the situation of increasing renewable energy penetration and the need to alleviate dependency on energy storage equipment. Different from the traditional manner of connection, this framework distinguishes itself by interconnecting the main grid and multiple microgrids via the energy router, and enabling flexible energy exchange among them.

The key idea is proposed based on the following considerations: (1) microgrids serving different purposes may differ substantially in terms of network architecture, type of RESs and loads within them; and therefore, their energy profiles of power production and consumption of the day may also differ; (2) this difference in the profiles of energy production and consumption usually leads to "complementarities" in their time periods of peak and valley demand. These complementarities can thus be utilized to compensate for each other's instant energy deficiency. For example, residential areas with rooftop solar energy will have an energy surplus during the day and peak demand in the evening, while commercial or office areas have different peak times of power consumption. If their locations are adjacent, it would definitely be beneficial for both communities if the surplus energy from the rooftop solar supply can be transferred to the areas in short energy supply. It is also beneficial to the main grid, since excessive solar power injections into the distribution network will lead to overvoltage in low voltage (LV) feeders [31]. Therefore, in comparison to the existing microgrid control methods, the work of this paper is focused on a novel framework to establish electrical interconnection and energy trade between microgrids. This work also presents a new paradigm to utilize direct energy consumption to suppress the impact of DER output fluctuations on the stability and reliability of the power system.

The rest of the paper will be arranged as follows: the analytical model is established for the whole system including the interconnected microgrids and the main grid, as well as the energy router, which is based on multiport bidirectional VSCs and a shared DC power line. Various operational modes of the interconnected microgrids facilitated by the energy router are analyzed, and corresponding control strategies are developed. Simulations are carried out on the Matlab/Simulink (R2014b, MathWorks, Natick, MA, USA) platform to demonstrate the validity and reliability of the presented interconnection framework and the corresponding energy control strategies.

2. Energy Router-Based Interconnecting Framework for Microgrids

The energy router-based framework for microgrid interconnection is demonstrated in Figure 1, where the energy router serves as an energy hub to establish electrical connections between the microgrids and the main grid. Its beneficial effects include: (1) the problem of instant energy surplus or deficiency can be resolved with "complementary" energy exchanges between the adjacent microgrids, thus relieving dependency on large-scale energy storage deployment. The impacts caused by the uncertainties and intermittencies of the DER outputs to the microgrid system, and also to the main grid, can further be alleviated to some extent; (2) the VSC converters and shared DC bus facilitate electrical isolation (in voltage, frequency and phase angle) between the microgrids and the main grid [32]. Each microgrid can have "stand-alone" voltage, frequency and phase angle according to its own operational situation, savingefforts for synchronization with the power grid or the other microgrids; (3) this isolation ensures that any voltage or frequency fluctuation at one end of the energy router (either on the microgrid side or the main grid side) will have no direct impact on the systems on other sides of the energy router, thus guaranteeing power quality and operation reliability; and (4) widespread implementation of the energy routers will promote the shift of the power system architecture from the traditional "tree-structure" hierarchical framework to a more connective and interactive framework. This final consequence will help to improve the power supply quality and the reliability of the whole system, which is an essential step in building the Internet of Energy of the future.

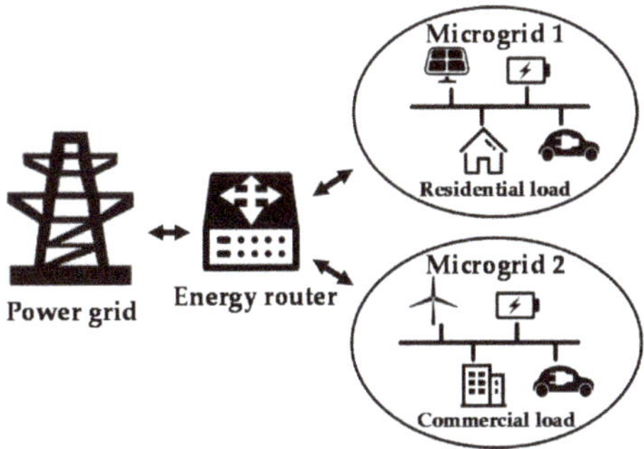

Figure 1. Energy router-based interconnecting framework for the microgrids system.

As shown in Figure 2, the schematic of the power router is based on multiport bidirectional voltage source converters (VSCs) and a shared DC power line. In this paper, we instigate the implementation of a three-port schematic—one port for connection to the main grid, and the other two for interconnections to adjacent microgrids. The VSCs in each port can operate in either rectifier or inverter mode, thus enabling bi-directional power flow control between the alternating current (AC) and shared DC power lines.

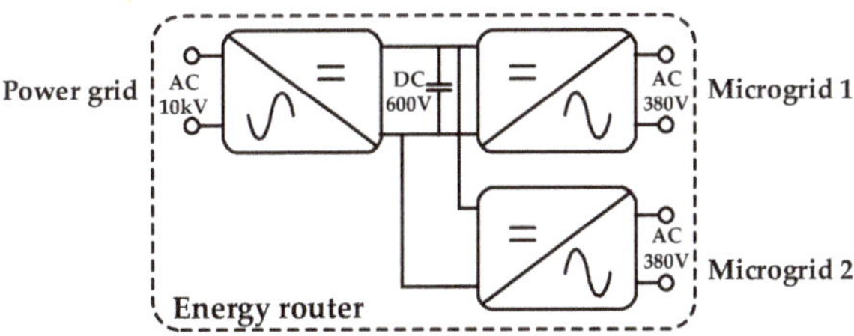

Figure 2. Schematic block diagram of the energy router.

With the DC power line serving as an "energy pool", autonomous energy trade can be performed between the interconnected microgrids without the intervention of the VSC on the main grid side. Energy flow management is easy and flexible, enabled by the multiport energy router. Besides the existing tied and islanded operational modes, there will be a new operational mode for the microgrids—the Parallel Mode. This will happen when a microgrid is having energy exchange (either input or output) only with the adjacent microgrid. For example, a microgrid providing excessive energy to its neighboring microgrid is working in the parallel mode; it does not have any energy exchange with the main grid, even when the VSC on the grid side is still in operation.

The various operational modes enabled by the energy router are demonstrated in Figure 3. According to the energy flow indicated in this figure, the operational modes of each microgrid can be summarized as:

(1)   as for the case in Figure 3a, both microgrid 1 and microgrid 2 are taking energy from the main grid, as the direction of power flow is from power grid to the microgrids. In this situation, both microgrid 1 and microgrid 2 are operating in the grid-connected mode;

(2)   as for the case in Figure 3b, microgrid 1 can be self-sufficient while microgrid 2 is taking energy from the main grid, as the direction of power flow is only from the power grid to microgrid 2. In this situation, microgrid 1 is operating in the islanded mode while microgrid 2 is operating in the grid-connected mode;

(3)   as for the case in Figure 3c, microgrid 1 is taking energy from both the main grid and microgrid 2, while microgrid 2 can not only operate on its own energy production, but provides surplus power to microgrid 1 as well. The direction of power flow is from both the main grid and microgrid 2 to microgrid 1. In this situation, microgrid 1 is operating in the grid-connected mode, while microgrid 2 is operating in the parallel mode;

(4)   as for the case in Figure 3d, microgrid 1 is providing surplus power to microgrid 2, while no power needs to be transferred from the main grid to the DC power line. The direction of power flow is from microgrid 1 to microgrid 2. In this situation, both microgrid 1 and microgrid 2 are operating in the parallel mode.

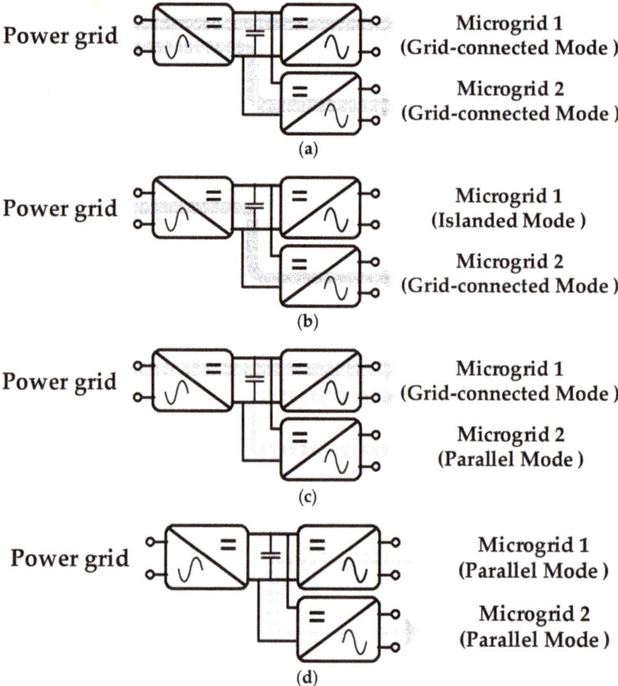

Figure 3. Operational modes enabled by the energy router: (**a**) case 1: grid-connected mode; (**b**) case 2: grid connected mode vs. islanded mode; (**c**) case 3: grid connected mode vs. parallel mode; and (**d**) case 4: parallel mode.

3. Topology and Mathematical Model of the Energy Router

3.1. Topology of the Energy Router

The topology of the energy router considered in this paper is shown in Figure 4. In practice, it can be regarded as a tri-port energy converter that comprises three VSCs with their RLC filters and a shared common DC power line.

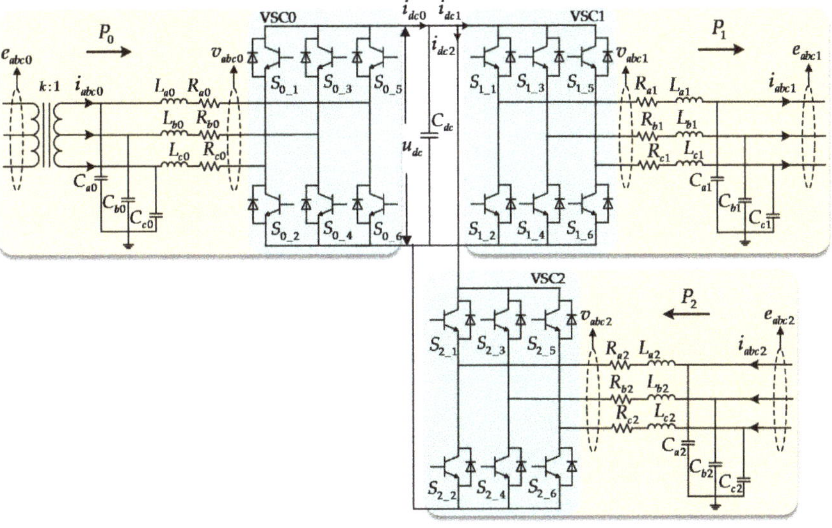

Figure 4. The detailed topology of the energy router.

Let VSCi (i = 0, 1, 2) represent the VSCs of the energy router, in which VSC0 is on the power grid side, while VSC1 and VSC2 are on the microgrid side. Assume that the VSCs and the corresponding RLC filters are three-phase symmetry circuits. As shown in Figure 4, e_{abci} and i_{abci} (i = 0, 1, 2) are the three-phase voltages and currents on the AC side, k is the transformer ratio, $C_i = C_{ai} = C_{bi} = C_{ci}$ are the single-phase capacitance of the RLC filters, $L_i = L_{ai} = L_{bi} = L_{ci}$ are the single-phase inductance of the RLC filters, $R_i = R_{ai} = R_{bi} = R_{ci}$ are the equivalent single-phase resistance of the RLC filters, v_{abci} are the three-phase output voltage of the VSCi, C_{dc} is the capacitance of the DC power line, u_{dc} is the DC-link voltage and i_{dci} is the DC current flowing in or out of the VSCi.

3.2. Analytical Model of the Energy Router

Because the VSC on the main grid side shall be working in a different pattern from the VSCs on the microgrid side, the model of the VSCs shall be described differently. According to the operation modes enabled by the energy router, as shown in Figure 3, VSC0 only works in the rectifier status, whereas VSC1 and VSC2 can work in either the rectifier or inverter statuses (which should be described by different model representations). Hence, the case in Figure 3c is taken as an example to formulate the state-space representations of the VSCs. In this situation, VSC0 and VSC2 work as rectifiers, and VSC1 works as an inverter.

The uni-polar logic switching function s_j is defined as:

$$s_j = \begin{cases} 1 & \text{upper bridge legs are on and lower off} \\ 0 & \text{upper bridge legs are off and lower on} \end{cases} \quad j = a, b, c. \tag{1}$$

Then, the three-phase circuit equations of the VSC0 are established and transformed into the dq rotating coordinate system, as shown in the following equations:

$$\begin{cases} L_\Sigma \frac{di_{d0}}{dt} = -R_0 i_{d0} + \omega_0 L_\Sigma i_{q0} + \frac{1}{k} e_{d0} - u_{dc} s_{d0} \\ L_\Sigma \frac{di_{q0}}{dt} = -R_0 i_{q0} - \omega_0 L_\Sigma i_{d0} + \frac{1}{k} e_{q0} - u_{dc} s_{q0} \end{cases}, \tag{2}$$

in which $L_\Sigma = L_T + L_0$ is the equivalent inductance on the AC side of VSC0, L_T is the equivalent inductance of the transformer, i_{di} and i_{qi} are the components of i_{abci} in the d and q axes, ω_i is the angular frequency of the AC system, e_{di} and e_{qi} are the components of e_{abci} in the d and q axes, k is the transformer ratio, and s_{di} and s_{qi} are the components of the switching function in the d and q axes of the dq rotating coordinate system.

The current flowing in or out of the VSCi can be calculated by:

$$i_{dci} = \frac{3}{2} \left(s_{di} i_{di} + s_{qi} i_{qi} \right). \tag{3}$$

The current balance equation of the DC power line is:

$$C_{dc} \frac{du_{dc}}{dt} = i_{dc0} - \left(i_{dc1} + i_{dc2} \right). \tag{4}$$

Substituting Equation (3) into Equation (4), Equation (5) can be derived:

$$C_{dc} \frac{du_{dc}}{dt} = \frac{3}{2} \left(s_{d0} i_{d0} + s_{q0} i_{q0} \right) - \frac{3}{2} \left(s_{d1} i_{d1} + s_{q1} i_{q1} + s_{d2} i_{d2} + s_{q2} i_{q2} \right). \tag{5}$$

Similarly, the mathematical models of the VSC1 working in the inverter status and VSC2 working in the rectifier status can be derived as:

$$\begin{cases} L_1 \frac{di_{d1}}{dt} = -R_1 i_{d1} + \omega_1 L_1 i_{q1} - e_{d1} + u_{dc} s_{d1} \\ L_1 \frac{di_{q1}}{dt} = -R_1 i_{q1} - \omega_1 L_1 i_{d1} - e_{q1} + u_{dc} s_{q1} \end{cases}, \text{ and} \tag{6}$$

$$\begin{cases} L_2 \frac{di_{d2}}{dt} = -R_2 i_{d2} + \omega_2 L_2 i_{q2} + e_{d2} - u_{dc} s_{d2} \\ L_2 \frac{di_{q2}}{dt} = -R_2 i_{q2} - \omega_2 L_2 i_{d2} + e_{q2} - u_{dc} s_{q2} \end{cases}. \tag{7}$$

Equations (2) and (5)–(7) constitute the analytical model of the energy router in the situation represented in Figure 3c, in which the microgrid 1 is in the grid-connected mode and microgrid 2 in the parallel mode. Models of the energy router operating in the other situations represented in Figure 3 can also be derived in the same way.

4. Energy Control Strategies of the Energy Router

4.1. Control Schematic of the VSCs

The VSCs in the energy router are under dual-loop control, and the controller structure of the VSCs is shown in the Figure 5.

The dual-loop controller mainly consists of four modules: (1) the inner-loop controller, also known as the current feedback loop; (2) the outer-loop controller, which can either be a DC voltage controller or an active/reactive power controller, depending on the control objective of each VSC; (3) the phase-locked loop (PLL), which can be used to obtain the phase angle of the AC voltage for Park's Transformation; and (4) the pulse width modulation (PWM) controller.

As shown in Figure 5, i_{idref} and i_{iqref} are the references of i_{di} and i_{qi}, while U_{dcref} is the reference of u_{dc}. As the VSC1 and VSC2 are identical in their structures, the detail of the controller of VSC2 is not shown in the Figure 5.

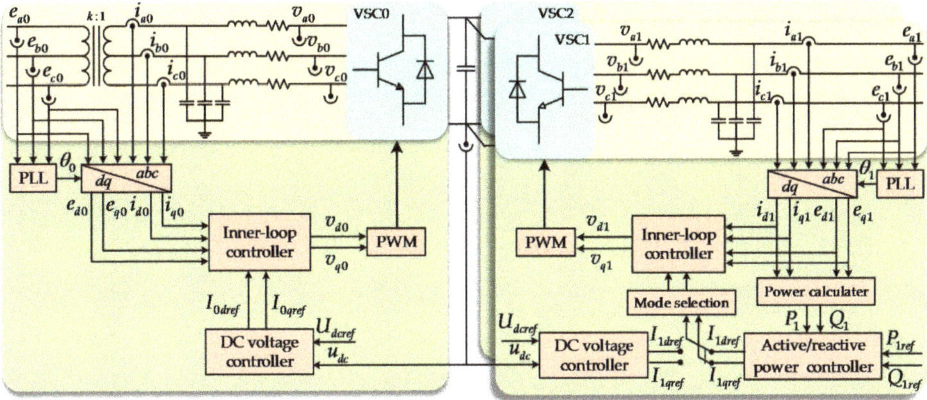

Figure 5. Control schematic of the VSCs.

4.2. Control Patterns for the VSCs

The VSCs of the energy router are working in the master/slave mode. For the master VSC, its outer-loop controller shall be working as the DC voltage controller, with its goal being to maintain stable and constant DC-link voltage. The rest of the VSCs shall be the slave converters, working in the manner of power transfer following up. Their out-loop controllers shall be working as the active/reactive power controllers, with the goal being to keep their power transfer following up the needs of the microgrids in connection.

The rules to choose the master converter are thus: (1) when the main grid is involved in power transfer, VSC0 is preferred as the master converter and the VSCs on the microgrid side are the slave converters; (2) when the main grid is not involved in energy exchange, the master converter shall be the VSC in connection with the microgrid which provides energy to its neighbor, while the VSC which intakes energy shall be the slave converter.

As defined in Table 1, there are three control patterns for the VSCs according to the operational status of their inner- and outer-loop controllers. The various operational modes of the microgrids are realized by choosing appropriate control patterns for the VSCs, which will also determine the roles of the VSCs (master or slave) in the process of power flow. For the master converter, the outer-loop controller shall be working as the DC voltage controller while the inner loop controller will keep the H-bridge of the converter working in the rectifier status. Thus, the master converter should be working in control pattern 1. Meanwhile, a slave converter can be selected to work in either pattern 2 or pattern 3, according to the role it plays in the process of power flow. In other words, the control pattern of a VSC is determined by the operational mode of the microgrid in connection with it.

Table 1. Definition of the control patterns for VSCs under different operational conditions.

Control Pattern	Inner-Loop Controller	Outer-Loop Controller
Pattern 1 (master control)	Rectifier	DC voltage loop control
Pattern 2	Rectifier	Active/reactive power control
Pattern 3	Inverter	Active/reactive power control

The control patterns of the VSCs can be listed as follows:

(1) once the main grid is involved in power transfer, VSC0 is chosen to be the master converter, working in pattern 1;

(2)   for a microgrid working in the grid-connected mode, the VSC beside the power grid must be working in pattern 3;

(3)   for a microgrid working in the parallel mode, the corresponding control pattern of its VSC is determined according to the other microgrid's operational mode. For example, when microgrid 2 works in parallel mode and microgrid 1 is in the grid-connected mode, which corresponds to the case represented in Figure 3c, VSC2 must be working in pattern 2, because the surplus energy of microgrid 2 is transferred to the DC power line through it;

(4)   when both microgrid 1 and microgrid 2 are in parallel mode, as in Figure 3d, VSC2 is working in pattern 3 and VSC1 is working in pattern 1, because the direction of energy flow is from microgrid 2 to microgrid 1.

The corresponding control patterns of each VSC, along with the operational modes of the microgrids, in the four cases of Figure 3 can thus be summarized as shown in Table 2.

Table 2. Operational modes of the microgrids vs. VSC control patterns in Figure 3a–d.

Operational Cases	Operational Mode of Microgrid 1	Operational Mode of Microgrid 2	VSC0	VSC1	VSC2
Case 1 (Figure 3a)	Grid-connected mode	Grid-connected mode	Pattern 1	Pattern 3	Pattern 3
Case 2 (Figure 3b)	Islanded mode	Grid-connected mode	Pattern 1	None	Pattern 3
Case 3 (Figure 3c)	Grid-connected mode	Parallel mode	Pattern 1	Pattern 3	Pattern 2
Case 4 (Figure 3d)	Parallel mode	Parallel mode	None	Pattern 1	Pattern 3

4.3. Control Patterns for the Energy Router

For the RESs, ESs and loads in the microgrids, the required active power of microgrid 1 and microgrid 2 can be calculated by:

$$P_1 = P_{DG1} + P_{bat1} - P_{load1}, \quad P_2 = P_{DG2} + P_{bat2} - P_{load2}, \tag{8}$$

where the *DG* outputs in microgrid 1 and microgrid 2 are denoted by P_{DG1} and P_{DG2}, the energy stored in ESs are P_{bat1} and P_{bat2}, and power demand from the loads are P_{load1} and P_{load2}, respectively.

$$P_{MG} = P_1 + P_2. \tag{9}$$

The total sum of active power for microgrid 1 and microgrid 2 is denoted by P_{MG}. When the power grid is involved in the energy exchange, the supplied power denoted by P_0 is:

$$P_0 = -P_{MG}. \tag{10}$$

Based on the above analysis, seven operational states can be summarized for the energy router, as shown in Table 3. The related state-transition diagram is demonstrated in Figure 6.

Table 3. All the states of the energy router.

State	Operation Case	Power Grid	Microgrid 1	Microgrid 2
1	Case 1		Grid-connected mode	Grid-connected mode
2	Case 2	Include	Grid-connected mode	Island mode
3	Case 2	Include	Island mode	Grid-connected mode
4	Case 3		Grid-connected mode	Parallel mode
5	Case 3		Parallel mode	Grid-connected mode
6	Case 4	Exclude	Parallel mode	Parallel mode
7	Case 4	Exclude	Parallel mode	Parallel mode
8	Downtime		Island mode	Island mode

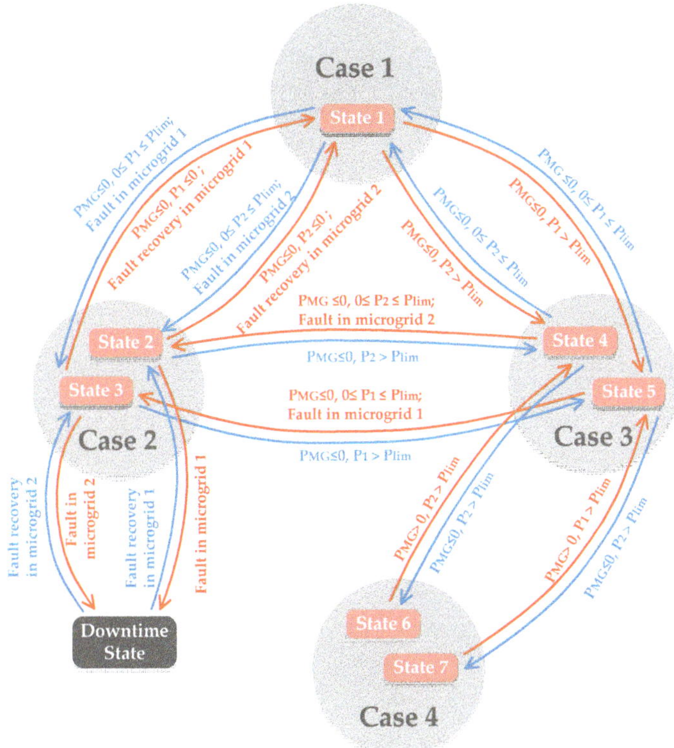

Figure 6. State-transition diagram of the energy router.

5. Dual-Loop Feedback Control

5.1. Inner Loop Controller

Based on Kirchhoff's law, the voltage equation of VSC0 in the synchronous dq coordinate system can be derived, and the items coupling between the d and q axes can be removed by the decoupling feed forward control method. Assuming that the energy router operates in case 3 (in Figure 3c), the voltage control equations of the VSCi inner-loop controller can be written as:

$$\begin{cases} v_{d0} = -(K_{P0} + \frac{K_{I0}}{s})(i_{0dref} - i_{d0}) + \omega_0 L_\Sigma i_{q0} + \frac{1}{k}e_{d0} \\ v_{q0} = -(K_{P0} + \frac{K_{I0}}{s})(i_{0qref} - i_{q0}) - \omega_0 L_\Sigma i_{d0} + \frac{1}{k}e_{q0} \end{cases} \tag{11}$$

where K_{Pi} and K_{Ii} are the proportional and integral regulation gain in the inner loop, respectively.

Similarly, the voltage equation of VSC1 working in the inverter status can be written as:

$$\begin{cases} v_{d1} = (K_{P1} + \frac{K_{I1}}{s})(i_{1dref} - i_{d1}) - \omega_1 L_1 i_{q1} + e_{d1} \\ v_{q1} = (K_{P1} + \frac{K_{I1}}{s})(i_{1qref} - i_{q1}) + \omega_1 L_1 i_{d1} + e_{q1} \end{cases}, \tag{12}$$

while the voltage equation of VSC2 working in the rectifier status can also be derived as:

$$\begin{cases} v_{d2} = -(K_{P2} + \frac{K_{I2}}{s})(i_{2dref} - i_{d2}) + \omega_2 L_2 i_{q2} + e_{d2} \\ v_{q2} = -(K_{P2} + \frac{K_{I2}}{s})(i_{2qref} - i_{q2}) - \omega_2 L_2 i_{d2} + e_{q2} \end{cases}. \tag{13}$$

5.2. Outer-Loop Controller

The outer-loop is used to generate the inner-loop current reference signal and input to the inner-loop control. According to different control objectives, the outer-loop controller can be classified as active/reactive power controller or DC voltage controller.

5.2.1. Active/Reactive Power Controller

The active/reactive power controller is used to keep the VSC working such that its active/reactive power output on the AC side shall follow the reference command value issued by the microgrid controller, with zero steady-state error. Classical proportional-integral (PI) regulators are used in the outer-loop control to calculate the current commands to the inner-loop with the power deviations.

Define the power direction symbol function sgn as:

$$\text{sgn} = \begin{cases} +1 \text{ power out AC system} \\ -1 \text{ power in AC system} \end{cases}. \tag{14}$$

The d-axis is in the same direction to the voltage vector, which leads to $e_{qi} = 0$. The active power P_i and reactive power Q_i, absorbed from the AC system to VSCi, can be expressed as:

$$\begin{cases} P_i = e_{di} i_{di} \times \text{sgn} \\ Q_i = -e_{di} i_{qi} \times \text{sgn} \end{cases}. \tag{15}$$

According to Equation (15), the control between active power and reactive power can be decoupled. Then, P_i and Q_i are compared with their reference inputs P_{iref} and Q_{iref}, respectively, and the result of the comparison is used by the PI controller to calculate the current command value. The control structure of the active power controller is shown in Figure 7.

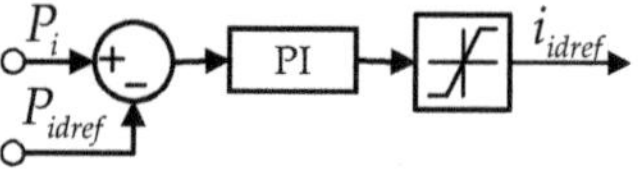

Figure 7. Control structure of the active power controller.

As a matter of fact, when the AC systems are injected with excessive power, the corresponding AC voltage e_i will rise. Conversely, the corresponding AC voltage will decline when the AC systems lack power energy. Hence, compared to the ordinary zero steady-state error tracking control, AC voltage-limiting controllers are added so that the internal stability of the AC systems can be maintained.

The purpose of the AC voltage-limiting controller is to keep the amplitude of AC voltage within a set range. When e_i exceeds the upper-limit value E_{imax}, the AC voltage-limiting controller outputs a negative value, which serves as the disturbance quantity and adds to the Q_{iref}. Thereby, the inner-loop current command value i_{iqref} decreases, so that the absorbed power in the AC system can be reduced. On the other hand, when e_i exceeds the lower-limit value E_{imin}, the AC voltage-limiting controller outputs a positive value and adds to the Q_{iref}. Thus, the inner-loop current command value i_{iqref} increases, so that the absorbed power in the AC system can be augmented. The control structure of reactive power control within the AC voltage controller is shown in Figure 8.

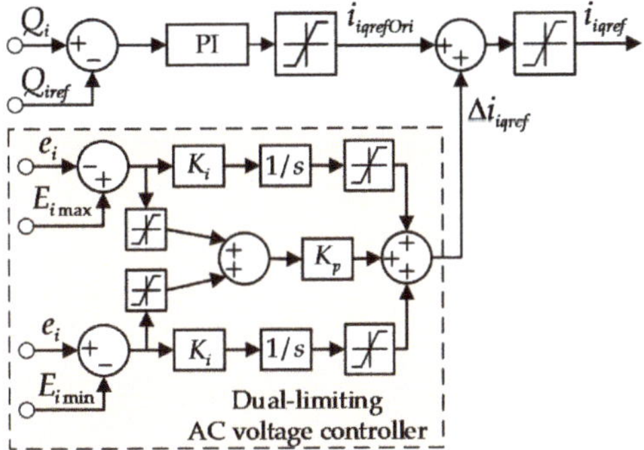

Figure 8. Control structures of reactive power controller.

5.2.2. DC Voltage Controller

The DC voltage controller is responsible for keeping the DC-link voltage within an adequate range around the command value issued by the microgrid controller. For VSCi (i = 1, 2) controlled by the DC voltage controller, the active power P_i exchanged with the AC systems is identical to the DC power P_{dc} stored in C_{dc}.

$$P_{dc} = \frac{1}{2} C_{dc} \frac{du_{dc}^2}{dt} = \sum_{i=1,2,3} P_i. \tag{16}$$

The current command input to the d-axis of the inner-loop current controller is

$$i_{idref} = \left(K_{Pdc} + \frac{K_{Idc}}{s} \right) \left(u_{dcref}^2 - u_{dc}^2 \right), \tag{17}$$

where K_{Pdc} and K_{Idc} are the proportional and integral regulation gains, respectively. Under this controller, the d-axis current command is derived from Equation (17), and the control structure is shown in Figure 9. Meanwhile, the q-axis current command is derived from Equation (15), while the control structure is shown in Figure 8.

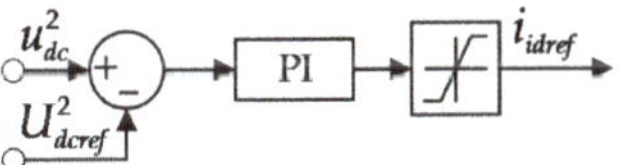

Figure 9. Control structure of the DC voltage controller.

6. Simulation Studies

The simulation model of the interconnected microgrid system was built on the MATLAB R2014b/Simulink platform. The simulation model of the overall system and the sub-models of VSC0 and its controller are demonstrated in Figure 10a–c, respectively. Apart from the transformer, the simulation models of VSC1, VSC2 and their controllers are similar to VSC0.

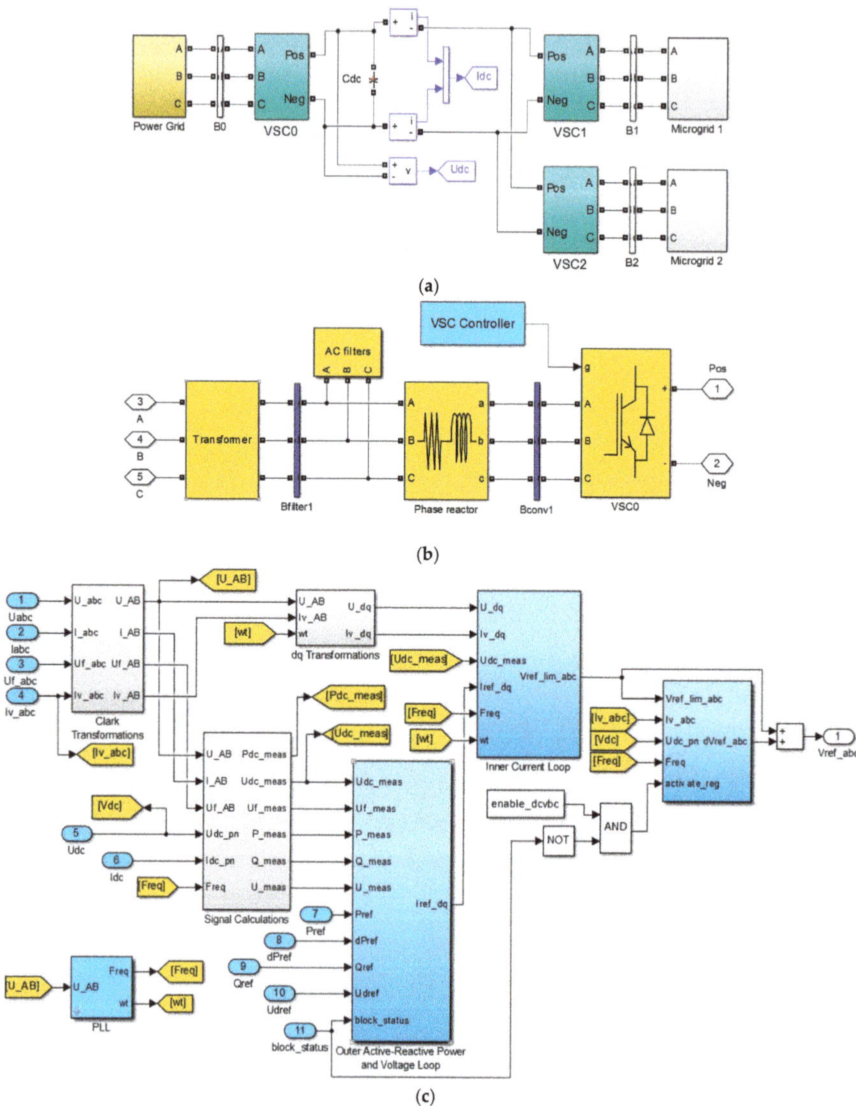

(a)

(b)

(c)

Figure 10. Simulation model of the interconnected microgrid system in MATLAB R2014b/Simulink platform: (**a**) overall simulation model; (**b**) VSC0 simulation model; and (**c**) VSC controller simulation model.

In this section, four simulation scenarios are presented to investigate the dynamic characteristics of the interconnected microgrids in all the operation modes defined previously, and to verify the feasibility of the proposed control strategies. The transient responses of the energy router for operational-mode shifting of the microgrids, the power balancing in Parallel Mode and the suppression of AC voltage disturbances are all demonstrated by these four simulated scenarios. The simulation parameters are listed in Table 4.

Table 4. Simulation parameters.

Parameters	Value
AC voltage of main grid	10 kV
AC voltage of microgrid	0.38 kV
DC-bus voltage	600 V
DC-bus capacitance	1 mF
Inductance of the filter	0.25 mH
Capacitance of the filter	250 µF
Switching frequency	10 k Hz

6.1. Scenario 1

In this paper, a new operational mode—the Parallel Mode—is defined for the interconnected microgrids. Microgrid 2 does not need to have any energy exchange with the power grid, because its power deficit can be met by the surplus power of the adjacent microgrid 1. In this scenario, both microgrid 1 and microgrid 2 are operating in the parallel mode, which the energy router is operating in case 4. Microgrid 1 is supplying its surplus energy to microgrid 2, so VSC1 is the master converter and VSC2 is the slave converter. The detail of the power flow from microgrid 1 to microgrid 2 and the DC-link voltage variation of the transient process are demonstrated in Figure 11.

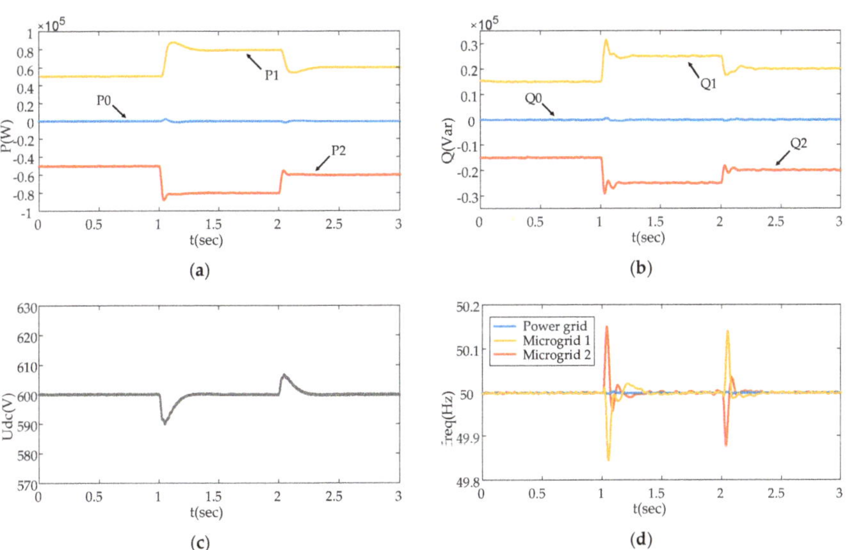

Figure 11. Simulation results of scenario 1: (a) active power curves; (b) reactive power curves; (c) DC-link voltage curves; and (d) frequency curves.

(1)   The process of active power flow within the interconnected microgrids is as follows:

$t = 0$ s, 50 kW active power is provided to microgrid 2 by microgrid 1; $t = 1$ s, the active power transfer from VSC1 to VSC2 increases from 50 to 80 kW; $t = 2$ s, the command value of the active power transfer changes from 80 to 60 kW. The active power variations of the VSCs and the command value curves are shown in Figure 11a.

(2)   The reactive power flow between the interconnected microgrids is as follows:

$t = 0$ s, 15 kVar reactive power is provided to microgrid 2 by microgrid 1; $t = 1$ s, the absorbed reactive power command value changes from 15 to 25 kVar; $t = 2$ s, the reactive power transfer changes from 25 to 20 kVar. The reactive power variations of the VSCs and the command value curves are shown in Figure 11b.

As demonstrated in Figure 11a,b, when the power transfer command value changes abruptly at $t = 1$ s and $t = 2$ s, there are only slight fluctuations in the output power curves, and the stable state is restored quickly. The active and reactive power transfer of VSC0 remains zero as the main grid is not involved in any power transfer through the energy router. As is shown in Figure 11c, the fluctuation range of the DC-link voltage is minor and can be ignored. Figure 11d shows that frequency fluctuates in an acceptable range for both microgrid 1 and microgrid 2 at the moments that energy exchange occurs. However, due to the electrical isolation of the energy router, the fluctuation does not have an impact on the frequency of the power grid, which remains stable at 50 Hz during the whole process. Note that the DC-link voltage is an indicator of instant power balance; the simulation results in Figure 11 have clearly verified the effectiveness of the control strategies in this scenario.

6.2. Scenario 2

This scenario demonstrates the dynamic characteristics of the interconnected microgrids during the transition process from the situation of case 1 (shown in Figure 3a) to case 2 (in Figure 3b), during which microgrid 1 switches from grid-connected mode to islanded mode, while microgrid 2 remains grid connected. The variations in power flow and AC frequency, as well as the DC-link voltage during the whole process, are demonstrated in Figure 12.

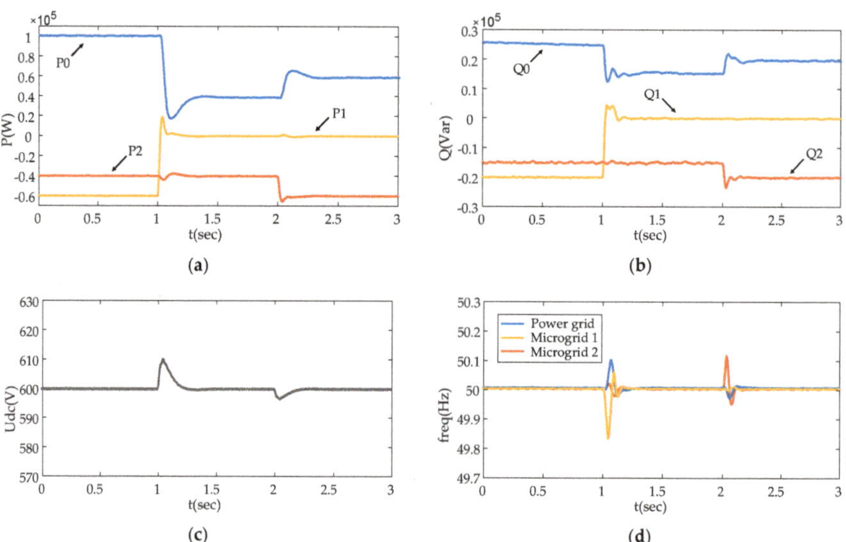

Figure 12. Simulation results of scenario 2: (**a**) active power curves; (**b**) reactive power curves; (**c**) DC-link voltage curves; and (**d**) frequency curves.

- $t = 0$–1 s

The energy router is working under the situation of case 1 (shown in Figure 3a). Both microgrid 1 and microgrid 2 are operating in the grid-connected mode, during which VSC0 serves as the master converter while VSC1 and VSC2 work as slave converters. During this period, the active power transferred from VSC0 to both VSC1 and VSC2 is 60 and 40 kW, respectively, and the reactive power transferred to VSC1 and VSC2 is 20 and 15 kVar, respectively.

- $t = 1$–2 s

The operational situation of the energy router switches to case 2 (shown in Figure 3b). During this period, the operational mode of microgrid 1 is changed from grid-connected to islanded, and VSC1 is not involved in any energy-exchange process. Meanwhile, the operational mode of microgrid 2 remains unchanged, with VSC0 serving as the master converter and VSC2 working as the slave converter. The active and reactive power transferred from the power grid to microgrid 2 remains unchanged.

- $t = 2$–3 s

In this period, the energy router is operating in the situation of case 2 (shown in Figure 3b), and the active power and reactive power imported from the power grid to microgrid 2 increases to 60 kW and 20 kVar, respectively.

As is shown in Figure 12a,b, the active and reactive power flows can track their command value accurately and rapidly after a small fluctuation. Instant power balance between the power grid and the interconnected microgrids is guaranteed in this case, which has verified the effectiveness of the presented control strategies. The active and reactive power transfer from VSC1 remains zero when microgrid 1 is not involved in the energy exchange. Shown in Figure 12c, the fluctuation range of the DC-link voltage at the moments of power transfer is minor and can be ignored. As for the frequency curves shown the Figure 12d, it can be seen that the frequencies of the power grid and microgrid 2 appear to be fluctuant at the moments that power exchange occurs. However, there is no frequency fluctuation of microgrid 1 at $t = 2$ s, since that microgrid was not involved in the energy exchange at that moment. The simulation results in Figure 12 have clearly verified the effectiveness of the control strategies in this scenario. Meanwhile, the feasibility of the control scheme in the island mode can also be verified.

6.3. Scenario 3

The microgrid with surplus power production (microgrid 2 in this scenario) may change from grid-connected to parallel mode, while the energy router needs to shift to the situation of case 3. In this scenario, operation of the energy router is switched from the situation of case 1 (shown in Figure 3a) to case 3 (in Figure 3c).The power flow from microgrid 1 to microgrid 2, the DC-link voltage and the AC frequency variations during the whole process are demonstrated in Figure 13.

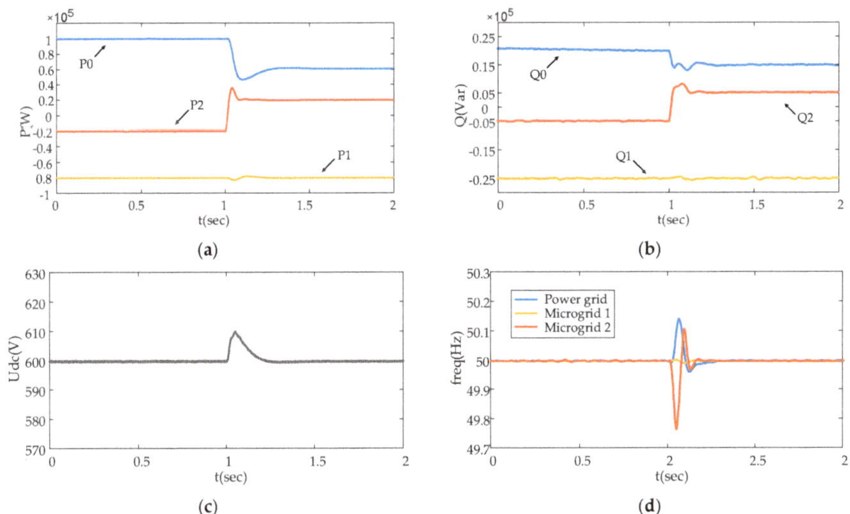

Figure 13. Simulation results of scenario 3: (**a**) active power curves; (**b**) reactive power curves; (**c**) DC-link voltage curves; and (**d**) frequency curves.

At $t = 1$ s, the status of microgrid 2 changes from energy consumption to energy production. Along with the power grid as energy provider, microgrid 2 transfers 20 kW active power and 5 kVar reactive power to microgrid 1. The situation of curve variations in Figure 13 is similar to that in Figure 12.

6.4. Scenario 4

In this scenario, the characteristics of the microgrids are investigated when a disturbance occurs on the grid side. While the energy router is operating in the situation of case 1, it is assumed that voltage variation occurs on the power-grid side.

As shown in Figure 14, the disturbance on the grid side starts from $t = 0.2$ s and continues for the next 0.1 s. The amplitude of the grid voltage decreases by 5%, due to the disturbance. The variations in active/reactive power, DC-link voltage and AC frequency on the microgrid side during the process are shown in Figure 15.

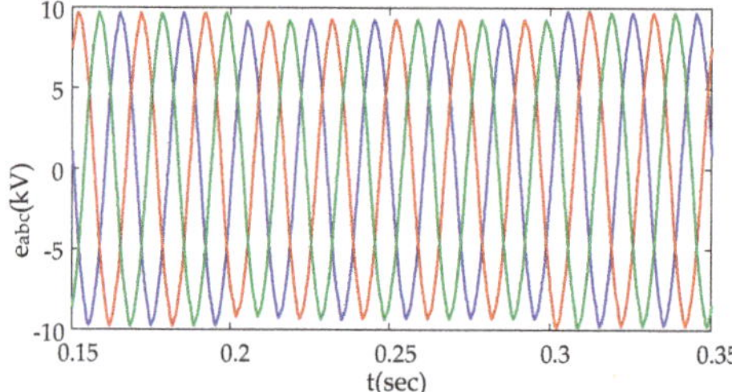

Figure 14. Three-phase voltage curves on the grid side at the moment of disturbance.

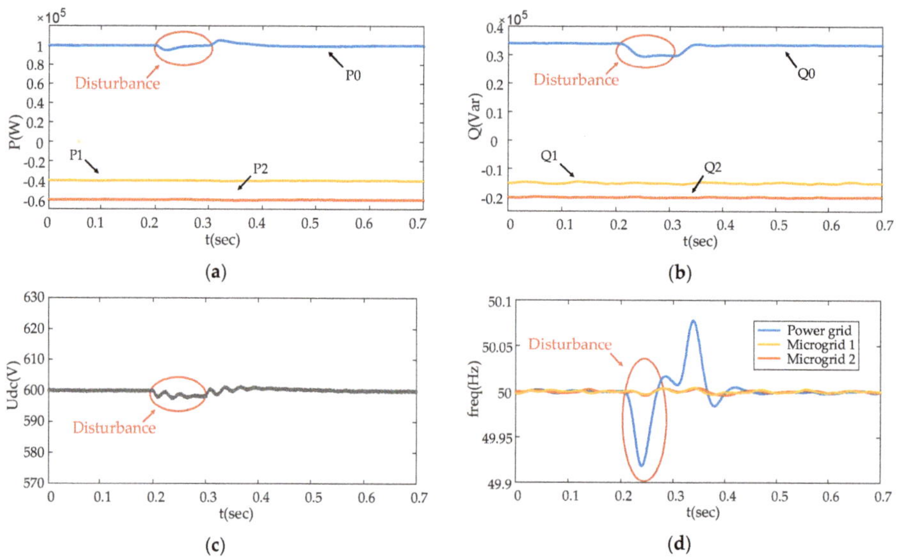

Figure 15. Simulation results of scenario 4: (**a**) active power curves; (**b**) reactive power curves; (**c**) DC-link voltage curves; and (**d**) frequency curves.

When the disturbance occurs at $t = 0.2$ s, the power supplied from the power grid fluctuates within a narrow range. The DC-link voltage also fluctuates along with it, but recovers from the disturbance rapidly. More importantly, due to the electrical isolation functionalities of the energy router, the AC frequency and voltage on the microgrid side are not affected by the disturbance, as shown in Figure 15d.

7. Conclusions

This paper focused on a novel and flexible interconnecting framework for microgrids and corresponding energy management strategies, with the aim to maintain the energy balance and operational stability of the power system, and also alleviate the dependency on energy storage equipment. The presented framework enabled direct energy exchange between microgrids via the multiport energy router, based on the thought that adjacent microgrids may have complementarity in terms of the pattern of energy production and consumption, which can be utilized to compensate for each other's instant energy deficiency. Facilitated by the energy router, the energy management patterns of the interconnected microgrids were expanded considerably. The operational modes were analyzed, and corresponding control strategies were developed for the energy router-based interconnected microgrids. Four scenarios were investigated based on the MATLAB R2014b/Simulink platform, whichdemonstrated that:

(1) apart from the grid-connected mode and the islanded mode, microgrids interconnected with the energy router can operate in the parallel mode when flexible energy flow among adjacent microgrids is enabled by the energy router. Therefore, energy surpluses or deficits can be compensated by complementary energy exchanges on the microgrid side without interference of the main power grid;

(2) facilitated by the energy router with the presented control strategies, the microgrids can switch freely and flexibly among the multiple operational modes, according to their energy demands and operational statuses;

(3) The VSC converters and shared DC bus of the energy router facilitate electrical isolation between the microgrids and the main grid, which ensures that any voltage or frequency fluctuation at one end of the energy router will have no direct impact on the systems on other sides of the energy router, thus guaranteeing power quality and operational reliability.

Widespread implementations of the presented energy router-based interconnected microgrid scheme will promote the shift of the power system architecture from the traditional hierarchical framework to a more connective and interactive framework, which will be an essential step to build the Internet of Energy of the future.

Acknowledgments: This work is supported by the "Science and Technology Project Plan of the Ministry of Housing and Urban-Rural Development of the People's Republic of China" (Project No.: 2016-K1-018).

Author Contributions: All the authors have contributed significantly. Yingshu Liu conceived and designed the model and corresponding control methods; Yue Fang performed the simulation; Yue Fang and Jun Li analyzed the results; Jun Li made all graphics; Yingshu Liu and Yue Fang wrote and revised the paper.

Conflicts of Interest: The authors declare no conflict of interest.

References

1. Huang, A.Q.; Crow, M.L.; Heydt, G.T. The Future Renewable Electric Energy Delivery and Management (FREEDM) System: The energy internet. *Proc. IEEE* **2011**, *99*, 133–148. [CrossRef]
2. Lasseter, R.H. Smart Distribution: Coupled Microgrids. *Proc. IEEE* **2011**, *99*, 1074–1082. [CrossRef]
3. Walling, R.A.; Saint, R.; Dugan, R.C. Summary of Distributed Resources Impact on Power Delivery Systems. *IEEE Trans. Power Deliv.* **2008**, *23*, 1636–1644. [CrossRef]
4. Barnes, M.; Kondoh, J.; Asano, H. Real-World MicroGrids-An Overview. In Proceedings of the 2007 IEEE International Conference on System of Systems Engineering, San Antonio, TX, USA, 16–18 April 2007; pp. 1–8.

5. Olivares, D.E.; Mehrizi-Sani, A.; Etemadi, A.H. Trends in Microgrid Control. *IEEE Trans. Smart Grid* **2014**, *5*, 1905–1919. [CrossRef]

6. Yang, Z.; Zhang, J.; Kintner-Meyer, M.C. Electrochemical Energy Storage for Green Grid. *Chem. Rev.* **2011**, *111*, 3577–3613. [CrossRef]

7. Díaz, N.L.; Luna, A.C.; Vasquez, J.C. Centralized Control Architecture for Coordination of Distributed Renewable Generation and Energy Storage in Islanded AC Microgrids. *IEEE Trans. Power Electron.* **2017**, *32*, 5202–5213. [CrossRef]

8. Bragard, M.; Soltau, N.; Thomas, S. The Balance of Renewable Sources and User Demands in Grids: Power Electronics for Modular Battery Energy Storage Systems. *IEEE Trans. Power Electron.* **2010**, *25*, 3049–3056. [CrossRef]

9. Divya, K.C.; Østergaard, J. Battery Energy Storage Technology for Power Systems—An Overview. *Electr. Power Syst. Res.* **2009**, *79*, 511–520. [CrossRef]

10. Cheng, S.; Sun, W.B.; Liu, W.L. Multi-Objective Configuration Optimization of a Hybrid Energy Storage System. *Appl. Sci.* **2017**, *7*, 163–174. [CrossRef]

11. Levelized Cost of Storage Analysis 2.0. Available online: https://www.lazard.com/perspective/levelized-cost-of-storage-analysis-20/ (accessed on 15 December 2016).

12. Lindley, D. The Energy Storage Problem. *Nature* **2010**, *7*, 18–20. [CrossRef]

13. Clement-Nyns, K.; Haesen, E.; Driesen, J. The Impact of Vehicle-to-grid on the Distribution Grid. *Power Syst. Res.* **2011**, *81*, 185–192. [CrossRef]

14. Baumann, C.; Schuster, R.; Moser, A. Economic Potential of Power-to-gas Energy Storages. In Proceedings of the International Conference on the European Energy Market(EEM), Stockholm, Sweden, 27–31 May 2013; pp. 1–6.

15. Rodrigues, E.M.G.; Godina, R.; Santos, S.F. Energy Storage Systems Supporting Increased Penetration of Renewables in Islanded Systems. *Energy* **2014**, *75*, 265–280. [CrossRef]

16. Díaz-González, F.; Sumper, A.; Gomis-Bellmunt, O. A Review of Energy Storage Technologies for Wind Power Applications. *Renew. Sustain. Energy Rev.* **2012**, *16*, 2154–2171. [CrossRef]

17. Liu, Z.; Chen, C.; Yuan, J. Hybrid Energy Scheduling in a Renewable Micro Grid. *Appl. Sci.* **2015**, *5*, 516–531. [CrossRef]

18. Huang, A.Q.; Baliga, J. FREEDM System: Role of Power Electronics and Power Semiconductors in Developing an Energy Internet. In Proceedings of the International Symposium on Power Semiconductor Devices and ICs, Barcelona, Spain, 14–18 June 2009; pp. 9–12.

19. Zhang, J.; Wang, W.; Bhattacharya, S. Architecture of Solid State Transformer-based Energy Router and Models of Energy Traffic. In Proceedings of the 2012 Innovative Smart Grid Technologies (ISGT), Washington, DC, USA, 16–20 January 2012; pp. 1–8.

20. Takahashi, R.; Kitamori, Y.; Hikihara, T. AC Power Local Network with Multiple Power Routers. *Energies* **2013**, *6*, 6293–6303. [CrossRef]

21. Kandula, R.P.; Iyer, A.; Moghe, R. Power Router for Meshed Systems Based on a Fractionally Rated Back-to-Back Converter. *IEEE Trans. Power Electron.* **2014**, *29*, 5172–5180. [CrossRef]

22. Xu, Y.; Zhang, J.; Wang, W. Energy Router: Architectures and Functionalities toward Energy Internet. In Proceedings of the 2011 IEEE International Conference on Smart Grid Communications (SmartGridComm 2011), Brussels, Belgium, 17–20 October 2011; pp. 31–36.

23. Takuno, T.; Kitamori, Y.; Takahashi, R. AC Power Routing System in Home Based on Demand and Supply Utilizing Distributed Power Sources. *Energies* **2011**, *4*, 717–726. [CrossRef]

24. Saponara, S.; Bacchillone, T. Network Architecture, Security Issues, and Hardware Implementation of a Home Area Network for Smart Grid. *J. Comput. Netw. Commun.* **2012**, *7*, 285–293. [CrossRef]

25. Nguyen, P.H.; Kling, W.L.; Ribeiro, P.F. Smart power router: A flexible Agent-based Converter Interface in Active Distribution Networks. *IEEE Trans. Smart Grid* **2011**, *2*, 487–495. [CrossRef]

26. Yu, X.; She, X.; Zhou, X. Power Management for DC Microgrid Enabled by Solid-state Transformer. *IEEE Trans. Smart Grid* **2014**, *5*, 954–965. [CrossRef]

27. Zhang, Y.; Umuhoza, J.; Liu, Y. Optimized control of isolated residential power router for photovoltaic applications. In Proceedings of the Energy Conversion Congress and Exposition, Pittsburgh, PA, USA, 14–18 September 2014; pp. 53–59.

28. Cao, J.W.; Meng, K.; Wang, J.Y. An energy internet and energy routers. *Sci. Sin. Inf.* **2014**, *44*, 714. [CrossRef]

29. Saponara, S. Distributed Measuring System for Predictive Diagnosis of Uninterruptible Power Supplies in Safety-Critical Applications. *Energies* **2016**, *9*, 327. [CrossRef]
30. Saponara, S.; Fanucci, L.; Bernardo, F. Predictive Diagnosis of High-Power Transformer Faults by Networking Vibration Measuring Nodes With Integrated Signal Processing. *IEEE Trans. Instrum. Meas.* **2016**, *65*, 1749–1760. [CrossRef]
31. Tonkoski, R.; Turcotte, D.; El-Fouly, T.H.M. Impact of High PV Penetration on Voltage Profiles in Residential Neighborhoods. *IEEE Trans. Sustain. Energy* **2012**, *3*, 518–527. [CrossRef]
32. Majumder, R.; Ghosh, A.; Ledwich, G. Power Management and Power Flow Control with Back-to-Back Converters in a Utility Connected Microgrid. *IEEE Trans. Power Syst.* **2010**, *25*, 821–834. [CrossRef]

Article

Modelling and Simulation of Electric Vehicle Fast Charging Stations Driven by High Speed Railway Systems

Morris Brenna [1], Michela Longo [1,*] and Wahiba Yaïci [2]

[1] Department of Energy, Politecnico di Milano, via La Masa, 34-20156 Milano, Italy; morris.brenna@polimi.it
[2] CanmetENERGY Research Centre, Natural Resources Canada, 1 Haanel Drive,
Ottawa, ON K1A 1M1, Canada; wahiba.yaici@canada.ca
* Correspondence: michela.longo@polimi.it; Tel.: +39-02-2399-3759

Academic Editor: Sergio Saponara
Received: 26 June 2017; Accepted: 10 August 2017; Published: 25 August 2017

Abstract: The aim of this investigation is the analysis of the opportunity introduced by the use of railway infrastructures for the power supply of fast charging stations located in highways. Actually, long highways are often located far from urban areas and electrical infrastructure, therefore the installations of high power charging areas can be difficult. Specifically, the aim of this investigation is the analysis of the opportunity introduced by the use of railway infrastructures for the power supply of fast charging stations located in highways. Specifically, this work concentrates on fast-charging electric cars in motorway service areas by using high-speed lines for supplying the required power. Economic, security, safety and environmental pressures are motivating and pushing countries around the globe to electrify transportation, which currently accounts for a significant amount, above 70 percent of total oil demand. Electric cars require fast-charging station networks to allowing owners to rapidly charge their batteries when they drive relatively long routes. In other words, this means about the infrastructure towards building charging stations in motorway service areas and addressing the problem of finding solutions for suitable electric power sources. A possible and promising solution is proposed in the study that involves using the high-speed railway line, because it allows not only powering a high load but also it can be located relatively near the motorway itself. This paper presents a detailed investigation on the modelling and simulation of a 2 × 25 kV system to feed the railway. A model has been developed and implemented using the SimPower systems tool in MATLAB/Simulink to simulate the railway itself. Then, the model has been applied to simulate the battery charger and the system as a whole in two successive steps. The results showed that the concept could work in a real situation. Nonetheless if more than twenty 100 kW charging bays are required in each direction or if the line topology is changed for whatever reason, it cannot be guaranteed that the railway system will be able to deliver the additional power that is necessary.

Keywords: electric vehicles; fast-charging stations; simulation; railway system

1. Introduction

In order to meet future mobility needs, decrease greenhouse gas and toxic emissions, and eliminate dependence on fossil fuels, nowadays automotive technologies require to be substituted by further effective, efficient and clean environmentally alternative energy sources. On the evolution to a sustainable society, proficient mobility technologies are mainly desirable worldwide. Electric vehicles have been recognised as being such a technology. In parallel, some nations, like Denmark, Germany and Sweden have agreed to replace electricity production from fossil fuel to renewable energy sources, hence further advancing sustainability and viability of electric vehicles when compared with internal combustion engine vehicles.

People realised in the 1950s and 1960s that trains could compete with commercial flights if travelling time by train was shorter than by plane. Two things have to be remembered when comparing trains and planes: the location of the airports and train stations; and the boarding procedures. Most airports are on the outskirts of the city and an additional trip by public transport or taxi is required to get to the city centre, while railway stations typically are located in the city centre [1]. Boarding procedures are also very different between trains and planes. Usually the boarding procedure for trains requires getting to the station in time to catch the train and proceeding through passport control, while flying requires getting to the airport at least one hour before the flight departure for national flights (two hours when flying internationally) to dropping off luggage and passing extensive security checks in addition to passport control [2]. Flying requires time after landing as well, especially for the luggage claim, which is not needed for train travel. This means train travelling can be much faster than flying for short distance travel. An example of this is the Tokaido Shinkansen line between Tokyo and Osaka. Built in 1964, today it carries, on a daily basis, more than 400,000 passengers on average, with 330,000 available seats on limited stop trains compared to 30,000 seats of airlines capacity. This is also due to the fact trains can carry far more passengers than any airliner, making trains more energy efficient. This led other developed countries to establish their own high-speed train services and to build a high-speed network. Because of the high-power demands of high-speed trains, it became clear that traditional electrification schemes were not very good at delivering that much power, so the 2×25 kV autotransformer system was used for new lines and new electrifications [3,4]. This system allows for having a relatively low number of substations while delivering a lot of power to the trains, with the advantage of minimising energy losses and voltage drops [5].

The issue of climate change and greenhouse effects led governments and regulating agencies to enforce stricter emission limits on traditional internal combustion engines [6]. This led to the rediscovery of the electric car. Invented in 1830, the electric car became a research topic when it was clear that internal combustion engines were the best solution available. Recent developments in battery technology, especially the lithium ion batteries, and in electronics led to the modern electric car, which is comparable with traditional cars in terms of performance. Despite the technological developments, the long charging times mean that electric vehicles are mostly used in cities and towns [6].

The increase in popularity of electric vehicles requires the construction of a suitable fast-charging infrastructure to allow driving longer distances. Usually high-speed lines are built (or designed to be built) near existing motorways [7–9]. Since the best place to install fast-charging facilities are motorways service areas because the necessary services are already there, it is interesting to study the possibility of connecting these charging facilities to the neighbouring high-speed line rather than the high-voltage transmission grid [10].

Electric vehicles need to have their batteries recharged to gain extra range, just as a traditional gasoline car needs to have its tank filled up at a gasoline station. There are three ways to get the batteries charged [11–13]. Battery swapping consists in swapping the depleted batteries with fully charged ones. In practice, almost no current electric car (except Tesla's) allows for it and has lost much of its appeal when a service provider went bankrupt in 2013. Wired charging brings electricity to the vehicle using cables and plugs. It is the most widely used technology. Induction charging transmits electricity through high frequency variable magnetic fields. It is still in development and it is only used in a few pilot locations [14].

The CENELEC EN 61851-1 standard on wired charging requires that the Electric Vehicle (EV) shall be connected to the EVSE (Electric Vehicle Supply Equipment) so that in normal conditions of use, the conductive energy transfer function operates safely [15,16]. In particular, the standard defines four charging modes but they are not all legal in every country. The four charging modes are: (a) Mode 1; (b) Mode 2; (c) Mode 3; and (d) Mode 4. Mode 1 charging requires connecting the EV to the Alternating Current (AC) supply network (mains) utilizing standardised socket-outlets not exceeding 16 A and not exceeding 250 V single-phase or 480 V three-phase, at the supply side, and utilizing the power and protective earth conductors [17–19]. Mode 2 charging requires a connection of the EV to the AC supply network (mains) not exceeding 32 A and not exceeding 250 V single-phase or 480 V three-phase utilising standardised single-phase or three-phase socket-outlets. It requires also utilising the power and protective earth conductors together with a control pilot function and system of personnel protection against electric shock (RCD) between the EV and the plug or as a part of the in-cable control box. The in-line control box needs to be located within 0:3 m of the plug or the EVSE or in the plug. Mode 3 charging requires connecting the EV to the AC supply network (mains) using dedicated EVSE where the control pilot function extends to control equipment in the EVSE, permanently connected to the AC supply network (mains). Mode 4 charging needs a connection of the EV to the AC supply network (mains) utilizing an on-board charger where the control pilot function extends to equipment permanently connected to the AC supply [20,21].

It is possible to classify the four charging modes according to the charging speed: Modes 1 and 2 are generally used for slow charging from the household plug, while Mode 4 is usually used for fast DC charging at a charging station. Mode 3, on the other hand, can be used both for slow charging and for fairly quick charging depending on the capability of the car and the infrastructure. The existence and the differences between Modes 1–3 are due to different factors: the consumer representatives want something cheap and feel that the stock household plugs are acceptable for home charging, while the industry feels that a purpose-built solution would be better thanks to higher safety and reliability.

There are three connectors standardised for AC charging: the US, Japan, and other countries that have a predominantly single-phase distribution network use the type 1 (or SAE) connector. Europe has settled on the Mennekes type 2 connector, which supports both single and three-phase power supplies up to 40 kW, and China uses its own standard. DC fast charging (Mode 4) sees a different situation, because there are four standards available on the market and the question has not been fully settled yet. The available standards are: (a) Tesla's Supercharger [22,23]; (b) the Chinese GB/T [24,25]; (c) the Combined Charging System (CCS) [26]; and (d) the CHAdeMO [27].

The power quality analyses carried out are also useful to predict the aging of the power devices as substation transformers [28,29].

A limited number of research studies propose different concepts for EV charging stations powered by renewable energies, including architectures, control strategies and infrastructure planning [30–37]. Despite the above efforts, the authors did not find detailed information on the modelling and simulation of electric vehicle fast charging stations powered by high-speed railway lines in Italy [38–40]. In order to further investigate the characteristics of the concept, a model of a 2 × 25 kV system to feed the railway was proposed and simulations of the battery charger were presented in detail. It is hoped that this contribution will lead to optimisation of the system design and characterisation and to encourage its widespread improvement and applications.

Moreover, many current researchers are focused on power electronic converters and devices for automotive applications [41,42].

The aim of this work is the analysis of the opportunity introduced by the use of railway infrastructures for the power supply of fast charging stations located in highways. In fact, long highways are often located far from urban areas and electrical infrastructure; hence, the installations of high power charging areas can be problematic.

This paper is organised as follows: Section 2 presents a detailed description of the 25 kV railway electrification, dividing the section in two different parts, a focus on the configuration and design of the 2 × 25 kV systems and another focus on the mathematical model of the 2 × 25 kV system, respectively. The specifications and configuration of the charging facility used in this analysis and the simulation model of the load are given in Section 3. Section 4 provides details of the system simulation of the charging system and results. Finally, the conclusions are reported in Section 5.

2. The 25 kV Railway Electrification

2.1. Configuration and Design of the 2 × 25 kV System

The 25 kV AC single phase is the most common electrification system for new high speed lines in the variant 2 × 25 kV autotransformer feed. This electrification system is commonly used with the industrial frequency (50 or 60 Hz) but it may be used for the railway frequency as well [43].

The 25 kV electrification schemes are single-phase electrification systems that are powered from the three phase industrial grid. This poses some problems due to the unbalance that a high power single-phase load creates on the grid's voltage. The degree of unbalance on the grid is limited to reduce the possibility of damage to generators and electric motors and it is measured though the unbalance coefficient, which is defined as follows:

$$K = \frac{V_i}{V_d} \tag{1}$$

where V_i is the negative sequence voltage and V_d is the positive sequence voltage. In practice, the unbalance coefficient is measured through the following Equation (2):

$$K = \frac{P_{sp}}{P_{sc}} \tag{2}$$

where P_{sp} is the power of the single-phase load and P_{sc} is the three-phase short circuit power of the connection node. International standards and transmission system operators set the maximum degree of unbalance between 1 and 2% that means that the load has to be connected to a high-voltage node with sufficiently high short-circuit-power. This later is not a problem where a high-power grid is available, but can be difficult on islands characterised by weak networks. The railway line is powered by several substations that are connected to different phase pairs in order to further reduce load unbalance on the industrial grid. Each substation is built with redundant apparatuses, such as two transformers and two connections to the industrial grid, in order to prevent the line grinding to a halt, in case of failure of any of these components. The connection between two different phase-couples of two successive ESS (energy storage system) implies that a neutral section has to be built when the power supply switches from one ESS to the next because of the voltage difference between two out-of-phase circuits. Usually, neutral sections are placed at the substations and halfway between them. Each neutral section is signalled to the engine driver to allow them to disconnect all the train loads (traction motors included) before entering the neutral section itself (Figure 1).

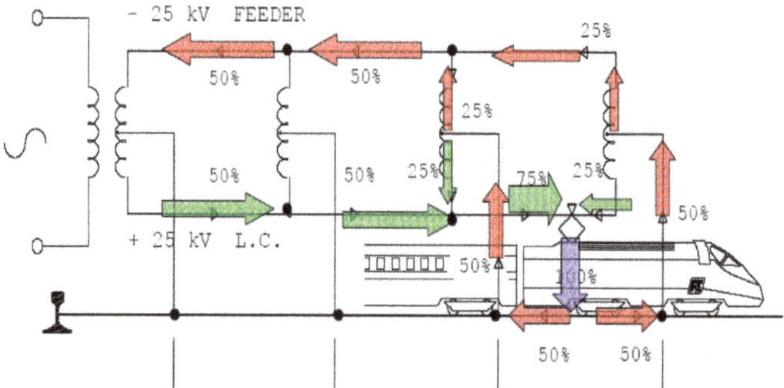

Figure 1. Power flow in an ideal 2 × 25 kV autotransformer system.

The neutral sections can be energised from either end when needed. International standards set the tolerance of the voltage in the system once the nominal value is chosen. The tolerances and the maximum duration allowed for a nominal voltage of 25 kV are reported in Table 1.

Table 1. The 25 kV system operating voltages [44].

U_{min2}	lowest non-permanent voltage (max. 10 min.)	17.5 kV
U_{min1}	lowest permanent voltage	19.0 kV
U_n	nominal voltage 25 kV	25.0 kV
U_{max1}	highest permanent voltage 27.5 kV	27.5 kV
U_{max2}	highest non-permanent voltage (max. 5 min.)	29.0 kV

Low traffic lines are usually fed directly from the transformer through the over-head line and the rail as return conductor. High traffic lines and high-speed lines are usually fed with an autotransformer system because it allows building the feeding stations further apart and, consequently, to contain the costs of building high-voltage lines to connect substations to the grid. Each electrical substation has two power transformers that supply the system from the high-voltage industrial network. The distance between substations is usually between 40 and 60 km. Autotransformers are installed regularly every 10 to 15 km between two consecutive substations [45].

The railway traction supply system has to be protected against faults so circuit breakers and switches are vital to isolate the faulted section and keep the system running. The heart of the protection system is a distance protection relay whose shape varies depending on the manufacturer and it is programmed to allow heavy loads without causing a trip. The same relay has an integrated overcurrent protection to clear close-in faults as fast as possible and a close-onto-fault protection [46].

2.2. Mathematical Model of the 2 × 25 kV System

Each high speed line is fed with a 2 × 25 kV autotransformer system capable of delivering 2 MW of power for each kilometre of line. This allows "the simultaneous presence of 12 MW trains travelling at 300 km/h (180 mph) 5 min apart with no limits but with margin" [47]. This has been achieved through two 60 MVA feeding transformers in each electrical substation that are located 50 km (30 miles) apart; during normal operation each transformer feeds a section that is 25 km long. The substations are fed through a dedicated power-line at the nominal voltage of 132 kV (in northern Italy) or 150 kV (in southern Italy) capable of delivering 200 MW that connects all the substations of the railway line. Both ends of the power-line are connected to the national grid in very high-voltage (400 kV) nodes through two 250 MVA autotransformers. This permits the system to feed the railway line from a single very high-voltage substation when needed [48].

Paralleling posts are located every 12.5 km on average and are equipped with two 15 MVA autotransformers; they allow for putting the two tracks electrically parallel and connecting them to the autotransformers. During normal operation, only one of the two autotransformers is in use, except when a phase break is present, in which case they are both energised for each phase. This normally happens halfway between two substations. The railway line comprises seven conductors of each track at three different voltage levels: contact wire and messenger wire are fed at +25 kV, the feeder is connected to the −25 kV bus, the two rails, the overhead earth wire and the buried-earth wire are connected to earth, i.e., they have zero volt potential.

The model of a two-track line is made of the following four elements: (a) wires and rails; (b) inductive couplers; (c) the feeder transformer; and (d) autotransformers.

The hypothesis the model is based on is that the current flow through the soil and the buried-earth conductor are negligible; therefore, it is possible to simulate the line as though all current flows through the rails themselves or the overhead earth-wire. Electrical lines are usually modelled with a series of resistances and inductances and with capacitances and conductances in parallel. In most cases, a loaded line can be modelled to exclude the parallel capacitances and conductances, as their effect is negligible. In the case of the high-speed railway line, the power-line is not symmetric on each track even if the two tracks are built symmetrically such that inductance and resistance are represented by a big matrix that, in our case, has dimensions twelve by twelve. The big advantage that the symmetry between the two tracks brings us is that the matrices are symmetric themselves, potentially reducing the amount of calculation needed. The reason why such big matrices are required is that the distance between the two track centrelines is 5 m, so the tracks cannot be considered independent in terms of magnetic coupling. Such a hypothesis would mean that the two tracks are very far apart, which is not viable, both in terms of environmental impact and economics. If it is assumed that all conductors are parallel and the messenger wire has a constant height, then self and mutual inductances of each conductor can be computed using Neumann formulas, which are:

$$L_{ij} = \frac{\mu_0}{2\pi} ln \frac{2 \times l}{D \times e} \tag{3}$$

$$L_{ij} = \frac{\mu_0}{2\pi} ln \frac{2 \times l}{D \times e} \tag{4}$$

$$L_{ij} = \frac{\mu_0}{2\pi} ln \frac{2 \times l}{K \times r_0 \times e} \tag{5}$$

where D is the distance between the wires, l is the wire length, r_0 is the wire radius and K is a coefficient used to represent how the current is distributed inside the wire. These equations can be simplified if the sum of the currents is taken into account; in this case, the sum of all currents is zero for each track, which means that the inductance equations can be written as following:

$$L_{ij} = \frac{\mu_0}{2\pi} ln \frac{l}{D} \tag{6}$$

$$L_{ij} = \frac{\mu_0}{2\pi} ln \frac{l}{K \times r_0} \tag{7}$$

If l is equal to 1 and all units are SI, the result is that the inductance per length unit is henry per meter. Exceptions to these formulas are the rails themselves: since steel is a magnetic material and the cross section is high and not entirely used, the preceding formula cannot be used, but the value for the self-inductance is available in the literature either measured or computed through finite elements simulation. In this case, the value 0.359 mH/km has been used. In opposition to the inductances, only the self-resistances of each conductor are meaningful, so the resistances matrix is diagonal. Another attractive feature of this matrix is due to how the line is built. As each track uses the same type of wires, the values can be computed for only one track, because the other track values are the same. The line resistances of each wire are computed with the usual formula:

$$R = \rho \times \frac{l}{S} \tag{8}$$

where ρ is the resistivity of the material, l is the wire length and S is the useful cross section of the wire itself. The rail resistances are again taken from the literature for the same reason it is not possible to compute the self-inductances easily: the value used is 0.116 Ω/km.

All calculations are done using MATLAB software for an Italian high-speed rail line built on an embankment. The wires and rail positions in Table 2 are given with coordinates on a plain whose origin is placed in the middle between the two tracks on rail level.

The line model itself has been built in MATLAB-Simulink using inductances and resistances. The blocks used to make the line model are mutual inductances, resistors and inductors. Each block represents a base line section that is 1.5 km long; this distance has been chosen because inductive couplers are installed every 1.5 km. The inductive couplers are inductors that are installed every 1.5 km and which connect the two running rails of each track together in order to allow the traction current to flow in the earth-wire without short-circuiting the rails themselves.

The inductance is 1.2 mH, which is low enough to let the 50 Hz traction current through and high enough to block the audio-frequency current and allow track circuits to function properly. The inductor is centre-tapped and has been modelled as two separate inductors in series to have the same total inductance and the tap to connect to the earth-wire. The installation requires the centre tap to be connected to the earth-wire and the two edges to be connected to each rail.

Table 2. Wires and rail positions on an Italian high-speed line built on an embankment.

Tracks	Wire	X Coordinate (m)	Y Coordinate (m)
Track 1	Contact wire	−2.5	5.3
	Messenger wire	−2.5	6.55
	Earth wire	−6.1	5.5
	Rail 1	−3.22	0
	Rail 2	−1.78	0
	Feeder	−6.6	8.0
Track 2	Contact wire	2.5	5.3
	Messenger wire	2.5	6.55
	Earth wire	6.1	5.5
	Rail 1	1.78	0
	Rail 2	3.22	0
	Feeder	6.6	8

The voltage drop along the transmission line is evaluated using a phasorial equation in steady-state conditions as follows:

$$-\left[\frac{dV}{dx}\right] = [Z'][I] \tag{9}$$

where $[V]$ and $[I]$ are the phasor vectors of the line-to-ground voltage and of the currents flowing in the conductor, respectively.

It is assumed that the ground is the node to which all the voltages are referred. Because of the ground presence, the resistive factors are introduced in the mutual couplings. The Z'_{ii} and Z'_{ik} values are expressed using the Carson equations, which are accurate for power systems if used with homogenous ground [14]. The impedance matrix elements are found from the conductor's placement geometry and their characteristics as shown in Figure 2. The self-impedance is then evaluated as follows:

$$Z'_{ii} = \left(R'_{i-int} + \Delta R'_{ii}\right) + j\left(\omega\frac{\mu_0}{2\pi}ln\frac{2h_i}{r_i} + X'_{i-int} + \Delta X'_{ii}\right) \tag{10}$$

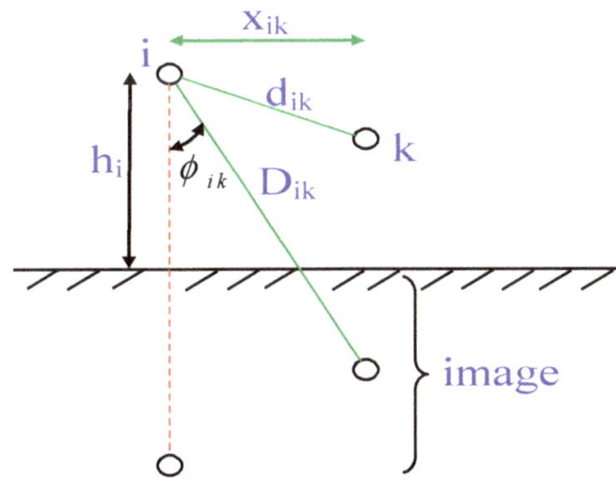

Figure 2. Geometry of the conductor's placement.

The mutual one is expressed as follows:

$$Z'_{ik} = Z'_{ki} = \Delta R'_{ik} + j(\omega \frac{\mu_0}{2\pi} ln \frac{D_{ik}}{d_{ik}} + \Delta X'_{ik}) \tag{11}$$

The conductor internal impedance $R'_{int} + jX'_{int}$ is also evaluated. The internal reactance is usually combined in a single equation with the external reactance $\omega \frac{\mu_0}{2\pi} \ln \frac{2h}{r}$, where the radius, r is replaced with the minor Geometric Mean Radius (GMR), which is available from the conductor datasheets, in order to take into account the internal magnetic field:

$$\omega \frac{\mu_0}{2\pi} ln \frac{2h}{r} + X'_{int} = \omega \frac{\mu_0}{2\pi} ln \frac{2h}{GMR} \tag{12}$$

The internal reactance can be evaluated as a part of the internal impedance. Since for non-magnetic conductors, the internal impedance represents a minor contribution of the total reactance, its accurate estimation is therefore not required. However, the evaluation of the internal resistance R'_{int} is more significant due to the increase of its rate with the frequency caused by the skin effect.

In this investigation, as the system is complex and comprises several wires, the traction line has been applied through an integrated parameter model, which can be considered rather precise. The resistance and the inductance can be held practically constant until 1 kHz. As the application of this study is on lower frequencies, therefore, this assumption is suitable. Furthermore, the traction line has been distributed into many multipoles in order to avoid using unnecessary approximations.

Additionally, all the joint connections between the 14 wires founding the system have been considered in each cell. There are actually 7 wires for each way, particularly:

- A copper messenger wire at 25 kV with a cross section of 150 mm^2;
- A copper contact line at 25 kV with a cross section of 120 mm^2;
- An aluminium steel feeder at -25 kV with a cross section of 307 mm^2;
- An aluminium alloy cable guard with a cross section of 147.1 mm^2;
- A copper or aluminium ground wire with a cross section of 95 mm^2;
- An external rail, and
- An internal rail.

The system is built differently depending on the number of tracks available. Each track has a complete set of wires if the rail is double tracked (Figure 3a), four track railways can be built in two different configurations: it is possible to feed the fast tracks separately, thus requiring two autotransformers at each AT site or feed all the tracks together that means only two feeders and one autotransformer are needed (Figure 3b).

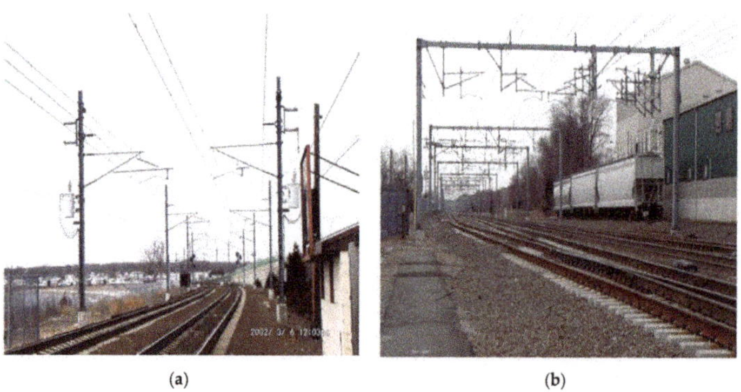

(a)   (b)

Figure 3. (**a**) Two track railway line; (**b**) Four track railway line.

The model details are shown in Figure 4. The feeding substation is fed through a high-voltage power-line connecting the various substations, so each substation has an incoming power-line connected.

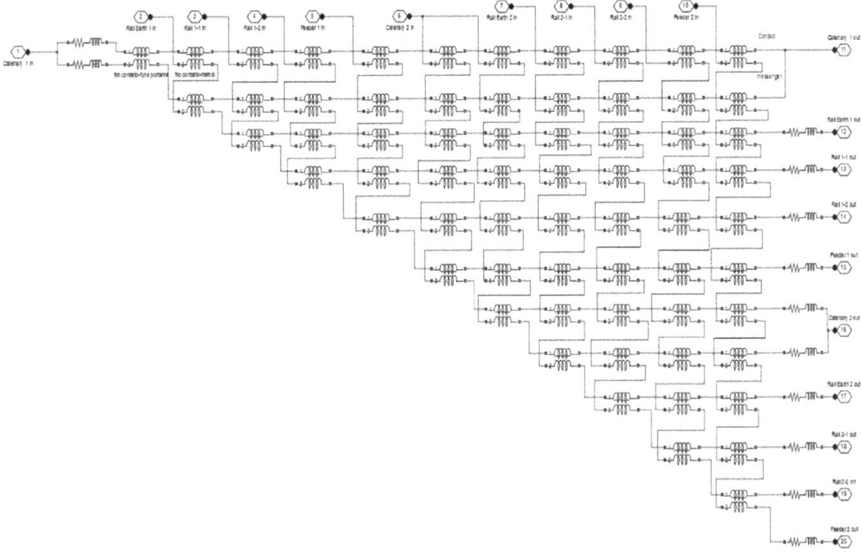

Figure 4. Electric model of the line including all mutual coupling among live and earthling conductors.

It is also possible to replace the three-phase generator with an AC ideal voltage source with the right voltage and frequency. There may be the need to represent a longer section of the high-speed line fed with multiple substations. In that case, it is possible to add a more detailed representation of the power-lines and the feeding 400 KV node. An appropriate power-line model can be made with the "distributed parameters line" Simulink block; the autotransformer can be replaced by an equivalent transformer and a suitable three-phase voltage source. It is also possible to skip the 400/132 kV autotransformers as long as the three-phase voltage source has the correct short-circuit power.

The feeder transformer is single phase with a centre tap on the secondary winding and it is modelled as a three-winding transformer whose low voltage windings are connected in series. Another important component is the autotransformer which is necessary to connect the 50 kV catenary to feeder transmission line to the 25 kV train feeding system. The real machine is made of two windings wound on each column of the steel core connected together so that both windings share the same magnetic flux. The model parameters are taken from the real machine and are:

- Nominal Power: 15 MVA;
- Nominal Voltage: 55/27.5 kV;
- Short-circuit voltage: 1%.

The main load of the line are going to be trains. The train model varies depending on the electronic converter used to rectify the AC current to DC. Old trains used diode or thyristor bridges, which make the train absorb highly distorted currents, i.e., the current absorbed is rich in low order harmonics especially the third, fifth, and seventh. This means that the fastest way to model these trains is by using a rectifier bridge feeding an appropriate load. When the use of GTOs and later insulated-gate bipolar transistors (IGBTs) became widespread, rectifier bridges were replaced by four quadrant converters in all newly constructed trains. This allows to convert AC to DC while absorbing a sinusoidal current in phase with the voltage (to be clear there are other harmonics, but their order is multiple of the switching frequency of the converter usually in the range of kilohertz). This means that, unless necessary, the load can be represented by a sinusoidal current injected through a current generator. This is achieved with a controlled generator; the control is very straightforward: the voltage is measured on site where the current will be injected and the signal is scaled with a gain block to the value of the current drawn by the load itself. If High Speed Trains (HST) is used as an example, the current drawn at the maximum power (8.8 MW) is 352 A at 25 KV; the gain block will scale the voltage to the current with the ratio 352/25,000 so that the current is in phase with the voltage and the power drawn is the nominal train power. The model can be used either to calculate the voltage profile along the line or to calculate the waveforms of voltages and currents wherever it is needed. As the line is built with 1.5 km long elementary line pieces, it is possible to simulate the case in which everything works correctly as well as the case with one or more faulty pieces of equipment.

3. Charging Facility

3.1. Specifications and Configuration of the Charging Facility

Fast-charging electric vehicles require a sufficiently powerful connection to the electrical grid, which may require connecting directly to the high-voltage transmission grid. It is quite expensive to connect it to the high-voltage mains because of the switchgear and the space required to build a substation. There are service areas distributed every 30 km on average in the Italian motorway system and, according to our survey, each service area has 13 fuel pumps on average in each direction. The survey data come from Google Earth and Google Street View imaging service; we counted the number of refuelling bays in some service stations on A4 Milan-Turin, A1 Milan-Bologna and A1 Rome-Naples motorway and then we calculated an average. If the refuelling process lasts 5 min and all the available fuel pumps are being used, it is possible to refuel 78 cars in half an hour, which means that 78 charging bays are needed to recharge the same number of vehicles in the same half-hour. This means that each direction needs 7.8 MW of power to recharge each vehicle with a 100 kW rating. The Italian high-speed rail network has been built near motorways (when possible) and is able to deliver high power at a relatively low voltage, so it makes sense to study the effects of such a solution on the 2×25 kV railway supply system to evaluate the possibility of connecting the motorway charging points to the nearby railway. This solution can be particularly advantageous in the countryside, where the high-voltage network node is relatively far away and the high-speed line is quite near the service station because it avoids the construction of high-voltage lines. It is important to note that the railway infrastructure owner has the right to disconnect the motorway car charging facility in case that power is needed to maintain a set quality of service on the railway line itself.

The hypothesis of simultaneous use of all the refuelling bays is not quite true, as usually only some are actually used simultaneously in real life. This can be expressed through a coefficient that indicates the percentage of bays used. This is actually very useful, because it allows for three options: reducing the number of available charging bays; or sharing the total 100 kW power rating between two bays; or both. If the power-sharing option is chosen, then it would be better to share the power dynamically between the two cars in order to give more power to the car with the lower state of charge rather than sharing the power fifty-fifty between the two users.

3.2. Simulation Model of the Load

Each charging station has a variable load with 100 kW maximum power. The load model has to potentially include each main rectifier topology for harmonics injection in the network and reactive power consumption.

The main topologies are Graetz bridges with both diodes and thyristors and the PWM (Pulse-width modulation) controlled switched AC/DC converter; these are very different in both harmonics injected and reactive power consumption. The Graetz bridge configuration is characterised by high harmonic currents at low frequency, i.e., bridge rectifiers inject high third, fifth and seventh harmonic order currents that normally have to be filtered out. On the other hand, they are easy to maintain because each valve is turned on and naturally, but the output voltage cannot be regulated without an additional switching DC/DC converter. Thyristors bridges are more complicated indeed requiring a controller, but they allow regulating the output voltage. Voltage regulation is possible by changing the ring angle, which causes the output voltage to drop; on the other hand, the ring angle is directly linked to the reactive power absorbed, which means that an appropriate power factor correction device is needed.

Switched AC/DC power converters constitute the most modern approach to rectification. Their main feature is that they achieve unity power factor and limit low order-harmonic-current injection through their high switching frequency, which, depending on the type of semiconductor used (MOSFETs, IGBTs, GTOs), varies from some kilohertz to over 20 kHz. Being directly linked to the switching frequency, the frequency of the current harmonics is high enough to make it possible for the current amplitude to be naturally dumped by cables and transformers inductances resulting in low THD without the need of expensive filters. What is more, if converters are grouped and controlled with a technique called interlacing, the net result is that each group of harmonic emissions is equivalent to that of a single converter operating at a frequency multiple of the converters number in each group and each converter's switching frequency.

Most simulations feature a switching AC/DC converter. The load itself is made of the battery, the converters and the interfacing transformer. The battery has been modelled using the battery block available in SimPower systems within MATLAB-Simulink. It is able to simulate every kind of battery form, from lead-acid to lithium-ion ones. The battery modelled in this case is the one available in the recent electric cars belonging to the category of Battery Electric vehicles (BEV) [40,41]. The battery's main features are summarised in Table 3.

Table 3. Main characteristics of the electric vehicle battery.

Description	Value
Nominal voltage (V)	364.8
Rated pack energy (kWh)	24
Rated pack capacity (Ah)	66.2

The four-quadrant (4Q) power converter used to rectify the AC voltage uses IGBTs as semiconductors and is controlled through a dedicated PWM controller. The main modifications needed to make it run consisted in disabling the maximum power point tracking (MPPT) system used to extract the maximum power available from the PV panel and adjusting the regulator parameters. This second step has been performed through a trial-and-error method until a functioning device was obtained. A further aspect that has been modified is the value of the DC bus capacitor, which has been changed to 40 mF in order to maintain the voltage ripple in a 5% band. The DC bus nominal voltage has been changed to 500 V from the previous value of 425 V. The new nominal value has been chosen because, actually, it is the maximum voltage that has CHAdeMO and CCS (Combined Charging System).

The converter and its control system are shown in Figure 5. It is necessary to install a DC/DC converter to adjust the current flow to the battery because of the difference between the battery voltage (about 400 V when fully charged) and the main DC bus (500 V). This converter is a two-quadrant converter that allows the power to flow in both directions, i.e., from the battery to the grid or vice-versa.

In this case, the battery charging function is more interesting. The control is done through a PI controller in order to obtain a voltage reference that is able to drive the PWM signal generator. The controller is again tuned through a trial-and-error process. The current reference signal has been made variable so that the regulator is enabled with a current reference equal to zero. The reference signal increases linearly to the maximum value of 250 A in 2 s. Battery, converter and filters are shown in Figure 6.

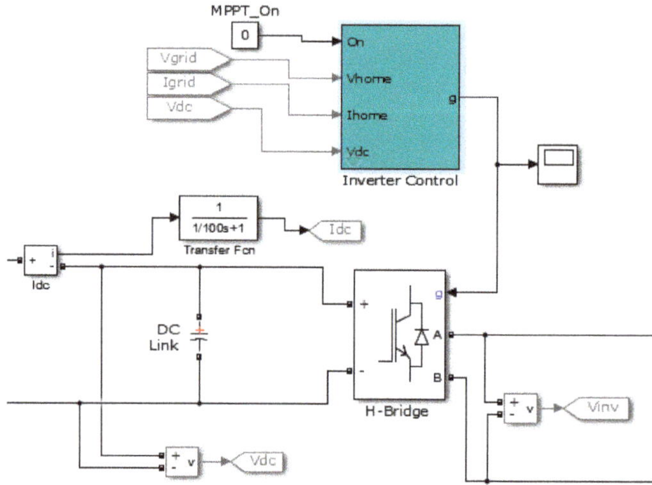

Figure 5. Model of the four-quadrant (4Q) converter and its control system.

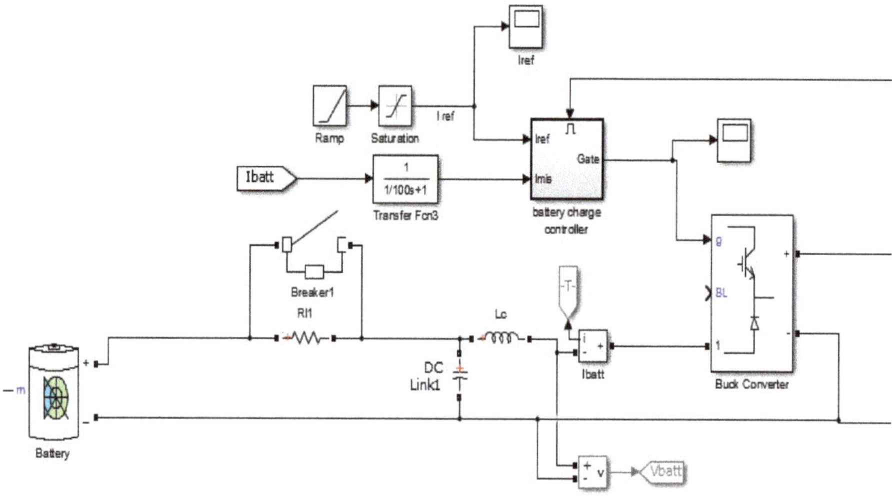

Figure 6. Model of the car battery and the converter.

The connection between the low-voltage systems that supply the power electronics and the medium voltage from the railway line is achieved through a short cable and a power transformer rated at 150 kVA. We decided to maintain the original topology of a centre-tapped two windings transformer used for the 120 V distribution system in the US for a couple of reasons; the most important of those is safety. As both lines are live, but their potential to earth is only 120 V, it is safer for people in case of an insulation failure. We hypothesised a cable of about 200 m length with a cross section of 180 mm^2 and a resistance of 0.106 km^{-1}. The transformer and line model is depicted in Figure 7. The resistor R_g represents the Earth system resistance.

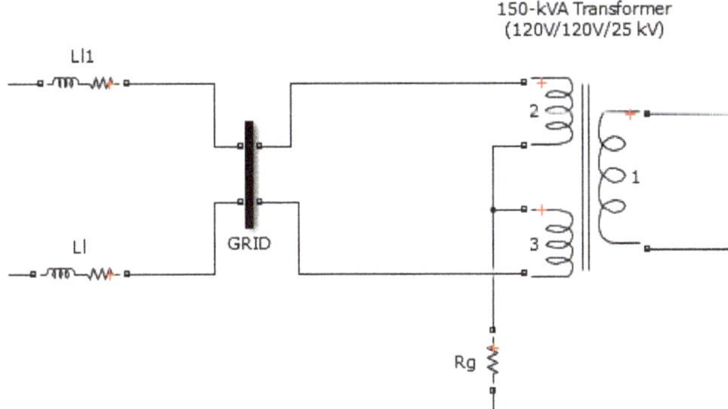

Figure 7. Model of the power transformer and the short low voltage line.

The 2 × 25 kV system is modelled as shown in Section 2.2. Four cases have been examined. These are: (a) the absence of trains; (b) the presence of one train; (c) the presence of two trains in the same cells; or (d) the presence of two trains in different cells. This is particularly important, as the main load continues to be the train traffic and the system is viable only if it is possible to power both the railway traffic and the charging facility at the same time. One of the possible scenarios simulated is the presence of two trains in the same cell to evaluate the voltage drop in the system. The considered trains are two modern High Speed Trains with power consumption of 8.8 MW and 9.8 MW, respectively. It is assumed that the two trains absorb a sinusoidal current with a unity power factor. Figure 8 shows the absorbed current by both trains without charging stations connected to the line. It can be seen that the system is capable of supplying both trains with no problem and maintains keeping the line voltage close to its nominal value.

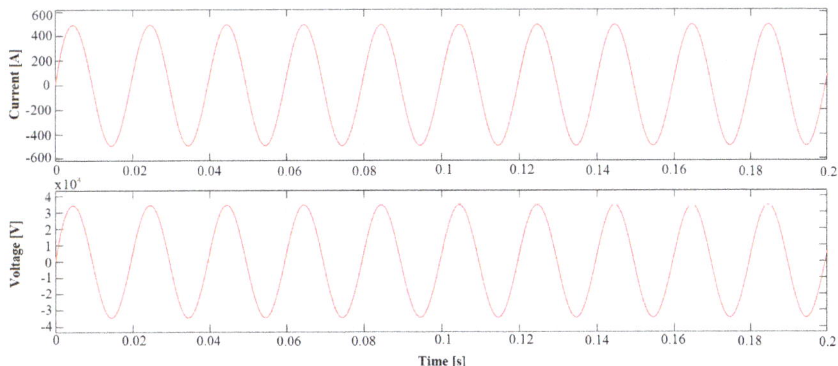

Figure 8. Voltage and current drawn by a High Speed Train (HST) on a high-speed line with no charging stations.

In order to verify that the actual current distribution is comparable to the ideal one, the current at the main transformer and at the autotransformer terminals have been measured (Figure 9). As can be seen, the actual current distribution is close to the ideal one even if some differences due to the real impedances of the line can be noted. In particular, it is possible to observe a small current in the rails and an imbalance between feeder and contact line (catenary).

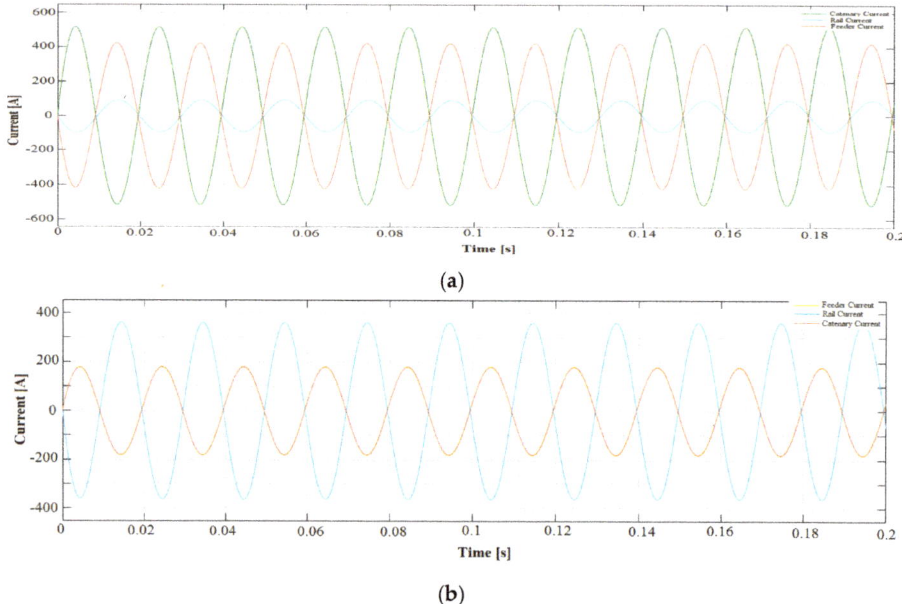

Figure 9. Currents measured on the (**a**) feeding transformer connections and (**b**) autotransformer connections with trains as the only load.

4. Simulation of the Charging System

The aim of these simulations is to assess how the quality of power of the railway systems is affected by the charging stations for road vehicles. Therefore, the analysis considers simulations of a very detailed model for a short period (max 3.5 s).

There are two simulation steps. The first step is about making the charging system work by itself. It is very important because it allows us to sort out the system problems before the two systems are made to work together. When both systems work as expected, it is then possible to connect them together and have the complete simulation of all the effects.

The first simulation phase has given the following results: the model was working with the following regulator values, 3000 and 1000 as proportional and integrator coefficients for the voltage regulator and 500 and 1000 as proportional and integrator coefficients for the current regulator. The current absorbed was not constant when the device was on full load as the DC bus voltage was still floating. The battery status is shown in Figure 10. It can be seen that the battery is actually recharged from the initial state of charge set to 10%. It shows current and voltage of the battery; current is represented with a negative value because it is charging. On the other end, the battery current would be positive if the battery was discharging. The battery is charged with a constant current using a soft inrush defended by ramp. As the current is constant, the state of charge increases linearly, instead the internal voltage drop is quite proportional to the current in a short time windows; it seems in this period, it is mainly due to the homic voltage drop. Moreover, it is possible to observe the effect of the switching of the power converters in contactors that produce a high-frequency ripple overlap to the main current.

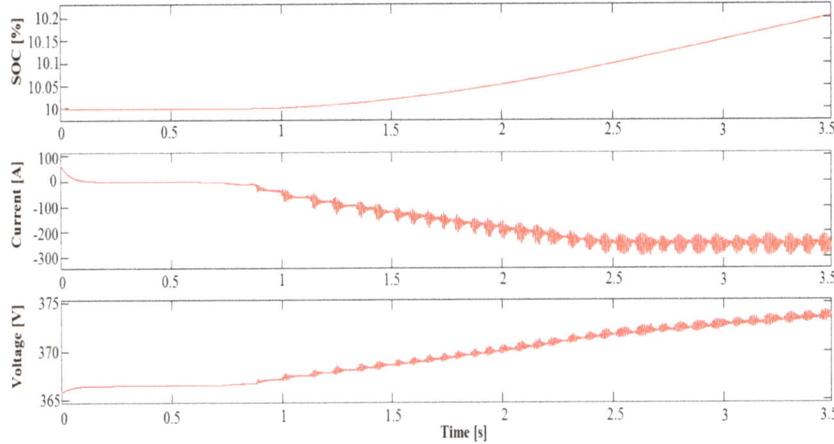

Figure 10. Battery main variables: state of charge, current and voltage.

The system is then tested inside the 2 × 25 kV railway line connected between the feeder and the grounding conductor. In this case, the model does not consider only one 100 kW charging spot, but twenty 100 kW charging bays for each direction that cause total power absorption of 4 MW from the railway systems. It is possible to replace the single bay with double bays sharing the same amount of power, thus allowing more people to charge simultaneously even if at a lower rate. The charging area is modelled with one charging bay and a current generator. The idea is to model one charging bay in order to see the current shape it absorbs and, at the same time, inject in the railway system the same current multiplied for the remaining charging bays. The assumption is that all charging bays are working at the same rate. The charging stations are connected from feeder to earth, rather than from catenary to earth, as trains are (Figure 11).

Figure 11. Section of a real High Voltage (HV) line with indication of the feeder and earth wires.

It has been decided to connect the charging facility in the middle of the cell. The first scenario is simulated with no train on the line and the results show that the system works quite well. The simulated time lapse is 3.5 s (Figure 12). In particular, the results are shown towards the end of the simulation because the charger is working at full load there. Figure 9 shows also the impact of all the charging infrastructures installed in two charging areas on the railway voltage. It is possible to remark that the voltage value near the charging area is still inside the limits set for the system on both tracks, despite the fact that the current is not always sinusoidal.

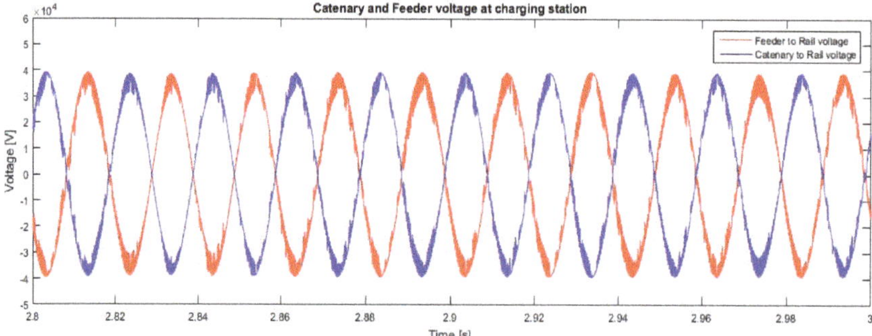

Figure 12. Feeder and catenary voltage due to the charging station presence.

Despite the shape of the current, Figure 13 shows that the autotransformers work as expected by shifting the earth current to the feeder and catenary system. This means that the feeding transformers see the most current coming from the feeder and the catenary with only a small current in the rail as Figure 14 shows. Actually, the two charging areas behave as a low-power HST, and it seems the power requested by the charging infrastructure is about one half of the power requested by a HST during acceleration. In fact, allowed connection between feeder and ground has the same behaviour of the train supplied from contact line and the rails.

The fact that the system works with no trains does not imply that it works while performing its main duty, i.e., powering high-speed trains. In order to verify this, a simulation has been produced in which the system has one or more trains on the tracks. The simulation represents the worst-case scenario with all the loads fed from the two autotransformers.

The first simulation involves the presence of one train on the line. The modelled train is an ETR 400 high-speed train with a maximum power of 9.8 MW located 3 km away from the charging stations. It has been computed that in 3.5 s the train travels less than 500 m, so it has been modelled to assume that the train is still. The total power of the three loads is about 14 MW, so it is expected that the system is able to power all the loads without an excessive voltage drop. The simulation results demonstrate exactly that the three loads can coexist. Figures 15 and 16 show that the voltage measured near the loads and the train is within the limits set by international standards with voltage value being around 23 kV.

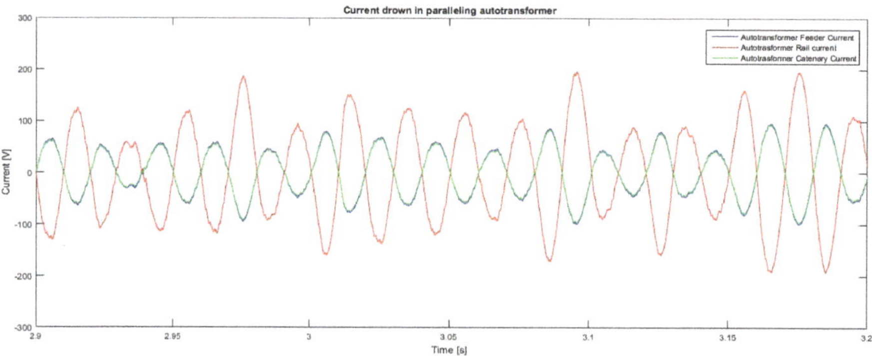

Figure 13. Current measured on the autotransformers connections due to the charging stations.

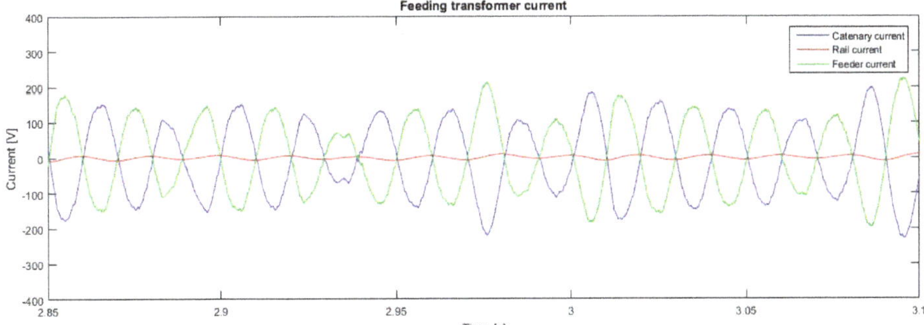

Figure 14. Current measured on the feeding transformer with the charging stations and no trains on the track.

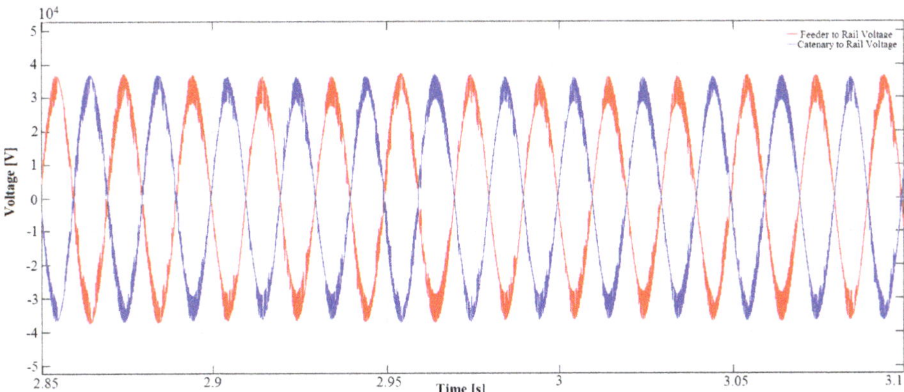

Figure 15. Catenary and feeder voltages on the charging facility connection point.

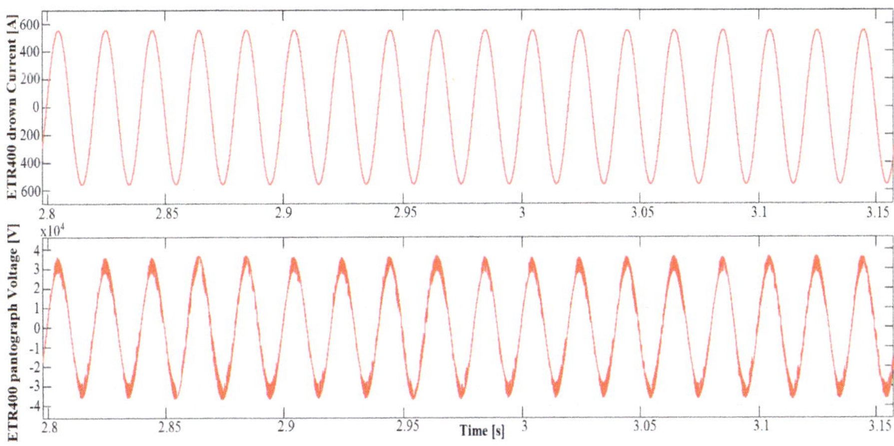

Figure 16. Voltage on the train position and current drawn by the train.

There are some differences between this and the preceding scenario; in particular, the currents have a different shape because of the sinusoidal train current superposed to the charger current. Figures 17 and 18 show the currents flowing in the autotransformers and in the main transformer.

The connection of the charging infrastructure between the feeder and the ground attempts to improve the balancing of the system as the current flows directly between the contact line and the feeder without involving the autotransformers and the sells not occupied by the train.

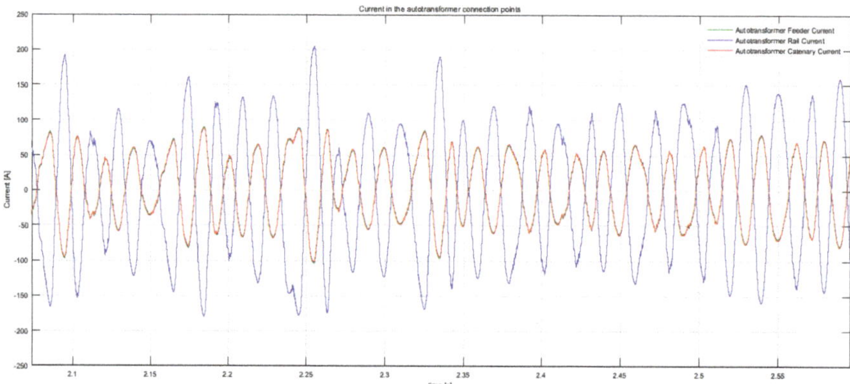

Figure 17. Current drawn by the balancing autotransformer with the charging stations and a train on the tracks.

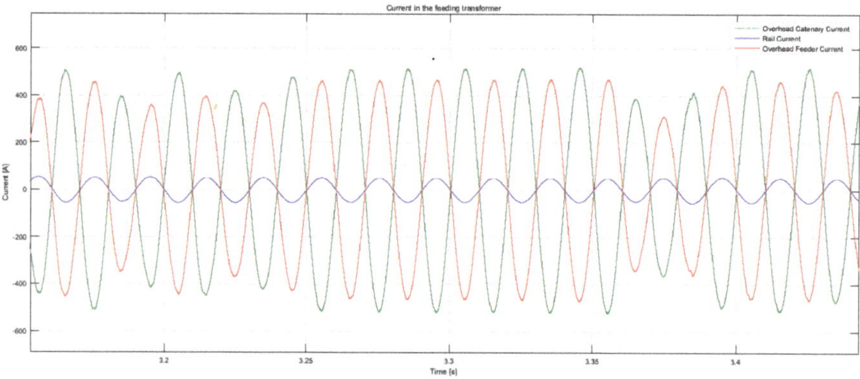

Figure 18. Current measured on the feeding transformer low voltage winding with the charging stations and a train on the tracks.

The last scenario evaluates the ability of the system to power two HST at 9.8 MW, each travelling in opposite directions and the same 40 charging bays. This case is very demanding and the system struggles to power the load. This is illustrated by the voltage value, which dropped to about 19 kV as shown in Figures 19 and 20.

The main similarity with the preceding scenario is that the currents in the feeding transformer and in the autotransformer are significantly influenced by the train rather than the relatively small load given by the charging bays. Figures 21 and 22 depict the current drawn by the autotransformers and the feeding transformer, respectively. The cause of this problem can be traced to two sources: the power converter of battery charging facility or the connection of such a big load between feeder and catenary. It is possible to speculate about the cause of this kind of problem being the connection of such a big load between feeder and earth. The system can cope with the load due to the line technical systems, such as signalling, switches and line diagnostics, but these loads have a maximum rating of 250 kVA each and each substation powers only a few of them.

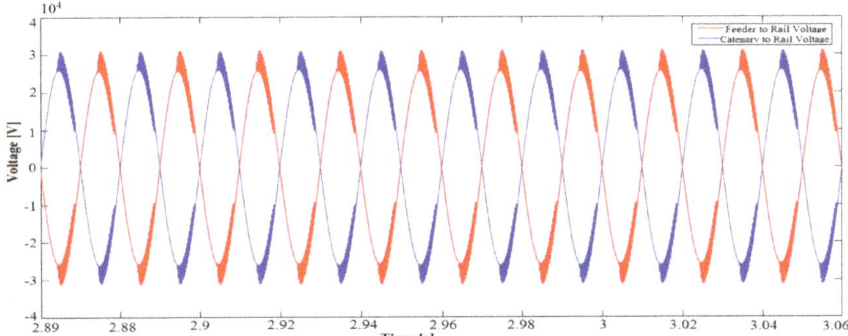

Figure 19. Voltage on the train position and current drawn by the train with two trains on the track.

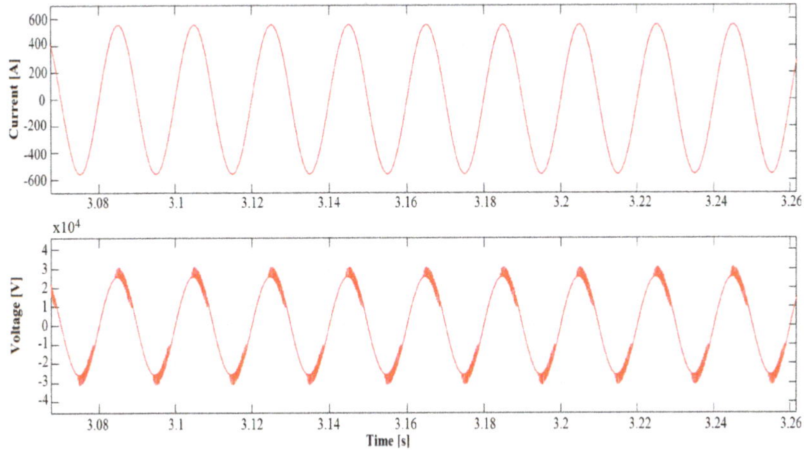

Figure 20. Voltage at the train pantograph and current drawn.

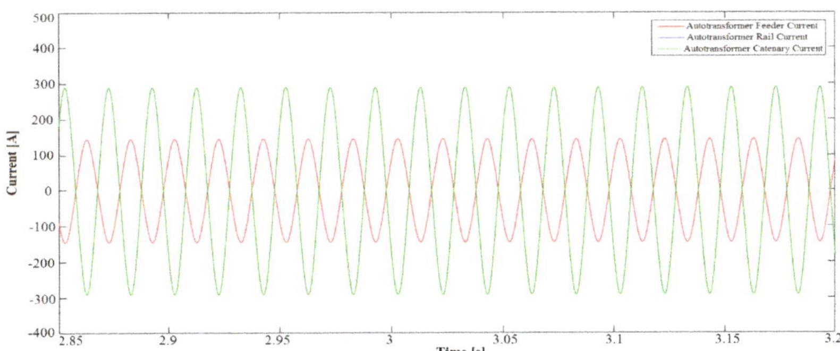

Figure 21. Current drawn by the autotransformer in the two-train scenario.

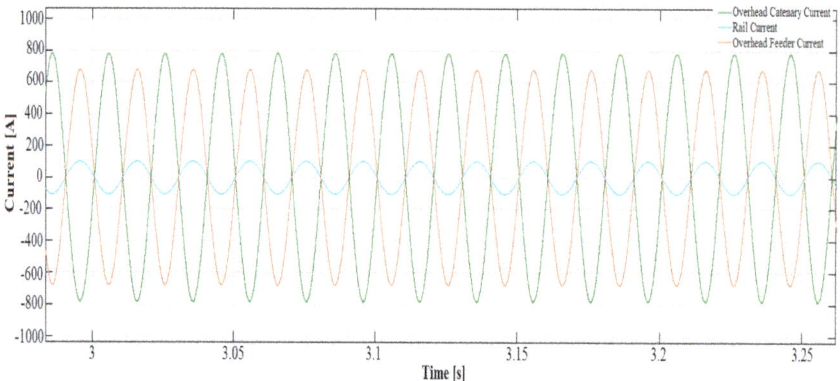

Figure 22. Current drawn by the feeding transformer in the two-train scenario.

5. Conclusions

The need to limit pollution enforces stricter emission standards that will increase the cost of producing traditional cars; moreover, increases in oil prices will result in fuel that is more expensive. In the meantime, a reduction in battery costs and government subsidies will lead to an increase in electric vehicle purchases. Another incentive to use electric vehicles comes from both battery manufacturers and the car industry: as battery technology develops further, it will be important to decrease the charging time for the battery pack while maintaining its performance for its entire life expectancy.

As electric vehicles are going to become even more popular, it is necessary to build fast charging infrastructures, especially on the highway network. One of the problems encountered is finding a suitable power source to charge many cars quickly despite the increase in battery capacity and in the number of vehicles in stock. Since the high-voltage grid is not always easily accessible, as the highway lines are usually far from the urban area and from the electricity grid but they are usually close to new high-speed railway lines, the authors have investigated the possibility to supply charging areas from the electric distribution system used for the railway service. These are promising solutions because the power absorbed by the charging areas is of the same order of the magnitude of the power absorbed by the train.

This solution has to compromise the needs of the rail operators, who want their trains adequately powered, and the service areas, who want to offer a service to car drivers.

To demonstrate the feasibility of the concept, a model of a 2×25 kV system to feed the railway has been developed. This latter has been implemented in MATLAB/Simulink/SimPower systems to simulate the railway. Then it has been applied to simulate the battery charger and the system.

The results through modelling and simulation disclosed that a compromise is possible but the charging station power has to be limited to allow trains to be properly powered. The simulations reveal that particular attention has to be paid to the quality of the power of the railway system in order to not be compromised by the high-power charging infrastructures. This means that it is not possible to achieve the same refuelling rate of traditional cars because the charging time is much greater than a traditional car refuelling time and the high-speed rail is not able to provide the extra power a larger charging facility needs. Nevertheless, it can be a good solution to begin building the fast charging infrastructure where the high-speed rail is easily accessible.

Author Contributions: Morris Brenna and Michela Longo proposed the core idea, developed the models. They performed the simulations, exported the results and analysed the data. Wahiba Yaïci revised the paper. Morris Brenna, Michela Longo and Wahiba Yaici contributed to the design of the models and the writing of this manuscript.

Conflicts of Interest: The authors declare no conflict of interest.

Nomenclature

V_i	Negative sequence voltage
V_d	Positive sequence voltage
P_{sp}	Power of the single-phase load
P_{sc}	Three-phase short circuit power of the connection node
U_{min2}	Lowest non-permanent voltage (max. 10 min)
U_{min1}	Lowest permanent voltage
U_n	Normal Voltage 25 kV
U_{max1}	Highest permanent voltage 27.5 kV
U_{max2}	Highest non-permanent voltage (max. 5 min)
D	Distance between the wires,
l	Wire length
r_0	Wire radius
K	Coefficient representing the current distribution inside the wire
ρ	Resistivity of the material,
S	Useful cross section of the wire itself
$[V]$	Phasor vectors
$[I]$	Line-to-ground voltage and of the currents flowing in the conductor.
Z'_{ii}, Z'_{ik}	values the Carson expressions
R_g	Earth system resistance

References

1. Barrero, R.; van Mierlo, J.; Tackoen, X. Energy saving in public transport. *IEEE Veh. Technol. Mag.* **2008**, *3*, 26–36. [CrossRef]
2. Franzitta, V.; Curto, D.; Milone, D.; Trapanese, M. Energy Saving in Public Transport Using Renewable Energy. *Sustainability* **2017**, *9*, 106. [CrossRef]
3. Gunselmann, W. Technologies for increased energy efficiency in railway system. *Proc. EPE Power Electron. Appl.* **2005**. [CrossRef]
4. Pengling, W.; Xuan, L.; Yuezong, L. Optimization analysis on the energy saving control for trains with adaptive genetic algorithm. In Proceedings of the International Conference on Systems and Informatics (ICSAI), Yantai, China, 19–20 May 2012.
5. González-Gil, A.; Palacin, R.; Batty, P. Optimal energy management of urban rail systems: Key performance indicators. *Energy Convers. Manag.* **2015**, *90*, 289–291. [CrossRef]
6. Zhang, D.; Zou, F.; Li, S.; Zhou, L. Green Supply Chain Network Design with Economies of Scale and Environmental Concerns. *J. Adv. Transp.* **2017**, *2017*, 1–14. [CrossRef]
7. Mozafar, M.R.; Moradi, M.H.; Amini, M.H. A simultaneous approach for optimal allocation of renewable energy sources and electric vehicle charging stations in smart grids based on improved GA-PSO algorithm. *Sustain. Cities Soc.* **2017**, *32*, 627–637. [CrossRef]
8. Adnan, N.; Nordin, S.M.; Rahman, I.; Amini, M.H. A market modeling review study on predicting Malaysian consumer behavior towards widespread adoption of PHEV/EV. *Environ. Sci. Pollut. Res. Int.* **2017**, 1–21. [CrossRef] [PubMed]
9. Amini, M.H.; Islam, A. Allocation of electric vehicles' parking lots in distribution network. In Proceedings of the IEEE PES Innovative Smart Grid Technologies Conference (ISGT), Washington, DC, USA, 19–22 February 2014.
10. Wang, J.; Wang, C.; Lv, J.; Zhang, Z.; Li, C. Modeling Travel Time Reliability of Road Network Considering Connected Vehicle Guidance Characteristics Indexes. *J. Adv. Transp.* **2017**, *2017*, 1–9. [CrossRef]
11. Bolla, V.; Pendolovska, V. *Driving Forces Behind EU-27 Greenhouse Gas Emissions over the Decade 1999–2008*; Statistics in Focus 10/2011; Eurostat: Luxembourg, 2011.
12. Faria, R.; Moura, P.; Delgado, J.; De Almeida, A.T. Managing the Charging of Electrical Vehicles: Impacts on the Electrical Grid and on the Environment. *IEEE Intell. Transp. Syst. Mag.* **2014**, *6*, 54–65. [CrossRef]

13. Wang, X.; Yuen, C.; Hassan, N.U.; An, N.; Wu, W. Electric Vehicle Charging Station Placement for Urban Public Bus Systems. *IEEE Trans. Intell. Transp. Syst.* **2017**, *18*, 128–139. [CrossRef]

14. European Commission. Taking Stock of the Europe 2020 Strategy for Smart, Sustainable and Inclusive Growth. COM (2014) 130 Final. Available online: http://ec.europa.eu/transparency/regdoc/rep/1/2014/EN/1-2014-130-EN-F2-1.Pdf (accessed on 1 August 2017).

15. Council Decision 406/2009/EC on the Effort of Member States to Reduce Their Greenhouse Gas Emissions to Meet the Community's Greenhouse Gas Emission Reduction Commitments Up to 2020. Available online: http://eur-lex.europa.eu/legal-content/EN/TXT/?uri=uriserv:OJ.L_.2009.140.01.0136.01.ENG (accessed on 1 August 2017).

16. Hess, A.; Malandrino, F.; Reinhardt, M.B.; Casetti, C.; Hummel, K.A.; Barceló-Ordinas, J.M. Optimal Deployment of Charging Stations for Electric Vehicular Networks. In Proceedings of the UrbaNe'12 First Workshop on Urban Networking, Nice, France, 10 December 2012; pp. 1–6.

17. Fox, G.H. Electric vehicle charging stations: Are we prepared? *IEEE Ind. Appl. Mag.* **2013**, *19*, 32–38. [CrossRef]

18. Falvo, M.C.; Sbordone, D.; Bayram, I.S.; Devetsikiotis, M. EV charging stations and modes: International standards. In Proceedings of the IEEE International Symposium on Power Electronics, Electrical Drivers, Automation and Motion (SPEEDAM), Ischia, Italy, 18–20 June 2014; pp. 1134–1139.

19. Baouche, F.; Billot, R.; Trigui, R.; El Faouzi, N.-E. Efficient allocation of electric vehicles charging stations: Optimization model and application to a dense urban network. *IEEE Intell. Transp. Syst. Mag.* **2014**, *6*, 33–34. [CrossRef]

20. Timpner, J.; Wolf, L. Design and evaluation of charging station scheduling strategies for electric vehicles. *IEEE Trans. Intell. Transp. Syst.* **2014**, *15*, 579–588. [CrossRef]

21. Alesiani, F.; Maslekar, N. Optimization of charging stops for fleet of electric vehicles: A genetic approach. *IEEE Intell. Transp. Syst. Mag.* **2014**, *6*, 10–21. [CrossRef]

22. Tesla Supercharger Deployment. Available online: http://www.teslamotors.com/supercharger (accessed on 1 August 2017).

23. Rajagopalan, S.; Maitra, A.; Halliwell, J.; Davis, M.; Duvall, M. Fast charging: An in-depth look at market penetration, charging characteristics, and advanced technologies. In Proceedings of the World Electric Symposium and Exhibition (EVS27), Barcelona, Spain, 17–20 November 2013; pp. 1–11.

24. China National Institute for Standardisation. China's Standard Proposal Adopted into IEC Standard. July 2014. Available online: http://en.cnis.gov.cn/xwdt/bzhdt/201407/t20140709_19405.shtml (accessed on 1 August 2017).

25. Han, Z.; Liu, S.; Gao, S.; Bo, Z. Protection scheme for china high-speed railway. In Proceedings of the 10th IET International Conference on Developments in Power System Protection (DPSP 2010), Managing the Change, Manchester, UK, 29 March–1 April 2010; pp. 1–5.

26. Kersting, K. Standardization Needs and Testing Methods for Multiple Outlet Chargers. Technical Report. 2014. Available online: http://www.egvi.eu/uploads/IDIADA%20standardisation%20needs%20MOC%20-%20published.pdf (accessed on 1 August 2017).

27. CHAdeMO Deployment. Available online: http://www.chademo.com/wp/usmap/ (accessed on 1 August 2017).

28. Saponara, S.; Fanucci, L.; Bernardo, F.; Falciani, A. Predictive Diagnosis of High-Power Transformer Faults by Networking Vibration Measuring Nodes with Integrated Signal Processing. *IEEE Trans. Instrum. Meas.* **2016**, *65*, 1749–1760. [CrossRef]

29. Saponara, S. Distributed Measuring System for Predictive Diagnosis of Uninterruptible Power Supplies in Safety-Critical Applications. *Energies* **2016**, *9*, 327. [CrossRef]

30. Ashique, R.H.; Salama, Z.; Abdul Aziz, M.J.B.; Bhattia, A.R. Integrated photovoltaic-grid dc fast charging system for electric vehicle: A review of the architecture and control Renewable and Sustainable. *Energy Rev.* **2017**, *69*, 1243–1257. [CrossRef]

31. Neumann, H.M.; Schär, D.; Baumgartner, F. The potential of photovoltaic carports to cover the energy demand of road passenger transport. *Prog. Photovolt. Res. Appl.* **2012**, *20*, 639–649. [CrossRef]

32. Tulpule, P.J.; Marano, V.; Yurkovich, S.; Rizzoni, G. Economic and environmental impacts of a PV powered workplace parking garage charging station. *Appl. Energy* **2013**, *108*, 323–332. [CrossRef]

33. Bhatti, A.R.; Salam, Z.; Aziz, M.J.B.A.; Yee, K.P. A comprehensive overview of electric vehicle charging using renewable energy. *Int. J. Power Electron. Drive Syst.* **2016**, *7*, 114–123. [CrossRef]
34. Bhatti, A.R.; Salam, Z.; Aziz, M.J.B.A.; Yee, K.P. A critical review of electric vehicle charging using solar photovoltaic. *Int. J. Energy Res.* **2016**, *40*, 439–461. [CrossRef]
35. Yagcitekin, B.; Uzunoglu, M. A double-layer smart charging strategy of electric vehicles taking routing and charge scheduling into account. *Appl. Energy* **2016**, *167*, 407–419. [CrossRef]
36. Mouli, C.; Bauer, G.R.; Zeman, P. System design for a solar powered electric vehicle charging station for workplaces. *Appl. Energy* **2016**, *168*, 434–443. [CrossRef]
37. Morrissey, P.; Weldon, P.; O'Mahony, M. Future standard and fast charging infrastructure planning: An analysis of electric vehicle charging behaviour. *Energy Policy* **2016**, *89*, 257–270. [CrossRef]
38. Wang, Y.; Shi, W.; Wang, B.; Chu, C.-C.; Gadh, R. Optimal operation of stationary and mobile batteries in distribution grids. *Appl. Energy* **2017**, *190*, 1289–1301. [CrossRef]
39. Wang, B.; Wang, Y.; Nazaripouya, H.; Qiu, C.; Chu, C.C.; Gadh, R. Predictive Scheduling Framework for Electric Vehicles With Uncertainties of User Behaviors. *IEEE Internet Things J.* **2017**, *4*, 52–63. [CrossRef]
40. Chen, N.; Gan, L.; Low, S.H.; Wierman, A. Distributional analysis for model predictive deferrable load control. In Proceedings of the IEEE 53rd Annual Conference on Decision and Control (CDC), Los Angeles, CA, USA, 15–17 December 2014; pp. 6433–6438.
41. Costantino, N.; Serventi, R.; Tinfena, F.; D'Abramo, P.; Chassard, P.; Tisserand, P.; Saponara, S.; Fanucci, L. Design and Test of an HV-CMOS Intelligent Power Switch With Integrated Protections and Self-Diagnostic for Harsh Automotive Applications. *IEEE Trans. Ind. Electron.* **2011**, *58*, 2715–2727. [CrossRef]
42. Saponara, S.; Pasetti, G.; Tinfena, F.; Fanucci, L.; D'Abramo, P. HV-CMOS design and characterization of a smart rotor coil driver for automotive alternators. *IEEE Trans. Ind. Electron.* **2013**, *60*, 2309–2317. [CrossRef]
43. Jensen, T.O.; Danmark, A. High Performance Railway Power Introduction to Autotransformer System (AT). May 2012. Available online: http://www.banekonference.dk/sites/default/files/AT-system%202012.pdf (accessed on 1 August 2017).
44. White, R.D. AC 25 kV 50 Hz electrification supply design. In Proceedings of the 4th IET Professional Course on Development Railway Electrification Infrastructure and Systems (REIS), London, UK, 1–5 June 2009; pp. 93–126.
45. Nardinocchi, A. Electrification and Power Supply. 2011. Available online: http://www.apta.com/mc/hsr/previous/2011/presentations/Presentations/Electrification-and-Power-Supply.pdf (accessed on 1 August 2017).
46. Sezi, T.; Menter, F.E. Protection scheme for a new ac railway traction power system. In Proceedings of the IEEE Transmission and Distribution Conference, New Orleans, LA, USA, 11–16 April 1999; Volume 1, pp. 388–393.
47. Shenoy, J.U.; Sheshadri, K.G.; Parthasarathy, K.; Khincha, H.P.; Thukaram, D. Matlab/psb based modeling and simulation of 25 kV AC railway traction system—A particular reference to loading and fault conditions. In Proceedings of the IEEE Region 10 Conference TENCON, Chiang Mai, Thailand, 24 November 2004; Volume 3, pp. 508–511.
48. US Department of Energy. Battery Test Results. 2011. Available online: http://media3.ev-tv.me/DOEleaftest.pd (accessed on 1 August 2017).

Article

Real-Time Analysis of a Modified State Observer for Sensorless Induction Motor Drive Used in Electric Vehicle Applications

Mohan Krishna S. [1], Febin Daya J. L. [2], Sanjeevikumar Padmanaban [3,*] and Lucian Mihet-Popa [4]

[1]   Department of Electrical and Electronics Engineering, MITS (Madanapalle Institute of Technology and Science), Madanapalle 517325, AP, India; smk87.genx@gmail.com

[2]   School of Electrical Engineering, VIT University—Chennai Campus, Chennai 600 048, India; febinresearch@gmail.com

[3]   Department of Electrical and Electronics Engineering Science, University of Johannesburg, Auckland Park, Johannesburg 2006, South Africa

[4]   Faculty of Engineering, Østfold University College, Kobberslagerstredet 5, 1671 Kråkeroy, Norway; lucian.mihet@hiof.no

*   Correspondence: sanjeevi_12@yahoo.co.in; Tel.: +27-79-219-9845

Academic Editor: Sergio Saponara

Received: 5 May 2017; Accepted: 3 July 2017; Published: 25 July 2017

Abstract: The purpose of this work is to present an adaptive sliding mode Luenberger state observer with improved disturbance rejection capability and better tracking performance under dynamic conditions. The sliding hyperplane is altered by incorporating the estimated disturbance torque with the stator currents. In addition, the effects of parameter detuning on the speed convergence are observed and compared with the conventional disturbance rejection mechanism. The entire drive system is first built in the Simulink environment. Then, the Simulink model is integrated with real-time (RT)-Lab blocksets and implemented in a relatively new real-time environment using OP4500 real-time simulator. Real-time simulation and testing platforms have succeeded offline simulation and testing tools due to their reduced development time. The real-time results validate the improvement in the proposed state observer and also correspond to the performance of the actual physical model.

Keywords: state estimation; model reference; sliding mode; adaptive; parameter detuning

---

1. Introduction

The utility of induction motors has risen considerably owing to its integration with power electronic converters, which made variable frequency operation realizable. This, in turn, made the induction motor the workhorse of the industry. Of all the variable frequency control strategies, the vector control or field oriented control principle was the most popular. It provided independent control of torque and flux resulting in fast torque response. The field orientation can be achieved by directly measuring the magnitude and direction of the flux by means of flux sensors or hall effect sensors in the machine (Direct Vector control) or it can be imposed indirectly by a slip frequency component from the rotor dynamics (Indirect Vector control). The latter was more feasible as it did not require the use of additional flux sensors that would occupy additional space and cost. The indirect vector control principle is shown in the phasor diagram expressed as steady state direct current (DC) quantities in Figure 1.

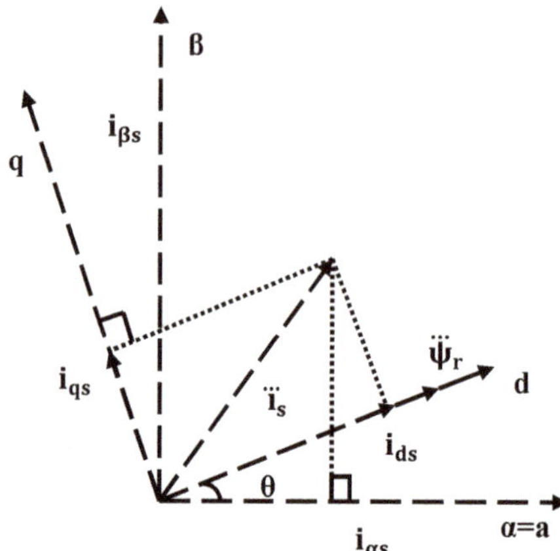

Figure 1. Indirect vector control principle.

By decoupling the induction motor at synchronously rotating reference frame, and forcing the direct axis stator current component (field producing) in phase with the rotor flux and orthogonal to the quadrature axis stator current' component (torque producing), independent control of torque and flux is obtained. However, the indirect vector control implementation required the utility of a shaft speed encoder to sense the rotor speed, which was processed along with the speed command to generate the reference torque request for vector control. The presence of the shaft speed encoder implies additional electronics, cost and mounting space. Therefore, to eliminate the shaft speed encoder, the speed estimation techniques were used.

The speed was estimated from either the terminal quantities of the machine or from its rotor saliency. However, speed estimation from the machine model was easier to implement, occupied less computational space and were most effective. Considerable research over the past two decades focused primarily on sensorless control of induction motor [1], with special emphasis on estimation from the machine model [2]. The state estimation schemes are shown in Figure 2.

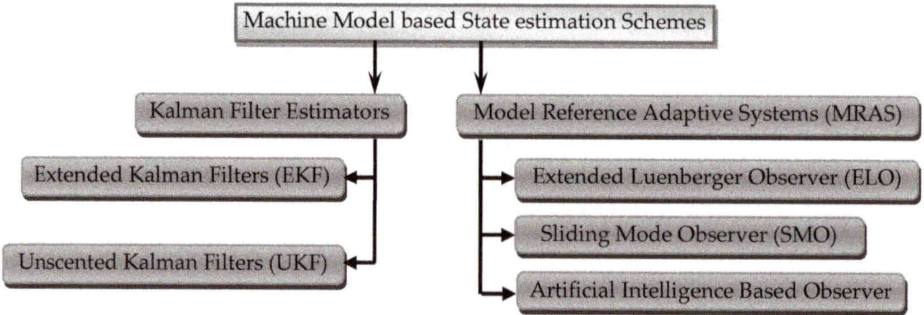

Figure 2. Machine model based state estimation schemes.

During the earlier stages, extended kalman filter (EKF) based estimators [3–5] were widely used for speed estimation, but they give accurate results only if the system dynamics are linearized and had an inherent disadvantage of a high sampling frequency and were computationally expensive. Estimators based on model reference adaptive systems (MRAS), extended luenberger observers (ELO) and sliding mode observers (SMO) [6,7] had a wider utility and were more extensively used owing to ease of use and flexibility. In addition, several configurations varying from lower order to higher order observers, as well as integration of variable structure or artificial intelligence could be developed from the MRAS. All the model based schemes were sensitive to variations or incorrect settings of parameters.

The MRAS and ELO were mainly used for simultaneous state estimation in order to prevent a mismatch between the actual and estimated values of the parameters under all speed ranges. Several studies focused on the performance analysis and parameter estimation [8–10] for the drive at low and zero speed regions [11–13]. As stated before, variable structure or sliding mode observers (SMO) based on MRAS have been implemented for wide speed bandwidth estimation and faster parameter convergence by constraining the states of the system to the sliding hyperplane [14–18] implements a Luenberger-SMO that estimates the critical parameters online. It is demonstrated by means of hardware in the loop (HIL) simulation setup with an field programmable gate array (FPGA) based controller and an induction motor, in order to verify the robustness of the algorithm. Reference [19] presents a Sliding Mode-MRAS observer based on a super twisting algorithm (STA), where the variations in critical parameters are intentionally considered. In the standard configuration of MRAS, the reference model is replaced by a stator current observer, which is designed based on STA. This, in turn, is insensitive to rotor resistance variations and disturbances when the states converge on the sliding hyperplane. The chattering phenomenon is eliminated and near zero speed operation is realized by means of a parallel identification of stator resistance. In [20], the concept of adaptive Luenberger flux observation is applied for the state estimation of a sensorless symmetrical six phase induction machine subjected to unbalanced operation. In addition to it, the efficacy and performance of the observer is tested by incorporating mechanical and electrical disturbances during the normal operation of the machine. Variation of the inertia up to the maximum value and the loss of one or more stator phases are considered for all the test cases of unbalanced conditions. There are also a certain class of load torque rejection observers that have been implemented. These disturbance observers, either comprised of a mechanical model of the motor or having error components that are dependent on the rotor speed, or gain coefficients dependent on the stator frequency [21–24]. In [24], the decoupling of current and subsequent control of the current components is applied to induction motor by employing a sliding mode controller and a disturbance observer. The coupled terms are modeled as disturbance, which, after observation, are utilized in the control law. In addition, the rotor speed is estimated based on the magnetizing current and the lyapunov stability criterion is used to ensure closed loop stability. Several of the above categories of observers have been implemented in many experimental platforms and also been verified by means of HIL testing.

The purpose of this paper is to demonstrate the improvement in the rejection of the external load by the proposed observer. The sliding hyperplane is altered and the disturbance estimated from the mechanical model is integrated into the sliding hyperplane along with the real and estimated stator currents [17]. In addition, to add to the nonlinearity of the observer, a Gaussian noise (measurement disturbance) is incorporated at the motor terminals, where the terminal voltages and currents are measured. The observer, along with the drive system, is first built using Matlab/Simulink blocksets and then validated in a comparatively new real-time simulation platform, RT-Lab, developed by Opal-RT (2011, Mathworks, Natick, MA, USA) and the real-time results add more credibility as compared to any other offline simulation platform.

2. General Configuration of Model Reference Adaptive Systems and System Modeling

Adaptive control is mainly used for parameter adaptation. The essence of an adaptive control mechanism is to adapt to the controlled system with parameters that need to be estimated. The concept of parameter adaptive MRAS along with the parallel disturbance torque estimation mechanism is illustrated in Figure 3. There is a reference motor model and the adaptive model as a function of the parameter to be estimated. The adaptive mechanism is used to ensure that the state of the observer (process) converges to the state of the motor (plant). Therefore, we have an optimization criterion X and the error to be constrained:

$$X = \int_0^T e^2 \mathrm{d}t, \tag{1}$$

$$e = X_{ref} - X_{adap}, \tag{2}$$

where X_{ref} and X_{adap} are outputs of respective models. The adaptation mechanism makes use of the classical Proportional-Integral theory to process the speed tuning signal. A Lyapunov function candidate is used for the speed derivation mechanism.

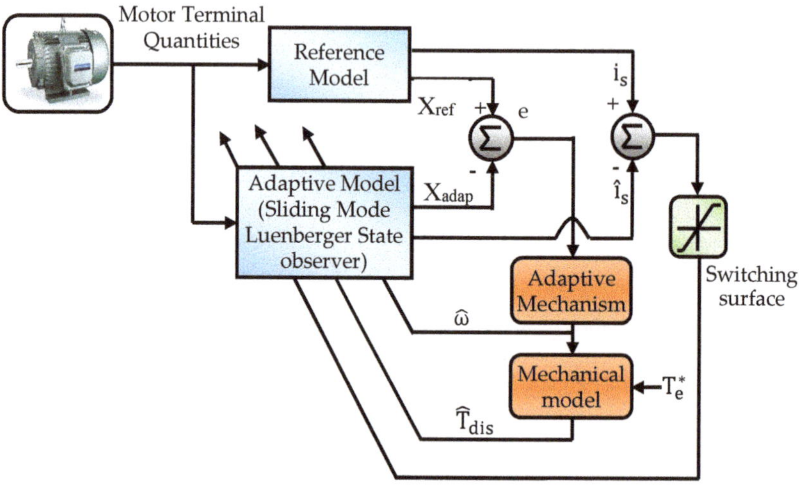

Figure 3. Basic configuration of parameter adaptive model reference adaptive system (MRAS) scheme with parallel disturbance torque estimation.

2.1. Structure of Sliding Mode Luenberger State Observer

The reduced order sliding mode luenberger observer (SMLO) with the modified switching surface is shown in Figure 4, where "A" is the parameter matrix, "$\hat{}$" is used for estimated parameters, "X" is the state variables comprised of the d and q-axes stator currents and rotor fluxes, "ksw" is the reduced order observer switching gain matrix, chosen in such a way that the eigenvalues of the observer and the machine are maintained proportional to ensure stability under normal operating conditions. "J" is the moment of inertia, "p" is the differential operator, "BV" is the viscous friction coefficient, "T_e^*" and "$\hat{T}_{dis}$" is the reference model electromagnetic torque and the estimated disturbance torque, "k" is an arbitrary positive gain [17]. The purpose of a sliding mode or variable structure strategy is to modify the dynamics of a non linear system state by means of a high frequency switching surface or a sliding hyperplane. The sliding hyperplane is selected in such a way that the Lyapunov function candidate "V" utilized for obtaining the convergence mechanism, and its derivative satisfies the Lyapunov stability criterion [17,25]. "V" is a scalar function of the sliding hyperplane "S". Therefore,

$$\dot{V}(S) = S(x)\dot{S}(x). \tag{3}$$

The control law is:

$$u(t) = u_{eq}(t) + u_{sw}(t), \tag{4}$$

where $u(t)$, $u_{eq}(t)$ and $u_{sw}(t)$ represent the control, equivalent control and the switching vector. For stability, the switching vector is obtained [17,26]:

$$u_{sw}(t) = \eta\,\text{sign}(S(x,t)), \tag{5}$$

where $\text{sign}(S) = \begin{cases} -1 \text{ for } S < 0 \\ 0 \text{ for } S = 0 \\ +1 \text{ for } S > 0 \end{cases}$. η is the switching control gain chosen such that (3) is negative definite, implying $S(x)\dot{S}(x) < 0$, thereby constraining the effect of the external disturbance. However, the high frequency switching plane increases non linearity of the observer, leading to chattering. Therefore, to eliminate the effect of this unwanted phenomenon, a saturation function having boundary layer of width (Φ) is used by replacing sign (S) with sat (S/Φ) and is given by [17]:

$$\text{sat}(S/\Phi) = \begin{cases} \text{sign}\left(\frac{S}{\Phi}\right) \text{ if } \left|\left(\frac{S}{\Phi}\right)\right| \geq 1 \\ \left(\frac{S}{\Phi}\right) \text{ if } \left|\left(\frac{S}{\Phi}\right)\right| < 1 \end{cases} . \tag{6}$$

Based on the theory of MRAS, the following equations depict the structure of the proposed observer scheme with the conventional and modified sliding hyperplane. The reference and the adaptive model are represented in state space form as they aid in the formulation of control and estimation problems [17,27].

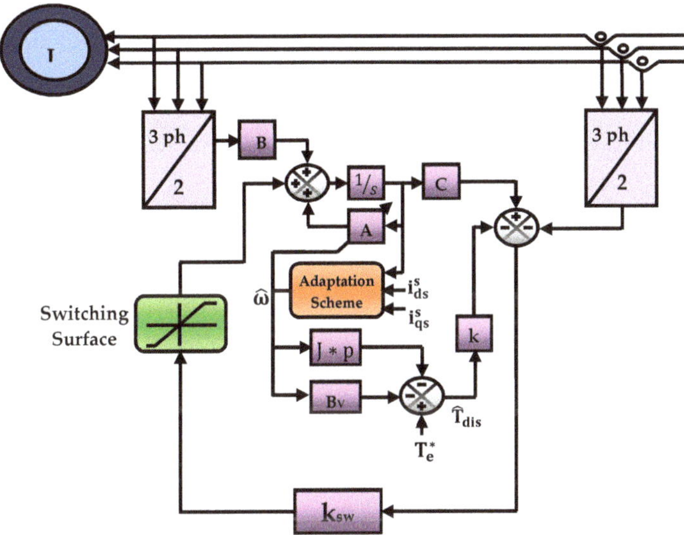

Figure 4. Proposed state observer with the modified switching hyperplane.

2.1.1. Reference Model (Motor)

$$\frac{dx}{dt} = [A]x + [B]u, \tag{7}$$

$$y = [C]x, \tag{8}$$

where:

$$x = \left[i_{ds}^s, i_{qs}^s, \psi_{dr}^s, \psi_{qr}^s\right]^T, A = \begin{bmatrix} A_{11} & A_{12} \\ A_{21} & A_{22} \end{bmatrix},$$

$$B = \left[\frac{1}{\sigma L_s}I\, 0\right]^T, C = [I, 0], u = \left[v_{ds}^s v_{qs}^s\right]^T,$$

$$I = \begin{bmatrix} 1 & 0 \\ 0 & 1 \end{bmatrix}, J = \begin{bmatrix} 0 & -1 \\ 1 & 0 \end{bmatrix},$$

$$A_{11} = -\left[\frac{R_s}{\sigma L_s} + \frac{1-\sigma}{\sigma T_r}\right]I = a_{r11}I, \quad A_{12} = \frac{L_m}{\sigma L_s L_r}\left[\frac{1}{T_r}I - \omega_r J\right] = a_{r12}I + a_{i12}J,$$

$$A_{21} = \frac{L_m}{T_r}I = a_{r21}I,$$

$$A_{22} = \frac{-1}{T_r}I + \omega_r J = a_{r22}I + a_{i22}J.$$

2.1.2. Estimation of Disturbance Torque from the Mechanical Model

By exploiting the machine model, the disturbance torque is estimated by utilizing the reference model electromagnetic torque and the estimated speed, respectively:

$$\hat{T}_{dis} = T_e^* - J\frac{d\,\hat{\omega}}{dt} - B_V \hat{\omega}. \tag{9}$$

2.1.3. SMLO 1—Observer with Conventional Disturbance Rejection Mechanism (Adaptive Model)

$$\frac{d\hat{x}}{dt} = [\hat{A}]\hat{x} + [B]u + k_{sw}\mathrm{sat}(\hat{i}_s - i_s) + \hat{d}, \tag{10}$$

where the sliding hyperplane, $s = \hat{i}_s - i_s$ and $\hat{d} = k\hat{T}_{dis}$ and

$$\hat{y} = [C]\hat{x}, \tag{11}$$

where $\hat{i}_s, i_s =$ estimated and measured value of stator current:

$$\hat{A} = \begin{bmatrix} \hat{A}_{11} & \hat{A}_{12} \\ \hat{A}_{21} & \hat{A}_{22} \end{bmatrix},$$

$$\hat{A}_{12} = \frac{L_m}{\sigma L_s L_r}\left[\frac{1}{T_r}I - \hat{\omega}_r J\right] = a_{r12}I + \hat{a}_{i12}J,$$

$$\hat{A}_{22} = \frac{-1}{T_r}I + \hat{\omega}_r J = a_{r22}I + \hat{a}_{i22}J.$$

The switching gain "ksw" is designed by the following lower order matrix given by

$$k_{sw} = \begin{bmatrix} k_1 & k_2 \\ -k_2 & k_1 \end{bmatrix}^T. \tag{12}$$

The switching gain matrix is designed appropriately to make Label (4) stable by means of pole placement. The eigenvalues are designed in such a way that, for the observer, they are comparatively more negative to that of the motor so that they ensure faster convergence of the desired performance to the process. Therefore,

$$k_1 = (m - 1)a_{r11}, \tag{13}$$

$$k_2 = k_p, k_p \geq -1, \tag{14}$$

where "m" and "k_2" are chosen in such a way that the eigenvalues of the observer are shifted more negative as compared to the eigenvalues of the motor. They also directly affect the dynamics and damping of the observer. "k_1" is dependent on the motor parameters.

2.1.4. SMLO 2—Observer with Modified Disturbance Rejection Mechanism (Adaptive Model)

The state dynamic equation is altered by changing the sliding hyperplane, i.e., by including the estimated disturbance torque [17]:

$$\frac{d\hat{x}}{dt} = [\hat{A}]\hat{x} + [B]u + k_{sw}\mathrm{sat}(\hat{i}_s - i_s - \hat{d}). \tag{15}$$

Therefore, the sliding hyperplane becomes, $s = \hat{i}_s - i_s - \hat{d}$ and $\hat{d} = k\hat{T}_{dis}$ and

$$\hat{y} = [C]\hat{x}. \tag{16}$$

2.1.5. Adaptive Mechanism

The Lyapunov function candidate used for speed derivation mechanism and to ensure stability is given by:

$$V = e^T e + \frac{(\hat{\omega}_r - \omega_r)^2}{\lambda}, \tag{17}$$

where λ is a positive constant.

We have:

$$\frac{dv}{dt} = e^T \left[(A + GC)^T + (A + GC) \right] e - \frac{2\Delta\omega_r \left(e_{ids} \hat{\varphi}_{qr}^s - e_{iqs} \hat{\varphi}_{dr}^s \right)}{c} + \frac{2\Delta\omega_r}{\lambda} \frac{d\hat{\omega}_r}{dt}, \tag{18}$$

where $e_{ids} = i_{ds}^s - \hat{i}_{ds}^s$, $e_{iqs} = i_{qs}^s - \hat{i}_{qs}^s$.

The second and third term of (18) is equalized to realize the expression for the estimated speed given by:

$$\frac{d\hat{\omega}_r}{dt} = \frac{\lambda}{c} \left(e_{ids} \hat{\varphi}_{qr}^s - e_{iqs} \hat{\varphi}_{dr}^s \right), \tag{19}$$

"c" being an arbitrary positive constant. The difference between SMLO 2 and SMLO 1 is the way in which the estimated disturbance is added and constrained in the sliding hyperplane along with the stator current error. The complexity of the observer increases due to the presence of the speed adaptation loop, the disturbance estimation and adaptation loop and the Luenberger observer gain loop, however, by tuning the feedback and switching gains, the dynamic performance and the stability of both the observers can be improved.

2.2. Stability Analysis of Both the Observers by Means of Pole Placement

For the conventional disturbance observer SMLO 1:

$$\left(A_{11} + k_{sw} + \hat{d} \right) = \begin{bmatrix} a_{r11} + k_1 + \hat{d} & -k_2 + \hat{d} \\ k_2 + \hat{d} & a_{r11} + k_1 + \hat{d} \end{bmatrix}. \tag{20}$$

The characteristic equation is:

$$SI - \left(A_{11} + k_{sw} + \hat{d} \right) = 0. \tag{21}$$

On solving:

$$S^2 - 2S \left(a_{r11} + k_1 + \hat{d} \right) + \left(a_{r11} + k_1 + \hat{d} \right)^2 + \left(k_2^2 - \hat{d}^2 \right) = 0. \tag{22}$$

The observer poles are:

$$S_1 = \left(a_{r11} + k_1 + \hat{d} \right) + j\left(k_2 - \hat{d} \right), \tag{23}$$

$$S_2 = \left(a_{r11} + k_1 + \hat{d} \right) - j\left(k_2 - \hat{d} \right). \tag{24}$$

For the modified disturbance observer SMLO 2:

$$\left(A_{11} + k_{sw} - k_{sw}\hat{d} \right) = \begin{bmatrix} a_{r11} + k_1 - k_{sw}\hat{d} & -k_2 - k_{sw}\hat{d} \\ k_2 - k_{sw}\hat{d} & a_{r11} + k_1 - k_{sw}\hat{d} \end{bmatrix}. \tag{25}$$

The characteristic equation is:

$$SI - \left(A_{11} + k_{sw} - k_{sw}\hat{d}\right) = 0. \tag{26}$$

On solving:

$$S^2 - 2S\left(a_{r11} + k_1 - k_{sw}\hat{d}\right) + \left(a_{r11} + k_1 - k_{sw}\hat{d}\right)^2 + \left(k_2^2 - k_{sw}^2\hat{d}^2\right) = 0. \tag{27}$$

The observer poles are:

$$S_1 = \left(a_{r11} + k_1 - k_{sw}\hat{d}\right) + j\left(k_2 - k_{sw}\hat{d}\right), \tag{28}$$

$$S_1 = \left(a_{r11} + k_1 - k_{sw}\hat{d}\right) - j\left(k_2 - k_{sw}\hat{d}\right). \tag{29}$$

2.3. Structure of Current Regulated Vector Controller

Current regulation or tolerance band current control has a fast torque response and is independent of load parameters. The speed error is processed by a PI controller whose output is the reference torque:

$$e_c = \hat{\omega}_r - \omega^*, \tag{30}$$

$$T_e^* = e_c\left[k_p + (k_i/s)*T_s\right], \tag{31}$$

where e_c is the speed error, kp and ki are the proportional and integral gains for tuning the speed error, and Ts is the sampling time. For operation in the motoring and flux weakening region, the rotor flux is constant for the former and as a function of the speed for the latter:

$$\psi_r = 0.96, \text{ If } \hat{\omega}_r < \omega_{bsync}, \tag{32}$$

$$\psi_r = 0.96 * \left(\frac{\hat{\omega}_r}{\omega_{bsync}}\right), \text{ If } \hat{\omega}_r > \omega_{bsync}. \tag{33}$$

The orthogonal direct and quadrature axes stator current components are [1]:

$$i_{ds}^* = \left(\frac{\psi_r}{L_m}\right)\left[1 + \frac{dT_r}{dT_s}\right], \tag{34}$$

$$i_{qs}^* = \left(\frac{2}{3}\right)\left(\frac{2}{P}\right)\left(\frac{L_r}{L_m}\right)\left(\frac{T_{ref}}{\psi_r}\right). \tag{35}$$

As the slip speed is used for imposing the field orientation, the field angle is determined from the slip speed, therefore:

$$\theta_f = \theta_{sl} + \theta_r. \tag{36}$$

The three-phase reference currents are obtained from the decoupled reference components of current by means of inverse transformation given as follows:

$$i_{as}^* = i_{ds}\sin\theta + i_{qs}\cos\theta, \tag{37}$$

$$i_{bs}^* = \left(\frac{1}{2}\right)\{-i_{ds}\cos\theta + \sqrt{3}\,i_{ds}\sin\theta\} + \left(\frac{1}{2}\right)\{i_{qs}\sin\theta + \sqrt{3}\,i_{qs}\cos\theta\}, \tag{38}$$

$$i_{cs}^* = -(i_{as}^* + i_{bs}^*). \tag{39}$$

The three-phase reference current components are compared with the actual sensed three-phase currents by means of hysteresis regulation and the gating pulses for the voltage source inverter (VSI) are generated. The hysteresis band is selected keeping the current and subsequent torque pulsation in mind. The entire sensorless drive scheme is illustrated in Figure 5 [17].

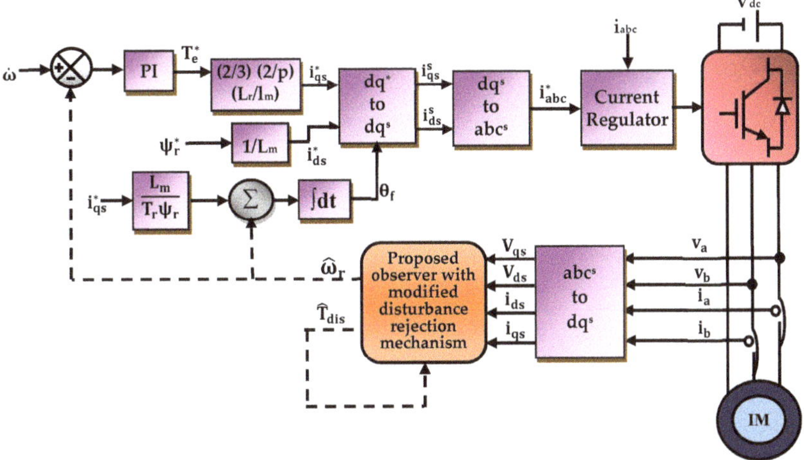

Figure 5. Voltage source inverter (VSI) fed speed sensorless induction motor drive system.

3. The Concept of Real-Time Simulation

Real-time simulation and test platform is one in which the computer model's performance corresponds to the performance of the actual physical system [28–36]. It would take the same amount of time like any real world application. Unlike many offline simulation platforms, where a variable step solver is used, in real time, a fixed step discrete solver is used. For a given computer model, in a real-time simulation, the processing of inputs, model calculations and the processing of outputs should be less than the fixed step. If it exceeds the fixed step, a phenomenon known as over run would occur. Therefore, it is imperative that the real-time simulator should produce the model calculations and output within the same time interval corresponding to its actual physical counterpart. The applications ranges from mechatronics, power electronic and power system based concepts to gaming and process control.

There are various real-time simulators available such as xPC Target (Mathworks, Natwick, United States), for power electronic system simulation there is eFPGAsim and eDRIVEsim (Opal-RT, Montreal, Canada) and for power system simulation, we have HYPERSIM (Opal-RT, Montreal, Canada) and RTDS (RTDS Technologies Inc., Winnipeg, MB, Canada).

RT-Lab is a distributed real-time platform with features ranging from virtual, control and plant prototyping, model based design, etc. It is flexible and has a fast execution time and can also be utilized for real-time processor-in-loop (PIL) and hardware in the loop (HIL) applications. The package is compatible with various offline platforms such as Matlab/Simulink, Labview, etc. The mathematical and dynamic model of the drive system is built in Simulink environment using sim-power systems toolbox. This acts as the front end interface after which the model is integrated with RT-Lab blocksets. RT-Lab generates the code to be simulated in a single or multiple targets. Here, the real-time simulation target used is OP4500 developed by Opal-RT. It is a multi core target, where the plant and controller can be placed in different cores.

It comprises of analog and digital input/output (I/O) channels with signal conditioning and is also integrated with powerful XILINX Kintex 7 FPGA, which has a very high processing power. The sensorless drive system is modeled and built offline using sim power systems toolbox in Simulink in the workstation. The offline simulink model uses a variable step solver. The RT-Lab integrated real-time platform uses a discretised fixed step time solver with a step size of 50 µs. The workstation is connected to the OP4500 real-time simulator through transmission control protocol/internet protocol (TCP/IP) protocol. The target executes the model and the results are viewed and recorded in the workstation, which is the front end interface. The model is executed and analyzed dynamically for different test cases as presented below.

4. Real-Time Simulation Results: Analysis and Discussion

The motor parameters and ratings used for the real-time simulation are given in Appendix A. In order to emphasize on the improvement in the performance of SMLO 2 over SMLO 1, some results are magnified to present a clearer picture. Parameter estimation of an observer at flux weakening regions is significant, as it indicates its robustness for a wider speed bandwidth as also its tracking performance at a constant power region.

4.1. Performance at Flux Weakening

Here, the observers are tested at a low flux weakening region for a given speed command of 165 rad/s. However, both reach the reference speed at almost identical time, and the estimated speed oscillations are comparatively very high for SMLO 1 as shown in Figure 6a. In the zoomed version shown in Figure 6b, the oscillation is as high as 70 rad/s for the speed command of 165 rad/s, which is almost 42%.

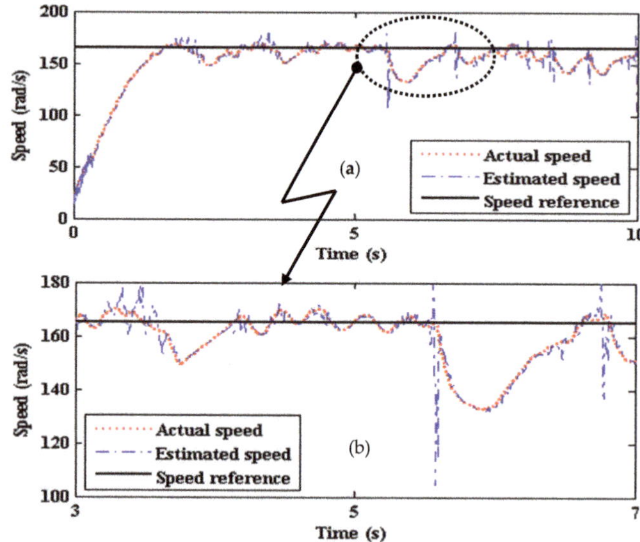

Figure 6. (a) Estimated speed of conventional sliding mode luenberger observer (SMLO) 1; and (b) zoomed version of (a).

The profiles of the estimated disturbance torque in Figure 7 and the electromagnetic torque in Figure 8 of SMLO1 are almost the same, since the viscous friction coefficient and the inertia constant are of a low value. The estimated rotor flux of SMLO 1 in Figure 9 has more oscillations and the magnitude is increasing with time. As compared to SMLO 1, speed performance of SMLO2 has relatively less oscillations shown in Figure 10a. In Figure 10b, although there is an initial overshoot and undershoot, the oscillation is initially around 20 rad/s, which is around 12%, and gradually dies out after 6 s.

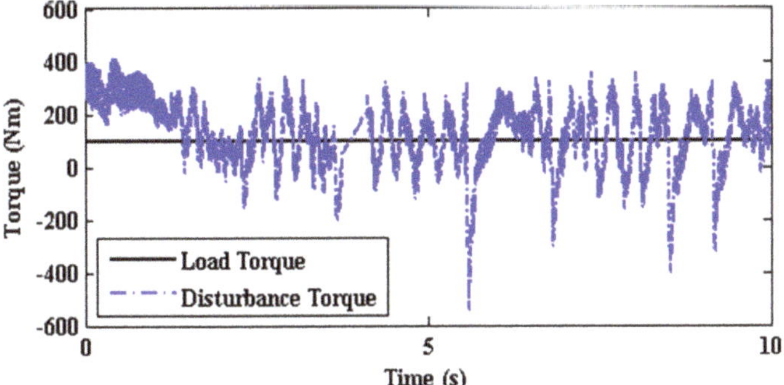

Figure 7. Estimated disturbance torque of SMLO 1.

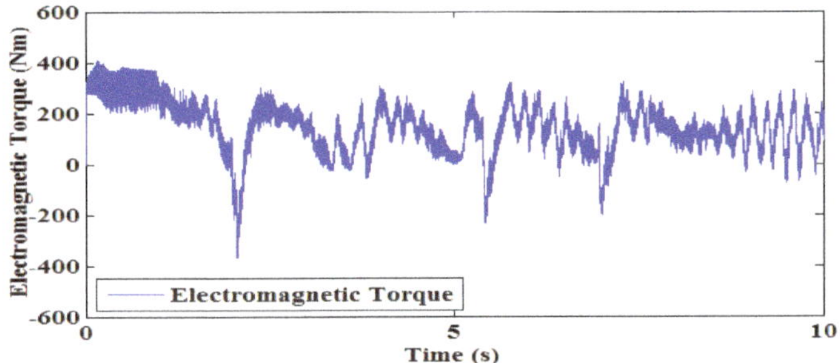

Figure 8. Electromagnetic torque of SMLO 1.

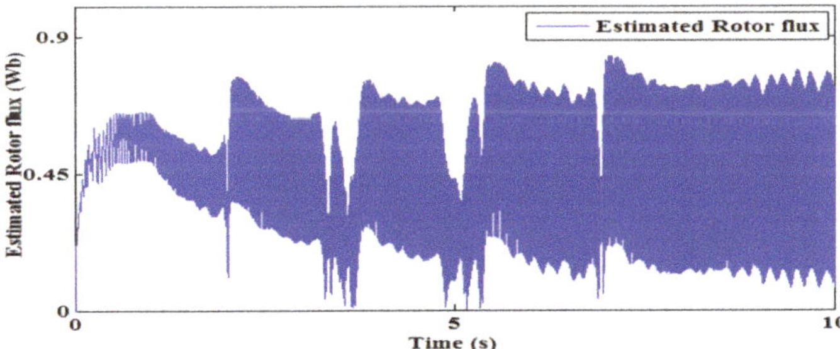

Figure 9. Estimated rotor flux of SMLO 1.

Estimated disturbance torque of SMLO2 shown in Figure 11 has relatively lesser ripples. Again, the electromagnetic torque profile of SMLO2 follows the estimated disturbance torque profile as shown in Figure 12. SMLO2 estimated rotor flux shown in Figure 13 has lesser oscillations. This shows that the SMLO 2 has better reception to flux weakening and operates well in this region.

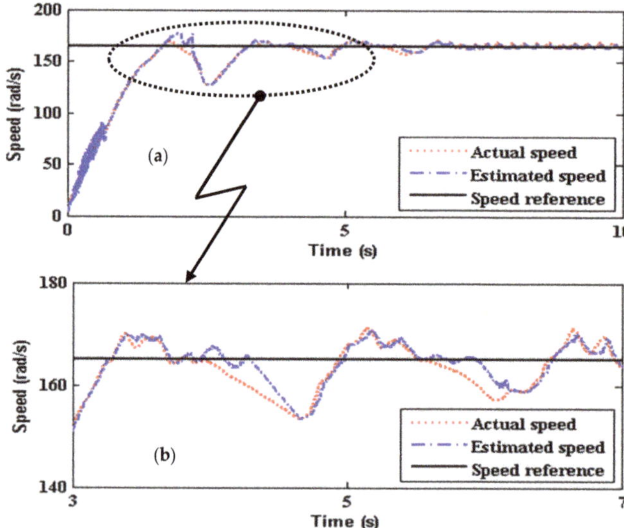

Figure 10. (**a**) Estimated speed of Modified Sliding Mode Luenberger Observer (SMLO 2); and (**b**) zoomed version of (**a**).

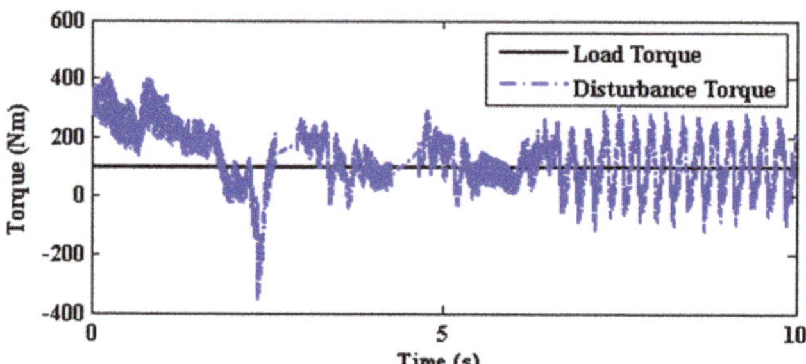

Figure 11. Estimated disturbance torque of SMLO 2.

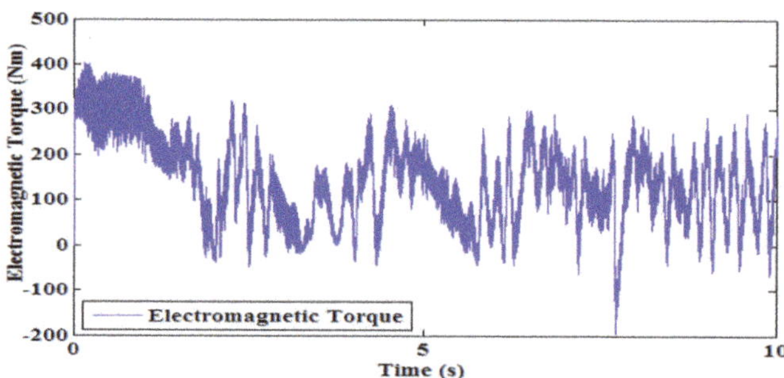

Figure 12. Electromagnetic torque of SMLO 2.

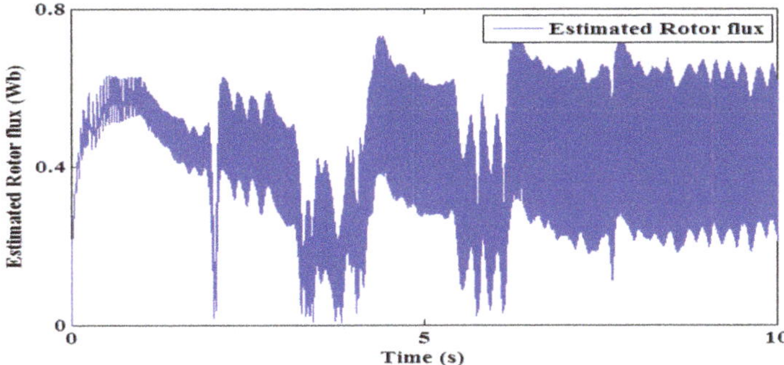

Figure 13. Estimated rotor flux of SMLO 2.

4.2. Performance at Step Speed Command

The difference is seen more distinctly at step speed command (initially 50 rad/s, after 2 s, stepped upto 150 rad/s, after 7 s, stepped down to 100 rad/s), which covers a wide speed bandwidth. SMLO 1 initially, tracks well for both 50 rad/s and 150 rad/s, but, during deceleration from 150 rad/s to 100 rad/s, there is a very high spike in the estimated speed (of almost 100 rad/s) shown in Figure 14a, which could prove detrimental to the drive system. This is mainly due to the sudden change in the flux level of the motor and SMLO 1, which directly affects its speed convergence. Even after settling down at 100 rad/s, after about 10 s, oscillations persist ranging between 98.5 and 101 rad/s, shown in Figure 14b. SMLO 2 in comparison tracks more smoothly and accurately, even during deceleration as shown in Figure 15a. The oscillations in the estimated speed are greatly reduced as shown in Figure 15b, ranging almost between 99.9 and 100.1 rad/s, which proves that it is near accurate and superior tracking as compared to SMLO 1.

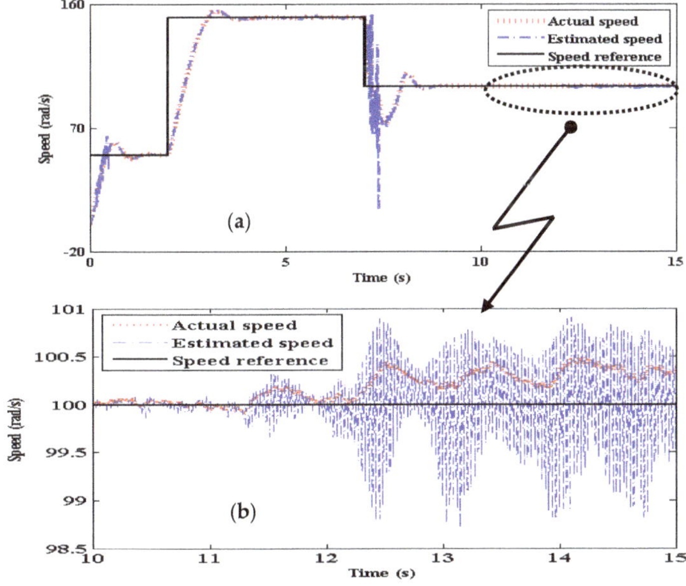

Figure 14. (**a**) Estimated speed of SMLO 1; and (**b**) zoomed version of (**a**).

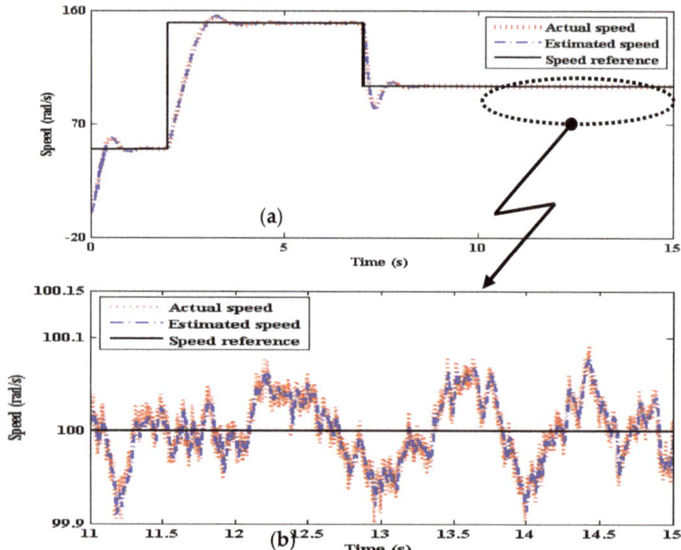

Figure 15. (**a**) Estimated speed of SMLO 2 (**b**); and zoomed version of (**a**).

Even the estimated disturbance torque profile of SMLO 1 shown in Figure 16a is comparatively less smoother with more pulsations as compared to SMLO 2 shown in Figure 16b. The estimated rotor flux performance for SMLO 1 in Figure 17a and SMLO 2 in Figure 17b can be observed. Here again, the latter gives smoother flux performance in spite of variations in the commanded speed at different time intervals, resulting in better torque holding capability.

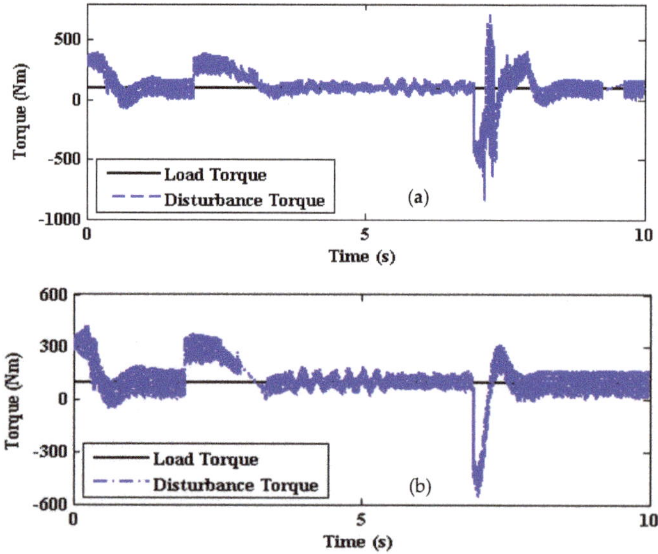

Figure 16. Estimated disturbance torque of (**a**) SMLO 1; and (**b**) SMLO 2.

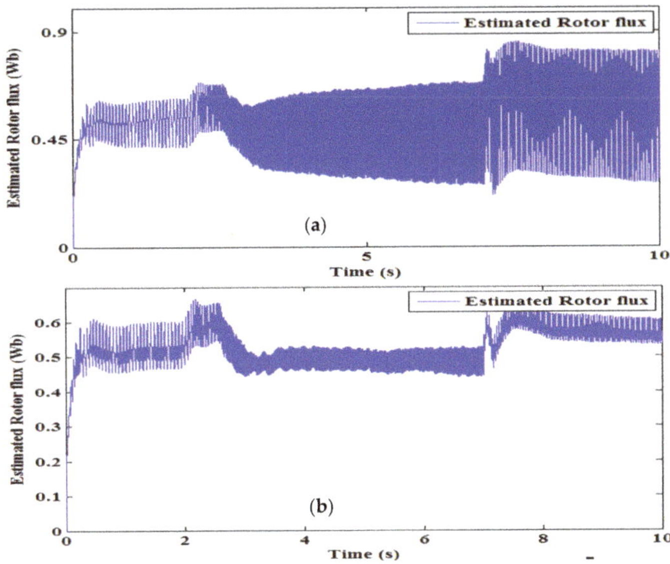

Figure 17. Estimated rotor flux of (a) SMLO 1; and (b) SMLO 2.

4.3. Performance at Low Speeds

At low speeds (for a speed command of 30 rad/s), SMLO 1 does not track becomes unstable and goes out of bounds. Therefore, the speed bandwidth of SMLO 1 is restricted to a range of 50–150 rad/s. However, the tracking performance of SMLO 2 is shown in Figure 18a, where, after an initial high overshoot for a very small amount of time, it tracks the command accurately, which is more distinct in the zoomed version in Figure 18b.

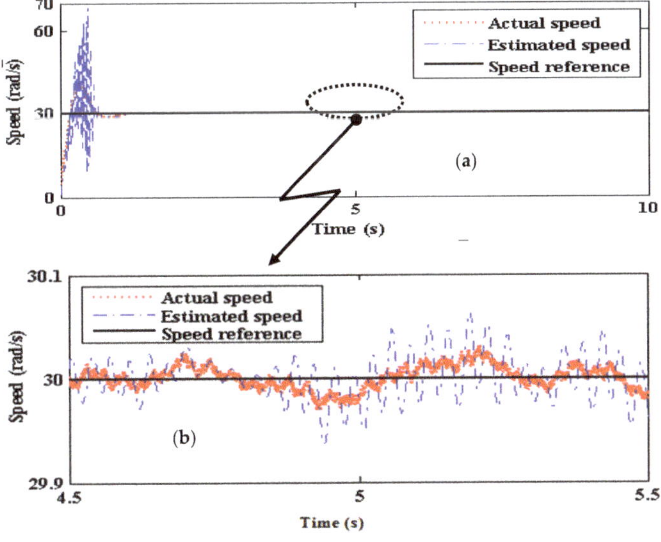

Figure 18. (a) estimated speed of SMLO 2 (b), zoomed version of (a).

The initial high disturbance torque pulsation shown in Figure 19 is responsible for the high overshoot in estimated speed.

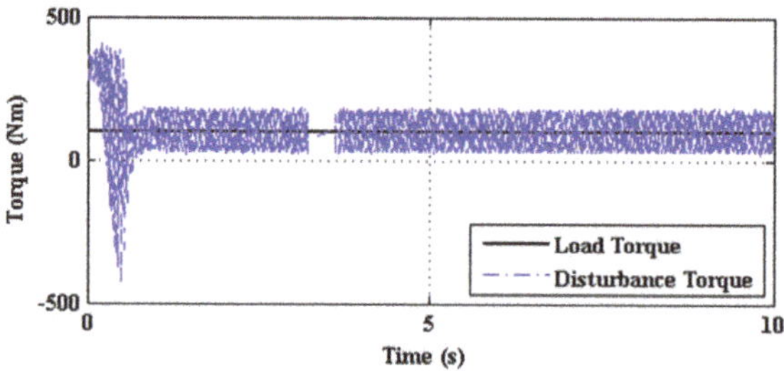

Figure 19. Estimated disturbance torque of SMLO 2.

4.4. Effect of Parameter Detuning on the Dynamic Performance

The detuning or incorrect setting of parameters plays a significant role in the ability of the adaptive mechanism to converge the states of the observer and motor. The stator resistance and rotor time constant play a critical role in the motor dynamics. Therefore, the speed convergence mechanisms of the observer are tested for incorrect settings of the said parameters. The speed tuning signal for the adaptive mechanism is derived from the difference between the products of the estimated d-axis flux linkages and the q-axis stator current error and q-axis flux linkages and the d-axis stator current error. For 50% incorrect setting in both, the convergence of SMLO 1 is shown in Figure 20a. The speed error gradually increases at the end, thereby affecting its ability to withstand parametric uncertainties, whereas, for SMLO 2, the speed error is approximately zero and consistent with time, as shown in Figure 20b.

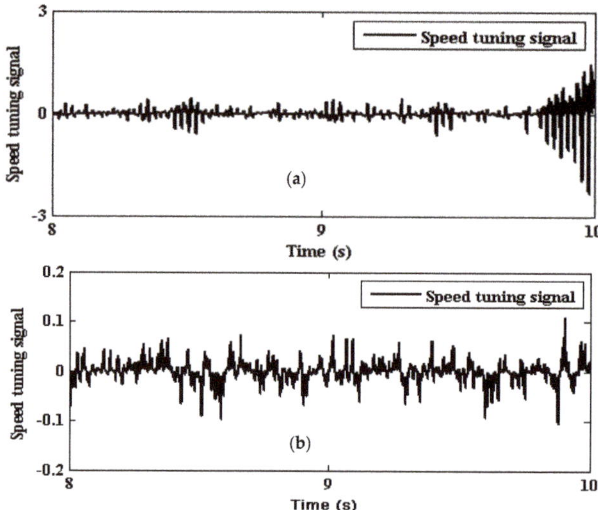

Figure 20. Speed tuning signal for 50% incorrect setting of stator resistance (R_s) and rotor time constant (T_r) for (**a**) SMLO 1, and (**b**) SMLO 2.

Parametric uncertainties can also be treated as model disturbances and this only reflects the ability of SMLO 2 to reject the effect of the disturbances. Again, for nominal setting of Rs and Tr, although the inconsistencies in the speed tuning signal of SMLO 1 has reduced, it is still pertinent, with speed error reaching almost 3 rad/s, as shown in Figure 21a. The speed tuning signal or speed error of SMLO 2 is almost confined to zero, as shown in Figure 21b, indicating faster convergence of the motor and the observer states.

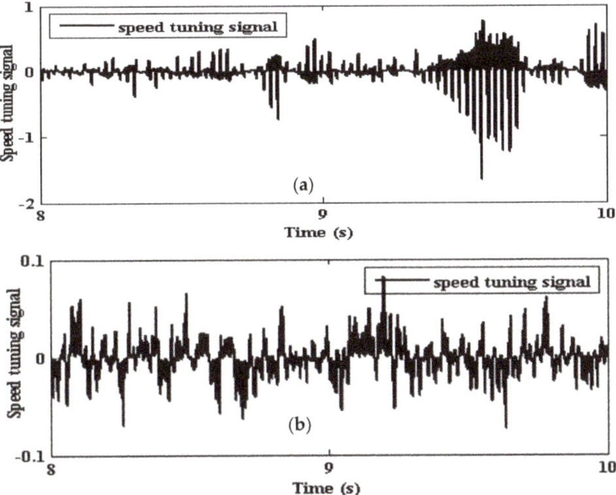

Figure 21. Speed tuning signal for nominal setting of stator resistance (R_s) and rotor time constant (T_r) for (**a**) SMLO 1; and (**b**) SMLO 2.

Now, for 150% incorrect setting in both the parameters, SMLO 1 exhibits a comparatively better convergence than previous cases, as shown in Figure 22a, but, here again, it is observed that SMLO 2 offers a greater and near accurate convergence of states, as shown in Figure 22b.

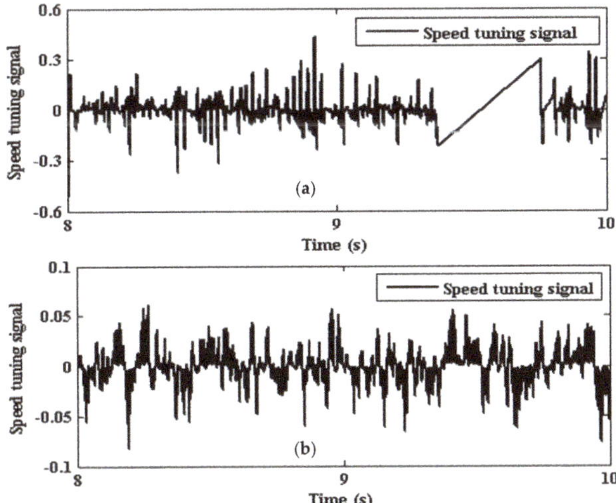

Figure 22. Speed tuning signal for 150% incorrect setting of stator resistance (R_s) and rotor time constant (T_r) for (**a**) SMLO 1; and (**b**) SMLO 2.

4.5. Switching Surface and Convergence of the Stator Current Error

The convergence of the stator current error and the sliding surface is also observed for both the observers. In both the cases, the profile of the sliding surface is identical to that of the stator current error, as both are dependent on each other. The profile of the sliding surface, direct and quadrature axes stator currents, stator current error of SMLO1 are shown in Figure 23a–d. However, as compared to SMLO 1, both the stator current error and the sliding surface of SMLO 2 are slightly displaced in the negative range. This can be primarily due to change in the configuration of the sliding surface, as, along with the stator current error, it also has to constrain the effect of the estimated disturbance torque. The direct and quadrature axes are also obtained. The performance of SMLO2 is shown in Figure 24a–d.

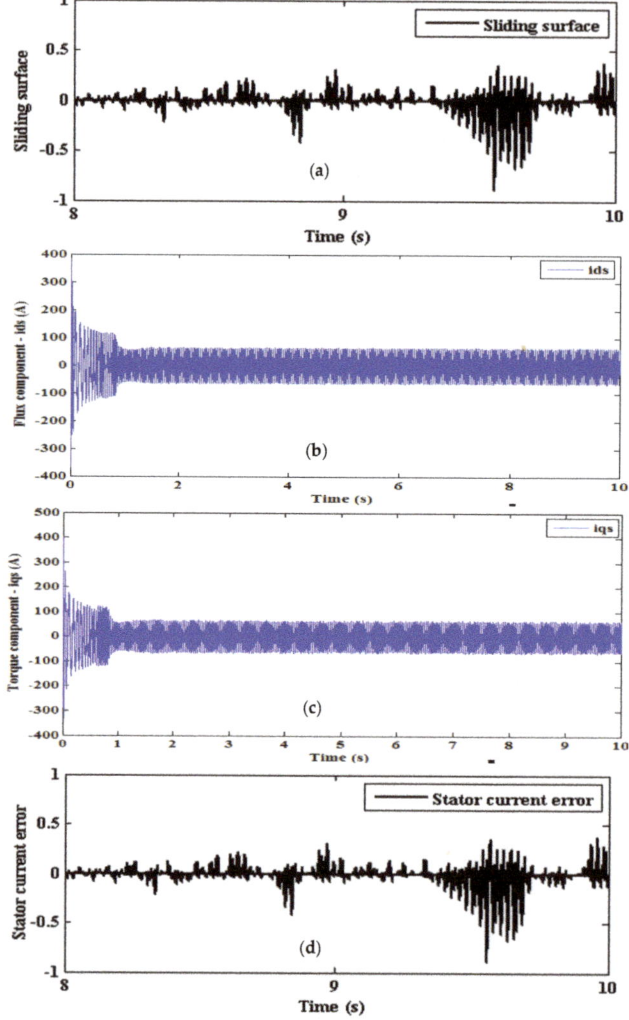

Figure 23. SMLO 1 (**a**) sliding surface; (**b**) flux component of stator current; (**c**) torque component of stator current; and (**d**) stator current error.

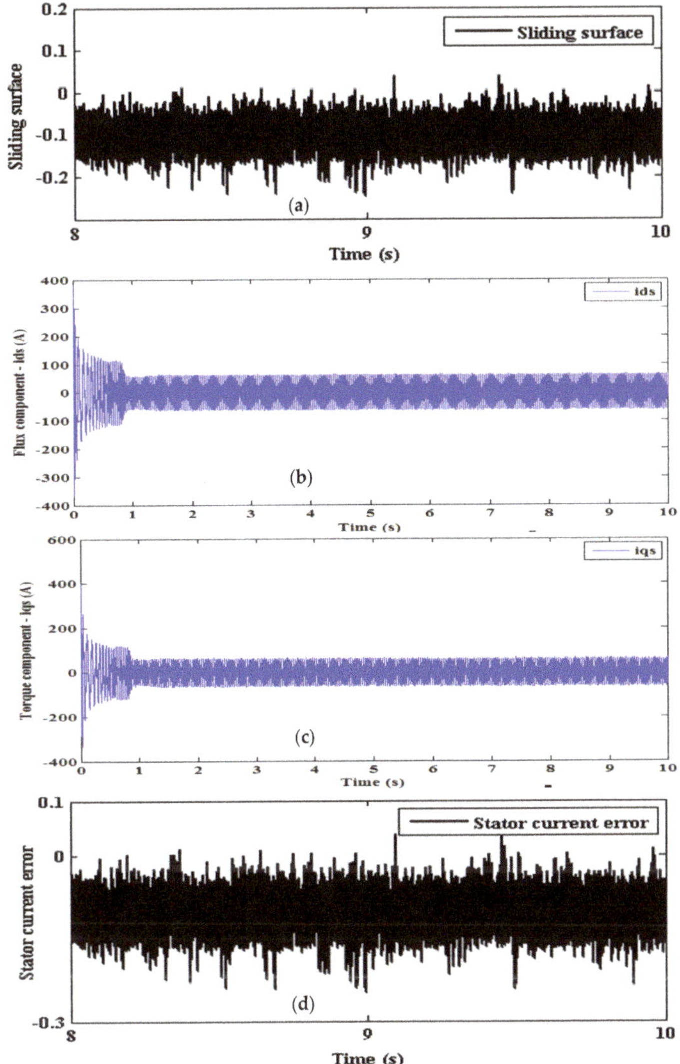

Figure 24. SMLO 2 (**a**) sliding surface; (**b**) flux component of stator current; (**c**) torque component of stator current; and (**d**) stator current error.

Although both are similar, the presence of pulsations in the stator current components, both during steady state and transient conditions, give rise to subsequent torque and flux pulsations in the observers. In addition, due to high power rating of the motor, the stator current dynamics play a major role in the torque performance. However, it is seen that the SMLO 2 comparatively displays better dynamic and steady state performance due to its ability to contain the stator current dynamics and ensure that it does not affect the tracking. For the entire study, the load torque was maintained constant at 100 Nm. For the last two test cases, along with the constant load torque, the speed command was also maintained constant at 100 rad/s, respectively.

4.6. Pole Placement Plot of the Modified nd Conventional Disturbance Observers

The pole plot shown in Figure 25 shows that the poles of SMLO 2 are shifted to left of SMLO 1, which indicates comparatively better stability performance of the same in low and flux weakening regions. However, the performance at medium speed regions remains more or less the same.

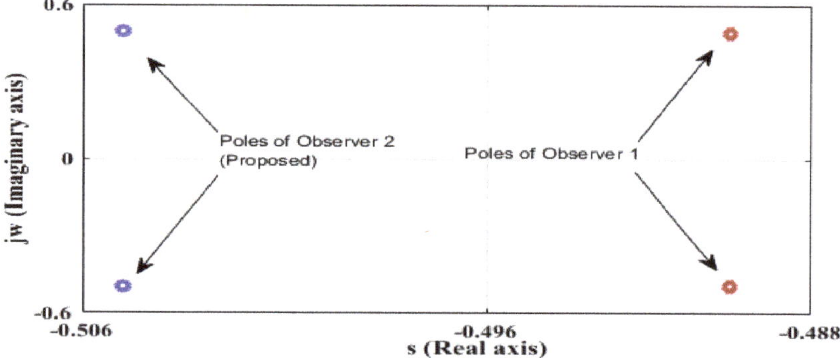

Figure 25. Pole placement of SMLO 1 and SMLO 2.

Since hysteresis band current regulation is used, the exact switching frequency cannot be predicted; however, the maximum switching frequency which is realizable by the OP45OO target is (1/(2*time step)), which is 10 kHz. As the model has been successfully executed by the target, it can be safely concluded that the equivalent switching frequency is within 10 kHz. The entire testing and analysis were performed in the motoring mode at low, medium speeds and low flux weakening regions. Although these real-time simulation results can also be considered to be equivalent to the experimental results while testing in model based design paradigm (provided all the model dynamics including uncertainties, plant disturbance etc. have been mathematically modeled), there are some problems which are encountered. While moving from simulations to real-time implementation of the model, in automotive and power applications, (EMI/EMC issues, over-voltages, overcurrent, temperature issues) arise which have to be managed with proper safe circuitry and diagnostic circuitry [30–32]. Table 1 summarizes the observations from the above analysis for both the observers.

Table 1. Salient features of both the observers.

Test Cases	SMLO 1	SMLO 2
Low flux weakening region	Maximum speed oscillation of around 70 rad/s (around 42% of the reference value). Speed oscillations do not die out.	Initial maximum speed oscillation of around 20 rad/s (around 12% of the reference value). Speed Oscillations gradually reduce with time.
Step speed command	Very high overshoot and undershoot observed at the instance of fast deceleration.	Smoother tracking during fast acceleration and deceleration.
Low speed operation	Does not track, becomes unstable and speed convergence goes out of bounds.	Tracks well, initial undershoot and overshoot, which results for a very small interval of time.
Disturbance torque	Higher torque pulsations as a result of high stator current pulsation.	Comparatively lower torque pulsation resulting in better torque holding capability.
Speed and Stator error convergence	Slower convergence, higher speed and stator current error	Faster convergence, resulting in smoother tracking

5. Conclusions

This paper presented an improved version of a sliding mode Luenberger observer with comparatively better tracking performance, robustness to the effect of external and model disturbances and a wider speed bandwidth than the conventional one. The drive system along with the proposed observer is executed in real-time using an RT-Lab package and an OP4500 real-time simulator. The real-time results validate the improvement in the disturbance rejection capability for the different test cases presented and also provide more credibility as compared to other offline simulated results. Some significant findings from the study are presented and summarized in a table to add more clarity. They present a realistic view of how the actual physical system would respond just like the virtual system present in the workstation. Furthermore, the plant or the controller can be made to interact in real-time PIL or HIL with the other components of the drive system in the workstation.

Acknowledgments: There were no funding resources.

Author Contributions: Mohan Krishna, Febin Daya, and Sanjeevikumar Padmanaban developed the proposed AC drives research work and implemented it with numerical simulation software and hardware implementation using RT-Lab solutions. Furthermore, Sanjeevikumar Padmanaban and Lucian Mihet-Popa extended the insight and technical expertise to make the work quality in its production. All authors were involved in validating the numerical simulation results with hardware implementation and experimental test results in accordance with the developed theoretical background. All authors were involved and contributed in each part of the article for its final depiction as a research paper.

Conflicts of Interest: The authors declare no conflict of interest.

Nomenclature

$i_{ds}{}^s, i_{qs}{}^s, i_{dr}{}^r, i_{qr}{}^r$	Direct and quadrature axes stator and rotor current components in stationary and rotating frame
$v_{ds}{}^s, v_{qs}{}^s$	Direct and quadrature stator voltages in stationary frame
T_r, R_s, R_r	Rotor time constant, stator and rotor resistance
σ, L_r, L_m, L_s	Leakage reactance, rotor, magnetizing and stator self inductance
L_{ls}, L_{lr}	Stator and rotor leakage inductances
$\omega_r, \hat{\omega}_r, \omega^*, \omega_{bsync}$	Actual, estimated, reference and base synchronous speed
$\psi_{ds}{}^s, \psi_{qs}{}^s, \psi_{dr}{}^s, \psi_{qr}{}^s$	Direct and quadrature axes stator and rotor flux linkages in stationary frame
φ_d, φ_q	Direct and quadrature axes estimated rotor flux linkages
$\theta_f, \theta_{sl}, \theta_r, T_e^*$	Field, slip and rotor angles and Torque reference
$i_{ds}{}^*, i_{qs}^*$	Direct and quadrature axes stator currents in synchronously rotating frame
$i_{as}^*, i_{bs}^*, i_{cs}^*$	Three-phase reference currents

Appendix A

The ratings of the model considered for the study are: A 50 HP, 415 V, 3Φ, 50 Hz, star connected, four-pole induction motor with equivalent parameters: R_s = 0.087 Ω, R_r = 0.228 Ω, L_{ls} = L_{lr} = 0.8 mH, L_m = 34.7 mH, Inertia, J = 1.662 kgm^2, friction factor = 0.1.

References

1. Anitha, P.; Badrul, H.C. Sensorless control of inverter-fed induction motor drives. *Electr. Power Syst. Res.* **2007**, *77*, 619–629.
2. Yang, Z.; Yue, Q.; Ye, Y. Induction Motor Speed Control Based on Model Reference. *Procedia Eng.* **2012**, *29*, 2376–2381.
3. Abd El-Halim, A.F.; Abdulla, M.M.; El-Arabawy, I.F. Simulation Aides in Comparison between Different Methodology of Field Oriented Control of Induction Motor Based on Flux and Speed Estimation. In Proceedings of the 22nd International Conference on Computer Theory and Applications, Alexandria, Egypt, 13–15 October 2012.

4. Sun, D.; Lin, W.; Diao, L.; Liu, Z. Speed Sensorless Induction Motor Drive Based on EKF and Γ-1 Model. In Proceedings of the IEEE International Conference on Computer Distributed Control and Intelligent Environmental Monitoring, Changsha, China, 19–20 February 2011.

5. Alonge, F.; D'Ippolito, F.; Sferlazza, A. Sensorless Control of Induction-Motor Drive Based on Robust Kalman Filter and Adaptive Speed Estimation. *IEEE Trans. Ind. Electron.* **2014**, *61*, 1444–1453. [CrossRef]

6. Rezgui, S.E.; Benalla, H. MRAS sensorless based control of IM combining sliding-mode, SVPWM, and Luenberger observer. In Proceedings of the International Conference on Computer as a Tool, Lisbon, Portugal, 27–29 April 2011.

7. Zheng, Y.; Loparo, K.A. Adaptive Flux Observer for Induction Motors. In Proceedings of the American Control Conference, Philadelphia, PA, USA, 24–26 June 1998.

8. Ticlea, A.; Besancon, G. Observer Scheme for State and Parameter Estimation in Asynchronous Motors with Application to Speed Control. *Eur. J. Control* **2006**, *12*, 400–412. [CrossRef]

9. Mikail, R.; Rahman, K.M. Sensorless Adaptive Rotor Parameter Estimation Method for Three Phase Induction Motor. In Proceedings of the IEEE 5th International Conference on Electrical and Computer Engineering, Dhaka, Bangladesh, 25–27 December 2008.

10. Levi, E.; Wang, M. Impact of Parameter Variations on Speed Estimation in Sensorless Rotor Flux Oriented Induction Machines. In Proceedings of the IEEE Power Electronics and Variable Speed Drives, London, UK, 21–23 September 1998.

11. Gadoue, S.M.; Giaouris, D.; Finch, J.W. Performance Evaluation of a Sensorless Induction Motor Drive at Very Low and Zero Speed Using a MRAS Speed Observer. In Proceedings of the IEEE 2008 Region 10 Colloquium and the Third International Conference on Industrial and Information Systems, Kharagpur, India, 8–10 December 2008.

12. Rashed, M.; Stronach, A.F. A stable back-EMF MRAS-based sensorless low-speed induction motor drive insensitive to stator resistance variation. *IEE Proc. Electr. Power Appl.* **2004**, *151*, 685–693. [CrossRef]

13. Beguenane, B.; Ouhrouche, M.A.; Trzynadlowski, A.M. A new scheme for sensorless induction motor control drives operating in low speed region. *Math. Comput. Simul.* **2006**, *71*, 109–120. [CrossRef]

14. Lascu, C.; Boldea, I.; Blaabjerg, F. A Class of Speed-Sensorless Sliding-Mode Observers for High-Performance Induction Motor Drives. *IEEE Trans. Ind. Electron.* **2009**, *56*, 3394–3403. [CrossRef]

15. Gadoue, S.M.; Giaouris, D.; Finch, J.W. MRAS Sensorless Vector Control of an Induction Motor Using New Sliding-Mode and Fuzzy-Logic Adaptation Mechanisms. *IEEE Trans. Energy Convers.* **2010**, *25*, 394–402. [CrossRef]

16. Zhang, X. Sensorless Induction Motor Drive Using Indirect Vector Controller and Sliding-Mode Observer for Electric Vehicles. *IEEE Trans. Veh. Technol.* **2013**, *62*, 3010–3018. [CrossRef]

17. Krishna, S.M.; Daya, J.L.F. A modified disturbance rejection mechanism in sliding mode state observer for sensorless induction motor drive. *Arab. J. Sci. Eng.* **2016**, *41*, 3571–3586. [CrossRef]

18. Nayeem Hasan, S.M.; Husain, I. A Luenberger-Sliding Mode Observer for Online Parameter Estimation and Adaptation in High-Performance Induction Motor Drives. *IEEE Trans. Ind. Appl.* **2009**, *45*, 772–781. [CrossRef]

19. Albu, M.; Horga, V.; Ratoi, M. Disturbance torque observers for the induction motor drives. *J. Electr. Eng.* **2006**, *6*, 1–6.

20. Krzeminski, Z. Observer of induction motor speed based on exact disturbance model. In Proceedings of the IEEE 13th International Power Electronics and Motion Control Conference, Poznan, Poland, 1–3 September 2008.

21. Krzeminski, Z. A new speed observer for control system of induction motor. In Proceedings of the IEEE International Conference on Power Electronics and Drive Systems, Hong Kong, China, 27–29 July 1999.

22. Vieira, R.P.; Gabbi, T.S.; Grundling, H.A. Sensorless decoupled IM current control by sliding mode control and disturbance observer. In Proceedings of the IEEE 40th Annual Conference, Dallas, TX, USA, 30 October–1 November 2014.

23. Comanescu, M. Design and analysis of a sensorless sliding mode flux observer for induction motor drives. In Proceedings of the IEEE International Electric Machines and Drives Conference, Niagara Falls, ON, Canada, 15–18 May 2011.

24. Krishna, S.M.; Daya, J.L.F. MRAS speed estimator with fuzzy and PI stator resistance adaptation for sesnorless induction motor drives using RT-Lab. *Perspect. Sci.* **2016**, *8*, 121–126. [CrossRef]

25. Krishna, S.M.; Daya, J.L.F. Adaptive Speed Observer with Disturbance Torque Compensation for Sensorless Induction Motor Drives using RT-Lab. *Turk. J. Electr. Eng. Comput. Sci.* **2016**, *24*, 3792–3806. [CrossRef]
26. Mikkili, S.; Prattipati, J.; Panda, A.K. Review of real-time simulator and the steps involved for implementation of a model from matlab/simulink to real-time. *J. Inst. Eng. India Ser. B.* **2014**, *96*, 179–196. [CrossRef]
27. Daya, J.L.F.; Sanjeevikumar, P.; Blaabjerg, F.; Wheeler, P.; Ojo, O.; Ertas, A.H. Analysis of Wavelet Controller for Robustness in Electronic Differential of Electric Vehicles—An Investigation and Numerical Developments. *J. Electr. Power Compon. Syst.* **2016**, *44*, 763–773. [CrossRef]
28. Sanjeevikumar, P.; Daya, J.L.F.; Blaabjerg, F.; Wheeler, P.; Oleschuk, V.; Ertas, A.H.; Mir-Nasiri, N. Wavelet-Fuzzy Speed Indirect Field Oriented Controller for Three-Phase AC Motor Drive-Investigation and Implementation. *Int. J. Eng. Sci. Technol.* **2016**, *19*, 1099–1107.
29. Daya, J.L.F.; Subbiah, V.; Iqbal, A.; Sanjeevikumar, P. A Novel Wavelet-Fuzzy based indirect field oriented control of Induction Motor Drives. *J. Power Elect.* **2013**, *13*, 656–668. [CrossRef]
30. Saponara, S.; Fanucci, L.; Bernardo, F.; Falciani, A. Predictive Diagnosis of High-Power Transformer Faults by Networking Vibration Measuring Nodes With Integrated Signal Processing. *IEEE. Trans. Instrum. Meas.* **2016**, *65*, 1749–1760. [CrossRef]
31. Costantino, N.; Serventi, R.; Tinfena, F.; D'Abramo, P.; Chassard, P.; Tisserand, P.; Saponara, S.; Fanucci, L. Design and Test of an HV-CMOS Intelligent Power Switch With Integrated Protections and Self-Diagnostic for Harsh Automotive Applications. *IEEE. Trans. Ind. Electron.* **2011**, *58*, 2715–2727. [CrossRef]
32. Saponara, S.; Petri, E.; Fanucci, L.; Terreni, P. Sensor Modeling, Low-Complexity Fusion Algorithms, and Mixed-Signal IC Prototyping for Gas Measures in Low-Emission Vehicles. *IEEE. Trans. Instrum. Meas.* **2011**, *60*, 372–384. [CrossRef]
33. Sanjeevikumar, P.; Daya, J.L.F.; Wheeler, P.; Blaabjerg, F.; Viliam, F.; Ojo, O. Wavelet Transform with Fuzzy Tuning Based Indirect Field Oriented Speed Control of Three-Phase Induction Motor Drive. Proceedings of 18th IEEE International Conference on Electrical Drives and Power Electronics, Tatranska Lomnica, Slovakia, 21–23 September 2015.
34. Daya, J.L.F.; Subbiah, V.; Sanjeevikumar, P. Robust Speed Control of an Induction Motor Drive using Wavelet-Fuzzy based Self-tuning Multiresolution Controller. *Int. J. Comput. Intell. Syst.* **2013**, *6*, 724–738. [CrossRef]
35. Sanjeevikumar, P.; Daya, J.L.F.; Blaabjerg, F.; Mir-Nasiri, N.; Ertas, A.H. Numerical Implementation of Wavelet and Fuzzy Transform IFOC for Three-Phase Induction Motor. *Int. J. Eng. Sci. Technol.* **2016**, *19*, 96–100.
36. Daya, J.L.F.; Sanjeevikumar, P.; Blaabjerg, F.; Wheeler, P.; Ojo, O. Implementation of Wavelet Based Robust Differential Control for Electric Vehicle Application. *IEEE Trans. Power Electron.* **2015**, *30*, 6510–6513. [CrossRef]

Article

Control Strategy for a Grid-Connected Inverter under Unbalanced Network Conditions—A Disturbance Observer-Based Decoupled Current Approach

Emre Ozsoy [1], Sanjeevikumar Padmanaban [1,*], Lucian Mihet-Popa [2], Viliam Fedák [3], Fiaz Ahmad [4], Rasool Akhtar [4] and Asif Sabanovic [4]

[1] Department of Electrical and Electronics Engineering, University of Johannesburg, Auckland Park 2006, South Africa; eemreozsoy@yahoo.co.uk

[2] Faculty of Engineering, Østfold University College, Kobberslagerstredet 5, 1671 Krakeroy-Fredrikstad, Norway; lucian.mihet@hiof.no

[3] Department of Electrical Engineering & Mechatronics, Technical University of Kosice, Rampová 1731/7, 040 01 Košice, Džung'a-Džung'a, Slovakia; viliam.fedak@tuke.sk

[4] Mechatronics Engineering, Faculty of Engineering and Natural Sciences, Sabancı University, Istanbul 34956, Turkey; fiazahmad@sabanciuniv.edu (F.A.); akhtar@sabanciuniv.edu (R.A.); asif@sabanciuniv.edu (A.S.)

* Correspondence: sanjeevi_12@yahoo.co.in; Tel.: +27-79-219-9845

Received: 17 June 2017; Accepted: 10 July 2017; Published: 22 July 2017

Abstract: This paper proposes a new approach on the novel current control strategy for grid-tied voltage-source inverters (VSIs) with circumstances of asymmetrical voltage conditions. A standard grid-connected inverter (GCI) allows the degree of freedom to integrate the renewable energy system to enhance the penetration of total utility power. However, restrictive grid codes require that renewable sources connected to the grid must support stability of the grid under grid faults. Conventional synchronously rotating frame dq current controllers are insufficient under grid faults due to the low bandwidth of proportional-integral (PI) controllers. Hence, this work proposes a proportional current controller with a first-order low-pass filter disturbance observer (DOb). The proposed controller establishes independent control on positive, as well as negative, sequence current components under asymmetrical grid voltage conditions. The approach is independent of parametric component values, as it estimates nonlinear feed-forward terms with the low-pass filter DOb. A numerical simulation model of the overall power system was implemented in a MATLAB/Simulink (2014B, MathWorks, Natick, MA, USA). Further, particular results show that double-frequency active power oscillations are suppressed by injecting appropriate negative sequence currents. Moreover, a set of simulation results provided in the article matches the developed theoretical background for its feasibility.

Keywords: power control; power electronics; pulse width modulation inverters; disturbance observer; grid connected system; grid stability; distorted voltage

---

1. Introduction

The rapid penetration of renewable energy sources (RESs) connected to the grid and distribution systems with power electronic converter topologies has changed the expected grid requirements to guarantee an appropriate performance under grid faults. In addition to the performance and reliability of the system under power electronic circuits in normal conditions, stability and grid support under grid faults are crucial due to restrictive grid code requirements [1,2]. Moreover, stability and reliability of the grid-connected inverter (GCI) under grid voltage faults must be considered for microgrid applications [3–8] with battery storage systems [9,10].

In particular, the most common fault type in electrical networks is unbalanced voltage conditions, which can easily occur in any voltage sags, and cause double-frequency power oscillations. In addition to requiring a positive sequence of active power (P) and reactive power (Q) injection by RESs through the GCI, these oscillations must be compensated for by injecting appropriate negative-sequence current sets. However, this aim cannot be realized by using conventional methods.

Proportional-integral (PI) controller-based vector control methods for GCI structures considering balanced voltage conditions are given in [11–13]. These methods decouple grid currents into P and Q generating components, and the PI current controllers achieve stable operation. However, this popular structure is fragile under voltage problems due to a low bandwidth of the PI controllers.

One of the first contributions related to the control of GCIs under unbalanced voltages is given in [14,15], by using decoupled PI control of positive- and negative-sequence dq frames. This structure is also known as the double synchronous reference frame (DSRF) method, and is used by many researchers [16,17]. Proportional-resonant (PR) [18,19] controllers are also extensively used for GCIs, which feed forward a resonant controller tuned at double the grid frequency. Direct power control methods [20,21] control the required power without additional inner current loops. The method given in [22] gives an enhanced operation of decoupled DSRF (DDSRF) operation by using feed-forwarded resonant controllers. Model-based predictive control [23,24] methods minimize the cost function by predicting the future current and power components of the GCI under an unbalanced voltage operation.

The decoupled control of synchronously rotating positive- and negative-sequence dq currents, as given in [14,15], is an effective method for the control of GCIs. However, this method suffers from simultaneously dissipating active and reactive power oscillating components. An instantaneous power theory calculations-based independent P and Q control strategy is given in [25], by proposing different current reference calculations depending on the power requirements. A robust power flow algorithm, which is based on the disturbance rejection control algorithm, is given in [26]. These methods given in [23–26] can independently dissipate P and Q double-frequency oscillations. However, the shape and magnitude of non-sinusoidal injected currents highly increase current harmonics in the system, which limits the effectiveness of these methods.

Three-phase four-leg inverters can generate sinusoidal voltage waveforms in a wide range of nonlinear operating conditions for more sensitive loads, such as for data transfer and military purposes, as they can also issue power quality requirements [27,28]. However, an additional phase-leg and inductance complicates the circuit and reduces the overall efficiency.

Grid synchronization is of great importance for robust control of GCIs; fast and accurate estimation of grid voltage parameters is essential to operate under grid faults. Different Phase Locked Loop (PLL) algorithms are available in the literature, aiming to operate under grid voltage problems [29–32]. It was assumed in this study that symmetrical positive- and negative-sequence component decomposition of the grid voltage was properly realized, such as is given in [33] under grid faults.

A disturbance observer (DOb)-based controller is a simple and robust structure that estimates external disturbances and uncertainties; thus the effect of disturbances and uncertainties are suppressed [34]. Estimated disturbances and system uncertainties are fed forward to the inner control loop; thus the robustness of the system is obtained. An additional external controller could be cascaded to achieve the desired performance goals, such as power and/or speed in electrical systems, as the DOb controls uncertain plant and removes the effect of external disturbances in the inner control loop.

Doubly fed induction generator (DFIG)-based wind turbines are also very fragile under grid voltage problems [7,8,35–38], and it can be considered that problem solution techniques applied to DFIG applications can be utilized in GCI applications. DOb-based current controllers are applied to DFIGs and GCIs in [39,40] by considering robustness against parameter variations under balanced voltage sets. However, this method must be carefully tuned to suppress double-frequency oscillations. This study modeled the grid dynamic model in synchronously rotating, symmetrical positive- and negative-sequence dq frames. Therefore, decoupled positive- and negative-sequence dq current components were independently controlled by achieving robust control under grid voltage faults.

In addition to the availability of simultaneous positive- and negative-sequence current injection, the proposed method was not affected by other external disturbances and uncertainties, such as grid impedance variations.

Integral terms in conventional PI controllers must be carefully tuned to prevent unwanted overshoots for a wide range of operations. In addition, windup effects of the integrator must be considered for real-time systems. Instead of conventional PI controllers and fed-forwarded parameter-dependent cross-coupling terms, proposed proportional controllers with a low-pass filter DOb are sufficient for robust operation, as the DOb accurately estimates and feeds forward uncertain terms. The control structure is simple and can be applied in real-time systems.

The main contribution of this study is a proportional decoupled current controller with a fed-forwarded low-pass filter DOb, which satisfies positive-sequence power requirements by independently controlling negative-sequence currents. The main advantage of this P + DOb current controller is to bring freedom from the sensitivity of the controllers with regard to variations in the grid parameters during operation for various reasons. Other methods outlined in [23–26] simultaneously control P and Q oscillations, as well as robustly satisfy positive-sequence power requirements. However, these methods inject non-sinusoidal currents to the grid at the instant of unbalanced voltage conditions. Conventional PI controllers are sensitive to parameter variations and anti-windup effects. This is the first reported study for a decoupled dq current control structure by using symmetrical component decomposition and estimating the disturbances with the DOb concept. The study was implemented on a Matlab/Simulink (2014B, MathWorks, Natick, MA, USA) simulation platform.

2. Dynamic Model

The equivalent circuit of the GCI is given in Figure 1 in the abc frame. The system was connected to the grid with respective grid resistance and inductance values. The dynamic model could be rewritten as either stationary or in the synchronously rotating dq frame, according to the given equivalent circuit.

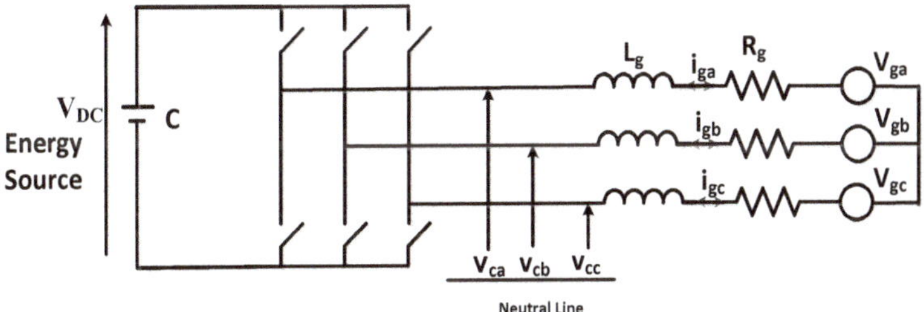

Figure 1. Equivalent circuit of the GCI in the abc frame.

The three-phase electrical variables, such as current, voltage, etc., could be indicated in several different types of reference frames [41,42]. Two orthogonal, synchronously rotating components in the dq frame are sufficient if a balanced system representation is required. However, they are insufficient in the case of an unbalanced system representation, and respective positive- and negative-sequence components must be presented.

The dynamical model could be arranged in the orthogonal frame of reference associated with positive and negative symmetrical components of the grid voltage, where positive sequence (dq)+ frames are composed of balanced voltages, while unbalanced voltage components generate negative sequence (dq)− frames, as is given in Figure 2.

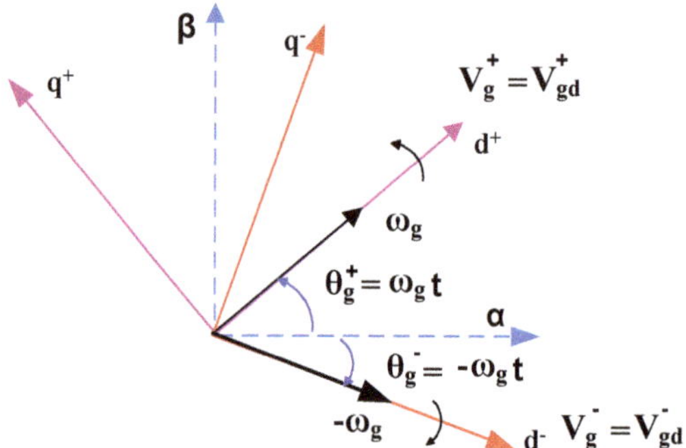

Figure 2. Orthogonal (dq)+ and (dq)− frames of references.

The current equation in symmetrical (dq)+ and (dq)− frames can be written as

$$L_g \frac{di_g}{dt} = v_s - R_g i_g + L i_g - v_g,\tag{1}$$

where

$$i_g^T = \begin{bmatrix} i_{gd}^+ & i_{gq}^+ & i_{gd}^- & i_{gq}^- \end{bmatrix},\, v_g^T = \begin{bmatrix} v_{gd}^+ & v_{gq}^+ & v_{gd}^- & v_{gq}^- \end{bmatrix},\, v_s^T = \begin{bmatrix} v_{sd}^+ & v_{sq}^+ & v_{sd}^- & v_{sq}^- \end{bmatrix},\tag{2}$$

$$L_g = \mathrm{diag}\begin{bmatrix} L_g & L_g & L_g & L_g \end{bmatrix},\, R_g = \mathrm{diag}\begin{bmatrix} R_g & R_g & R_g & R_g \end{bmatrix},\tag{3}$$

$$L = \begin{bmatrix} 0 & \omega_g L_g & 0 & 0 \\ \omega_g L_g & 0 & 0 & 0 \\ 0 & 0 & 0 & \omega_g L_g \\ 0 & 0 & \omega_g L_g & 0 \end{bmatrix},\tag{4}$$

The terms i_g and v_g, represent the grid currents and voltages in the synchronously rotating dq frame. The term v_s is the GCI output voltage. The terms R_g and L_g represent the grid resistances and inductances. All diagonal elements of the L_g and R_g matrix for the symmetrical systems are equal. The meaning of the +/− superscripts are for (dq)+ and (dq)− rotating frames, respectively. The d/q subscript refers to dq rotating frames. The term ω_g is the grid electrical speed. The rotating frame is aligned with the d axis, and $v_q = 0$. The line currents are assumed to be measured, and the GCI-output generated voltage is known. The GCI circuit can be written as is given below:

$$\frac{di_g}{dt} = L_g^{-1} v_s - L_g^{-1} R_g i_g + L_g^{-1} L i_g - L_g^{-1} v_g,\tag{5}$$

$$\varepsilon_g = i_g^{\mathrm{ref}} - i_g,\tag{6}$$

where $\varepsilon_g^T = \begin{bmatrix} \varepsilon_{gd}^+ & \varepsilon_{gq}^+ & \varepsilon_{gd}^- & \varepsilon_{gq}^- \end{bmatrix}$ is the error of control performance. If Equation (5) is inserted into the derivative of Equation (6), the error dynamics can be given as

$$\frac{d\varepsilon_g}{dt} = \frac{di_g^{\mathrm{ref}}}{dt} - L_g^{-1} v_s + L_g^{-1} R_g i_g - L_g^{-1} L i_g + L_g^{-1} v_g,\tag{7}$$

The closed-loop error equation is given as follows:

$$\frac{d\varepsilon_g}{dt} + k_g \varepsilon_g = 0,\tag{8}$$

The term $k_g{}^T = \text{diag}\begin{bmatrix} k_{gd}^+ & k_{gq}^+ & k_{gd}^- & k_{gq}^- \end{bmatrix}$ is a positive controller gain. The error of control performance ε_g is defined by asymptotic convergence to zero. The definition of convergence speed is dependent on the value of k_g coefficients. If Equation (7) is inserted into Equation (8), applied generated voltages to the GCI are written as follows:

$$L_g^{-1}v_s = \frac{di_g^{\text{ref}}}{dt} + L_g^{-1}R_g i_g - L_g^{-1}L i_g + L_g^{-1}v_g + k_g\varepsilon_g, \tag{9}$$

The grid inductance base value, L_g, is insensitive to disturbances. Thus, voltages applied for the GCI are written as below:

$$v_s{}^{\text{ref}} = \underbrace{L_g\Big(\frac{di_g^{\text{ref}}}{dt} + L_g^{-1}R_g i_g - L_g^{-1}L i_g + L_g^{-1}v_g\Big)}_{f_g} + L_g k_g\varepsilon_g, \tag{10}$$

The terms $f_g{}^T = \begin{bmatrix} f_{gd}^+ & f_{gq}^+ & f_{gd}^- & f_{gq}^- \end{bmatrix}$ are nonlinear, and an accurate determination of grid and GCI parameters is required to define these terms; this is impractical and f_g is considered as a disturbance.

Necessary and sufficient conditions for asymptotic stability of the control structure must satisfy the following conditions of the Lyapunov candidate function:

$$V(0) = 0, \; V > 0 \text{ and } \dot{V} < 0, \tag{11}$$

The term V is the Lyapunov candidate function. The Lyapunov function and time derivative of the Lyapunov function can be selected, as given below, to prove the asymptotic stability:

$$V = \frac{1}{2}\varepsilon_g^2, \; \frac{dV}{dt} = \varepsilon_g\frac{d\varepsilon_g}{dt}, \tag{12}$$

The first condition for Lyapunov stability is satisfied for $V(0) = 0$ The second condition for Lyapunov stability ($V > 0$) is valid for all real ε values. Finally, the third condition ($\dot{V} < 0$) can be satisfied by inserting Equation (8) into the time derivative of the Lyapunov candidate function.

$$\frac{dV}{dt} = -\varepsilon_g k_g\varepsilon_g, \tag{13}$$

It is obvious from Equation (13) that the time derivative of the Lyapunov candidate function is negative for positive, definite k_g values. Thus, necessary and sufficient conditions for the asymptotic stability of the controller structure are satisfied.

2.1. First-Order Low-Pass Filter Disturbance Observer

The term f_g can be estimated by modifying the voltage equations. If Equation (8) is inserted into Equation (9), determination of the grid voltage is possible to enforce the desired control performance in the current loop. The disturbance terms are considered as bounded, and are defined by $\dot{f}_g = 0$ with unknown initial conditions [43]. System inputs and outputs (v_s and i_g) are considered to be known or measured.

$$f_g = v_s - L_g\frac{di_g}{dt}, \tag{14}$$

The first-order low-pass filter DOb is applied to Equation (14) in the s domain, as is given below:

$$\hat{f}_g = T(v_s - sL_g i_g), \tag{15}$$

where $T^T = \text{diag}\left[\dfrac{g_d^+}{s+g_d^+} \quad \dfrac{g_q^+}{s+g_q^+} \quad \dfrac{g_d^-}{s+g_d^-} \quad \dfrac{g_q^-}{s+g_q^-} \right]$.

The term s is the Laplace operator. The coefficients g_d and g_q are the cut-off frequency gains. To simplify the implementation of the DOb, Equation (15) can be rewritten as is given below.

$$\hat{f}_g = T(v_s - L_g i_g) + gL_g i_g \tag{16}$$

where $g = \text{diag}\left[g_d^+ \quad g_q^+ \quad g_d^- \quad g_q^- \right]$. The block diagram of the DOb could be drawn as is given in Figure 3.

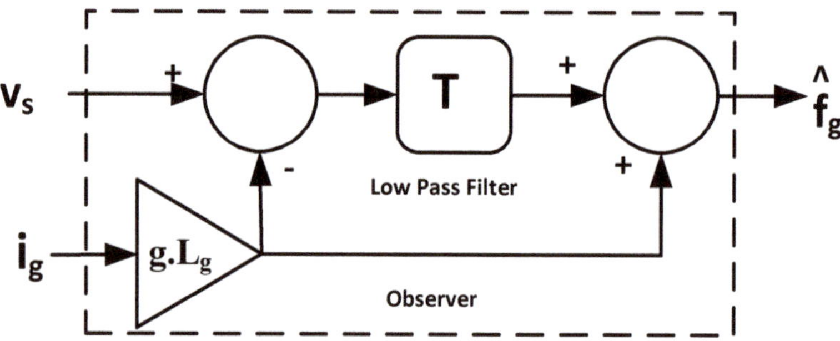

Figure 3. Disturbance observer (DOb) block diagram.

The final grid current error equations are given by

$$\frac{d\varepsilon_g}{dt} + k_g \varepsilon_g = f_g - \hat{f}_g \tag{17}$$

It can be stated from Equation (17) that the right-hand-side tends towards zero, as is given below. The optimal selection of the low-pass filter parameter is to set $[T] = \text{diag}[1]$ in the frequency range in which disturbance is expected. The bandwidth of the DOb should be as high as possible, so the disturbance error can converge to zero in a wide range of frequencies. The DOb compensation error will converge to zero in practical terms with a proper selection of the cut-off frequency [43]. This estimated disturbance plays a very critical role in the controller structure as a feed-forward term, and does not influence the stability of the closed-loop controller structure with the properly selected cut-off frequencies. Because of the effectiveness of the feed-forward disturbance term, the integral action is not required in the closed-loop structure. Therefore, the proportional controller with a positive definite k_g value is sufficient for the controller error to converge to zero in a finite time. As a result, the proposed controller structure is more robust and simple, compared to conventional PI controllers, as it estimates and feeds forward the disturbance terms without the integral part of the controller.

2.2. Instantaneous Power Equations

The instantaneous powers associated with unbalanced current and voltage components can be written in the following form [44], with multiplication of the double-frequency oscillating components.

$$\begin{bmatrix} P(t) \\ Q(t) \end{bmatrix} = \begin{bmatrix} P_{g0} \\ Q_{g0} \end{bmatrix} + \begin{bmatrix} P_{sc2} \\ Q_{sc2} \end{bmatrix} \cos(2\omega_g t) + \begin{bmatrix} P_{ss2} \\ Q_{ss2} \end{bmatrix} \sin(2\omega_g t), \tag{18}$$

where

$$\begin{bmatrix} P_{g0} \\ Q_{g0} \end{bmatrix} = 1.5 \begin{bmatrix} v_{gd}^{+} & v_{gq}^{+} & v_{gd}^{-} & v_{gd}^{-} \\ v_{gq}^{+} & -v_{gd}^{+} & v_{gq}^{-} & -v_{gd}^{-} \end{bmatrix} \begin{bmatrix} i_{gd}^{+} \\ i_{gq}^{+} \\ i_{gd}^{-} \\ i_{gq}^{-} \end{bmatrix}, \tag{19}$$

$$\begin{bmatrix} P_{sc2} \\ Q_{sc2} \end{bmatrix} = 1.5 \begin{bmatrix} v_{gd}^{-} & v_{gq}^{-} & v_{gd}^{+} & v_{gq}^{+} \\ v_{gq}^{-} & -v_{gd}^{-} & v_{gq}^{+} & -v_{gd}^{+} \end{bmatrix} \begin{bmatrix} i_{gd}^{+} \\ i_{gq}^{+} \\ i_{gd}^{-} \\ i_{gq}^{-} \end{bmatrix}, \tag{20}$$

$$\begin{bmatrix} P_{ss2} \\ Q_{ss2} \end{bmatrix} = 1.5 \begin{bmatrix} v_{gq}^{-} & -v_{gd}^{-} & -v_{gq}^{+} & v_{gd}^{+} \\ -v_{gd}^{-} & -v_{gq}^{-} & v_{gd}^{+} & v_{gq}^{+} \end{bmatrix} \begin{bmatrix} i_{gd}^{+} \\ i_{gq}^{+} \\ i_{gd}^{-} \\ i_{gq}^{-} \end{bmatrix}, \tag{21}$$

The terms, P_{g0} and Q_{g0} are fundamental instantaneous P and Q components, which consist of positive- and negative-sequence power equations, while the terms P_{sc2}-P_{ss2} and Q_{sc2}-Q_{ss2} are four pulsating terms, which are the result of asymmetrical network conditions. The maximum four variables (i_{gd}^{+} i_{gq}^{+} i_{gd}^{-} i_{gq}^{-}) could be controlled to achieve the P_{g0} and Q_{g0} requirements and compensate for the P_{sc2}-P_{ss2} and Q_{sc2}-Q_{ss2} oscillating components. Thus, P and Q oscillations cannot be compensated for simultaneously in positive- and negative-sequence dq frames [44]. It is necessary to calculate an appropriate set of current references to ensure a constant value of P is absorbed or injected by the GCI under balanced and unbalanced voltage conditions. These P_{g0} and Q_{g0} requirements and the P_{sc2}-P_{ss2} oscillation compensation can be addressed by using the following expression:

$$\begin{bmatrix} P_{g0} \\ Q_{g0} \\ P_{sc2} \\ P_{ss2} \end{bmatrix} = 1.5 \begin{bmatrix} v_{gd}^{+} & v_{gq}^{+} & v_{gd}^{-} & v_{gq}^{-} \\ v_{gq}^{+} & -v_{gd}^{+} & v_{gq}^{-} & -v_{gd}^{-} \\ -v_{gd}^{-} & v_{gq}^{-} & v_{gd}^{+} & v_{gq}^{+} \\ v_{gq}^{-} & -v_{gd}^{-} & -v_{gq}^{+} & v_{gd}^{+} \end{bmatrix} \begin{bmatrix} i_{gd}^{+} \\ i_{gq}^{+} \\ i_{gd}^{-} \\ i_{gq}^{-} \end{bmatrix}, \tag{22}$$

Equation (22) defines how positive-sequence grid current controllers achieve P and Q requirements, while negative-sequence current controllers can compensate for the P oscillations depending on the negative-sequence current injection strategy.

The proposed scheme is depicted in Figure 4. If zero i_{gd}^{-} and i_{gd}^{-} references are chosen, injected currents towards the grid are sinusoidal; this supports power quality requirements. If a zero P_{sc2}-P_{ss2} reference selection is selected, double-frequency oscillating power components can be compensated for by injecting negative-sequence currents towards the grid. The proportional current controllers are sufficient to track the desired current requirements with accurately estimated disturbance terms. The block diagram in Figure 3 is used to estimate disturbance terms. An online Second Order Generalized Integrator (SOGI)-based symmetrical component estimation is achieved with the method given in [33]. PLL structures separately calculate the symmetrical voltage phase and angle. It is assumed that symmetrical component decomposition of the voltage and currents is perfectly estimated, and an accurate PLL voltage phase and angle estimation is achieved.

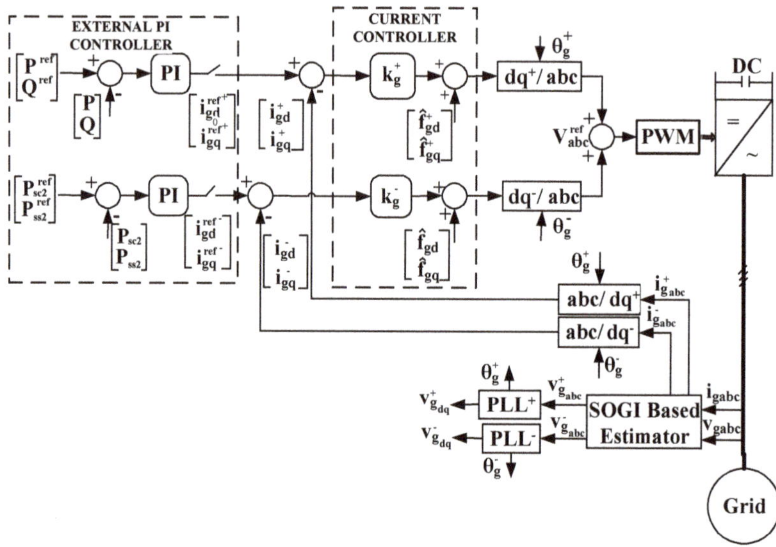

Figure 4. Proposed controller structure.

3. Simulation Results

Figure 5 depicts the simulation circuit implemented in MATLAB/Simulink using the SimPowerSystem tool. The GCI was connected to a transmission system, and all necessary parameters for the simulation are given in Table 1. Four different simulations were implemented to validate the proposed controller structure. The first simulation demonstrated the deteriorated current and power waveforms under unbalanced voltage conditions with the positive-sequence controller, without enabling the negative-sequence controller (Simulation A). The dual-current controller with the enabled negative-sequence current controller enforced negative-sequence currents to zero in the second simulation (Simulation B). The third simulation enforced double-frequency P_{sc2}-P_{ss2} power oscillations to zero. In addition, the dynamic performance of positive-sequence controllers was demonstrated by applying appropriate dq current steps (Simulation C). Finally, the fourth simulation compared the performance of conventional PI controllers to DOb-based current controllers (Simulation D).

Table 1. Parameters used in simulations.

Symbol	Quantity	Unit
Grid Connnected Inverter (GCI) DC Voltage	750	V
Nominal GCI Current	500	A
Nominal GCI Power	350	KVA
Switching Frequency	10	kHz
L_g Filter of GCI	0.25	mH
X/R Ratio of Grid	7	-
K_P (+)/K_P (−)	20	-
g_d	500	rad

The DC voltage was kept constant at 750 V to reduce the harmonic stress in the currents, which meant RESs were connected to the DC bus, and could inject required power to the grid at any instant of the simulation. Reference of i_{gd}^+ was kept at 75 A, meaning that the injection of currents were applied towards the grid. Reference of i_{gq}^+ was kept at 0 A to ensure a zero reactive power injection. The applied steps at different instants of Simulation A and B were given as follows:

0.20–0.27 s: 30% unbalanced voltage condition was generated on phase-A in the grid.

0.30–0.35 s: i_{gd}^{+} reference step was applied from 75 to 150 A.

0.38–0.43 s: i_{gq}^{+} reference step was applied from 0 to 50 A.

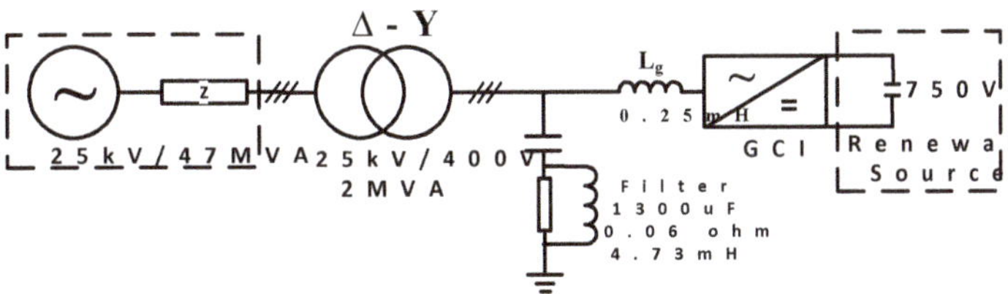

Figure 5. Simulation circuit.

Figure 6 shows the first simulation results (Simulation A) with the disabled negative-sequence controller. A 30% unbalanced voltage on phase-A between 0.20 and 0.27 s was applied, which is shown in Figure 6a. Sinusoidal grid currents (Figure 6b) show that the sinusoidal shape deteriorated without the negative-sequence current controller. Figure 6c shows the respective dq axis current references that changed P_g and Q_g properly. Respective i_{gd}^{+} and i_{gq}^{+} step-response tests are shown in Figure 6d. The performance criteria was satisfied with the DOb-based current controllers without any steady-state error or overshoot. Double-grid frequency power oscillations exist under unbalanced voltages, and could be dissipated by injecting negative-sequence currents. Figure 7 shows the uncontrolled i_{gd}^{-} and i_{gq}^{-} currents and the resultant oscillation P_{ss2} and P_{sc2} components that were the root cause of double-frequency oscillations.

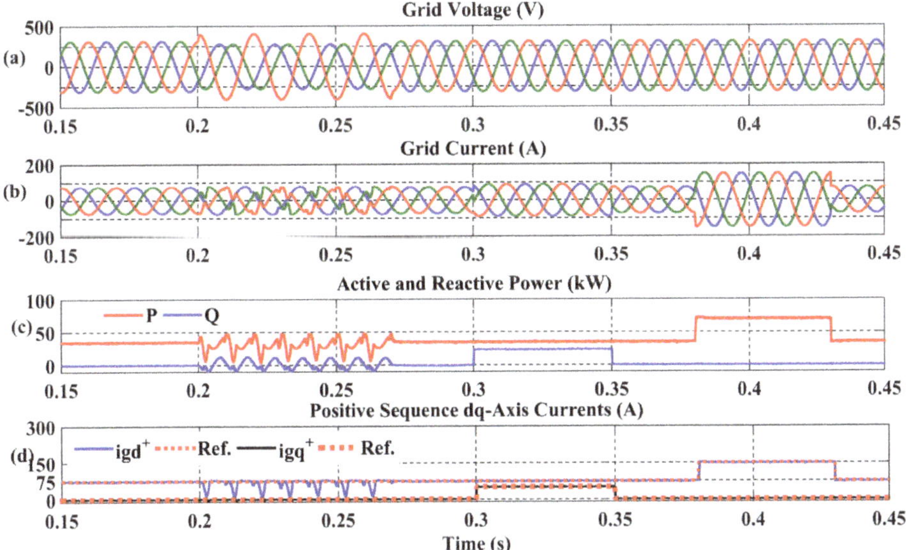

Figure 6. Simulation A results without dual positive- and negative-sequence controllers: (**a**) grid voltage (V) (red: phase-A, blue: phase-B, green: phase-C); (**b**) grid currents (A) (red: phase-A, blue: phase-B, green: phase-C); (**c**) P_g (kW) and Q_g (kVAr); and (**d**) positive-sequence dq axis currents (A).

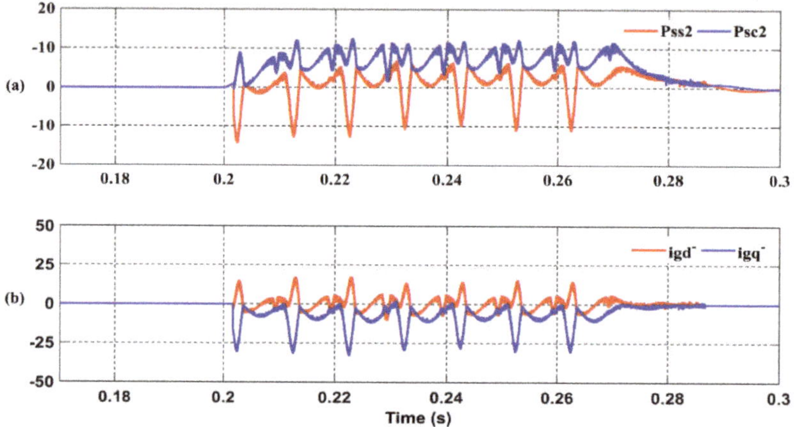

Figure 7. Simulation A results without dual positive- and negative-sequence controllers: (**a**) P_{ss2} (kW) and P_{sc2} (V) components, and (**b**) i_{gd}^{-} (A) and i_{gq}^{-} (A).

Similarly, Figure 8 shows the second simulation results (Simulation B) with dual positive- and negative-sequence current controller results. The negative-sequence controller was only enabled when an unbalanced voltage existed in the grid because it was observed in simulations that enabling the negative-sequence controller in balanced voltage conditions unnecessarily deteriorated the dynamic performance of the overall system [45]. Thus, a simple logic condition was added in the simulation to enable or disable the negative-sequence controller, depending on the negative-sequence voltage level. Figure 8b shows the balanced grid current sets with a zero negative-sequence grid current injection. Figure 8c shows that double-frequency P oscillations still existed due to the P_{ss2} and P_{sc2} components.

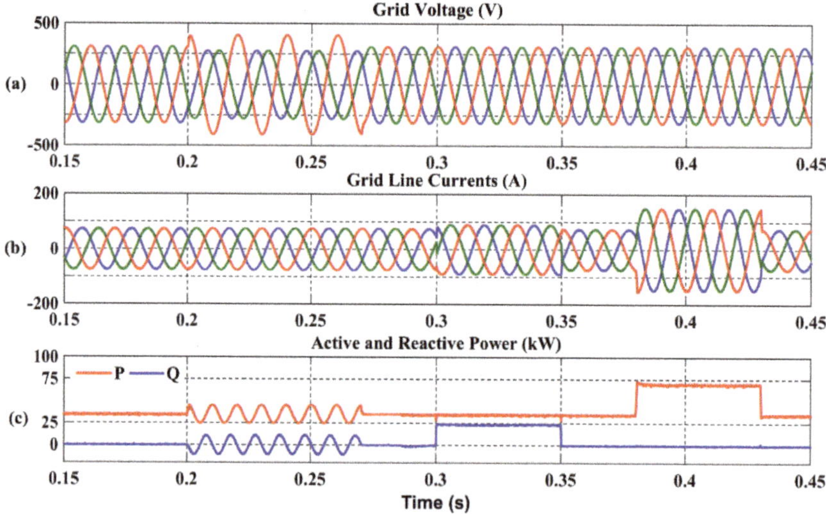

Figure 8. Simulation B results with dual positive- and negative-sequence controllers: (**a**) grid voltage (V) (red: phase-A, blue: phase-B, green: phase-C); (**b**) grid currents (A) (red: phase-A, blue: phase-B, green: phase-C); and (**c**) P_g (kW) and Q_g (kVAr).

Figure 9a shows the positive-sequence currents, where the performance of the current trajectory was satisfied. Figure 9b shows f_{gd}^+ and the estimated $\hat{f}_{gd}^+$. The term f_{gd}^+ was calculated with dynamic equations and compared with $\hat{f}_{gd}^+$. The mean value of $\hat{f}_{gd}^+$ was equal to f_{gd}^+, which proved that the term $\hat{f}_{gd}^+$ was accurately estimated. Similarly, Figure 9c shows both the parameters f_{gq}^+ and the estimated $\hat{f}_{gq}^+$, pointing out that the term $\hat{f}_{gq}^+$ was accurately estimated.

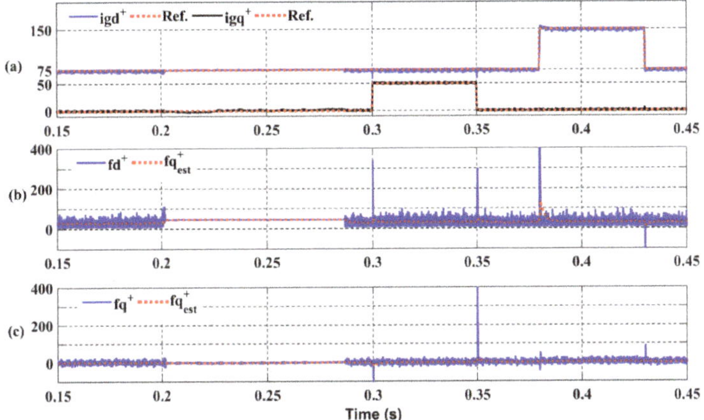

Figure 9. Simulation B results with dual positive- and and negative-sequence controllers: (a) positive-sequence dq axis currents (A), (b) f_{gd}^+ and $\hat{f}_{gd}^+$ (V), and (c) f_{gq}^+ and $\hat{f}_{gq}^+$ (V).

Negative-sequence current components were enforced to zero at the instant of unbalanced voltage conditions, and deteriorated grid current waveforms were dissipated, as shown in Figure 8b. Double-frequency power oscillations could not be removed without an appropriate injection of negative-sequence currents. Figure 10a shows that the negative-sequence currents could be controlled at zero references. Figure 10b,c shows the negative sequence of the parameters $f_{gd}^- - f_{gq}^-$ and the estimated $\hat{f}_{gd}^- - \hat{f}_{gq}^-$. The mean values of the estimated components were equal to the calculated terms, which proved that the estimated terms were accurately estimated.

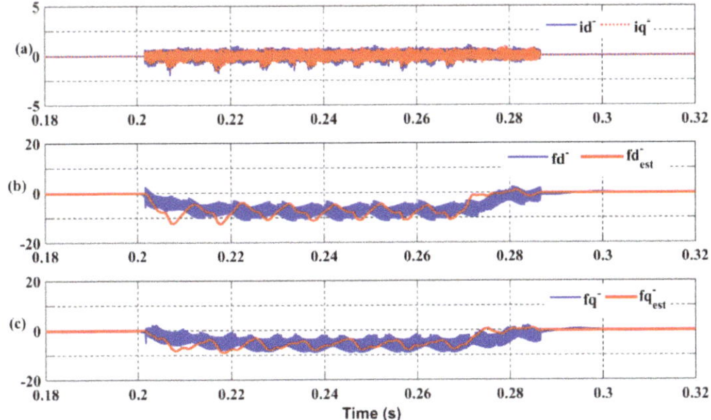

Figure 10. Simulation B results with dual positive- and negative-sequence controllers: (a) negative sequence dq axis currents (A), (b) f_{gd}^- and $\hat{f}_{gd}^-$ (V), and (c) f_{gq}^- and $\hat{f}_{gq}^-$ (V).

Figures 11–13 show the third simulation results for dissipating the P_{sc2}-P_{ss2} power oscillations (Simulation C). External PI controllers with a reference of $P_{sc2} - P_{ss2} = 0$ were enabled, as shown in Figure 4. A 30% unbalanced voltage was generated between 0.2 and 0.27 s (Figure 11a). Similar to for Simulation B, the negative-sequence controller was enabled at predefined unbalanced voltage levels. The reference value of i_{gd}^+ was kept at 75 A, and i_{gq}^+ was kept at zero. Figure 11c shows that double-frequency P oscillations were dissipated under the unbalanced voltage operation, and that P and Q could be independently controlled under balanced conditions.

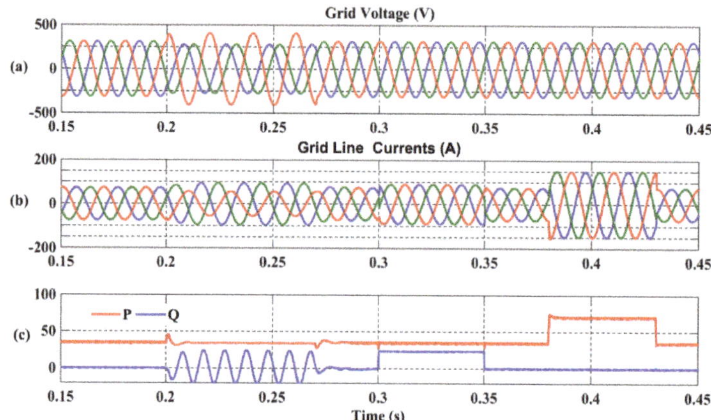

Figure 11. Simulation C results with external P_{ss2}-P_{sc2} controllers: (**a**) grid voltage (V) (red: phase-A, blue: phase-B, green: phase-C); (**b**) grid currents (A) (red: phase-A, blue: phase-B, green: phase-C); and (**c**) P_g (kW) and Q_g (kVAr).

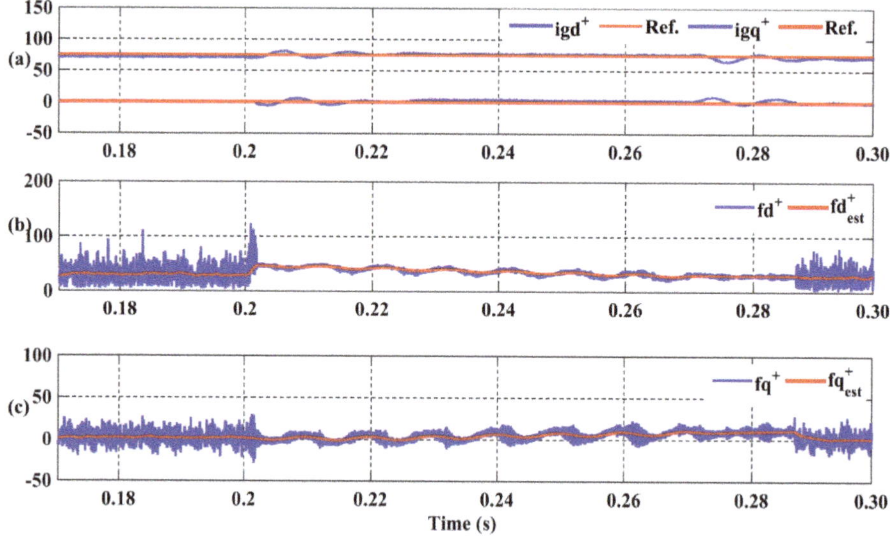

Figure 12. Simulation C results with external P_{ss2}-P_{sc2} controllers: (**a**) grid voltage (V); (**b**) grid currents (A); and (**c**) positive-sequence dq axis currents (A).

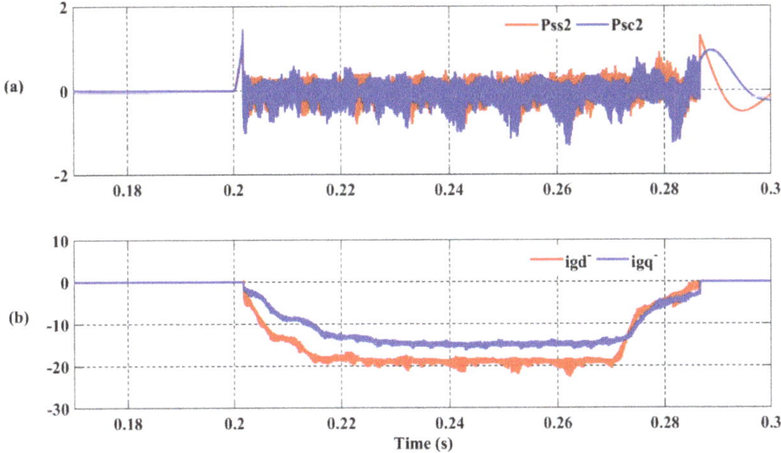

Figure 13. Simulation C results with external P_{ss2}-P_{sc2} controllers: (**a**) P (kW) and Q (kVAr), (**b**) grid currents (A).

Figure 12 shows that positive-sequence currents could follow their respective references at the instant of the unbalanced voltage generation. Figure 12b,c shows that the calculated f_g and estimated $\hat{f}_g^+$ terms were equal, which meant that $\hat{f}_{gd}^+$ and $\hat{f}_{gq}^+$ were accurately estimated.

The oscillating components P_{sc2}-P_{ss2} were enforced to zero with external PI controllers. PI controller gains could easily be determined with trial and error methods ($K_p = 1.1$ and $K_I = 5.2$). It can be noted from Equation (22) that oscillating P_{sc2} and P_{ss2} components could be compensated for by internal i_{gd}^- and i_{gq}^- controllers, respectively. Double-frequency oscillations were removed on P by enforcing the P_{sc2}-P_{ss2} components to zero, as can be seen in Figure 13a, and Figure 13b shows the resultant injected i_{gd}^- and i_{gq}^- components.

Finally, the proposed DOb-based current controller was compared with the conventional PI controller. The performance comparison seemed to be equivalent under a balanced operation, depending on the controller's proportional and integral gains. In addition, it is difficult to comment whether either the conventional PI or proposed DOb-based current controller was better in performance under balanced voltage conditions. However, the DOb-based current controller showed a better dynamic performance, and did not cause any steady-state error in positive-sequence currents under unbalanced voltage conditions; this is shown in previous plots (Figures 11 and 12). The conventional PI controller resulted in steady-state current and power errors under unbalanced voltage conditions. This problem is also stated in [45], in that symmetrical decomposition methods degrade dynamic performance and may cause steady-state errors in PI controllers. It was shown in Simulation D that the aforementioned problem exists in constructed simulation platforms with $k_p = 20$ and $k_i = 5$ values. The dynamic performance seemed equivalent under a balanced operation; steady-state error plots under an unbalanced voltage operation are demonstrated in Figure 14. Figure 14a,b shows that the i_{gd}^+ and i_{gq}^+ components could not follow the respective trajectories under an unbalanced operation, and if a longer unbalanced voltage operation was applied, the steady-state error would have slowly increased to unacceptable values. A similar behavior also existed in the P component, as shown in Figure 14c.

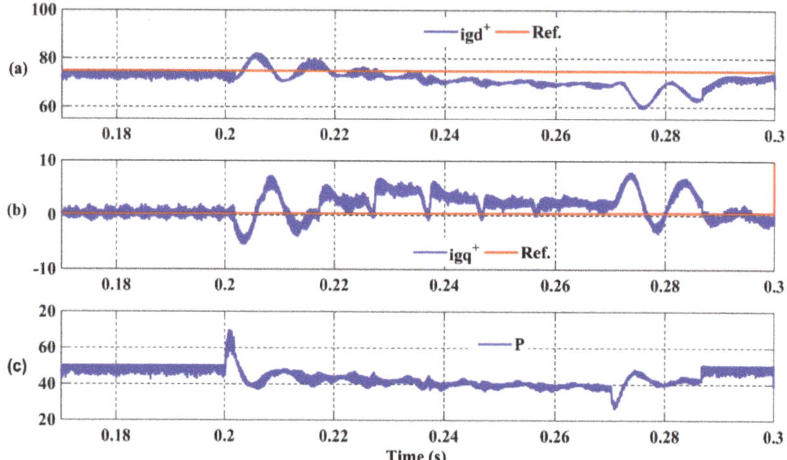

Figure 14. Simulation D results with the conventional PI controller: (a) i_{gd}^+ (A), (b) i_{gq}^+ (A), and (c) P (kW).

4. Conclusions

The objective of this paper was to investigate a novel current controller that was based on a low-pass filter DOb, to provide a precise control of currents under unbalanced grid voltage conditions for a grid-tied inverter. The GCI was modeled in the symmetrical synchronous reference frames, and estimated disturbance parameters were fed to current controllers. P and Q were defined by using the instantaneous power theory, and double-frequency P_{ss2} and P_{sc2} pulsations were removed under a full propagation cycle. PI and proposed DOb-based proportional current controllers were compared, and it was demonstrated that conventional PI controllers may cause steady-state errors under an asymmetrical grid voltage operation. Numerical simulation results also proved that the methods applied were able to compensate for the double-frequency power oscillations for the grid-tied inverter application, which means that the objective was achieved. The proposed current controller seems to be a valid alternative solution for GCIs under unbalanced conditions. Hopefully, the results presented will form a basis for diagnosis methods regarding the control techniques of GCIs under unbalanced network conditions. Due to the fact that the study was limited to the simulation results, instead the effect of real components, more research is certainly needed.

Author Contributions: Emre Ozsoy and Sanjeevikumar Padmanaban developed the proposed control strategy for the grid-connected system under distorted conditions. Emre Ozsoy, Sanjeevikumar Padmanaban, Lucian Mihet-Popa, Fiaz Ahmad, Rasool Akhtar and Asif Sabanovic were involved in further development of the study and in the implementation strategy. Sanjeevikumar Padmanaban, Lucian Mihet-Popa, Viliam Fedák and Asif Sabanovic contributed expertise in the grid-connected system for the verification of theoretical concepts and validation of obtained results. All authors contributed to and were involved in framing the final version of the full research article in its current form.

Conflicts of Interest: The authors declare no conflict of interest.

References

1. Basso, T.S.; DeBlasio, R. IEEE 1547 series of standards: interconnection issues. *IEEE Trans. Power Electron.* **2004**, *19*, 1159–1162. [CrossRef]
2. Markiewicz, H.; Klajn, A. *Voltage Disturbances Standard EN50160—Voltage Characteristics in Public Distribution Systems*; Copper Development Association: Hertfordshire, UK, 2004.

3. Sridhar, V.; Umashankar, S.; Sanjeevikumar, P. Decoupled Active and reactive power control of cascaded H-Bridge PV-Inverter for grid connected applications. In *Lecture Notes in Electrical Engineering*; Springer: Berlin/Heidelberg, Germany, 2017.

4. Swaminathan, G.; Ramesh, V.; Umashankar, S.; Sanjeevikumar, P. Fuzzy Based micro grid energy management system using interleaved boost converter and three level NPC inverter with improved grid voltage quality. In *Lecture Notes in Electrical Engineering*; Springer: Berlin/Heidelberg, Germany, 2017.

5. Swaminathan, G.; Ramesh, V.; Umashankar, S.; Sanjeevikumar, P. Investigations of microgrid stability and optimum power sharing using Robust Control of grid tie PV inverter. In *Lecture Notes in Electrical Engineering*; Springer: Berlin/Heidelberg, Germany, 2017.

6. Awasthi, A.; Karthikeyan, V.; Rajasekar, S.; Sanjeevikumar, P.; Siano, P.; Ertas, A. Dual mode control of inverter to integrate solar-wind hybrid fed DC-Grid with distributed AC grid. In Proceedings of the IEEE Conference on Environment and Electrical Engineering, Florence, Italy, 7–10 June 2016.

7. Mihet-Popa, L. Current Signature Analysis as diagnosis media for incipient fault detection. *J. Adv. Electr. Comput. Eng.* **2007**, *7*, 11–16. [CrossRef]

8. Mihet-Popa, L.; Prostean, O.; Szeidert, I.; Filip, I.; Vasar, C. Fault detection methods for frequency converters fed induction machines. In Proceedings of the 12th IEEE Conference on Emerging Technologies and Factory Automation—ETFA, Patras, Greece, 25–28 September 2007; pp. 161–168.

9. Das, V.; Sanjeevikumar, P.; Karthikeyan, V.; Rajasekar, S.; Blaabjerg, F.; Pierluigi, S. Recent advances and challenges of fuel cell based power system architectures and control—A review. *Renew. Sustain. Energy* **2017**, *73*, 10–18. [CrossRef]

10. Vavilapalli, S.; Sanjeevikumar, P.; Umashankar, S.; Mihet-Popa, L. Power balancing control for grid energy storage system in PV spplications—Real time digital simulation implementation. *Energies* **2017**, *10*, 928. [CrossRef]

11. Pena, R.; Clare, J.C.; Asher, G.M. A doubly-fed induction generator using two back-to-back PWM converters and its application to variable speed wind energy system. *Proc. Inst. Elect. Eng. B* **1996**, *143*, 231–241.

12. Mihet-Popa, L.; Groza, V.; Prostean, G.; Filip, I.; Szeidert, I. Variable Speed Wind Turbines Using Cage Rotor Induction Generators Connected to the Grid. Proceeding of the IEEE Canada Electrical Power Conference, EPC 2007, Montreal, QC, Canada, 25–26 October 2007.

13. Mihet-Popa, L.; Groza, V.; Prostean, O.; Szeidert, I. Modeling and design of a grid connection control mode for a small variable-speed wind turbine system. In Proceedings of the IEEE I2MTC-International Instrumentation & Measurement Technology Conference, Vancouver Island, BC, Canada, 12–15 May 2008; pp. 288–293.

14. Rioual, P.; Pouliquen, H.; Louis, H.P. Control of a PWM rectifier in the unbalanced state by robust voltage regulation. In Proceedings of the 5th European Conference Power Electronics Applications, Brighton, UK, 13–16 September 1993; Volume 4, pp. 8–14.

15. Song, H.S.; Nam, K. Dual current control scheme for PWM converter under unbalanced input voltage conditions. *IEEE Trans. Ind. Electron.* **1999**, *46*, 953–959. [CrossRef]

16. Suh, Y.; Tijeras, V.; Lipo, T.A. Control scheme in hybrid synchronous stationary frame for PWMAC/DC converter under generalized unbalanced operating conditions. *IEEE Trans. Ind. Appl.* **2006**, *42*, 825–835.

17. Suh, Y.; Tijeras, V.; Lipo, T.A. A control method in dq synchronous frame for PWM boost rectifier under generalized unbalanced operating conditions. In Proceedings of the IEEE 33rd Annual Power Electronics Specialists Conference, Cairns, Australia, 23–27 June 2002; Volume 3, pp. 1425–1430.

18. Teodorescu, R.; Blaabjerg, F.M.; Liserre, M.; Loh, P.C. Proportional Resonant controllers and filters for grid-connected voltage-source converters. *IEE Proc.-Electr. Power Appl.* **2006**, *153*, 750–762. [CrossRef]

19. Lascu, C.; Asiminoaei, L.; Boldea, I.; Blaabjerg, F. High performance current controller for selective harmonic compensation in active power filters. *IEEE Trans. Power Electron.* **2007**, *22*, 1826–1835. [CrossRef]

20. Serpa, L.A.; Ponnaluri, S.; Barbosa, P.M.; Kolar, J.W. A modified direct power control strategy allowing the connection of three-phase inverters to the grid through LCL filters. *IEEE Trans. Ind. Appl.* **2007**, *43*, 1388–1400. [CrossRef]

21. Vazquez, S.; Sanchez, J.A.; Carrasco, J.M.; Leon, J.I.; Galvan, E. A model-based direct power control for three-phase power converters. *IEEE Trans. Ind. Electron.* **2008**, *55*, 1647–1657. [CrossRef]

22. Reyes, M.; Rodriguez, P.; Vazquez, S.; Luna, A.; Teodorescu, R.; Carrasco, J.M. Enhanced decoupled double synchronous reference frame current controller for unbalanced grid-voltage conditions. *IEEE Trans. Power Electron.* **2012**, *27*, 3934–3943. [CrossRef]

23. Dawei, Z.; Lie, X.; Williams, B.W. Model-based predictive direct power control of doubly fed induction generators. *IEEE Trans. Power Electron.* **2010**, *25*, 341–351. [CrossRef]

24. Cort'es, P.; Rodriguez, J.; Antoniewicz, P.; Kazmierkowski, M. Direct power control of an AFE using predictive control. *IEEE Trans. Power Electron.* **2008**, *23*, 2516–2553. [CrossRef]

25. Rodriguez, C.P.; Timbus, A.V.; Teodorescu, R.; Liserre, M.; Blaabjerg, F. Independent PQ control for distributed power generation systems under grid faults. In Proceedings of the IEEE Industrial Electronics 32nd Annual Conference IECON, Paris, France, 6–10 November 2006; pp. 5185–5190.

26. Wang, Y.; Beibei, R.; Zhong, Q.C. Robust power flow control of grid-connected inverters. *IEEE Trans. Ind. Electron.* **2016**, *63*, 6687–6897. [CrossRef]

27. Ortjohann, E.; Arturo, A.; Morton, D.; Mohd, A.; Hamsic, N.; Omari, O. Grid-forming three-phase inverters for unbalanced loads in hybrid power systems. In Proceedings of the 2006 IEEE 4th World Conference on Photovoltaic Energy Conference, Waikoloa, HI, USA, 7–12 May 2006.

28. Vechium, I.; Camblong, H.; Tapia, G.; Curea, O.; Dakyo, B. Modelling and control of four-wire voltage source inverter under unbalanced voltage condition for hybrid power system applications. In Proceedings of the 2005 European Conference on Power Electronics and Applications, Dresden, Germany, 11–14 September 2005.

29. Chung, S.K. A phase tracking system for three phase utility interface inverters. *IEEE Trans. Power Electron.* **2000**, *15*, 431–438. [CrossRef]

30. Rodriguez, P.; Pou, J.; Bergas, J.; Candela, J.; Burgos, R.P.; Boroyevich, D. Decoupled double synchronous reference frame PLL for power converters control. *IEEE Trans. Power Electron.* **2007**, *22*, 584–592. [CrossRef]

31. Rodriguez, P.; Luna, A.; Teodorescu, R.; Iov, F.; Blaabjerg, F. Fault ride-through capability implementation in wind turbine converters using a decoupled double synchronous reference frame PLL. In Proceedings of the Power Electronics and Applications European Conference, Aalborg, Denmark, 2–5 September 2007; pp. 1–10.

32. Ciobotaru, M.; Teodorescu, R.; Blaabjerg, F. A new single-phase PLL structure based on second order generalized integrator. In Proceedings of the Power Electronics Specialists Conference, Jeju, Korea, 18–22 June 2006; pp. 1–6.

33. Iravani, M.R.; Karimi-Ghartemani, M. Online estimation of steady state and instantaneous symmetrical components. *IEEE Proc. Gener. Transm. Distrib.* **2003**, *150*, 616–622. [CrossRef]

34. Ohnishi, K.; Shibata, M.; Murakami, T. Motion control for advanced mechatronics. *IEEE/ASME Trans. Mechatron.* **1996**, *1*, 56–67. [CrossRef]

35. Tamvada, K.; Umashankar, S.; Sanjeevikumar, P. Investigation of double fed induction generator behavior under Symmetrical and Asymmetrical Fault Conditions. In *Lecture Notes in Electrical Engineering*; Springer: Berlin/Heidelberg, Germany, 2017.

36. Tamvada, K.; Umashankar, S.; Sanjeevikumar, P. Impact of Power Quality Disturbances on Grid Connected Double Fed Induction Generator. In *Lecture Notes in Electrical Engineering*; Springer: Berlin/Heidelberg, Germany, 2017.

37. Tiwaria, R.; Babu, N.R.; Sanjeevikumar, P. A Review on GRID CODES—Reactive power management in power grids for Doubly-Fed Induction Generator in Wind Power application. In *Lecture Notes in Electrical Engineering*; Springer: Berlin/Heidelberg, Germany, 2017.

38. Kalaivani, C.; Sanjeevikumar, P.; Rajambal, K.; Bhaskar, M.S.; Mihet-Popa, L. Grid Synchronization of Seven-phase Wind Electric Generator using d-q PLL. *Energies* **2017**, *10*, 926. [CrossRef]

39. Ozsoy, E.E.; Golubovic, E.; Sabanovic, A.; Gokasan, M.; Bogosyan, S. A novel current controller scheme for doubly fed induction generators. *Autom. J. Control Meas. Electron Comput. Commun.* **2015**, *56*, 186–195. [CrossRef]

40. Ozsoy, E.E.; Golubovic, E.; Sabanovic, A.; Gokasan, M.; Bogosyan, S. Modeling and control of a doubly fed induction generator with a disturbance observer: A stator voltage oriented approach. *Turk. J. Electr. Eng. Comput. Sci.* **2016**, *24*, 961–972. [CrossRef]

41. Akagi, H.; Watanabe, E.H.; Aredes, M. *Instantaneous Power Theory and Applications to Power Conditioning*; John Wiley & Sons: New York, NY, USA, 2007.

42. Krause, P.C.; Wasynczuk, O.; Sudhoff, S.D. *Analysis of Electrical Machinery and Drive Systems*, 2nd ed.; John Wiley & Sons: New York, NY, USA, 2002.

43. Sabanovic, A.; Ohnishi, K. *Motion Control Systems*; John Wiley & Sons: New York, NY, USA, 2011.

44. Teodorescu, R.; Liserre, M.; Rodriguez, P. *Grid Converters for Photovoltaic and Wind Power Systems*; John Wiley & Sons: New York, NY, USA, 2011.

45. Xu, L. Coordinated control of DFIG's rotor and grid side converters during network unbalance. *IEEE Trans. Power Electron.* **2008**, *23*, 1041–1049.

Article

Power Balancing Control for Grid Energy Storage System in Photovoltaic Applications—Real Time Digital Simulation Implementation

Sridhar Vavilapalli [1], Sanjeevikumar Padmanaban [2,*], Umashankar Subramaniam [1] and Lucian Mihet-Popa [3]

[1] Department of Energy and Power Electronics, School of Electrical Engineering, VIT University, Vellore 632014, India; sridhar.spark@gmail.com (S.V.); shankarums@gmail.com (U.S.)
[2] Department of Electrical and Electronics Engineering, University of Johannesburg, Auckland, Johannesburg 2006, South Africa
[3] Faculty of Engineering, Østfold University College, Kobberslagerstredet 5, 1671 Kråkeroy, Fredrikstad, Norway; lucian.mihet@hiof.no
* Correspondence: sanjeevi_12@yahoo.co.in; Tel.: +27-79-219-9845

Academic Editor: Sergio Saponara
Received: 19 May 2017; Accepted: 30 June 2017; Published: 5 July 2017

Abstract: A grid energy storage system for photo voltaic (PV) applications contains three different power sources i.e., PV array, battery storage system and the grid. It is advisable to isolate these three different sources to ensure the equipment safety. The configuration proposed in this paper provides complete isolation between the three sources. A Power Balancing Control (PBC) method for this configuration is proposed to operate the system in three different modes of operation. Control of a dual active bridge (DAB)-based battery charger which provides a galvanic isolation between batteries and other sources is explained briefly. Various modes of operation of a grid energy storage system are also presented in this paper. Hardware-In-the-Loop (HIL) simulation is carried out to check the performance of the system and the PBC algorithm. A power circuit (comprised of the inverter, dual active bridge based battery charger, grid, PV cell, batteries, contactors, and switches) is simulated and the controller hardware and user interface panel are connected as HIL with the simulated power circuit through Real Time Digital Simulator (RTDS). HIL simulation results are presented to explain the control operation, steady-state performance in different modes of operation and the dynamic response of the system.

Keywords: active power control; battery charging; dual active bridge; energy storage system; hardware-in-the-loop; LCL filter

---

1. Introduction

In solar power plants, active power transfers from the photo voltaic (PV) array to the grid during daytime and the array loses its power generating capability during nighttime or when the solar irradiation is weak. To also supply power to the grid during nighttime, energy storage is required. Since the power requirements during nighttime are usually much lower than those during the daytime, energy storage with 25% of the PV array rated power may be selected for 24-h operation. A block diagram of a grid energy storage system in a solar PV power plant is shown in Figure 1.

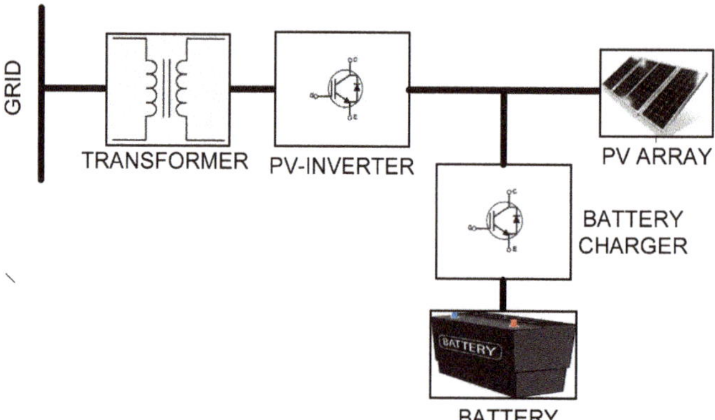

Figure 1. Generalized block diagram of a grid energy storage system in photo voltaic applications.

The different modes of operation of the above system are explained below:

Mode1: During the daytime, the PV array feeds active power to the grid through an inverter and provides charging current to the battery through the battery charger.

Mode2: The battery is in charged condition and the PV array cannot feed full power to the grid i.e., during partial cloudiness or during nighttime or when the solar irradiation is weak. In this mode of operation, PV array feeds the power to the grid based on maximum power point (MPP) and the batteries also feed active power to the grid.

Mode3: The battery is in fully discharged state and the PV array cannot provide the charging current to the battery i.e., during nighttime. In this mode of operation, the grid provides the charging current to the batteries through the inverter and battery charger.

With such systems, it is also possible to charge the batteries from the grid during non-peak load hours and the batteries along with PV array feed power to the grid during peak load hours [1]. Since the system is connected to three different power sources i.e., PV array, battery storage system and the grid, these three power sources need to be isolated to ensure the safety of the equipment. Existing energy storage systems for PV applications using a buck-boost chopper-based battery charger are briefly explained below.

In the configuration presented in [2], a DC-DC converter is connected between the PV array and PV inverter and the battery is connected across the DC link as shown in Figure 2a. In such systems, the DC/DC converter needs to be designed for the maximum capacity of the PV array even though the battery capacity is much less when the system operates in Mode3, the inverter should act like an active rectifier to charge the batteries and there is no isolation between the PV array and the batteries. In the configuration shown in Figure 2b, the PV array and PV inverter are connected to the DC link and the battery is connected to the DC link through a buck-boost chopper. In this case, the charger needs to be rated only for the rating of the battery. In the configuration presented in [3,4], independent DC-DC converters are required to connect the battery and PV array to the DC link as shown in Figure 2c.

An optimized operation of a dual active bridge (DAB) converter feeding a PV inverter connected to the grid is presented in [5,6]. Isolation between the grid and DC side is provided through a high-frequency transformer used in the DAB as shown in Figure 2d. In such a configuration, the DAB needs to be designed for the full capacity of the PV array. Since the design of a DAB is complex for high power ratings, this configuration is more suitable for low power applications. There is also no isolation between the DC link and the power bank with this configuration.

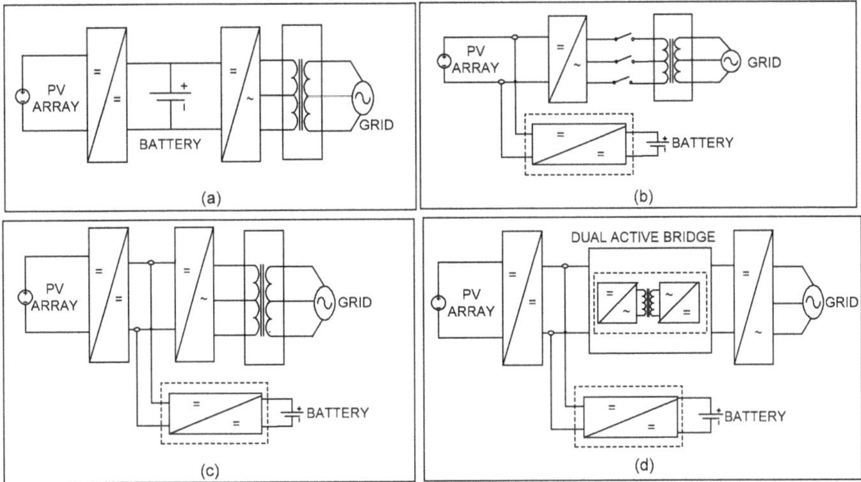

Figure 2. Buck-boost chopper-based energy storage system configurations for photo voltaic applications (**a**); two stage conversion with battery directly connected to DC Link (**b**); single stage conversion with chopper based Battery charger (**c**); two stage conversion with chopper based battery charger (**d**); dual active bridge based photo voltaic inverter with chopper based battery charger.

From the above discussions, it is observed that buck-boost chopper-based ESS cannot provide complete isolation. In this paper, a DAB-based energy storage system (ESS) for PV applications is proposed which can mitigate the drawbacks of buck-boost chopper-based systems. In the DAB-based ESS configuration, PV array and the PV inverter are directly connected to the DC link and the battery is connected to the DC link through a DAB-based bi-directional battery charger as shown in Figure 3. A high-frequency transformer in the DAB provides isolation between the DC link and the power bank. A transformer connected between the inverter and grid provides isolation between the DC sources and AC grid.

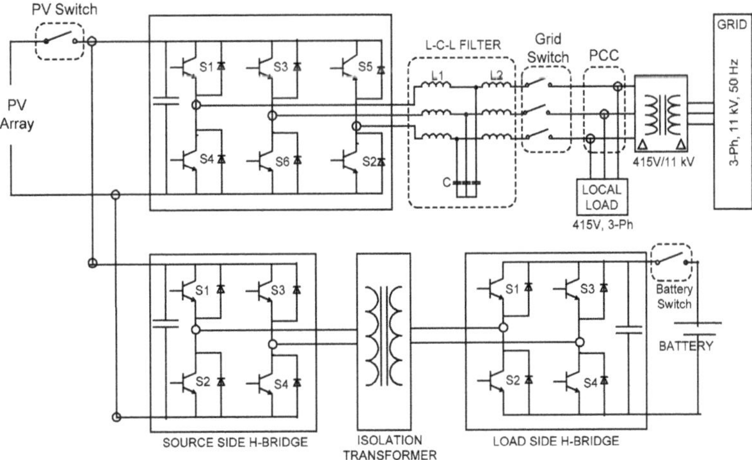

Figure 3. Dual active bridge-based energy storage system for a grid connected PV system.

The following technical features are the main advantages of the proposed system:

- The battery charger only needs to be designed for the battery capacity.
- Independent controls for the battery charger and inverter are possible.
- When the system operates in Mode3, the inverter acts like a simple diode rectifier and the battery charger takes care of the charging current.
- The high-frequency transformer in the DAB provides isolation between the PV array and the battery.

In this paper, a power balancing control for the DAB-based energy storage system is proposed and validated through real-time simulations. Detailed discussions on the proposed system, design calculations, and the control structure are presented in Section 2. The proposed power balancing control algorithm is explained in Section 3. In Section 4, a hardware-in-the-loop (HIL) simulation setup to validate the control algorithm is explained. HIL results are presented in Section 5.

2. Dual Active Bridge Based Energy Storage System for Photo Voltaic Applications

As shown in Figure 3, in a dual active bridge-based energy storage system, the PV array is connected to the DC link directly and the battery is connected to the DC link through a DAB-based bidirectional DC-DC converter. The PV inverter is connected to the grid through an isolation transformer. The transformer secondary is the point of common coupling (PCC) i.e., the coupling point of the grid, PV inverter output and the local load. A sine filter is used at the output terminals of the PV-inverter to smoothen the inverter output voltage. In this paper, design, and control of the proposed system with a 100 kVA PV-inverter and energy storage of 25% capacity for 4 h minimum backup time are presented. An overview of the electrical requirements of the system is shown in Table 1. Since the local load is rated for a 415 V, 50 Hz, 3-phase, the PCC voltage is selected to be the same as the rated voltage of the local load to avoid an additional transformer across the load and the PCC. Design calculations for the system to meet the electrical specifications are presented in the next subsection.

Table 1. Electrical specifications/requirements of the system.

SL. NO	Parameter	Value	Units	Remarks
1	Grid Voltage	11	kV	3-Phase
2	Grid Frequency (F)	50	Hz	
3	PCC Voltage	415	V	Rated Voltage of Local load
4	Maximum PV-Inverter Power	100	kW	
5	Backup Power	25	kW	25% of PV-Inverter Power
6	Minimum Back Up Time	4	h	

2.1. Design Calculations for the Photo Voltaic-Inverter

The design calculations for the inverter and the filter are presented in Table 2. An LCL filter (2 inductances-L connected in series with a capacitor-C in parallel) is often used to interconnect an inverter to the utility grid in order to filter the harmonics produced by the inverter. Since the PCC voltage is 415 V and the grid voltage is 11 kV, the grid side transformer with a transformation ratio of 11 kV/415 V with minimum 100 kVA rating is selected.

Due to the L-C-L filter used at the output side of the inverter, there is an voltage drop across the filter, so the inverter output voltage should be the sum of PCC voltage and the voltage drop across the filter. Considering the sinusoidal PWM (SPWM) technique for pulse generation, the minimum DC link voltage required is calculated based on the inverter output voltage. Since the inverter needs to be designed to handle the filter capacitor current in addition to the rated load current, an optimal filter capacitor is selected which draw less than 5% of rated current. The Inverter side inductor L1 is derived from the value of the capacitor C and the corner frequency Fc. After selecting the inductor L1, the grid side inductor L2 can be selected based on the maximum filter drop allowed. Inductance L2 can also be made part of the transformer on the grid side to eliminate the physical inductor L2 in the system.

Table 2. Design calculations for the photo voltaic-inverter.

SL. NO	Parameter	Value	Units	Remarks
1	Power Rating	100	kW	
2	PCC Voltage (Vpcc)	415	V	
3	Inverter RMS Current	140	A	Power/(1.732 ∗ PCC Voltage)
4	Maximum Filter Drop	6	%	Drop across L-C-L Filter
5	Inverter Voltage (Vinv)	440	V	PCC Voltage + filter Drop
6	Minimum DC Link Voltage	620	V	Vdc = (Vinv/0.71) With SPWM
	Selection of Filter Capacitor (C)			
7	Maximum Reactive Power(Qc)	5	%	5% of 500 kVA i.e., 25 kVAR
8	Current Rating of Capacitor (Ic)	11.5	A	Qc/PCC Voltage
9	Maximum Capacitance	85	uF	Ic/(2 ∗ pi ∗ F ∗ Vpcc)
10	Selected Value of Capacitance	80	uF	<Maximum capacitance
	Selection of Inverter Side Filter Inductor (L1)			
11	Corner Frequency selected (Fc)	1.25	kHZ	Switching Frequency/4
12	Inductance of Inductor L1	203	uH	Fc = 1/[2 ∗ pi ∗ √(LC)]
13	% Voltage Drop in Inductor L1	2.1	%	[Irms × (2 ∗ pi ∗ F ∗ L1)]/Vpcc
	Selection of Grid Side Filter Inductor (L2)			
14	Maximum Drop allowed across L2	3.9	%	Max Drop-% Drop across L1
15	Maximum Inductance of L2	382	uH	(3.9% ∗ Vpcc)/[Irms ∗ 2 ∗ pi ∗ F]

2.2. Selection of Battery Type

The procedure for the calculation of PV power requirement for battery charging is explained in Table 3. As mentioned earlier, since the power requirement during nighttime is much lower than that during the daytime, an energy storage with 25% of the rated power of the PV array is selected. A Lithium-ion battery with a nominal voltage of 350 V is selected as an energy storage in this system. The ampere-hour rating of the battery is decided based on the minimum backup time required and the battery discharging current. Similarly, the charging current of the battery is calculated based on the charging time and ampere-hour rating of the battery. The power required from the PV array for charging the battery is determined from the battery nominal voltage and the charging current.

Table 3. Electrical parameters of the battery.

SL. NO	Parameter	Value	Units	Remarks
	Selection of Battery			
1	Nominal Voltage of Battery (Vnom)	350	V	
2	Maximum Battery Voltage	406	V	116% of Vnom for Li-Ion battery
3	Minimum battery Voltage	306	V	>87.5% for Safe operation
4	Battery Rated Power	25	kW	25% of PV-Inverter rating
5	Maximum Battery Current	72	A	Battery Power/Vnom
5	Minimum Backup time	4	h	
6	Ah Rating of Battery	288	Ah	Current X Backup Time
7	Battery Charging time	8	H	PV Power availability time
8	Charging Current (I_Charging)	36	A	Ah Rating/Charging Time
9	PV Power Required for Charging	12.5	kW	Vnom × I_charging

The battery selected for this system can be modeled as a voltage source [1,7,8] and the model was implemented based on the equation for the battery voltage expressed as below [9].

Battery Voltage:

$$V_{Batt} = E_0 - K\,[Q/(Q\text{-}I*T)] + Ae^{(-B*I*T)} - [I_{Batt}*R] \tag{1}$$

where V_{Batt} is the battery voltage (V), I_{Batt} is the battery current (A), E_0 is the nominal voltage (V), Q is ampere-hour rating of the battery (Ah), B is the nominal discharge current $(Ah)^{-1}$, K is the fully charged voltage (V)which is around 116%, A is the exponential voltage (V), which is around 105%, R is the internal resistance of the battery, and "I * T" is the discharged capacity (Ah) which is determined by integrating the battery current.

State of charge of the battery can be obtained after integrating the battery current:

$$\% \, SOC = 100 \times \{1 - [(I * T)/Q]\} \tag{2}$$

2.3. Selection of the PV Array

From the above discussions, a PV array for a minimum power rating of 113 kW is required, since a power of 100 kW for the grid and 13 kW for battery charging is required from the PV array. Design calculations for the PV array using 435 watt PV module (Make: M/s Sunpower, Model: SPR-435NE-WHT-D) is explained and the procedure for selecting a number of series, parallel PV modules in a PV array is also presented in Table 4. An operating temperature range of 25 to 55 °C is considered for the calculations. PV module parameters such as MPP voltage, MPP current, open circuit voltage and short circuit current at 25 °C are obtained from the data sheet and the values at 55 °C are derived using the temperature coefficients of the PV modules. The operating range of PV module voltages and currents are tabulated.

Table 4. Photo voltaic array selection.

SL. NO	Parameter	Value	Units	Remarks
	PV Array Requirement			
1	Minimum Power Requirement	113	kW	PV Inverter + Charging Power
2	Minimum PV Voltage (V_PV_Min)	620	V	Vdc Minimum Refer Table 2
3	Maximum PV Current	182	A	Power/V_PV_Min
	Details of Selected PV Module			
4	Make			Sunpower
5	Type Number			SPR-435NE-WHT-D
6	Operating Temperature Range			25–55 °C
	Electrical Ratings of Selected PV Module at 25 °C			
7	Power Rating of Each Module	435	W	
8	Open Circuit Voltage (Voc)	85.6	V	
9	Short Circuit Current (Isc)	6.43	A	From Datasheet
10	MPP Voltage (Vmpp	72.9	V	
11	MPP Current (Impp)	5.97	A	
	Temperature Coefficients of Selected PV Module			
12	Temperature Coefficient for power	−0.38	%/K	
13	Temperature Coefficient for Voltage	−233.5	mV/K	From Datasheet
14	Temperature Coefficient for Voltage	3.5	mA/K	
	Electrical Ratings of Selected PV Module at 55 °C			
15	Open Circuit Voltage (Voc)	78.59	V	
16	Short Circuit Current (Isc)	6.535	A	Derived From Values at 25 °C and
17	MPP Voltage (Vmpp)	65.89	V	the Temperature Coefficients
18	MPP Current (Impp)	6.08	A	
	Selected PV Module Electrical Ratings in the Operating Temperature Range			
19	Minimum Voltage (Vmod_min)	65.89	V	Vmpp at 55 °C
20	Maximum Voltage (Vmod_max)	85.6	V	Voc at 25 °C
21	Maximum Current (Imod_max)	6.535	A	Isc at 55 °C
22	Maximum power (Pmod_max)	435	W	From Datasheet

Table 4. *Cont.*

SL. NO	Parameter	Value	Units	Remarks
	Electrical Ratings of PV Array			
23	Minimum No. of Modules required (N)	260	No's	PV power/Pmod_max
24	Minimum No. of Modules in Series (Nse)	10	No's	V_PV_Min/Vmod_min
25	No. of Modules in parallel (Np)	26	No's	N/Nse
26	Minimum Voltage of PV Array	658.9	V	Vmod_min × Nse
27	Maximum Voltage of PV Array	856	V	Vmod_max × Nse
28	Maximum power from PV Array	113	kW	Nse × Np × Pmod_max

The total number of PV modules required in the PV array is calculated based on the total power requirement and the power rating of each PV module. The number of series PV modules in a PV array is selected based on the minimum DC link voltage requirement for the PV-inverter. For this system, the minimum DC link voltage required is 620 Volts to match the inverter voltage with the PCC voltage. Hence the PV array minimum voltage should always be more than 620 V in the operating temperature range. The minimum number of parallel PV modules in the PV array is calculated from a total number of PV modules and the number of series PV modules selected.

The PV array can be modeled as a current source [1,7,10] and the mathematical expression for the PV array model is as given below [11].

PV Current:

$$I = I_{ph} - [I_s \times (e^{(V + I * Rs)/N * Vt - 1)}] - [(V + I * R_s)/R_p] \tag{3}$$

where I is PV current and V is the PV voltage. I_{ph} is the photon current and it is expressed as:

$$I_{ph} = \text{Irradiance} * (I_{sc}/I_{ro}) \tag{4}$$

where I_{sc} is the short circuit current of the PV array = I_{sc} of each module, N represents the number of parallel modules, I_{ro} is the measured irradiance = 1000 W/m^2 (from the datasheet), I_s is the diode saturation current and expressed as:

$$I_s = I_{sc}/(\exp(Voc/(n * V_t)) - 1) \tag{5}$$

where I_{sc} is the short circuit current of the PV array, Voc is the open-circuit voltage= Voc of each module, multiply by the number of series modules, n is the quality factor and Vt is the thermal voltage and expressed as:

$$V_t = k * T/q \tag{6}$$

where k is Boltzmann's constant = 1.3806×10^{-23}, T is the operating temperature = 25 °C, q is charge of an electron = 1.602×10^{-19}, R_S is the series resistance of the PV array, R_p is the parallel resistance of the PV array.

2.4. Design Calculations for the Battery Charger

Battery charger ratings are decided after selecting the PV array. Since the PV array and the battery charger are connected to the common DC link, the battery charger input voltage range is the PV array operating voltage range. The battery charger output voltage range is the battery operating voltage range. Based on the input and output voltage range of the battery charger, an isolation transformer with a transformation ratio of 1:2 is selected. Electrical parameters for the battery charger system are listed in Table 5.

Design calculations for the proposed system are explained briefly in this subsection. The control methodology for the battery charger and the PV-inverter are explained in next subsections.

Table 5. Electrical parameters of the battery charger system.

SL. NO	Parameter	Value	Units	Remarks
	Selection of Battery			
1	Battery Charger Rated Power	25	kW	Battery discharging capacity
2	Battery Charger Input Voltage	658–856	V	PV Operating Range
3	Battery Charger Output Voltage	306–406	V	Battery Operating Voltage
4	Isolation Transformer Turns Ratio	1:2		
5	Transformer Primary Current	38	A	Rated Power/Min Input Voltage
6	Transformer Secondary Current	82	A	Rated Power/Min Output Voltage
7	Minimum kVA of Primary	32.5	kVA	Primary Max Voltage × Current
8	Minimum kVA of Secondary	33.3	kVA	Secondary Max Voltage × Current
9	Selected Transformer KVA Rating	35	kVA	More than minimum kVA

2.5. Control of the Dual Active Bridge-Based Battery Charger

The DAB-based DC-DC converter consists of two H-bridges and a high-frequency transformer. The source side H-bridge is connected to the DC link and the load side H-bridge is connected to the battery, as shown in Figure 3. A high-frequency transformer is required to match the battery voltage with the DC link voltage and also to provide isolation between the PV array and the battery. The transformer's leakage inductance helps in boost operation mode [12,13]. The transformer winding connected to the source side H-bridge is considered as the primary and the winding connected to the load side H-bridge is considered as the secondary. A control block diagram of the DAB-based battery charger is shown in Figure 4. Source side and load side H-bridges act like a simple square wave inverter. Square pulses with 50% duty cycle are provided to the load side and source side H-bridges. Power flow through the DAB is controlled using phase shift control. The pulse generator provides the gate pulses for the source side and load side H-bridges based on the phase shift obtained through a PI controller.

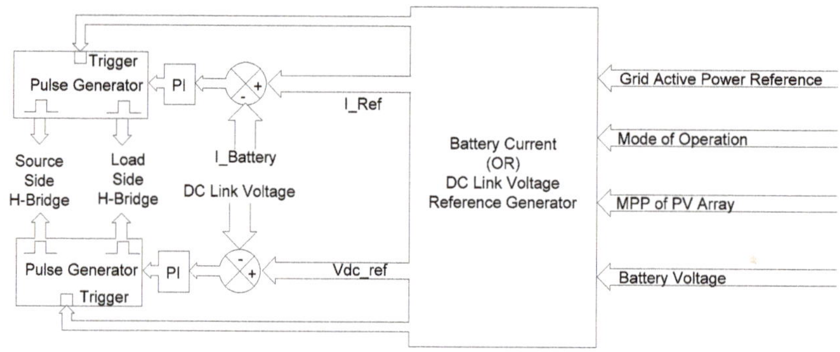

Figure 4. Control block diagram of the DAB-based battery charger.

During current control mode, when the battery current reference is zero, then the gate pulses for source side and load side H-bridges will be in phase with each other. During forward power flow i.e., for charging the battery, the gate pulse of the source side H-bridge will be in leading to the gate pulse of the load side H-bridge. Similarly, during reverse power flow i.e., during battery discharging mode, the load side gate pulse will be leading. The amount of power transfer depends on the phase angle between the gate pulses for the source side bridge and load side bridge. As the size and cost of a high-frequency transformer are much less than those of a high-frequency transformer for the same power rating, battery charger size and cost can be reduced by using a high-frequency transformer in a DAB.

During voltage control mode, the battery side H-bridge receives 50% duty cycle gate pulses and the DC link side H-bridge is controlled to maintain the DC link equal to the reference DC link voltage.

The detailed control philosophy of the DAB-based battery charging system during various modes of operation is explained below:

(1)   Inputs to the reference generator block are the mode of operation, MPP of the PV array, active power reference, and battery voltage signals.

(2)   When the system is operating in either Mode1 or Mode3:

- The battery is in charging mode of operation hence the battery current reference is taken as positive.
- Based on the battery SOC, the reference charging current is obtained through a look-up-table.
- When the battery SOC is in the range of 80 to 115%, then the battery charging current is maintained at 0.12 C i.e., 36 A (0.12 * amp-hour rating of battery) as shown in Table 3.
- When the battery is fully charged, then the battery voltage will reach the maximum voltage, then the reference battery current is made zero to avoid overcharging. In this case, the operation can be transferred to Mode2, if the load requirement is more than the PV power.
- When the battery is fully discharged, then the battery voltage will be less than 0.9 times the nominal voltage, then the charging current is adjusted to 0.2 C i.e., 57 A for fast charging.

(3)   When the system is operating in Mode2:

- In this mode, the battery is in discharging mode of operation hence the battery current reference is taken as negative.
- In this case, if the PV array is in an inactive state i.e., PV voltage is more than the minimum DC link voltage required (i.e., 620 V) but the MPP of the PV array is less than the critical load requirement:
  - The battery discharging current is obtained from the Amp-hour rating of the battery and the discharging time.
  - Since the minimum backup time in this system is 4 h, the user can adjust the backup time to be more than 4 h.
- In case the PV voltage is less than the minimum DC link voltage required then the battery charger needs to provide the required voltage to the DC link:
  - In this case, the DC link voltage reference (Vdc_ref) is generated by the reference generator.
  - Vdc_ref is always maintained at more than the minimum required DC link voltage (620 V).

2.6. Control of the Grid-Connected Photo Voltaic Inverter

A typical grid connected solar power conditioning system consists of a three-phase two level PV-inverter for converting DC power to AC power, a sine filter to smoothen the AC output and a transformer to couple the inverter and the grid. The transformer also provides isolation between AC side and DC side. Figure 5 shows a control block diagram for a grid connected PV-inverter. In this system, the PV array voltage and currents are to be monitored for MPP tracking and the grid voltage is to be monitored for the phase-locked loop (PLL). The controller senses the charging current or discharging current of the battery and the MPP of the PV array and then calculates the maximum possible power that can be fed to the grid. The current reference is generated based on the maximum possible power and the PLL output. Three phase grid voltage is applied to the PLL to find out the angle wt. Angle wt obtained through the PLL is used to generate Id and Iq components from three phase grid currents. After comparing the reference Idq currents and actual Idq currents, the error signals are given to the PI controllers for the active and reactive power control. The PI controller outputs are converted back to three Phase modulating signals and given to the PWM generator to generate inverter gate pulses [14,15].

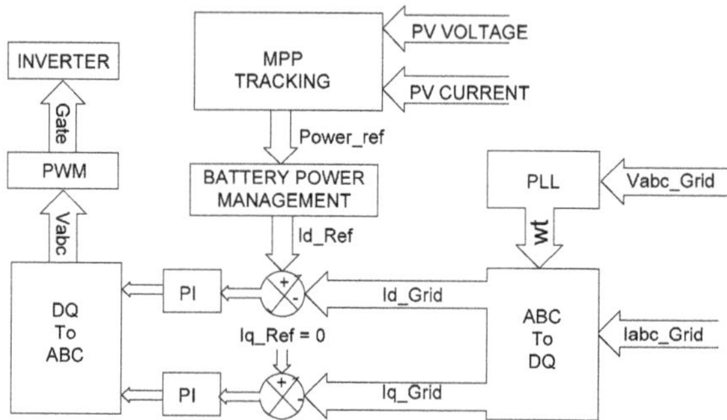

Figure 5. Control block diagram of a grid-connected photo voltaic inverter.

The perturb and observe method is used for maximum power point tracking (MPPT). In this method, the following activities are carried out:

(a) Initially, When the system starts the PV power (Ppv) = 0

- In this state, PV voltage = Voc and the PV current = 0
- Initialize PV power reference MPP = 0
- Minimum PV power reference (MPP_Ref_Min) is limited to 0.
- Maximum PV power reference (MPP_Ref_Max) is limited to 113 kW.

(b) Now increase the PV power reference (MPP) in 500 Watt steps:

- Measure PV voltage and current and calculate the new PV power (Ppv_New)
- If Ppv_New is more than Ppv

  - Ppv == Ppv_New
  - MPP == MPP + 500 Watt

- If Ppv_New is less than Ppv

  - Ppv == Ppv_New
  - MPP == MPP − 500 Watt

(c) Based on MPP_Ref and battery current, the reference inverter current is obtained.
(d) Steps 'b' and 'c' operate in a continuous loop.
(e) Since this activity is not required when the solar irradiation is weak, this loop can be bypassed during the nighttime.

2.7. Operation Sequence of the System

The normal operation sequence of the system is explained below:

(a) During nighttime, the PV voltage is less than the minimum required DC link voltage. Considering that the battery is in discharged mode and being charged from the grid, hence the system is in Mode3.
(b) Now when the irradiation improves in the morning, MPP tracking is started and when the MPP becomes more than the minimum power required for system operations and other critical requirements then the system switches to Mode1.
(c) As the irradiance improves during the daytime, since the battery charging power is almost constant, power transfer to the grid increases.
(d) Again when the irradiation is getting reduced, the power transferred to the grid also reduces.

(e)   In case the power requirement for the grid is more than the available PV power, then the system can be transferred manually to theMode2 operation to meet the power demand.

(f)   When the irradiation reduces further and becomes zero, then the battery stays in Mode2 till the battery gets discharged and the operation shifts to Mode3 and the process loops back to step (a).

In this work, a new power balancing control algorithm for the proposed configuration is developed to meet the operational requirements of the system. The proposed algorithm is explained in the next section.

3. Power Balancing Control of the Grid Energy Storage System in Photo Voltaic Applications

Power control of PV with ESS for off-grid applications is presented in [16]. In the system presented, the battery and PV arrays are connected to the common AC load through independent converters i.e., as an AC-centric system. During charging of the battery, the PV array supplies power to the AC load and battery. When the battery is fully charged, the battery and PV arrays supply power to the common AC load. Conditions for battery charging and discharging and control of power converters during Mode1 and Mode2 operation are explained briefly. Experimental results were presented to show the dynamic response of the system during mode changeover.

Control for high power PV + a fuel cell plant with hybrid energy storage consisting of a battery and the supercapacitor is presented in [17]. Each energy source and energy storing element are connected to a common DC link through independent converters in this configuration. Energy management among the different sources, control for charging the super-capacitor and batteries is explained briefly. Experimental results were presented for explaining the dynamic characteristics of the system and plant performance during long load and short load cycles. Since the system presented is for off-grid applications, Mode3 operation i.e., charging the battery from the grid supply is not covered in this work. Real-time simulation of the hybrid energy system with wind-PV-battery storage is presented in [18]. In the presented system, independent converters for battery, PV modules, and the wind are used. Based on the power availability of all the sources, an algorithm is developed for battery charging, discharging and load shedding.

The systems presented in [16–18] are for off-grid applications hence the control during Mode3 of operation is not covered. System configurations presented in the above works require independent converters for each source, which may increase the cost of the system and also may increase the complexity of the control algorithm. The above mentioned drawbacks can be mitigated with the proposed system configuration and with the power balancing control algorithm explained in the next subsection.

The proposed system is the combination of three phase PV inverter and a DAB-based battery charger explained in Section 2. Power flow through the inverter can be controlled over a wide range through current control. Battery current can be controlled in both directions through DAB using a phase angle control. Power balancing among the three sources in the presently proposed system is achieved by controlling the power flow through the battery charger and inverter. The power balance control algorithm shown in Figure 6 is explained below:

(1)   Once the system is ready and the start command is given by the user, the controller reads the grid voltages for determining wt through PLL.

(2)   The controller initializes the value of the inverter reference current (Id_Inv_Ref) and battery reference currents (I_Batt_Ref) as 0.

(3)   The controller reads the PV voltage (Vpv), PV current(Ipv), battery voltage(VBatt) and battery current (I Batt)

- If the PV voltage (Vpv) is less than the minimum PV voltage required (Vpv_Min) then MPP of the PV array is zero. Vpv_Min is the minimum DC required to match the inverter output voltage with the transformer secondary voltage. In this system, 620 V is the minimum DC link voltage required, as shown in Table 2.

- ○ In this case, if the battery voltage is also less than the nominal battery voltage Vb_Nominal then the system is in Mode3.
- ○ In this mode, the battery needs to be charged but the PV array cannot provide any power for battery charging, so the grid shall supply the power required for battery charging.
- ○ In this mode of operation, the inverter acts like a simple diode rectifier to provide DC input to the battery charger.
- ○ Based on the SOC of battery, the reference battery charging current I_Batt_Ref is obtained.

- If the PV voltage (Vpv) is higherthanthe minimum PV voltage (Vpv_Min) then the controller tracks the MPP of the PV array by monitoring the PV voltage and current.

  - ○ In case the MPP is higher than minimum value i.e., P_PV_Min then the system is in Mode1.
  - ○ In this mode of operation, the battery will be in charging state and the PV array provides the power for battery charging.
  - ○ The remaining power after battery charging will be transferred to the grid.
  - ○ Based on the SOC of the battery, the reference battery charging current I_Batt_Ref is obtained.
  - ○ Through the power balancing equation, the inverter reference current Id_Inv_Ref is calculated based on the MPP and battery current:
    Inverter power reference = MPP − battery power reference
    (the battery Power reference is positive during charging mode and negative in discharging mode)

    - ⇨ Inverter Power Reference = MPP − (I_Batt_ref * V_Battery)
    - ⇨ $\sqrt{3}$ * Vabc_rms * Iabc_rms_Ref = MPP − (I_Batt_ref * V_Battery)
      where Vabc_rms is the RMS value of line voltage of the grid/inverter
      and Iabc_rms_ref is the reference RMS value of line current of the inverter
    - ⇨ Iabc_rms_ref = (MPP − [I_Batt_ref * V_Battery])/($\sqrt{3}$ * Vabc_rms)

  - ○ Id_Inv_Ref can be calculated through the abc to dq transformation. Since in this system Iq_reference is always maintained at zero, the magnitude of Id_Inv_Ref can also be obtained as given below:

    - ⇨ Id_Inv_Ref = $\sqrt{2}$ * Iabc_rms_ref
    - ⇨ Id_Inv_Ref = $\sqrt{2}$ * (MPP − [I_Batt_ref * V_Battery])/($\sqrt{3}$ * Vabc_rms)

- If the MPP is less than the minimum value (P_PV_Min) but the battery is in charged condition then the system is in Mode2.

  - ○ In this case based on the backup time adjusted by the user, the reference battery current I_Batt_ref is calculated. In case the backup time adjusted by the user is 6 h, then the battery current reference is calculated as follows:

    - ⇨ I_Batt_ref = rated Amp-hour rating of the battery/backup time
    - ⇨ I_Batt_ref = 288 Ah/6 h = 48 A

  - ○ Through the power balancing equation, the inverter reference current Id_Inv_Ref is calculated based on the MPP and battery current:
    Id_Inv_Ref = $\sqrt{2}$ * (MPP − [I_Batt_ref * V_Battery])/($\sqrt{3}$ * Vabc_rms)
  - ○ When the PV voltage is more than the minimum DC link voltage, then the battery charger is operated with closed loop current control to maintain I_Batt = I_Batt_Ref.
  - ○ When the PV voltage is less than the minimum DC link voltage then the battery charger operates with closed loop voltage control to maintain a constant DC link voltage i.e., Vdc_Link = Vdc_Link_Ref.

(4)   After determining the battery reference current I_Batt_Ref and inverter reference current Id_Inv_ref, the controller implements the closed loop current control through PI controllers and releases the gate pulses to the inverter stack and battery charger stack.

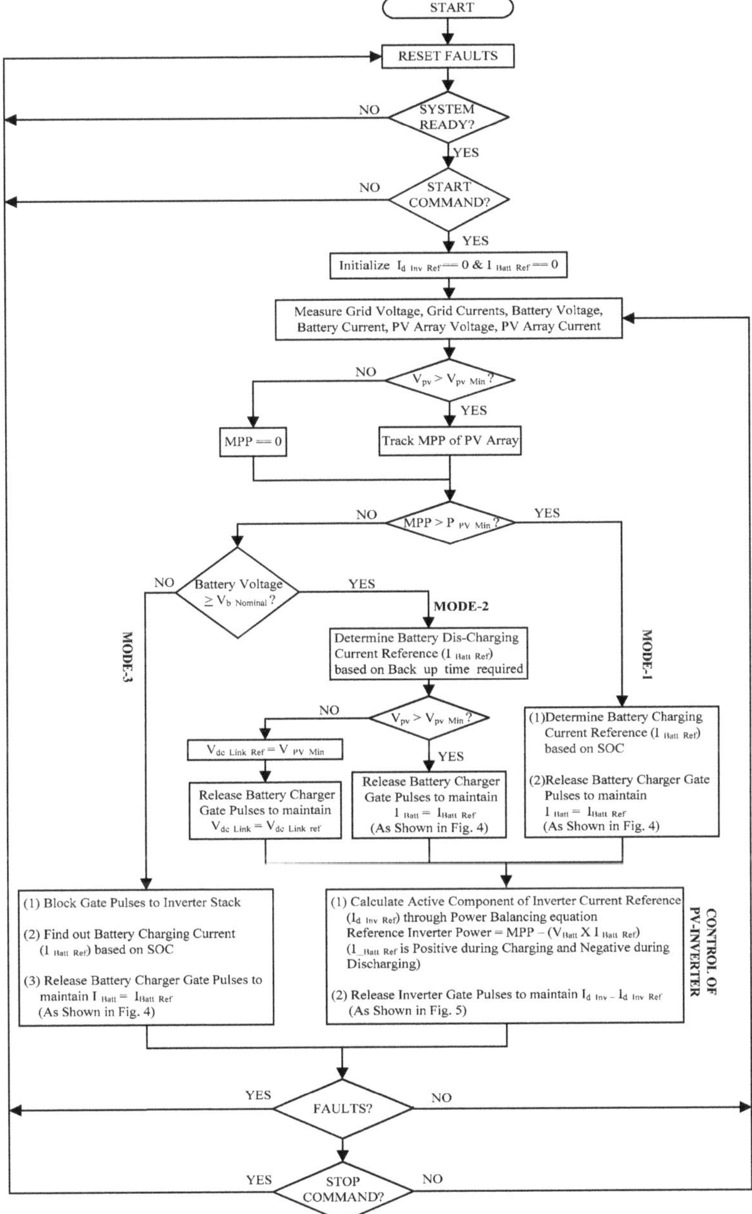

Figure 6.   Algorithm for power balancing control of grid energy storage system in photo voltaic applications.

The proposed algorithm was tested on the real controller with the help of Hardware-In-Loop simulations. The need for HIL simulations, features of the real-time digital simulator and the setup built for HIL simulation for the proposed configuration are explained in the next sections.

4. Hardware-in-the-Loop Simulation Setup for the Proposed System

In general, controller and control software are validated by integrating the controller with actual plant hardware. However, in the case of any error in the control system, there are risks of personal injuries, damage to the equipment and delays. Hardware-in-the-Loop (HIL) simulation is a useful tool to avoid such issues. In HIL simulation, instead of a real plant a mathematical model representing the plant loaded in the real-time simulator to act like an actual plant. Through HIL simulations, the response of a controller in real time operation can be validated [19]. The following are the advantages with the HIL simulation: (a) prototype controller software can be developed with minor assumptions about the plant parameters; (b) once the control parameters are calculated with the HIL simulation, it is easy to tune the parameters of the actual system; (c) it saves design cost and time (d) system protections in real time can be analyzed by simulating faults. To validate the control software for the proposed system, HIL simulations were carried out. A plant consisting of a PV array, inverter, battery charger, isolation transformers, grid, battery and contactors was simulated using the Matlab-Simulink software. The simulated model is compiled and loaded into the processor of a real-time digital simulator (RTDS). The simulated plant can be accessed by the external controller cards and other hardware through the I/O channels available in the RTDS. The DSP-based controller is connected as hardware in the loop as shown in Figure 7.

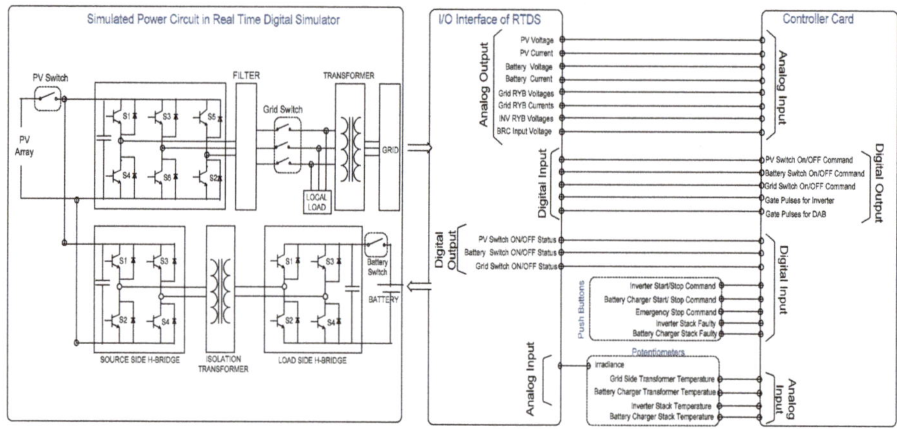

Figure 7. Block diagram for hardware-in-the-loop simulation of the proposed grid energy storage system.

The RTDS used for the HIL simulations is the Opal-RT Simulator and the controller hardware is based on a Texas Instruments TMS320F2812 DSP-based controller card. The user interface panel consisting of pushbuttons and potentiometers (POT) is used for user commands and for simulating the faults. Figure 8 shows the hardware setup for the HIL simulations.

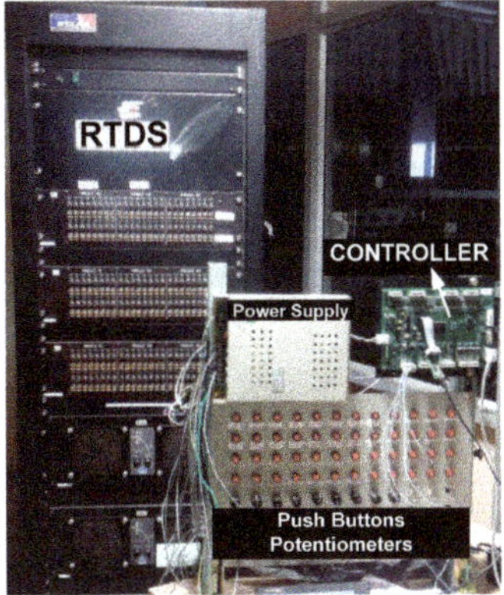

Figure 8. Hardware-in-the-loop simulation setup for the proposed grid energy storage system for photo voltaic applications.

4.1. Input-Output Channels of Real-Time Digital Simulator

The Opal-RT RTDS is equipped with analog and digital input-output modules. The voltage range for analog signals is ±15 V whereas the voltage levels for digital signals is 0 and +15 V.

4.2. Input-Output Channels of Controller Card

The controller used in this work is a TMS320F2812 DSP processor-based controller card. The voltage range for analog signals is ±10 V whereas the voltage levels for digital signals is 0 and +15 V.

4.3. User Interface Panel Signals

The input to the simulated PV array is solar irradiance, which can be provided to the simulated plant through the analog input channel of RTDS from a POT mounted on a user interface panel. The minimum value of the POT output refers to an irradiance of 0 and the maximum value of the POT refers to 1000 W/m^2. From the user interface panel, start/stop commands, and emergency stop commands are given to the controller card for the plant operations. The controller card receives temperature signals of from the isolation transformers and power stacks from the user interface panel. The possible fault signals are also sent to the controller card hence different faults can be simulated to check the functionality of the controller and control algorithm.

4.4. Signals from Simulated Plant to Controller

The controller receives the analog signals of the plant through the analog output channels of the RTDS. The controller receives PV voltage and current signals which are required for tracking MPP. Battery voltage is required for finding out the charging current reference and battery current signal is required for closed loop current control of the battery charger. Three phase grid voltage signals are required for the PLL and inverter side voltages are monitored for synchronization purposes. Three phase inverter currents are required for the closed loop current control of the PV inverter. Based on the start-stop commands received from the user interface panel, the controller gives the ON/OFF commands to the PV switch, grid switch and battery switches through the digital input channels of RTDS. Switch status outputs to the controller are given to the controller through the digital output channels of the RTDS.

4.5. Online Plant Parameter Modifications

Online modification of simulated plant parameters i.e., transformer parameters, filter parameters, battery SOC, DC link capacitor values and load parameters, etc. during the real-time digital simulation is also possible through RT lab main controller. Through online modifications, the optimum values of the plant components can also be obtained.

5. Results and Discussions

5.1. Inverter, Grid and Load Currents in Different Modes of Operation

Figure 9 shows the current waveforms in Mode1 of operation. Since the PV array can produce more than the minimum power required for charging the battery and feeding the internal loads connected to the plant, the additional power produced by the PV array is supplied to the grid. An irradiance of 1000 W/m^2 is adjusted on the user interface panel, hence the PV array is producing the maximum possible power. From the presented result, it can be observed that the grid current is phase displaced by 180 degree with respect to the inverter current.

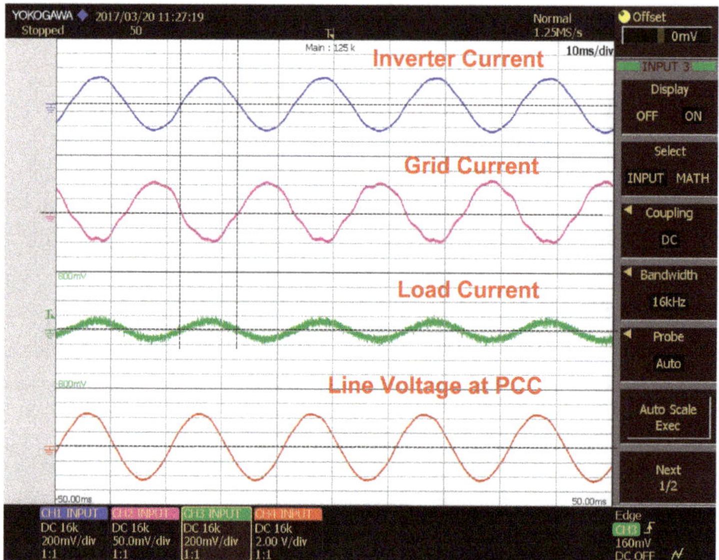

Figure 9. Currents of inverter, grid, load, and voltage at PCC in Mode1 of operation.

In Mode2 of operation, an irradiance of 0 W/m^2 is adjusted on the user interface panel; hence the PV array cannot produce any power. The battery is in charged condition and supplies the power to the load based on the Amp-hour rating of the battery and the discharging time or backup time adjusted by the user.

Since the local loads consume more than the inverter supplied current, the remaining current is drawn from the grid as shown in Figure 10. Since the load is drawing current from both the sources, the grid current and the inverter current are in phase with each other.

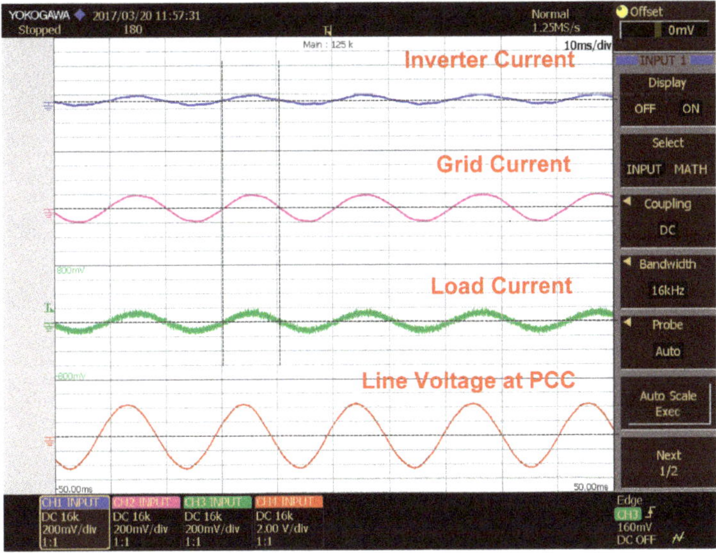

Figure 10. Currents of inverter, grid, load , andvoltage at PCC in Mode2 of operation.

In Mode3 Operation, the PV array cannot produce any power and the battery is in discharged condition. Since the battery is to be charged, the grid supplies the necessary charging current to the battery and the current required for the local loads as shown in Figure 11.

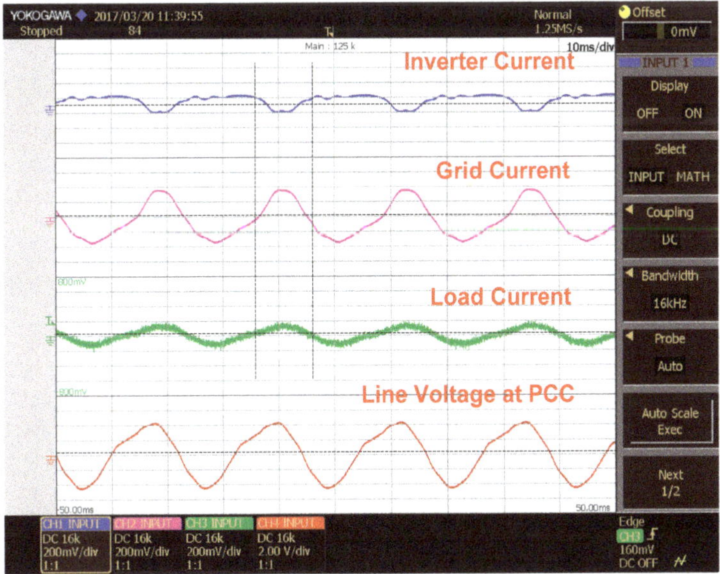

Figure 11. Currents of inverter, grid, load, and voltage at PCC in Mode3 of operation.

The dynamic response of the system to a step change in inverter reference power is shown in Figure 12.

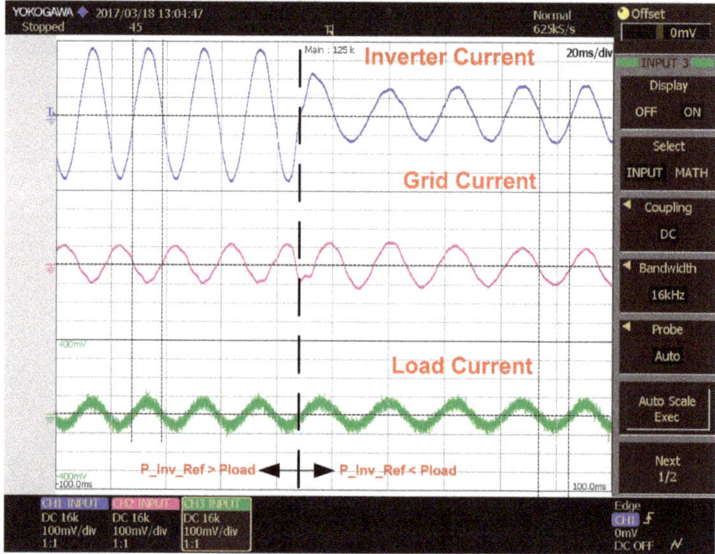

Figure 12. Currents of inverter, grid, load for a step change in reference power.

When the inverter power reference is more than the local load requirement then the inverter is supplying current to grid and load. Since the grid is receiving the current, the phase displacement between grid and inverter currents is 180 degrees. After a step change in the reference power, since the reference power is less than the load requirement, the load current is supplied from both the inverter and the grid, hence the both the currents are in phase with each other. With the present controls, the system reaches the steady state within one cycle time.

5.2. Battery Charger Input and Output Currents in Different Modes of Operation

Battery current is considered as positive during charging and negative during discharging of the battery. The battery will be in charged condition in Mode1 and Mode3 of operations as explained earlier. During charging, depending on the SOC of the battery, the charging current reference is obtained and the controller carries the closed loop current control of the battery charger. The battery charger input and output currents for different modes of operation are discussed below. In Mode1 operation, since the battery is in charging condition, battery current and the average value of battery charger input current are positive, as shown in Figure 13.

In Mode2 of operation, when the PV array voltage is less than the minimum DC link voltage then the battery charger operates with closed loop voltage control and maintains a constant DC link voltage. The current through the battery depends on the Id Reference of the inverter which is obtained through the Amp-hour rating of the battery and the backup time required for the user. Since the battery is in discharging condition, battery current and the average value of the battery charger input current are negative, as shown in Figure 14.

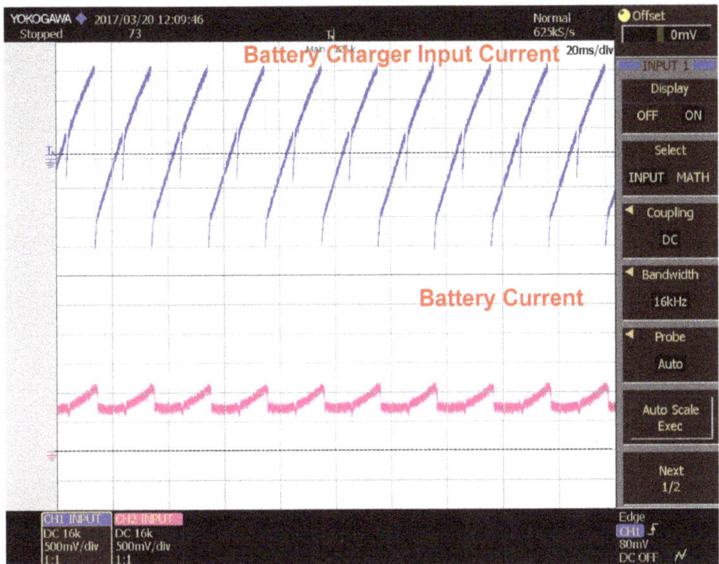

Figure 13. Battery charger input and output currents in mode1 operation.

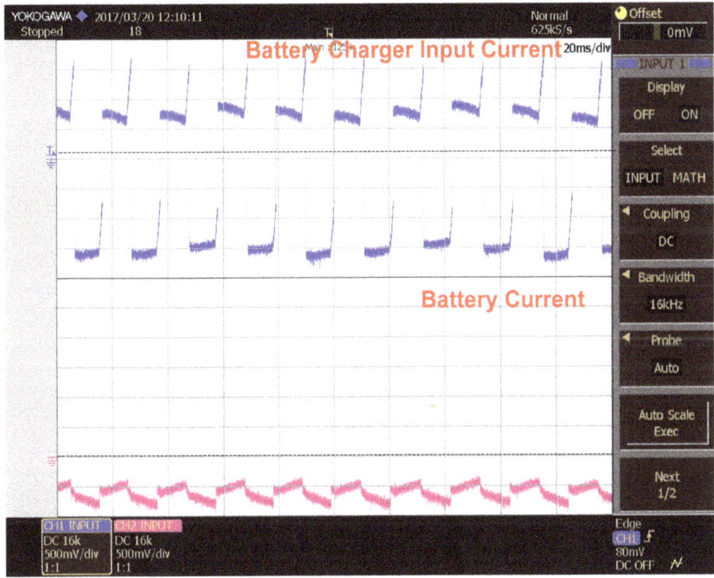

Figure 14. Battery charger input and output currents in mode-2 operation.

Similar to Mode1, in Mode3 operation battery current and the average value of the battery charger input current are also positive as shown in Figure 15. The charging current required for the battery is provided from the grid supply in this case.

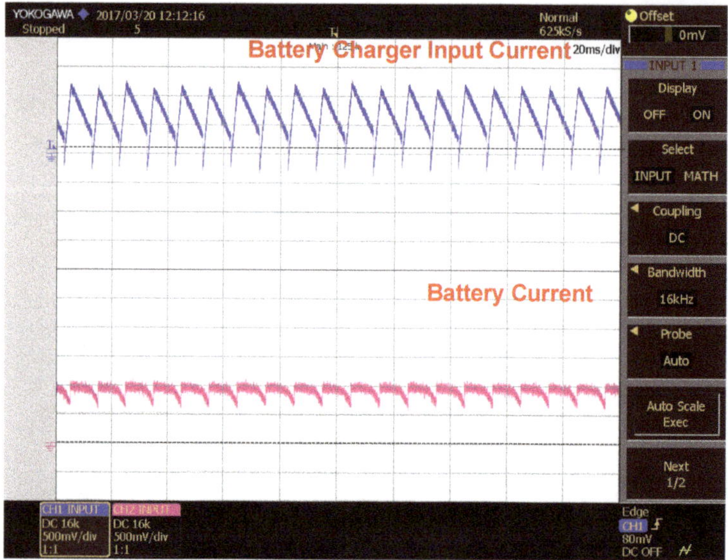

Figure 15. Battery charger input and output currents in mode3 operation.

The dynamic response of the battery charger system is observed by applying a step change in the battery current reference. The system takes approximately 250 milliseconds to come to the steady state as shown in Figure 16.

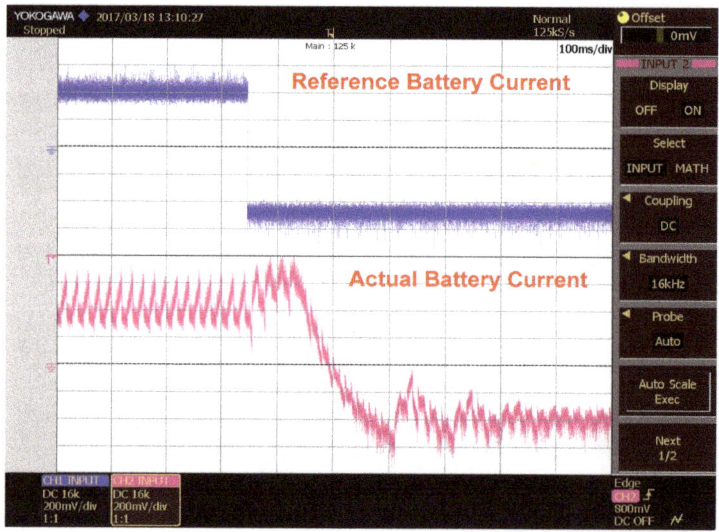

Figure 16. Battery charger reference and actual currents for a step change in reference current.

6. Future Scope

The results presented in this paper are obtained through real-time digital simulations. A scaled down model of the plant i.e., power circuit can also be made to test the controller and control algorithm.

A predictive diagnostic in high-power transformers used in traction grade uninterruptible power supplies is presented in [20]. In similar lines, predictive diagnosis of the system components in a grid energy storage system can also be addressed as future work.

The presented system can also be extended for smart grid applications by incorporating additional energy sources along with the PV and battery, then as future work, the security and privacy problems in this smart grid can be addressed, as discussed in [21].

In this paper, design and control of grid energy storage system with a conventional PV inverter is explained in detail. Additional feature reactive power compensation can be incorporated to make the system work as PV-STATCOM. This system can also be extended for high power applications by using multilevel configurations for the PV inverter [22,23].

7. Conclusions

In this paper, a power balancing control for a grid energy storage system is presented. The PBC technique is implemented on a TMS320F2812 processor-based controller card and tested. Dynamic responses of the inverter and battery charger system are verified by applying a step change in the reference values. From the presented HIL results, it is observed that the performance of PBC control is satisfactory in all three modes of operation and good dynamic performance is also achieved using this technique. As the controls are tested through HIL simulations in this work, plant parameters are considered as ideal whereas the plant parameters vary with operating temperatures in real time operation. Hence, minor modifications in plant parameters and tuning of control parameters are required to implement the proposal on areal system. The same system can be extended further to have the feature of reactive power compensation through the modified PBC control.

Acknowledgments: No funding resources.

Author Contributions: Sridhar Vavilapalli, Sanjeevikumar Padmanaban, Umashankar Subramaniam had developed the originally proposed research work and implemented with numerical simulation software and real time RTDS system for investigation and performance validation. Sanjeevikumar Padmanaban, Umashankar Subramaniam, Lucian Mihet-Popa contributed their expertise in the proposed subject of research and verification of the obtained results based on theoretical concepts and insight background. All authors involved to articulate the research work for its final depiction as a research paper.

Conflicts of Interest: The authors declare no conflict of interest.

References

1. Mihet-Popa, L.; Bindner, H. Simulation models developed for voltage control in a distribution network using energy storage systems for PV penetration. In Proceedings of the 39th Annual Conference of the IEEE Industrial Electronics Society—IECON'13, Vienna, Austria, 10–13 November 2013; pp. 7487–7492.
2. El Khateb, A.; Rahim, N.A.; Selvaraj, J. Ćuk-Buck Converter for Standalone Photovoltaic System. *J. Clean Energy Technol.* **2013**, *1*, 69–74. [CrossRef]
3. Wu, T.; Xiao, Q.; Wu, L.; Zhang, J.; Wang, M. Study and implementation on batteries charging method of Micro-Grid photovoltaic systems. *Smart Grid Renew. Energy* **2011**, *204*, 324–329. [CrossRef]
4. Choi, H.; Jang, M.; Ciobotaru, M.; Agelidis, V.G. Hybrid energy storage for large PV systems using bidirectional high-gain converters. In Proceedings of the 2016 IEEE International Conference on Industrial Technology (ICIT), Taipei, Taiwan, 14–17 March 2016; pp. 425–430.
5. Shi, Y.; Li, R.; Xue, Y.; Li, H. Optimized operation of current-fed dual active bridge DC–DC converter for PV applications. *IEEE Trans. Ind. Electron.* **2015**, *62*, 6986–6995. [CrossRef]
6. Bharathi, K.; Sasikumar, M. Voltage Compensation of Smart Grid using Bidirectional Intelligent Semiconductor Transformer and PV Cell. *Indian J. Sci. Technol.* **2016**, *9*, 1–8. [CrossRef]
7. Mihet-Popa, L.; Koch-Ciobotaru, C.; Isleifsson, F.; Bindner, H. Development of tools for DER Components in a distribution network. In Proceedings of the 2012 XXth International Conference on Electrical Machines (ICEM), Marseille, France, 2–5 September 2012; pp. 1022–1031.
8. Chen, M.; Rincon-Mora, G.A. Accurate electrical battery model capable of predicting runtime and I-V performance. *IEEE Trans. Energy Convers.* **2006**, *21*, 504–511. [CrossRef]

9. MathWorks. Available online: https://in.mathworks.com/help/physmod/sps/powersys/ref/battery.html;
    jsessionid=96072057f2374167c734b7a8e92d (accessed on 15 June 2017).
10. Koch-Ciobotaru, C.; Mihet-Popa, L.; Isleifsson, F.; Bindner, H. Simulation Model developed for a Small-Scale
    PV-System in a Distribution Network. In Proceedings of the 7th International Symposium on Applied
    Computational Intelligence and Informatics—IEEE SACI 2012, Timisoara, Romania, 24–26 May 2012;
    pp. 257–261.
11. Rahman, S.A.; Varma, R.K.; Vanderheide, T. Generalised model of a photovoltaic panel. *IET Renew.
    Power Gener.* **2014**, *8*, 217–229. [CrossRef]
12. Jeong, D.K.; Kim, H.S.; Baek, J.W.; Kim, J.Y.; Kim, H.J. Dual active bridge converter for Energy Storage
    System in DC microgrid. In Proceedings of the 2016 IEEE Conference and Expo Transportation Electrification
    Asia-Pacific (ITEC Asia-Pacific), Busan, Korea, 1–4 June 2016; pp. 152–156.
13. Dutta, S.; Hazra, S.; Bhattacharya, S. A Digital Predictive Current-Mode Controller for a Single-Phase
    High-Frequency Transformer-Isolated Dual-Active Bridge DC-to-DC Converter. *IEEE Trans. Ind. Electron.*
    2016, *63*, 5943–5952. [CrossRef]
14. Kumar, N.; Saha, T.K.; Dey, J. Sliding-mode control of PWM dual inverter-based grid-connected PV system:
    Modeling and performance analysis. *IEEE J. Emerg. Sel. Top. Power Electron.* **2016**, *4*, 435–444. [CrossRef]
15. Toodeji, H.; Farokhnia, N.; Riahy, G.H. Integration of PV module and STATCOM to extract maximum power
    from PV. In Proceedings of the International Conference on Electric Power and Energy Conversion Systems,
    Sharjah, UAE, 10–12 November 2009; pp. 1–6.
16. Serban, E.; Ordonez, M.; Pondiche, C.; Feng, K.; Anun, M.; Servati, P. Power management control strategy in
    photovoltaic and energy storage for off-grid power systems. In Proceedings of the 2016 IEEE 7th International
    Symposium on Power Electronics for Distributed Generation Systems (PEDG), Vancouver, BC, Canada,
    27–30 June 2016; pp. 1–8.
17. Sikkabut, S.; Mungporn, P.; Ekkaravarodome, C.; Bizon, N.; Tricoli, P.; Nahid-Mobarakeh, B.; Thounthong, P.
    Control of High-Energy High-Power Densities Storage Devices by Li-ion Battery and Supercapacitor for
    Fuel Cell/Photovoltaic Hybrid Power Plant for Autonomous System Applications. *IEEE Trans. Ind. Appl.*
    2016, *52*, 4395–4407. [CrossRef]
18. Merabet, A.; Ahmed, K.T.; Ibrahim, H.; Beguenane, R.; Ghias, A.M. Energy Management and Control
    System for Laboratory Scale Microgrid Based Wind-PV-Battery. *IEEE Trans. Sustain. Energy* **2017**, *8*, 145–154.
    [CrossRef]
19. Lemaire, M.; Sicard, P.; Belanger, J. Prototyping and Testing Power Electronics Systems Using Controller
    Hardware-In-the-Loop (HIL) and Power Hardware-In-the-Loop (PHIL) Simulations. In Proceedings of the
    Vehicle Power and Propulsion Conference (VPPC), Montreal, QC, Canada, 19–22 October 2015; pp. 1–6.
20. Saponara, S.; Fanucci, L.; Bernardo, F.; Falciani, A. Predictive diagnosis of high-power transformer faults
    by networking vibration measuring nodes with integrated signal processing. *IEEE Trans. Instrum. Measure.*
    2016, *65*, 1749–1760. [CrossRef]
21. Saponara, S.; Bacchillone, T. Network architecture, security issues, and hardware implementation of a home
    area network for smart grid. *J. Comput. Netw. Commun.* **2012**, *2012*, 534512. [CrossRef]
22. Sridhar, V.; Umashankar, S. A comprehensive review on CHB MLI based PV inverter and feasibility study of
    CHB MLI based PV-STATCOM. *Renew. Sustain. Energy Rev.* **2017**, *78*, 138–156. [CrossRef]
23. Das, V.; Sanjeevikumar, P.; Karthikeyan, V.; Rajasekar, S.; Blaabjerg, F.; Pierluigi, S. Recent Advances and
    Challenges of Fuel Cell Based Power System Architectures and Control—A Review. *Renew. Sustain. Energy*
    2017, *73*, 10–18. [CrossRef]

Article

Grid Synchronization of a Seven-Phase Wind Electric Generator Using d-q PLL

Kalaivani Chandramohan [1], Sanjeevikumar Padmanaban [2,*], Rajambal Kalyanasundaram [1], Mahajan Sagar Bhaskar [2] and Lucian Mihet-Popa [3]

[1] Department of Electrical and Electronics Engineering, Pondicherry Engineering College, Kalapet, Puducherry 605014, India; kalaivani46@pec.edu (K.C.); rajambalk@pec.edu (R.K.)
[2] Department of Electrical and Electronics Engineering, University of Johannesburg, Auckland Park 2006, South Africa; sagar25.mahajan@gmail.com
[3] Faculty of Engineering, Østfold University College, Kobberslagerstredet 5, 1671 Kråkeroy-Fredrikstad, Norway; lucian.mihet@hiof.no
* Correspondence: sanjeevi_12@yahoo.co.in; Tel.: +27-79-219-9845

Received: 10 May 2017; Accepted: 26 June 2017; Published: 4 July 2017

Abstract: The evolving multiphase induction generators (MPIGs) with more than three phases are receiving prominence in high power generation systems. This paper aims at the development of a comprehensive model of the wind turbine driven seven-phase induction generator (7PIG) along with the necessary power electronic converters and the controller for grid interface. The dynamic model of the system is developed in MATLAB/Simulink (R2015b, The MathWorks, Inc., Natick, MA, USA). A synchronous reference frame phase-locked loop (SRFPLL) system is incorporated for grid synchronization. The modeling aspects are detailed and the system response is observed for various wind velocities. The effectiveness of the seven phase induction generator is demonstrated with the fault tolerant capability and high output power with reduced phase current when compared to the conventional 3-phase wind generation scheme. The response of the PLL is analysed and the results are presented.

Keywords: multi-phase induction machine; synchronous reference frame; induction generator; PWM inverter; seven phase rectifier; PLL; grid

---

1. Introduction

Electric power generation gained by exploring the use of renewable energy sources is a viable solution for reducing the dependency on fast depleting fossil fuels and to adhere to environmentally friendly conditions [1]. Among all existing non-conventional sources, wind has latent qualities that can be utilized to meet the heaping energy demand [2]. Self-excited induction generators (SEIGs) are usually deployed for wind energy conversion systems in standalone applications with their inherent characteristics as mentioned in [3,4]. Later they also operated in a grid connected mode for distributed power generation in hybrid micro grids [5]. However, they are suitable for low and medium power applications [4]. Multiphase induction generator (MPIG) with more than three phases is a potential contender which combines the advantages of MPIG with SEIG technologies to yield an efficient, reliable, and fault tolerant machine that has diverse applications [6–9]. Multiphase systems can be employed for different applications, such as offshore energy harvesting, electrical vehicles, electric ship propulsion, and aircrafts. The earlier proposed research works describe the supremacy of multiphase machines for obtaining a better reliable performance [10–22].

As a consequence, MPIG research has evoked interest among researchers in the recent past which has culminated into gradual but steady progress in this field. However, the available literature suggests that finite modeling approaches should be implemented for MPIG analysis.

The *d-q* model of the six phase induction generator with a dual stator and single rotor has been presented in many papers [23–26]. The performance of the six phase dual stator induction generator has been investigated in [25–27]. Dynamic analysis of the six phase induction generator for standalone wind power generation was investigated in [28], while steady state performance analysis of the machine, and its experimental validation have been carried out in [29–31].

The most challenging requirement for wind electric systems is the low voltage ride through capability that requires generators to remain connected during grid faults and to contribute to the system recovery. Unbalanced voltage conditions and dips in the grid can have significant negative effects on the performance of induction generators. These effects can decrease the lifetime of sensitive components in the wind energy converter in the long term and in extreme cases, they can cause damage and tripping of the system, leading to violation of the grid code requirements [32–35].

The grid integration of wind electric generators (WEG) is a critical aspect in the planning of a wind power generation system. The variation in production and higher intermittency of wind generation makes it difficult for grid integration. Hence it is necessary to provide the appropriate synchronization techniques such that the system maintains constant frequency and voltage to ensure stable and reliable operation of the grid [36,37]. A good synchronization method must detect the frequency and phase angle variations proficiently in order to reduce the harmonics and disturbances for safe operation of the grid. Further simple implementation and cost decides the reliability of the synchronization schemes [38]. The power transfer between distributed generation and the grid is enhanced by a good synchronization method. Earlier known zero crossing detectors have adverse power quality issues in a weak grid. Nowadays, phase-locked loop (PLL) is one of the generally used techniques and it controls the distributed power generation system and other applications. Several types of PLL are analysed in [37,38]. This paper aims to develop a PLL based grid connected seven phase WEG where PLL enables the frequency and voltage synthesis.

A d-q model of the seven-phase induction generator (7PIG) with the stator windings phase shifted by 51.42° is developed. A simulation is carried out to study the performance under varying wind velocities. The voltage build up process is shown. The generator voltage, current, and power output is presented under varying load conditions. The reliability of the machine under a fault condition is examined with one or two phases open. The results are compared with the three phase generator. The power electronic interface, namely the seven phase rectifier, boost converter, and the three phase neutral point clamped (NPC) inverter are simulated for varying modulation indices and the results are explored. The synchronous reference frame (SRF) PLL is designed to track the phase angle and frequency. The PLL response analyses various grid conditions such as unbalanced grid voltages, voltage sags, line to line (LL) faults, and line to line ground faults (LLG), and the results are explored.

2. Proposed System Description

Figure 1 shows the proposed multiphase AC power system for wind power application. The grid connected seven phase wind generation system considered for the study, consists of a wind turbines that is driven by a 7PIG through a gearbox. The generated seven phase AC output is rectified by the seven phase rectifier and filtered by an LC filter. The filtered and boosted DC output voltage is injected through a three phase inverter to the grid with proper synchronization through SRF PLL. The inverter is controlled using the synchronous *d-q* reference frame approach. The phase lock loop technique that is incorporated synchronizes the inverter and the grid. The high frequency ripple at the inverter is filtered. The filtered output of the inverter is fed into the grid through a step-up transformer.

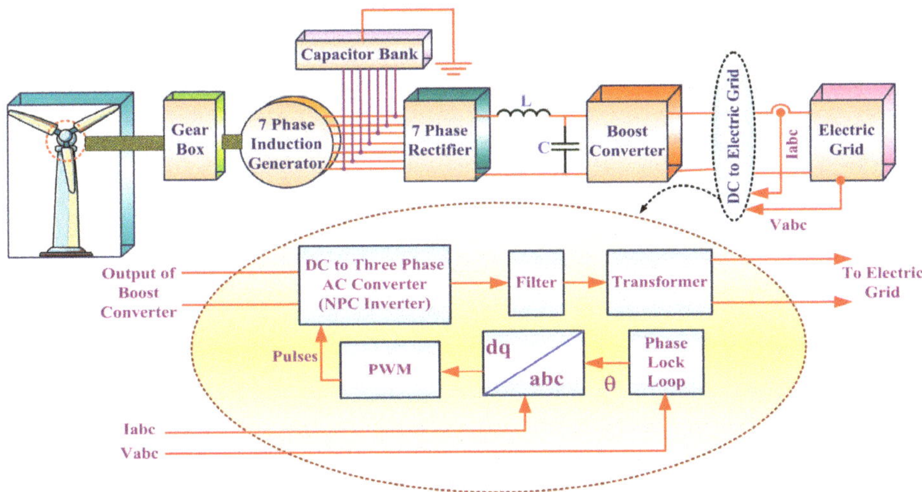

Figure 1. Seven Phase Grid Connected Wind Electric Generator.

3. Modeling of System Components

The mathematical modeling of the seven phase wind generator components, namely the wind turbine, seven phase induction generator, seven phase rectifier, and three phase inverter and PLL are discussed in the following sections.

3.1. Wind Turbine

The following equation defines the power output of the wind turbine, which is the aerodynamic power developed on the main shaft of the wind turbine:

$$P_{tur} = 0.5\rho A C_p(\lambda) V_w{}^3, \tag{1}$$

C_p is a dimensionless power coefficient that depends on the wind speed and constructional characteristics of the wind turbine. For the wind turbine used in this study, the following form approximates C_p as a function of λ known as the tip-speed ratio, which depends on the rotor speed of the turbine and the wind speed.

$$C_p = 0.5\left(\frac{116}{\lambda_1} - 0.4\beta - 5\right)e^{\frac{-16.5}{\lambda_1}}, \tag{2}$$

$$\lambda = \frac{R\omega_{tur}}{V_w}, \tag{3}$$

$$\lambda_1 = \frac{1}{\frac{1}{(\lambda+0.089)} - \frac{0.035}{(\beta^3+1)}}, \tag{4}$$

where, ρ_{tur}: Air density (kg/m^3); V_w—wind speed (m/s); R—Radius of the wind turbine rotor (m); A—Area swept out by the turbine blades (m^2); C_p: power coefficient defined by Equation (2); λ—Tip speed ratio given by Equation (3); ω_{tur}: angular rotor speed of the turbine (rad/s); β: The blade pitch angle (degree).

3.2. 7PIG Model

A 7PIG has seven stator windings sinusoidally distributed with a phase displacement of 51.4° (360°/7) and the rotor is short circuited for the squirrel cage induction machine. The 7P induction machine operating as a generator is represented as a two phase equivalent circuit. The *ds-qs* represent the stator direct and quadrature axes and *dr-qr* represents the rotor direct and quadrature axes. The transformation of the seven phase stationary reference frame variables to a two phase stationary reference frame is given by Equation (5). The assumptions made in modeling 7PIG are the same as those given in [8,39–41]. The modeling of 7PIG is carried out using a *d-q* equivalent circuit, as shown in Figure 2 [39–41].

$$
\begin{bmatrix} V_{qs} \\ V_{ds} \\ V_{xs} \\ V_{ys} \\ \cdot \\ V_{os} \end{bmatrix} = \begin{bmatrix} 1 & \cos\alpha & \cos 2\alpha & \cos 3\alpha & \cdot & \cos n\alpha \\ 0 & \sin\alpha & \sin 2\alpha & \sin 3\alpha & \cdot & \sin n\alpha \\ 1 & \cos 2\alpha & \cos 4\alpha & \cos 6\alpha & \cdot & \cos 2n\alpha \\ 0 & \sin 2\alpha & \sin 4\alpha & \sin 6\alpha & \cdot & \sin 2n\alpha \\ \cdot & \cdot & \cdot & \cdot & \cdot & \cdot \\ \frac{1}{\sqrt{2}} & \frac{1}{\sqrt{2}} & \frac{1}{\sqrt{2}} & \frac{1}{\sqrt{2}} & \cdot & \frac{1}{\sqrt{2}} \end{bmatrix} X \begin{bmatrix} V_a \\ V_b \\ V_c \\ V_d \\ \cdot \\ V_n \end{bmatrix}, \tag{5}
$$

where $\alpha = 2\pi/n$; n = number of phases; s and r represent stator and rotor quantities, respectively; d-q represents a direct and quadrature axis.

Equations (5) and (6) define the stator side voltages

$$
V_{qs} = -R_s i_{qs} + \omega\lambda_{ds} + p\lambda_{qs}, \tag{6}
$$

$$
V_{ds} = -R_s i_{ds} - \omega\lambda_{qs} + p\lambda_{ds}, \tag{7}
$$

Equations (7) and (8) define the rotor side voltages

$$
V_{qr} = R_r i_{qr} + (\omega - \omega_r)\lambda_{dr} + p\lambda_{qr}, \tag{8}
$$

$$
V_{dr} = R_r i_{dr} - (\omega - \omega_r)\lambda_{qr} + p\lambda_{dr}, \tag{9}
$$

The voltage equations for dynamic performance analysis under balanced conditions are represented in a stationary reference frame ($\omega = 0$). The rotor side voltages V_{qr} and V_{dr} are zero for the squirrel cage induction generators. The rotor side quantities are referred to as stator reference frame. The flux linkage expression as a function of the current is given by Equations (10)–(15).

$$
\lambda_{qs} = -L_{ls} i_{qs} + L_m (i_{qr} - i_{qs}), \tag{10}
$$

$$
\lambda_{ds} = -L_{ls} i_{ds} + L_m (i_{dr} - i_{ds}), \tag{11}
$$

$$
\lambda_{qr} = L_{lr} i_{qr} + L_m (i_{qr} - i_{qs}), \tag{12}
$$

$$
\lambda_{dr} = L_{lr} i_{dr} + L_m (i_{dr} - i_{ds}), \tag{13}
$$

$$
\lambda_{dm} = L_m (i_{ds} + i_{dr}), \tag{14}
$$

$$
\lambda_{qm} = L_m (i_{qs} + i_{qr}), \tag{15}
$$

The leakage inductance of the stator and rotor are assumed to be constant. The degree of magnetic saturation decides the magnetizing inductance L_m and it is a non-linear function of the magnetizing current, which is given by the following equation

$$
I_m = \sqrt{(i_{qr} + i_{qs})^2 + (i_{dr} + i_{ds})^2}, \tag{16}
$$

The non-linear piecewise relationship between the magnetizing inductance and the current (L_m, i_m) is given by

$$L_m = \begin{cases} 0.012726, 0 \le i_m < 25.944 \\ 1.94597/(i_m + 117.6), 25.944 \le i_m < 51.512 \\ 1.79031/(i_m + 61.2), 52.512 \le i_m < 73.8 \\ 1.41566/(i_m + 46.296), 73.8 \le i_m < 85.872 \\ 2.67838/(i_m + 31.608), i_m \ge 85.872 \end{cases} \qquad (17)$$

The developed electromagnetic torque of the 7PIG is defined by

$$T_g = -\frac{7}{2}\left(\frac{P}{2}\right) L_m \left(i_{qs}i_{dr} - i_{ds}i_{qr}\right), \qquad (18)$$

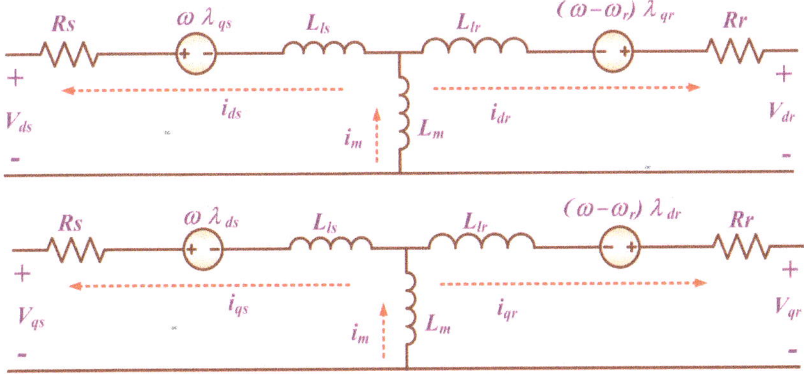

Figure 2. *d-q*-axis Equivalent Circuit of Seven-Phase Induction Generator (7PIG).

A negative (-ve) sign indicates generation action.

$$L_r = L_{lr} + L_m, \qquad (19)$$

$$L_s = L_{ls} + L_m, \qquad (20)$$

3.3. Modeling of the Shunt Capacitor and Load

The modeling equations of the voltage and current of the excitation capacitor and the load in the *d-q*-axis are given by Equations (21)–(26)

$$pV_{qs} = \left(\frac{1}{C}\right) i_{cqs} - \omega V_{ds}, \qquad (21)$$

$$pV_{ds} = \left(\frac{1}{C}\right) i_{cds} + \omega V_{qs}, \qquad (22)$$

$$i_{cqs} = i_{qs} - i_{Rqs}, \qquad (23)$$

$$i_{cds} = i_{ds} - i_{Rds}, \qquad (24)$$

$$i_{Rqs} = \frac{V_{qs}}{R}, \qquad (25)$$

$$i_{Rds} = \frac{V_{ds}}{R}, \qquad (26)$$

The 7P voltages are transformed to 2P using Equation (27).

$$\left.\begin{aligned}
V_a &= V_{qs}\cos\theta_e + V_{ds}\sin\theta_e \\
V_b &= V_{qs}\cos(\theta_e - \alpha) + V_{ds}\sin(\theta_e - \alpha) \\
V_c &= V_{qs}\cos(\theta_e - 2\alpha) + V_{ds}\sin(\theta_e - 2\alpha) \\
V_d &= V_{qs}\cos(\theta_e - 3\alpha) + V_{ds}\sin(\theta_e - 3\alpha) \\
V_e &= V_{qs}\cos(\theta_e - 4\alpha) + V_{ds}\sin(\theta_e - 4\alpha) \\
V_f &= V_{qs}\cos(\theta_e - 5\alpha) + V_{ds}\sin(\theta_e - 5\alpha) \\
V_g &= V_{qs}\cos(\theta_e - 6\alpha) + V_{ds}\sin(\theta_e - 6\alpha)
\end{aligned}\right\}, \tag{27}$$

4. DC Link Converter

The power electronics based interface system, namely the DC link converter, involves a seven phase rectifier, three phase inverter, and a DC-DC boost converter. The uncontrolled seven phase rectifier converts the seven phase AC output of the generator to DC and is boosted by the boost converter.

4.1. Seven Phase Diode Bridge Rectifier

A variable magnitude, the variable frequency voltage at the seven phase induction generator terminal, is converted to DC using a seven-phase diode bridge rectifier [42–45]. The voltage V_{rec} at the output is given by Equation (28) in terms of the peak phase voltage V_{ds} of the generator. The LC filter reduces the output voltage ripple of the seven phase rectifier.

$$V_{rec} = \frac{1}{(2\pi/14)} \int_{-\pi/14}^{\pi/14} 1.949 V_{ds}\cos(\omega t)d(\omega t), \tag{28}$$

$$V_{rec} = 1.932 V_{ds}, \tag{29}$$

4.2. DC-DC Boost Converter

A DC-DC boost converter (Figure 3) steps up the input voltage depending on the duty ratio, inductor, and capacitor values [40]. The output voltage of the boost converter is given by

$$V_{dc} = \frac{V_{rec}}{1 - \delta}, \tag{30}$$

where, V_{rec}—Input voltage from the seven phase rectifier; δ—Duty cycle of the switch. The inductance and capacitance are determined using Equations (31) and (32).

$$\text{Inductance, } L = \frac{R * \delta(1 - \delta)^2}{2 * f_S} \text{ (Henry)}, \tag{31}$$

$$\text{Capacitance, } C \geq \frac{V_O * \delta}{f_S * \Delta V_O * R} \text{ (Farad)}, \tag{32}$$

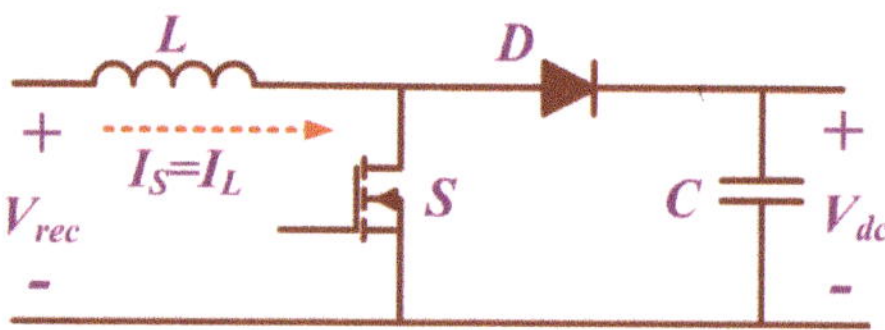

Figure 3. Power Circuit of Boost Converter.

4.3. Three Level Neutral Point Clamped Inverter

The DC input is given to this inverter from the DC/DC converter (Figure 3) and the three phase, three level output obtained is given to the grid through a step-up transformer. The modulation index of the reference signal is varied to control the output voltage of the inverter and is given by Equation (33)

$$\text{Modulation Index, } M_a = \frac{V_m}{\frac{V_{dc}}{2}},\tag{33}$$

where, V_m—Peak value of the Phase voltage (V); V_{dc}—Input Input DC voltage/Output of the Boost converter.

5. Grid Interface Using PLL

The effective power transfer between the grid and the source can be realized by the efficient synchronization technique. The most familiar method is tracking of the phase angle using the PLL which synchronizes the voltage and frequency of a given reference and output signal. A phase detector, loop filter, and voltage controlled oscillator (VCO) together make a basic PLL system, wherein the phase detector generates an error signal by comparing the reference and output signal. The harmonics of the error signal are eliminated by the loop filter. Depending on the output of the loop filter, the VCO generates the output signal. The basic structure of the PLL circuit is shown in Figure 4. A linear PLL is usually used in a single phase system, whereas a three phase system employs an SRF PLL or otherwise a *d-q* PLL.

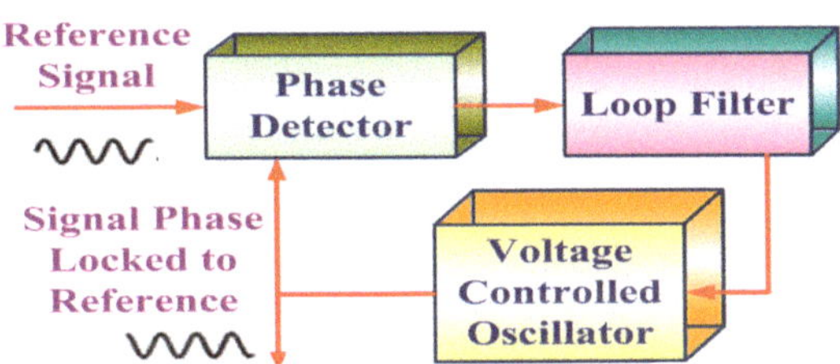

Figure 4. Basic Phase Locked Loop (PLL) Structure.

Synchronous Reference Frame (SRF/d-q) PLL

In the synchronous frame PLL, Clarke's transformation [39] is applied to the three-phase voltage vector to transform *abc* to the $\alpha\beta$ stationary reference frame. Park's transformation changes $\alpha\beta$ to the *d-q* rotating frame, as shown in Figure 5. The feedback loop controls the angular position of the *d-q* reference, making the *q*-axis component zero in the steady state. The *d*-axis will be the voltage amplitude during steady state conditions.

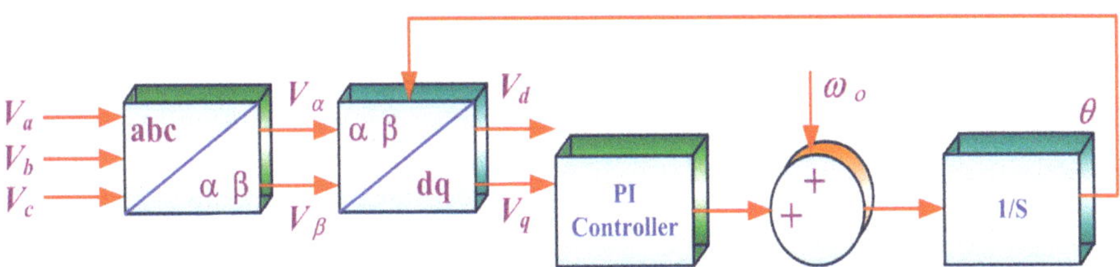

Figure 5. Synchronous Reference Frame (SRF)/*d-q* PLL Structure.

The d- and q-axis components are defined by the following equation under balanced conditions.

$$\begin{bmatrix} V_d \\ V_q \end{bmatrix} = \begin{bmatrix} \cos\hat{\theta} & \sin\hat{\theta} \\ -\sin\hat{\theta} & \cos\hat{\theta} \end{bmatrix} \begin{bmatrix} U\cos\theta \\ U\sin\theta \end{bmatrix} = \begin{bmatrix} U\cos(\theta - \hat{\theta}) \\ U\sin(\theta - \hat{\theta}) \end{bmatrix}, \tag{34}$$

where U, θ—amplitude and phase of the input signal; $\hat{\theta}$—PLL output; and V_d, V_q are the d- and q-axis components.

The phase is denoted by the q-axis and the amplitude in steady state is denoted by the d-axis error. The generalized voltage vector under unbalanced utility conditions (without voltage harmonics) is represented by

$$V = V_+ + V_- + V_0, \tag{35}$$

The positive, negative, and zero sequence components are represented by subscripts $+$, $-$, and 0. The $\alpha\beta$ component using Clarke's transformation is given by

$$V_{\alpha\beta\gamma} = \begin{bmatrix} V_\alpha \\ V_\beta \\ V_\gamma \end{bmatrix} = T_{\alpha\beta/abc} \begin{bmatrix} V_a \\ V_b \\ V_c \end{bmatrix}, \tag{36}$$

$$T_{\alpha\beta/abc} = \frac{2}{3} \begin{bmatrix} 1 & \frac{-1}{2} & \frac{-1}{2} \\ 0 & \frac{\sqrt{3}}{2} & \frac{\sqrt{3}}{2} \\ \frac{1}{2} & \frac{1}{2} & \frac{1}{2} \end{bmatrix}, \tag{37}$$

The zero-sequence component is neglected as it is on the γ-axis. The expression of the voltage vector on the $\alpha\beta$-plane is:

$$V_{\alpha\beta} = T_{\alpha\beta/abc}(V_+ + V_-) = \begin{bmatrix} U_+ \cos\theta_+ + U_- \cos\theta_- \\ U_+ \sin\theta_+ + U_- \sin\theta_- \end{bmatrix}, \tag{38}$$

The $\alpha\beta$ frame is transformed to the d-q frame using Park's transformation.

$$V_{dq} = T_{dq/\alpha\beta} V_{\alpha\beta} = \begin{bmatrix} U_+ \cos(\theta_+ - \hat{\theta}) + U_- \cos(\theta_- - \hat{\theta}) \\ U_+ \sin(\theta_+ - \hat{\theta}) + U_- \sin(\theta_- - \hat{\theta}) \end{bmatrix} = \begin{bmatrix} U_+ + U_- \cos(2\omega t) \\ U_+ - U_- \sin(2\omega t) \end{bmatrix}, \tag{39}$$

$$T_{dq/\alpha\beta} = \begin{bmatrix} \cos\hat{\theta} & \sin\hat{\theta} \\ -\sin\hat{\theta} & \cos\hat{\theta} \end{bmatrix}, \tag{40}$$

ω is the angular frequency of the voltage vector and $\hat{\theta} = \theta_+ = \theta_- = \omega t$.

6. Simulation Result

The modeling equations of the various system components are simulated with the parameters given in Appendix A. The individual component models are analysed and integrated to study the performance of the seven phase wind electric generator. The simulation results are discussed in the following sections.

6.1. Wind Turbine

A 250 kW wind turbine is simulated using the Equations (1)–(4) for various wind speeds and rotational speeds. Figure 6 shows the power curves of the wind turbine at various wind speeds. The rated power of 250 kW is achieved at the rated wind speed of 15 m/s and 40 rpm. The wind turbine produces the maximum power at various rotational speeds for the different wind speeds.

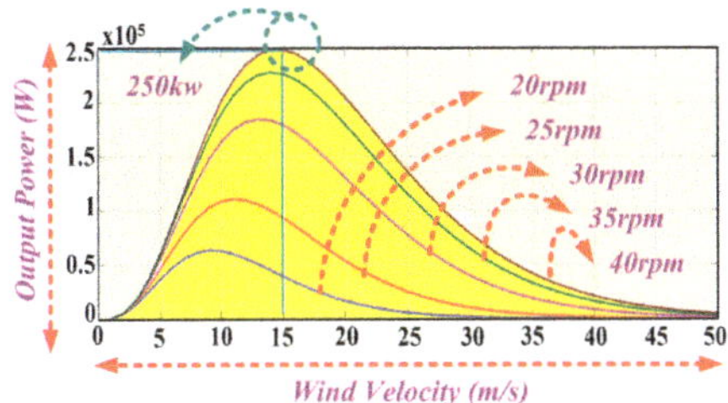

Figure 6. Wind Turbine Power Output vs. Wind Velocity (Wind Speed).

6.2. Seven Phase Induction Generator

The mathematical equations represented by Equations (5)–(27) are used to develop a mathematical model of the seven phase induction generator from the *d-q* equivalent circuit shown in Figure 2. The performance of 7PIG is investigated under various operating conditions.

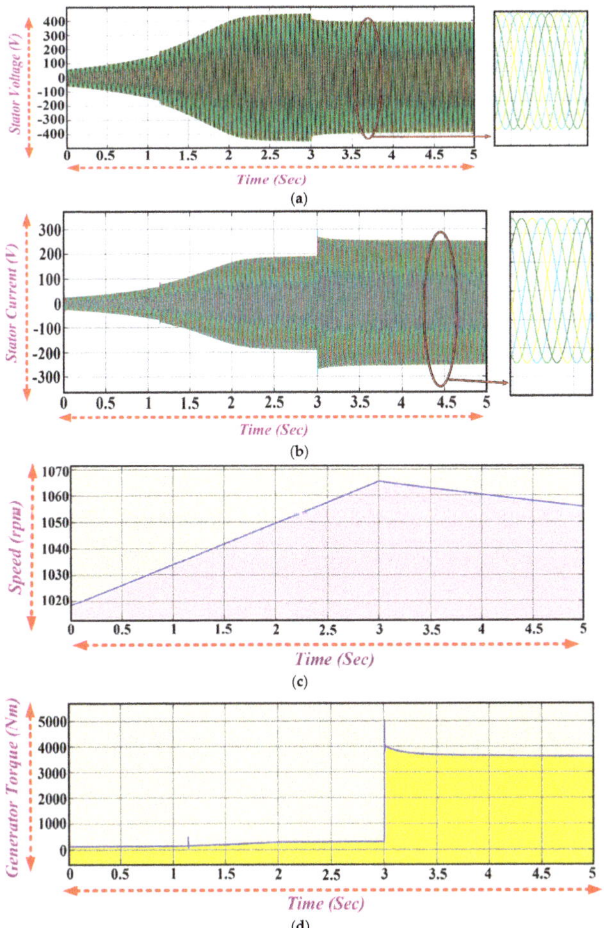

Figure 7. 7PIG (**a**) Output Voltage; (**b**) Current; (**c**) Generator speed; (**d**) Generator Torque.

The rated speed of 1018 rpm with the excitation capacitance of 2332 µF is given as the input to the generator and the voltage and current of the 7PIG are obtained at no load and are presented in Figure 7. The self-excitation process begins at time $t = 0$, and the stator voltage builds and the steady state value of 419 V (peak) and a current of about 165 A is reached at $t = 2.2$ s with the phases mutually displaced by 51.4° $(2\pi/n)$.

The 7PIG is loaded at $t = 3$ s with the excitation capacitance held constant at 2332 µF. At $t = 3$ s, the terminal voltage of the stator is reduced from 419 volts to 386 volts and the current increases from 165 A to 241 A. The steady state is reached at $t = 2$ s as shown in Figure 7a,b. The generated torque and speed of the generator are represented in Figure 7c,d which show that for increasing the load, the speed of the generator decreases with an increase in the torque. The line voltages of the seven phase induction generator varies for the adjacent ($V_{ab} = 0.8676\ V_m$) and non-adjacent sides ($V_{ac} = 1.5629\ V_m$) and ($V_{ad} = 1.949\ V_m$), which is clearly illustrated using the results shown in the Figure 8a–c.

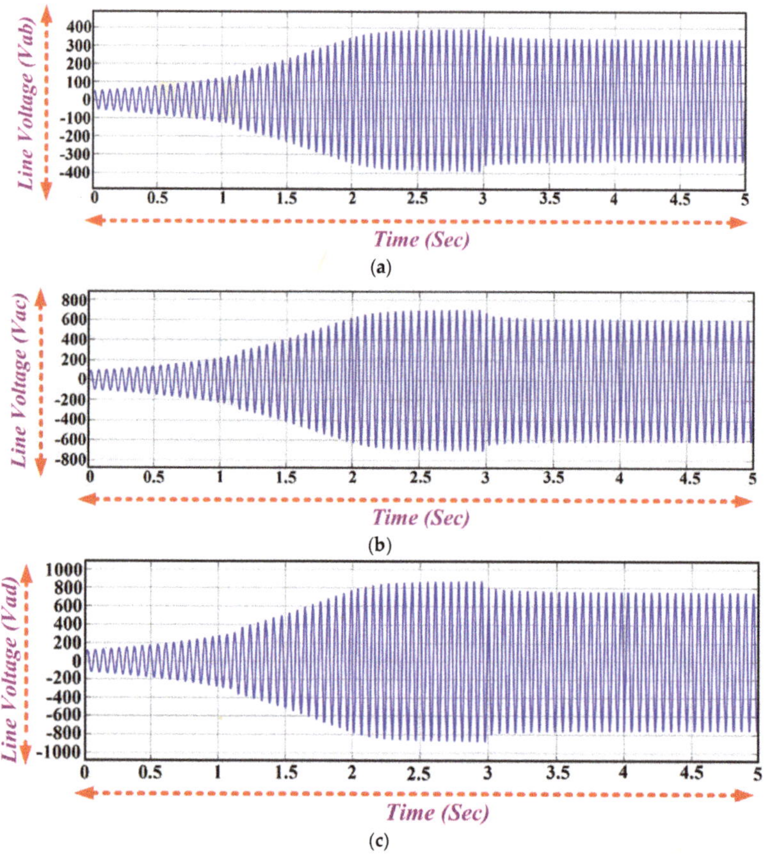

Figure 8. Generated Line Voltage of 7PIG Adjacent Side and Non Adjacent Side (**a**) V_{ab}; (**b**) V_{ac}; (**c**) V_{ad}.

6.3. Fault Tolerant Operation of 7PIG

The most important ability of the multiphase phase generator is that it continues to operate even after the fault occur in one (or more) phase(s), whereas three-phase machines can hardly continue their operation. Under the faulty conditions, the additional degrees of freedom available in MPIG are efficiently used for the post fault operating strategy. One-, two-phases are open circuited V_c, V_c and V_d

at t = 3 s for the seven phase induction generator for the investigation test. It is clear from the obtained numerical results by Figures 9 and 10 that the generator continues to operate with the reduced phase current in amplitude for continuous propagation under open circuit faulty conditions.

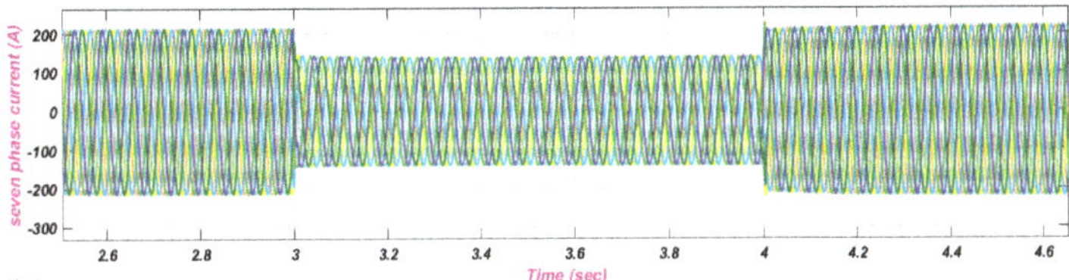

Figure 9. Fault Current of 7PIG with One Phase Open (V_c).

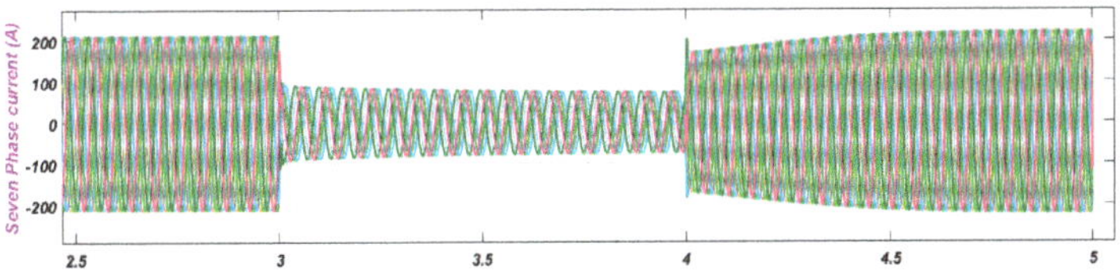

Figure 10. Fault Current of 7PIG with Two Phases Open (V_c and V_d).

The performance of the seven phase induction generator is compared with the three phase generator in terms of generating voltage and current. The per phase voltage of the three phase and seven phase system remains the same whereas the current per phase is reduced, as clearly shown in Figure 11a,b.

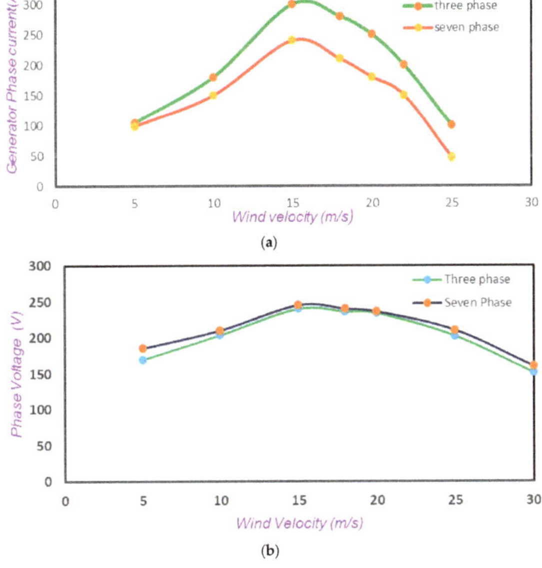

Figure 11. (**a**) Seven Phase Current vs. Three Phase Current for Varying Wind Velocity; (**b**) Three Phase Voltage vs. Seven Phase Voltage.

6.4. DC Link Converter

The generated seven phase AC output is fed as an input to the seven phase rectifier which converts AC to DC. The rectified DC output voltage feeds the boost converter. The boost converter output voltage and current of 845 V and 279 A is achieved, as shown in Figure 12a,b.

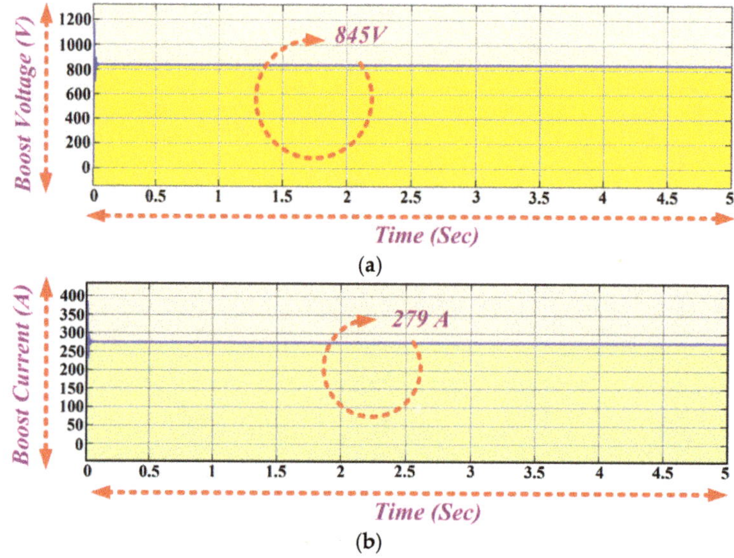

Figure 12. (a) Boost Voltage; (b) Boost Current.

6.5. Grid Integration

The grid tied inverter is the power electronic converter that converts the DC signal into AC, but with the appropriate synchronizing techniques. It is basically used in the integration of renewable energy to the utility line. The magnitude and phase of the inverter voltage should be the same as that of the grid and its output frequency should be equal to the grid frequency for proper grid synchronization. The output phase voltage of the inverter is 365 V (peak) and a current of about 508 V (peak) is achieved at a 0.85 modulation index with a DC input of 845 V as shown in Figures 13 and 14. The line voltage of the inverter is given by Figure 15.

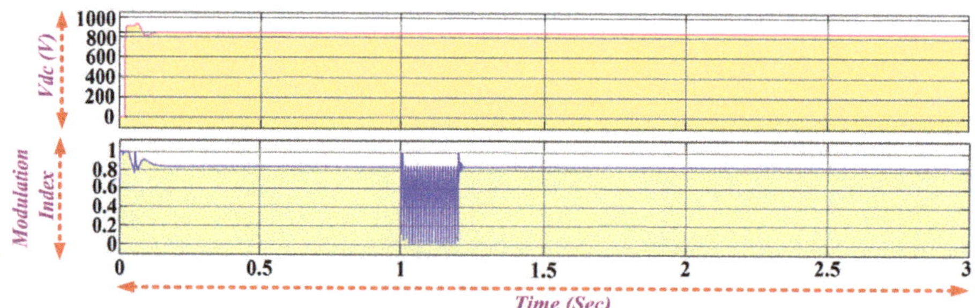

Figure 13. Vdc and Modulation Index.

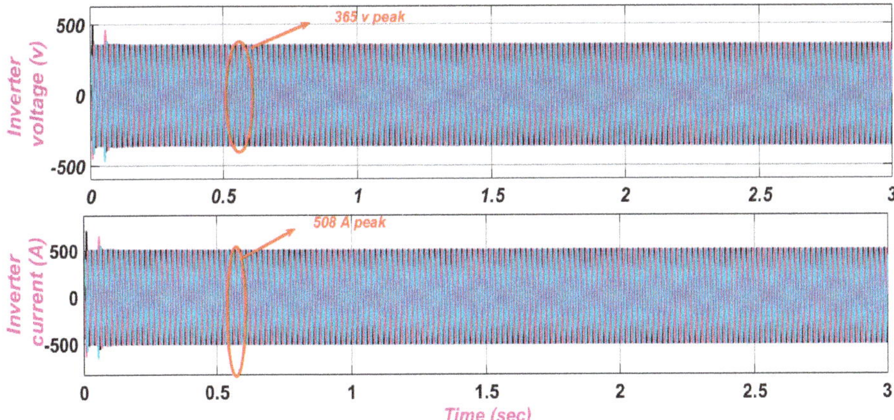

Figure 14. Inverter Output Voltage and Inverter Current.

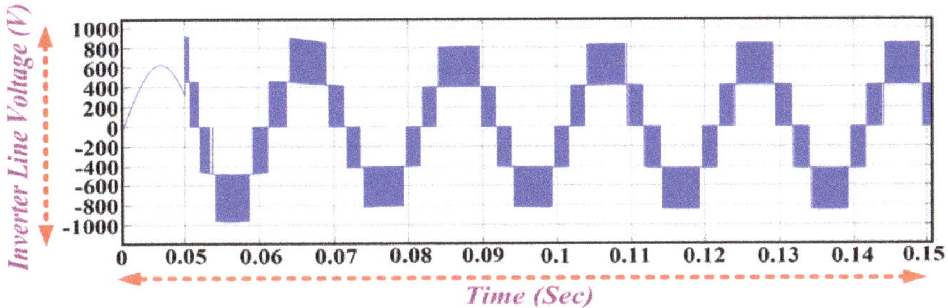

Figure 15. Inverter Line Voltage.

The d- and q-axis voltage of the d-q PLL and frequency tracking is shown in Figure 16. The voltage and current drawn by the load connected at the point of common coupling is shown in Figure 17. The grid voltage and current are shown in Figure 18. The power injected into the grid is about 196 kW, which is shown in Figure 19.

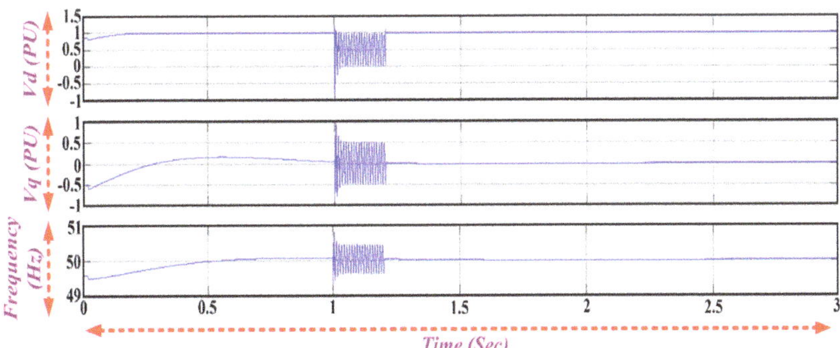

Figure 16. Graph of V_d, V_q, and Frequency.

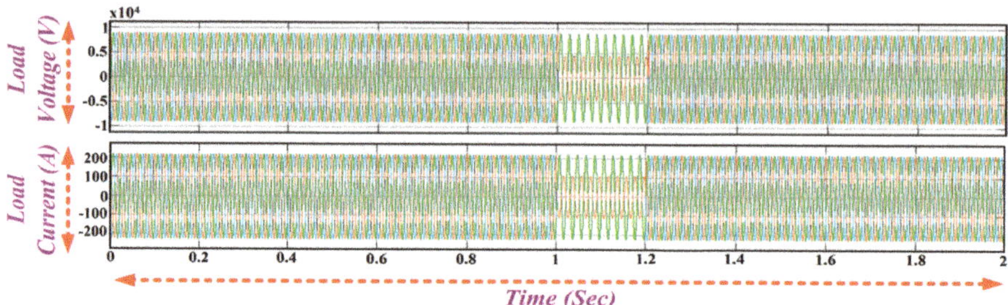

Figure 17. Load Voltage and Load Current.

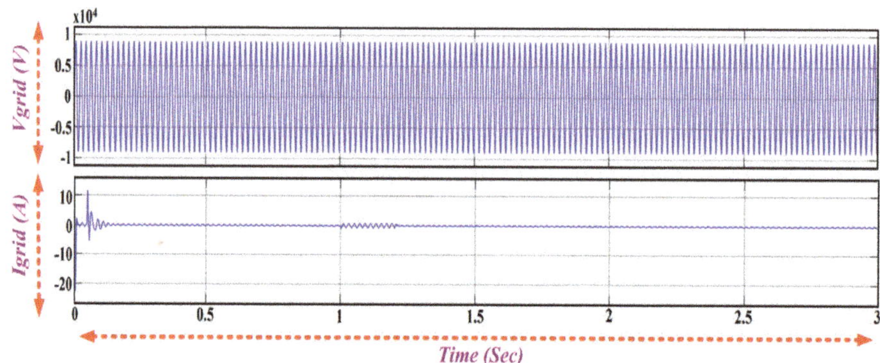

Figure 18. Grid Voltage (Vgrid) and Grid Current (Igrid).

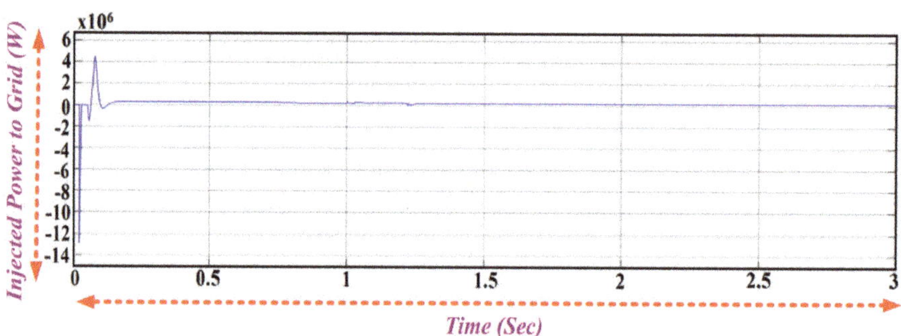

Figure 19. Power injected into the Grid.

6.6. SRF PLL Performance under Various Grid Conditions

The grid is subjected to different fault conditions to investigate the performance of the SRF PLL. Figure 20a shows the frequency and phase detection variation during a line to line fault. It is clear from the figure that phases B and C are in phase with each other and their magnitude is less than phase A, whereas the magnitude of phase of B and C are zero during a line to line ground fault as shown in Figure 21a. The voltages of the *d-q*-axis also vary, as it contains second harmonic ripples as given by Equation (35), which is illustrated by Figures 20b and 21b.

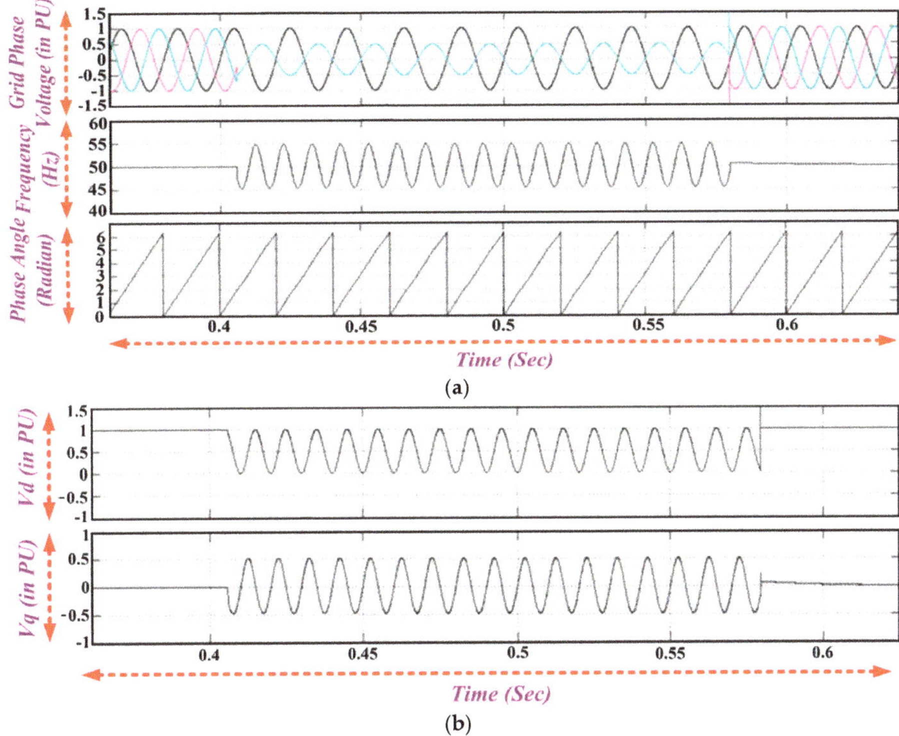

(a)

(b)

Figure 20. (**a**) Frequency and Phase Angle Variation during a Line to Line Fault; (**b**) *q*-axis and *d*-axis Voltage Magnitude during a Line to Line Fault.

During unbalanced grid voltage condition, the sinusoidal nature of the *q*-axis voltage component affects the output of the PI controller. Therefore, the PI controller generates a sinusoidal error signal, angular frequency is shown in Figure 22a,b, which is similar to that of the line to line fault.

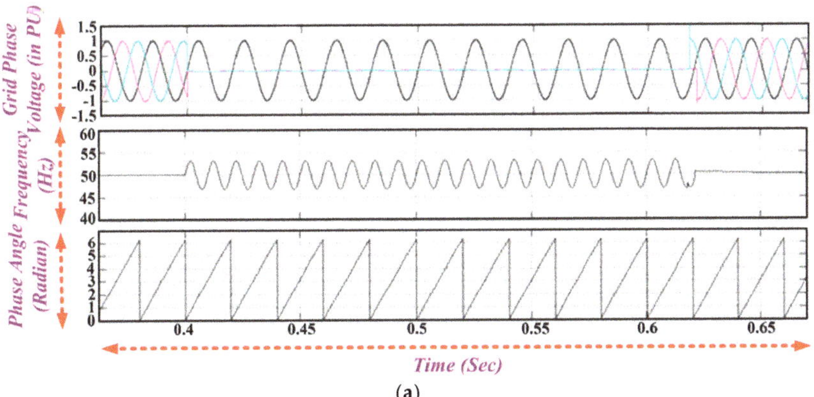

(a)

Figure 21. *Cont.*

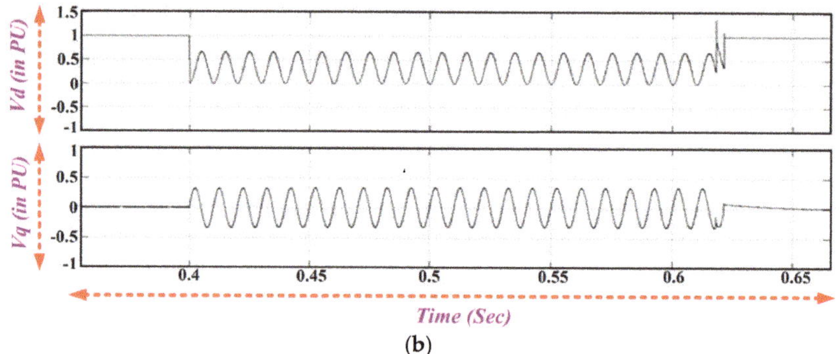

(b)

Figure 21. (**a**) Frequency and Phase Angle Variation during a Line-Line ground LLG Fault; (**b**) *q*-axis and *d*-axis Voltage Magnitude during a LLG Fault.

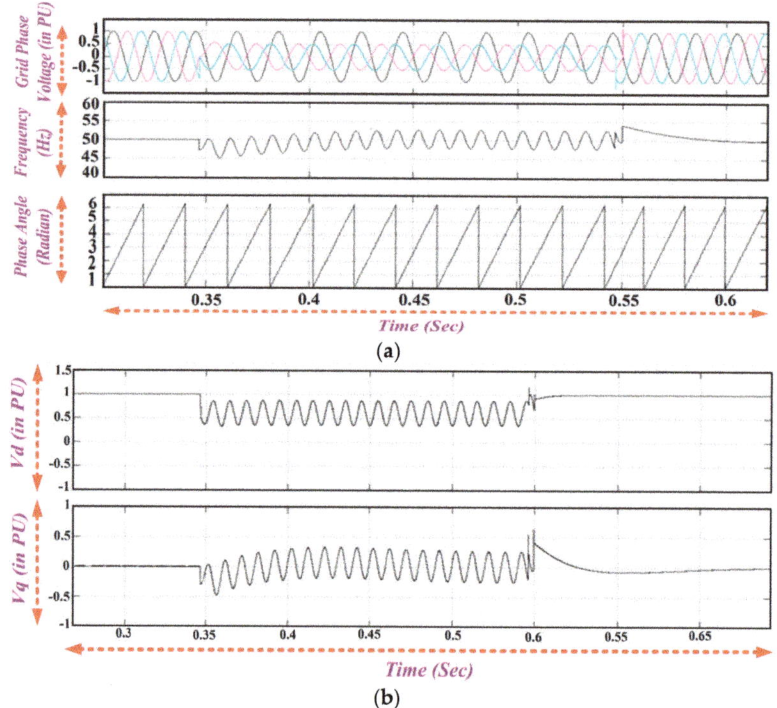

(a)

(b)

Figure 22. (**a**) Frequency and Phase Detection Variation during Unbalanced Grid Voltages; (**b**) *q*-axis and *d*-axis Voltage Magnitude during Unbalanced Grid Voltages.

The SRF PLL performance during voltage sag is shown in Figure 23a,b. Voltage sag occurs in the grid such that the magnitude of all phase voltages are equal and their magnitudes are 50% of the nominal voltage. It is noticed that it does not cause any oscillations in the frequency and the *d-q* voltages. Balanced voltage sag does not affect PLL tracking. However, a sudden change in the magnitude causes a dip in the estimated frequency of *d-q* PLL, and later it tracks the phase angle of the grid voltages.

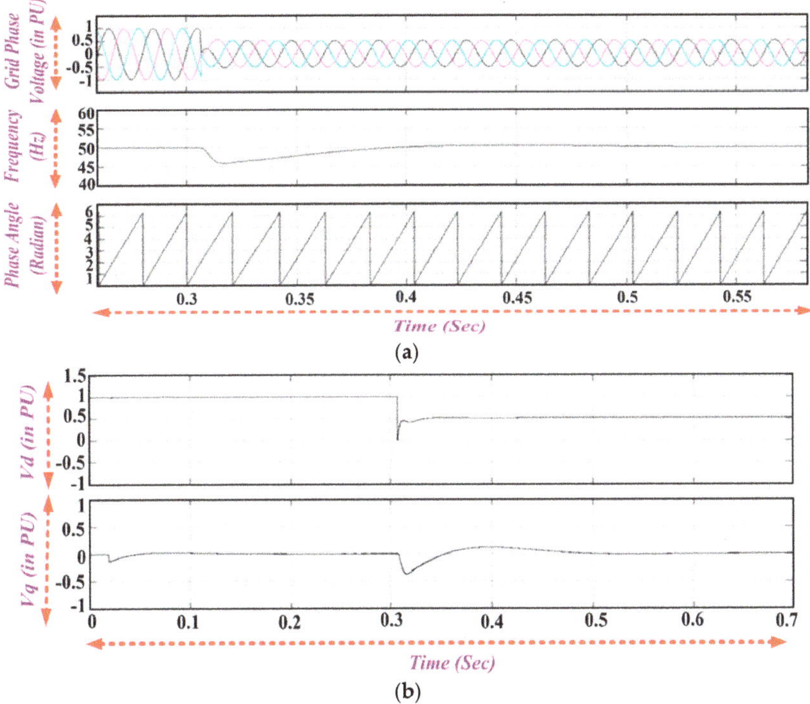

Figure 23. (**a**) Frequency and Phase Detection Variation during Voltage Sag; (**b**) *q*-axis and *d*-axis Voltage Magnitude during Voltage Sag.

7. Conclusions

In this article, a comprehensive model of a wind driven 7PIG in grid connected mode was developed using the two axis *d-q* equivalent circuit. A seven phase wind electric generator is integrated using the individual system components and the performance of the seven phase wind electric generator is analysed for varying wind speed [46]. A synchronous reference frame PLL incorporated for the grid interface is simulated and analysed. The enhanced performance of 7PIG is evaluated through the fault tolerant capability and high output power with reduced current per phase when compared with the three phase model. The performance of SRF-PLL incorporated in the grid connected seven phase wind electric generator was analysed for various operating grid conditions. The use of multiphase machines along with the PLL synchronization of the grid increases the reliability of the WEG. Notably by the possibility of achieving post-fault disturbance free operation provided by the seven phase machine, as well as the constant voltage and frequency operation enabled by the *d-q* PLL.

Acknowledgments: No funding resources.

Author Contributions: Kalaivani Chandramohan, Sanjeevikumar Padmanaban, and Rajambal Kalyanasundaram, has developed the concept of the research proposed and developed the numerical background; Mahajan Sagar Bhaskar involved in the implementation of numerical simulation along with other authors for its depiction in quality of the work with predicted output results. Lucian Mihet-Popa has contributed his experience in AC drives and Wind Energy Conversion for further development and verification of theoretical concepts. All authors involved in articulating the paper work in its current form in each part their contribution to research investigation.

Conflicts of Interest: The authors declare no conflict of interest.

Nomenclature

MPIG	Multiphase Induction Generator
7PIG	Seven Phase Induction Generator
WEG	Wind Electric Generator
d-q	Direct-Quadrature axis
R_s, R_r	Stator, Rotor resistance (Ω)
L_s, L_r	Stator. Rotor leakage inductance (mH)
L_m	Mutual inductance (mH)
i_{ds}, i_{qs}	Stator d-q-axis currents (Amps)
i_{dr}, i_{qr}	Rotor d-q-axis currents (Amps)
V_{ds}, V_{qs}	Stator d-q-axis voltage (V)
V_{dr}, V_{qr}	Rotor d-q-axis voltage (V)
$\lambda_{ds}, \lambda_{qs}$	Stator d-q-axis flux linkage
SRFPLL	Synchronous Reference Frame Phase Locked Loop
$\lambda_{dr}, \lambda_{qr}$	λ_{qr} Stator d-q-axis flux linkage
P	Numbers of poles
p	Differential operator with respect to t
Δ	Tip Speed ratio
B	Blade Pitch Angle
ω_{tur}	Angular speed of turbine
T_g	Electromagnetic Torque
f_s	Switching frequency (Hz)
δ	Duty ratio
V_m	Peak value of phase voltage (V)
U, θ	Amplitude and phase of input

Appendix A

Table A1. Wind Turbine and Seven Phase Induction Generator Parameters taken for Investigation.

Wind Turbine		7PIG	
Rated power	250 kW	Rated power	210 kW
No. of blades	3	Rated voltage	240 V
Rated speed	40 rpm	Rated current	240 A
Rotor Diameter	29.8 m	Rated frequency	50 Hz
Air density	1.2 kg/m^3	Rated power factor	0.82
Blade pitch angle	-1.1	Rated speed	1018 rpm
Gear Ratio	1:24.52	No. of poles	6
Cut-in wind speed	3 m/s	Stator resistance	0.12 ohms
Cut-out wind speed	25 m/s	Stator leakage inductance	0.017197 mH
Rated wind speed	15 m/s	Rotor resistance referred to stator	0.0047 ohms
Equivalent inertia	1542 kg·m^2	Rotor leakage inductance referred to stator	0.015605 mH

References

1. Jain, S.; Ramulu, C.; Padmanaban, S.; Ojo, J.O.; Ertas, A.H. Dual MPPT algorithm for dual PV source fed open-end winding induction motor drive for pumping application. *Int. J. Eng. Sci. Technol.* **2016**, *19*, 1771–1780. [CrossRef]
2. Yaramasu, V.; Wu, B.; Sen, P.C.; Kouro, S.; Narimani, M. High-power wind energy con-version system: State-of-the-art and emerging technologies. *IEEE Proc.* **2015**, *103*, 740–788. [CrossRef]
3. Singh, G.K. Self-excited induction generator research—A survey. *Electr. Power Syst. Res.* **2004**, *69*, 107–114. [CrossRef]
4. Bansal, R.C. Three phase self-excited induction generator-an overview. *IEEE Trans. Energy Convers.* **2015**, *20*, 292–299. [CrossRef]

5.   Thomsen, B.; Guerrero, J.; Thogersen, P. Faroe islands wind-powered space heating microgrid using self-excited 220-kW induction generator. *IEEE Trans. Sustain. Energy* **2014**, *5*, 1361–1366. [CrossRef]

6.   Khan, M.F.; Khan, M.R.; Iqbal, A. Modeling, implementation and analysis of a high (six) phase self-excited induction generator. *J. Electr. Syst. Inf. Technol.* **2017**. [CrossRef]

7.   Levy, D. Analysis of double stator induction machine used for a variable speed constant frequency small scale hydro/wind electric generator. *Electr. Power Syst. Res.* **1986**, *11*, 205–223. [CrossRef]

8.   Levi, E.; Bojoi, R.; Profumo, F.; Toliyat, H.A.; Williamson, S. Multiphase induction motor drives—A technology status review. *IET Electr. Power Appl.* **2007**, *1*, 489–516. [CrossRef]

9.   Singh, G.K. Multiphase Induction Machine drive research. *Electr. Power Syst. Res.* **2002**, *61*, 139–147. [CrossRef]

10. Jones, M.; Levi, E. A literature survey of state-of-the-art in multi-phase ac drives. In Proceedings of the 37th University Power Engineering Conference (UPEC), Stafford, UK, 9–11 September 2002; pp. 505–510.

11. Apsley, J.M.; Williamson, S.; Smith, A.; Barnes, M. Induction machine performance as a function of phase number. *IEE Proc. Electr. Power Appl.* **2006**, *153*, 898–904. [CrossRef]

12. Apsley, J.; Williamson, S. Analysis of multiphase induction machines with winding faults. *IEEE Trans. Ind. Appl.* **2006**, *42*, 465–472. [CrossRef]

13. Wang, T.; Fang, F.; Wu, X.; Jiang, X. Novel Filter for Stator Harmonic Currents Reduction in Six-Step Converter Fed Multiphase Induction Motor Drives. *IEEE Trans. Power Electr.* **2013**, *28*, 498–506. [CrossRef]

14. Ayman, S.; Khalik, A.; Ahmed, S. Performance Evaluation of a Five-Phase Modular Winding Induction Machine. *IEEE Trans. Ind. Electr.* **2012**, *59*, 2654–2699.

15. Sanjeevikumar, P.; Grandi, G.; Blaabjerg, F.; Wheeler, P.W; Ojo, J.O. Analysis and implementation of power management and control strategy for six-phase multilevel AC drive system in fault condition. *Int. J. Eng. Sci. Technol.* **2016**, *19*, 31–39.

16. Sanjeevikumar, P.; Grandi, G.; Blaabjerg, F.; Ojo, J.O.; Wheeler, P.W. Power sharing algorithm for vector controlled six-phase AC motor with four customary three-phase voltage source inverter drive. *Int. J. Eng. Sci. Technol.* **2015**, *18*, 408–415.

17. Sanjeevikumar, P.; Pecht, M. An isolated/non-isolated novel multilevel inverter configuration for dual three-phase symmetrical/asymmetrical converter. *Int. J. Eng. Sci. Technol.* **2016**, *19*, 1763–1770.

18. Sanjeevikumar, P; Bhaskar, M.S.; Blaabjerg, F.; Norum, L.; Seshagiri, S.; Hajizadeh, A. Nine-phase hex-tuple inverter for five-level output based on double carrier PWM technique. In Proceedings of the 4th IET International Conference on Clean Energy and Technology, IET-CEAT'16, Kuala Lumpur, Malaysia, 14–15 November 2016.

19. Dragonas, F.A; Nerrati, G.; Sanjeevikumar, P; Grandi, G. High-voltage high-frequency arbitrary waveform multilevel generator for DBD plasma actuators. *IEEE Trans. Ind. Appl.* **2015**, *51*, 3334–3342. [CrossRef]

20. Sanjeevikumar, P.; Blaabjerg, F.; Wheeler, P.; Lee, K.; Mahajan, S.B.; Dwivedi, S. Five-phase five-level open-winding/star-winding inverter drive for low-voltage/high-current applications. In Proceedings of the 2016 IEEE Transportation Electrification Conference and Expo, Asia Pacific, Busan, Korea, 1–4 June 2016; pp. 66–71.

21. Sanjeevikumar, P.; Blaabjerg, F.; Wheeler, P.; Siano, P.; Martirano, L.; Szcześniak, P. A novel multilevel quad-inverter configuration for quasi six-phase open-winding converter. In Proceedings of the 2016 10th International Conference on Compatibility, Power Electronics and Power Engineering, Bydgoszcz, Poland, 29 June–1 July 2016; pp. 325–330.

22. Ayman, S.; Khalik, A.; Masoud, M. Effect of Current Harmonic Injection on Constant Rotor Volume Multiphase Induction Machine Stators: A Comparative Study. *IEEE Trans. Ind. Appl.* **2012**, *48*, 2002–2013.

23. Wang, L.; Jian, Y.-S. Dynamic Performance of isolated self-Excited Induction Generator under various Loading conditions. *IEEE Trans. Energy Convers.* **1999**, *14*, 93–100. [CrossRef]

24. Mihet-Popa, L.; Blaabjerg, F.; Boldea, I. Wind Turbine Generator Modeling and Simulation where Rotational Speed is the Controlled Variable. *IEEE Transac. Ind. Appl.* **2004**, *40*, 3–10. [CrossRef]

25. Mihet-Popa, L.; Proştean, O.; Szeidert, I. The soft-starters modeling, simulations and control implementation for 2 MW constant-speed wind turbines. *Int. Rev. Electric. Eng.* **2008**, *3*, 129–135.

26. Mihet-Popa, L.; Groza, V. Modeling and simulations of a 12 MW wind farm. *J. Advan. Electric. Comput. Eng.* **2010**, *10*, 141–144. [CrossRef]

27. Singh, G.K.; Yadav, K.B.; Saini, R.P. Modeling and analysis of multi-phase (six phase) self-excited induction generator. In Proceedings of the Eighth International Conference on Electrical Machines and Systems, Nanjing, China, 29 September 2005; Volume 3.

28. Singh, G.K.; Yadav, K.B.; Saini, R.P. Analysis of a saturated multi-phase (six-phase) self-excited induction generator. *Int. J. Emerg. Electr. Power Syst.* **2006**, *7*. [CrossRef]

29. Singh, G.K. Modeling and experimental analysis of a self-excited six-phase induction generator for stand-alone renewable energy generation. *Int. J. Renew Energy* **2008**, *33*, 1605–1621. [CrossRef]

30. Singh, G.K.; Yadav, K.B.; Saini, R.P. Capacitive self-excitation in six-phase induction generator for small hydro power—An experimental investigation. In Proceedings of the International Conference on Power Electronics, Drives and Energy Systems (PEDES), New Delhi, India, 12–15 December 2006.

31. Singh, G.K. Steady-state performance analysis of six-phase self-excited induction generator for renewable energy generation. In Proceedings of the 11th International Conference on Electrical Machines and Systems (ICEMS), Wuhan, China, 17–20 October 2008.

32. Mittal, R.; Sandhu, K.S.; Jain, D.K. An overview of some important issues related to wind energy conversion system (WECS). *Int. J. Environ. Sci. Dev.* **2010**, *1*. [CrossRef]

33. Blaabjerg, F.; Liserre, M.; Ma, K. Power Electronics Converters for Wind Turbine Systems. *IEEE Trans. Ind. Appl.* **2012**, *48*, 708–719. [CrossRef]

34. Pavlos, T.; Sourkounis, C. Review of control strategies for DFIG-based wind turbines under unsymmetrical grid faults. In Proceedings of the 2014 Ninth International Conference on Ecological Vehicles and Renewable Energies (EVER), Monte-Carlo, Monaco, 25–27 March 2014.

35. Haniotis, A.E.; Soutis, K.S.; Kladas, A.G.; Tegopoulos, J.A. Grid connected variable speed wind turbine modeling, dynamic performance and control. In Proceedings of the 2004 IEEE PES Power Systems Conference and Exposition, New York, NY, USA, 10–13 October 2004.

36. Ramesh, M.; Jyothsna, T.R. A Concise Review on different ascepts of wind energy systems. In Proceedings of the 2016 3rd international conference on Electrical Energy systems (ICEES), Chennai, India, 17–19 March 2016.

37. Limongi, L.R.; Bojoi, R.; Pica, C.; Profumo, F.; Tenconi, A. Analysis and Comparison of Phase Locked Loop Techniques for Grid Utility Applications. In Proceedings of the IEEE Power Conversion Conference PCC, Nagoya, Japan, 2–5 April 2007; pp. 674–681.

38. Karimi-Ghartemani, M.; Iravani, M. A method for synchronization of power electronic converters in polluted and variable-frequency environments. *IEEE Trans. Power Syst.* **2004**, *19*, 1263–1270. [CrossRef]

39. Krause, P.; Wasynczuk, O.; Sudhoff, S.D.; Pekarek, S. *Analysis of Electric Machinery and Drive Systems*, 3rd ed.; John Wiley & Sons: Hoboken, NJ, USA, 2013.

40. Bimal, K.B. *Modern Power Electronics and AC Drives*; Prentice Hall: Upper Saddle River, NJ, USA, 2002.

41. Renukadevi, G.; Rajambal, K. Generalized model of multi-phase induction motor drive using Matlab/Simulink. In Proceedings of the 2011 IEEE PES Innovative Smart Grid Technologies, Kerala, India, 1–3 December 2011.

42. Renukadevi, G.; Rajambal, K. Novel carrier-based PWM technique for n-phase VSI. *Int. J. Energy Technol.* **2011**, *1*, 1–9.

43. Renukadevi, G.; Rajambal, K. Comparison of different PWM schemes for n-phase VSI. In Proceedings of the 2012 International Conference on Advances in Engineering, Science and Management (ICAESM), Nagapattinam, Tamil Nadu, India, 30–31 March 2012; pp. 559–564.

44. Renukadevi, G.; Rajambal, K. Field programmable gate array implementation of space-vector pulse-width modulation technique for five-phase voltage source inverter. *IET Power Electr.* **2014**, *7*, 376–389. [CrossRef]

45. Masoud, M. Five-phase Uncontrolled Line Commutated Rectifier: AC Side Compensation using Shunt Active Power Filter. In Proceedings of the 8th IEEE GCC Conference and Exhibition, Muscat, Oman, 1–4 February 2015.

46. Ranjana, M.S.B.; Sanjeevikumar, P.; Siano, P.; Fedák, V.; Vaidya, H.; Aishwarya, S.T. On The structural implementation of magnetic levitation windmill. In Proceedings of the IEEE 1st Industrial and Commercial Power System Europe, 17th International Conference on Environment and Electrical Engineering, Milan, Italy, 6–9 June 2017; pp. 2326–2330.

Review

A Comprehensive Study of Key Electric Vehicle (EV) Components, Technologies, Challenges, Impacts, and Future Direction of Development

Fuad Un-Noor [1], Sanjeevikumar Padmanaban [2,*], Lucian Mihet-Popa [3],
Mohammad Nurunnabi Mollah [1] and Eklas Hossain [4,*]

[1] Department of Electrical and Electronic Engineering, Khulna University of Engineering and Technology, Khulna 9203, Bangladesh; fuad9304@gmail.com (F.U.-N.); nurunnabim12@gmail.com (M.N.M.)
[2] Department of Electrical and Electronics Engineering, University of Johannesburg, Auckland Park 2006, South Africa
[3] Faculty of Engineering, Østfold University College, Kobberslagerstredet 5, 1671 Kråkeroy-Fredrikstad, Norway; lucian.mihet@hiof.no
[4] Department of Electrical Engineering & Renewable Energy, Oregon Tech, Klamath Falls, OR 97601, USA
*   Correspondence: sanjeevi_12@yahoo.co.in (S.P.); eklas.hossain@oit.edu (E.H.);
    Tel.: +27-79-219-9845 (S.P.); +1-541-885-1516 (E.H.)

Academic Editor: Sergio Saponara
Received: 8 May 2017; Accepted: 21 July 2017; Published: 17 August 2017

Abstract: Electric vehicles (EV), including Battery Electric Vehicle (BEV), Hybrid Electric Vehicle (HEV), Plug-in Hybrid Electric Vehicle (PHEV), Fuel Cell Electric Vehicle (FCEV), are becoming more commonplace in the transportation sector in recent times. As the present trend suggests, this mode of transport is likely to replace internal combustion engine (ICE) vehicles in the near future. Each of the main EV components has a number of technologies that are currently in use or can become prominent in the future. EVs can cause significant impacts on the environment, power system, and other related sectors. The present power system could face huge instabilities with enough EV penetration, but with proper management and coordination, EVs can be turned into a major contributor to the successful implementation of the smart grid concept. There are possibilities of immense environmental benefits as well, as the EVs can extensively reduce the greenhouse gas emissions produced by the transportation sector. However, there are some major obstacles for EVs to overcome before totally replacing ICE vehicles. This paper is focused on reviewing all the useful data available on EV configurations, battery energy sources, electrical machines, charging techniques, optimization techniques, impacts, trends, and possible directions of future developments. Its objective is to provide an overall picture of the current EV technology and ways of future development to assist in future researches in this sector.

Keywords: electric vehicle; energy sources; motors; charging technologies; effects of EVs; limitations of EVs; energy management; control algorithms; global EV sales; trends and future developments

1. Introduction

In recent times, electric vehicles (EV) are gaining popularity, and the reasons behind this are many. The most eminent one is their contribution in reducing greenhouse gas (GHG) emissions. In 2009, the transportation sector emitted 25% of the GHGs produced by energy related sectors [1]. EVs, with enough penetration in the transportation sector, are expected to reduce that figure, but this is not the only reason bringing this century old and once dead concept back to life, this time as a commercially viable and available product. As a vehicle, an EV is quiet, easy to operate, and does not have the fuel costs associated with conventional vehicles. As an urban transport mode, it is highly useful. It does not use any stored energy or cause any emission while idling, is capable of frequent start-stop driving, provides the total torque from the startup, and does not require trips to the gas station. It does not contribute either to any of the smog making the city air highly polluted. The instant torque makes it highly preferable for motor sports. The quietness and low infrared signature makes it useful in military use as well. The power sector is going through a changing phase where renewable sources are gaining momentum. The next generation power grid, called 'smart grid' is also being developed. EVs are being considered a major contributor to this new power system comprised of renewable generating facilities and advanced grid systems [2,3]. All these have led to a renewed interest and development in this mode of transport.

The idea to employ electric motors to drive a vehicle surfaced after the innovation of the motor itself. From 1897 to 1900, EVs became 28% of the total vehicles and were preferred over the internal combustion engine (ICE) ones [1]. But the ICE types gained momentum afterwards, and with very low oil prices, they soon conquered the market, became much more mature and advanced, and EVs got lost into oblivion. A chance of resurrection appeared in the form of the EV1 concept from General Motors, which was launched in 1996, and quickly became very popular. Other leading carmakers, including Ford, Toyota, and Honda brought out their own EVs as well. Toyota's highly successful Prius, the first commercial hybrid electric vehicle (HEV), was launched in Japan in 1997, with 18,000 units sold in the first year of production [1]. Today, almost none of those twentieth century EVs exist; an exception can be Toyota Prius, still going strong in a better and evolved form. Now the market is dominated by Nissan Leaf, Chevrolet Volt, and Tesla Model S; whereas the Chinese market is in the grip of BYD Auto Co., Ltd. (X'an National Hi-tech Industrial Development Zone, Xi'an, China).

EVs can be considered as a combination of different subsystems. Each of these systems interact with each other to make the EV work, and there are multiple technologies that can be employed to operate the subsystems. In Figure 1, key parts of these subsystems and their contribution to the total system is demonstrated. Some of these parts have to work extensively with some of the others, whereas some have to interact very less. Whatever the case may be, it is the combined work of all these systems that make an EV operate.

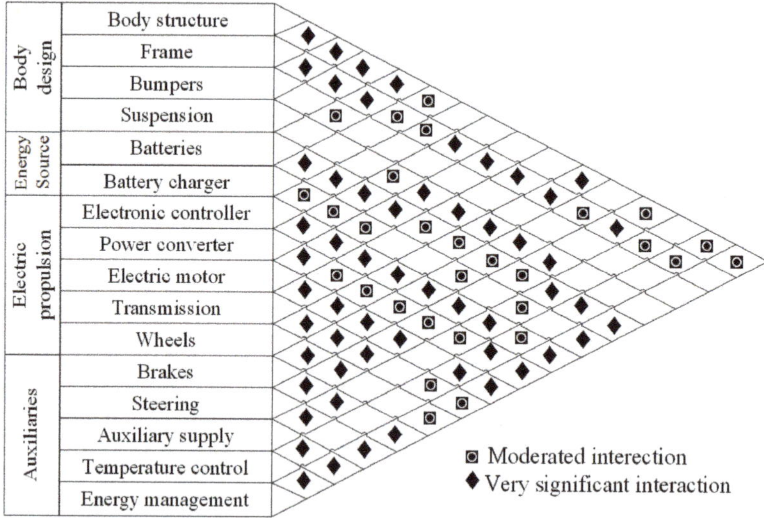

Figure 1. Major EV subsystems and their interactions. Some of the subsystems are very closely related while some others have moderated interactions. Data from [4].

There are quite a few configurations and options to build an EV with. EVs can be solely driven with stored electrical power, some can generate this energy from an ICE, and there are also some vehicles that employ both the ICE and the electrical motors together. The general classification is discussed in Section 2, whereas different configurations are described in Section 3. EVs use different types of energy storage to store their power. Though batteries are the most used ones, ultracapacitors, flywheels and fuel cells are also up and coming as potential energy storage systems (ESS). Section 4 is dedicated to these energy sources. The types of motors that have been used in EVs and can be used in future are discussed in Section 5. Different charging voltages and charger configurations can be used in charging the vehicles. Wireless charging is also being examined and experimented with to increase convenience. These charger standards, configurations and power conversion systems are demonstrated in Sections 6–8 discusses the effects EVs create in different sectors. Being a developing technology, EVs still have many limitations that have to be overcome to enable them to penetrate deeper into the market. These limitations are pointed out in Section 9 along with probable solutions. Section 10 summed up some strategies used in EVs to enable proper use of the available power. Section 11 presented different types of control algorithms used for better driving assistance, energy management, and charging. The current state of the global EV market is briefly presented in Section 12, followed by Section 13 containing the trends and sectors that may get developed in the future. Finally, the ultimate outcomes of this paper is presented in Section 14. The topics covered in this paper have been discussed in different literatures. Over the years, a number of publications have been made discussing different aspects of EV technology. This paper was created as an effort to sum up all these works to demonstrate the state-of-the art of the system and to position different technologies side by side to find out their merits and demerits, and in some cases, which one of them can make its way to the future EVs.

2. EV Types

EVs can run solely on electric propulsion or they can have an ICE working alongside it. Having only batteries as energy source constitutes the basic kind of EV, but there are kinds that can employ other energy source modes. These can be called hybrid EVs (HEVs). The International Electrotechnical Commission's Technical Committee 69 (Electric Road Vehicles) proposed that vehicles using two or more types of energy source, storage or converters can be called as an HEV as long as at least one of those provide electrical energy [4]. This definition makes a lot of combinations possible for HEVs like ICE and battery, battery and flywheel, battery and capacitor, battery and fuel cell, etc. Therefore, the common population and specialists both started calling vehicles with an ICE and electric motor combination HEVs, battery and capacitor ones as ultra-capacitor-assisted EVs, and the ones with battery and fuel cell FCEVs [2–4]. These terminologies have become widely accepted and according to this norm, EVs can be categorized as follows:

(1)   Battery Electric Vehicle (BEV)
(2)   Hybrid Electric Vehicle (HEV)
(3)   Plug-in Hybrid Electric Vehicle (PHEV)
(4)   Fuel Cell Electric Vehicle (FCEV)

2.1. Battery Electric Vehicle (BEV)

EVs with only batteries to provide power to the drive train are known as BEVs. BEVs have to rely solely on the energy stored in their battery packs; therefore the range of such vehicles depends directly on the battery capacity. Typically they can cover 100 km–250 km on one charge [5], whereas the top-tier models can go a lot further, from 300 km to 500 km [5]. These ranges depend on driving condition and style, vehicle configurations, road conditions, climate, battery type and age. Once depleted, charging the battery pack takes quite a lot of time compared to refueling a conventional ICE vehicle. It can take as long as 36 h completely replenish the batteries [6,7], there are far less time consuming ones as well, but none is comparable to the little time required to refill a fuel tank. Charging time depends on the charger configuration, its infrastructure and operating power level. Advantages of BEVs are their simple construction, operation and convenience. These do not produce any greenhouse gas (GHG), do not create any noise and therefore beneficial to the environment. Electric propulsion provides instant and high torques, even at low speeds. These advantages, coupled with their limitation of range, makes them the perfect vehicle to use in urban areas; as depicted in Figure 2, urban driving requires running at slow or medium speeds, and these ranges demand a lot of torque. Nissan Leaf and Teslas are some high-selling BEVs these days, along with some Chinese vehicles. Figure 3 shows basic configuration for BEVs: the wheels are driven by electric motor(s) which is run by batteries through a power converter circuit.

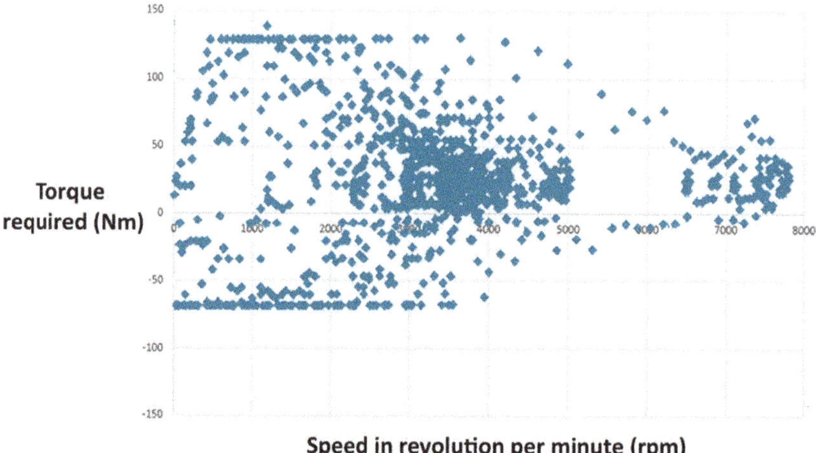

Figure 2. Federal Urban Driving Schedule torque-speed requirements. Most of the driving is done in the 2200 to 4800 rpm range with significant amount of torque. Lower rpms require torques as high as 125 Nm; urban vehicles have to operate in this region regularly as they face frequent start-stops. Data from [4].

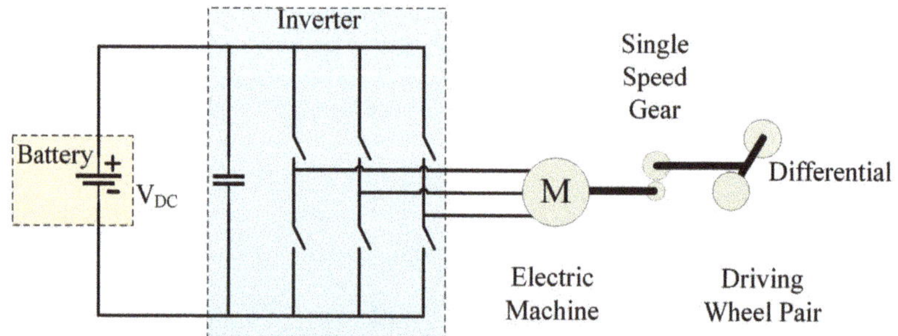

Figure 3. BEV configuration. The battery's DC power is converted to AC by the inverter to run the motor. Adapted from [5].

2.2. Hybrid Electric Vehicle (HEV)

HEVs employ both an ICE and an electrical power train to power the vehicle. The combination of these two can come in different forms which are discussed later. An HEV uses the electric propulsion system when the power demand is low. It is a great advantage in low speed conditions like urban areas;

it also reduces the fuel consumption as the engine stays totally off during idling periods, for example, traffic jams. This feature also reduces the GHG emission. When higher speed is needed, the HEV switches to the ICE. The two drive trains can also work together to improve the performance. Hybrid power systems are used extensively to reduce or to completely remove turbo lag in turbocharged cars, like the Acura NSX. It also enhances performance by filling the gaps between gear shifts and providing speed boosts when required. The ICE can charge up the batteries, HEVs can also retrieve energy by means of regenerative braking. Therefore, HEVs are primarily ICE driven cars that use an electrical drive train to improve mileage or for performance enhancement. To attain these features, HEV configurations are being widely adopted by car manufacturers. Figure 4 shows the energy flows in a basic HEV. While starting the vehicle, the ICE may run the motor as a generator to produce some power and store it in the battery. Passing needs a boost in speed, therefore the ICE and the motor both drives the power train. During braking the power train runs the motor as generator to charge the battery by regenerative braking. While cruising, ICE runs the both the vehicle and the motor as generator, which charges the battery. The power flow is stopped once the vehicle stops. Figure 5 shows an example of energy management systems used in HEVs. The one demonstrated here splits power between the ICE and the electric motor (EM) by considering the vehicle speed, driver's input, state of charge (SOC) of battery, and the motor speed to attain maximum fuel efficiency.

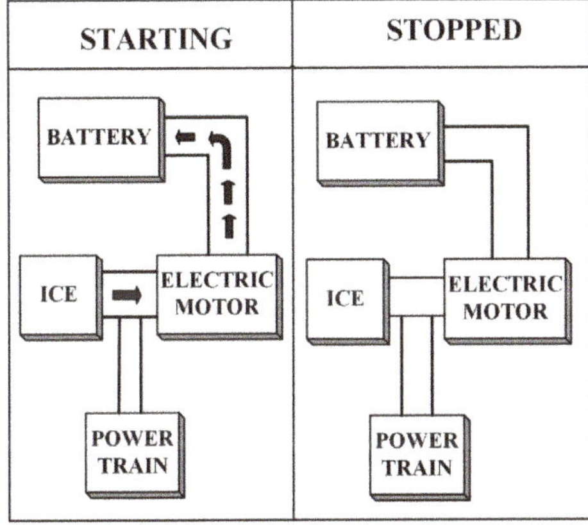

(a) Direction of power flow during starting and when stopped.

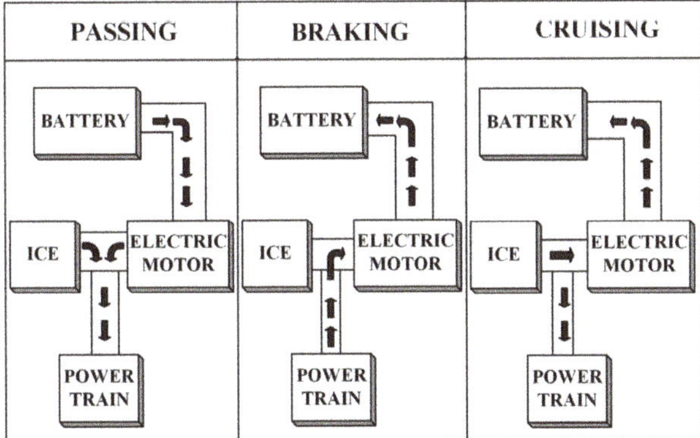

(b) Direction of power flow during passing, braking and cruising.

Figure 4. Power flow among the basic building blocks of an HEV during various stages of a drive cycle. Adapted from [8].

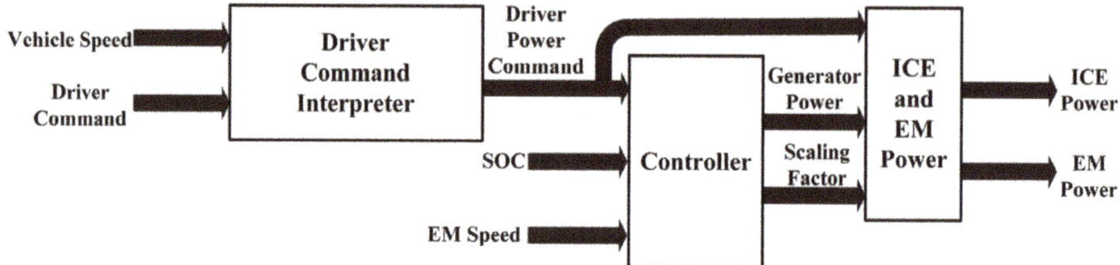

Figure 5. Example of energy management strategy used in HEV. The controller splits power between the ICE and the motor by considering different input parameters. Adapted from [8].

2.3. Plug-In Hybrid Electric Vehicle (PHEV)

The PHEV concept arose to extend the all-electric range of HEVs [9–14]. It uses both an ICE and an electrical power train, like a HEV, but the difference between them is that the PHEV uses electric propulsion as the main driving force, so these vehicles require a bigger battery capacity than HEVs. PHEVs start in 'all electric' mode, runs on electricity and when the batteries are low in charge, it calls on the ICE to provide a boost or to charge up the battery pack. The ICE is used here to extend the range. PHEVs can charge their batteries directly from the grid (which HEVs cannot); they also have the facility to utilize regenerative braking. PHEVs' ability to run solely on electricity for most of the time makes its carbon footprint smaller than the HEVs. They consume less fuel as well and thus reduce the associated cost. The vehicle market is now quite populated with these, Chevrolet Volt and Toyota Prius sales show their popularity as well.

2.4. Fuel Cell Electric Vehicle (FCEV)

FCEVs also go by the name Fuel Cell Vehicle (FCV). They got the name because the heart of such vehicles is fuel cells that use chemical reactions to produce electricity [15]. Hydrogen is the fuel of choice for FCVs to carry out this reaction, so they are often called 'hydrogen fuel cell vehicles'. FCVs carry the hydrogen in special high pressure tanks, another ingredient for the power generating process is oxygen, which it acquires from the air sucked in from the environment. Electricity generated from the fuel cells goes to an electric motor which drives the wheels. Excess energy is stored in storage systems like batteries or supercapacitors [2,3,16–18]. Commercially available FCVs like the Toyota Mirai or Honda Clarity use batteries for this purpose. FCVs only produce water as a byproduct of its power generating process which is ejected out of the car through the tailpipes. The configuration of an FCV is shown in Figure 6. An advantage of such vehicles is they can produce their own electricity which emits no carbon, enabling it to reduce its carbon footprint further than any other EV. Another major advantage of these are, and maybe the most important one right now, refilling these vehicles takes the same amount of time required to fill a conventional vehicle at a gas pump. This makes adoption of these vehicles more likely in the near future [2–4,19]. A major current obstacle in adopting this technology is the scarcity of hydrogen fuel stations, but then again, BEV or PHEV charging stations were not a common scenario even a few years back. A report to the U.S. Department of Energy (DOE) pointed to another disadvantage which is the high cost of fuel cells, that cost more than $200 per kW, which is far greater than ICE (less than $50 per kW) [20,21]. There are also concerns regarding safety in case of flammable hydrogen leaking out of the tanks. If these obstacles were eliminated, FCVs could really represent the future of cars. The possibilities of using this technology in supercars is shown by Pininfarina's H2 Speed (Figure 7). Reference [22] compared BEVs and FCEVs in different aspects, where FCEVs appeared to be better than BEVs in many ways; this comparison is shown in Figure 8. In this figure, different costs and cost associated issues of BEV and FCEV: weight, required storage volume, initial GHG emission, required natural gas energy, required wind energy, incremental costs, fueling infrastructure cost per car, fuel cost per kilometer, and incremental life cycle cost are all compared for 320 km (colored blue) and 480 km (colored green) ranges. The horizontal axis shows the attribute ratio of BEV to FCEV. As having a less value in these attributes indicates an advantage, any value higher than one in the horizontal axis will declare FCEVs superior to BEVs in that attribute. That being said, BEVs only appear better in the fields of required wind energy and fuel cost per kilometer. Fuel cost still appears to be one of the major drawbacks of FCEVs, as a cheap, sustainable and environment-friendly way of producing hydrogen is still lacking, and the refueling infrastructure lags behind that of BEVs; but these problems may no longer prevail in the near future.

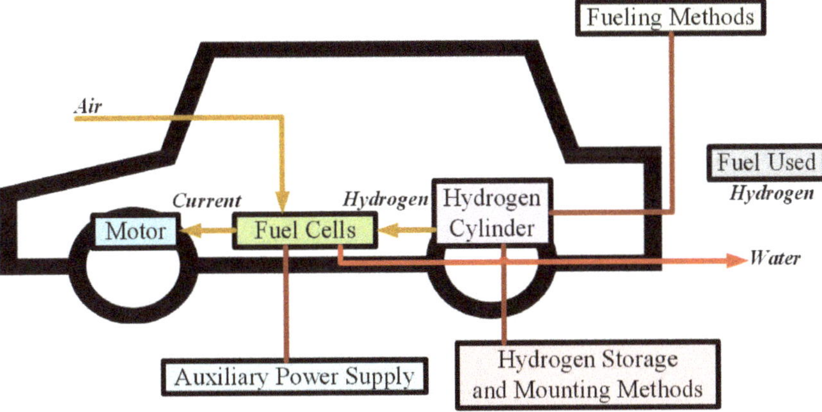

Figure 6. FCEV configuration. Oxygen from air and hydrogen from the cylinders react in fuel cells to produce electricity that runs the motor. Only water is produced as by-product which is released in the environment.

Figure 7. Pininfarina H2 Speed, a supercar employing hydrogen fuel cells.

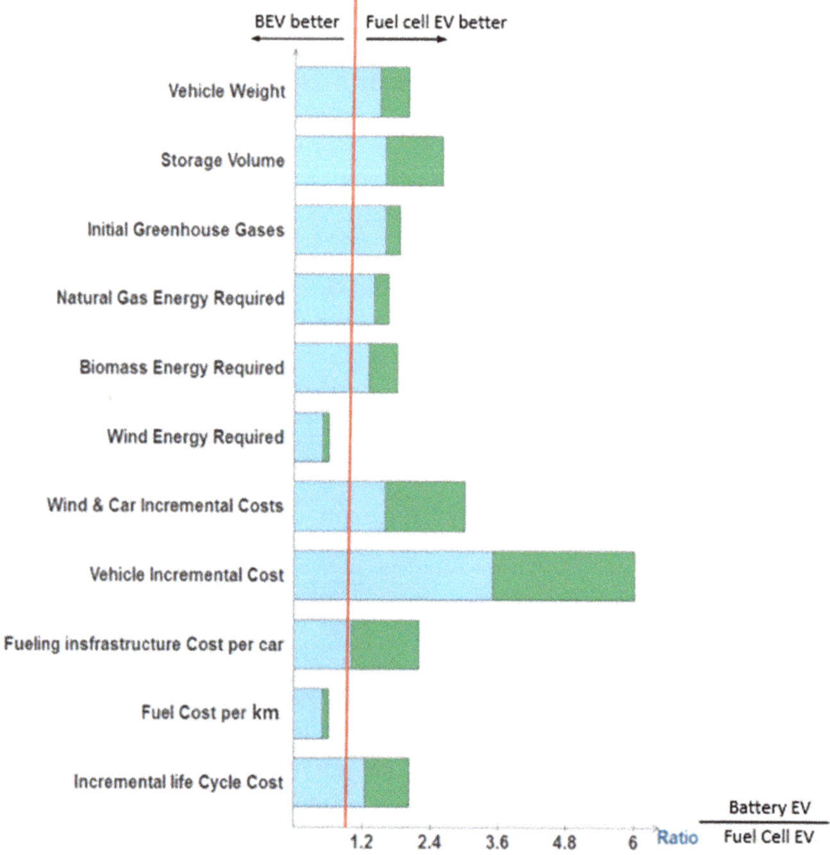

Figure 8. Advanced battery EV attribute and fuel cell EV attribute ratio for 320 km (colored blue) and 480 km (colored green) ranges, with assumptions of average US grid mix in 2010–2020 time-range and all hydrogen made from natural gas (values greater than one indicate a fuel cell EV advantage over the battery EV). Data from [22].

Rajashekara predicted a slightly different future for FCVs in [23]. He showed a plug-in fuel cell vehicle (PFCV) with a larger battery and smaller fuel cell, which makes it battery-dominant car. According to [23], if hydrogen for such vehicles can be made from renewable sources to run the fuel cells and the energy to charge the batteries comes from green sources as well, these PFCVs will be the future of vehicles. The FCVs we see today will not have much appeal other than some niche markets. Figure 9 shows a basic PFCV configuration. Table 1 compares the different vehicle types in terms of driving component, energy source, features, and limitations.

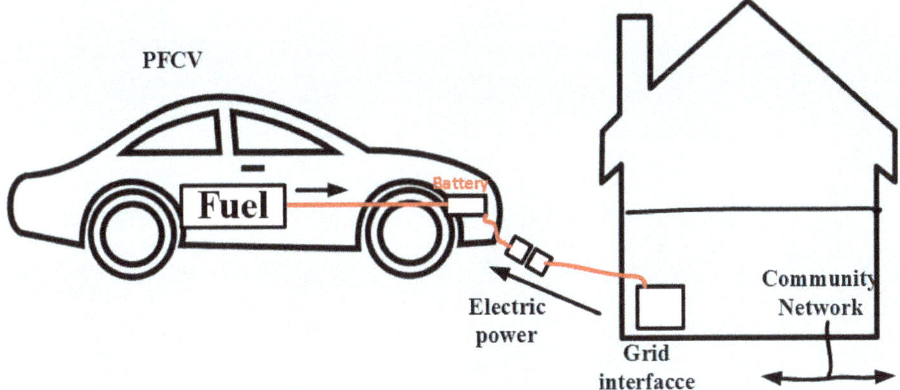

Figure 9. PFCV configuration. In addition to the fuel cells, this arrangement can directly charge the battery from a power outlet.

Table 1. Comparison of different vehicle types. Adapted from [4].

EV Type	Driving Component	Energy Source	Features	Problems
BEV	• Electric motor	• Battery • Ultracapacitor	• No emission • Not dependent on oil • Range depends largely on the type of battery used • Available commercially	• Battery price and capacity • Range • Charging time • Availability of charging stations • High price
HEV	• Electric motor • ICE	• Battery • Ultracapacitor • ICE	• Very little emission • Long range • Can get power from both electric supply and fuel • Complex structure having both electrical and mechanical drivetrains • Available commercially	• Management of the energy sources • Battery and engine size optimization
FCEV	• Electric motor	• Fuel cell	• Very little or no emission • High efficiency • Not dependent on supply of electricity • High price • Available commercially	• Cost of fuel cell • Feasible way to produce fuel • Availability of fueling facilities

3. EV Configurations

An electric vehicle, unlike its ICE counterparts, is quite flexible [4]. This is because of the absence of intricate mechanical arrangements that are required to run a conventional vehicle. In an EV, there is only one moving part, the motor. It can be controlled by different control arrangements and techniques. The motor needs a power supply to run which can be from an array of sources. These two components can be placed at different locations on the vehicle and as long as they are connected through electrical wires, the vehicle will work. Then again, an EV can run solely on electricity, but an ICE and electric motor can also work in conjunction to turn the wheels. Because of such flexibility, different configurations emerged which are adopted according to the type of vehicle. An EV can be considered as a system incorporating three different subsystems [4]: energy source, propulsion and auxiliary. The energy source subsystem includes the source, its refueling system and energy management system. The propulsion subsystem has the electric motor, power converter, controller, transmission and the driving wheels as its components. The auxiliary subsystem is comprised of auxiliary power supply, temperature control system and the power steering unit. These subsystems are shown in Figure 10.

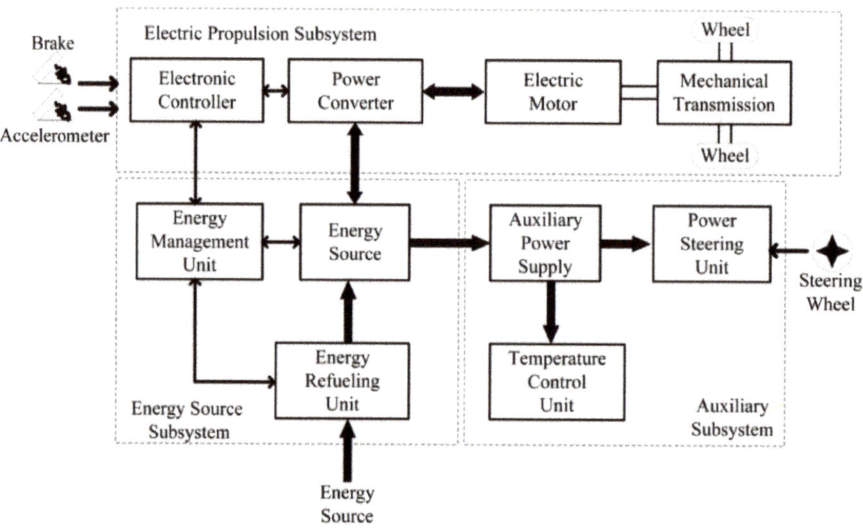

Figure 10. EV subsystems. Adapted from [4].

The arrows indicate the flow of the entities in question. A backward flow of power can be created by regenerative actions like regenerative braking. The energy source has to be receptive to store the energy sent back by regenerative actions. Most of the EV batteries along with capacitors/flywheels (CFs) are compatible with such energy regeneration techniques [4].

3.1. General EV Setup

EVs can have different configurations as shown in [4]. Figure 11a shows a front-engine front-wheel drive vehicle with just the ICE replaced by an electric motor. It has a gearbox and clutch that allows high torque at low speeds and low torque at high speeds. There is a differential as well that allows the wheels to rotate at different speeds. Figure 11b shows a configuration with the clutch omitted. It has a fixed gear in place of the gearbox which removes the chance of getting the desired torque-speed characteristics. The configuration of Figure 11c has the motor, gear and differential as a single unit that drives both the wheels. The Nissan Leaf, as well as the Chevrolet Spark, uses an electric motor mounted at the front to drive the front axle. In Figure 11d,e, configurations to obtain differential action by using two motors for the two wheels are shown. Mechanical interaction can be further reduced by placing the motors inside the wheels to produce an 'in-wheel drive'. A planetary gear system is employed here because advantages like high speed reduction ratio and inline arrangement of input and output shafts. Mechanical gear system is totally removed in the last configuration (Figure 11f) by mounting a low-speed motor with an outer rotor configuration on the wheel rim. Controlling the motor speed thus controls the wheel speed and the vehicle speed.

EVs can be built with rear wheel drive configuration as well. The single motor version of the Tesla Model S uses this configuration (Figure 12). The Nissan Blade Glider is a rear wheel drive EV with in-wheel motor arrangement. The use of in-wheel motors enables it to apply different amount of torques at each of the two rear wheels to allow better cornering.

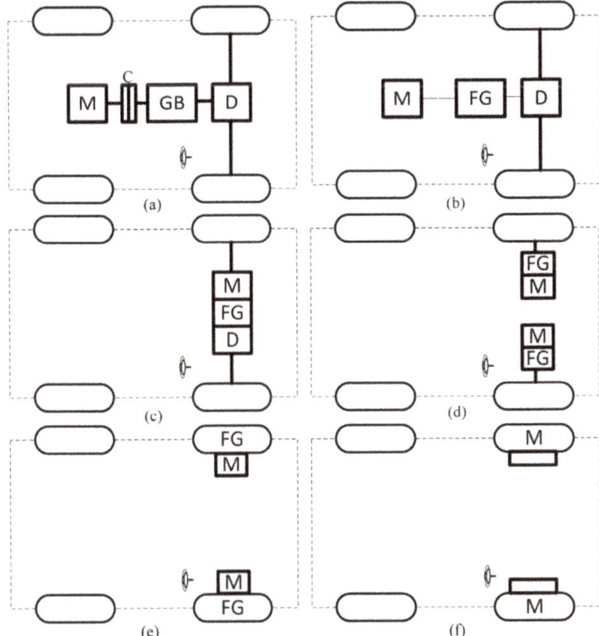

Figure 11. Different front wheel drive EV configurations. (**a**) Front-wheel drive vehicle with the ICE replaced by an electric motor; (**b**) Vehicle configuration with the clutch omitted; (**c**) Configuration with motor, gear and differential combined as a single unit to drive the front wheels; (**d**) Configuration with individual motors with fixed fearing for the front wheels to obtain differential action; (**e**) Modified configuration of Figure 11d with the fixed gearing arrangement placed within the wheels; (**f**) Configuration with the mechanical gear system removed by mounting a low-speed motor on the wheel rim. Adapted from [4].

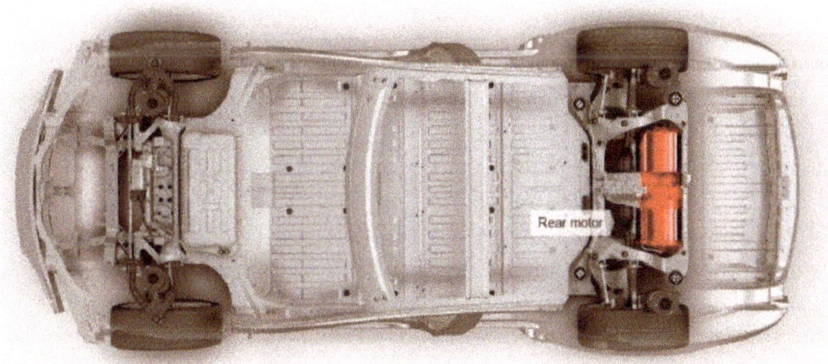

Figure 12. Tesla Model S, rear wheel drive configuration [22,24]. (Reprint with permission [24]; 2017, Tesla).

For more control and power, all-wheel drive (AWD) configurations can also be used, though it comes with added cost, weight and complexity. In this case, two motors can be used to drive the front and the rear axles. An all-wheel drive configuration is shown in Figure 13. AWD configurations are useful to provide better traction in slippery conditions, they can also use torque vectoring for better cornering performance and handling. AWD configuration can also be realized for in-wheel motor systems. It can prove quite useful for city cars like the Hiriko Fold (Figure 14) which has steering actuator, suspension, brakes and a motor all integrated in each wheel. Such arrangements can provide efficient all wheel driving, all wheel steering along with ease of parking and cornering.

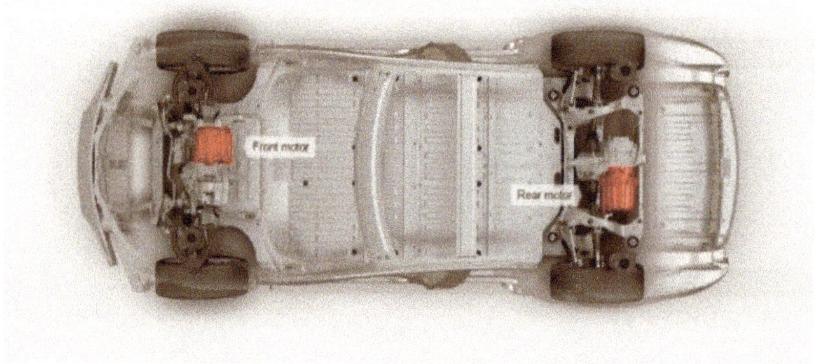

Figure 13. Tesla Model S, all-wheel drive configuration [24]. (Reprint with permission [24]; 2017, Tesla.)

Figure 14. Hiriko Fold—a vehicle employing in-wheel motors.

In-wheel motor configurations are quite convenient in the sense that they reduce the weight of the drive train by removing the central motor, related transmission, differential, universal joints and drive shaft [25]. They also provide more control, better turning capabilities and more space for batteries, fuel cells or cargo, but in this case the motor is connected to the power and control systems through wires that can get damaged because of the harsh environment, vibration and acceleration, thus causing serious trouble. Sato et al., proposed a wireless in-wheel motor system (W-IWM) in [26] which they had implemented in an experimental vehicle (shown in Figure 15). Simply put, the wires are replaced by two coils which are able to transfer power in-between them. Because of vibrations caused by road conditions, the motor and the vehicle can be misaligned and can cause variation in the secondary side voltage. In-wheel motor configurations are shown in Figure 16, whereas the efficiencies at different stages of such a system are shown in Figure 17. In conditions like this, magnetic resonance coupling is preferred for wireless power transfer [27] as it can overcome the problems associated with such misalignments [28]. The use of a hysteresis comparator and applying the secondary inverter power to a controller to counter the change in secondary voltage was also proposed in [28]. Wireless power transfer (WPT) employing magnetic resonance coupling in a series-parallel arrangement can provide a transmitting efficiency of 90% in both directions at 2 kW [29]. Therefore, W-IWM is compliant with regenerative braking as well.

Figure 15. Experimental vehicle with W-IWM system by Sato et al. [26]. (Reprint with permission [26]; 2015, IEEE.)

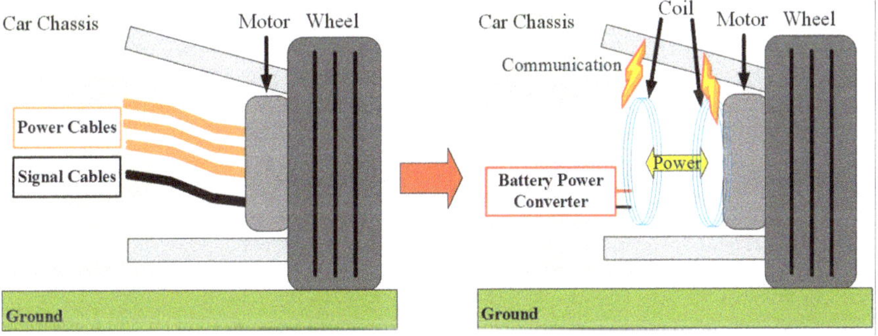

Figure 16. Conventional and wireless IWM. In the wireless setup, coils are used instead of wires to transfer power from battery to the motor. Adapted from [26].

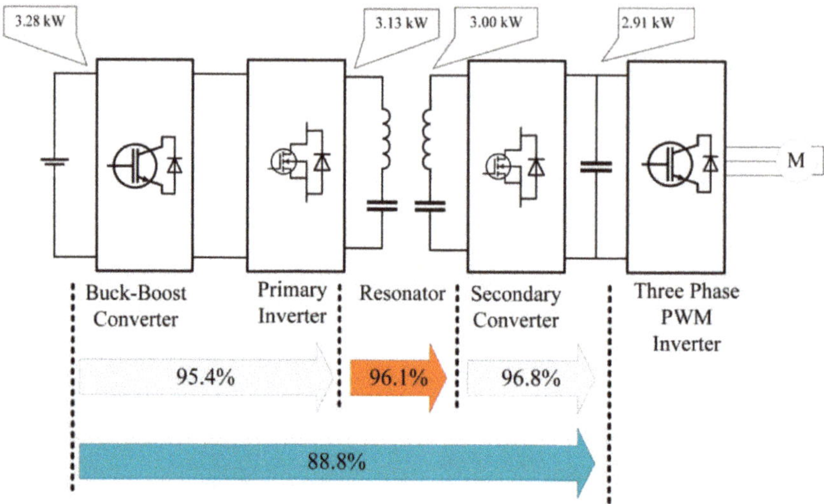

Figure 17. W-IWM setup showing efficiency at 100% torque reference. Adapted from [26].

3.2. HEV Setup

HEVs use both an electrical propulsion system and an ICE. Various ways in which these two can be set up to spin the wheels creates different configurations that can be summed up in four categories [4]:

(1)   Series hybrid
(2)   Parallel hybrid
(3)   Series-parallel hybrid
(4)   Complex hybrid

3.2.1. Series Hybrid

This configuration is the simplest one to make an HEV. Only the motor is connected to the wheels here, the engine is used to run a generator which provides the electrical power. It can be put as an EV that is assisted by an ICE generator [4]. Series hybrid drive train is shown in Figure 18. Table 2 shows the merits and demerits of this configuration.

Table 2. Advantages and limitations of series hybrid configuration. Adapted from [8].

Advantages	Efficient and optimized power-plant Possibilities for modular power-plant Optimized drive line Possibility of swift 'black box' service exchange Long lifetime Mature technology Fast response Capable of attaining zero emission
Limitations	Large traction drive system Requirement of proper algorithms Multiple energy conversion steps

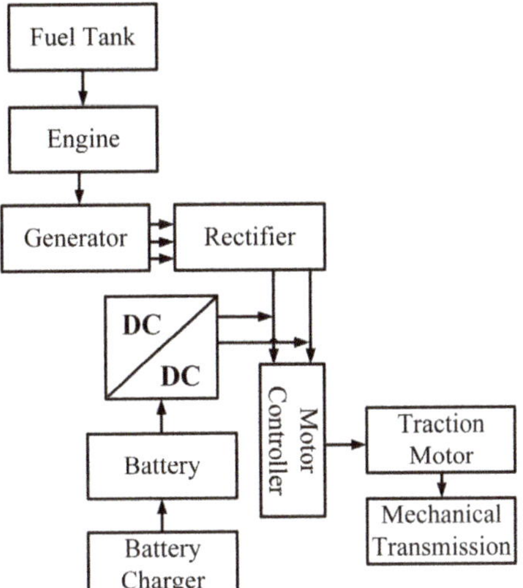

Figure 18. Drive train of series hybrid system. The engine is used to generate electricity only and supply to the motor through a rectifier. Power from the battery goes to the motor through a DC-DC converter [30].

3.2.2. Parallel Hybrid

This configuration connects both the ICE and the motor in parallel to the wheels. Either one of them or both take part in delivering the power. It can be considered as an IC engine vehicle with electric assistance [4]. The energy storages in such a vehicle can be charged by the electric motor by means of regenerative braking or by the ICE when it produces more than the power required to drive the wheels. Parallel hybrid drive train is shown in Figure 19. Table 3 shows the merits and demerits of this configuration, while Table 4 compares the series and the parallel systems.

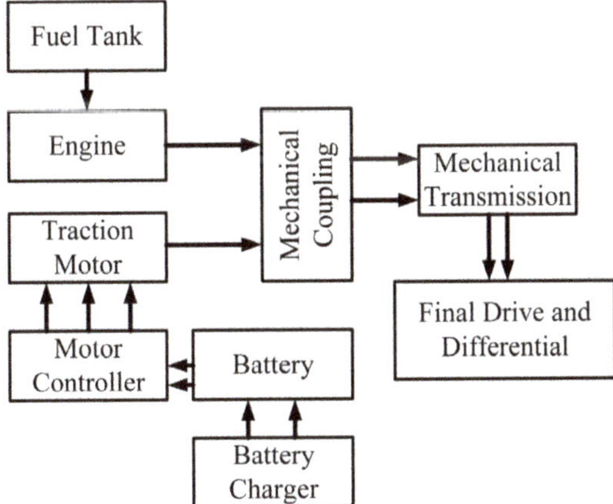

Figure 19. Drive train of parallel hybrid system. The engine and the motor both can run the can through the mechanical coupling [30].

Table 3. Advantages and limitations of parallel hybrid configuration. Adapted from [30].

Advantages	Capable of attaining zero emission Economic gain More flexibility
Limitations	Expensive Complex control Requirement of proper algorithms Need of high voltage to ensure efficiency

Table 4. Comparison of parallel and series hybrid configurations. Data from [8].

Parameters	Parallel HEV	Series HEV
Voltage	14 V, 42 V, 144 V, 300 V	216 V, 274 V, 300 V, 350 V, 550 V, 900 V
Power requirement	3 KW–40 KW	>50 KW
Relative gain in fuel economy (%)	5–40	>75

3.2.3. Series-Parallel Hybrid

In an effort to combine the series and the parallel configuration, this system acquires an additional mechanical link compared to the series type, or an extra generator when compared to the parallel type. It provides the advantages of both the systems but is more costly and complicated nonetheless. Complications in drive train are caused to some extent by the presence of a planetary gear unit [30]. Figure 20 shows a planetary gear arrangement: the sun gear is connected to the generator, the output shaft of the motor is connected to the ring gear, the ICE is coupled to the planetary carrier, and the pinion gears keep the whole system connected. A less complex alternative to this system is to use a transmotor, which is a floating-stator electric machine. In this system the engine is attached to the stator, and the rotor stays connected to the drive train wheel through the gears. The motor speed is the relative speed between the rotor and the stator and controlling it adjusts the engine speed for any particular vehicle speed [30]. Series-parallel hybrid drive train with planetary gear system is shown in Figure 21; Figure 22 shows the system with a transmotor.

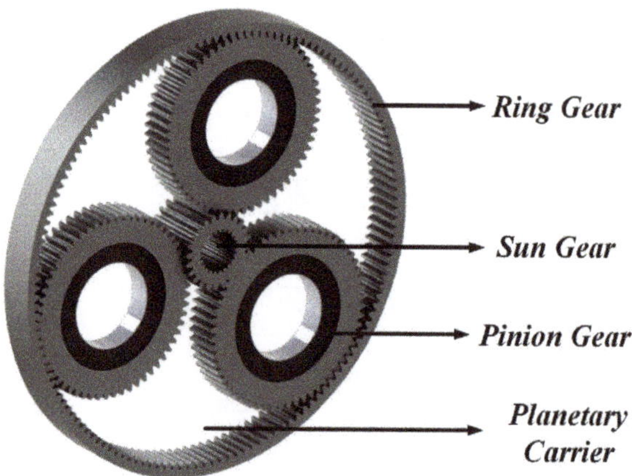

Figure 20. Planetary gear system [31].

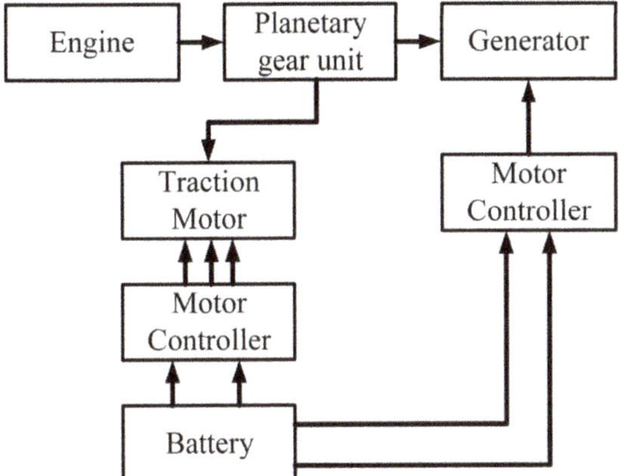

Figure 21. Drive train of series-parallel hybrid system using planetary gear unit. The planetary gear unit combines the engine, the generator and the motor [30].

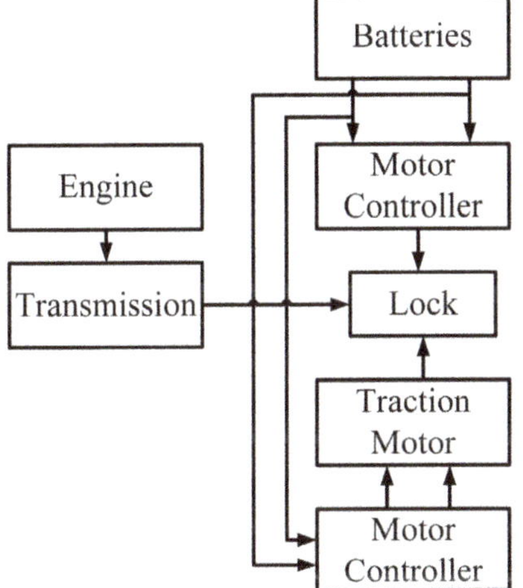

Figure 22. Drive train of series-parallel hybrid system using transmotor. The planetary gear system is absent in this arrangement [30].

3.2.4. Complex Hybrid

This system has one major difference with the series-parallel system, that is, it allows bidirectional flow of power whereas the series-parallel can provide only unidirectional power flow. However, using current market terminologies, this configuration is denoted as series-parallel system too. High complexity and cost are drawbacks of this system, but it is adopted by some vehicles to use dual-axle propulsion [4]. Constantly variable transmission (CVT) can be used for power splitting in a complex hybrid system or choosing between the power sources to drive the wheels. Electric arrangements can be used for such processes and this is dubbed as e-CVT, which has been developed and introduced by Toyota Motor Co. (Toyota City, Aichi Prefecture 471-8571, Japan). CVTs can be implemented hydraulically, mechanically, hydro-mechanically or electromechanically [32]. Two methods of power splitting—input splitting and complex splitting are shown in [32]. Input splitting got the name as it has a power split device placed at the transmission input. This system is used by certain Toyota and Ford models [32]. Reference [32] also showed different modes of these two splitting mechanisms and provided descriptions of e-CVT systems adopted by different manufacturers which are shown in Figures 23 and 24. Such power-split HEVs require two electric machines, wheels, an engine and a planetary gear (PG), combining all of them can be done in twenty-four different ways. If another PG is used, that number gets greater than one thousand. An optimal design incorporating a single PG is proposed in [31]. Four-wheel drive (4WD) configurations can benefit from using a two-motor hybrid configuration as it nullifies the need of a power transmission system to the back wheels (as they get their own motor) and provides the advantage of energy reproduction by means of regenerative braking [33]. Four-wheel drive HEV structure is shown in Figure 25. A stability enhancement scheme for such a configuration by controlling the rear motor is shown in [33].

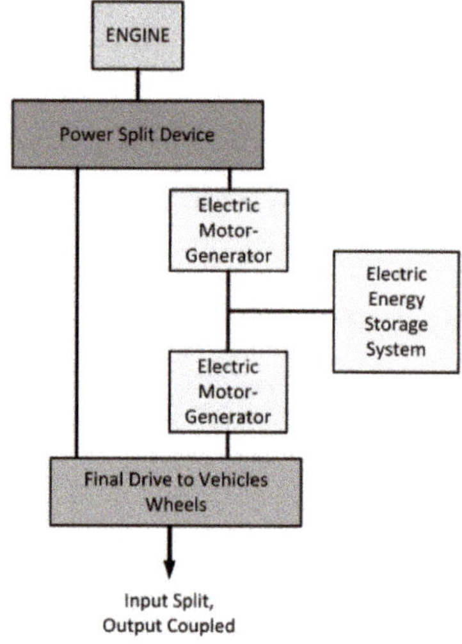

Figure 23. Input split e-CVT system. Adapted from [32].

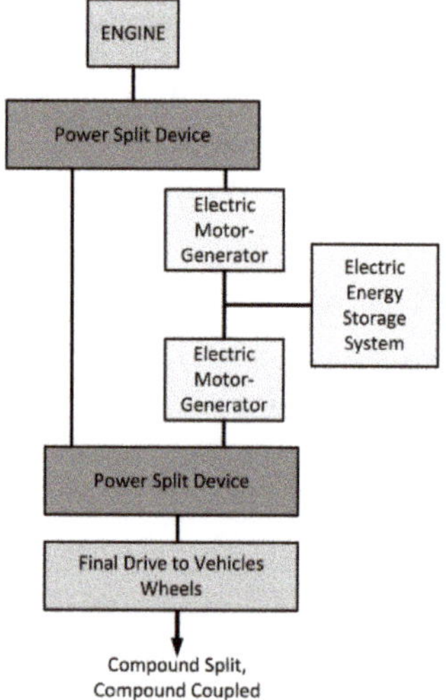

Figure 24. Compound split e-CVT system. Adapted from [32].

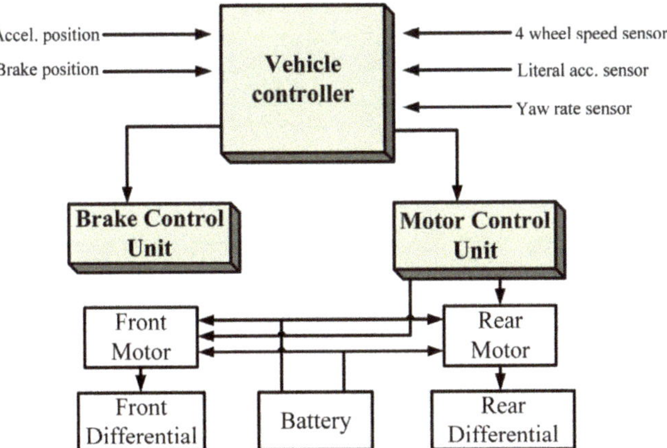

Figure 25. Structure for four-wheel drive HEV [32]. This particular system uses a vehicle controller which employs a number of sensors to perceive the driving condition and keeps the vehicle stable by controlling the brake control and the motor control units.

4. Energy Sources

EVs can get the energy required to run from different sources. The criteria such sources have to satisfy are mentioned in [4], high energy density and high power density being two of the most important ones [30]. There are other characteristics that are sought after to make a perfect energy source, fast charging, long service and cycle life, less cost and maintenance being a few of them. High specific energy is required from a source to provide a long driving range whereas high specific power helps to increase the acceleration. Because of the diverse characteristics that are required for the perfect source, quite a few sources or energy storage systems (ESS) come into discussion; they are also used in different combinations to provide desired power and energy requirements [4].

4.1. Battery

Batteries have been the major energy source for EVs for a long time; though of course, was time has gone by, different battery technologies have been invented and adopted and this process is still going on to attain the desired performance goals. Table 5 shows the desired performance for EV batteries set by the U.S. Advanced Battery Consortium (USABC).

Table 5. Performance goal of EV batteries as set by USABC. Data from [4].

	Parameters	Mid-Term	Long-Term
Primary goals	Energy density (C/3 discharge rate) (Wh/L)	135	300
	Specific energy (C/3 discharge rate) (Wh/kg)	80 (Desired: 100)	200
	Power density (W/l)	250	600
	Specific power (80% DOD/30 s) (W/kg)	150 (Desired: 200)	400
	Lifetime (year)	5	10
	Cycle life (80% DOD) (cycles)	600	1000
	Price (USD/kWh)	<150	<100
	Operating temperature (°C)	−30 to 65	−40 to 84
	Recharging time (hour)	<6	3 to 6
	Fast recharging time (40% to 80% SOC) (hour)	0.25	
Secondary goals	Self-discharge (%)	<15 (48 h)	<15 (month)
	Efficiency (C/3 discharge, 6 h charge) (%)	75	80
	Maintenance	No maintenance	No maintenance
	Resistance to abuse	Tolerance	Tolerance
	Thermal loss	3.2 W/kWh	3.2 W/kWh

Some of the prominent battery types are: lead-acid, Ni-Cd, Ni-Zn, Zn/air, Ni-MH, Na/S, Li-polymer and Li-ion batteries. Yong et al., also showed a battery made out of graphene for EV use whose advantages, structural model and application is described in [34]. Different battery types have their own pros and cons, and while selecting one, these things have to be kept in mind. In [35], Khaligh et al., provided key features of some known batteries which are demonstrated in Table 6. In Table 7, common battery types are juxtaposed to relative advantage of one battery type over the others.

Table 6. Common battery types, their basic construction components, advantages and disadvantages. Data from [35–44].

Battery Type	Components	Advantage	Disadvantage
Lead-acid	• Negative active material: spongy lead • Positive active material: lead oxide • Electrolyte: diluted sulfuric acid	• Available in production volume • Comparatively low in cost • Mature technology as used for over fifty years	• Cannot discharge more than 20% of its capacity • Has a limited life cycle if operated on a deep rate of SOC (state of charge) • Low energy and power density • Heavier • May need maintenance
NiMH (Nickel-Metal Hydride)	• Electrolyte: alkaline solution • Positive electrode: nickel hydroxide • Negative electrode: alloy of nickel, titanium, vanadium and other metals.	• Double energy density compared to lead-acid • Harmless to the environment • Recyclable • Safe operation at high voltage • Can store volumetric power and energy • Cycle life is longer • Operating temperature range is long • Resistant to over-charge and discharge	• Reduced lifetime of around 200–300 cycles if discharged rapidly on high load currents • Reduced usable power because of memory effect
Li-Ion (Lithium-Ion)	• Positive electrode: oxidized cobalt material • Negative electrode: carbon material • Electrolyte: lithium salt solution in an organic solvent	• High energy density, twice of NiMH • Good performance at high temperature • Recyclable • Low memory effect • High specific power • High specific energy • Long battery life, around 1000 cycles	• High cost • Recharging still takes quite a long time, though better than most batteries
Ni-Zn (Nickel-Zinc)	• Positive electrode: nickel oxyhydroxide • Negative electrode: zinc	• High energy density • High power density • Uses low cost material • Capable of deep cycle • Friendly to environment • Usable in a wide temperature range from −10 °C to 50 °C	• Fast growth of dendrite, preventing use in vehicles
Ni-Cd (Nickel-Cadmium)	• Positive electrode: nickel hydroxide • Negative electrode: cadmium	• Long lifetime • Can discharge fully without being damaged • Recyclable	• Cadmium can cause pollution in case of not being properly disposed of • Costly for vehicular application

Table 7. Cross comparison of different battery types to show relative advantages. Adapted from [45].

Advantages Over	Lead-Acid	Ni-Cd (Nickel-Cadmium)	NiMH (Nickel–Metal Hydride)	Li-Ion (Lithium-Ion)	
				Conventional	Polymer
Lead-acid		• Volumetric energy density • Gravimetric energy density • Range of operating temperature • Rate of self-discharge reliability	• Volumetric energy density • Gravimetric energy density • Rate of self-discharge	• Volumetric energy density • Gravimetric energy density • Rate of self-discharge	• Volumetric energy density • Gravimetric energy density • Rate of self-discharge • Design features
Ni-Cd (Nickel-Cadmium)	• Output voltage • Cost • Higher cyclability		• Volumetric energy density • Gravimetric energy density	• Volumetric energy density • Gravimetric energy density • Rate of self-discharge • Output voltage	• Volumetric energy density • Gravimetric energy density • Rate of self-discharge • Design features
NiMH (Nickel-Metal Hydride)	• Output voltage • Cost • Higher cyclability	• Range of operating temperature • Cost • Higher cyclability • Rate of self-discharge		• Volumetric energy density • Gravimetric energy density • Range of operating temperature	• Volumetric energy density • Gravimetric energy density • Range of operating temperature • Rate of self-discharge • Design features
Li-Ion (conventional)	• Cost • Safety • Higher cyclability • Re-cyclability	• Range of operating temperature • Cost • Safety • Higher cyclability • Recyclability	• Cost • Safety • Rate of discharge • Re-cyclability		• Volumetric energy density • Gravimetric energy density (potential) • Cost • Design features • Safety
Li-Ion (polymer)	• Cost • Higher cyclability	• Range of operating temperature • Higher cyclability • Cost	• Volumetric energy density • Cost • Higher cyclability	• Range of operating temperature • Higher cyclability	
Absolute advantages	• Cost • Higher cyclability	• Cost • Range of operating temperature	• Volumetric energy density	• Volumetric energy density • Gravimetric energy density • Range of operating temperature • Rate of self-discharge • Output voltage	• Volumetric energy density • Gravimetric energy density • Range of operating temperature • Rate of self-discharge • Output voltage • Design features

The battery packs used in EVs are made of numerous battery cells (Figure 26). The Tesla Model S, for example, has 7104 Li-Ion cells in the 85 kWh pack. All these cells are desired to have the same SOC at all times to have the same degradation rate and same capacity over the lifetime, preventing premature end of life (EOL) [46]. A power electronic control device, called a cell voltage equalizer, can achieve this feat by taking active measures to equalize the SOC and voltage of each cell. The equalizers can be of different types according to their construction and working principle. Resistive equalizers keep all the cells at the same voltage level by burning up the extra power at cells with higher voltages. Capacitive equalizers, on the other hand, transfers energy from the higher energy cells to the lower energy ones by switching capacitors. Inductive capacitors can be of different configurations: basic, Cuk, and single of multiple transformer based; but all of them transfer energy from higher energy cells to the ones with lower energy by using inductors [46–52]. All these configurations have their own merits and demerits, which are shown in Table 8; the schematics are shown in Figures 27 and 28. Table 9 shows comparisons between the equalizer types.

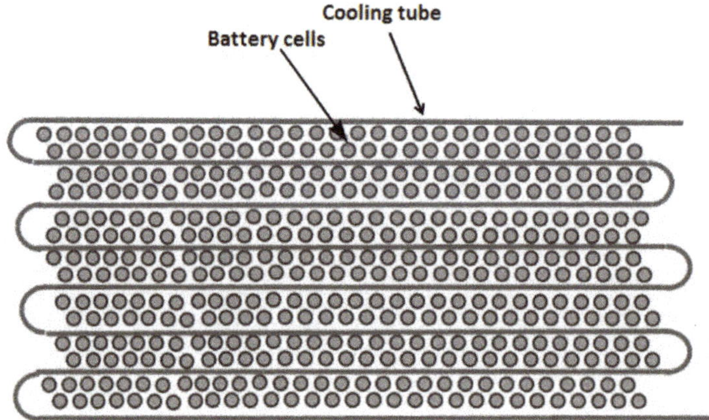

Figure 26. Battery cell arrangement in a battery pack. Cooling tubes are used to dissipate the heat generated in the battery cells.

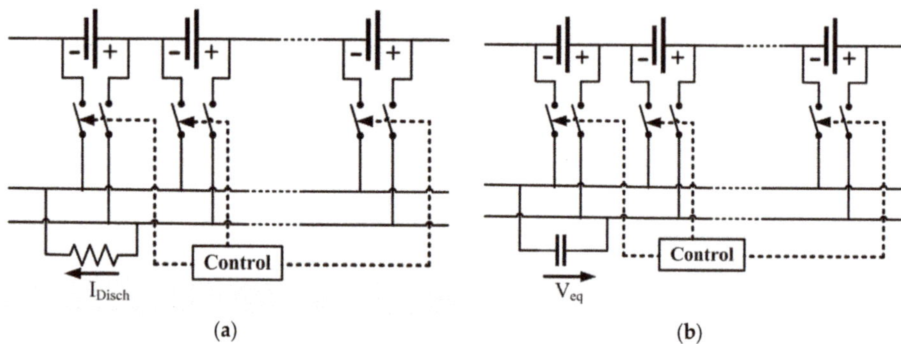

Figure 27. Equalizer configurations: (**a**) Resistive equalizer, extra power from any cell is burned up in the resistance; (**b**) Capacitive equalizer, excess energy is transferred to lower energy cells by switching of capacitors.

Table 8. Advantages and disadvantages of different equalizer types. Data from [46–52].

Equalizer Type	Advantage	Disadvantage
Resistive	• Cheapest, widely utilized for laptop batteries	• Inherent heating problem • Low equalizing current (300–500) mA • Only usable in the last stages of charging and flotation • Efficiency is almost 0% • All equalizing current transforms into heat for EV application, therefore not recommended
Capacitive	• Better current capabilities than resistive equalizers • No control issue • Simple implementation	• Unable to control inrush current • Potentially harmful current ripples can flow for big cell voltage differences • Cannot provide any required voltage difference which is essential for SOC equalization
Basic Inductive	• Relatively simple • Capable of transporting high amount of energy • Can handle complex control schemes like voltage difference control and current limitation • Can compensate for internal resistance of cells • Increased equalizing current • Not dependent on cell voltage	• Requires additional components to prevent ripple currents • Needs two switches in addition to drivers and controls in each cell • Current distribution is highly concentrated in neighboring cells because of switching loss
Cuk Inductive	• Has all the advantages of inductive equalizres • Can accommodate complex control and withstand high current	• Additional cost of higher voltage and current rated switches, power capacitors • Subjected to loss caused by series capacitor • A little less efficient than typical inductive equalizers • Faces problems during distributing equalizing currents all over the cell string • May need additional processing power
Transformer based Inductive	• Theoretically permits proper current distribution in all cells without addition control or loss	• Complex transformer with multiple secondary, which is very much challenging to mass produce • Not an option for EV packs • Cannot handle complex control algorithms
Multiple transformer based Inductive	• Separate transformers are used which are easier for mass production	• Still difficult to build with commercial inductors without facing voltage and current imbalance

Table 9. Comparison of equalizers; a ↑ sign indicates an advantage whereas the ↓ signs indicate drawbacks. Adapted from [46].

Equalizer Type	Equalizer Current	Current Distribution	Current Control	Current Ripple	Manufacture	Cost	Control
Resistive	↓↓	N/A	↑	↑↑↑	↑↑↑	↑↑↑	↑↑↑
Capacitive	↓	↑	↓↓	↓↓	↑↑	↑↑	↑↑
Basic Inductive	↑↑	↑	↑	↑↑	↑	↓	↓
Cuk	↑↑	↑	↑	↑↑↑	↓	↓↓	↓
Transformer	↑	↑↑↑	↓↓	↓↓	↓↓	↓↓	↑↑

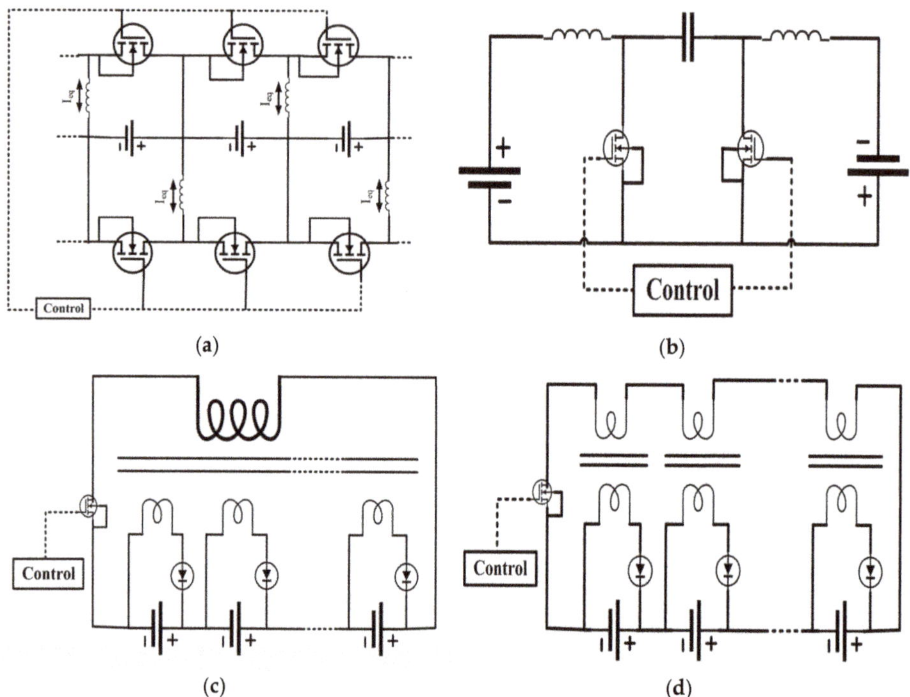

Figure 28. Inductive equalizer configurations: (**a**) Basic; (**b**) Cuk; (**c**) Transformer based; (**d**) Multiple transformers based. Excess energy is transferred to lower energy cells by using inductors.

Lithium-ion batteries are being used everywhere these days. It has replaced the lead-acid counterpart and became a mature technology itself. Their popularity can be justified by the fact that best-selling EVs, for example, Nissan Leaf and Tesla Model S—all use these batteries [53,54]. Battery parameters of some current EVs are shown in Table 10. Lithium batteries also have lots of scope to improve [55]. Better battery technologies have been discovered already, but they are not being pursued because of the exorbitant costs associated with their research and development, so it can be said that, lithium batteries will dominate the EV scene for quite some time to come.

Table 10. Battery parameters of some current EVs. Data from [5].

Model	Total Energy (kWh)	Usable Energy (kWh)	Usable Energy (%)
i3	22	18.8	85
C30	24	22.7	95
B-Class	36	28	78
e6	61.4	57	93
RAV4	41.8	35	84

4.2. Ultracapacitors (UCs)

UCs have two electrodes separated by an ion-enriched liquid dielectric. When a potential is applied, the positive electrode attracts the negative ions and the negative electrode gathers the positive ones. The charges get stored physically stored on electrodes this way and provide a considerably high power density. As no chemical reactions take place on the electrodes, ultra- capacitors tend to have a long cycle life; but the absence of any chemical reaction also makes them low in energy density [35].

The internal resistance is low too, making it highly efficient, but it also causes high output current if charged at a state of extremely low SOC [56,57]. A UC's terminal voltage is directly proportional to its SOC; so it can also operate all through its voltage range [35]. Basic construction of an UC cell is shown in Figure 29. EVs go through start/stop conditions quite a lot, especially in urban driving situations. This makes the battery discharge rate highly changeable. The average power required from batteries is low, but during acceleration or conditions like hill-climb a high power is required in a short duration of time [4,35]. The peak power required in a high-performance electric vehicle can be up to sixteen times the average power [4]. UCs fit in perfectly in such a scenario as it can provide high power for short durations. It is also fast in capturing the energy generated by regenerative braking [2,35]. A combined battery-UC system (as shown in Figure 30) negates each other's shortcomings and provides an efficient and reliable energy system. The low cost, load leveling capability, temperature adaptability and long service life of UCs make them a likable option as well [4,30].

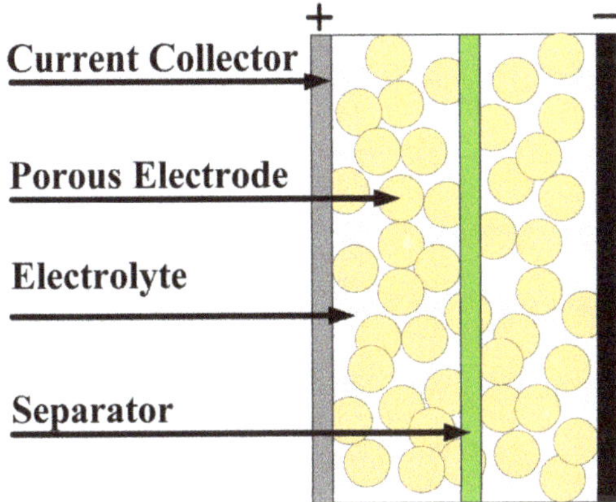

Figure 29. An UC cell; a separator keeps the two electrodes apart [58].

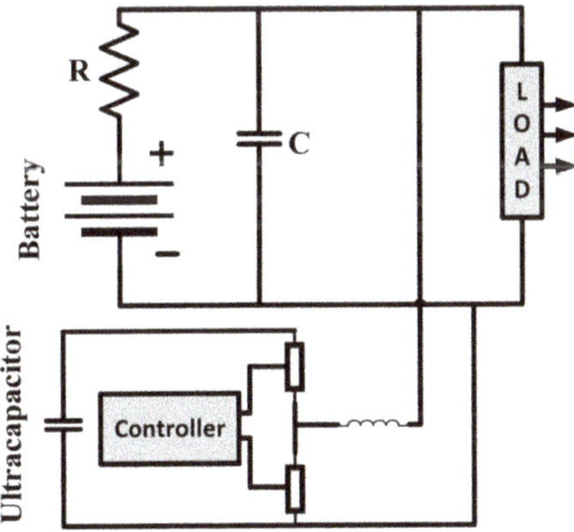

Figure 30. Combination of battery and UC to complement each-other's shortcomings [59].

4.3. Fuel Cell (FC)

Fuel cells generate electricity by electrochemical reaction. An FC has an anode (A), a cathode (C) and an electrolyte (E) between them. Fuel is introduced to the anode, gets oxidized there, the ions created travel through the electrolyte to the cathode and combine with the other reactant introduced there. The electrons produced by oxidation at the anode produce the electricity. Hydrogen is used in FCEVs because of its high energy content, and the facts it is non-polluting (producing only water as exhaust) and abundant in Nature in the form of different compounds such as hydrocarbons [4]. Hydrogen can be stored in different methods for use in EVs [4]; commercially available FCVs like the Toyota Mirai use cylinders to store it. The operating principle of a general fuel cell is demonstrated in Figure 31, while Figure 32 shows a hydrogen fuel cell. According to the material used, fuel cells can be classified into different types. A comparison among them is shown in Table 11. The chemical reaction governing the working of a fuel cell is stated below:

$$2H_2 + O_2 = 2H_2O \tag{1}$$

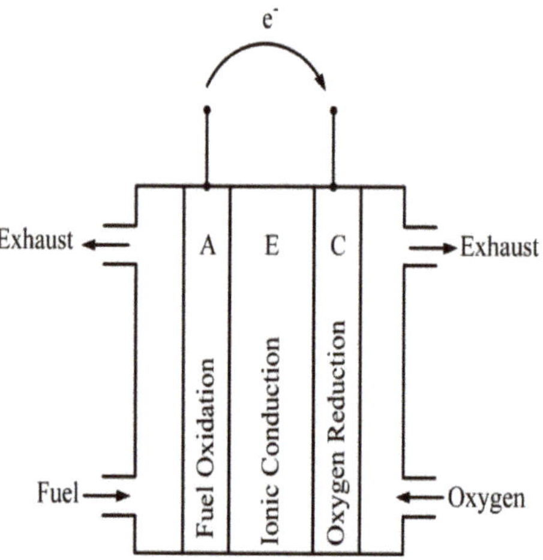

Figure 31. Working principle of fuel cell. Fuel and oxygen is taken in, exhaust and current is generated as the products of chemical reaction. Adapted from [4].

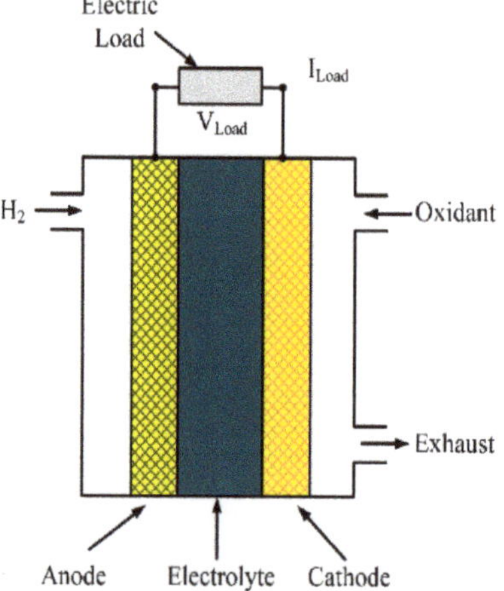

Figure 32. Hydrogen fuel cell configuration. Hydrogen is used as the fuel which reacts with oxygen and produces water and current as products. Adapted from [35].

Table 11. Comparison of different fuel cell configurations. Data from [2].

	PAFC	AFC	MCFC	SOFC	SPFC	DMFC
Working temp. (°C)	150–210	60–100	600–700	900–1000	50–100	50–100
Power density (W/cm^2)	0.2–0.25	0.2–0.3	0.1–0.2	0.24–0.3	0.35–0.6	0.04–0.25
Estimated life (kh)	40	10	40	40	40	10
Estimated cost (USD/kW)	1000	200	1000	1500	200	200

PAFC: Phosphoric acid fuel cell; AFC: Alkaline fuel cell; SOFC: Solid oxide fuel cell; SPFC: Solid polymer fuel cell, also known as proton exchange membrane fuel cell.

Fuel cells have many advantages for EV use like efficient production of electricity from fuel, noiseless operation, fast refueling, no or low emissions, durability and the ability to provide high density current output [24,60]. A main drawback of this technology is the high price. Hydrogen also have lower energy density compared to petroleum derived fuel, therefore larger fuel tanks are required for FCEVs, these tanks also have to capable enough to contain the hydrogen properly and to minimize risk of any explosion in case of an accident. FC's efficiency depends on the power it is supplying; efficiency generally decreases if more power is drawn. Voltage drop in internal resistances cause most of the losses. Response time of FCs is comparatively higher to UCs or batteries [35]. Because of these reasons, storage like batteries or UCs is used alongside FCs. The Toyota Mirai uses batteries to power its motor and the FC is used to charge the batteries. The batteries receive the power reproduced by regenerative braking as well. This combination provides more flexibility as the batteries do not need to be charged, only the fuel for the FC has to be replenished and it takes far less time than recharging the batteries.

4.4. Flywheel

Flywheels are used as energy storage by using the energy to spin the flywheel which keeps on spinning because of inertia. The flywheel acts as a motor during the storage stage. When the energy is needed to be recovered, the flywheel's kinetic energy can be used to rotate a generator to produce power. Advanced flywheels can have their rotors made out of sophisticated materials like carbon composites and are placed in a vacuum chamber suspended by magnetic bearings. Figure 33 shows a flywheel used in the Formula One (F1) racing kinetic energy recovery system (KERS). The major components of a flywheel are demonstrated in Figure 34. Flywheels offer a lot of advantages over other storage forms for EV use as they are lighter, faster and more efficient at absorbing power from regenerative braking, faster at supplying a huge amount of power in a short time when rapid acceleration is needed and can go through a lot of charge-discharge cycles over their lifetime. They are especially favored for hybrid racecars which go through a lot of abrupt braking and acceleration, which are also at much higher g-force than normal commuter cars. Storage systems like batteries or UCs cannot capture the energy generated by regenerative braking in situations like this properly. Flywheels, on the other hand, because of their fast response, have a better efficiency in similar scenarios, by making use of regenerative braking more effectively; it reduces pressure on the brake pads as well. The Porsche 911GT3R hybrid made use of this technology. Flywheels can be made with different materials, each with their own merits and demerits. Characteristics of some these materials are shown in Table 12; among the ones displayed in the table, carbon T1000 offers the highest amount of energy density, but it is much costlier than the others. Therefore, there remains a trade-off between cost and performance.

Figure 33. A flywheel used in the Formula One racing kinetic energy recovery system (KERS).

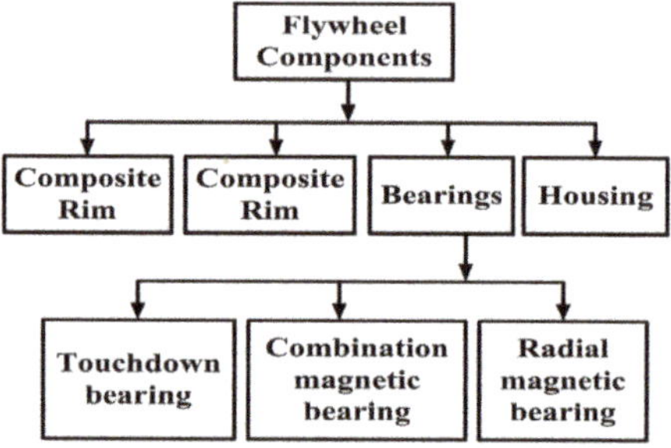

Figure 34. Basic flywheel components. The flywheel is suspended in tis hosing by bearings, and is connected to a motor-generator to store and supply energy [61].

Table 12. Characteristics of different materials used for flywheels [62].

Material		Density (kg/m^3)	Tensile Strength (mpa)	Max Energy Density (mj/kg)	Cost (USD/kg)
Monolithic material	4340 steel	7700	1520	0.19	1
Composites	E-glass	2000	100	0.05	11
	S2-glass	1920	1470	0.76	24.6
	Carbon T1000	1520	1950	1.28	101.8
	Carbon AS4C	1510	1650	1.1	31.3

Currently, no single energy source can provide the ideal characteristics, i.e., high value of both power and energy density. Table 13 shows a relative comparison of the energy storages to demonstrate this fact. Hybrid energy storages can be used to counter this problem by employing one source for high energy density and another for high power density. Different combinations are possible to create this hybrid system. It can be a combination of battery and ultracapacitor, battery and flywheel, or fuel cell and battery [4]. Table 14 shows the storage systems used by some current vehicles.

Table 13. Relative energy and power densities of different energy storage systems [63].

Storage	Energy Density	Power Density
Battery	High	Low
Ultracapacitor	Low	High
Fuel cell	High	Low
Flywheel	Low	High

Table 14. Vehicles using different storage systems.

Storage System	Vehicles Using the System
Battery	Tesla Model S, Nissan Leaf
Fuel cell + battery	Toyota Mirai, Honda Clarity
Flywheel	Porsche 911GT3R Hybrid

5. Motors Used

The propulsion system is the heart of an EV [64–69], and the electric motor sits right in the core of the system. The motor converts electrical energy that it gets from the battery into mechanical energy which enables the vehicle to move. It also acts as a generator during regenerative action which sends energy back to the energy source. Based on their requirement, EVs can have different numbers of motors: the Toyota Prius has one, the Acura NSX has three—the choice depends on the type of the vehicle and the functions it is supposed to provide. References [4,23] listed the requirements for a motor for EV use which includes high power, high torque, wide speed range, high efficiency, reliability, robustness, reasonable cost, low noise and small size. Direct current (DC) motor drives demonstrate some required properties needed for EV application, but their lack in efficiency, bulky structure, lack in reliability because of the commutator or brushes present in them and associated maintenance requirement made them less attractive [4,30]. With the advance of power electronics and control systems, different motor types emerged to meet the needs of the automotive sector, induction and permanent magnet (PM) types being the most favored ones [23,30,70].

5.1. Brushed DC Motor

These motors have permanent magnets (PM) to make the stator; rotors have brushes to provide supply to the stator. Advantages of these motors can be the ability to provide maximum torque in low speed. The disadvantages, on the other hand, are its bulky structure, low efficiency, heat generated because of the brushes and associated drop in efficiency. The heat is also difficult to remove as it is generated in the center of the rotor. Because of these reasons, brushed DC motors are not used in EVs any more [70].

5.2. Permanent Magnet Brushless DC Motor (BLDC)

The rotor of this motor is made of PM (most commonly NdFeB [4]), the stator is provided an alternating current (AC) supply from a DC source through an inverter. As there are no windings in the rotor, there is no rotor copper loss, which makes it more efficient than induction motors. This motor is also lighter, smaller, better at dissipating heat (as it is generated in the stator), more reliable, has more torque density and specific power [4]. But because of its restrained field-weakening ability, the constant power range is quite short. The torque also decreases with increased speed because of back EMF generated in the stator windings. The use of PM increases the cost as well [30,70]. However, enhancement of speed range and better overall efficiency is possible with additional field windings [4,71]. Such arrangements are often dubbed PM hybrid motors because of the presence of both PM and field windings. But such arrangements too are restrained by complexity of structure; the speed ratio is not enough to meet the needs of EV use, specifically in off-roaders [30]. PM hybrid motors can also be constructed using a combination of reluctance motor and PM motor. Controlling the conduction angle of the power converter can improve the efficiency of PM BLDCs as well as speed range, reaching as high as four times the base speed, though the efficiency may decrease at very high speed resulting from demagnetization of PM [4]. Other than the PM hybrid configurations, PM BLDCs can be buried magnet mounted—which can provide more air gap flux density, or surface magnet mounted—which require less amount of magnet. BLDCs are useful for use in small cars requiring a maximum 60 kW of power [72]. The characteristics of PM BLDCs are shown in Figure 35.

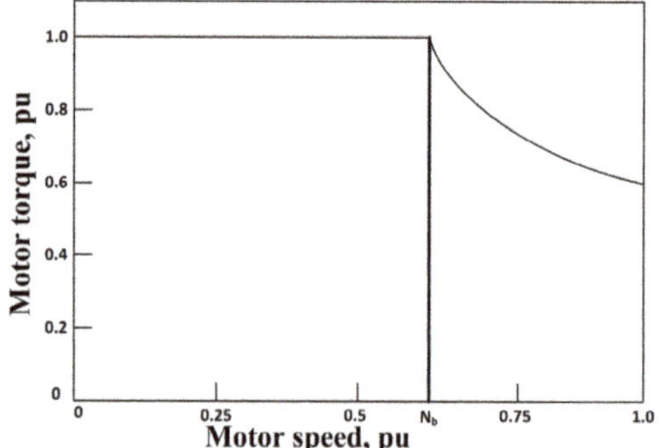

Figure 35. Characteristics of a Permanent Magnet Brushless DC Motor. The torque remains constant at the maximum right from the start, but starts to decrease exponentially for speeds over the base speed.

5.3. Permanent Magnet Synchronous Motor (PMSM)

These machines are one of the most advanced ones, capable of being operated at a range of speeds without the need of any gear system. This feature makes these motors more efficient and compact. This configuration is also very suitable for in-wheel applications, as it is capable of providing high torque, even at very low speeds. PMSMs with an outer rotor are also possible to construct without the need of bearings for the rotor. But these machines' only notable disadvantage also comes in during in-wheel operations where a huge iron loss is faced at high speeds, making the system unstable [73]. NdFeB PMs are used for PMSMs for high energy density. The flux linkages in the air-gap are sinusoidal in nature; therefore, these motors are controllable by sinusoidal voltage supplies and vector control [70]. PMSM is the most used motor in the BEVs available currently; at least 26 vehicle models use this motor technology [5].

5.4. Induction Motor (IM)

Induction motors are used in early EVs like the GM EV1 [23] as well as current models like the Teslas [54,74]. Among the different commutatorless motor drive systems, this is the most mature one [2]. Vector control is useful to make IM drives capable of meeting the needs of EV systems. Such a system with the ability to minimize loss at any load condition is demonstrated in [75]. Field orientation control can make an IM act like a separately excited DC motor by decoupling its field control and torque control. Flux weakening can extend the speed range over the base speed while keeping the power constant [30], field orientation control can achieve a range three to five times the base speed with an IM that is properly designed [76]. Three phase, four pole AC motors with copper rotors are seen to be employed in current EVs. Characteristics of IM are shown in Figure 36.

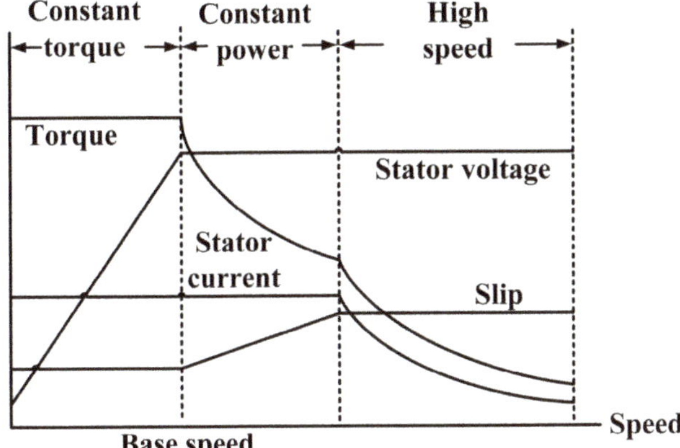

Figure 36. Induction motor drive characteristics. Maximum torque is maintained till base speed, and then decreases exponentially. Adapted from [4].

5.5. Switched Reluctance Motor (SRM)

SRMs, also known as doubly salient motor (because of having salient poles both in the stator and the rotor) are synchronous motors driven by unipolar inverter-generated current. They demonstrate simple and robust mechanical construction, low cost, high-speed, less chance of hazards, inherent long constant power range and high power density useful for EV applications. PM is not required for such motors and that facilitates enhanced reliability along with fault tolerance. On the downside, they are very noisy because of the variable torque nature, have low efficiency, and are larger in size and weight when compared to PM machines. Though such machines have a simple construction, their design and control are not easy resulting from fringe effect of slots and poles and high saturation of the pole-tips [4,23,30,70]. Because of such drawbacks, these machines did not advance as much as the PM or induction machines. However, because of the high cost rare-rare earth materials needed in PM machines, interest in SRMs are increasing. Advanced SRMs like the one demonstrated by Nidec in 2012 had almost interior permanent machine (IPM)-like performance, with a low cost. Reducing the noise and torque ripple are the main concerns in researches associated with SRMs [23]. One of the configurations that came out of these researches uses a dual stator system, which provides low inertia and noise, superior torque density and increased speed-range compared to conventional SRMs [77,78]. Design by finite element analysis can be employed to reduce the total loss [79], control by fuzzy sliding mode can also be employed to reduce control chattering and motor nonlinearity management [80].

5.6. Synchronous Reluctance Motor (SynRM)

A Synchronous Reluctance Motor runs at a synchronous speed while combining the advantages of both PM and induction motors. They are robust and fault tolerant like an IM, efficient and small like a PM motor, and do not have the drawbacks of PM systems. They have a control strategy similar to that of PM motors. The problems with SynRM can be pointed as the ones associated with controllability, manufacturing and low power factor which hinder its use in EVs. However, researches have been going on and some progress is made as well, the main area of concern being the rotor design. One way to improve this motor is by increasing the saliency which provides a higher power factor. It can be achieved by axially or transversally laminated rotor structures, such an arrangement is shown in Figure 37. Improved design techniques, control systems and advanced manufacturing can help it make its way into EV applications [23].

Figure 37. SynRM with axially laminated rotor [23].

5.7. PM Assisted Synchronous Reluctance Motor

Greater power factors can be achieved from SynRMs by integrating some PMs in the rotor, creating a PM assisted Synchronous Reluctance Motor. Though it is similar to an IPM, the PMs used are fewer in amount and the flux linkages from them are less too. PMs added in the right amount to the core of the rotor increase the efficiency with negligible back EMF and little change to the stator. This concept is free from the problems associated with demagnetization resulting from overloading and high temperature observed in IPMs. With a proper efficiency optimization technique, this motor can have the performance similar to IPM motors. A PM-assisted SynRM suitable for EV use was demonstrated by BRUSA Elektronik AG (Sennwald, Switzerland). Like the SynRM, PM-assisted SynRMs can also get better with improved design techniques, control systems and advanced manufacturing systems [23]. A demonstration of the rotor of PM-assisted SynRM is shown in Figure 38.

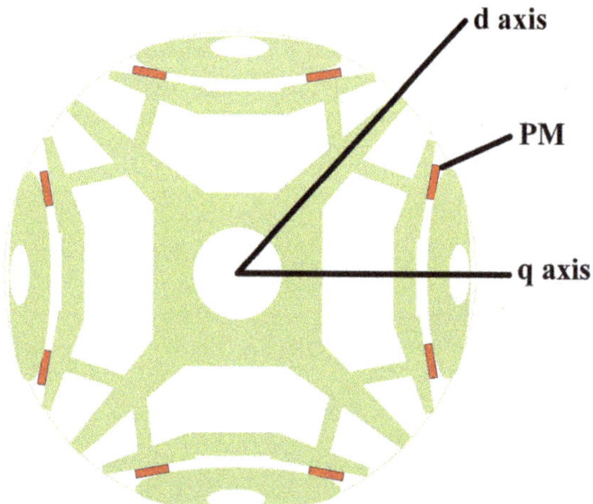

Figure 38. Permanent magnet (PM) assisted SynRM. Permanent magnets are embedded in the rotor [23].

5.8. Axial Flux Ironless Permanent Magnet Motor

According to [70], this motor is the most advanced one to be used in EVs. It has an outer rotor with no slot; use of iron is avoided here as well. The stator core is absent too, reducing the weight of the machine. The air gap here is radial field type, providing better power density. This motor is a variable speed one too. One noteworthy advantage of this machine is that the rotors can be fitted on lateral sides of wheels, placing the stator windings on the axle centrally. The slot-less design also improves the efficiency by minimizing copper loss as there is more space available [70].

Power comparison of three different motor types is conducted in Table 15. Table 16 compares torque densities of three motors. Table 17 summarizes the advantages and disadvantages of different motor types, and shows some vehicles using different motor technologies.

Table 15. Power comparison of different motors having the same size. Data from [72].

Motor Type	Power (kW)		Base Speed	Maximum Speed
	HEV	**BEV**		
IM	57	93	3000	12,000
SRM	42	77	2000	12,000
BLDC	75	110	4000	9000

Table 16. Typical torque density values of some motors. Data from [30].

Motor Type	Torque/Volume (Nm/m^3)	Torque/Cu Mass (Nm/kg Cu)
PM motor	28,860	28.7–48
IM	4170	6.6
SRM	6780	6.1

Table 17. Advantages, disadvantages and usage of different motor types.

Motor Type	Advantage	Disadvantage	Vehicles Used In
Brushed DC Motor	• Maximum torque at low speed	• Bulky structure • Low efficiency • Heat generation at brushes	Fiat Panda Elettra (Series DC motor), Conceptor G-Van (Separately excited DC motor)
Permanent Magnet Brushless DC Motor (BLDC)	• No rotor copper loss • More efficiency than induction motors • Lighter • Smaller • Better heat dissipation • More reliability • More torque density • More specific power	• Short constant power range • Decreased torque with increase in speed • High cost because of PM	Toyota Prius (2005)
Permanent Magnet Synchronous Motor (PMSM)	• Operable in different speed ranges without using gear systems • Efficient • Compact • Suitable for in-wheel application • High torque even at very low speeds	• Huge iron loss at high speeds during in-wheel operation	Toyota Prius, Nissan Leaf, Soul EV
Induction Motor (IM)	• The most mature commutatorless motor drive system • Can be operated like a separately excited DC motor by employing field orientation control		Tesla Model S, Tesla Model X, Toyota RAV4, GM EV1

Table 17. *Cont.*

Motor Type	Advantage	Disadvantage	Vehicles Used In
Switched Reluctance Motor (SRM)	• Simple and robust construction • Low cost • High speed • Less chance of hazard • Long constant power range • High power density	• Very noisy • Low efficiency • Larger and heavier than PM machines • Complex design and control	Chloride Lucas
Synchronous Reluctance Motor (SynRM)	• Robust • Fault tolerant • Efficient • Small	• Problems in controllability and manufacturing • Low power factor	
PM assisted Synchronous Reluctance Motor	• Greater power factor than SynRMs • Free from demagnetizing problems observed in IPM		BMW i3
Axial Flux Ironless Permanent Magnet Motor	• No iron used in outer rotor • No stator core • Lightweight • Better power density • Minimized copper loss • Better efficiency • Variable speed machine • Rotor is capable of being fitted to the lateral side of the wheel		Renovo Coupe

6. Charging Systems

For charging of EVs, DC or AC systems can be used. There are different current and voltage configurations for charging, generally denoted as 'levels'. The time required for a full charge depends on the level being employed. Wireless charging has also been tested and researched for quite a long time. It has different configurations as well. The charging standards are shown in Table 18. The safety standards that should be complied by the chargers are the following [46]:

- SAE J2929: Electric and Hybrid Vehicle Propulsion Battery System Safety Standard
- ISO 26262: Road Vehicles—Functional safety
- ISO 6469-3: Electric Road Vehicles—Safety Specifications—Part 3: Protection of Persons Against Electric Hazards
- ECE R100: Protection against Electric Shock
- IEC 61000: Electromagnetic Compatibility (EMC)
- IEC 61851-21: Electric Vehicle Conductive Charging system—Part 21: Electric Vehicle Requirements for Conductive Connection to an AC/DC Supply
- IEC 60950: Safety of Information Technology Equipment
- UL 2202: Electric Vehicle (EV) Charging System Equipment
- FCC Part 15 Class B: The Federal Code of Regulation (CFR) FCC Part 15 for EMC Emission Measurement Services for Information Technology Equipment.
- IP6K9K, IP6K7 protection class
- $-40\,°C$ to $105\,°C$ ambient air temperature

6.1. AC Charging

AC charging system provides an AC supply that is converted into DC to charge the batteries. This system needs an AC-DC converter. According to the SAE EV AC Charging Power Levels, they can be classified as below:

- Level 1: The maximum voltage is 120 V, the current can be 12 A or 16 A depending on the circuit ratings. This system can be used with standard 110 V household outlets without requiring any special arrangement, using on-board chargers. Charging a small EV with this arrangement can take 0.5–12.5 h. These characteristics make this system suitable for overnight charging [5,46,81].
- Level 2: Level 2 charging uses a direct connection to the grid through an Electric Vehicle Service Equipment (EVSE). On-board charger is used for this system. Maximum system ratings are 240 V, 60 A and 14.4 kW. This system is used as a primary charging method for EVs [46,81].
- Level 3: This system uses a permanently wired supply dedicated for EV charging, with power ratings greater than 14.4 kW. 'Fast chargers'—which recharge an average EV battery pack in no more than 30 min, can be considered level 3 chargers. All level 3 chargers are not fast chargers though [46,82]. Table 19 shows the AC charging characteristics defined by Society of Automotive Engineers (SAE).

Table 18. Charging standards. Data from [81].

Standard		Scope
IEC 61851: Conductive charging system	IEC 61851-1	Defines plugs and cables setup
	IEC 61851-23	Explains electrical safety, grid connection, harmonics, and communication architecture for DCFC station (DCFCS)
	IEC 61851-24	Describes digital communication for controlling DC charging
IEC 62196: Socket outlets, plugs, vehicle inlets and connectors	IEC 62196-1	Defines general requirements of EV connectors
	IEC 62196-2	Explains coupler classifications for different modes of charging
	IEC 62196-3	Describes inlets and connectors for DCFCS
IEC 60309: Socket outlets, plugs, and couplers	IEC 60309-1	Describes CS general requirements
	IEC 60309-2	Explains sockets and plugs sizes having different number of pins determined by current supply and number of phases, defines connector color codes according to voltage range and frequency.
IEC 60364		Explains electrical installations for buildings
SAE J1772: Conductive charging systems		Defines AC charging connectors and new Combo connector for DCFCS
SAE J2847: Communication	SAE J2847-1	Explains communication medium and criteria for connecting EV to utility for AC level 1&2 charging
	SAE J2847-2	Defines messages for DC charging
SAE J2293	SAE J2293-1	Explains total EV energy transfer system, defines requirements for EVSE for different system architectures
SAE J2344		Defines EV safety guidelines
SAE J2954: Inductive charging		Being developed

Table 19. SAE (Society of Automotive Engineers) AC charging characteristics. Data from [44,80].

AC Charging System	Supply Voltage (V)	Maximum Current (A)	Branch Circuit Breaker Rating (A)	Output Power Level (kW)
Level 1	120 V, 1-phase	12	15	1.08
	120 V, 1-phase	16	20	1.44
Level 2	208 to 240 V, 1-phase	16	20	3.3
	208 to 240 V, 1-phase	32	40	6.6
	208 to 240 V, 1-phase	≤80	Per NEC 635	≤14.4
Level 3	208/480/600 V	150–400	150	3

6.2. DC Charging

DC systems require dedicated wiring and installations and can be mounted at garages or charging stations. They have more power than the AC systems and can charge EVs faster. As the output is DC, the voltage has to be changed for different vehicles to suit the battery packs. Modern stations have the capability to do it automatically [46]. All DC charging systems has a permanently connected Electric Vehicle Service Equipment (EVSE) that incorporates the charger. Their classification is done depending on the power levels they supply to the battery:

- Level 1: The rated voltage is 450 V with 80 A of current. The system is capable of providing power up to 36 kW.

- Level 2: It has the same voltage rating as the level 1 system; the current rating is increased to 200 A and the power to 90 kW.

- Level 3: Voltage in this system is rated to 600 V. Maximum current is 400 A with a power rating of 240 kW. Table 20 shows the DC charging characteristics defined by Society of Automotive Engineers (SAE).

Table 20. SAE (Society of Automotive Engineers) DC charging characteristics. Data from [46].

DC Charging System	DC Voltage Range (V)	Maximum Current (A)	Power (kW)
Level 1	200–450	$\leq$80	$\leq$36
Level 2	200–450	$\leq$200	$\leq$90
Level 3	200–600	$\leq$400	$\leq$240

6.3. Wireless Charging

Wireless charging or wireless power transfer (WPT) enjoys significant interest because of the conveniences it offers. This system does not require the plugs and cables required in wired charging systems, there is no need of attaching the cable to the car, low risk of sparks and shocks in dirty or wet environment and less chance of vandalism. Forerunners in WPT research include R&D centers and government organizations like Phillips Research Europe, Energy Dynamic Laboratory (EDL), US DOT, DOE; universities including the University of Tennessee, the University of British Columbia, Korea Advance Institute of Science and Technology (KAIST); automobile manufacturers including Daimler, Toyota, BMW, GM and Chrysler. The suppliers of such technology include Witricity, LG, Evatran, HaloIPT (owned by Qualcomm), Momentum Dynamics and Conductix-Wampfler [27]. However, this technology is not currently available for commercial EVs because of the health and safety concerns associated with the current technology. The specifications are determined by different standardization organizations in different countries: Canadian Safety Code 6 in Canada [83], IEEE C95.1 in the USA [84], ICNIRP in Europe [85] and ARPANSA in Australia [86]. There are different technologies that are being considered to provide WPT facilities. They differ in the operating frequency, efficiency, associated electromagnetic interference (EMI), and other factors.

Inductive power transfer (IPT) is a mature technology, but it is only contactless, not wireless. Capacitive power transfer (CPT) has significant advantage at lower power levels because of low cost and size, but not suitable for higher power applications like EV charging. Permanent magnet coupling power transfer (PMPT) is low in efficiency, other factors are not favorable as well. Resonant inductive power transfer (RIPT) as well as On-line inductive power transfer (OLPT) appears to be the most promising ones, but their infrastructuret may not allow them to be a viable solution. Resonant antennae power transfer (RAPT) is made on a similar concept as RIPT, but the resonant frequency in this case is in MHz range, which is capable of damage to humans if not shielded properly. The shielding is likely to hinder range and performance; generation of such high frequencies is also a challenge for power electronics [87]. Table 21 compares different wireless charging systems in terms of performance, cost, size, complexity, and power level. Wireless charging for personal vehicles is unlikely to be available soon because of health, fire and safety hazards, misalignment problems and range. Roads with WPT systems embedded into them for charging passing vehicles also face major cost issues [27]. Only a few wireless systems are available now, and those too are in trial stage. WiTricity is working with Delphi Electronics, Toyota, Honda and Mitsubishi Motors. Evatran is collaborating with Nissan and GM for providing wireless facilities for Nissan Leaf and Chevrolet Volt models. However, with significant advance in the technology, wireless charging is likely to be integrated in the EV scenario, the conveniences it offers are too appealing to overlook.

Table 21. Comparison of wireless charging systems.

Wireless Charging System	Performance			Cost	Volume/Size	Complexity	Power Level
	Efficiency	EMI	Frequency				
Inductive power transfer (IPT)	Medium	Medium	10–50 kHz	Medium	Medium	Medium	Medium/High
Capacitive power transfer (CPT)	Low	Medium	100–500 kHz	Low	Low	Medium	Low
Permanent magnet coupling power transfer (PMPT)	Low	High	100–500 kHz	High	High	High	Medium/Low
Resonant inductive power transfer (RIPT)	Medium	Low	1–20 MHz	Medium	Medium	Medium	Medium/Low
On-line inductive power transfer (OLPT)	Medium	Medium	10–50 kHz	High	High	Medium	High
Resonant antennae power transfer (RAPT)	Medium	Medium	100–500 kHz	Medium	Medium	Medium	Medium/Low

For the current EV systems, on-board AC systems are used for the lowest power levels, for higher power, DC systems are used. DC systems currently have three existing standards [16]:

- Combined Charging System (CCS)
- CHAdeMO (CHArge de MOve, meaning: 'move by charge')
- Supercharger (for Tesla vehicles)

The powers offered by CCS and CHAdeMO are 50 kW and 120 kW for the Supercharger system [88,89]. CCS and CHAdeMO are also capable of providing fast charging, dynamic charging and vehicle to infrastructure (V2X) facilities [6,90]. Most of the EV charging stations at this time provides level 2 AC charging facilities. Level 3 DC charging network, which is being increased rapidly, is also available for Tesla cars. The stations may provide the CHAdeMO standard or the CCS, therefore, a vehicle has to be compatible with the configuration provided to be charged from the station. The CHAdeMO system is favored by the Japanese manufacturers like Nissan, Toyota and Honda whereas the European and US automakers, including Volkswagen, BMW, General Motors and Ford, prefer the CCS standard. Reference [5] discusses the charging systems used by current EVs along with the time required to get them fully charged.

7. Power Conversion Techniques

Batteries or ultracapacitors (UC) store energy as a DC charge. Normally they have to obtain that energy from AC lines connected to the grid, and this process can be wired or wireless. To deliver this energy to the motors, it has to be converted back again. These processes work in the reverse direction as well i.e., power being fed back to the batteries (regenerative braking) or getting supplied to grid when the vehicle in idle (V2G) [91]. Typical placement of different converters in an EV is shown in Figure 39 along with the power flow directions. This conversion can be DC-DC or DC-AC. For all this conversion work required to fill up the energy storage of EVs and then to use them to propel the vehicle, power converters are required [72], and they come in different forms. A detailed description of power electronics converters is provided in [92]. Further classification of AC-AC converters is shown in [93]. A detailed classification of converters is shown in Figure 40.

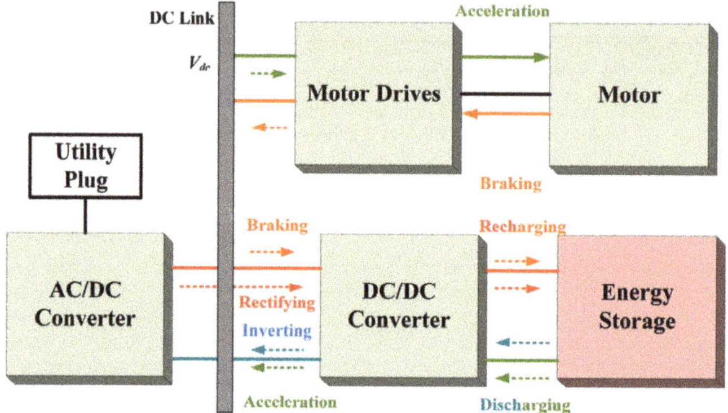

Figure 39. Typical placements of different converters in an EV. AC-DC converter transforms the power from grid to be stored in the storage through another stage of DC-DC conversion. Power is supplied to the motor from the storage through the DC-DC converter and the motor drives [72].

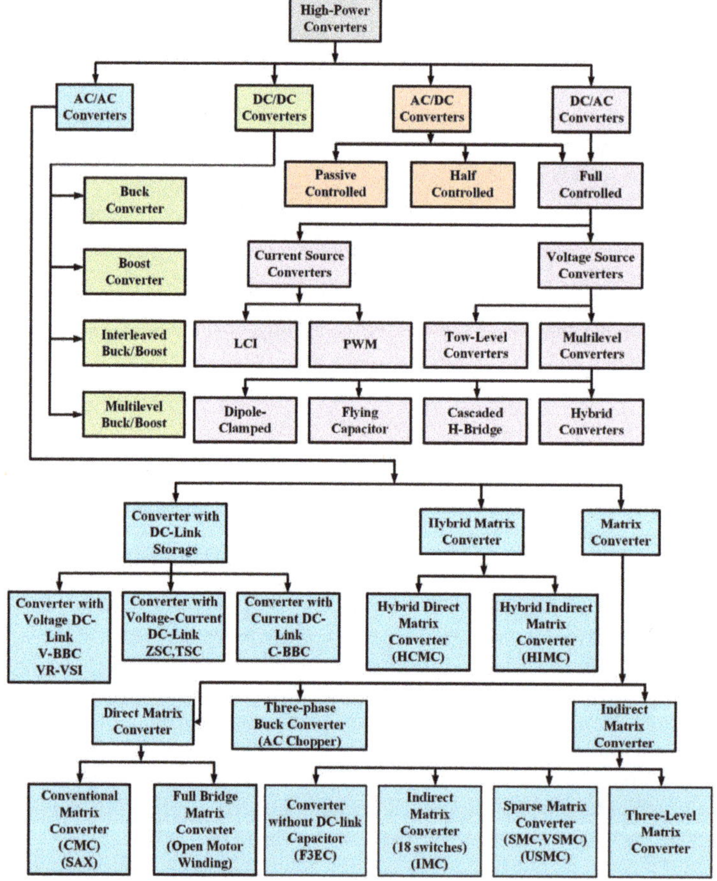

Figure 40. Detailed classification of converters. Data from [92,93].

7.1. Converters for Wired Charging

DC-DC boost converter is used to drive DC motors by increasing the battery voltage up to the operating level [72]. DC-DC converters are useful to combine a power source with a complementing energy source [94]. Figure 41 shows a universal DC-DC converter used for DC-DC conversion. It can be used as a boost converter for battery to DC link power flow and as a buck converter when the flow is reversed. The operating conditions and associated switching configuration is presented in Table 22. DC-DC boost converters can also use a digital signal processor [95].

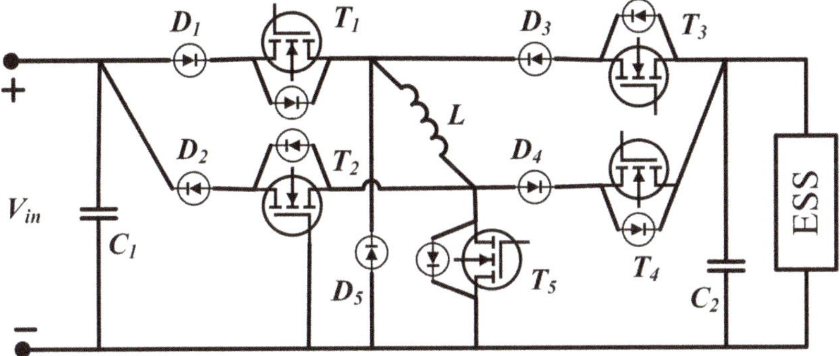

Figure 41. Universal DC-DC converter [72].

Table 22. Operating conditions for universal DC-DC converter. Adapted from [88].

Direction	Mode	T_1	T_2	T_3	T_4	T_5
V_{dc} to V_{batt}	Boost	On	Off	Off	On	PWM
V_{dc} to V_{batt}	Buck	PWM	Off	Off	On	Off
V_{batt} to V_{dc}	Boost	Off	On	On	Off	PWM
V_{batt} to V_{dc}	Buck	Off	On	PWM	Off	Off

According to [72], dual inverter is the most updated technology to drive AC motors like permanent magnet synchronous motors (PMSMs), shown in Figure 42. For dual voltage source applications, the system of Figure 43 is used [96]. These inverters operate on space vector PWM. For use on both PMSMs and induction motors (IMs), a bidirectional stacked matrix inverter can be used; such a system is shown in Figure 44.

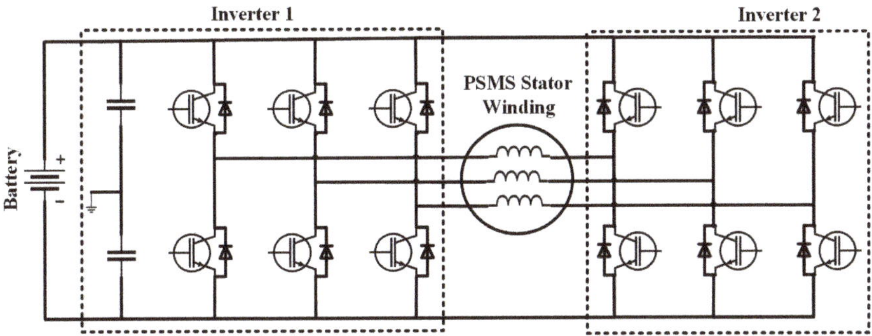

Figure 42. Dual inverter for single source [72].

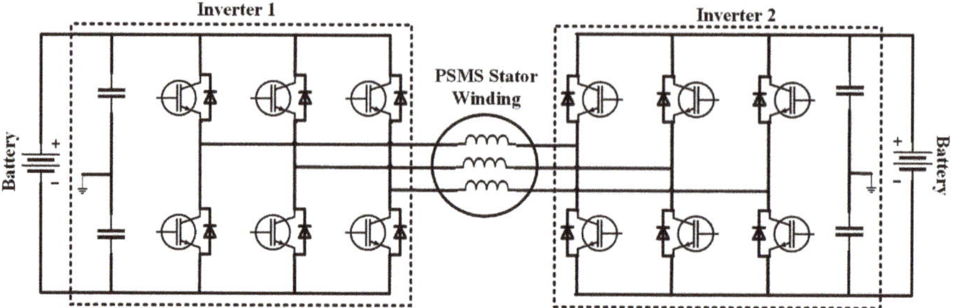

Figure 43. Dual inverter with dual sources [72].

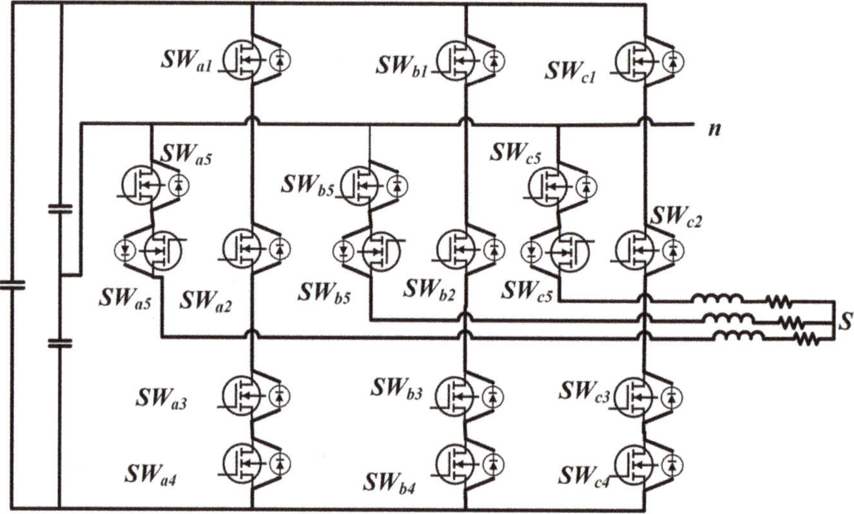

Figure 44. Novel stacked matrix inverter as shown in [97].

Some notable conventional DC-DC converters are: phase-shift full-bridge (PSFB), inductor-inductor-capacitor (LLC), and series resonant converter (SRC). A comparison of components used in these three converters is presented in [98], which is demonstrated here in Table 23. The DC-DC converters used are required to have low cost, weight and size for being used in automobiles [99]. Interleaved converters are a preferable option regarding these considerations, it offers some other advantages as well [100–103], though using it may increase the weight and volume of the inductors compared to the customary single-phase boost converters [99]. To solve this problem, Close-Coupled Inductor (CCI) and Loosely-Coupled Inductor (LCI) integrated interleaved converters have been proposed in [99]. In [48] converters for AC level-1 and level-2 chargers are shown by Williamson et al., who stated that Power Factor Correction (PFC) is a must to acquire high power density and efficiency. Two types of PFC technique are shown here: single-stage approach and two-stage approach. The first one suits for low-power use and charge only lead-acid batteries because of high low frequency ripple. To avoid these problems, the second technique is used.

Table 23. Comparison of components used in PSFB, LLC and SRC converter. Adapted from [98].

Item	PSFB	LLC	SRC
Number of switch blocks	4	4	4
Number of diode blocks	4	4	4
Number of transformers	1	1	2
Number of inductors	1	0	0
Additional capacitor	Blocking capacitor	-	-
Output filter size	Small	-	Large

In [34], Yong et al., presented the front end AC-DC converters. The Interleaved Boost PFC Converter (Figure 45) has a couple of boost converters connected in parallel and working in 180° out of phase [104–106]. The ripple currents of the inductors cancel each other. This configuration also provides twice the effective switching frequency and provides a lower ripple in input current, resulting in a relatively small EMI filter [103,107]. In Bridgeless/Dual Boost PFC Converter (Figure 46), the gating signals are made identical here by tying the power-train switches. The MOSFET gates are not made decoupled. Rectifier input bridge is not needed here. The Bridgeless Interleaved Boost PFC Converter (Figure 47) is proposed to operate above the 3.5 kW level. It has two MOSFETS and uses two fast diodes; the gating signals have a phase difference of 180°.

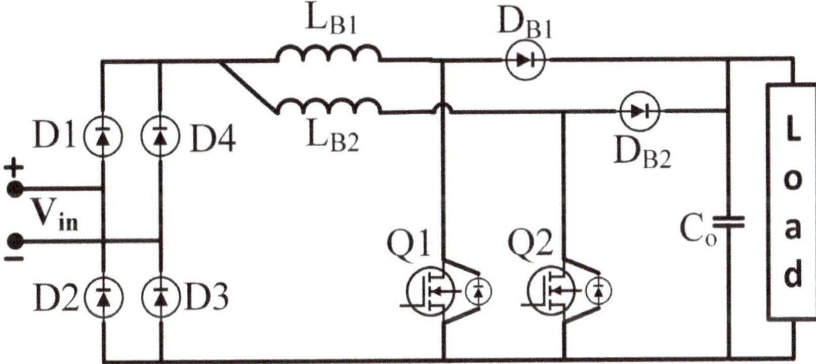

Figure 45. Interleaved Boost PFC Converter [46].

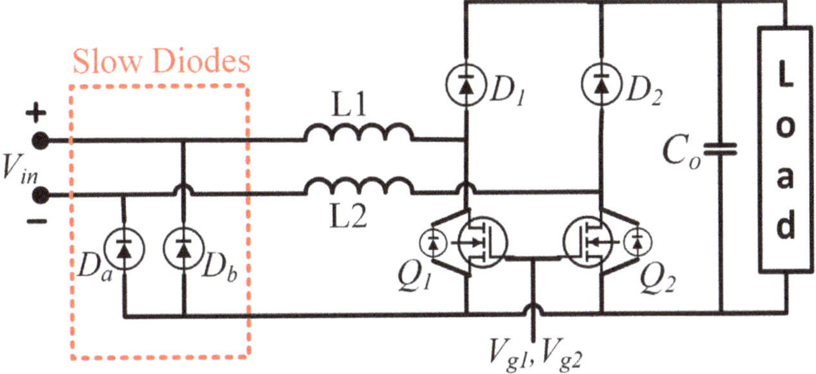

Figure 46. Bridgeless/Dual Boost PFC Converter. Adapted from [46].

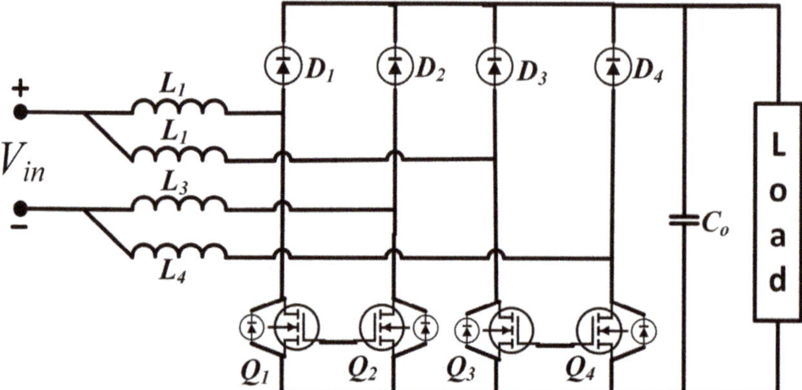

Figure 47. Bridgeless Interleaved Boost PFC Converter [46].

Williamson et al., presented some isolated DC-DC converter topologies in [44]. The ZVS FB Converter with Capacitive Output Filter (Figure 48) can achieve high efficiency as it uses zero voltage switching (ZVS) along with the capacitive output filters which reduces the ringing of diode rectifiers. The trailing edge PWM full-bridge system proposed in [107]. The Interleaved ZVS FB Converter with Voltage Doubler (Figure 49) further reduces the voltage stress and ripple current on the capacitive output filter, it reduces the cost too. Interleaving allows equal power and thermal loss distribution in each cell. The number of secondary diodes is reduced significantly by the voltage doubler rectifier at the output [34]. Among its operating modes, DCM (discontinuous conduction mode) and BCM (boundary conduction mode) are preferable. The Full Bridge LLC Resonant Converter (Figure 50) is widely used in telecom industry for the benefits like high efficiency at resonant frequency. But unlike the telecom sector, EV applications require a wide operating range. Reference [41] shows a design procedure for such configurations for these applications.

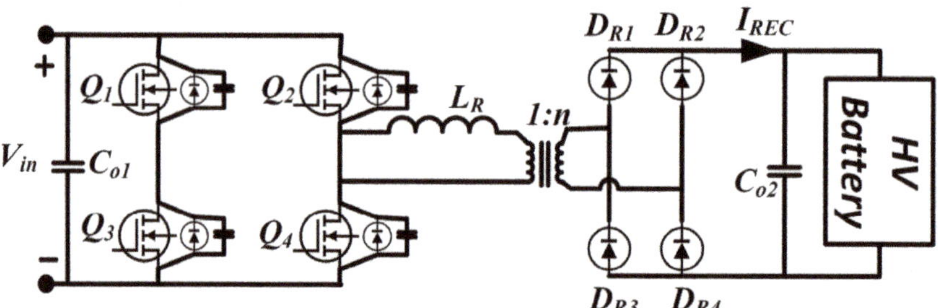

Figure 48. ZVS FB Converter with Capacitive Output Filter [46].

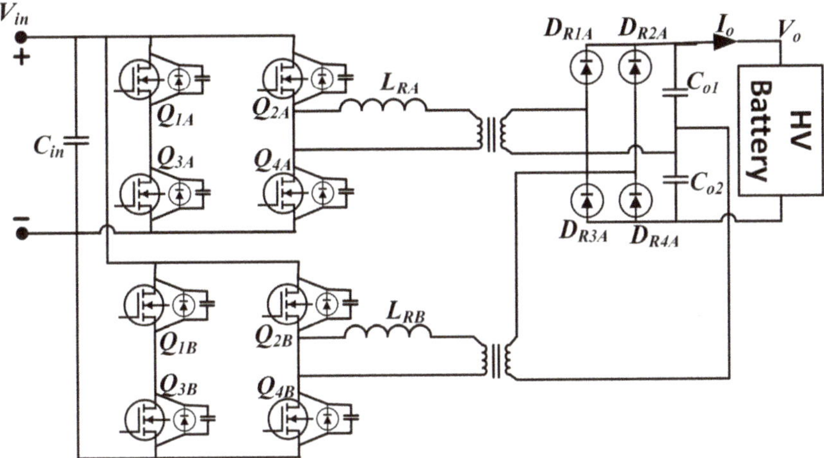

Figure 49. Interleaved ZVS FB Converter with Voltage Doubler [46].

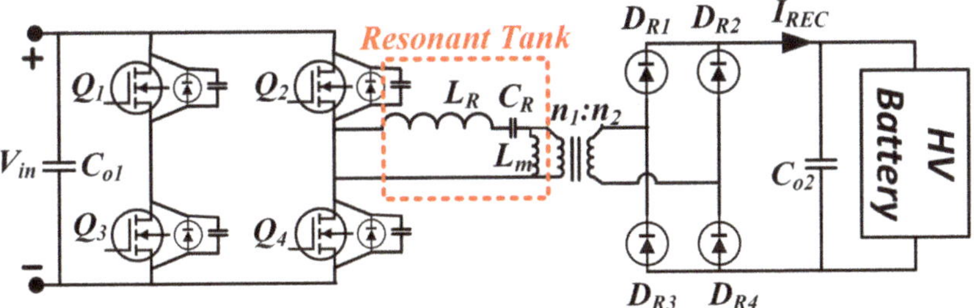

Figure 50. Full Bridge LLC Resonant Converter. Adapted from [46].

Balch et al., showed converter configurations that are used in different types of EVs in [42]. In Figure 51, a converter arrangement for a BEV is shown. An AC-DC charger is used for charging the battery pack here while a two-quadrant DC-DC converter is used for power delivery to the DC bus form the battery pack. This particular example included an ultracapacitor as well. An almost similar arrangement was shown in [42] for PHEVs (Figure 52) where a bidirectional DC-DC converter was used between the DC bus and the battery pack to facilitate regeneration. Use of integrated converter in PHEV is shown in Figure 53. Figure 54 shows converter arrangement for a PFCV; this configuration is quite similar to one shown for BEV, but it contains an additional boost converter to adjust the power produced by the fuel cell stack to be sent to the DC bus.

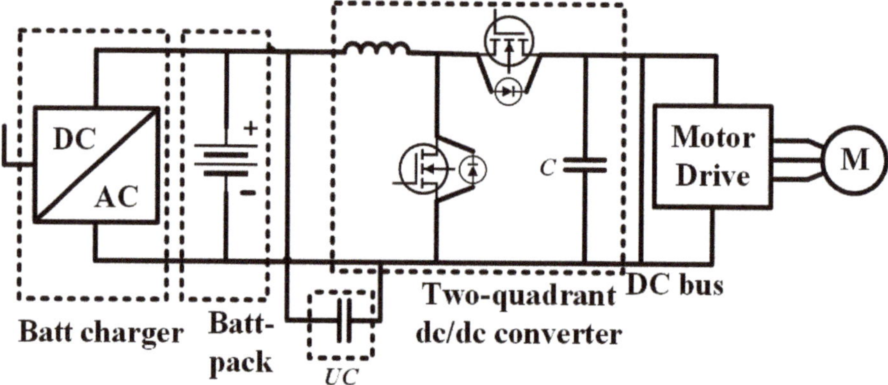

Figure 51. Converter placement in a pure EV [35]. The charger has an AC-DC converter to supply DC to the battery from the grid, whereas the DC-DC converter converts the battery voltage into a value required to drive the motor.

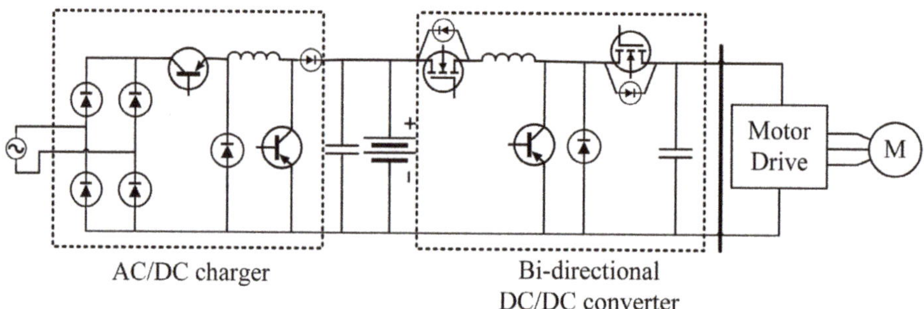

Figure 52. Cascaded converter to use in PHEV. Adapted from [35]. A bidirectional DC-DC converter is used between the DC bus and the battery pack to allow regenerated energy to flow back to the battery from the motor.

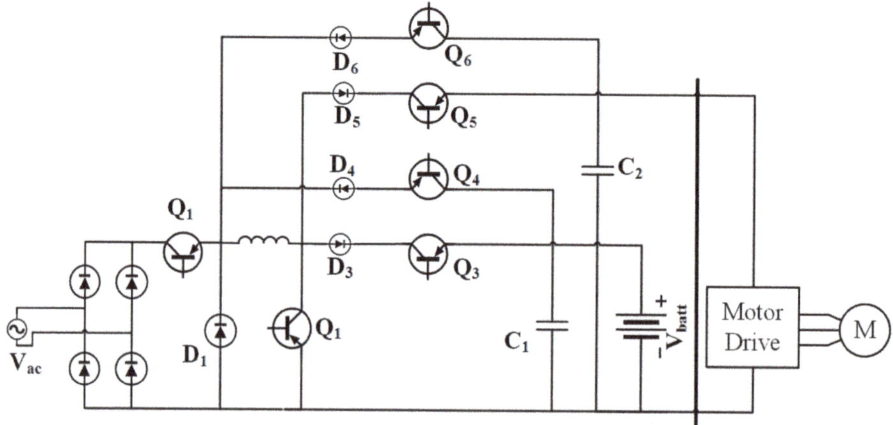

Figure 53. Integrated converter used in PHEV [35].

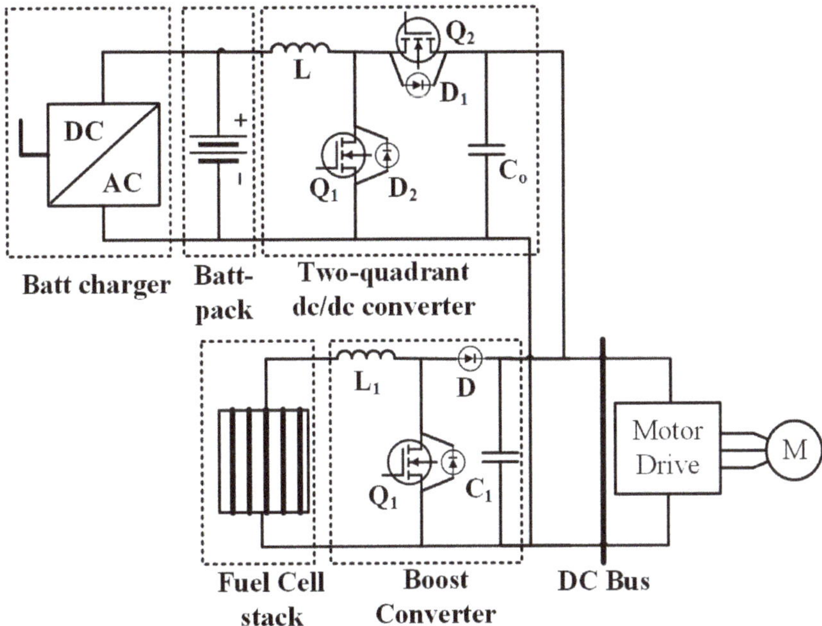

Figure 54. Converter arrangement in PFCV. Adapted from [35]. An AC-DC converter is used to convert the power from the grid; DC-DC converter is used for power exchange between the DC bus and battery; boost converter is used to make the voltage generated from the fuel cell stack suitable for the DC bus.

Bidirectional converters allow transmission of power from the motors to the energy sources and also from vehicle to grid. Novel topologies for bidirectional AC/DC-DC/DC converters to be used in PHEVs are being researched [103,108–112], such a configuration in shown in Figure 55. Kok et al., showed different DC-DC converter arrangements for EVs using multiple energy sources in [94] which are presented in Figure 56. The first system has both battery and ultracapacitor added in cascade, while the second one has them connected in parallel. The third one shows a system employing fuel cells, and battery for backup. In [113], Koushki et al., classified bidirectional AC-DC converters into two main groups: Low frequency AC-High frequency AC-DC (Figure 57), and Low frequency AC-DC- High frequency AC-DC (Figure 58). The first kind can also be called single-stage converters where the latter may be described as two-stage, which can be justified from their topologies. Converters employed for EV application are compiled in Table 24. From this table, it is evident that step down converters are required for charging the batteries from a higher voltage grid voltage, bidirectional converters are needed for providing power flow in both directions, and specialized converters such as the last three, are needed for better charging performances.

AC-DC converters are used to charge the batteries from AC supply-lines; DC-DC converters are required for sending power to the motors from the batteries. The power flow can be reversed in case of regenerative actions or V2G. Bidirectional converters are required in such cases. Different converter configurations have different advantages and shortcomings which engendered a lot of research and proliferation of hybrid converter topologies.

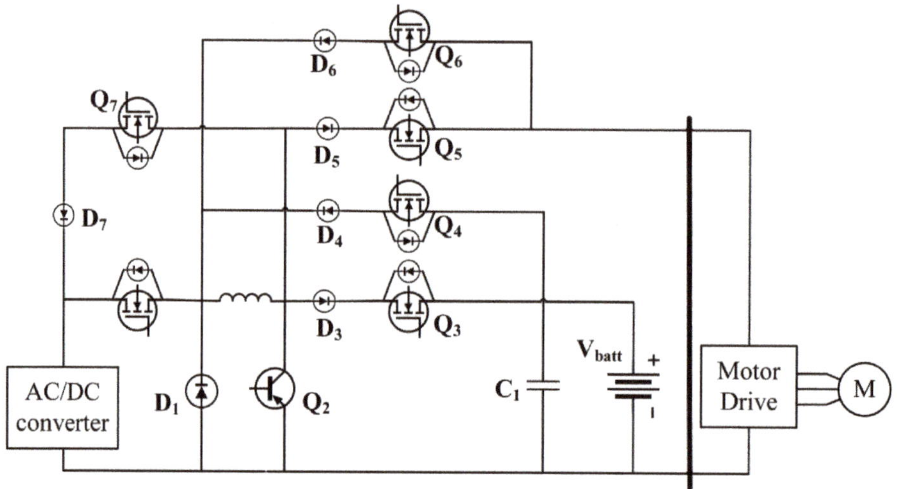

Figure 55. Integrated bidirectional AC/DC-DC/DC converter [33].

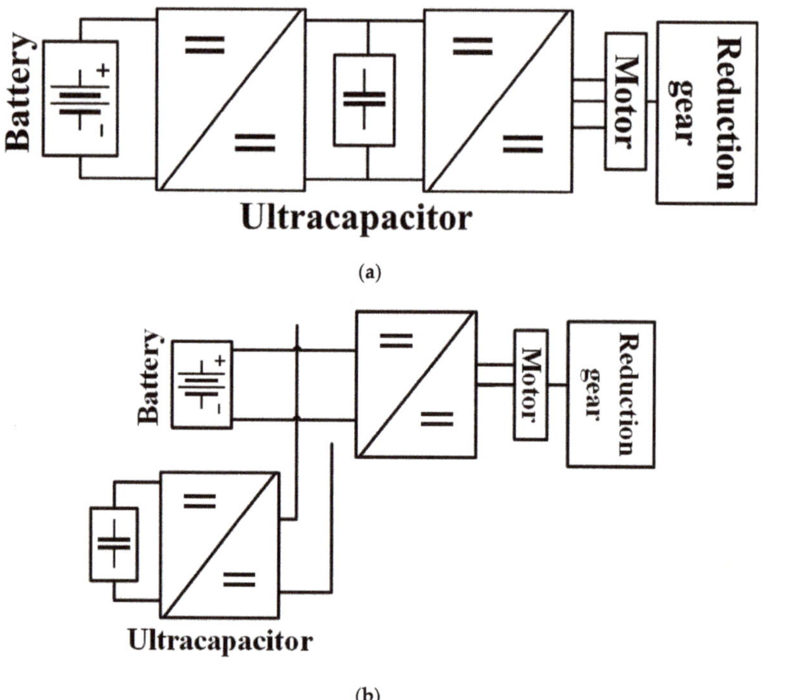

(a)

(b)

Figure 56. *Cont.*

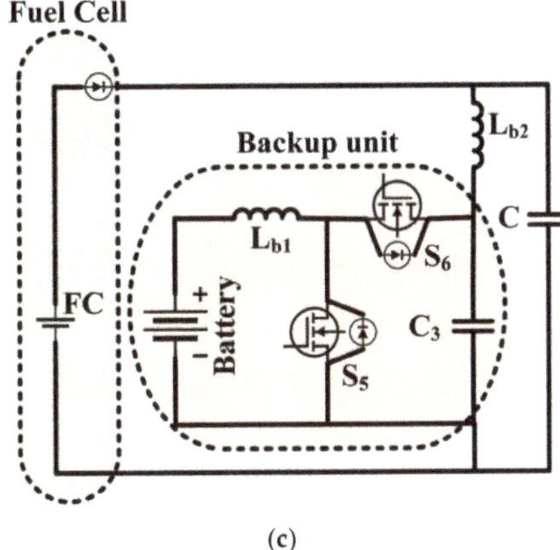

(c)

Figure 56. Converter arrangements as shown in [94]: (**a**) Cascaded connection; (**b**) Parallel connection; (**c**) Fuel cell with battery backup. Adapted from [94].

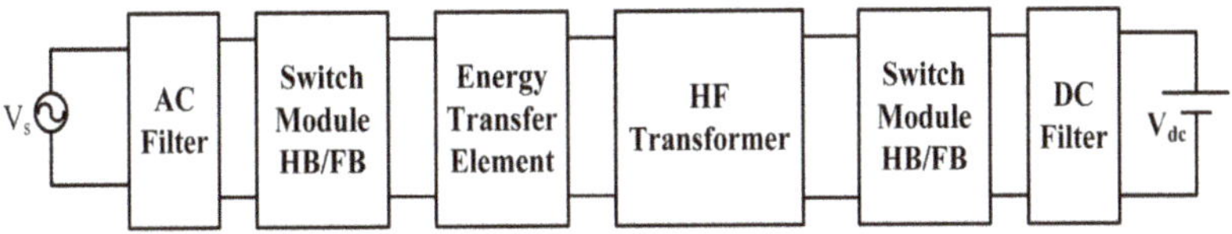

Figure 57. Low frequency AC-High frequency AC-DC converter, also called single-stage converter [113].

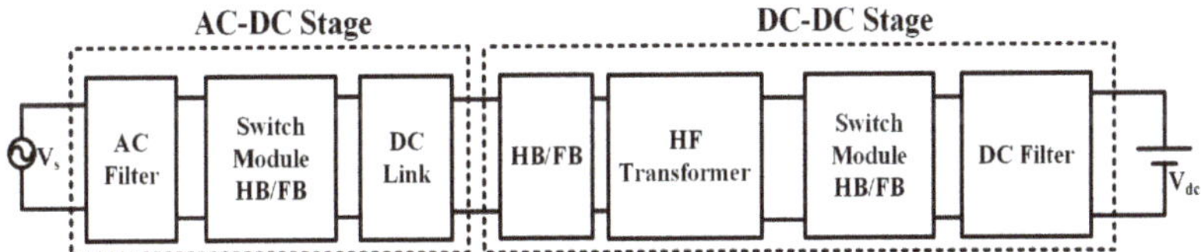

Figure 58. Low frequency AC-DC-High frequency AC-DC converter, also called two-stage converter. Adapted from [113].

Table 24. Converters with EV application displaying their key features and uses in EVs.

Configuration	Reference	Operation	Key Features	Application in EV
Buck converter	Bose [92]	Step down	Can operate in continuous or discontinuous mode	Sending power to the battery
Buck-Boost converter	Bose [92]	Step up and step down	Two quadrant operation of chopper	Regenerative action
Interleaved Boost PFC converter	Williamson et al. [46]	Step up with power factor correction	Relatively small input EMI filter	Charging
Bridgeless/Dual Boost PFC Converter	Williamson et al. [46]	Step up with power factor correction	Does not require rectifier input bridge	Charging
ZVS FB Converter with Capacitive Output Filter	Williamson et al. [46]	AC-DC conversion	Zero voltage switching	Charging

7.2. Systems for Wireless Charging

Wireless charging or wireless power transfer (WPT) uses a principle similar to transformer. There is a primary circuit at the charger end, from where the energy is transferred to the secondary circuit located at the vehicle. In case of inductive coupling, the voltage obtained at the secondary side is:

$$v_2 = L_2(di_2/dt) + M(di_1/dt) \tag{2}$$

M is the mutual inductance and can be calculated by:

$$M = k\sqrt{(L_1 L_2)} \tag{3}$$

The term k here is the coupling co-efficient; L_1 and L_2 are the inductances of primary and secondary circuit. Figure 59 shows the 'double D' arrangement for WPT which demonstrates the basic principle of wireless power transfer by means of flux linkages. A variety of configurations can be employed for wireless power transfer; some of them meet a few desired properties to charge vehicles. Inductive WPT, shown in Figure 60a, is the most rudimentary type, transfer power from one coil to another just like the double D system. Capacitive WPT (Figure 60b) uses a similar structure as the inductive system, but it has two coupling transformers at its core. Low frequency permanent magnet coupling power transfer (PMPT) is shown in Figure 60c; it uses a permanent magnet rotor to transmit power, another rotor placed in the vehicle acts as the receiver. Resonant antennae power transfer (RAPT) (Figure 60d) uses resonant antennas for wireless transfer of power. Resonant inductive power transfer (RIPT), shown in Figure 60e, uses resonance circuits for power transfer. Online power transfer (OLPT) has a similar working principle as RIPT, it can be used in realizing roadways that can charge vehicles wirelessly by integrating the transmitter with the roadway (pilot projects using similar technology placed them just beneath the road surface), and equipping vehicles with receivers to collect power from there. Schematic for this system is shown in Figure 60f. Characteristics of these systems are shown in Table 25.

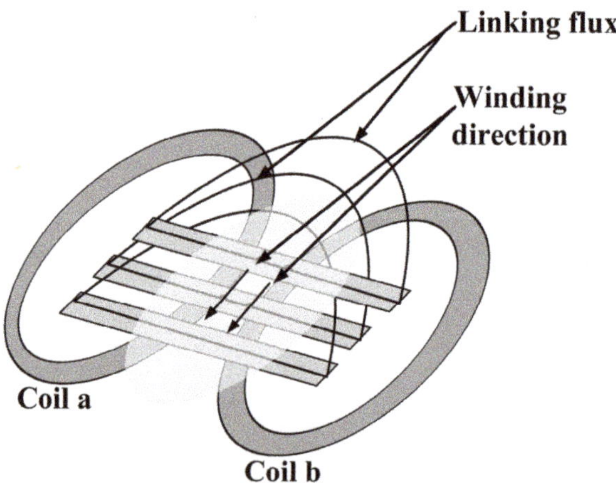

Figure 59. Double D arrangement for WPT. Fluxes generated in one coil cut the other one and induces a voltage there, enabling power transfer between the coils without any wired connection [27].

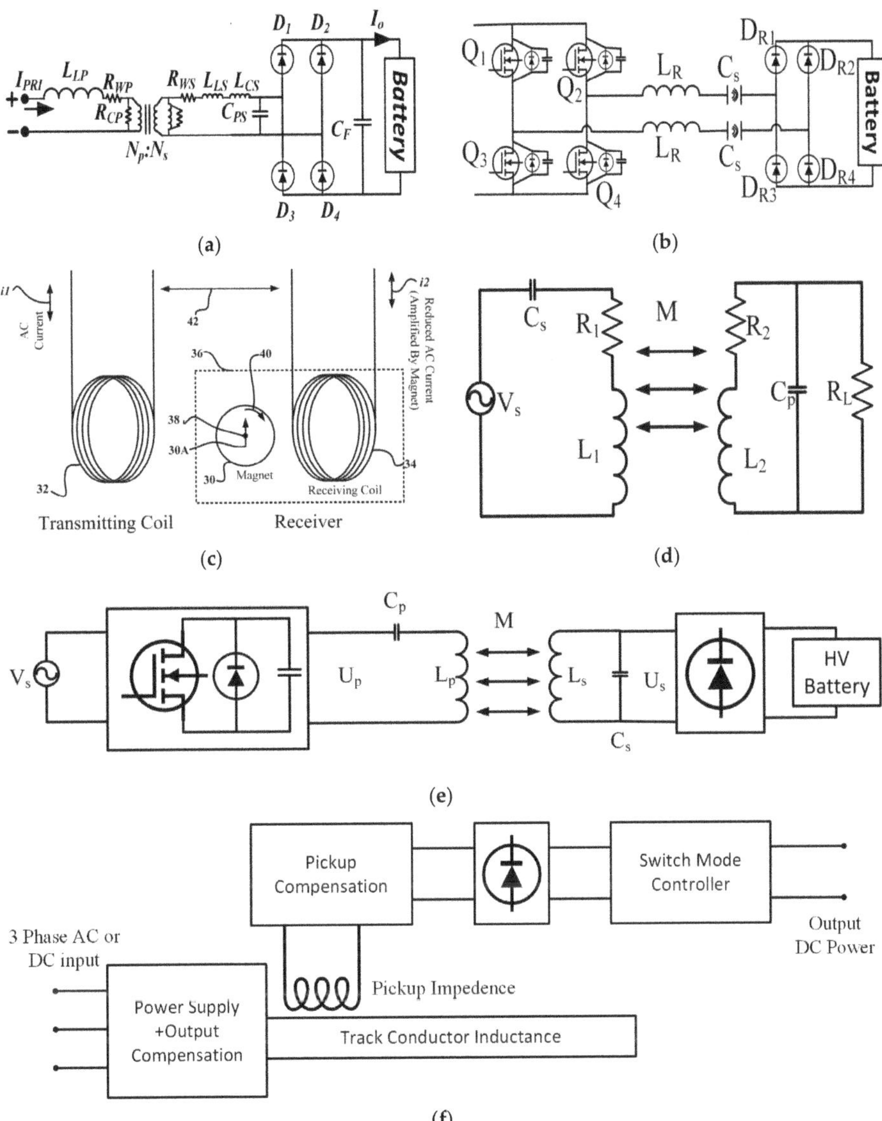

Figure 60. Different configurations used for wireless power transfer over the years: (**a**) Inductive WPT; (**b**) Capacitive WPT; (**c**) Low frequency permanent magnet coupling power transfer (PMPT); (**d**) Resonant antennae power transfer (RAPT); (**e**) Resonant inductive power transfer (RIPT); (**f**) Online power transfer (OLPT).

Table 25. Characteristics of wireless charging systems [87].

Technology	Characteristics
Inductive WPT	• It is not actually wireless, just does not require any connection. • Primary and secondary coils are sealed in epoxy. • Can provide power of either 6.6 kW or 50 kW. • Coaxial winding transformer can be used to place all the transformer core materials off-board. • Losses including geometric effects, eddy current loss, EMI are mainly caused by nonlinear flux distribution. • A piecewise assembly of ferrite core and dividing the secondary winding symmetrically can help minimizing the losses.
Capacitive WPT	• Capacitive power transfer or CPT interface is built with two coupling transformers at the center; the rest of the system is similar to inductive WPT. • Capacitive interface is helpful in reducing the size and cost of the required galvanic isolating parts. • Cheaper and smaller for lower power applications, but not preferred for high power usage. • Useful in consumer electronics, may not be sufficient for EV charging.
Low frequency permanent magnet coupling power transfer (PMPT)	• The transmitter is a cylinder-shaped, permanent magnet rotor driven by static windings placed on the rotor, inside it if the rotor is hollow, or outside the motor, separated by an air-gap. • The receiver is placed on the vehicle, similar to the transmitter in construction. • Transmitter and receiver have to be within 150 mm for charging. • Because of magnetic gear effect, the receiver rotor rotates at the same speed as the transmitter and energy is transferred. • The disadvantages may be the vibration, noise and lifetime associated with the mechanical components used.
Resonant inductive power transfer (RIPT)	• Most popular WPT system. • Uses two tuned resonant tanks or more, operating in the same frequency in resonance. • Resonant circuits enable maximum transfer of power, efficiency optimization, impedance matching, compensation of magnetic coupling and magnetizing current variation. • Can couple power for a distance of up to 40 cm. • Advantages include extended range, reduced EMI, operation at high frequency and high efficiency.
Online power transfer (OLPT)	• Has a similar concept like RIPT, but uses a lower resonant frequency. • Can be used for high power applications. • This system is proposed to be applied in public transport system in [87]. • The primary circuit—a combination of the input of resonant converter and distributed primary windings is integrated in the roadway. This primary side is called the 'track'. • The secondary is placed in vehicles and is called the 'pickup coil'. • Supply of this system is high voltage DC or 3-phase AC. • It can provide frequent charging of the vehicles while they are on the move, reducing the required battery capacity, which will reduce the cost and weight of the cars. • The costs associated with such arrangement may also make its implementation unlikely.
Resonant antennae power transfer (RAPT)	• This system uses two resonant antennas, or more, with integrated resonant inductances and capacitances. The antennas are tuned to identical frequencies. • Large WPT coils are often used as antennas; resonant capacitance is obtained there by controlled separation in the helical structure. • The frequencies used are in MHz range. • Can transfer power efficiently for distances up to 10 m. • The radiations emitted by most of such systems exceed the basic limits on human exposure and are difficult to shield without affecting the range and performance. • Generating frequencies in the MHz range is also challenging and costly with present power electronics technologies.

8. Effects of EVs

Vehicles may serve the purpose of transportation, but they affect a lot of other areas. Therefore, the shift in the vehicle world created by EVs impacts the environment, the economy, and being electric, the electrical systems to a great extent. EVs are gaining popularity because of the benefits they provide in all these areas, but with them, there come some problems as well. Figure 61 illustrates the impacts of EVs on the power grid, environment and economy.

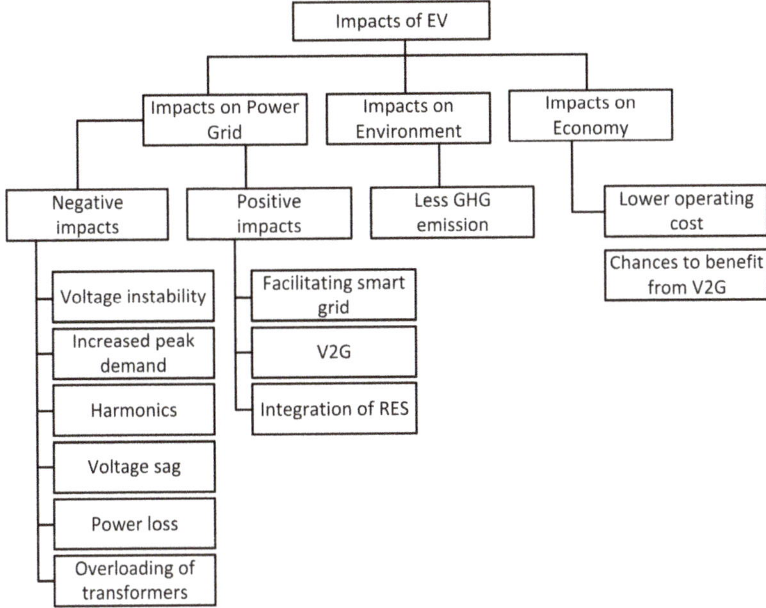

Figure 61. A short list of the impacts of EVs on the power grid, environment and economy.

8.1. Impact on the Power Grid

8.1.1. Negative Impacts

EVs are considered to be high power loads [114] and they affect the power distribution system directly; the distribution transformers, cables and fuses are affected by it the most [115,116]. A Nissan Leaf with a 24 kWh battery pack can consume power similar to a single European household. A 3.3 kW charger in a 220 V, 15 A system can raise the current demand by 17% to 25% [117]. The situation gets quite alarming if charging is done during peak hours, leading to overload on the system, damage of the system equipment, tripping of protection relays, and subsequently, an increase in the infrastructure cost [117]. Charging without any concern to the time of drawing power from the grid is denoted as uncoordinated charging, uncontrolled charging or dumb charging [117,118]. This can lead to the addition of EV load in peak hours which can cause load unbalance, shortage of energy, instability, and decrease in reliability and degradation of power quality [116,119]. In case of the modified IEEE 23 kV distribution system, penetration of EVs can deviate voltage below the 0.9 p.u. level up to 0.83 p.u., with increased power losses and generation cost [118]. Level 1 charging from an 110 V outlet does not affect the power system much, but problems arise as the charging voltage increases. Adding an EV for fast charging can be equivalent to adding several households to the grid. The grid is likely to be capable of withstanding it, but distribution networks are designed with specific numbers of households kept into mind, sudden addition of such huge loads can often lead to problems. Reducing the charging time to distinguish their vehicles in the EV market has become the current norm among the manufacturers, and it requires higher voltages than ever. Therefore, mitigating the adverse effects is not likely by employing low charging voltages.

To avoid these effects, and to provide efficient charging with the available infrastructure, coordinated charging (also called controlled or smart charging) has to be adopted. In this scheme, the EVs are charged during the time periods when the demand is low, for example, after midnight. Such schemes are beneficial in a lot of ways. It not only prevents addition of extra load during peak hours, but also increases the load in valley areas of the load curve, facilitating proper use of the power plants with better efficiency. In [116], Richardson et al., showed that a controlled charging rate can make high EV penetration possible in the current residential power network with only a few upgrades in the infrastructure. Geng et al., proposed a charging strategy in [120] comprising of two stages aimed at providing satisfactory charging for all connected EVs while shifting the loads on the transformers. On the consumer side, it can reduce the electricity bill as the electricity is consumed by the EVs during off peak hours, which generally have a cheaper unit rate than peak hours. According to [121], smart charging systems can reduce the increase investment cost in distribution system by 60–70%. The major problems that are faced in the power systems because of EVs can be charted as following:

- Voltage instability: Normally power systems are operated close to their stability limit. Voltage instabilities in such systems can occur because of load characteristics, and that instability can lead to blackouts. EV loads have nonlinear characteristics, which are different than the general industrial or domestic loads, and draw large quantities power in a short time period [81,122]. Reference [123] corroborated to the fact that EVs cause serious voltage instability in power systems. If the EVs have constant impedance load characteristics, then it is possible for the grid to support a lot of vehicles without facing any instability [81]. However, the EV loads cannot be assumed beforehand and thus their power consumptions stay unpredictable; addition of a lot of EVs at a time therefore can lead to violation of distribution constraints. To anticipate these loads properly, appropriate modeling methods are required. Reference [124] suggested tackling the instabilities by damping the oscillations caused by charging and discharging of EV batteries using a wide area control method. The situation can also be handled by changing the tap settings of transformers [125], by a properly planned charging system, and also by using control systems like fuzzy logic controllers to calculate voltages and SOCs of batteries [81].
- Harmonics: The EV charger characteristics, being nonlinear, gives raise high frequency components of current and voltage, known as harmonics. The amount of harmonics in a system can be expressed by the parameters total current harmonic distortion (THD_i) and total voltage harmonic distortion (THD_v):

$$THD_i = \frac{\sqrt{\sum_{h=2}^{H} I_h^2}}{I_1} \times 100\% \tag{4}$$

$$THD_v = \frac{\sqrt{\sum_{h=2}^{H} V_h^2}}{V_1} \times 100\% \tag{5}$$

Harmonics distort the voltage and current waveforms, thus can reduce the power quality. It also causes stress in the power system equipment like cables and fuses [122]. The present cabling is capable of withstanding 25% EV penetration if slow charging is used, in case of rapid charging, the amount comes down to 15% [126]. Voltage imbalance and harmonics can also give rise to current flow in the neutral wire [127,128]. Different approaches have been adopted to determine the effects of harmonics due to EV penetration. Reference [127] simulated the effects of harmonics using Monte Carlo analysis to determine the power quality. In [129] the authors showed that THD_v can reach 11.4% if a few number of EVs are fast charging. This is alarming as the safety limit of THD_v is 8%. According to

Melo et al. [130], THD_i also becomes high, in the range of 12% to 14%, in case of fast charging, though it remains in the safe limit during times of slow charging. Studies conducted in [131] show the modern EVs generate less THD_i than the conventional ones, though their THD_v values are higher. However, with increased number of EVs, there are chances of harmonics cancellation because of different load patterns [132,133]. Different EV chargers can produce different phase angles and magnitudes which can lead to such cancellations [133]. It is also possible to reduce, even eliminate harmonics by applying pulse width modulation in the EV chargers [132]. High THD_i can be avoided by using filtering equipment at the supply system [134].

- Voltage sag: A decrease in the RMS value of voltage for half a cycle or 1 min is denoted as voltage sag. It can be caused by overload or during the starting of electric machines. Simulation modeled with an EV charger and a power converter in [135] stated 20% EV penetration can exceed the voltage sag limit. Reference [136] stated that 60% EV penetration is possible without any negative impact is possible if controlled charging is employed. The amount, however, plummets to 10% in case of uncontrolled charging. Leemput et al., conducted a test employing voltage droop charging and peak shaving by EV charging [137]. This study exhibited considerable decrease in voltage sag with application of voltage droop charging. Application of smart grid can help in great extents in mitigating the sag [138].
- Power loss: The extra loss of power caused by EV charging can be formulated as:

$$PL_E = PL_{EV} - PL_{original} \tag{6}$$

PL_{original} is the loss occurred when the EVs are not connected to the grid and PLEV is the loss with EVs connected. Reference [121] charted the increased power loss as high as 40% in off peak hours considering 60% of the UK PEVs to be connected to distribution system. Uncoordinated charging, therefore, can increase the amount of loss furthermore. Taking that into account, a coordinated charging scheme, based on objective function, to mitigate the losses was proposed in [139]. Coordinated charging is also favored by [140,141] to reduce power losses significantly. Power generated in the near vicinity can also help minimizing the losses [142], and distributed generation can be quite helpful in this prospect, with the vehicle owners using energy generated at their home (by PV cells, CHP plants, etc.) to charge the vehicles.

- Overloading of transformers: EV charging directly affects the distribution transformers [81]. The extra heat generated by EV loads can lead to increased aging rate of the transformers, but it also depends on the ambient temperature. In places with generally cold weather like Vermont, the aging due to temperature is negligible [81]. Estimation of the lifetime of a transformer is done in [143], where factors taken into account are the rate of EV penetration, starting time of charging and the ambient temperature. It stated that transformers can withstand 10% EV penetration without getting any decrease in lifetime. The effect of level 1 charging, is in fact, has negligible effect on this lifetime, but significant increase in level 2 charging can lead to the failure of transformers [144]. Elnozahy et al., stated that overloading of transformer can happen with 20% PHEV penetration for level 1 charging, whereas level 2 does it with 10% penetration [145]. According to [122], charging that takes place right after an EV being plugged in can be detrimental to the transformers.
- Power quality degradation: The increased amount of harmonics and imbalance in voltage will degrade the power quality in case of massive scale EV penetration to the grid.

8.1.2. Positive Impacts

On the plus side, EVs can prove to be quite useful to the power systems in a number of ways:

- Smart grid: In the smart grid system, intelligent communication and decision making is incorporated with the grid architecture. Smart grid is highly regarded as the future of power grids and offers a vast array of advantages to offer reliable power supply and advanced control. In such a system, the much coveted coordinated charging is easily achievable as interaction with the grid system becomes very much convenient even from the user end. The interaction of EVs and smart grid can facilitate opportunities like V2G and better integration of renewable energy. In fact, EV is one the eight priorities listed to create an efficient smart grid [117].

- V2G: V2G or vehicle to grid is a method where the EV can provide power to the grid. In this system, the vehicles act as loads when they are drawing energy, and then can become dynamic energy storages by feeding back the energy to the grid. In coordinated charging, the EV loads are applied in the valley points of the load curve, in V2G; EVs can act as power sources to provide during peak hours. V2G is realizable with the smart grid system. By making use of the functionalities of smart grid, EVs can be used as dynamic loads or dynamic storage systems. The power flow in this system can be unidirectional or bidirectional. The unidirectional system is analogous to the coordinated charging scheme, the vehicles are charged when the load is low, but the time to charge the vehicles is decided automatically by the system. Vehicles using this scheme can simply be plugged in anytime and put there; the system will choose a suitable time and charge it. Smart meters are required for enabling this system. With a driver variable charging scheme, the peak power demand can be reduced by 56% [117]. Sortomme et al., found this system particularly attractive as it required little up gradation of the existing infrastructure; creating a communication system in-between the grid and the EVs is all that is needed [146]. The bidirectional system allows vehicles to provide power back to the grid. In this scenario, vehicles using this scheme will supply energy to the grid from their storage when it is required. This method has several appealing aspects. With ever increasing integration of renewable energy sources (RES) to the grid, energy storages are becoming essential to overcome their intermittency, but the storages have a very high price. EVs have energy storages, and in many cases, they are not used for a long time. Example for this point can be the cars in the parking lots of an office block, where they stay unused till the office hour is over, or vehicles that are used in a specific time of the year, like a beach buggy. Studies also revealed that, vehicles stay parked 95% of the time [117]. These potential storages can be used when there is excess generation or low demand and when the energy is needed, it is taken back to the grid. The vehicle owners can also get economically beneficial by selling this energy to the grid. In [147], Clement-Nyns et al., concluded that a combination of PHEVs can prove beneficial to distributed generation sources by providing storage for the excess generation, and releasing that to the grid later. Bidirectional charging, however, needs chargers capable of providing power flow in both directions. It also needs smart meters to keep track of the units consumed and sold, and advanced metering architecture (AMI) to learn about the unit charges in real time to get actual cost associated with the charging or discharging at the exact time of the day. The AMI system can shift 54% of the demand to off-peak periods, and can reduce peak consumption by 36% [117]. The bidirectional system, in fact, can provide 12.3% more annual revenue than the unidirectional one. But taking the metering and protections systems required in the bidirectional method, this revenue is nullified and indicates the unidirectional system is more practical. Frequent charging and discharging caused by bidirectional charging can also reduce battery life and increase energy losses from the conversion processes [81,117]. In a V2G scenario, operators with a vehicle fleet are likely to reduce their cost of operation by 26.5% [117]. Another concept is produced using the smart grid and the EVs, called virtual power plant (VPP), where a cluster of vehicles is considered as a power plant and dealt like one in the system. VPP architecture and control is shown in Figure 62. Table 26 shows the characteristics of unidirectional and bidirectional V2G.

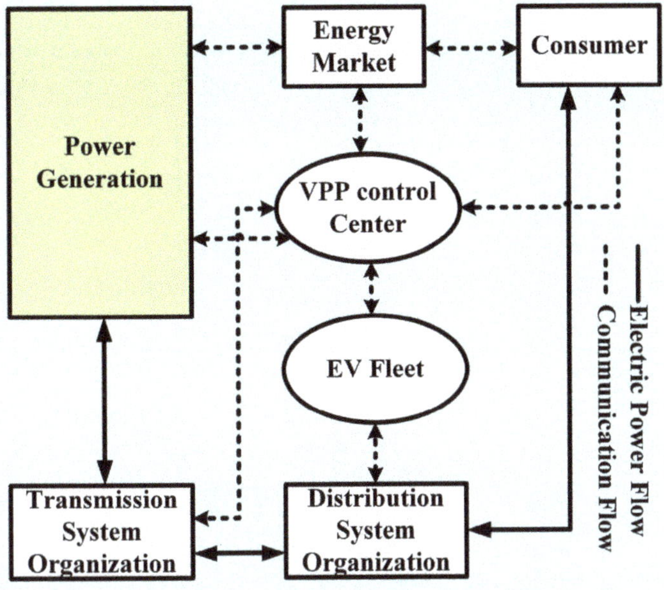

Figure 62. VPP architecture and control [117].

Table 26. Unidirectional and bidirectional V2G characteristics. Adapted from [1].

V2G System	Description	Services	Advantages	Limitations
Unidirectional	Controls EV charging rate with a unidirectional power flow directed from grid to EV based on incentive systems and energy scheduling	• Ancillary service—load levelling	• Maximized profit • Minimized power loss • Minimized operation cost • Minimized emission	• Limited service range
Bidirectional	Bidirectional power flow between grid and EV to attain a range of benefits	• Ancillary service—spinning reserve • Load leveling • Peak power shaving • Active power support • Reactive power support/Power factor correction • Voltage regulation • Harmonic filtering • Support for integration of renewable	• Maximized profit • Minimized power loss • Minimized operation cost • Minimized emission • Prevention of grid overloading • Failure recovery • Improved load profile • Maximization of renewable energy generation	• Fast battery degradation • Complex hardware • High capital cost • Social barriers

- Integration of renewable energy sources: Renewable energy usage becomes more promising with EVs integrated into the picture. EV owners can use RES to generate power locally to charge their EVs. Parking lot roofs have high potential for the placement of PV panels which can charge the vehicles parked underneath as well as supplying the grid in case of excess generation [148–150], thus serving the increase of commercial RES deployment. The V2G structure is further helpful to integrate RES for charging of EVs, and to the grid as well, as it enables the selling of energy to the grid when there is surplus, for example, when vehicles are parked and the system knows the user will not need the vehicle before a certain time. V2G can also enable increased penetration of wind energy (41%–59%) in the grid in an isolated system [121]. References [151–154] worked with different architectures to observe the integration scenario of wind energy with EV assistance. Figure 63 demonstrates integration of wind and solar farm with conventional coal and nuclear power grid with EV charging station employing bidirectional V2G. Table 27 shows the types of assistance EVs can provide for integrating renewable energy sources to the grid.

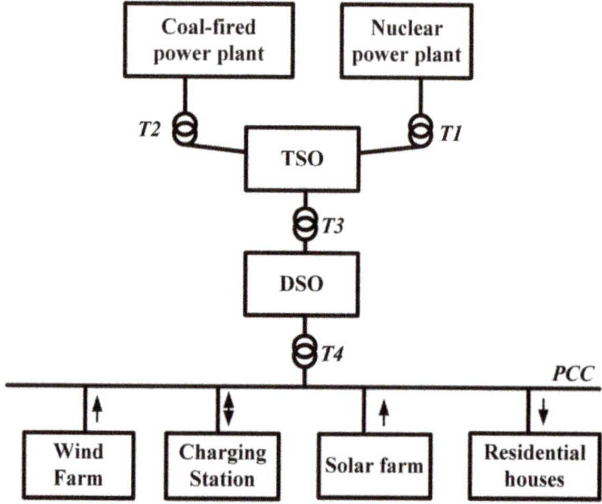

Figure 63. Wind and solar integration in the grid with the help of EV in V2G system. TSO stands for transmission system organization; DSO for distribution system organization; T1 to T4 represent the transformers coupling the generation, transmission, and distribution stages [117].

Table 27. Scopes of assisting renewable energy source (RES) integration using EV. Adapted from [1].

Interaction with RES	Field of Application	Contribution
Solar PV	Smart home	• Implementation of PV and EV in smart home to reduce emission • Development of stand-alone home EV charger based on solar PV system • Development of future home with uninterruptable power by implementing V2G with solar PV
	Parking lot	• Analysis of EV charging using solar PV at parking lots • Scheduling of charging and discharging for intelligent parking lot
	Grid distribution network	• Assessment of power system performance with integration of grid connected EV and solar PV • Development of EV charging control strategy for grid connected solar PV based charging station • Development of optimization algorithm to coordinate V2G services
	Micro grid	• Development of generation scheduling for micro grid consisting of EV and solar PV
Wind turbine	Grid distribution network	• Determination of EV interaction potential with wind energy generation • Development of V2G systems to overcome wind intermittency problems
	Micro grid	• Development of coordinating algorithm for energy dispatching of V2G and wind generation

Table 27. *Cont.*

Interaction with RES	Field of Application	Contribution
Solar PV and wind turbine	Smart home	• Development of control strategy for smart homes with grid-interactive EV and renewable sources
	Parking lot	• Design of intelligent optimization framework for integrating renewable sources and EVs
	Grid distribution network	• Potential analysis of grid connected EVs for balancing intermittency of renewable sources • Emission analysis of EVs associated with renewable generation • Development of optimized algorithm to integrate EVs and renewable sources to the grid
	Micro grid	• Development of V2G control for maximized renewable integration in micro grid

8.2. Impact on Environment

One of the main factors that propelled the increase of EVs' popularity is their contribution to reduce the greenhouse gas (GHG) emissions. Conventional internal combustion engine (ICE) vehicles burn fuels directly and thus produce harmful gases, including carbon dioxide and carbon monoxide. Though HEVs and PHEVs have IC engines, their emissions are less than the conventional vehicles. But there are also theories that the electrical energy consumed by the EVs can give rise to GHG emission from the power plants which have to produce more because of the extra load added in form of EVs. This theory can be justified by the fact that the peak load power plants are likely to be ICE type, or can use gas or coal for power generation. If EVs add excess load during peak hours, it will lead to the operation of such plants and will give rise to CO_2 emission [155]. Reference [156] also stated that power generation from coal and natural gas will produce more CO_2 from EV penetration than ICEs. However, all the power is not generated from such resources. There are many other power generating technologies that produce less GHG. With those considered, the GHG production from power plants because of EV penetration is less than the amount produced by equivalent power generation from ICE vehicles. The power plants also produce energy in bulk, thus minimizing the per unit emission. With renewable sources integrated properly, which the EVs can support strongly, the emission from both power generation and transportation sector can be reduced [115]. Over the lifetime, EVs cause less emission than conventional vehicles. This parameter can be denoted as well-to-wheel emission and it has a lower value for EVs [157]. In [158], well-to-wheel and production phases are taken into account to calculate the impact of EVs on the environment. This approach stated the EVs to be the least carbon intensive among the vehicles. Denmark managed to reduce 85% CO_2 emission from transportation by combining EVs and electric power. EVs also produce far less noise, which can highly reduce sound pollution, mostly in urban areas. The recycling of the batteries raises serious concerns though, as there are few organizations capable of recycling the lithium-ion batteries fully. However, like the previous nickel-metal and lead-acid ones, lithium-ion cells are not made of caustic chemicals, and their reuse can reduce 'peak lithium' or 'peak oil' demands [81].

8.3. Impact on Economy

From the perspective of the EV owners, EVs provide less operating cost because of their superior efficiency [22]; it can be up to 70% where ICE vehicles have efficiencies in the range of 60% to 70% [159]. The current high cost of EVs is likely to come down from mass production and better energy policies [3] which will further increase the economic gains of the owners. V2G also allows the owners to obtain a financial benefit from their vehicles by providing service to the grid [160]. The power service providers benefit from EV integration mainly by implementing coordinated charging and V2G. It allows them to adopt better peak shaving strategies as well as to integrate renewable sources. EV fleets can lead to $200 to $300 savings in cost per vehicle per year [161,162].

8.4. Impacts on Motor Sports

Hybrid technologies are not used extensively in motor sports to enhance the performance of the vehicles. Electric vehicles now have their own formula racing series named 'Formula E' [163] which started in Beijing in September 2014. Autonomous EVs are also being planned to take part in a segment of this series called 'Roborace'.

9. Barriers to EV Adoption

Although electric vehicles offer a lot of promises, they are still not widely adopted, and the reasons behind that are quite serious as well.

9.1. Technological Problems

The main obstacles that have frustrated EVs' domination are the drawbacks of the related technology. Batteries are the main area of concern as their contribution to the weight of the car is significant. Range and charging period also depend on the battery. These factors, along with a few others, are demonstrated below:

9.1.1. Limited Range

EVs are held back by the capacity of their batteries [4]. They have a certain amount of energy stored there, and can travel a distance that the stored energy allows. The range also depends on the speed of the vehicle, driving style, cargo the vehicle is carrying, the terrain it is being driven on, and the energy consuming services running in the car, for example air conditioning. This causes 'range anxiety' among the users [81], which indicates the concern about finding a charging station before the battery drains out. People are found to be willing to spend up to $75 extra for an extra range of one mile [164]. Though even the current BEVs are capable of traversing equivalent or more distance than a conventional vehicle can travel with a full tank (Tesla Model S 100D has a range of almost 564 km on 19″ wheels when the temperature is 70 °C and the air conditioning is off [24], the Chevrolet Bolt's range is 238 miles or 383 km [165]), range anxiety remains a major obstacle for EVs to overcome. This does not affect the use of EVs for urban areas though, as in most cases this range is enough for daily commutation inside city limits. Range extenders, which produce electricity from fuel, are also available with models like BMW i3 as an option. Vehicles with such facilities are currently being called as Extended Range Electric Vehicles (EREV).

9.1.2. Long Charging Period

Another major downside of EVs is the long time they need to get charged. Depending on the type of charger and battery pack, charging can take from a few minutes to hours; this truly makes EVs incompetent against the ICE vehicles which only take a few minutes to get refueled. Hidrue et al., found out that, to have an hour decreased from the charging time; people are willing to pay $425–$3250 [164]. A way to make the charging time faster is to increase the voltage level and employment of better chargers. Some fast charging facilities are available at present, and more are being studied. There are also the fuel cell vehicles that do not require charging like other EVs. Filling up the hydrogen tank is all that has to be done in case of these vehicles, which is as convenient as filling up a fuel tank, but FCVs need sufficient hydrogen refueling stations and a feasible way to produce the hydrogen in order to thrive.

9.1.3. Safety Concerns

The concerns about safety are rising mainly about the FCVs nowadays. There are speculations that, if hydrogen escapes the tanks it is kept into, can cause serious harm, as it is highly flammable. It has no color either, making a leak hard to notice. There is also the chance of the tanks to explode in case of a collision. To counter these problems, the automakers have taken measures to ensure the integrity of the tanks; they are wrapped with carbon fibers in case of the Toyota Mirai. In this car, the hydrogen handling parts are placed outside the cabin, allowing the gas to disperse easily in case of any leak, there are also arrangements to seal the tank outlet in case of high-speed collision [166].

9.2. Social Problems

9.2.1. Social Acceptance

The acceptance of a new and immature technology, along with its consequences, takes some time in the society as it means change of certain habits [167]. Using an EV instead of a conventional vehicle means change of driving patters, refueling habits, preparedness to use an alternative transport in case of low battery, and these are not easy to adopt.

9.2.2. Insufficient Charging Stations

Though public charging stations have increased a lot in number, still they are not enough. Coupled with the lengthy charging time, this acts as a major deterrent against EV penetration. Not all the public charging stations are compatible with every car as well; therefore it also becomes a challenge to find a proper charging point when it is required to replete the battery. There is also the risk of getting a fully occupied charging station with no room for another car. But, the manufacturers are working on to mitigate this problem. Tesla and Nissan have been expanding their own charging networks, as it, in turn means they can sell more of their EVs. Hydrogen refueling stations are not abundant yet as well. It is necessary as well to increase the adoption of FCVs. In [168], a placement strategy for hydrogen refueling stations in California is discussed. It stated that a total of sixty-eight such stations will be sufficient to provide service to FCVs in the area. To get the better out of the remaining stations, there are different trip planning applications, both web based and manufacturer provided, which helps to obtain a route so that there are enough charging facilities to reach the destination.

9.3. Economic Problems

High Price

The price of the EVs is quite high compared to their ICE counterparts. This is because of the high cost of batteries [81] and fuel cells. To make people overlook this factor, governments in different countries including the UK and Germany, have provided incentives and tax breaks which provide the buyers of EVs with subsidies. Mass production and technological advancements will lead to a decrease in the prices of batteries as well as fuel cells. Affordable EVs with a long range like the Chevrolet Bolt has already appeared in the market, while another vehicle with the same promises (the Tesla Model 3) is anticipated to arrive soon. Figure 64 shows the limitations of EVs in the three sectors. Table 29 demonstrates the drawbacks in key factors, while Table 28 suggests some solutions for the existing limitations.

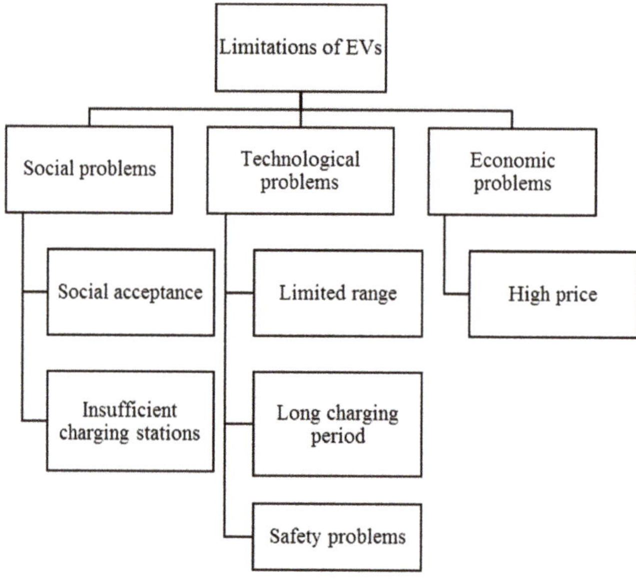

Figure 64. Social, technological, and economic problems faced by EVs.

Table 28. Tentative solutions of current limitations of EVs.

Limitation	Probable Solution
Limited range	Better energy source and energy management technology
Long charging period	Better charging technology
Safety problems	Advanced manufacturing scheme and build quality
Insufficient charging stations	Placement of sufficient stations capable of providing services to all kinds of vehicles
High price	Mass production, advanced technology, government incentives

10. Optimization Techniques

To make the best out of the available energy, EVs apply various aerodynamics and mass reduction techniques, lightweight materials are used to decrease the body weight as well. Regenerative braking is used to restore energy lost in braking. The restored energy can be stored in different ways. It can be stored directly in the ESS, or it can be stored by compressing air by means of hydraulic motor, springs can also be employed to store this energy in form of gravitational energy [169].

Table 29. Hurdles in key EV factors. Adapted from [170].

Factor	Hurdles
Recharging	Weight of charger, durability, cost, recycling, size, charging time
Hybrid EV	Battery, durability, weight, cost
Hydrogen fuel cell	Cost, hydrogen production, infrastructure, storage, durability, reliability
Auxiliary power unit	Size, cost, weight, durability, safety, reliability, cooling, efficiency

Formula One vehicles employ kinetic energy recovery systems (KERSs) to use the energy gathered during braking to provide extra power during accelerating. The Porsche 911 GT3R hybrid uses a flywheel energy storage system to store this energy. The energy consuming accessories on a car include power steering, air conditioning, lights, infotainment systems etc. Operating these in an energy efficient way or turning some of these off can increase the range of a vehicle. LEDs can be used for lighting because of their high efficiency [169]. Table 30 shows different methods of recovering the energy lost during braking.

Table 30. Different methods of recovering energy during braking [169].

Storage System	Energy Converter	Recovered Energy	Application
Electric storage	Electric motor/generator	~50%	BEV, HEV
Compressed gas storage	Hydraulic motor	>70%	Heavy-duty vehicles
Flywheel	Rotational kinetic energy	>70%	Formula One (F1) racing
Gravitational energy storage	Spring storage system	-	Train

Aerodynamic techniques are used in vehicles to reduce the drag coefficient, which reduces the required power. Power needed to overcome the drag force is:

$$P_d = \frac{1}{2}\rho v^3 A C_d \tag{7}$$

Here C_d is the drag coefficient, the power to overcome the drag increases if the drag coefficient's value increases. The Toyota Prius claims a drag coefficient of 0.24 for the 2017 model, the same as the Tesla Model S. The 2012 Nissan Leaf SL had this value set at 0.28 [171].

To ensure efficient use of the available energy, different energy management schemes can be employed [6]. Presented different control strategies for energy management which included systems using fuzzy logic, deterministic rule and optimization based schemes. Geng et al., worked on a plug-in series hybrid FCV. The objective of their control system was to consume the minimum amount of hydrogen while preserving the health of the proton exchange membrane fuel cell (PEMFC) [172]. The control system was comprised of two stages; the first stage determined the SOC and control references, whereas the second stage determined the PEMFC health parameters. This method proved to be capable of reducing the hydrogen consumption while increasing the life-time to the fuel cell. Another intelligent management system is examined in [173] by Murphey et al., which used machine learning combined with dynamic programming to determine energy optimization strategies for roadway and traffic-congestion scenarios for real-time energy flow control of a hybrid EV. Their system is simulated using a Ford Escape Hybrid model; it revealed the system was effective in finding out congestion level, optimal battery power and optimal speed. Geng et al., proposed a control mechanism for energy management for a PHEV employing batteries and a micro turbine in [174]. In this work, they introduced a new parameter, named the "energy ratio", to produce the equivalent factor (EF) which was used in the popular Equivalent Consumption Minimization Strategy (ECMS) to deduce the minimum driving cost by applying Pontryagin's minimum principle. This method claimed to reduce the cost by 7.7–21.6%. In [175], Moura et al., explored efficient ways to split power demand among different power sources of mid-sized sedan PHEVs. They used a number of drive cycles, rather than a single one, assessed the potential of depleting charge in a controlled manner, and considered relative pricing of fuel and electricity for optimal power management of the vehicle.

11. Control Algorithms

Control systems are crucial for proper functioning of EVs and associated systems. Sophisticated control mechanisms are required for providing a smooth and satisfactory ride quality, for providing the enough power when required, estimating the energy available from the on-board sources and using them properly to cover the maximum distance, charging in a satisfactory time without causing burden on the grid, and associated tasks. Different algorithms are used in these areas, and as the EV culture is becoming more mainstream, need for better algorithms are on the rise.

Driving control systems are required to assist the driver in keeping the vehicle in control, especially at high speeds and in adverse conditions such as slippery surfaces caused by rain or snow. Driving control systems such as traction control, cruise control, and different driving modes have been being applied in conventional vehicles for a long time. Application of such systems appeared more efficient in EVs as the driving forces of EVs can be controlled with more ease, with less conversion required in-between the mechanical and the electrical domains. In any condition, forces act on a vehicle at different directions; for a driving control system, if is essential to perfectly perceive these forces, along with other sensory inputs, and provide torques to the wheels to maintain desired stability. In Figure 65, the forces in different direction acting on each wheel of a car is shown in a horizontal plane. In [176], Magallan et al., proposed and simulated a control system to utilize the maximum torque in a rear-wheel-drive EV without causing the tires to skid. The model they worked on had independent driving systems for the two rear wheels. A sliding mode system, based on a *LuGre* dynamic friction model, was used to estimate the vehicle's velocity and wheel slip on unknown road surfaces. Utilizing these data, the control algorithm determined the maximum allowable traction force, which was applied to the road by torque controlling of the two rear motors. Juyong Kang et al., presented an algorithm aimed at driving control systems for four-wheel-drive EVs in [177]. Their vehicle model had two motors driving the front and the rear shafts. The algorithm had three parts: a supervisory level for determine the desirable dynamics and control mode, an upper level computing the yaw moment and traction force inputs, and a lower level determining the motor and braking commands. This system proved useful for enhancing lateral stability, maneuverability, and reducing rollover. Figure 66 shows the acting components of this system on a vehicle model while Figure 67 shows a detailed diagram of the system with the inputs, controller levels, and actuators. Tahami et al., introduced a stability system for driving assistance for all-wheel drive EVs in [25]. They trained a neural network to produce a reference yaw rate. A fuzzy logic controller dictated independent wheel torques; a similar controller was used for controlling wheel slip. This system is shown in Figure 68. In [178], Wang et al., showed a system to assist steering using differential drive for in-wheel drive system. A proportional integral (PI) closed loop control system was used here to monitor the reference steering position. It was achieved by distributing torque at the front wheels. Direct yaw moment control and traction control were also employed to make the differential drive system better. This approach maintained the lateral stability of the vehicle, and improved stability at high speeds. The structure of this system is shown in Figure 69. In a separate study conducted by Nam et al., lateral stability of an in-wheel drive EV was attained by estimating the sideslip angle of the vehicle employing sensors to measure lateral tire forces [179]. In this study, a state observer was proposed which was derived from extended-Kalman-filtering (EKF) method and was evaluated by implementing in an experimental EV alongside Matlab/Simulink-Carsim simulations.

Energy management is a big issue for EVs. Proper measurement of the available energy is crucial for calculating the range and plans the driving strategy thereafter. For vehicles with multiple energy sources (e.g., HEVs), efficient energy management algorithms are required to make proper use of the energy on-board. Zhou et al., proposed a battery state-of-charge (SOC) measuring algorithm for lithium polymer batteries which made use of a combination of particle filter and multi-model data fusion technique to produce results real time and is not affected by measurement noise [180]. They used different battery models and presented the tuning strategies for each model as well. Their multi-model approach proved to be more effective than single model methods for providing real time results. Working principle of this system is shown in Figure 70. Moura et al., explored efficient ways to split power demand among different power sources of mid-sized sedan PHEVs in [175], which can be used for other vehicle configurations as well. Their method made use of different drive cycles, rather than using a single one; assessed the potential of depleting charge in a controlled manner; and considered relative pricing of fuel and electricity to optimally manage the power of the vehicle. In [181], Hui et al., presented a novel hybrid vehicle using parallel hybrid architecture which employed a hydraulic/electric synergy configuration to mitigate the drawbacks faced by heavy hybrid vehicles using a single energy source. Transition among the operating modes of such a vehicle is shown in Figure 71. They developed an algorithm to optimize the key parameters and adopted a logic threshold approach to attain desired performance, stable SOC at the rational operating range constantly, and maximized fuel economy. The operating principle of this system is shown in Figure 72. Chen et al., proposed an energy management algorithm in [182] to effectively control battery current, and thus reduce fuel usage by allowing the engine operate more effectively. Quadratic programming was used here to calculate the optimum battery current. In [183], Li et al., used fuzzy logic to create a new quantity: battery working state or BWS which was used in an energy management system run by fuzzy logic to provide proper power division between the engine and the battery. Simulation results proved this approach to be effective in making the engine operate in the region of maximum fuel efficiency while keeping the battery away from excess discharging. Yuan et al., compared Dynamic Programming and Pontryagin's Minimum Principle (PMP) for energy management in parallel HEVs using Automatic Manual Transmission. The PMP method proved better as it was more efficient to implement, required considerably less computational time, and both of the systems provided almost similar results [184]. In [185], Bernard et al., proposed a real time control system to reduce hydrogen consumption in FCEVs by efficiently sharing power between the fuel cell arrangement and the energy buffer (ultracapacitor or battery). This control system was created from an optimal control theory based non-causal optimization algorithm. It was eventually implemented in a hardware arrangement built around a 600 W fuel cell arrangement. In an attempt to create an energy management system for a still-not-commercialized PHEV employing a micro turbine, Geng et al., used an equivalent consumption minimization strategy (ECMS) in [174] to estimate the optimum driving cost. Their system used the battery SOC and the vehicle telemetry to produce the results, which were available in real time and provided driving cost reductions of up to 21.6%.

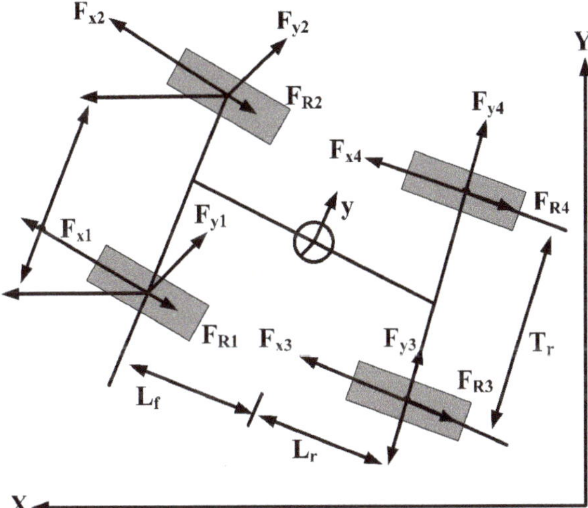

Figure 65. Forces acting on the wheels of a car. Each of the wheels experience forces in all three directions, marked with the 'F' vectors. L_f and L_r show the distances of front and rear axles from the center of the vehicle, while T_r shows the distance between the wheels of an axle. Adapted from [25].

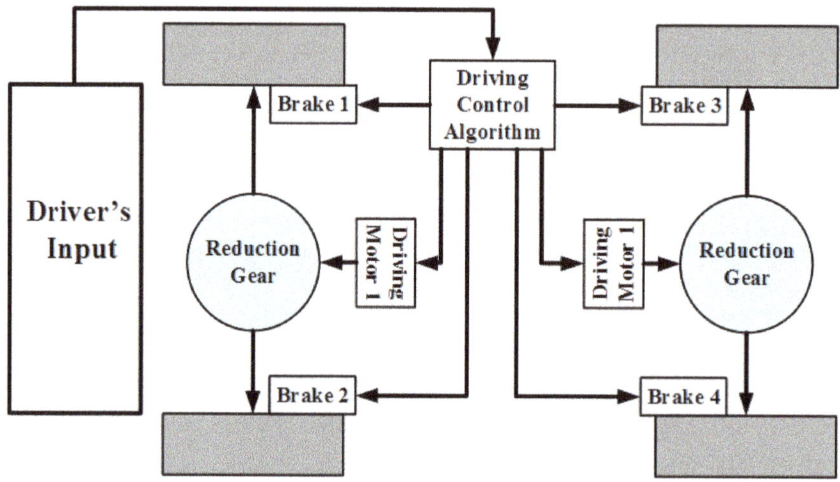

Figure 66. Main working components of the driving control system for four-wheel-drive EVs proposed by Juyong Kang et al. The driving control algorithm takes the driver's inputs, and then determines the actions of the brakes and the motors according to the control mode [177].

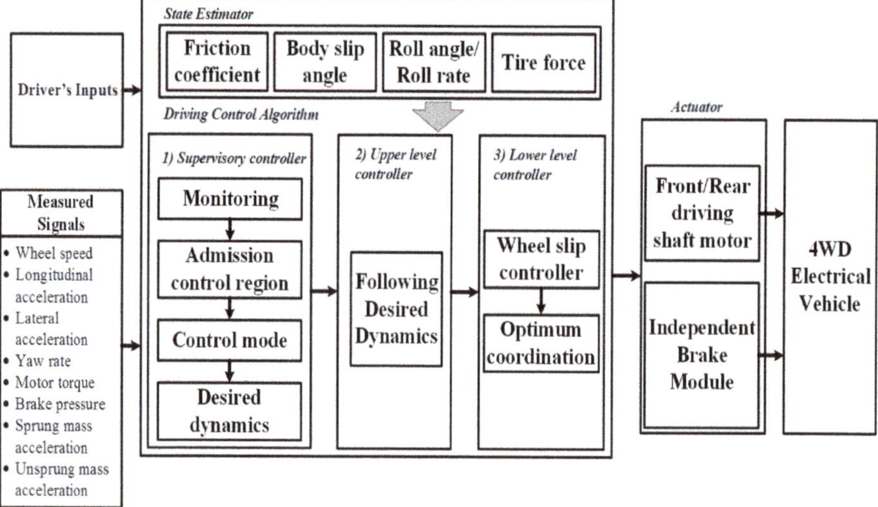

Figure 67. Working principle of the control system proposed by Kang et al. The system uses both the driver's commands and sensor measurements as inputs, and then drives the actuators as determined by the three level control algorithms. Adapted from [177].

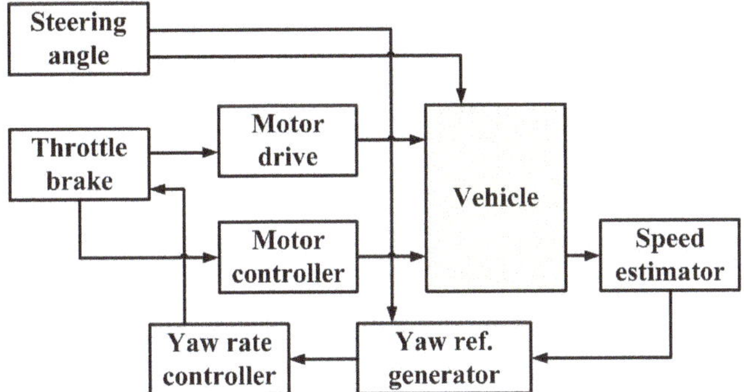

Figure 68. Working principle of vehicle stability system proposed by Tahami et al. A neural network was used in the yaw reference generator [25].

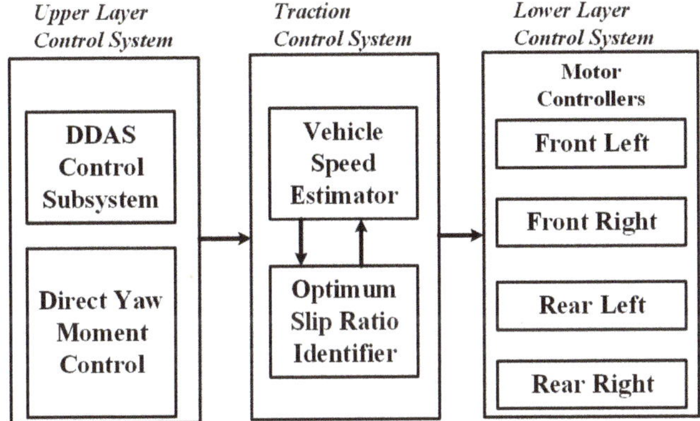

Figure 69. Independent torque control system proposed by Wang et al., Differential drive assisted steering (DDAS) subsystem and direct yaw moment control subsystem creates the upper layer. The traction control subsystem processes the inputs, and the controlling is done through the lower layer [178].

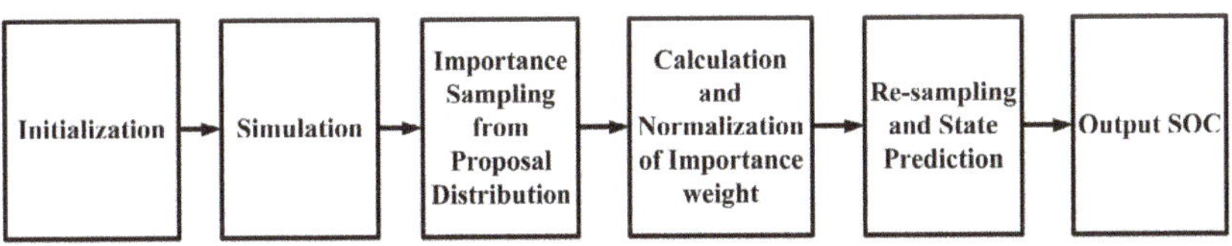

Figure 70. Working principle of the SOC measuring algorithm proposed by Zhou et al. [180].

As pointed out in Section 8, the grid is facing some serious problems with the current rise in EV penetration. Reducing the charging time of the vehicles while creating minimal pressure on the grid has become difficult goal to achieve. However, ample research has already been done on this matter and a number of charging system algorithms have been proposed to attain satisfactory charging performance.

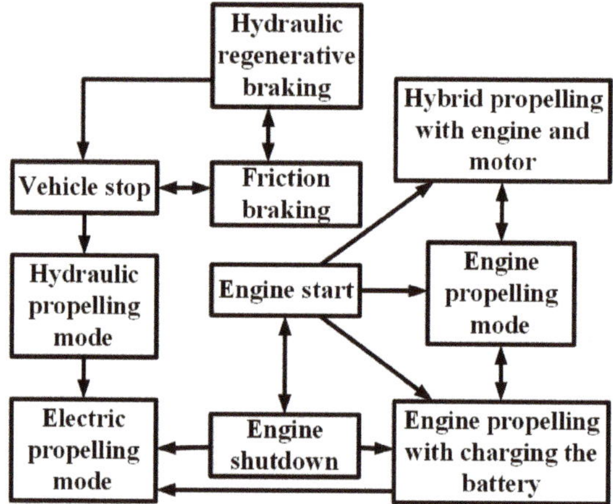

Figure 71. Transition of the operating modes of the vehicle used in [181] by Hui et al. From engine start to shutdown through stops, the vehicle can use either the hydraulic or the electric system, or it can use both.

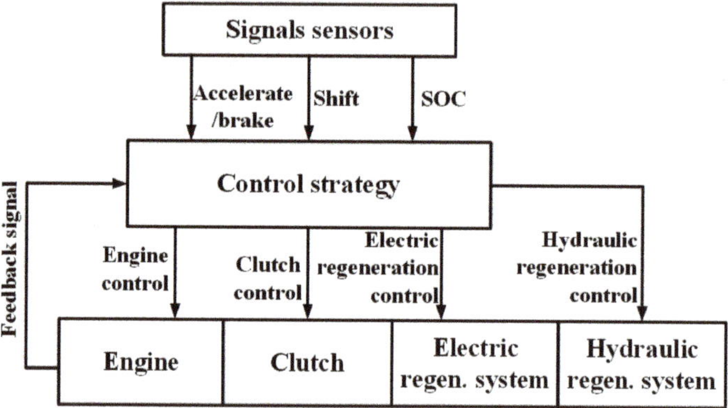

Figure 72. Operating principle of the control system proposed by Hui et al. The control strategy drives the actuating systems according to the decisions made from the sensor inputs. Adapted from [181].

In [186], Su et al., presented an algorithm (shown in Figure 73) capable of providing charge intelligently to a large fleet of PHEVs docked at a municipal charging station. This algorithm—which used the estimation of distribution (EDA) algorithm—considered real-world factors such as remaining charging time, remaining battery capacity, and energy price. The load management system proposed by Deilami et al., in [140] considered market energy prices that vary with time, time zones preferred by EV owners by priority selection, and random plugging-in of EVs—for providing coordinated charging in a smart grid system. It then used the maximum sensitivities selection (MSS) optimization technique to enable EVs charge as soon as possible depending on the priority time zones while maintaining the operation criteria of the grid such as voltage profile, limits of generation, and losses. This system was simulated using an IEEE 23 kV distribution system modified for this purpose. Mohamed et al., designed an energy management algorithm to be applied in EV charging parks incorporating renewable generation such as PV systems [187]. The system they developed used a fuzzy controller to manage the charging/discharging times of the connected EVs, power sharing among them, and V2G services.

The goal of this system was to minimize the charging cost while reducing the impact on the grid as well as contributing to peak shaving. The flowchart associated to this system is shown in Figure 74.

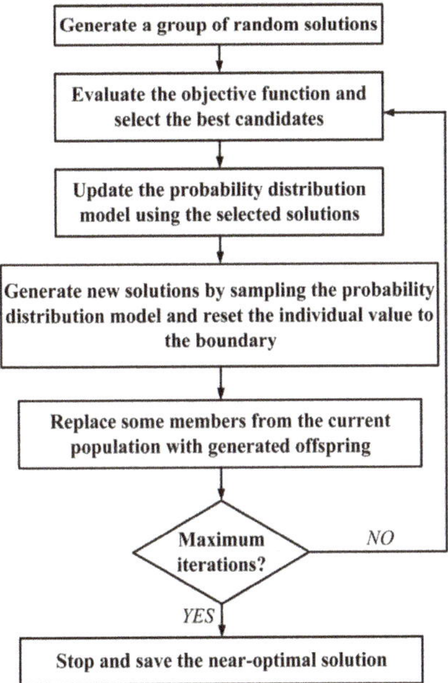

Figure 73. Intelligent charging algorithm proposed by Su et al., for a municipal charging station [186].

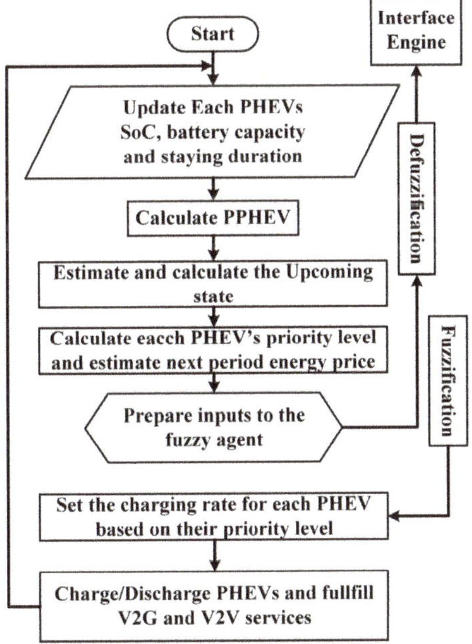

Figure 74. Flowchart of the management system proposed by Mohamed et al. [187].

To alleviate the problems at the distribution stage of the grid—which is highly affected by EV penetration—Geng et al., proposed a charging strategy comprising of two stages aimed at providing satisfactory charging for all connected EVs while shifting the loads on the transformers [120]. The first stage utilized Pontryagin's minimum principle and was based on the concept of dynamic aggregator; it derived the optimal charging power for all the EVs in the system. The second stage used fuzzy logic to distribute the power calculated in the first stage among the EVs. According to the authors, the system was feasible to be implemented practically [120]. In [116], Richardson et al., employed a linear programming based technique to calculate the optimal rate of charging for each EV connected in a distribution network to enable maximized power delivery to the vehicles while maintaining the network limits. This approach can provide high EV penetration possible in existing residential power systems with no or a little upgrade. Sortomme et al., developed an algorithm to maximize profit from EV charging in a unidirectional V2G system where an aggregator is present to manage the charging [146]. Table 31 summarizes the algorithms presented in this section.

Table 31. Summary of the control algorithms presented.

References	Algorithm Based on	Application
Magallan et al. [176]	*LuGre* dynamic friction model	Driving control system in rear-wheel-drive EV
Kang et al. [177]	Optimization-based control allocation strategy	Driving control system in four-wheel-drive EV
Tahami et al. [25]	Fuzzy logic	Driving control system in all-wheel-drive EV
Wang et al. [178]	Proportional-integral (PI) closed loop control system	Driving control system in in-wheel-drive EV
Nam et al. [179]	Extended Kalman filtering (EKF) method	Driving control system in in-wheel-drive EV
Zhou et al. [180]	Particle filter and multi-model data fusion	SOC measurement for lithium polymer batteries
Moura et al. [175]	Markov process	Power splitting in mid-sized sedan PHEV
Hui et al. [181]	Torque control strategy	Heavy hybrid vehicles using a single energy source
Chen et al. [182]	Quadratic programming	Reduction of fuel consumption by effective battery current control
Li et al. [183]	Fuzzy logic	Attaining maximum fuel efficiency without excess discharging of battery
Yuan et al. [184]	Dynamic Programming and Pontryagin's Minimum Principle	Efficient energy management in parallel HEV using Automatic Manual Transmission or AMT
Bernard et al. [185]	Non-causal optimization algorithm	Reduction of hydrogen consumption in FCEV
Geng et al. [174]	Equivalent consumption minimization strategy (ECMS)	Energy management in PHEV employing microturbine
Su et al. [186]	Estimation of distribution (EDA) algorithm	Intelligent charging of large fleet of PHEVs docked at a municipal charging station
Deilami et al. [140]	Maximum sensitivities selection (MSS) optimization	Load management system for intelligent charging
Mohamed et al. [187]	Fuzzy controller	V2G system for EV charging parks incorporating renewable generation
Geng et al. [120]	Pontryagin's minimum principle, fuzzy logic	Load shifting while charging EVs in the distribution network
Richardson et al. [116]	Linear programming	Enabling high EV penetration in existing residential power system network
Sortomme et al [146]	Preferred operating point (POP) algorithm	Maximizing profit from EV charging through an aggregator

12. Global EV Sales Figures

The electric vehicle market is growing much faster than the conventional vehicle market, and in some regions EVs are catching up with ICE vehicles in terms of the number of units sold. China has become the largest market for EVs, its market claiming 35.4% of the worldwide EV scene in 2017, an exorbitant rise from the mere 6.3% in 2013 [188]. Chinese consumers bought a world-topping 24.38 million passenger electric vehicles in 2016. China has the greatest number of manufacturers, led by BYD autos, which sold 96,000 EVs in 2016. This drive in China is fueled by government initiatives adopted to promote EV use to mitigate the country's serious air pollution. However, the majority of Chinese vehicles are in the $36,000 range and offers limited range, but high-end vehicles manufacturing is on the rise in China too. This huge market has attracted major carmakers all over the world—Ford, Volkswagen, Volvo, and General Motors—who have their own EVs in the Chinese market and are poised to introduce more models in the coming years [189]. Figure 75 shows the ten highest selling EVs in China in 2016.

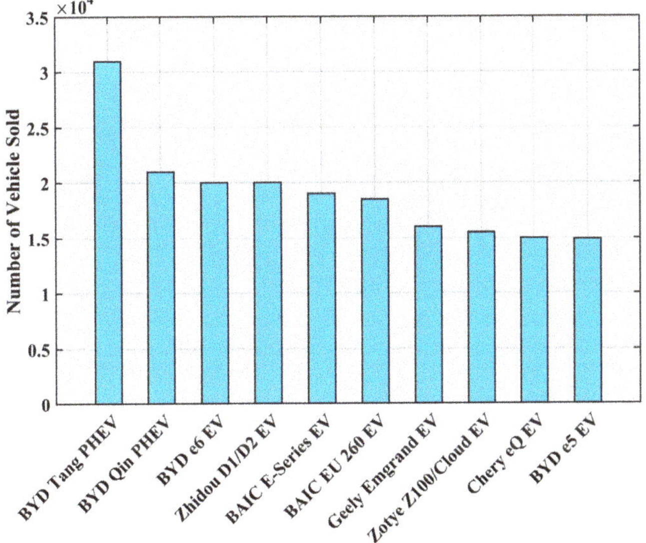

Figure 75. Top ten EVs in China in 2016 according to the number of units sold. Data from [190].

From a global perspective, sales of EV grew by 36% in the USA; Europe saw a growth of 13%, while Japan observed a decrease of 11% in the same period. BYD dominated the global market with a 13.2% share, followed by Tesla in second place (9.9%); the other major contributors can be listed as Volkswagen Group, BMW Group, Nissan, BAIC, and Zoyte. However, the Tesla Model S remained the best-selling EV in 2016 with 50,935 units sold, followed by the Nissan Leaf EV with 49,818 units [191]. The top ten best-selling vehicles around the globe in shown in Figure 76.

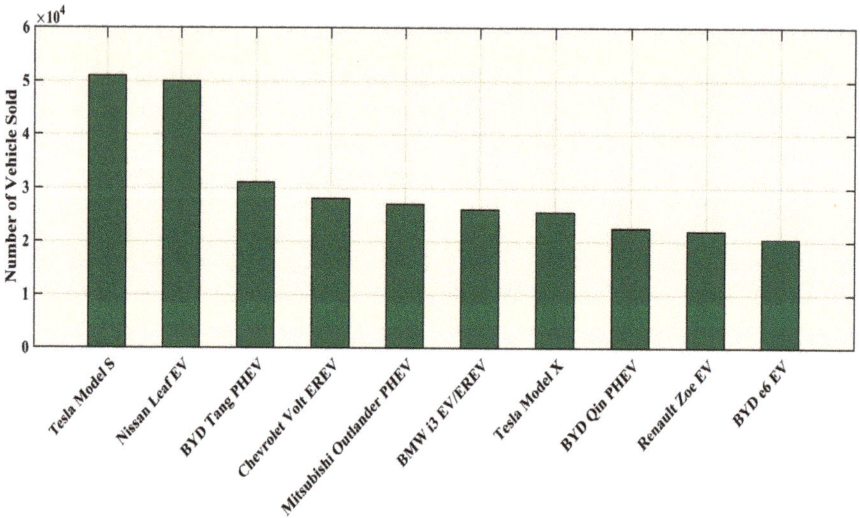

Figure 76. Top ten best-selling EVs globally in 2016. Data from [191].

The American market was dominated predictably by the Tesla Model S in 2016, 28,821 of these were sold; Chevrolet Volt EREV sold 24,739 units, thus securing the second place. The third place was achieved by another Tesla, the Model X; 18,192 of these SUVs were sold in 2016 [192]. The ten best-selling EVs in the USA in 2016 are shown in Figure 77.

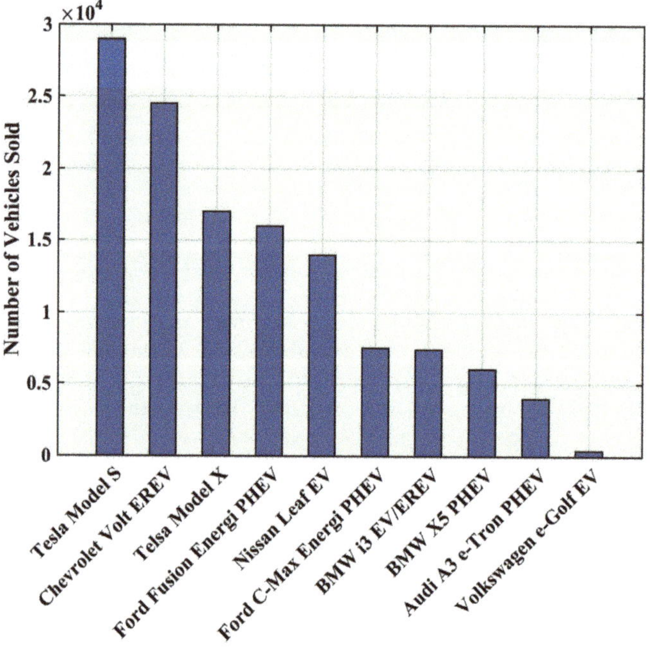

Figure 77. Top ten best-selling EVs in the USA in 2016. Data from [192].

The Renault Zoe was the best-selling BEV in Europe in 2016, with 21,338 units sold, followed by the Nissan Leaf with 18,614 units. In the PHEV segment, the Mitsubishi Outlander PHEV was the market leader in Europe in 2016, with 21,333 units sold; the Volkswagen Passat GTE held the second position with 13,330 units [193]. Figures 78 and 79 shows the BEV and PHEV market shares in Europe in 2016.

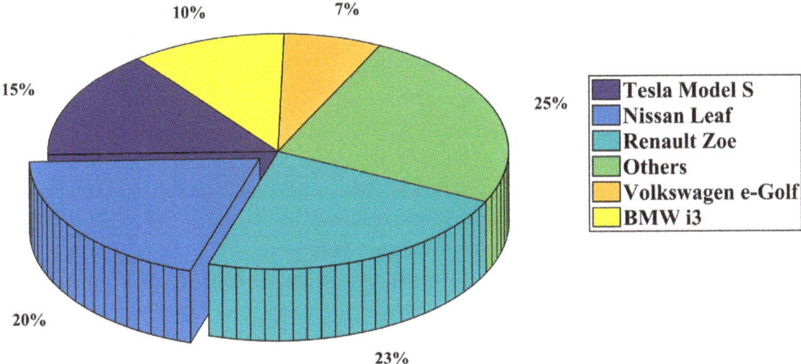

Figure 78. BEV market shares in Europe in 2016. Data from [193].

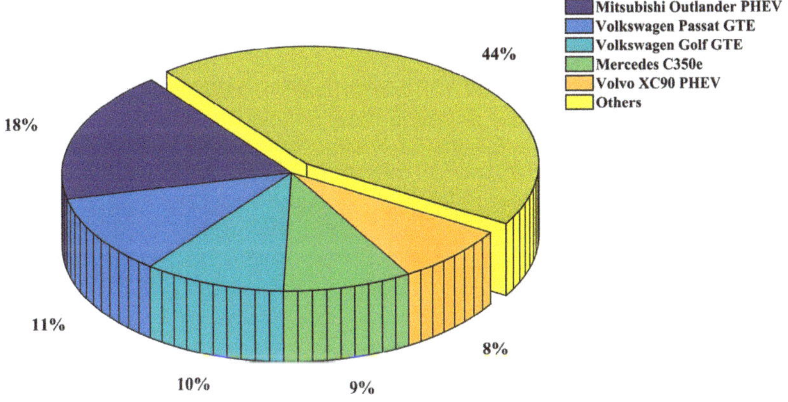

Figure 79. PHEV market shares in Europe in 2016. Data from [193].

13. Trends and Future Developments

The adoption of EVs has opened doors for new possibilities and ways to improve both the vehicles and the systems associated with it, the power system, for example. EVs are being considered as the future of vehicles, whereas the smart grid appears to be the grid of the future [194,195]. V2G is the link between these two technologies and both get benefitted from it. With V2G comes other essential systems required for a sustainable EV scenario—charge scheduling, VPP, smart metering etc. The existing charging technologies have to improve a lot to make EVs widely accepted. The charging time has to be decreased extensively for making EVs more flexible. At the same time, chargers and EVSEs have to able to communicate with the grid for facilitating V2G, smart metering, and if needed, bidirectional charging [23]. Better batteries are a must to take the EV technology further. There is a need for batteries that use non-toxic materials and have higher power density, less cost and weight, more capacity, and needs less time to recharge. Though technologies better than Li-ion have been discovered already, they are not being pursued industrially because of the huge costs associated with creating a working version. Besides, Li-ion technology has the potential to be improved a lot more. Li-air batteries could be a good option to increase the range of EVs [23]. EVs are likely to move away from using permanent magnet motors which use rare-earth materials. The motors of choice can be induction motor, synchronous reluctance motor, and switched reluctance motor [23]. Tesla is using an induction motor in its models at present. Motors with internal permanent magnet may stay in use [23]. Wireless power transfer systems are likely to replace the current cabled charging system. Concepts revealed by major automakers adopted this feature to highlight their usefulness and convenience. The Rolls-Royce 103EX and the Vision Mercedes-Maybach 6 can be taken as example for that. Electric roads for wireless charging of vehicles may appear as well. Though this is not still viable, the situation may change in the future. Recent works in this sector includes the work of Electrode, an Israeli startup, which claims to be able to achieve this feat in an economic way. Vehicles that follow a designated route along the highway, like trucks, can get their power from overhead lines like trains or trams. It will allow them to gather energy as long as their route resides with the power lines, then carry on with energy from on-board sources. Such a system has been tested by Siemens using diesel-hybrid trucks from Scania on a highway in Sweden [196]. New ways of recovering energy from the vehicle may appear. Goodyear has demonstrated a tire that can harvest energy from the heat generated there using thermo-piezoelectric material. There are also chances of solar-powered vehicles. Until now, these have not appeared useful as installed solar cells only manage to convert up to 20% of the input power [70]. Much research is going on to make the electronics and sensors in EVs more compact, rugged and cheaper—which in many cases are leading to advanced solid state devices that can achieve these goals with promises of cheaper products if they can be mass-produced. Some examples can be the works on gas sensors [197], smart LED drivers [198], smart drivers for automotive alternators [199], advanced gearboxes [200], and compact and smart power switches to weather harsh conditions [201]. The findings of [202–208] may prove helpful for studies regarding fail-proof on-board power supplies for EVs. The future research topics will of course, revolve around making the EV technology more efficient, affordable, and convenient. A great deal of research has already been conducted on making EVs more affordable and capable of covering more distance: energy management, materials used for construction, different energy sources etc. More of such researches are likely to go on emphasizing on better battery technologies, ultracapacitors, fuel cells, flywheels, turbines, and other individual and hybrid configurations. FCVs may get significant attention in military and utility-based studies, whereas the in-wheel drive configuration for BEVs may be appealing to researchers focusing on better urban transport systems. Better charging technologies will remain a crucial research topic in near future. This is one of the areas the EV technology is lacking very badly; wireless charging technologies are very likely to attract more researchers' attention. A lot of research has already been done incorporating EVs and the grid: the challenges and possibilities that the EVs bring with them to the existing grid and also to the grid of the future. With more implementation of smart grids, distributed generation, and renewable energy sources, researches in these fields are likely to increase. And as researches in the entire aforementioned field's increase, exploration for better algorithms to run the systems is bound to rise. Figure 80 shows the major trends and sectors for future developments for EVs.

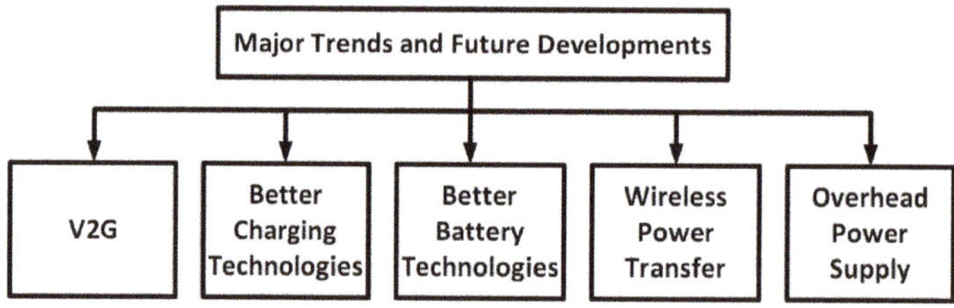

Figure 80. Major trends and sectors for future developments for EV.

14. Outcomes

The goal of this paper is to focus on the key components of EV. Major technologies in different sections are reviewed and the future trends of these sectors are speculated. The key findings of this paper can be summarized as follows:

- EVs can be classified as BEV, HEV, PHEV, and FCEV. BEVs and PHEVs are the current trends. FCEVs can become mainstream in future. Low cost fuel cells are the main prerequisite for that and there is need of more research to make that happen. There are also strong chances for BEVs to be the market dominators with ample advancement in key technologies; energy storage and charging systems being two main factors. Currently FCVs appear to have little chance to become ubiquitous, these may find popularity in niche markets, for example, the military and utility vehicles.

- EVs can be front wheel drive, rear wheel drive, even all-wheel drive. Different configurations are applied depending on the application of the vehicle. The motor can also be placed inside the wheel of the vehicle which offers distinct advantages. This configuration is not commercially abundant now, and has scopes for more study to turn it into a viable product.

- The main HEV configurations are classified as series, parallel, and series-parallel. Current vehicles are using the series-parallel system mainly as it can operate in both battery-only and ICE-only modes, providing more efficiency and less fuel consumption than the other two systems.

- Currently EVs use batteries as the main energy source. Battery technology has gone through significant changes, the lead-acid technology is long gone, as is the NiMH type. Li-ion batteries are currently in use, but even they are not capable enough to provide the amount of energy required to appease the consumers suffering from 'range anxiety' in most cases. Therefore the main focus of research in this area has to be creating batteries with more capacity, and also with better power densities. Metal-air batteries can be the direction where the EV makers will head towards. Lithium-sulfur battery and advanced rechargeable zinc batteries also have potential provide better EVs. Nevertheless, low cost energy sources will be sought after always as ESS cost is one of the major contributors to high EV cost.

- Ultracapacitors are considered as auxiliary power sources because of their high power densities. If coupled with batteries, ultracapacitors produce a hybrid ESS that can satisfy some requirements demanded from an ideal source. Flywheels are also being used, especially because of their compact build and capability to store and discharge power on demand. Fuel cells can also be used more in the future if FCVs gain popularity.

- Different types of motors can be employed for EV use. The prominent ones can be listed as induction motor, permanent magnet synchronous motor, and synchronous reluctance motor. Induction motors are being extensively these days, they can also dominate in future because of their independence on rare-earth material permanent magnets.

- EVs can be charged with AC or DC supply. There are different voltage levels and they are designated accordingly. Higher voltage levels provide faster charging. DC supplies negate the need of rectification from AC, which reduces delay and loss. However, with increased voltage level, the pressure on the grid increases and can give rise to harmonics as well as voltage imbalance in an unsupervised system. Therefore, there are ample chances of research in the field of mitigating the problems associated with high-voltage charging.

- Two charger configurations are mainly available now: CCS and CHAdeMO. These two systems are not compatible with each other and each has a number of automakers supporting them. Tesla also brought their own 'supercharger' system, which provides a faster charging facility. It is not possible to determine now which one of these will prevail, or if both will co-exist, technical study is needed to find out the most useful one of these configurations or ways to make them compatible with each other.

- Whatever the charging system is, the charging time is still very long. This is a major disadvantage that is thwarting the growth of the EV market. Extensive research is needed in this sector to provide better technologies that can provide much faster charging and can be compatible with the small time required to refill an ICE vehicle. Wireless charging is also something in need of research. With all the conveniences it promises, it is still not in a viable form to commercialize.

- EV impacts the environment, power system, and economy alongside the transportation sector. It shows promises to reduce the GHG emissions as well as efficient and economical transport solutions. At the same time, it can cause serious problems in the power system including voltage instability, harmonics, and voltage sag, but these shortcomings may be short-lived if smart grid technologies are employed. There are prospects of research in the areas of V2G, smart metering, integration of RES, and system stability associated with EV penetration.

- EVs employ different techniques to reduce energy loss and increase efficiency. Reducing the drag coefficient, weight reduction, regenerative braking, and intelligent energy management are some of these optimization techniques. Further research directions can be better aerodynamic body designs, new materials with less weight and desired strength, ways to generate and restore the lost energy.

- Different control algorithms have been developed for driving assist, energy management, and charging. There is lots of room left for more research into charging and energy management algorithms. With increased EV penetration in the future, demands for efficient algorithms are bound to increase.

15. Conclusions

EVs have great potential of becoming the future of transport while saving this planet from imminent calamities caused by global warming. They are a viable alternative to conventional vehicles that depend directly on the diminishing fossil fuel reserves. The EV types, configurations, energy sources, motors, power conversion and charging technologies for EVs have been discussed in detail in this paper. The key technologies of each section have been reviewed and their characteristics have been presented. The impacts EVs cause in different sectors have been discussed as well, along with the huge possibilities they hold to promote a better and greener energy system by collaborating with smart grid and facilitating the integration of renewable sources. Limitations of current EVs have been listed along with probable solutions to overcome these shortcomings. The current optimization techniques and control algorithms have also been included. A brief overview of the current EV market has been presented. Finally, trends and ways of future developments have been assessed followed by the outcomes of this paper to summarize the whole text, providing a clear picture of this sector and the areas in need of further research.

Acknowledgments: No funding has been received in support of this research work.

Author Contributions: All authors contributed for bringing the manuscript in its current state. Their contributions include detailed survey of the literatures and state of art which were essential for the completion of this review paper.

Conflicts of Interest: The authors declare no conflict of interest.

References

1. Yong, J.Y.; Ramachandaramurthy, V.K.; Tan, K.M.; Mithulananthan, N. A review on the state-of-the-art technologies of electric vehicle, its impacts and prospects. *Renew. Sustain. Energy Rev.* **2015**, *49*, 365–385. [CrossRef]

2. Camacho, O.M.F.; Nørgård, P.B.; Rao, N.; Mihet-Popa, L. Electrical Vehicle Batteries Testing in a Distribution Network using Sustainable Energy. *IEEE Trans. Smart Grid* **2014**, *5*, 1033–1042. [CrossRef]

3. Camacho, O.M.F.; Mihet-Popa, L. Fast Charging and Smart Charging Tests for Electric Vehicles Batteries using Renewable Energy. *Oil Gas Sci. Technol.* **2016**, *71*, 13–25. [CrossRef]

4. Chan, C.C. The state of the art of electric and hybrid vehicles. *Proc. IEEE* **2002**, *90*, 247–275. [CrossRef]

5. Grunditz, E.A.; Thiringer, T. Performance Analysis of Current BEVs Based on a Comprehensive Review of Specifications. *IEEE Trans. Transp. Electr.* **2016**, *2*, 270–289. [CrossRef]

6. SAE International. SAE Electric Vehicle and Plug-in Hybrid Electric Vehicle Conductive Charge Coupler. In *SAE Standard J1772*; Society of Automotive Engineers (SAE): Warrendale, PA, USA, 2010.

7. Yilmaz, M.; Krein, P.T. Review of battery charger topologies, charging power levels, and infrastructure for plug-in electric and hybrid vehicles. *IEEE Trans. Power Electr.* **2013**, *28*, 2151–2169. [CrossRef]

8. Bayindir, K.Ç.; Gözüküçük, M.A.; Teke, A. A comprehensive overview of hybrid electric vehicle: Powertrain configurations, powertrain control techniques and electronic control units. *Energy Convers. Manag.* **2011**, *52*, 1305–1313. [CrossRef]

9. Marchesoni, M.; Vacca, C. New DC–DC converter for energy storage system interfacing in fuel cell hybrid electric vehicles. *IEEE Trans. Power Electron.* **2007**, *22*, 301–308. [CrossRef]

10. Schaltz, E.; Khaligh, A.; Rasmussen, P.O. Influence of battery/ultracapacitor energy-storage sizing on battery lifetime in a fuel cell hybrid electric vehicle. *IEEE Trans. Veh. Technol.* **2009**, *58*, 3882–3891. [CrossRef]

11. Kramer, B.; Chakraborty, S.; Kroposki, B. A review of plug-in vehicles and vehicle-to-grid capability. In Proceedings of the 34th IEEE Industrial Electronics Annual Conference, Orlando, FL, USA, 10–13 November 2008; pp. 2278–2283.

12. Williamson, S.S. Electric drive train efficiency analysis based on varied energy storage system usage for plug-in hybrid electric vehicle applications. In Proceedings of the IEEE Power Electronics Specialists Conference, Orlando, FL, USA, 17–21 June 2007; pp. 1515–1520.

13. Wirasingha, S.G.; Schofield, N.; Emadi, A. Plug-in hybrid electric vehicle developments in the US: Trends, barriers, and economic feasibility. In Proceedings of the IEEE Vehicle Power and Propulsion Conference, Harbin, China, 3–5 September 2008; pp. 1–8.

14. Gao, Y.; Ehsani, M. Design and control methodology of plug-in hybrid electric vehicles. *IEEE Trans. Ind. Electron.* **2010**, *57*, 633–640.

15. EG&G Technical Services, Inc. *The Fuel Cell Handbook*, 6th ed.; U.S. Department of Energy: Morgantown, WV, USA, 2002.

16. Miller, J.F.; Webster, C.E.; Tummillo, A.F.; DeLuca, W.H. Testing and evaluation of batteries for a fuel cell powered hybrid bus. In Proceedings of the Energy Conversion Engineering Conference, Honolulu, HI, USA, 27 July–1 August 1997; Volume 2, pp. 894–898.

17. Rodatz, P.; Garcia, O.; Guzzella, L.; Büchi, F.; Bärtschi, M.; Tsukada, A.; Dietrich, P.; Kötz, R.; Scherer, G.; Wokaun, A. Performance and operational characteristics of a hybrid vehicle powered by fuel cells and supercapacitors. In Proceedings of the SAE 2003 World Congress and Exhibition, Detroit, MI, USA, 3 March 2003; Volume 112, pp. 692–703.

18. Thounthong, P.; Raël, S.; Davat, B. Utilizing fuel cell and supercapacitors for automotive hybrid electrical system. In Proceedings of the Applied Power Electronics Conference and Exposition, Austin, TX, USA, 6–10 March 2005; Volume 1, pp. 90–96.

19. Why the Automotive Future Will Be Dominated by Fuel Cells—IEEE Spectrum. Available online: http://spectrum.ieee.org/green-tech/fuel-cells/why-the-automotive-future-will-be-dominated-by-fuel-cells (accessed on 8 May 2017).

20. Rose, R. *Questions and Answers about Hydrogen and Fuel Cells; Report Style*; U.S. Department of Energy: Washington, DC, USA, 2005.

21. *U.S. Climate Technology Program: Technology Options for the Near and Long Term (Report Style)*; U.S. Climate Change Technology Program: Washington, DC, USA, 2005.

22. Thomas, C.E. Fuel cell and battery electric vehicles compared. *Int. J. Hydrogen Energy* **2009**, *34*, 6005–6020. [CrossRef]

23. Rajashekara, K. Present status and future trends in electric vehicle propulsion technologies. *IEEE J. Emerg. Sel. Top. Power Electron.* **2013**, *1*, 3–10. [CrossRef]

24. Model S | Tesla. Available online: https://www.tesla.com/models (accessed on 8 May 2017).

25. Tahami, F.; Kazemi, R.; Farhanghi, S. A novel driver assist stability system for all-wheel-drive electric vehicles. *IEEE Trans. Veh. Technol.* **2003**, *52*, 683–692. [CrossRef]

26. Sato, M.; Yamamoto, G.; Gunji, D.; Imura, T.; Fujimoto, H. Development of Wireless In-Wheel Motor Using Magnetic Resonance Coupling. *IEEE Trans. Power Electron.* **2016**, *31*, 5270–5278. [CrossRef]

27. Kurs, A.; Karalis, A.; Moffatt, R.; Joannopoulos, J.D.; Fisher, P.; Soljačić, M. Wireless power transfer via strongly coupled magnetic resonances. *Science* **2007**, *317*, 83–86. [CrossRef] [PubMed]

28. Imura, I.; Uchida, T.; Hori, Y. Flexibility of contactless power transfer using magnetic resonance coupling to air gap and misalignment for EV. *World Electr. Veh. J.* **2009**, *3*, 24–34.

29. Nakadachi, S.; Mochizuki, S.; Sakaino, S.; Kaneko, Y.; Abe, S.; Yasuda, T. Bidirectional contactless power transfer system expandable from unidirectional system. In Proceedings of the 2013 IEEE Energy Conversion Congress and Exposition, Denver, CO, USA, 15–19 September 2013; pp. 3651–3657.

30. Gao, Y.; Ehsani, M.; Miller, J.M. Hybrid Electric Vehicle: Overview and State of the Art. In Proceedings of the IEEE International Symposium on Industrial Electronics, Dubrovnik, Croatia, 20-23 June 2005; pp. 307–316.

31. Kim, H.; Kum, D. Comprehensive Design Methodology of Input- and Output-Split Hybrid Electric Vehicles: In Search of Optimal Configuration. *IEEE/ASME Trans. Mechatron.* **2016**, *21*, 2912–2923. [CrossRef]

32. Miller, J.M. Hybrid electric vehicle propulsion system architectures of the e-CVT type. *IEEE Trans. Power Electron.* **2006**, *21*, 756–767. [CrossRef]

33. Kim, D.; Hwang, S.; Kim, H. Vehicle Stability Enhancement of Four-Wheel-Drive Hybrid Electric Vehicle Using Rear Motor Control. *IEEE Trans. Veh. Technol.* **2008**, *57*, 727–735.

34. Li, Y.; Yang, J.; Song, J. Nano energy system model and nanoscale effect of graphene battery in renewable energy electric vehicle. *Renew. Sustain. Energy Rev.* **2017**, *69*, 652–663. [CrossRef]

35. Khaligh, A.; Li, Z. Battery, ultracapacitor, fuel cell, and hybrid energy storage systems for electric, hybrid electric, fuel cell, and plug-in hybrid electric vehicles: State of the art. *IEEE Trans. Veh. Technol.* **2010**, *59*, 2806–2814. [CrossRef]

36. Olson, J.B.; Sexton, E.D. Operation of lead–acid batteries for HEV applications. In Proceedings of the 15th Battery Conference on Applications and Advances, Long Beach, CA, USA, 11–14 January 2000; pp. 205–210.

37. Edwards, D.B.; Kinney, C. Advanced lead acid battery designs for hybrid electric vehicles. In Proceedings of the 16th Battery Conference on Applications and Advances, Long Beach, CA, USA, 12 January 2001; pp. 207–212.

38. Cooper, A.; Moseley, P. Progress in the development of lead–acid batteries for hybrid electric vehicles. In Proceedings of the IEEE Vehicle Power and Propulsion Conference, Windsor, UK, 6–8 September 2006; pp. 1–6.

39. Fetcenko, M.A.; Fetcenko, M.A.; Ovshinsky, S.R.; Reichman, B.; Young, K.; Fierro, C.; Koch, J.; Zallen, A.; Mays, W.; Ouchi, T. Recent advances in NiMH battery technology. *J. Power Sources* **2007**, *165*, 544–551. [CrossRef]

40. Li, H.; Liao, C.; Wang, L. Research on state-of-charge estimation of battery pack used on hybrid electric vehicle. In Proceedings of the Asia-Pacific Power and Energy Engineering Conference, Wuhan, China, 27–31 March 2009; pp. 1–4.

41. Chalk, S.G.; Miller, J.F. Key challenges and recent progress in batteries, fuel cells, and hydrogen storage for clean energy systems. *J. Power Sources* **2006**, *159*, 73–80. [CrossRef]

42. Balch, R.C.; Burke, A.; Frank, A.A. The affect of battery pack technology and size choices on hybrid electric vehicle performance and fuel economy. In Proceedings of the 16th IEEE Annual Battery Conference on Applications and Advances, Long Beach, CA, USA, 12 January 2001; pp. 31–36.

43. Viera, J.C.; Gonzalez, M.; Anton, J.C.; Campo, J.C.; Ferrero, F.J.; Valledor, M. NiMH vs. NiCd batteries under high charging rates. In Proceedings of the 28th Annual Telecommunications Energy Conference, Providence, RI, USA, 10–14 September 2006; pp. 1–6.

44. Gao, Y.; Ehsani, M. Investigation of battery technologies for the army's hybrid vehicle application. In Proceedings of the 56th IEEE Vehicular Technology Conference, Vancouver, BC, Canada, 24–28 September 2002; pp. 1505–1509.

45. Pilot, C. The Rechargeable Battery Market and Main Trends 2014–2025. Available online: http://www.avicenne.com/pdf/Fort_Lauderdale_Tutorial_C_Pillot_March2015.pdf (accessed on 29 July 2017).

46. Williamson, S.S.; Rathore, A.K.; Musavi, F. Industrial electronics for electric transportation: Current state-of-the-art and future challenges. *IEEE Trans. Ind. Electron.* **2015**, *62*, 3021–3032. [CrossRef]

47. Cassani, P.A.; Williamson, S.S. Feasibility analysis of a novel cell equalizer topology for plug-in hybrid electric vehicle energy-storage systems. *IEEE Trans. Veh. Technol.* **2009**, *58*, 3938–3946. [CrossRef]

48. Baughman, A.C.; Ferdowsi, M. Double-tiered switched-capacitor battery charge equalization technique. *IEEE Trans. Ind. Electron.* **2008**, *55*, 2277–2285. [CrossRef]

49. Nishijima, K.; Sakamoto, H.; Harada, K. A PWM controlled simple and high performance battery balancing system. In Proceedings of the IEEE Power Electronics Specialists Conference, Galway, Ireland, 23 June 2000; Volume 1, pp. 517–520.

50. Cassani, P.A.; Williamson, S.S. Design, testing, and validation of a simplified control scheme for a novel plug-in hybrid electric vehicle battery cell equalizer. *IEEE Trans. Ind. Electron.* **2010**, *57*, 3956–3962. [CrossRef]

51. Lee, Y.S.; Cheng, M.W. Intelligent control battery equalization for series connected lithium-ion battery strings. *IEEE Trans. Ind. Electron.* **2005**, *52*, 1297–1307. [CrossRef]

52. Lee, Y.S.; Cheng, M.W.; Yang, S.C.; Hsu, C.L. Individual cell equalization for series connected lithium-ion batteries. *IEICE Trans. Commun.* **2006**, *E89-B*, 2596–2607. [CrossRef]

53. 2017 Nissan LEAF® Electric Car Specs. Available online: https://www.nissanusa.com/electric-cars/leaf/versions-specs/ (accessed on 8 May 2017).

54. Model S Specifications | Tesla. Available online: https://www.tesla.com/support/model-s-specifications (accessed on 8 May 2017).

55. Why We Still Don't Have Better Batteries—MIT Technology Review. Available online: https://www.technologyreview.com/s/602245/why-we-still-dont-have-better-batteries/ (accessed on 8 May 2017).

56. Ribeiro, P.F.; Johnson, B.K.; Crow, M.L.; Arsoy, A.; Liu, Y. Energy storage systems for advanced power applications. *Proc. IEEE* **2001**, *89*, 1744–1756. [CrossRef]

57. Bartley, T. Ultracapacitors and batteries for energy storage in heavy-duty hybrid-electric vehicles. In Proceedings of the 22nd International Battery Seminar & Exhibit, Fort Lauderdale, FL, USA, 14–17 March 2005.

58. Gigaom | How Ultracapacitors Work (and Why They Fall Short). Available online: https://gigaom.com/2011/07/12/how-ultracapacitors-work-and-why-they-fall-short/ (accessed on 8 May 2017).

59. Singh, A.; Karandikar, P.B. A broad review on desulfation of lead-acid battery for electric hybrid vehicle. *Microsyst. Technol.* **2017**, *23*, 1–11. [CrossRef]

60. Chiu, H.J.; Lin, L.W. A bidirectional DC-DC converter for fuel cell electric vehicle driving system. *IEEE Trans. Power Electron.* **2006**, *21*, 950–958. [CrossRef]

61. Mahlia, T.M.I.; Saktisahdan, T.J.; Jannifar, A.; Hasan, M.H.; Matseelar, H.S.C. A review of available methods and development on energy storage; technology update. *Renew. Sustain. Energy Rev.* **2014**, *33*, 532–545. [CrossRef]

62. Bolund, B.; Bernhoff, H.; Leijon, M. Flywheel energy and power storage systems. *Renew. Sustain. Energy Rev.* **2007**, *11*, 235–258. [CrossRef]

63. Luo, X.; Wang, J.; Dooner, M.; Clarke, J. Overview of current development in electrical energy storage technologies and the application potential in power system operation. *Appl. Energy* **2015**, *137*, 511–536. [CrossRef]

64. Chan, C.C.; Chau, K.T. An overview of power electronics in electric vehicles. *IEEE Trans. Ind. Electron.* **1997**, *44*, 3–13. [CrossRef]

65. Chan, C.C.; Chau, K.T.; Jiang, J.Z.; Xia, W.A.X.W.; Zhu, M.; Zhang, R. Novel permanent magnet motor drives for electric vehicles. *IEEE Trans. Ind. Electron.* **1996**, *43*, 331–339. [CrossRef]

66. Chan, C.C.; Chau, K.T.; Yao, J. Soft-switching vector control for resonant snubber based inverters. In Proceedings of the IEEE International Conference Industrial Electronics, New Orleans, LA, USA, 14 November 1997; pp. 605–610.

67. Chan, C.C.; Jiang, J.Z.; Chen, G.H.; Chau, K.T. Computer simulation and analysis of a new polyphase multipole motor drive. *IEEE Trans. Ind. Electron.* **1993**, *40*, 570–576. [CrossRef]

68. Chan, C.C.; Jiang, J.Z.; Chen, G.H.; Wang, X.Y.; Chau, K.T. A novel polyphase multipole square-wave permanent magnet motor drive for electric vehicles. *IEEE Trans. Ind. Appl.* **1994**, *30*, 1258–1266. [CrossRef]

69. Chan, C.C.; Jiang, J.Z.; Xia, W.; Chan, K.T. Novel wide range speed control of permanent magnet brushless motor drives. *IEEE Trans. Power Electron.* **1995**, *10*, 539–546. [CrossRef]

70. Jose, C.P.; Meikandasivam, S. A Review on the Trends and Developments in Hybrid Electric Vehicles. In *Innovative Design and Development Practices in Aerospace and Automotive Engineering*; Springer: Singapore, 2017; pp. 211–229.

71. Chan, C.C.; Chau, K.T. *Morden Elcetric Vehicle Technology*; Oxford University Press, Inc.: New York, NY, USA, 2001; pp. 122–133.

72. Lulhe, A.M.; Date, T.N. A technology review paper for drives used in electrical vehicle (EV) & hybrid electrical vehicles (HEV). In Proceedings of the 2015 International Conference on Control, Instrumentation, Communication and Computational Technologies (ICCICCT), Kumaracoil, India, 18–19 December 2015.

73. Magnussen, F. On design and analysis of synchronous permanent magnet for field—Weakening operation. Ph.D. Thesis, KTH Royal Institute of Technology, Sweden, 2004.

74. Model X Specifications | Tesla. Available online: https://www.tesla.com/support/model-x-specifications (accessed on 8 May 2017).

75. Yamada, K.; Watanabe, K.; Kodama, T.; Matsuda, I.; Kobayashi, T. An efficiency maximizing induction motor drive system for transmissionless electric vehicle. In Proceedings of the 13th International Electric Vehicle Symposium, Osaka, Japan, 13–16 Octobor 1996; Volume II, pp. 529–536.

76. Boglietti, A.; Ferraris, P.; Lazzari, M.; Profumo, F. A new design criteria for spindles induction motors controlled by field oriented technique. *Electr. Mach. Power Syst.* **1993**, *21*, 171–182. [CrossRef]

77. Abbasian, M.; Moallem, M.; Fahimi, B. Double-stator switched reluctance machines (DSSRM): Fundamentals and magnetic force analysis. *IEEE Trans. Energy Convers.* **2010**, *25*, 589–597. [CrossRef]

78. Cameron, D.E.; Lang, J.H.; Umans, S.D. The origin and reduction of acoustic noise in doubly salient variable-reluctance motors. *IEEE Trans. Ind. Appl.* **1992**, *28*, 1250–1255. [CrossRef]

79. Chan, C.C.; Jiang, Q.; Zhan, Y.J.; Chau, K.T. A high-performance switched reluctance drive for P-star EV project. In Proceedings of the 13th International Electric Vehicle Symposium, Osaka, Japan, 13–16 Octobor 1996; Volume II, pp. 78–83.

80. Zhan, Y.J.; Chan, C.C.; Chau, K.T. A novel sliding-mode observer for indirect position sensing of switched reluctance motor drives. *IEEE Trans. Ind. Electron.* **1999**, *46*, 390–397. [CrossRef]

81. Shareef, H.; Islam, M.M.; Mohamed, A. A review of the stage-of-the-art charging technologies, placement methodologies, and impacts of electric vehicles. *Renew. Sustain. Energy Rev.* **2016**, *64*, 403–420. [CrossRef]

82. Yu, X.E.; Xue, Y.; Sirouspour, S.; Emadi, A. Microgrid and transportation electrification: A review. In Proceedings of the 2012 IEEE Transportation Electrification Conference and Expo (ITEC), Dearborn, MI, USA, 18–20 June 2012.

83. Consumer and Clinical Radiation Protection Bureau; Environmental and Radiation Health Sciences Directorate; Healthy Environments and Consumer Safety Branch; Health Canada. Limits of human exposure to radiofrequency electromagnetic energy in the frequency range from 3 kHz to 300 GHz. *Health Can. Saf. Code* **2009**, *6*, 10–11.

84. *IEEE Standard for Safety Levels with Respect to Human Exposure to Radio Frequency Electromagnetic Fields, 3 kHz to 300 GHz*; IEEE Std C95.1; IEEE: New York, NY, USA, 1999.

85. Ahlbom, A.; Bergqvist, U.; Bernhardt, J.H.; Cesarini, J.P.; Court, L.A.; Grandolfo, M.; Hietanen, M.; McKinlay, A.F.; Repacholi, M.H.; Sliney, D.H. Guidelines: For limiting exposure to time-varying electric, magnetic and electromagnetic fields (up to 300 GHz). *Health Phys.* **1998**, *74*, 494–521.

86. Australian Radiation Protection and Nuclear Safety Agency (ARPANSA). *Radiation Protection Standard: Maximum Exposure Levels to Radiofrequency Fields—3 kHz to 300 GHz*; Radiation Protection Series Publication No. 3; ARPANSA: Melbourne, Australia, 2002.

87. Musavi, F.; Eberle, W. Overview of wireless power transfer technologies for electric vehicle battery charging. *IET Power Electron.* **2014**, *7*, 60–66. [CrossRef]

88. Chademo-Ceritifed Chrager List. Available online: www.chademo.com (accessed on 6 July 2015).

89. Supercharger. Available online: www.chademo.com (accessed on 7 July 2015).

90. International Electrotechnical Commission. *Standard IEC 62196—Plugs, Socket-Outlets, Vehicle Couplers and Vehicle Inlets—Conductive Charging of Electric Vehicles*; The International Electrotechnical Commission (IEC): Geneva, Switzerland, 2003.

91. Onar, O.C.; Kobayashi, J.; Khaligh, A. A Fully Directional Universal Power Electronic Interface for EV, HEV, and PHEV Applications. *IEEE Trans. Power Electron.* **2013**, *28*, 5489–5498. [CrossRef]

92. Bose, B.K. Power electronics—A technology review. *Proc. IEEE* **1992**, *80*, 1303–1334. [CrossRef]

93. Yaramasu, V.; Wu, B.; Sen, P.C.; Kouro, S.; Narimani, M. High-power wind energy conversion systems: State-of-the-art and emerging technologies. *Proc. IEEE* **2015**, *103*, 740–788. [CrossRef]

94. Kok, D.; Morris, A.; Knowles, M. Novel EV drive train topology—A review of the current topologies and proposal for a model for improved drivability. In Proceedings of the 2013 15th European Conference on Power Electronics and Applications (EPE), Lille, France, 2–6 September 2013.

95. Hegazy, O.; Van Mierlo, J.; Lataire, P. Analysis, control and comparison of DC/DC boost converter topologies for fuel cell hybrid electric vehicle applications. In Proceedings of the 14th European Conference on Power Electronics and Applications (EPE 2011), Birmingham, UK, 30 August–1 September 2011.

96. Hong, J.; Lee, H.; Nam, K. Charging Method for the Secondary Battery in Dual-Inverter Drive Systems for Electric Vehicles. *IEEE Trans. Power Electron.* **2015**, *30*, 909–921. [CrossRef]

97. Sangdehi, S.M.M.; Hamidifar, S.; Kar, N.C. A novel bidirectional DC/AC stacked matrix converter design for electrified vehicle applications. *IEEE Trans. Veh. Technol.* **2014**, *63*, 3038–3050. [CrossRef]

98. Kim, Y.J.; Lee, J.Y. Full-Bridge+ SRT Hybrid DC/DC Converter for a 6.6-kW EV On-Board Charger. *IEEE Trans. Veh. Technol.* **2016**, *65*, 4419–4428. [CrossRef]

99. Kimura, S.; Itoh, Y.; Martinez, W.; Yamamoto, M.; Imaoka, J. Downsizing Effects of Integrated Magnetic Components in High Power Density DC–DC Converters for EV and HEV Applications. *IEEE Trans. Ind. Appl.* **2016**, *52*, 3294–3305. [CrossRef]

100. Schroeder, J.C.; Fuchs, F.W. Detailed Characterization of Coupled Inductors in Interleaved Converters Regarding the Demand for Additional Filtering. In Proceedings of the 2012 IEEE Energy Energy Conversion Congress and Exposition (ECCE), Raleigh, NC, USA, 15–20 September 2012; pp. 759–766.

101. Imaoka, J.; Yamamoto, M.; Nakamura, Y.; Kawashima, T. Analysis of output capacitor voltage ripple in multi-phase transformer-linked boost chopper circuit. *IEEE J. Ind. Appl.* **2013**, *2*, 252–260. [CrossRef]

102. Zhu, J.; Pratt, A. Capacitor Ripple Current in an interleaved PFC Converter. *IEEE Trans. Power Electron.* **2009**, *24*, 1506–1514. [CrossRef]

103. Wang, C.; Xu, M.; Lee, F.C.; Lu, B. EMI Study for the Interleaved Multi-Channel PFC. In Proceedings of the IEEE Power Electronics Specialists Conference (PESC), Orlando, FL, USA, 17–21 June 2007; pp. 1336–1342.

104. O'Loughlin, M. An Interleaved PFC Preregulator for High-Power Converters. Available online: http://www.ti.com/download/trng/docs/seminar/Topic5MO.pdf (accessed on 7 August 2017).

105. Balogh, L.; Redl, R. Power-factor correction with interleaved boost converters in continuous-inductor-current mode. In Proceedings of the IEEE Applied Power Electronics Conference and Exposition, San Diego, CA, USA, 7–11 March 1993; pp. 168–174.

106. Jang, Y.; Jovanovic, M.M. Interleaved boost converter with intrinsic voltage-doubler characteristic for universal-line PFC front end. *IEEE Trans. Power Electron.* **2007**, *22*, 1394–1401. [CrossRef]

107. Kong, P.; Wang, S.; Lee, F.C.; Wang, C. Common-mode EMI study and reduction technique for the interleaved multichannel PFC converter. *IEEE Trans. Power Electron.* **2008**, *23*, 2576–2584. [CrossRef]

108. Gautam, D.S.; Musavi, F.; Eberle, W.; Dunford, W.G. A zero voltage switching full-bridge dc-dc converter with capacitive output filter for a plug-in-hybrid electric vehicle battery charger. In Proceedings of the IEEE Applied Power Electronics Conference and Exposition, Orlando, FL, USA, 5–9 February 2012; pp. 1381–1386.

109. Musavi, F.; Craciun, M.; Gautam, D.S.; Eberle, W.; Dunford, W.G. An LLC resonant DC-DC Converter for wide output voltage range battery charging applications. *IEEE Trans. Power Electron.* **2013**, *28*, 5437–5445. [CrossRef]

110. Gurkaynak, Y.; Li, Z.; Khaligh, A. A novel grid-tied, solar powered residential home with plug-in hybrid electric vehicle (PHEV) loads. In Proceedings of the 5th Annual IEEE Vehicle Power and Propulsion Conference, Dearborn, MI, USA, 7–10 September 2009; pp. 813–816.

111. Onar, O. *Bi-Directional Rectifier/Inverter and Bi-Directional DC/DC Converters Integration for Plug-in Hybrid Electric Vehicles with Hybrid Battery/Ultra-Capacitors Energy Storage Systems*; Illinois Institute of Technology: Chicago, IL, USA, 2009.

112. Rashid, M.H. *Power Electronics Handbook: Devices, Circuits and Applications*; Elsevier: Amsterdam, The Netherlands, 2010.

113. Koushki, B.; Safaee, A.; Jain, P.; Bakhshai, A. Review and comparison of bi-directional AC-DC converters with V2G capability for on-board EV and HEV. In Proceedings of the 2014 IEEE Transportation Electrification Conference and Expo (ITEC), Dearborn, MI, USA, 15–18 June 2014.

114. Yao, L.; Lim, W.H.; Tsai, T.S. A Real-Time Charging Scheme for Demand Response in Electric Vehicle Parking Station. *IEEE Trans. Smart Grid* **2017**, *8*, 52–62. [CrossRef]

115. Kütt, L.; Saarijärvi, E.; Lehtonen, M.; Mõlder, H.; Niitsoo, J. A review of the harmonic and unbalance effects in electrical distribution networks due to EV charging. In Proceedings of the 2013 12th International Conference on Environment and Electrical Engineering (EEEIC), Wroclaw, Poland, 5–8 May 2013.

116. Richardson, P.; Flynn, D.; Keane, A. Optimal charging of electric vehicles in low-voltage distribution systems. *IEEE Trans. Power Syst.* **2012**, *27*, 268–279. [CrossRef]

117. Mwasilu, F.; Justo, J.J.; Kim, E.K.; Do, T.D.; Jung, J.W. Electric vehicles and smart grid interaction: A review on vehicle to grid and renewable energy sources integration. *Renew. Sustain. Energy Rev.* **2014**, *34*, 501–516. [CrossRef]

118. Green, R.C.; Wang, L.; Alam, M. The impact of plug-in hybrid electric vehicles on distribution networks: A review and outlook. *Renew. Sustain. Energy Rev.* **2011**, *15*, 544–553. [CrossRef]

119. Qian, K.; Zhou, C.; Allan, M.; Yuan, Y. Modeling of load demand due to EV battery charging in distribution systems. *IEEE Trans. Power Syst.* **2011**, *26*, 802–810. [CrossRef]

120. Geng, B.; Mills, J.K.; Sun, D. Two-stage charging strategy for plug-in electric vehicles at the residential transformer level. *IEEE Trans. Smart Grid* **2013**, *4*, 1442–1452. [CrossRef]

121. Fernandez, L.P.; San Román, T.G.; Cossent, R.; Domingo, C.M.; Frias, P. Assessment of the impact of plug-in electric vehicles on distribution networks. *IEEE Trans. Power Syst.* **2011**, *26*, 206–213. [CrossRef]

122. Gómez, J.C.; Morcos, M.M. Impact of EV battery chargers on the power quality of distribution systems. *IEEE Trans. Power Deliv.* **2003**, *18*, 975–981. [CrossRef]

123. Dharmakeerthi, C.H.; Mithulananthan, N.; Saha, T.K. Impact of electric vehicle fast charging on power system voltage stability. *Int. J. Electr. Power Energy Syst.* **2014**, *57*, 241–249. [CrossRef]

124. Mitra, P.; Venayagamoorthy, G.K. Wide area control for improving stability of a power system with plug-in electric vehicles. *IET Gener Transm. Distrib.* **2010**, *4*, 1151–1163. [CrossRef]

125. Rajakaruna, S.; Shahnia, F.; Ghosh, A. *Plug in Electric Vehicles in Smart Grids*, 1st ed.; Springer Science and Business Media Singapore Pte Ltd.: Singapore, 2015.

126. Akhavan-Rezai, E.; Shaaban, M.F.; El-Saadany, E.F.; Zidan, A. Uncoordinated charging impacts of electric vehicles on electric distribution grids: Normal and fast charging comparison. In Proceedings of the IEEE Power and Energy Society General Meeting, San Diego, CA, USA, 22–26 July 2012; pp. 1–7.

127. Jiang, C.; Torquato, R.; Salles, D.; Xu, W. Method to assess the power-quality impact of plug-in electric vehicles. *IEEE Trans. Power Deliv.* **2014**, *29*, 958–965.

128. Desmet, J.J.M.; Sweertvaegher, I.; Vanalme, G.; Stockman, K.; Belmans, R.J.M. Analysis of the neutral conductor current in a three- phase supplied network with nonlinear single-phase loads. *IEEE Trans. Ind. Appl.* **2003**, *39*, 587–593. [CrossRef]

129. Nguyen, V.L.; Tuan, T.Q.; Bacha, S. Harmonic distortion mitigation for electric vehicle fast charging systems. In Proceedings of the 2013 IEEE Grenoble PowerTech (POWERTECH), Grenoble, France, 16–20 June 2013; pp. 1–6.

130. Melo, N.; Mira, F.; De Almeida, A.; Delgado, J. Integration of PEV in Portuguese distribution grid: Analysis of harmonic current emissions in charging points. In Proceedings of the International Conference on Electrical Power Quality and Utilization, Lisbon, Portugal, 17–19 October 2011; pp. 791–796.

131. Zamri, M.; Wanik, C.; Siam, M.F.; Ayob, A.; Mohamed, A.; Hanifahazit, A.; Sulaiman, S.; Ali, M.A.M.; Hussein, Z.F.; MatHussin, A.K. Harmonic measurement and analysis during electric vehicle charging. *Engineering* **2013**, *5*, 215–220.

132. Bentley, E.C.; Suwanapingkarl, P.; Weerasinghe, S.; Jiang, T.; Putrus, G.A.; Johnston, D. The interactive effects of multiple EV chargers within a distribution network. In Proceedings of the IEEE Vehicle Power and Propulsion Conference (VPPC), Lille, France, 1–3 September 2010; pp. 1–6.

133. Staats, P.T.; Grady, W.M.; Arapostathis, A.; Thallam, R.S. A statistical analysis of the effect of electric vehicle battery charging on distribution system harmonic voltages. *IEEE Trans. Power Deliv.* **1998**, *13*, 640–646. [CrossRef]

134. Balcells, J.; García, J. Impact of plug-in electric vehicles on the supply grid. In Proceedings of the IEEE Vehicle Power and Propulsion Conference (VPPC), Lille, France, 1–3 September 2010; pp. 1–4.

135. Lee, S.J.; Kim, J.H.; Kim, D.U.; Go, H.S.; Kim, C.H.; Kim, E.S.; Kim, S.K. Evaluation of voltage sag and unbalance due to the system connection of electric vehicles on distribution system. *J. Electr. Eng. Technol.* **2014**, *9*, 452–460. [CrossRef]

136. Tie, C.H.; Gan, C.K.; Ibrahim, K.A. The impact of electric vehicle charging on a residential low voltage distribution network in Malaysia. In Proceedings of the 2014 IEEE Innovative Smart Grid Technologies—Asia (ISGT Asia), Kuala Lumpur, Malaysia, 20–23 May 2014; pp. 272–277.

137. Leemput, N.; Geth, F.; Van Roy, J.; Delnooz, A.; Buscher, J.; Driesen, J. Impact of electric vehicle on board single-phase charging strategies on a Flemish residential grid. *IEEE Trans. Smart Grid* **2014**, *5*, 1815–1822. [CrossRef]

138. Masoum, M.A.S.; Moses, P.S.; Deilami, S. Load management in smart grids considering harmonic distortion and transformer derating. In Proceedings of the IEEE Innovative Smart Grid Technologies Europe (ISGT Europe), Gaithersburg, MD, USA, 19–21 January 2010; pp. 1–7.

139. Nyns, K.C.; Haesen, E.; Driesen, J. The impact of charging plug-in hybrid electric vehicles on a residential distribution grid. *IEEE Trans. Power Syst.* **2010**, *25*, 371–380. [CrossRef]

140. Deilami, S.; Masoum, A.S.; Moses, P.S.; Masoum, M.A.S. Real-time coordination of plug-in electric vehicle charging in smart grids to minimize power losses and improve voltage profile. *IEEE Trans. Smart Grid* **2011**, *2*, 456–467. [CrossRef]

141. Sortomme, E.; Hindi, E.M.M.; MacPherson, S.D.J.; Venkata, S.S. Coordinated charging of plug-in hybrid electric vehicles to minimize distribution system losses. *IEEE Trans. Smart Grid* **2011**, *2*, 198–205. [CrossRef]

142. Sadeghi-Barzani, P.; Rajabi-Ghahnavieh, A.; Kazemi-Karegar, H. Optimal fast charging station placing and sizing. *Appl. Energy* **2014**, *125*, 289–299. [CrossRef]

143. Qian, K.; Zhou, C.; Yuan, Y. Impacts of high penetration level of fully electric vehicles charging loads on the thermal ageing of power transformers. *Int. J. Electr. Power Energy Syst.* **2015**, *65*, 102–112. [CrossRef]

144. Razeghi, G.; Zhang, L.; Brown, T.; Samuelsen, S. Impacts of plug-in hybrid electric vehicles on a residential transformer using stochastic and empirical analysis. *J. Power Sources* **2014**, *252*, 277–285. [CrossRef]

145. Elnozahy, M.S.; Salama, M.M. A comprehensive study of the impacts of PHEVs on residential distribution networks. *IEEE Trans. Sustain. Energy* **2014**, *5*, 332–342. [CrossRef]

146. Sortomme, E.; El-Sharkawi, M.A. Optimal charging strategies for unidirectional vehicle-to-grid. *IEEE Trans. Smart Grid* **2011**, *2*, 131–138. [CrossRef]

147. Clement-Nyns, K.; Haesen, E.; Driesen, J. The impact of vehicle-to-grid on the distribution grid. *Electr. Power Syst. Res.* **2011**, *81*, 185–192. [CrossRef]

148. Tulpule, P.; Marano, V.; Yurkovich, S.; Rizzoni, G. Economic and environmental impacts of a PV powered workplace parking garage charging station. *Appl. Energy* **2013**, *108*, 323–332. [CrossRef]

149. Birnie, D.P. Solar-to-vehicle (S2V) systems for powering commuters of the future. *J. Power Sources* **2009**, *186*, 539–542. [CrossRef]

150. Derakhshandeh, S.Y.; Masoum, A.S.; Deilami, S.; Masoum, M.A.; Golshan, M.H. Coordination of generation scheduling with PEVs charging in industrial microgrids. *IEEE Trans. Power Syst.* **2013**, *28*, 3451–3461. [CrossRef]

151. Pillai, R.J.; Heussen, K.; Østergaard, P.A. Comparative analysis of hourly and dynamic power balancing models for validating future energy scenarios. *Energy* **2011**, *36*, 3233–3243. [CrossRef]

152. Borba, B.S.M.; Szklo, A.; Schaeffer, R. Plug-in hybrid electric vehicles as a way to maximize the integration of variable renewable energy in power systems: The case of wind generation in northeastern Brazil. *Energy* **2012**, *37*, 469–481. [CrossRef]

153. Wu, T.; Yang, Q.; Bao, Z.; Yan, W. Coordinated energy dispatching in microgrid with wind power generation and plug-in electric vehicles. *IEEE Trans. Smart Grid* **2013**, *4*, 1453–1463. [CrossRef]

154. Liu, C.; Wang, J.; Botterud, A.; Zhou, Y.; Vyas, A. Assessment of impacts of PHEV charging patterns on wind-thermal scheduling by stochastic unit commitment. *IEEE Trans. Smart Grid* **2012**, *3*, 675–683. [CrossRef]

155. Ma, H.; Balthser, F.; Tait, N.; Riera-Palou, X.; Harrison, A. A new comparison between the life cycle greenhouse gas emissions of battery electric vehicles and internal combustion vehicles. *Energy Policy* **2012**, *44*, 160–173. [CrossRef]

156. Sioshansi, R.; Miller, J. Plug-in hybrid electric vehicles can be clean and economical in dirty power systems. *Energy Policy* **2011**, *39*, 6151–6161. [CrossRef]

157. Donateo, T.; Ingrosso, F.; Licci, F.; Laforgia, D. A method to estimate the environmental impact of an electric city car during six months of testing in an Italian city. *J. Power Sources* **2014**, *270*, 487–498. [CrossRef]

158. Onat, N.C.; Kucukvar, M.; Tatari, O. Conventional, hybrid, plug-in hybrid or electric vehicles? State-based comparative carbon and energy footprint analysis in the United States. *Appl. Energy* **2015**, *150*, 36–49. [CrossRef]

159. Jorgensen, K. Technologies for electric, hybrid and hydrogen vehicles: Electricity from renewable energy sources in transport. *Util. Policy* **2008**, *16*, 72–79. [CrossRef]

160. Kempton, W.; Letendrem, S. Electric vehicles as a new power source for electric utilities. *Transp. Res. Part D* **1997**, *2*, 157–175. [CrossRef]

161. Peterson, S.; Whitacre, J.; Apt, J. The economics of using plug-in hybrid electric vehicles battery packs for grid storage. *J. Power Sources* **2010**, *195*, 2377–2384. [CrossRef]

162. Sioshansi, R.; Denholm, P. The value of plug-in hybrid electric vehicles as grid resources. *Energy J.* **2010**, *31*, 1–16. [CrossRef]

163. Formula, E. Available online: http://www.fiaformulae.com/en (accessed on 8 May 2017).

164. Hidrue, M.K.; Parsons, G.R.; Kempton, W.; Gardner, M.P. Willingness to pay for electric vehicles and their attributes. *Resour. Energy Econ.* **2011**, *33*, 686–705. [CrossRef]

165. 2017 Bolt EV: All-Electric Vehicle | Chevrolet. Available online: http://www.chevrolet.com/bolt-ev-electric-vehicle.html (accessed on 8 May 2017).

166. Hydrogen Fuel Cell Car | Toyota Mirai. Available online: https://ssl.toyota.com/mirai/fcv.html (accessed on 8 May 2017).

167. Wolsink, M. The research agenda on social acceptance of distributed generation in smart grids: Renewable as common pool resources. *Renew. Sustain. Energy Rev.* **2012**, *16*, 822–835. [CrossRef]

168. Kang, J.E.; Brown, T.; Recker, W.W.; Samuelsen, G.S. Refueling hydrogen fuel cell vehicles with 68 proposed refueling stations in California: Measuring deviations from daily travel patterns. *Int. J. Hydrogen Energy* **2014**, *39*, 3444–3449. [CrossRef]

169. Tie, S.F.; Tan, C.W. A review of energy sources and energy management system in electric vehicles. *Renew. Sustain. Energy Rev.* **2013**, *20*, 82–102. [CrossRef]

170. Chan, C.C.; Wong, Y.S. Electric vehicles charge forward. *IEEE Power Energy Mag.* **2004**, *2*, 24–33. [CrossRef]

171. Five Slippery Cars Enter a Wind Tunnel, One Slinks Out a Winner. Available online: https://www.tesla.com/sites/default/files/blog_attachments/the-slipperiest-car-on-the-road.pdf (accessed on 8 May 2017).

172. Geng, B.; Mills, J.K.; Sun, D. Two-stage energy management control of fuel cell plug-in hybrid electric vehicles considering fuel cell longevity. *IEEE Trans. Veh. Technol.* **2012**, *61*, 498–508. [CrossRef]

173. Murphey, Y.L.; Park, J.; Chen, Z.; Kuang, M.L.; Masrur, M.A.; Phillips, A.M. Intelligent hybrid vehicle power control—Part I: Machine learning of optimal vehicle power. *IEEE Trans. Veh. Technol.* **2012**, *61*, 3519–3530. [CrossRef]

174. Geng, B.; Mills, J.K.; Sun, D. Energy management control of microturbine powered plug-in hybrid electric vehicles using telemetry equivalent consumption minimization strategy. *IEEE Trans. Veh. Technol.* **2011**, *60*, 4238–4248. [CrossRef]

175. Moura, S.J.; Fathy, H.K.; Callaway, D.S.; Stein, J.L. A stochastic optimal control approach for power management in plug-in hybrid electric vehicles. *IEEE Trans. Control Syst. Technol.* **2011**, *19*, 545–555. [CrossRef]

176. Magallan, G.A.; De Angelo, C.H.; Garcia, G.O. Maximization of the traction forces in a 2WD electric vehicle. *IEEE Trans. Veh. Technol.* **2011**, *60*, 369–380. [CrossRef]

177. Kang, J.; Yoo, J.; Yi, K. Driving control algorithm for maneuverability, lateral stability, and rollover prevention of 4WD electric vehicles with independently driven front and rear wheels. *IEEE Trans. Veh. Technol.* **2011**, *60*, 2987–3001. [CrossRef]

178. Wang, J.N.; Wang, Q.N.; Jin, L.Q.; Song, C.X. Independent wheel torque control of 4WD electric vehicle for differential drive assisted steering. *Mechatronics* **2011**, *21*, 63–76. [CrossRef]

179. Nam, K.; Fujimoto, H.; Hori, Y. Lateral stability control of in-wheel-motor-driven electric vehicles based on sideslip angle estimation using lateral tire force sensors. *IEEE Trans. Veh. Technol.* **2012**, *61*, 1972–1985.

180. Zhou, D.; Ravey, A.; Gao, F.; Miraoui, A.; Zhang, K. Online Estimation of Lithium Polymer Batteries State-of-Charge Using Particle Filter-Based Data Fusion with Multimodels Approach. *IEEE Trans. Ind. Appl.* **2016**, *52*, 2582–2595. [CrossRef]

181. Hui, S.; Lifu, Y.; Junqing, J.; Yanling, L. Control strategy of hydraulic/electric synergy system in heavy hybrid vehicles. *Energy Convers. Manag.* **2011**, *52*, 668–674. [CrossRef]

182. Chen, Z.; Mi, C.C.; Xiong, R.; Xu, J.; You, C. Energy management of a power-split plug-in hybrid electric vehicle based on genetic algorithm and quadratic programming. *J. Power Sources* **2014**, *248*, 416–426. [CrossRef]

183. Li, S.G.; Sharkh, S.M.; Walsh, F.C.; Zhang, C.N. Energy and battery management of a plug-in series hybrid electric vehicle using fuzzy logic. *IEEE Trans. Veh. Technol.* **2011**, *60*, 3571–3585. [CrossRef]

184. Yuan, Z.; Teng, L.; Fengchun, S.; Peng, H. Comparative study of dynamic programming and Pontryagin's minimum principle on energy management for a parallel hybrid electric vehicle. *Energies* **2013**, *6*, 2305–2318. [CrossRef]
185. Bernard, J.; Delprat, S.; Guerra, T.M.; Büchi, F.N. Fuel efficient power management strategy for fuel cell hybrid powertrains. *Control Eng. Pract.* **2010**, *18*, 408–417. [CrossRef]
186. Su, W.; Chow, M.Y. Performance evaluation of an EDA-based large-scale plug-in hybrid electric vehicle charging algorithm. *IEEE Trans. Smart Grid* **2012**, *3*, 308–315. [CrossRef]
187. Mohamed, A.; Salehi, V.; Ma, T.; Mohammed, O. Real-time energy management algorithm for plug-in hybrid electric vehicle charging parks involving sustainable energy. *IEEE Trans. Sustain. Energy* **2014**, *5*, 577–586. [CrossRef]
188. Worldwide EV Sales Are on The Move. Available online: http://evercharge.net/blog/infographic-worldwide-ev-sales-are-on-the-move/ (accessed on 8 May 2017).
189. China's Quota Threat Charges Up Electric Car Market | The Daily Star. Available online: http://www.thedailystar.net/business/chinas-quota-threat-charges-electric-car-market-1396066 (accessed on 8 May 2017).
190. EV-Volumes—The Electric Vehicle World Sales Database. Available online: http://www.ev-volumes.com/news/china-plug-in-sales-2016-q4-and-full-year/ (accessed on 8 May 2017).
191. EV-Volumes—The Electric Vehicle World Sales Database. Available online: http://www.ev-volumes.com/country/total-world-plug-in-vehicle-volumes/ (accessed on 8 May 2017).
192. EV-Volumes—The Electric Vehicle World Sales Database. Available online: http://www.ev-volumes.com/country/usa/ (accessed on 8 May 2017).
193. EAFO. Available online: http://www.eafo.eu/vehicle-statistics/m1 (accessed on 8 May 2017).
194. Hossain, E.; Kabalci, E.; Bayindir, R.; Perez, R. Microgrid testbeds around the world: State of art. *Energy Convers. Manag.* **2014**, *86*, 132–153. [CrossRef]
195. Bayindir, R.; Hossain, E.; Kabalci, E.; Perez, R. A comprehensive study on microgrid technology. *Int. J. Renew. Energy Res.* **2014**, *4*, 1094–1107.
196. eHighway—Siemens Global Website. Available online: https://www.siemens.com/press/en/feature/2015/mobility/2015-06-eHighway.php?content[]=MO (accessed on 8 May 2017).
197. Saponara, S.; Petri, E.; Fanucci, L.; Terreni, P. Sensor modeling, low-complexity fusion algorithms, and mixed-signal IC prototyping for gas measures in low-emission vehicles. *IEEE Trans. Instrum. Meas.* **2011**, *60*, 372–384. [CrossRef]
198. Saponara, S.; Pasetti, G.; Costantino, N.; Tinfena, F.; D'Abramo, P.; Fanucci, L. A flexible LED driver for automotive lighting applications: IC design and experimental characterization. *IEEE Trans. Power Electron.* **2012**, *27*, 1071–1075. [CrossRef]
199. Saponara, S.; Pasetti, G.; Tinfena, F.; Fanucci, L.; D'Abramo, P. HV-CMOS design and characterization of a smart rotor coil driver for automotive alternators. *IEEE Trans. Ind. Electron.* **2013**, *60*, 2309 2317. [CrossRef]
200. Baronti, F.; Lazzeri, A.; Roncella, R.; Saletti, R.; Saponara, S. Design and characterization of a robotized gearbox system based on voice coil actuators for a Formula SAE Race Car. *IEEE/ASME Trans. Mechatron.* **2013**, *18*, 53–61. [CrossRef]
201. Costantino, N.; Serventi, R.; Tinfena, F.; D'Abramo, P.; Chassard, P.; Tisserand, P.; Saponara, S.; Fanucci, L. Design and test of an HV-CMOS intelligent power switch with integrated protections and self-diagnostic for harsh automotive applications. *IEEE Trans. Ind. Electron.* **2011**, *58*, 2715–2727. [CrossRef]
202. Saponara, S.; Fanucci, L.; Bernardo, F.; Falciani, A. Predictive diagnosis of high-power transformer faults by networking vibration measuring nodes with integrated signal processing. *IEEE Trans. Instrum. Meas.* **2016**, *65*, 1749–1760. [CrossRef]
203. Abhishek, A.; Karthikeyan, V.; Sanjeevikumar, P.; Rajasekar, S.; Blaabjerg, F.; Asheesh, K.S. Optimal Planning of Electric Vehicle Charging Station at the Distribution System Using Hybrid Optimization Algorithm. *Energy* **2017**, *133*, 70–78.
204. Febin Daya, J.L.; Sanjeevikumar, P.; Blaabjerg, F.; Wheeler, P.; Ojo, O.; Ahmet, H.E. Analysis of Wavelet Controller for Robustness in Electronic Differential of Electric Vehicles—An Investigation and Numerical Implementation. *Electr. Power Compon. Syst.* **2016**, *44*, 763–773. [CrossRef]

205. Febin Daya, J.L.; Sanjeevikumar, P.; Blaabjerg, F.; Wheeler, P.; Ojo, O. Implementation of Wavelet Based Robust Differential Control for Electric Vehicle Application. *IEEE Trans. Power Electron.* **2015**, *30*, 6510–6513. [CrossRef]

206. Sanjeevikumar, P.; Febin Daya, J.L.; Blaabjerg, F.; Mir-Nasiri, N.; Ahmet, H.E. Numerical Implementation of Wavelet and Fuzzy Transform IFOC for Three-Phase Induction Motor. *Eng. Sci. Technol. Int. J.* **2016**, *19*, 96–100.

207. Dragonas, F.A.; Nerrati, G.; Sanjeevikumar, P.; Grandi, G. High-Voltage High-Frequency Arbitrary Waveform Multilevel Generator for DBD Plasma Actuators. *IEEE Trans. Ind. Appl.* **2015**, *51*, 3334–3342. [CrossRef]

208. Mohan, K.; Febin Daya, J.L.; Sanjeevikumar, P.; Mihet-Popa, L. Real-time Analysis of a Modified State Observer for Sensorless Induction Motor Drive used in Electric Vehicle Applications. *Energies* **2017**, *10*, 1077.

Review

Toward Green Vehicles Digitalization for the Next Generation of Connected and Electrified Transport Systems

Lucian Mihet-Popa [1,*] and Sergio Saponara [2,*]

[1] Faculty of Engineering, Østfold University College, Kobberslagstredet 5, 1671 Kråkeroy, Norway
[2] Dipartimento Ingegneria dell'Informazione (DII), University of Pisa, via G. Caruso 16, 56122 Pisa, Italy
* Correspondence: lucian.mihet@hiof.no (L.M.-P.); sergio.saponara@unipi.it (S.S.)

Received: 15 October 2018; Accepted: 8 November 2018; Published: 12 November 2018

Abstract: This survey paper reviews recent trends in green vehicle electrification and digitalization, as part of a special section on "Energy Storage Systems and Power Conversion Electronics for E-Transportation and Smart Grid", led by the authors. First, the energy demand and emissions of electric vehicles (EVs) are reviewed, including the analysis of the trends of battery technology and of the recharging issues considering the characteristics of the power grid. Solutions to integrate EV electricity demand in power grids are also proposed. Integrated electric/electronic (E/E) architectures for hybrid EVs (HEVs) and full EVs are discussed, detailing innovations emerging for all components (power converters, electric machines, batteries, and battery-management-systems). 48 V HEVs are emerging as the most promising solution for the short-term electrification of current vehicles based on internal combustion engines. The increased digitalization and connectivity of electrified cars is posing cyber-security issues that are discussed in detail, together with some countermeasures to mitigate them, thus tracing the path for future on-board computing and control platforms.

Keywords: green vehicles (GV); electrified vehicles (EV); smart grids; hybrid electric vehicles (HEV); integrated starter-generators

1. Introduction

Today, about 20% of the global primary energy usage and about 25% of energy-related emissions of CO_2, are due to the contribution of the transportation field [1]. Moreover, half of the emissions for transportation originate from passenger vehicles, mostly based on the internal combustion engine (ICE) propulsion [1]. One of the most promising technologies to solving the above issues is represented by electric vehicles (EVs), including hybrid EVs (HEVs), thanks to the cut of oil usage and of CO_2 emissions, on a per-km basis [2].

EVs will play a major role in meetings Europe's need for clean and efficiency mobility. The objectives set in the European Green Vehicles Initiative (EGVI) report [3] are quite ambitious; an overall efficiency improvement of the transport system by 50% in 20 years (i.e., in 2030 compared to 2010). The major EV car manufacturers have started the production and sale of EVs, with projections estimating one million EVs circulating in Europe by 2020, anticipating a significant expansion by 2030 and beyond. More than that, the sales of new electric vehicles worldwide exceeded one million cars in 2017 [4]. This market volume represents a huge growth in the sales of new electric cars, by 54% compared with 2016 [5]. In some countries, the market of electric cars is no more a niche market; for example, in Norway, the market share of electric cars, in terms of new car sales, was 39% in 2017 [6]. More than half of the global sales were in China, more than double the amount delivered in the United States [7,8]. The global stock of electric vehicles is growing rapidly and was already beyond the threshold of three million vehicles in 2017.

To reduce the emission of pollutions (e.g., CO_2 and NOx), the main trend in new vehicle design is electrifying the propulsion [9–19]. The evolution towards electric/hybrid mobility has been accelerated by the so-called "diesel gate" in Europe and the United States [20,21], as well as by the high economic cost for cars equipped with conventional ICE, to face restrictive regulations regarding greenhouse gases. Beside low-volume premium car brands, like Tesla, several large-volume car makers announced a cut in all petrol/diesel vehicles (e.g., by 2019, all new car models from Volvo and those produced in Europe by Toyota will be fully electric and/or hybrid). By 2030, a ban of diesel car sales has been announced by several countries.

Together with the increased connectivity of the car with the smart grid, thanks to V2SG/V2I (vehicle to smart-grid/infrastructure) wireless technology, the cars of the future will be electrified, connected, shared, and autonomous. The major impacts of the current R&D activities in academia and industry will be as follows:

(i)   to promote the development of a new generation of EVs with a minimum autonomy of 400 km to meet customer expectations;

(ii) to support the production of batteries, power components, and electrified vehicles at a low-cost and to be standardized as much as possible all over the world; and

(iii) to stimulate the accessibility and ease of use of all types of charging infrastructure—public, private, or mixed.

Unfortunately, full EVs are expensive, mainly because of the cost of the battery-based energy storage. Moreover, the autonomy of full EVs is still not comparable to ICE-based ones. A full EV needs also a rethinking of the drivetrain architecture, which now should involve high voltage power buses (up to 300 V or 400 V), with extra shock protection and insulating costs, power electronic converters, battery energy storage with battery management systems, new electric machines, and new transmission schemes. Indeed, as an alternative to the classic solution of a single-engine plus a complex transmission sub-system, constructions with a dedicated motor per driving wheel (two or four) are proposed. Electrical machines on-board electrical vehicles should have power levels ranging from tens of kWs, in cases of light vehicles or in cases of one motor for each driving wheel, to hundreds of kWs. Because of large time and development costs, the widespread diffusion of electric vehicles on the market is still to come.

Hence, the hybridization of the propulsion is the most viable solution to ensure, in the short term, a smooth transition from classic petrol/diesel cars to fully electric ones. Hybrid vehicles still need the evolution of current power electronic/electric parts of ICE cars; classic alternators should evolve towards an integrated starter-generator, capable of providing torque assistance to the ICE at low rpm conditions (motor mode), or to generate electric energy when the vehicle is braking or the ICE is working at a high rpm (generator mode). The classic 12 V power bus should evolve to a 48 V bus, in order to increase the power delivered with the same current levels. Several Original Equipment Manufacturers (OEMs) already succeeded in launching/going onto the mass market with 48 V topologies, with an acceptable cost level reaching the CO_2 emission targets and solving the technical challenges in areas such as battery technology, power electronics, EMC, and the electrification of further features.

The development of new tools, functionalities, and methods integrated with the controlled development of a vehicle-centralized controller will also be part of the future solutions for the next generation of EVs. For improving on the safety analysis and reduction costs, the solutions will be based on flexible user-friendly interfaces and specialized software tools, as well as on Ethernet, according to the existing ISO/IEC standards. The current trends is integrating Ethernet with other communication domains using already established technologies in the automotive industry, such as Controller Area Network (CAN), Local Interconnect Network (LIN) protocol, and Flex-Ray. These new technologies, based on digitalization and connectivity, will enable new ways to structure and design electric and electronic (E/E) architectures, such as seamless hierarchical architectures, for the next generation of cars.

As a matter of fact, digital technologies are already used in energy end-use sectors, and new technologies with the potential for widespread deployment are appearing, such as autonomous driving, intelligent domotic systems and machine learning. These platforms offer a high performance and fulfil the highest security and safety requirements [4,22].

The implementation and validation of tools under real-driving conditions based on the Internet of Things (IoT) paradigm for the over-the-air (OTA) diagnostic and flashing, and for V2SG issues, will also be part of future solutions. OTA will cut the time and cost for electronic control unit (ECU) diagnostic and reprogramming by at least 50% in case of software (SW) bugs.

To address the above issues, this paper reviews recent trends in green vehicle electrification. Particularly, Section 2 refers to the energy demand and emissions of EVs, and reviews the battery technology trends, including recharging issues and solutions, to integrate the EV electricity demand in power grids. Sections 3 and 4 deal with the integrated electric/electronic architectures and control systems of new generations of hybrid and full electric vehicles. The increased digitalization and connectivity of cars is posing cyber-security issues, which are discussed in Section 5. Conclusions are drawn in Section 6.

2. Global EV Outlook towards GV Electrification and the Impact of Digitalization on Energy Demand

Supportive policies and cost reductions are probably going to bring important growth in the market assimilation of EVs from the outlook period to 2030. In the New Policies scenario, which adopts existing and notified policies, the number of electric light-duty vehicles (LDVs) on the road will reach 125 million by 2030 [5]. The market volume of electric LDVs on the road can grow up to 220 million in 2030, if the policy ambitions continue to rise to meet more challenging climate goals and sustainability targets, as reported in Figure 1, in the case of the EV30@30 scenario. More particularly, it is estimated that there will be 130 million battery electric vehicles (BEVs) and 90 million plug-in hybrid vehicles (PHEVs). The EV30@30 campaign, started at the Eighth Clean Energy Ministerial in 2017, redefined the electrical vehicle initiative (EVI) ambition by setting the collective ambition target for all EVI members, as a 30% market share for electric vehicles for all vehicles (except two-wheelers) by 2030. The dispatching campaign contains various implementing actions to help achieve the target, which are in agreement with the priorities and programs of each EVI country.

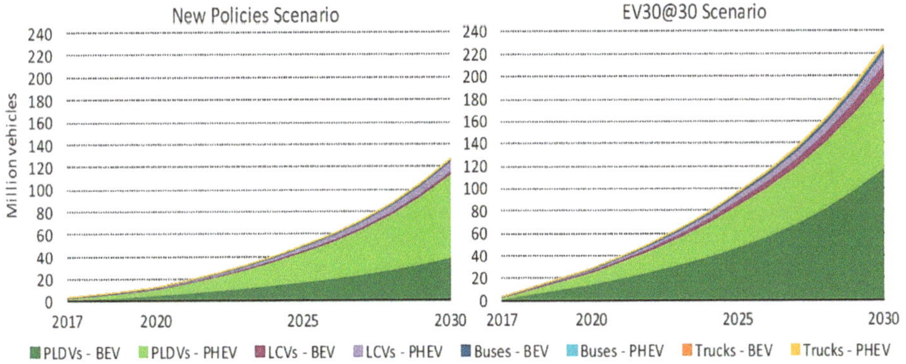

Notes: PLDVs = passenger light duty vehicles; LCVs = light commercial vehicles; BEVs = battery electric vehicles; PHEV = plug-in hybrid electric vehicles.

Figure 1. Global electric vehicle (EV) supply in the New Policies and EV30@30 scenarios, 2017–2030.

Fast developments in sizing battery production and reducing costs, allowed by the increasing sale of EVs, primarily driven by policies targeting LDVs, have positive disperse effects across other transport modes. Eventually, electric two-wheelers, which are not at present a prime policy focus in most regions, are planned to experience a significant increase. As matter of fact, in terms of stock share, about 40% of the world's two-wheelers will be electric by 2030 in the New Policies scenario. China has worldwide leadership of the two-wheeler electrification market, with a continuing commitment to the electrification of mobility; however India, whose population is predicted to be as numerous as that of China in the coming years, has the ambition to electrify its two-wheelers. Europe, where fuel taxes cause a faster cost recovery over the vehicle life, is also at the top of this transition. As shown in Figure 1, if regulatory pressure is applied to better harness the full economic and environmental benefits, then a 50% global stock share can be achieved in 2030 in the EV30@30 scenario [1].

2.1. Impact of Digitalization on Energy Demand in Electrified Vehicles

Digitalization is having a major impact on the transport and automotive industry, especially on EVs and HEVs. How significant this revert will be in the future will vary for each sector, and particular, application. The transport sector is becoming smarter and more connected, improving on safety and efficiency. In electrified vehicles, connectivity is empowering new mobility dividing services. Digitalization depicts the increasing application of information and communications technologies (ICT) across the economy, including energy systems. Advances in big data analysis, analytics, and connectivity enable the trend toward greater digitalization, as follows:

- increasing volumes of data on the strength of decreasing costs of sensors and data storage;
- fast progress in advanced analytics, such as artificial intelligence (AI) and machine learning (ML);
- major connectivity of people and devices as well as faster and cheaper data transmission (4G, 5G);
- digitalization encompasses a range of digital technologies, concepts, and trends, such as AI, IoT, OTA update of SW, and the fourth industrial revolution (industry 4.0).

Mixed with the advances in vehicle automation and electrification, digitalization will result in significant but uncertain energy and emissions impacts.

2.2. Current Impact of EVs on Energy Demand

In 2017, the estimated global electricity demand from all EVs was 54 terawatt-hours (TWh) (see Figure 2), an amount corresponding to little more than the electricity demand of Greece [5]. China has worldwide leadership of this demand (91%), and its energy consumption is mainly due to two-wheelers and buses. These two modes combined account for 87% of EV electricity demand worldwide. Yet, the electricity demand for LDVs has increased the fastest since 2015 (143%), followed by buses (110%) and two-wheelers (13%).

Figure 2 shows that the approximated electricity demand from EVs in 2017 increased by 21% compared with 2016. With reference to last year (2017), the electricity demand of the EVs corresponds to 0.2% of the total global electricity consumption [5,6,19]. In countries like China and Norway, which have the largest fleet and market share of EVs, the electricity demand of EVs is still below 1% of the total demand—0.45% in China and 0.78% in Norway. As yet, the expanding numbers of EVs have had a limited impact on the electricity demand, thus providing support of confidence for the transition to greater electric mobility. As electric vehicles are started growing, they will increase electricity demand and with that will affect transmission and distribution grids.

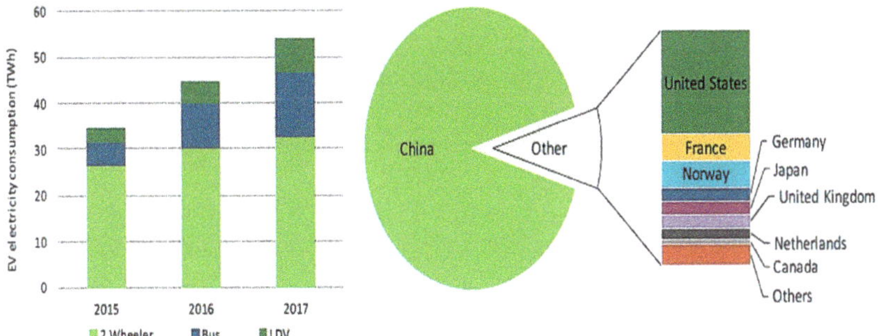

Notes: TWh = terawatt-hours. The pie chart refers to 2017 data. The assumptions are: passenger vehicle consumption 20-27 kWh/100 km, annual mileage 8 500-18 800 km; two-wheelers consumption 3-5 kWh/100 km, annual mileage 5 900-7 500 km; electric urban bus consumption 135-170 kWh/100 km, annual mileage 28 000-47 000 km, (the range indicates the variation across countries). The share of electric driving for PHEVs is assumed to be 36% of the annual mileage. Charging is assumed to have an efficiency of 90%.

Figure 2. EV electricity consumption by country in 2017 from EVs.

2.3. Possible Issues and Solutions to Integrate EV Electricity Demand in Power Networks

Strong peaks characterize the demand for energy in the transportation sector in the morning and in the evening on weekdays, with limited variations across modes. Power demand also displays a morning and an evening peak in most regions, while the demand is lower during the night and in the afternoon. The low period of power demand during daytime is less visible for the summer days in warm climates, where there is a high electricity demand from cooling appliances, or in winter days in cold climates, because of the higher power demand for heating purposes.

Figure 3 shows a scenario with day changes in traffic flow in three different cities from three continents, Hong Kong (China), Long Beach (CA, USA), and Manchester (UK). Figure 3 also shows the electricity load curve of each region. The power demand and road mobility demand are both characterised by two peaks during the morning and evening hours, and a period of low demand during night time.

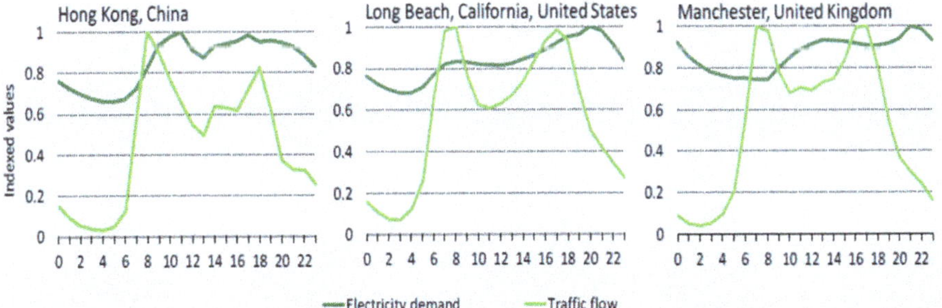

Notes: The two lines aim to describe the variation of road traffic, using a 0 to 1 index profile, and power demand during an average weekday, as a fraction of the maximum daily road traffic and power demand respectively. This figure uses data from three studies for which road traffic measurements were conducted and load curves which are representative for the power demand of the respective city. Power demand curves refer to April for Long Beach and Manchester, and are yearly averages for Hong Kong.

Figure 3. Road traffic and power demand profile during an average weekday for three cities from three different continents.

In all three cities, analyzed in Figure 3, as it is expected, there is a peak in traffic activity in the morning after a period of low electricity demand during the night. These specific features of electricity demand and transport activity point out that the overnight charging of EVs is well timed, before they are used in the morning. Moreover, the overnight charging of EVs has the added benefit of minimizing both the need for incremental electricity generation capacity and investment in distribution infrastructure upgrades.

The peak in electricity demand often follows the traffic peak in the evenings. Plugging EVs into the grid after the evening traffic peak may exacerbate the peak power draw. This couples with a higher risk of overloading of the power distribution network, requiring grid upgrades such as the replacement of distribution transformers and cables [23]. If not properly managed, the increased power draw at peak times could also require additional generation capacity. To avoid the economic and environmental effects of an increased peak load demand in the evenings, transferring the load to the night is advised.

Demand-side management (DSM), also called demand-side response, is an important tool that can significantly reduce the need for grid upgrades and additional generation capacity, because of the electrification of road transport, as well as facilitating the integration of renewable energy sources (RES) [5,6].

Regulators, utilities, transmission system operators, distribution system operators, and retailers are already taking DSM measures and designing policy mechanisms to ensure that the EV uptake will not overload the power grid. For EVs, DSM largely consists of the optimization of the charging time of the vehicles, shifting the loads to ensure a good match between the power supply and demand, with the aim of moving the bulk of the EV charging related power demand from the evening peak to the night. In addition to relieving the load on the distribution grid and reducing the investment needs for grid reinforcements, achieving this has the capacity to deliver a number of potential benefits, such as:

- The need for additional generation capacity is decreased by shifting EVs' charging loads to periods where the energy demand is lower. This can lead to lower electricity prices thanks to the possibility of relying on the power produced by the generation of assets with a lower marginal price.
- Optimizing the utilization of the grid assets during the day, increasing their utilization factor and maximizing their profitability, therefore reducing their cost per kWh.
- In a more efficient way exploiting the energy produced with RES by shifting the EVs' charging loads to periods of high output from RES. For example, exploiting the nighttime charging when the generation from wind generators is often highest, or mid-day when there are peaks from photovoltaic generation [5].

Realizing these benefits with DSM is facilitated by the implementation of a dynamic tariff policy such as time-of-use (TOU) pricing and/or real-time pricing (RTP) [24]. TOU pricing incentivizes consumers to charge EVs in a way that maximizes the power draw when electricity prices are low and minimizes it when they are high.

Dynamic pricing aims to discourage EV owners from charging their vehicle at peak times. However, it can also be used to shift the demand towards times when electricity production from RES is abundant, or to get all these benefits concurrently.

The charging process should be assisted by smart charging applications. Manufacturers such as BMW already have developed products to optimize the home charging process in an automatic way to benefit from low electricity prices [25]. DSM products may also be used to optimize the usage patterns of other residential appliances (e.g., heating and cooling) that contribute to electricity peak loads. Integrated systems may enable consumers to prioritize appliances, for instance, by temporarily reducing the electric heating to offset any additional load from charging an EV during the peak load.

DSM can also provide valuable ancillary services to the power grid, including frequency regulation, voltage support, and power factor correction, as well as the possibility to balance loads across the distribution network. The effectiveness of DSM measures could be further enhanced by a bidirectional "vehicle-to-grid" (V2G) capability. V2G is a bi-directional connection between the EV and the grid through which power can flow from the grid to the vehicle and vice-versa [25,26].

2.4. New Trends in Battery Technology and the Implication for EVs

Lithium-ion (Li-ion) storage technology prices have decreased, while the manufacturing volumes have increased. Experience in manufacturing batteries for consumer electronics has driven cost reductions to the benefit of EV packs as well as stationary storage.

Figure 4 illustrates the cost reductions comparative to the cumulative manufactured capacity across Li-ion storage technologies used in various applications. It also shows that Li-ion batteries have proved significant cost reductions since their market introduction in the 1990s.

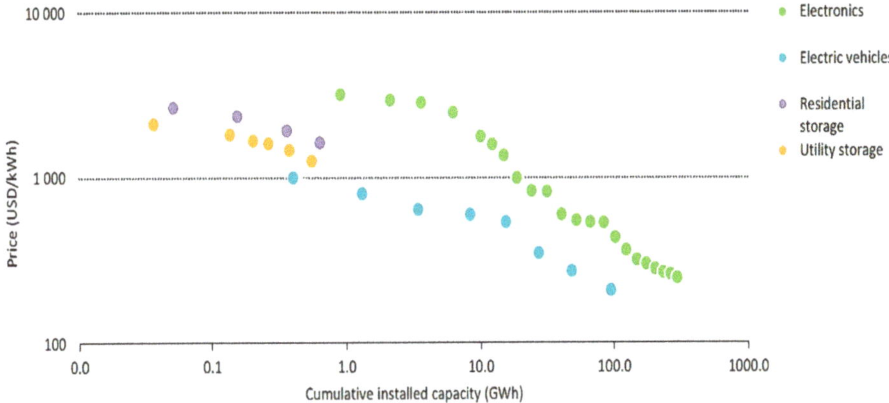

Notes: Axes are on a logarithmic scale. Electronics refer to power electronic batteries (only cells); electric vehicles refer to battery packs for EVs; utility and residential storage refer to Li-ion battery packs plus power conversion system and includes costs for engineering, procurement and construction.

Figure 4. Evolution of Lithium-ion storage technology prices vs. installed capacity adapted from [27].

The early development of batteries for consumer electronics (e.g., smartphones, laptops, etc.) provided invaluable experience in the production of Li-ion cells, underpinning the attainment of the cumulative production capacity of 100 gigawatt-hours (GWh) by 2010 [27], enabling the achievement of very significant cost reductions and performance improvements over the past decade. These same developments made the development of Li-ion battery packs for EVs increasingly viable.

At the state-of-art, most of battery packs used in EVs exploit the lithium-ion technology. This technology is reaching a maturity level that is enabling the design of EVs with a performance comparable to ICE vehicles. Current battery packs for light-duty applications have gravimetric energy densities of 200 Watt-hours per kilogram (Wh/kg) [28], and volumetric pack energy densities of 200–300 Watt-hours per liter (Wh/L) [29]. The lifetime of the battery is another important parameter. For EV batteries, a good proxy is the expected mileage associated with a battery's lifetime, as well as its ability to retain a good share of its initial capacity (usually 80%).

According to the recent literature, modern Li-ion chemistry for EV batteries is able to withstand 1000-cycle degradation [30]. To compare this value to the lifetime of an ICE-based vehicle, let us assume a vehicle with a battery capacity of 35 kWh and an average consumption of 0.2 kWh/km. Withstanding 1000-cycle degradation suggests that the cycle life threshold would not be reached over the first 175,000 km of driving. Considering a car used mainly in an urban scenario for about 12,500 km/year, then 175,000 km and 1000-cycle recharging degradation means a lifetime of 14 years for the battery pack. Therefore, the lifetime of the battery is compatible with the expected lifetime for a car.

Notwithstanding the complexity of battery design and manufacturing, four key cost and performance drivers have been identified for Li-ion batteries—chemistry, capacity, manufacturing capacity, and charging speeds. The cost per kWh of the currently available battery chemistries varies because of the different energy densities and material needs.

The size of the battery packs used at the state-of-art in EVs vary considerably. For BEVs, the size of the battery packs ranges from about 20 kWh to about 100 kWh. In China, which is, as discussed before, the worldwide market leader for EVs, the three bestselling EVs have battery sizes ranging between 18.3 kWh and 23 kWh, mainly because the market in China is focused on small vehicles, and their design is focused on affordability. Instead, in Europe and North America, for mid-sized cars, the capacity of battery packs ranges between 23 kWh and 60 kWh, whereas larger cars and SUVs have battery capacities ranging from 75 kWh and 100 kWh.

As far as the current charging speed is concerned, at the state-of-art, fast chargers enable 80% recharging in less than 1 h. Such a charging speed does not constitute a challenge for current battery design. Further increasing the maximum speed of charging to ultra-fast charging (which implies working at power levels as high as 300 kW or 400 kW) is a desirable feature that would decrease the performance gap of EVs compared to ICE vehicles. However, designing batteries for ultra-fast charging has the negative effects of increasing the complexity of their design and of shortening their lifetime. Accommodating fast charging requires specific battery design considerations, such as decreasing the thickness of the electrodes. These added design constraints tend to increase the cost of the battery and to decrease its energy density. With an appropriate design and appropriately sized thermal management system, the increases in fast charging are not expected to affect the battery's lifetime. On the other hand, an analysis conducted for the United States Department of Energy suggests that the change in battery design to accommodate 400 kW charging would nearly double the cell costs [31]. Indications from the recent assessments of battery technologies suggest that lithium-ion is expected to remain the technology of choice for the next decade (see Figure 5). The main developments in cell technology that are likely to be deployed in the next few years include the following:

- For the cathode, the reduction of cobalt content in existing cathode chemistries, aiming to reduce cost and increase energy density (i.e., from today's Nickel Manganese Cobalt (NMC) 111 to NMC 622 by 2020, or from the 80% nickel and 15% cobalt of current Lithium Nickel Cobalt Aluminium Oxide (NCA) batteries to higher shares of nickel) [28,31,32].
- For the anode, further improvement to the graphite structure, enabling faster charging rates [28].
- For the electrolyte, the development of a gel-like electrolyte material [28].

The next generation of Li-ion batteries entering the mass production market around 2025 is expected to have a low cobalt content, high energy density, and NMC 811 cathodes. To increase the energy density up to 50% silicon can be added in small quantities to the graphite anode [28]. To contribute, better performance electrolyte salts that are able to withstand higher voltages, can be used. As reported in Figure 5, lithium-ion is assumed to remain the technology of choice for the next decade, when it is expected to take advantage of a number of improvements to increase the battery performance. According to Figure 5, other technology options are expected to become available after 2030.

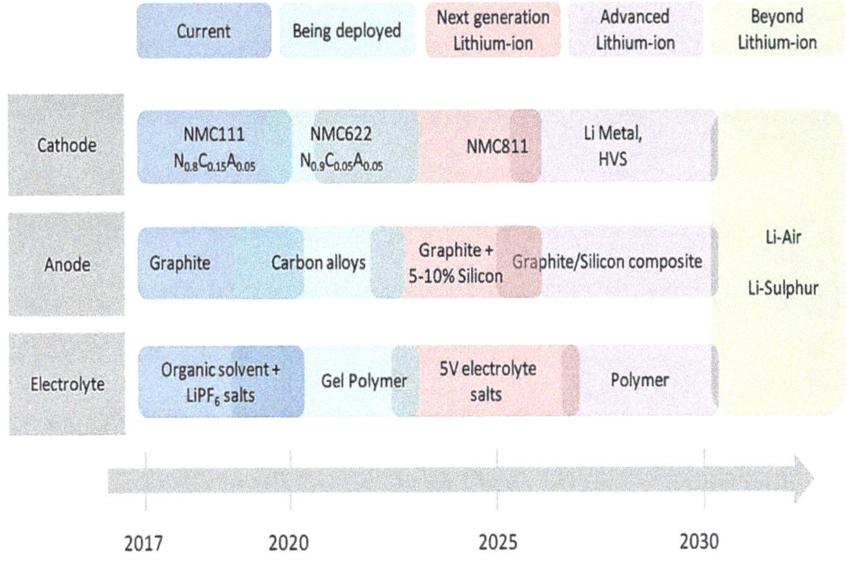

Figure 5. Expected battery technology commercialization timeline.

3. Hybrid-Electric 48 V Vehicles

Hybrid vehicles may ensure a rapid and smooth transition from the current generation of petrol/diesel-based propulsion to new electrified mobility. HEVs may be realized by combining a downsized ICE with an electric motor, usually within 10 kW and with power DC buses up to 48 V nominal. In such cases, the on-board energy storage to provide extra energy for torque assist, or to store recovered energy when braking, can be limited to less than 1 kWh. The immediate benefit vs. current vehicle generations is fuel saving and CO_2 reduction. According to this approach, an electrical machine able to work in both motor and generator modes, with power levels to 10 or 15 kW [12,17,21], can implement several "green" functions, aimed at fuel saving and/or a reduction of pollutant emissions. Among these "green" functions, it is worth mentioning the following: start-and-stop to cut ICE emissions in urban scenarios, torque assistance to improve efficiency at low-rpm regimes where an ICE is not working in optimal conditions, and regenerative braking to avoid just dissipating energy when a deceleration is needed. Besides assisting with an electric machine, the ICE for the propulsion, a saving of energy and of pollution emissions can be achieved through the electrifications of comfort and/or chassis control functions. The latter include, as an example, the electrically controlled air-conditioning compressors or the electric steering.

In this context, a belt-driven starter generator (BSG), or an integrated starter generator (ISG), replaces the conventional alternator, thus creating a parallel hybrid architecture. Figure 6 shows the layout of a BSG. The main idea is that the starter-generator electrical machines replaces the conventional alternator. In this way, there is a low impact on the layout of the engine compartment, and just the redesign of the belt tensioner is needed. Differently from the BSG solution, in the ISG solution, the starter-generator electric machine should be inserted between the ICE and the gearbox. Hence, the ISG approach requires a complete revision of the engine compartment. This is why the authors focused their work, as reported in this survey paper, on the design of a BSG electrical machine.

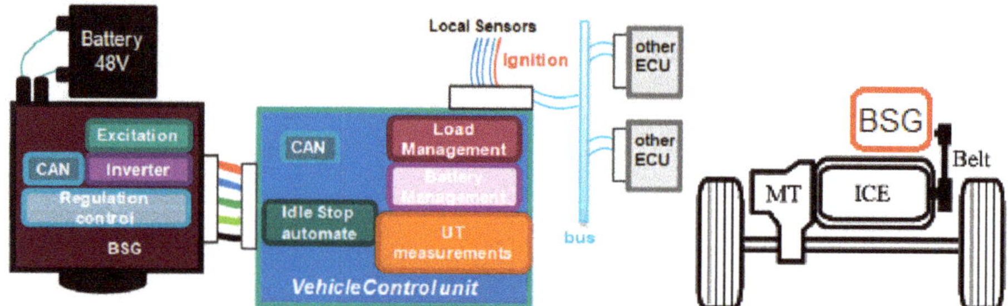

Figure 6. 48 V belt-driven starter generator (BSG) interfaced to the electronics (**left**) and to the mechanical-transmission (MT) and internal combustion engine (ICE) (**right**).

For hybrid vehicles using BSG or ISG machines, the classic 12 V automotive DC bus reaches its limit as follows: at 12 V, the supplied current would be above 600 A in the case of cranking (e.g., peak current at cars start). This will require cables with a high cross section, with a too high increase in wiring cost and size. To solve this issue, there is a trend toward the adoption of a 48 V DC bus technology (52 V in generator mode) for hybrid vehicles. With respect to a classic 12 V solution, the drawn current at equivalent power is reduced by a factor of four. As DC voltages below 60 V do not require electrical shock protection, adopting a 48 V DC bus does not increase the cost of the on-board electrical implant by too much. In the short term, to reduce the cost of migrating from ICE-based vehicle generation using 12 V electrical systems to a new 48 V HEV generation, most of the 12 V automotive components will be reused. As a consequence, new 48 V HEVs require the development of energy-efficient and compact DC/DC converters [20,21]. These DC/DC converters should be able to interconnect the 48 V DC domain to the low voltage supply domains required by low-power components (such as ECUs and memories and sensors). In the case of a fault and/or malfunction, the 48 V domain should be isolated from the low-voltage domain. In emerging 48 V HEVs, another issue is the design of an integrated H-bridge for rotor excitation. The integration is important for reducing the use of discrete devices, thus reducing the size, weight, and cost. In summary, the main issues of hybrid vs. ICE vehicles are the introduction of (i) BSG/ISG machines and related power drives; (ii) 48 V architecture; and (iii) DC/DC converters among the 48 V and 12 V buses to reuse most of the 12 V components.

Figure 6 shows the vehicle electrical architecture for a hybrid vehicle, including a BSG or ISG unit. The BSG or ISG unit is a mechatronic sub-system; it receives the electric power from the battery pack (at 48 V in Figure 6) plus proper command/configuration signals from the vehicle control unit (VCU). The VCU is interfaced also to local sensors, including the ignition signal, and through the main car bus it is connected to all of the other ECUs. At the state-of-art, the controller area network (CAN) is the de facto standard to connect vehicle ECUs with smart actuator and sensing units. As it will be discussed in Section 5, the fact that the BSG or ISG machine (providing torque assistance, start and stop, and electric energy generation functionalities) is connected to the CAN poses several cybersecurity issues to be addressed, as they can lead to safety issues. The core of the BSG proposed in our research work is a six-phase wounded rotor synchronous machine, showed with its dedicated electronics in Figure 7. As we detail in the literature [21], the architecture of the proposed BSG electrical machine includes a wounded rotor and a double three-phase stator. The prototype of the BSG electrical machine is shown in Figure 7. The power level of the proposed BSG electrical machine is up to 8 kW in operating conditions, with a peak of 15 kW. The adoption of two stators instead of a classic solution with two stators allows for the use of lower current levels in each stator and of smaller copper winding. The two stators in the BSG of Figure 7 are electrically shifted by 30 degrees. This solution allows for reducing the ripple on the output rectification. The new BSG electrical machine of Figure 7, within the EU project ATHENIS-3D with the Valeo company, has been integrated and tested on a commercial car—a Peugeot308 hybrid [21]. The proposed drive reaches a max speed of 20,000 rpm (no load condition), and a max torque at a low-rpm of 70 Nm, and is able to start a 1.5 L diesel engine in less than 400 ms, going from 0 to more than 1000 rpm.

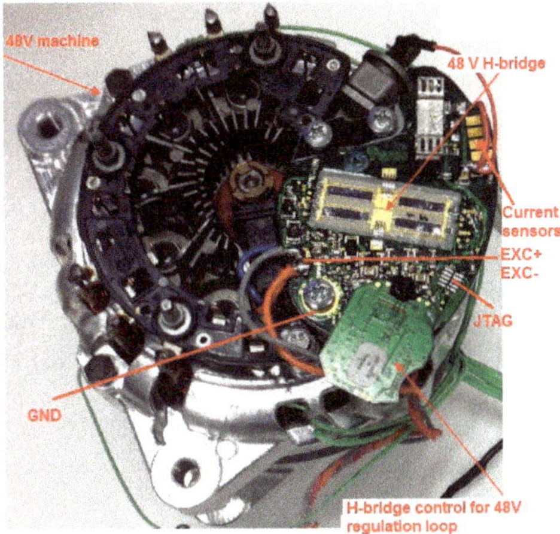

Figure 7. Prototype of the 48 V BSG plus control and sensing electronics.

The electronics of the BSG electrical machine in Figure 6 include the following two main power sections:

-   The stators connected to power modules, where each three-phase stator includes six MOSFETs transistors that can be controlled to work both in inverter mode or rectifier mode. The synchronous rectification mode is adopted when the power system is working in generator mode. The six MOSFETs are controlled in the DC/AC inverter mode, using a pulse width modulation/full wave (PWM/FW) technique, when the power system is working in motor mode.
-   The rotor current control that is managed by an H-bridge using four power MOSFETs is controlled by PWM signals. The full H-bridge approach (instead of using just a half bridge with two MOSFETs) is needed to ensure fast demagnetization in cases of load dump, and to increase safety vs. overvoltage phenomena in cases of component failures.

The control electronics of the BSG electrical machine include the following two power domains:

-   The 48 V domain, which is dedicated to the control of the wounded rotor and the two three-phase stators. The controller in the 48 V domain generates all of the signals needed to drive the power MOSFETs of the rectifier/inverter stages of the two three-phase stators, and to drive the power MOSFETs of the H-bridge used to control the rotor current. The 48 V domain also includes analog and digital circuits that are used to interface all of the sensors in the stator and rotor control. Examples of such sensors are phase current sensors, phase voltage sensors, rotor position sensors, stator thermal sensors, and power module thermal sensors.
-   The 12 V domain, which is used for the ECU. Concerning the control algorithm used for the BSG electrical machine, seen in Figure 7, a closed-loop flux vector control running in real-time on a field programmable gate array (FPGA) is the preferred solution. Differently from the classic 32 bit microcontrollers, the use of FPGA as a computing platform for BSG control ensures a better trade-off in terms of computational capability, flexibility, size, and power consumption. It is worth noting that at the state-of-art, FPGA devices are available as automotive qualified components (e.g., AECQ-100).

Another innovation discussed in this work is the use of direct copper bonding (DCB) technology for the 48 V H-bridge. The aim of the DCB usage is the reduction of the ON-resistance ("drain-source on resistance", is the total resistance between the drain and source in a Metal Oxide Field Effect Transistor, or MOSFET when the MOSFET is "on") of the H-bridge switches. Indeed, low ON-resistance values are needed to minimize power dissipation when the current is flowing through the H-bridge switch. The ON-resistance is due not only to the RDS_{ON} of the power MOSFETs (that can be reduced by increasing the silicon area), but also to the resistance of the connections between the power MOSFETs and the electronic board on which the devices are assembled. To achieve low ohmic transistor connections in the 48 V H-bridge, DCB technology can be adopted when designing the electronic board and assembling the devices. In the DCB approach, an interconnect plate with Al_2O_3 ceramic substrate and electroplated copper is used. The thickness of the electronic board is 0.38 mm. The thickness of the copper is about 0.3 mm, with a surface plating for die stacking with a 5 µm Ni and 0.03 µm Au layer. Each transistor chip is connected with four lines (two source lines and two drain lines) to the electronic board. A stress analysis has been carried out within the ATHENIS-3D EU project [21] with an AMS partner on a 48 V H-bridge, using transistors implemented in High Voltage Power MOSFET (HV-MOS) technology, with a DCB connection to the board. This stress analysis has demonstrated that the DCB technology can allow for operations at high temperatures (up to 175 °C).

The supply voltage of the H-bridge, used in Figure 7, is nominal 48 V (in generator mode the voltage is 52 V). The rotor currents are up to 12 A in nominal conditions and up to 17 A in transient conditions. The four transistors of the H-bridge are realized in AMS HV-MOS technology, with DCB on ceramic substrate, as two P-channel MOSFETs for the high side devices and two N-channel MOSFETs for the low side devices. Using P-channel MOSFETs as high side devices in the H-bridge allows for avoiding the use of complex charge-pump devices, although at the price of a higher ON-resistance vs. N-channel MOSFETs. The minimum ON-resistance is 8 mΩ for N-channel MOSFETs and 11 mΩ for P-channel MOSFETs when using a 5 V gate-source supply for the transistors. Table 1 shows a comparison of the performance parameters of the integrated H-bridge vs. the state-of-art of the integrated devices for rotor coil driving. With reference to the H-bridge in the literature [33], implemented using the same HV-MOS technology, the solution proposed in this work sustains a current 2.125 times higher and a voltage 4 times higher, whereas the ON-resistance is reduced by 6 times. The improved performance of the H-bridge proposed in this work vs. that in the literature [33] ($\times$2.125 higher current, $\times$4 higher voltage, and $\times$6 lower ON-resistance) is justified by both the increase of the transistor area and by the adoption of the DCB approach, missing in the literature [33]. Thanks to the DCB technology, the transistors are assembled on top of the ceramic substrate. The DCB technology also allows for a reduction of the printed circuit board (PCB) size with respect to the layouts of electronic boards, where large conductive plates are placed around the chip to achieve low-ohmic contacts, as in the literature [33]. Since in this work, the electronics is coupled with the 48 V BSG machine, and the adopted technology is a low-cost one, and the DCB approach allows for a reduction of the PCB size, then the increased area of the transistors is not an issue.

Table 1. 48 V H-bridge vs. state-of-art, rotor current control.

	Max. Current	**Voltage**	**ON-Resistance (High/Low-Side)**
Current work	17 A	48 V	11 mΩ/8 mΩ
[33]	8 A	12 V	62 mΩ/47 mΩ

4. Full Electric Vehicles

In recent years, EVs or BEVs have been gaining popularity, and one of the most important reason behind this is their contribution to reducing greenhouse gas (GHG) emissions.

EVs, including the different types of BEVs, HEVs, PHEVs, and fuel-cell EVs (FCEVs), have been becoming more commonplace in the transportation sector in the last five years. A block diagram with the main EV's components is depicted in Figure 8. Comparing this figure for EV with that of Figure 6 (right), it is clear how full EV needs a major change in current car architectures with lot of new electronic and electrical sub-systems. This is why many leading OEM's and suppliers are developing 48 Volt architecture, as discussed in Section 3, on the basis that it can achieve large efficiency gains at lower costs in the medium term rather than full electrification, as discussed in this section.

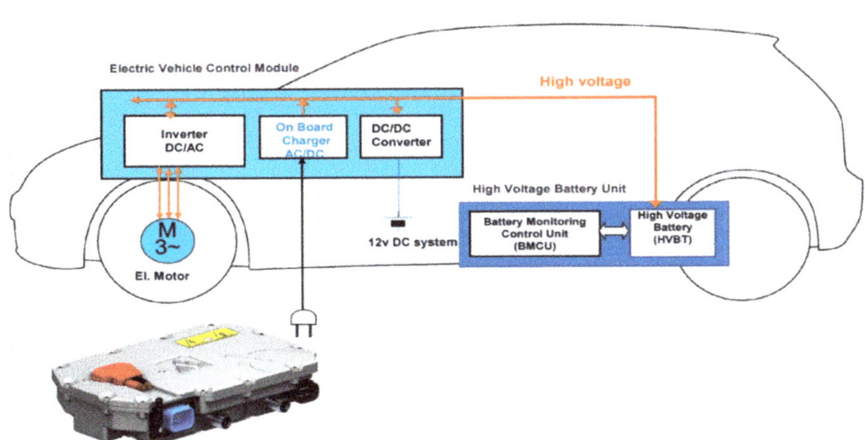

Figure 8. Car's block diagram with the main EV's components.

The main energy source of BEV is the high voltage battery, while the most important components of the EV are the electrical motors and the high voltage battery, which, together with the transmission system, builds the vehicle drivetrain architecture. These main components are assisted by a number of auxiliary subsystems, such as an ECU, a battery management system (BMS), and power electronics converters. An ECU, which can be called "e-motor control", is liable for the electric motors operation mode. BMS, also called the battery monitoring and control unit (BMCU), supervises the storage operations, including charging and discharging states. Most EVs contain on-board single/phase or three/phase unidirectional or bidirectional chargers; hence, an AC/DC power electronic converter is required. Power electronic converters, for the charging operation mode are able to transform energy from AC to DC, while for the discharging operation mode, they should be able to transform energy from DC to AC. At the same time, the EV system operation requires other energy conversions in order to supply other subsystems. Therefore, several additional converters are needed, such as the following:

- DC/DC switch/mode converters between internal LV and HV battery, to charge LV battery;
- DC/DC switch/mode converters between electrical motors and HV battery system in order to provide braking energy regeneration;
- AC/DC switch/mode converters and DC/DC chopper or single DC/DC converter between the alternator and the HV battery.

EVs can produce significant impacts on the environment, on the network for power system distribution, and on other related sectors. The actual power system could face huge instabilities with enough EV penetration, but with a suitable management and coordination system, EVs can be turned into a major contributor to the successful implementation of the smart grid concept.

4.1. E/E-Architecture for EVs

The physical architecture as well as the electrical and electronics (E/E) architecture will be the keys to manage the increased complexity of the third generation of EVs, which will require a faster increase in electronics, software, and communication capabilities.

The new generation of electric and autonomous driving vehicles is evolving towards a distributed connection of smart sensors, ECUs, and actuator control units (ACUs), up to one hundred devices for a premium car, with stringent requirements in terms of bandwidth, functional safety, and security. In addition, the complexity of the software, stored in the non-volatile memories of the ECUs, is continuously growing [34]; today, the number of software code lines on a premium car is reaching 100 million. Security against cyberattacks is a recent but very important issue in the transportation world. As vehicles are evolving towards autonomous driving, and the vehicle-to-X (vehicle, infrastructure, road, or pedestrian) connections and on-board vehicles networking allow for access to every ECU in a car, then a remote cyberattack can force failure in any of the key functions of a vehicle, like the control of propulsion, braking, steering, and so on. To illustrate the importance of automotive security, recent studies forecast investments in this field of up to 11 billion by 2021. By 2020, some estimates forecast that 40% of the cost of a car will be in wiring and connections. Cybersecurity issues of digitized and electrified vehicles are discussed in Section 5.

A seamless hierarchical E/E architecture for the next generation of cars is shown in Figure 9. In this architecture, the central computing platform partition the main software functions offering high performance and fulfil the highest security and safety requirements.

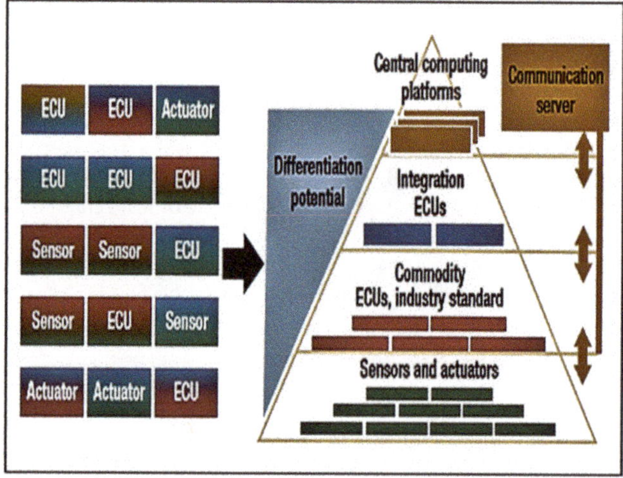

Figure 9. A powerful integration platform enable a seamless hierarchical electrical and electronics (E/E) architecture.

The advantages of this platform, shown in Figure 9, are that each ECU class has specific requirements, such that the classification is requirements-based and the system level optimization is the focus. BMW has introduced a service-oriented architecture (SOA), similar to the one in Figure 9, for the next generation of E/E architecture, with hierarchy enabled testing against interfaces using agile methods for system complexity reduction.

4.2. Power Train Design and Component Trends of EVs

The majority of European automotive manufacturers have already decided to introduce 48 V technology to reduce the consumption of their fleets and to meet the new European CO_2 boundaries, which will came into force from 2021. In the near future, plug and play or start–stop systems, 48 V systems, and high-voltage electrification will all occur together in most fleets. A strong worldwide global trend towards plug-in hybrids has emerged. For instance, in China, all-electric vehicles are particularly in demand.

In Europe, 48 V systems still have their development core, although many automotive manufacturers have now recognized their advantages, and global programs have been introduced immediately after that. An analysis of the extra costs of HEVs proves that 48 V mild HEVs are only 30–50% as cost-intensive as high-voltage HEVs [35]. As such, the 48 V system represents an intelligent and, in particular, affordable supplement to full and plug-in hybrids vehicles. In addition, 48 V systems can be integrated more easily into existing vehicle powertrains and architectures, but a fewer extensive modifications are also required. Consequently, it can be anticipated that the 48 V voltage level will very quickly become established in the market.

Furthermore, to make use of hybridization, the additional 48 V system also makes it possible to have a choice of electrical components in the vehicle at higher voltages. This is significant, because the number of electrical components is continuing to expand dramatically, especially in the mid-size and luxury car markets [1,35]. In addition to that, high-power components work more efficiently at higher voltages, and converting them to the 48 V on-board power system also decreases the load on the 12 V system. Market predictions highlight the fact that 25% of newly registered cars will have an electrified powertrain by 2025, as can be seen in Figure 10, and that almost half of these vehicles will use 48 V technology.

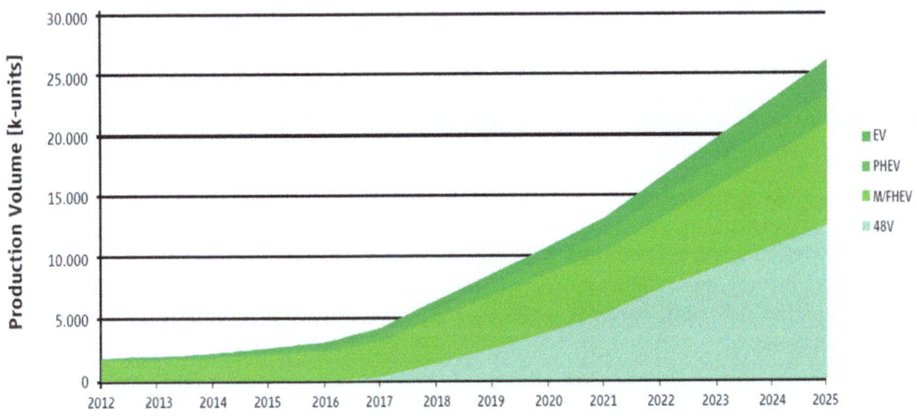

Figure 10. Market forecast electrified powertrain by 2025 (source: Continental).

As can also be seen in Figure 10, beginning with the year 2020 onwards, a global potential of up to four million of 48 V systems could be deployed.

Because of the different options to mechanically connect and integrate the 48 V electrical machines (usually as a starter generator in 48 V technology) into the drivetrain, and the different types of 48 V electrical machines available for selection, several powertrain topologies are achievable. The powertrain topology chosen significantly influences the performance and characteristics with which the aforementioned functions can be implemented, as well as the costs involved. Carmakers and automotive suppliers are currently analyzing and evaluating four major powertrain topologies [36–38]. Depending on the e-machine architecture, the topologies vary in relation to their potential for energy recovery and electrical boost capacity.

In the next-generation of EVs, power and efficiency are critical parameters, as the number of engine control units and ECUs within the cars grows exponentially. The electrical and electronic component's models, such as electrical machines, power electronic converters, and controllers, represent the e-drivetrain system components for closed-loop tests (HIL).

4.3. High Voltage Battery Systems Technology and Battery Management

The high voltage battery system is in fact the EV energy source subsystem. As it is one of the most important components of BEV, it may require special care and handling. For instance, in the case of DC fast charging, EV's BMS system and charging station should establish the bi-directional communication to exchange information such as the battery charging pattern, state of charge (SOC), temperature, and so on [39]. At present, the battery system is one of the most demandable and critical components of the EV system. Main issues include their weight, price, capacity, energy density, lifetime or degradation, electrical parameters, and dimensions [39–43]. For all of this, EV batteries' market segment is rapidly developing and increasing, and manufacturers and e-mobility stakeholders are conscious that batteries are crucial for the sector to be further development [1,39].

The most used battery technology for EV is Lithium-ion, as already highlighted in Section 2. This battery type can be designed using various cathode and anode materials such as lithium titanate, lithium–cobalt, oxide-based, or lithium–iron–phosphorus. Li-ion promises a high energy density, lifetime, and charging cycles [1,40–44]. Li-ion batteries are characterized by an energy density of 130 Wh/kg, a cell voltage of 3.7 V, and their expected number of cycles (e.g., 3000), assuming that the depth of discharge (DoD) is at 80% [39]. The highest practical energy density can be achieved within the cobalt cathode (120–180 Wh/kg). Battery development is crucial for further e-mobility development. The most urgent issues include increasing the energy storage capacity, allowing for high current charging, and extending cycle lifetime, as well as improving safety and reducing cost.

The BMS handles operation of the EV battery and its functionalities are very important for the optimal use and handling of high voltage batteries. It controls the charging and discharging cycles together with the on-board or off-board charger. The control strategy is in most cases arranged for extending the battery lifetime. BMS impedes also from deep discharge and wrong charging parameters. The main important BMS tasks can be summarized as follows:

-   controlling/giving assistance to charging and discharging cycles, including fast charging and smart charging;
-   protecting battery from operating outside the admissible range;
-   monitoring the battery SOC and battery condition (BMS includes control of voltage, power level, temperature, SOC, state of health, current, and coolant flow);
-   reporting data.

The BMS system/unit contains three main subsystems, centralized, distributed, and modular parts. Centralized subsystem is based on wire connections and is a single controller connected to the battery cells. The distributed components of the BMS implies BMS on/off board battery chargers, in which each cell is equipped. Modular subsystem involves a few controllers, which carry out the operation of a certain number of cells. BMS needs to send the required data to other devices in the EV system. The main measures used for this purpose include CAN BUS, FlexRay, or direct wiring, although data transmission over the power bus or fast wireless communication based on IoT may be used.

4.4. Electric Motors and Control

It is imagined that the 48 V powertrain systems of all types will operate without the traditional 12 V generator, as the 48 V electrical machine takes over the generator function. Thanks to the higher voltage, the electrical machine's performance and efficiency will be improved. Differently from classic 12 V generators, electrical machines in new EV generations fulfil two different functions, as they work in both regimes, as generators and starters, or rather electrical motors, to support propulsion.

Key parameters of the powertrain system include the total system load, the charging status (SOC and SOH), and the dimensioning/size of the batteries. The energy management system controls the activation of individual functions and features, such as the charge, boost, or recovery modes, in the circumstances of the specific driving situations and conditions. While 12 V electric generators are employed, the claw-pole machines, because of their system design, the introduction of 48 V powertrain systems will lead to the coexistence of different technologies. Two electrical machine technologies are employed—synchronous and asynchronous/induction machines, as can be seen in Figure 11.

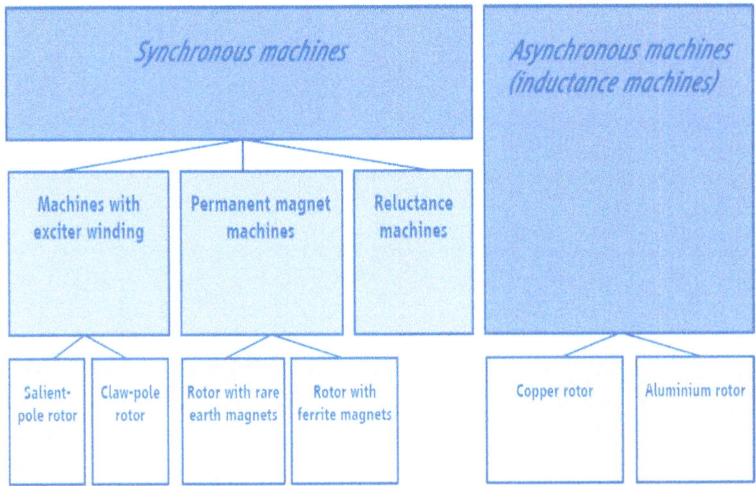

Figure 11. A block diagram with a short classification of electrical machines used in EV applications.

Synchronous machines are divided into different machines categories with excitation winding, of either the salient-pole or claw-pole rotor type, permanent magnet synchronous machines, and reluctance machines. The rotor of asynchronous or induction machines consists of a laminated core with uniform slotting accommodating either aluminum (copper) bars short-circuited by end-rings (the squirrel cage), or a three-phase winding (as in the stator) connected to some copper rings and fixed brushes—the wound rotor. Induction machines with squirrel-cage rotors are mostly used because of their robustness technology. The speed, efficiency, and power density of the machines may vary as a function of the active power and maximum current of their respective rectifiers. It is therefore difficult to classify any single machine as the best type. Furthermore, the automotive industry also requires additional important factors such as battery package space, costs, robustness, and standardization that need to be considered. This explains why different technologies will be used in the 48 V powertrain system.

4.5. Inverters and DC/DC Converters Used in EV Applications

An inverter with a bidirectional power flow is required for operating as a starter/generator. This power electronic inverter is used to convert the battery's direct current into a three-phase alternating current, supplying the individual windings of the electrical machine with electric energy. The energy flow is reversed for the regenerative operation mode. In this case, the power electronic inverter converts the alternating current generated into direct current to charge the battery. In terms of its functional configuration, the power electronic inverter of the 48 V powertrain systems is similar to that of the high-voltage inverters used in full hybrids or all-electric vehicles, as can be seen in Figure 12. One of the main differences lies in the power of the semiconductors used. Unlike the high-voltage systems using primarily, insulated gate bipolar transistors (IGBTs), MOSFETs are predestined for use as switching elements in power electronic inverters for starter generators, because of their lower voltages and switching losses. To control a three-phase machine, MOSFETs are mainly used as three half-bridge converters.

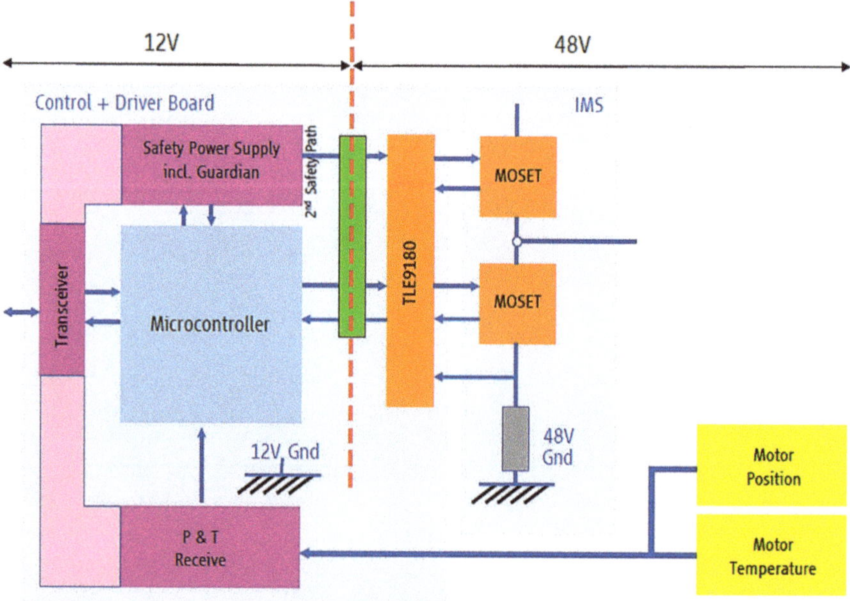

Figure 12. Block diagram of inverter electronics (source: Infineon Technologies).

A switch-mode converter (DC/DC converter/chopper) is used to transfer the energy between the two subsystems of a dual voltage system. As it mostly transfers energy from the 48 V level system to the 12 V level, it is primarily working as a step-down (buck) converter. In this energy transfer direction, the DC/DC switch-mode DC/DC converter replaces the generator for the traditional 12 V system. Scenarios with switch-mode converters operating in a step-up operation mode usually only involve partial load requirements so as to ensure 48 V operation. Alternatively, they can be stand-alone system solutions without a 48 V generator. The evaluation of various application scenarios of 48 V implementation, with and without 48 V components, reveals that the switching-mode DC–DC converter needs to be implemented in different power classes from one to three kW. To ensure the cost-efficient production, a modular and scalable DC–DC switch-mode converter architecture is necessary to support the different power classes and levels. The scalable converter must be configured with multi-phase half-bridges, polarity reversal protection for 14 V, anti-touch protection and short-circuit protection, and it must be protected against single-point failures in the components carrying current [39]. Several active or passive air-cooling or water-cooling concepts can be used.

4.6. Active Electronic Components

The electronic control units of 48 V powertrain systems always follow the same fundamental/basic topologies as those used in 12 V or high-voltage systems. Anyway, the semiconductor devices/switches selected for 48 V power electronic converter applications must take the different voltage level and different loads into account. They are mainly used in 48 V systems to control the electrical motors and other electric loads, in addition to connecting the 48 V and 12 V system levels by means of a DC/DC switch-mode converter [45,46].

The semiconductor devices to be used in EVs can be classified into the following topologies: sensors, microcontrollers, power supply and power management integrated circuits (ICs), communication, and driver ICs. In 48 V power systems, MOSFETs are often used as power output stage ICs, as, compared to IGBT devices, their performance regarding switching and conduction losses and cost is better [39]. A key issue for semiconductor devices/switches managing the current levels of more than 100 amperes, is the selection of their "case style", to ensure a sufficiently good dissipation of lost heat (cooling). Different case styles are possible subject to the configuration of the control unit, ranging from standard-organic field-effect transistors (OFET)-cases for single transistors and the integration of output stages (one or all stages) into one power module to the direct integration of power components into the motor. Driver ICs are another important element. Their primary purpose is to adjust the (PWM) signals generated by the microcontroller to control the motor to the level required for the power output stages. It may sometimes be necessary to use several drivers to ensure that this is achieved.

Multiple options are possible for the design of a DC/DC switch-mode converter, depending on individual requirements. The most important requirements, as discussed in the literature [39], are the converter performance (i.e., line regulation, load regulation, etc.), the efficiency (and hence the power losses), the package weight and volume, whether only unidirectional or also bidirectional power transfer is supported, whether power sources and load have galvanic separation or are coupled, and which classifications are reached in terms of the safety integrity level (SIL). These multiple options result in different circuit topologies (single-phase, two-phase, or multi-phase). The choice of converter switching frequency is also an important parameter.

5. In-Vehicle Cyber-Security for New GV Generations: Threats and Countermeasures

The new electronics control architecture of hybrid and electric vehicles, discussed in Sections 3 and 4, where control units are interconnected through in-vehicle networks, like CAN, will create new cyber-security issues. The latter can lead to severe safety issues as, as discussed in Sections 3 and 4, cyber-attacks propagating through the in-vehicle network can compromise the control of electrical components, providing key functionalities like energy generation, braking, and torque generation. To mitigate these issues, some countermeasures will be discussed in this section.

Some key features, such as integrity of the data, privacy, identification, and availability, should characterize a secure on-board communication system. However, a lot of security threats [46–54] characterize state-of-art technologies for on-board networking. The state-of-art is based on the use of CAN, and its evolutions time-triggered CAN (TTCAN) and flexible data-rate CAN (CAN-FD), as the backbone of the in-vehicle network. Then, several types of communication technologies are used for specific tasks; for example, the local interconnect network (LIN) is used for low data-rate nodes, FlexRay is used for high throughput control tasks, multimedia oriented system transport (MOST), or IDB-1394 (Automotive Firewire) are used for infotainment applications, where the human–machine interfaces exploit Bluetooth and/or USB connectivity

At the state-of-art, on-board gateway units are used to interconnect each other the different networking domains. This approach, which allows access to any vehicular bus from every other existing bus system, is a severe source of cybersecurity threats. An ECU that is interconnected on a low security-level and non-safe multimedia bus like MOST can transmit information (data packets) to other ECUs, even those operating in safety domains and interconnected with CAN or FlexRay [55,56]. As a consequence, a security-violation of a single subsystem can propagate and lead to the failure of the whole communication infrastructure. The challenge is when the propagation reaches safety-related domains, like those related to propulsion, braking, and navigation control, which are typically based on CAN or FlexRay technologies. The trends towards autonomous driving, towards a vehicle's connectivity with V2I/V2V technologies, and towards the electrification of propulsion, braking, and energy storage through electric components (machines, converters, and batteries) controlled by digital ECU, are exacerbating the above security risks.

In the state-of-art of on-board networking technologies, the requirement of data integrity is satisfied thanks to the use of error detection techniques, like, for example, cyclic redundancy check (CRC) codes. Instead, security features like data confidentiality, data authentication, and data availability are not guaranteed with the current in-vehicle technologies.

The broadcast nature of CAN, and its evolutions CAN-FD and TTCAN, is a source of threats for data confidentiality. A security-compromised ECU, which is under the control of a hacker, because of the broadcasting communication approach, can monitor all of the messages that are transferred through the bus (generated from or transmitted to all the other non-compromised ECUs).

At the state-of-art, to ensure a car's E/E scalability, ECUs can be added or removed with a plug-and-play approach. For example, if a new ECU is added to a CAN bus, then a new CAN identifier is assigned to it, without any change to the other installed ECUs. Combining this with the fact that signature mechanisms are currently missing, it is easy to understand that at the state-of-art there is a high risk of correct authentication. As consequence, it is possible for a hacker to attack an ECU and to emulate protocol-compliant behavior, as for all the other ECUs, is difficult to understand if a received packet has been transmitted by an authorized or unauthorized (and hence malicious) ECU.

Another feature of the current in-vehicle technologies that is a source of security threats is the arbitration among multi-masters based on identifier priority. This feature creates a risk for data availability. Indeed, a hacked ECU can use a high priority identifier to generate false packets. Because of the arbitration based on the identifier priority, the hacked controller using a fake high priority identifier can continuously send false packets, thus jamming the whole communication infrastructure, which will be no more available to the other ECUs. The jamming attack will cause a denial of service, with a high severity in terms of safety issues when the subsystem under attack is related to propulsion, braking, or navigation functionalities.

To address the above cybersecurity threats, several countermeasures are under analysis in the current R&D activities of the automotive industry and academia.

Cybersecurity hardware accelerators should be integrated in new automotive controllers to implement in real-time the encryption of data packets. To this aim, cybersecurity hardware accelerators to implement in real-time advanced encryption standard (AES), at the core of symmetric cryptography, or elliptic curve cryptography (ECC) for asymmetric cryptography, are under development [56–60].

Moreover, the trusted zone concept can be exploited. This means that the several network domains, and the relevant subsystems, should be clustered in separated security zones. Such zones should each be separated by gateways with integrated cybersecurity features. Thanks to this countermeasure, an attack on a non-safety domain, for example MOST, will be blocked when trying to propagate to a safety-related domain, like CAN. As consequence, a cybersecurity failure of the infotainment system (not related to driver and passenger safety) will not be the first step for a failure of the braking or propulsion systems, which instead are safety-critical.

Besides the adoption of the trusted zone concept, it is important to cluster the ECUs in different classes with different trustability levels. Such clustering should take into account both "how easy an ECU can be attacked" and "which are the safety consequences in an ECU is attacked".

Anomaly detection and intrusion detection mechanisms should be also implemented, exploiting both physical and packet layer features. In this way it will be possible to identify malicious ECUs.

6. Conclusions

This work has reviewed the recent trends for the electrification and digitalization of GVs. The current and foreseen, until 2030, market penetration of different types of EVs have been discussed, as well as their energy demand and their pollutant emissions. These trends have been compared to the evolution trends of battery technology, mainly based on lithium technology, and of the recharging issues considering the characteristics of the power grid. Real power grid scenarios in three cities, Hong Kong (China), Long Beach (CA, USA), and Manchester (UK) are considered. As result, from the vehicle point of view, light BEVs and 48 V HEVs are seen as the most promising technologies in terms of penetration and market acceptance. From the power grid point of view, demand-side management is the key technology to face the energy demand of transport electrification and to overcome the limits of current power grid. Solutions to integrate EV electricity demand in power grids have been also discussed and proposed. Integrated E/E architectures for HEVs and full EVs have been analyzed, detailing the innovations emerging for all components—inverter and DC/DC converters, new electric drives, battery-packs, and related BMS. 48 V HEVs are emerging as the most promising solution for a short-term electrification of vehicles. To this aim, a new integrated starter generator e-drive has been proposed; a power rating up to 10 kW is capable of ensuring the hybridization of small and medium cars, with a limited impact of the car architecture. This approach will ensure a smooth and rapid transition from the current ICE-based car generation to full EV. The increased digitalization and connectivity of electrified cars is also posing cyber-security issues. The main limits of state-of-art on-board vehicles, particularly CAN, have been discussed, and a list of countermeasures to mitigate them has been proposed.

Author Contributions: Both authors contributed equally to this survey paper.

Funding: This work was partially supported by PRA2017 project from University of Pisa.

Acknowledgments: With reference to Section 3, discussions with Valeo and AMS partners in the framework of the ATHENIS-3D EU projects are gratefully acknowledged.

Conflicts of Interest: The authors declare no conflict of interest.

References

1. International Energy Agency Key Word Energy Statistics. Available online: https://www.iea.org/statistics/kwes/ (accessed on 10 September 2018).
2. OECD/IEA. *Energy Technology Perspectives 2010*; OECD: Paris, France; IEA: Paris, France, 2010; ISBN 978 92-64-08597-8. Available online: http://www.iea.org/ (accessed on 12 September 2018).
3. Estefan, J. *Survey of Model-Based Systems Engineering (MBSE) Methodologies*; Technical Report; INCOSE: San Diego, CA, USA, 2008.
4. International Energy Agency-IEA. *Global EV Outlook 2018-Towards Cross-Modal Electrification*; OECD: Paris, France; IEA: Paris, France, 2018; Available online: www.iea.org/t&c/ (accessed on 10 September 2018).
5. Electric Cars Report, BMW Targets 140,000 Plug-in Vehicle Sales in 2018. Available online: https://electriccarsreport.com/2018/02/bmw-targets-140000-plug-vehicle-sales-2018/?utm_source=dlvr.it&utm_medium=twitter (accessed on 5 April 2018).
6. IEA. *Nordic EV Outlook 2018: Insights from Leaders in Electric Mobility*; IEA: Paris, France, 2018; Available online: www.iea.org/publications/freepublications/publication/nordic-ev-outlook-2018.html (accessed on 14 September 2018).
7. ICCT. *China's New Energy Vehicle Mandate Policy (Final Rule)*; ICCT: Brussels, Belgium, 2018; Available online: www.theicct.org/publications/china-nev-mandate-final-policy-update-20180111 (accessed on 14 September 2018).
8. ICCT. *Expanding the Electric Vehicle Market in US Cities*; ICCT: Brussels, Belgium, 2017; Available online: www.theicct.org/sites/default/files/publications/US-Cities-EVs_ICCT-White-Paper_25072017_vF.pdf (accessed on 14 October 2018).
9. Han, P.; Cheng, M.; Chen, Z. Dual-electrical-port control of cascaded doubly-fed induction machine for EV/HEV applications. *IEEE Trans. Ind. Appl.* **2017**, *53*, 1390–1398. [CrossRef]
10. Jurkovic, S.; Rahman, K.M.; Morgante, J.C.; Savagian, P.J. Induction machine design and analysis for general motors e-assist electrification technology. *IEEE Trans. Ind. Appl.* **2015**, *51*, 631–639. [CrossRef]
11. Tenconi, A.; Tenconi, A.; Vaschetto, S. Experimental characterization of a belt-driven multiphase induction machine for 48-V automotive applications: Losses and temperatures assessments. *IEEE Trans. Ind. Appl.* **2016**, *52*, 1321–1330.
12. Bojoi, R.; Cavagnino, A.; Cossale, M.; Tenconi, A. Multiphase starter generator for a 48-V mini-hybrid powertrain: Design and testing. *IEEE Trans. Ind. Appl.* **2016**, *52*, 1750–1758.
13. Chen, S.; Lequesne, B.; Henry, R.R.; Xue, Y.; Ronning, J.J. Design and testing of a belt-driven induction starter–generator. *IEEE Trans. Ind. Appl.* **2002**, *38*, 1750–1758.
14. Ala, G.; Giaconia, G.C.; Giglia, G.; Piazza, M.C.D.; Vitale, G. Design and performance evaluation of a high power-density EMI filter for PWM inverter-fed induction-motor drives. *IEEE Trans. Ind. Appl.* **2016**, *52*, 2397–2404. [CrossRef]

15. Niu, S.; Chau, K.T.; Jiang, J.Z. A permanent-magnet double-stator integrated-starter-generator for hybrid electric vehicles. In Proceedings of the IEEE Vehicle Power and Propulsion Conference (VPPC), Harbin, China, 3–5 September 2008; pp. 1–6.

16. Bounadja, M.; Belarbi, A.W.; Belmadani, B. A high performance svm-dtc scheme for induction machine as integrated starter generator in hybrid electric vehicles. *Nat. Technol.* **2010**, *2*, 41–47.

17. Richard, D.; Dubel, Y. Valeo stars technology: A competitive solution for hybridization. In Proceedings of the IEEE 2007 Power Conversion Conference, Nagoya, Japan, 2–5 April 2007; pp. 1601–1605.

18. Vint, M. Powertrain electrification for the 21st Century. In Proceedings of the UMTRI Conference, Ann Arbor, MI, USA, 23 July 2014.

19. Yang, L.; Franco, V.; Campestrini, A.; German, J.; Mock, P. *NOx Control Technologies for Euro 6 Diesel Passenger Cars*; White Paper; International Council Clean Transportation: Brussels, Belgium, 2015; pp. 1–22.

20. Saponara, S.; Ciarpi, G.; Groza, V.Z. Design and Experimental Measurement of EMI Reduction Techniques for Integrated Switching DC/DC Converters. *IEEE Can. J. Electr. Comput. Eng.* **2017**, *40*, 116–127.

21. Saponara, S.; Tisserand, P.; Chassard, P.; Ton, D.-M. Design and measurement of integrated converters for belt-driven starter-generator in 48 V micro/mild hybrid vehicles. *IEEE Trans. Ind. Appl.* **2017**, *53*, 3936–3949. [CrossRef]

22. International Energy Agency (IEA). *Digitalization and Energy*; IEA: Paris, France, 2017; Available online: www.iea.org/digital (accessed on 1 November 2018).

23. Muratori, M. Impact of uncoordinated plug-in electric vehicle. *Nat. Energy* **2018**, *3*, 193–201. [CrossRef]

24. Eurelectric. *Dynamic Pricing in Electricity Supply—A Eurelectric Position Paper*; Eurelectric: Brussels, Belgium, 2017; Available online: www3.eurelectric.org/media/309103/dynamic_pricing_in_electricity_supply-2017-2520-0003-01-e.pdf (accessed on 24 August 2018).

25. BMW Group. *BMW Group Announces Next Step in Electrification Strategy*; BMW Group: Munich, Germany, 2017; Available online: www.press.bmwgroup.com/global/article/detail/T0273122EN/bmwgroup-announces-next-step-in--lectrification-strategy?language=en (accessed on 24 August 2018).

26. BMW Group. *BMW Launches First App to Automate the Home Charging Process for BMW iElectric Vehicles*; BMW Group: Munich, Germany, 2018; Available online: www.press.bmwgroup.com/usa/article/detail/T0183262EN_US/bmw-launches-first-appto-automate-the-home-charging-process-for-bmw-i-electric-vehicles?language=en_US (accessed on 26 April 2018).

27. Schmidt, O.; Hawkes, A.; Gambhir, A.; Staffell, I. The future cost of electrical energy storage based on experience rates. *Nat. Energy* **2017**, *2*, 17110. [CrossRef]

28. Meeus, M. Review of Status of the Main Chemistries for the EV Market. 2018. Available online: www.iea.org/media/Workshops/2018/Session1MeeusSustesco.pdf (accessed on 1 November 2018).

29. ANL. *BatPaC: A Lithium-Ion Battery Performance and Cost Model for Electric-Drive Vehicles*; Argonne National Laboratory: Lemont, IL, USA, 2018. Available online: www.cse.anl.gov/batpac/index.html (accessed on 1 November 2018).

30. Warner, J. *The Handbook of Lithium-Ion Battery Pack Design*; Elsevier Science: Amsterdam, The Netherlands; Oxford, UK; Waltham, MA, USA, 2015.

31. Nitta, N.; Wu, F.; Lee, J.T.; Yushin, G. Li-ion battery materials: Present and future. *Mater. Today* **2015**, *18*, 252–264. [CrossRef]

32. Chung, J.; Lee, J. Asian Battery Makers Eye Nickel Top-up as Cobalt Price Bites. 2017. Available online: www.reuters.com/article/us-southkorea-battery-cobalt/asianbattery-makers-eye-nickel-top-up-as-cobalt-price-bites-idUSKBN1AJ0S8 (accessed on 25 September 2018).

33. Saponara, S.; Pasetti, G.; Tinfena, F.; Fanucci, L.; D'Abramo, P. HV-CMOS design and characterization of a smart rotor coil driver for automotive alternators. *IEEE Trans. Ind. Electron.* **2013**, *60*, 2309–2317. [CrossRef]

34. Pieri, F.; Zambelli, C.; Nannini, A.; Olivo, P.; Saponara, S. Is Consumer Electronics Redesigning Our Cars?: Challenges of Integrated Technologies for Sensing, Computing, and Storage. *IEEE Consum. Electron. Mag.* **2018**, *7*, 8–17. [CrossRef]

35. The German Electrical & Electronic Industry—Facts & Figures. 2018. Available online: www.zvei.org (accessed on 6 October 2018).

36. Power Electronics, Control Unit for Electric Drive Systems. Available online: https://www.bosch-mobility-solutions.com/en/products-and-services/passenger-cars-and-light-commercial-vehicles/powertrain-systems/high-voltage-hybrid-systems/power-electronics/ (accessed on 6 October 2018).

37. Volkswagen Introduces 48 V Mild-Hybrid System to the Golf. Available online: https://www.enginetechnologyinternational.com/news/hybrid-powertrain-technologies/volkswagen-48v.html (accessed on 1 November 2018).

38. Timmann, M.; Inderka, R.; Eder, T. Development of 48V powertrain systems at Mercedes-Benz. In *Internationales Stuttgarter Symposium*; Springer: Braunschweig, Germany, 2018; pp. 567–577.

39. Available online: www.gridinnovation-on-line.eu (accessed on 14 October 2018).

40. Camacho, O.M.F.; Nørgård, P.B.; Rao, N.; Mihet-Popa, L. Electrical Vehicle Batteries Testing in a Distribution Network using Sustainable Energy. *IEEE Trans. Smart Grid* **2014**, *5*, 1033–1042. [CrossRef]

41. Camacho, O.M.F.; Mihet-Popa, L. Fast Charging and Smart Charging Tests for Electric Vehicles Batteries using Renewable Energy. *Oil Gas Sci. Technol. Rev. IFP Energies Nouv. OGST J.* **2014**, *71*. [CrossRef]

42. Un-Noor, F.; Padmanaban, S.; Mihet-Popa, L.; Mollah, M.N.; Hossain, E. A Comprehensive Study of Key Electric Vehicle (EV) Components, Technologies, Challenges, Impacts, and Future Direction of Development. *Energies* **2017**, *10*, 1217. [CrossRef]

43. Harighi, T.; Bayindir, R.; Padmanaban, S.; Mihet-Popa, L.; Hossain, E. An Overview of Energy Scenarios, Storage systems and the Infrastructure for Vehicle-to-Grid Technology. *Energies* **2018**, *11*, 2174. [CrossRef]

44. Mihet-Popa, L.; Camacho, O.M.F.; Nørgård, P.B. Charging and discharging tests for obtaining an accurate dynamic electro-thermal model of high power lithium-ion pack system for hybrid and EV applications. In Proceedings of the IEEE PES Power Tech Conference, Grenoble, France, 16–20 June 2013; ISBN 978-146735669-5.

45. Saponara, S.; Ciarpi, G. IC Design and Measurement of an Inductorless 48 V DC/DC Converter in Low-Cost CMOS Technology Facing Harsh Environments. *IEEE Trans. Circuits Syst. I* **2018**, *65*, 380–393. [CrossRef]

46. Saponara, S.; Ciarpi, G. Electrical, Electromagnetic, and Thermal Measurements of 2-D and 3-D Integrated DC/DC Converters. *IEEE Trans. Instrum. Meas.* **2018**, *67*, 1070–1080. [CrossRef]

47. Nilsson, D.K.; Larson, U.E.; Picasso, F.; Jonsson, E. A first simulation of attacks in the automotive network communications protocol flexray. In Proceedings of the International Workshop on Computational Intelligence in Security for Information Systems CISIS'08, Burgos, Spain, 23–26 September 2009; pp. 84–91.

48. Lin, C.W.; Sangiovanni-Vincentelli, A. Cyber-security for the controller area network (can) communication protocol. In Proceedings of the 2012 International Conference on Cyber Security, Washington, DC, USA, 14–16 December 2012; pp. 1–7.

49. Wolf, M.; Weimerskirch, A.; Paar, C. *Secure In-Vehicle Communication*; Springer: Berlin/Heidelberg, Germany, 2006; pp. 95–109.

50. Avatefipour, O.; Malik, H. State-of-the-art survey on in-vehicle network communication can-bus security and vulnerabilities. *Int. J. Comput. Sci. Netw.* **2017**, *6*, 720–727.

51. Cho, K.-T.; Shin, K.G. Fingerprinting electronic control units for vehicle intrusion detection. In Proceedings of the 25th USENIX Security Symposium, Austin, TX, USA, 10–12 August 2016; pp. 911–927.

52. Santos, E.D.; Simpson, A.; Schoop, D. A formal model to facilitate security testing in modern automotive systems. *EPTCS* **2018**, *271*, 95–104. [CrossRef]

53. Hoppe, T.; Kiltz, S.; Dittmann, J. Security threats to automotive can networks–practical examples and selected short-term countermeasures. *Reliab. Eng. Syst. Saf.* **2011**, *96*, 11–25. [CrossRef]

54. Lukasiewycz, M.; Mundhenk, P.; Steinhorst, S. Security-aware obfuscated priority assignment for automotive can platforms. *ACM Trans. Des. Autom. Electron. Syst.* **2016**, *21*. [CrossRef]

55. Eisenbarth, T.; Kasper, T.; Moradi, A.; Paar, C.; Salmasizadeh, M.; Shalmani, M.T.M. On the power of power analysis in the real world: A complete break of the keeloq code hopping scheme. In Proceedings of the Advances in Cryptology–CRYPTO 2008, Barbara, CA, USA, 17–21 August 2008; Wagner, D., Ed.; Springer: Berlin/Heidelberg, Germany, 2008; pp. 203–220.

56. Koscher, K.; Czeskis, A.; Roesner, F.; Patel, S.; Kohno, T.; Checkoway, S.; McCoy, D.; Kantor, B.; Anderson, D.; Shacham, H.; et al. Experimental security analysis of a modern automobile. In Proceedings of the IEEE Symposium on Security and Privacy, Berkeley, CA, USA, 16–19 May 2010; pp. 447–462.

57. Patsakis, C.; Dellios, K.; Bouroche, M. Towards a distributed secure in-vehicle communication architecture for modern vehicles. *Comput. Secur.* **2014**, *40*, 60–74. [CrossRef]

58. Sghaier, A.; Zeghid, M.; Machhout, M. Fast hardware implementation of ECDSA signature scheme. In Proceedings of the 2016 International Symposium on Signal, Image, Video and Communications, Tunis, Tunisia, 21–23 November 2016; pp. 343–348.

59. Baldanzi, L.; Crocetti, L.; Bartolucci, M.; Fanucci, L.; Saponara, S. Analysis of Cybersecurity Weakness in Automotive In-Vehicle Networking and Hardware Accelerators for Real-time Cryptography. In Proceedings of the APPLEPIES2018, Pisa, Italy, 26–27 September 2018.

60. Lo Bello, L.; Mariani, R.; Mubeen, S.; Saponara, S. Recent Advances and Trends in On-board Embedded and Networked Automotive Systems. *IEEE Trans. Ind. Inform.* **2018**. [CrossRef]